T0142296

Advances in Intelligent Systems and Computing

Volume 814

Series editor

Janusz Kacprzyk, Polish Academy of Sciences, Warsaw, Poland
e-mail: kacprzyk@ibspan.waw.pl

The series "Advances in Intelligent Systems and Computing" contains publications on theory, applications, and design methods of Intelligent Systems and Intelligent Computing. Virtually all disciplines such as engineering, natural sciences, computer and information science, ICT, economics, business, e-commerce, environment, healthcare, life science are covered. The list of topics spans all the areas of modern intelligent systems and computing such as: computational intelligence, soft computing including neural networks, fuzzy systems, evolutionary computing and the fusion of these paradigms, social intelligence, ambient intelligence, computational neuroscience, artificial life, virtual worlds and society, cognitive science and systems, Perception and Vision, DNA and immune based systems, self-organizing and adaptive systems, e-Learning and teaching, human-centered and human-centric computing, recommender systems, intelligent control, robotics and mechatronics including human-machine teaming, knowledge-based paradigms, learning paradigms, machine ethics, intelligent data analysis, knowledge management, intelligent agents, intelligent decision making and support, intelligent network security, trust management, interactive entertainment, Web intelligence and multimedia.

The publications within "Advances in Intelligent Systems and Computing" are primarily proceedings of important conferences, symposia and congresses. They cover significant recent developments in the field, both of a foundational and applicable character. An important characteristic feature of the series is the short publication time and world-wide distribution. This permits a rapid and broad dissemination of research results.

More information about this series at http://www.springer.com/series/11156

Ajith Abraham · Paramartha Dutta
Jyotsna Kumar Mandal · Abhishek Bhattacharya
Soumi Dutta
Editors

Emerging Technologies in Data Mining and Information Security

Proceedings of IEMIS 2018, Volume 3

 Springer

Editors
Ajith Abraham
Machine Intelligence Research Labs
Auburn, WA, USA

Abhishek Bhattacharya
Institute of Engineering and Management
Kolkata, West Bengal, India

Paramartha Dutta
Department of Computer and Systems
   Sciences
Visva-Bharati University
Santiniketan, West Bengal, India

Soumi Dutta
Institute of Engineering and Management
Kolkata, West Bengal, India

Jyotsna Kumar Mandal
Department of Computer Science
   and Engineering
University of Kalyani
Kalyani, India

ISSN 2194-5357   ISSN 2194-5365   (electronic)
Advances in Intelligent Systems and Computing
ISBN 978-981-13-1500-8   ISBN 978-981-13-1501-5   (eBook)
https://doi.org/10.1007/978-981-13-1501-5

Library of Congress Control Number: 2018947481

This Springer imprint is published by the registered company Springer Nature Singapore Pte Ltd.
The registered company address is: 152 Beach Road, #21-01/04 Gateway East, Singapore 189721, Singapore

Foreword

Welcome to the Springer International Conference on Emerging Technologies in Data Mining and Information Security (IEMIS 2018) held from February 23 to 25, 2018, in Kolkata, India. As a premier conference in the field, IEMIS 2018 provides a highly competitive forum for reporting the latest developments in the research and application of information security and data mining. We are pleased to present the proceedings of the conference as its published record. The theme of this year is Crossroad of Data Mining and Information Security, a topic that is quickly gaining traction in both academic and industrial discussions because of the relevance of privacy-preserving data mining model (PPDM).

IEMIS is a young conference for research in the areas of information and network security, data sciences, big data, and data mining. Although 2018 is the debut year for IEMIS, it has already witnessed a significant growth. As evidence of that, IEMIS received a record of 532 submissions. The authors of submitted papers come from 35 different countries several authors contributed their researcher work in this volume. The authors of accepted papers are from 11 countries.

We hope that this program will further stimulate research in information security and data mining and provide the practitioners with better techniques, algorithms, and tools for deployment. We feel honored and privileged to serve the best recent developments in the field of WSDM to the readers through this exciting program.

Kolkata, India

Bimal Kumar Roy
General Chair, IEMIS 2018
Indian Statistical Institute

Preface

This volume presents the proceedings of the International Conference on Emerging Technologies in Data Mining and Information Security, IEMIS 2018, which took place at the University of Engineering and Management in Kolkata, India, from February 23 to 25, 2018. The volume appears in the series "Advances in Intelligent Systems and Computing" (AISC) published by Springer Nature, one of the largest and most prestigious scientific publishers. It is one of the fastest growing book series in their program. AISC series publication is meant to include various high-quality and timely publications, primarily proceedings of relevant conferences, congresses, and symposia but also monographs, on the theory, applications, and implementations of broadly perceived modern intelligent systems and intelligent computing, in their modern understanding, i.e., including not only tools and techniques of artificial intelligence (AI), and computational intelligence (CI)—which includes data mining, information security, neural networks, fuzzy systems, evolutionary computing, as well as hybrid approaches that synergistically combine these areas—but also topics such as multiagent systems, social intelligence, ambient intelligence, Web intelligence, computational neuroscience, artificial life, virtual worlds and societies, cognitive science and systems, perception and vision, DNA- and immune-based systems, self-organizing and adaptive systems, e-learning and teaching, human-centered and human-centric computing, autonomous robotics, knowledge-based paradigms, learning paradigms, machine ethics, intelligent data analysis, and various issues related to "big data," security, and trust management. These areas are at the forefront of science and technology and have been found useful and powerful in a wide variety of disciplines such as engineering, natural sciences, computer, computation and information sciences, ICT, economics, business, e-commerce, environment, health care, life science, and social sciences. The AISC book series is submitted for indexing to ISI Conference Proceedings Citation Index (now run by Clarivate), Ei Compendex, DBLP, SCOPUS, Google Scholar, and SpringerLink, and many other indexing services around the world. IEMIS 2018 is a debut annual conference series organized by the School of Information Technology, under the aegis of Institute of Engineering and Management, India. Its idea came from the heritage of the other two cycles of the events: IEMCON and UEMCON, which were organized by the Institute of Engineering and Management under the leadership of Prof. (Dr.) Satyajit Chakrabarti.

In this volume of "Advances in Intelligent Systems and Computing," we would like to present the results of studies on the selected problems of data mining and information security. Security implementation is the contemporary answer to the new challenges in threat evaluation of complex systems. Security approach in theory and engineering of complex systems (not only computer systems and networks) is based on multidisciplinary attitude to information theory, technology, and maintenance of the systems working in real (and very often unfriendly) environments. Such a transformation has shaped the natural evolution in the topical range of subsequent IEMIS conferences, which can be seen over the recent years. Human factors likewise infest the best digital dangers. Workforce administration and digital mindfulness are fundamental for accomplishing all-encompassing cybersecurity. This book will be of an extraordinary incentive to a huge assortment of experts, scientists, and understudies concentrating on the human part of the Internet, and for the compelling assessment of safety efforts, interfaces, client-focused outline, and plan for unique populaces, especially the elderly. We trust this book is instructive yet much more than it is provocative. We trust it moves, driving peruser to examine different inquiries, applications, and potential arrangements in making sheltered and secure plans for all.

The Program Committee of IEMIS 2018, its organizers, and the editors of this proceedings would like to gratefully acknowledge the participation of all the reviewers who helped to refine the contents of this volume and evaluated the conference submissions. Our thanks go to Prof. Bimal Kumar Roy, Dr. Ajith Abraham, Dr. Sheng Lung peng, Dr. Detlef Streitferdt, Dr. Shaikh Fattah, Dr. Celia Shahnaz, Dr. Swagatam Das, Dr. Niloy Ganguly, Dr. K. K. Shukla, Dr. Nilanjan Dey, Dr. Florin PopentiuVladicescu, Dr. Dewan Md. Farid, Dr. Saptarshi Ghosh, Dr. Rita Choudhury, Dr. Asit Kumar Das, Prof. Tanupriya Choudhury, Prof. Arijit Ghosal, Prof. Rahul Saxena, Prof. Monika Jain, Dr. Aakanksha Sharaff, Prof. Dr. Sajal Dasgupta, Prof. Rajiv Ganguly, and Prof. Sukalyan Goswami.

Thanking all the authors who have chosen IEMIS 2018 as the publication platform for their research, we would like to express our hope that their papers will help in further developments in design and analysis of engineering aspects of complex systems, being a valuable source material for scientists, researchers, practitioners, and students who work in these areas.

Auburn, USA Ajith Abraham
Santiniketan, India Paramartha Dutta
Kalyani, India Jyotsna Kumar Mandal
Kolkata, India Abhishek Bhattacharya
Kolkata, India Soumi Dutta

Organizing Committee

Patron

Prof. (Dr.) Satyajit Chakrabarti, Institute of Engineering and Management, India

Conference General Chair

Dr. Bimal Kumar Roy, Indian Statistical Institute, Kolkata, India

Convener

Dr. Subrata Saha, Institute of Engineering and Management, India
Abhishek Bhattacharya, Institute of Engineering and Management, India

Co-convener

Sukalyan Goswami, University of Engineering and Management, India
Krishnendu Rarhi, Institute of Engineering and Management, India
Soumi Dutta, Institute of Engineering and Management, India
Sujata Ghatak, Institute of Engineering and Management, India
Dr. Abir Chatterjee, University of Engineering and Management, India

Keynote Speakers

Dr. Ajith Abraham, Machine Intelligence Research Labs (MIR Labs), USA
Dr. Fredric M. Ham, IEEE Life Fellow, SPIE Fellow, and INNS Fellow,
    Melbourne, USA

Dr. Sheng-Lung Peng, National Dong Hwa University, Hualien, Taiwan
Dr. Shaikh Fattah, Editor, IEEE Access & CSSP (Springer), Bangladesh
Dr. Detlef Streitferdt, Technische Universität Ilmenau, Germany
Dr. Swagatam Das, Indian Statistical Institute, Kolkata, India
Dr. Niloy Ganguly, Indian Institute of Technology Kharagpur, India
Dr. K K Shukla, IIT (B.H.U.), Varanasi, India
Dr. Nilanjan Dey, Techno India College of Technology, Kolkata, India
Dr. Florin Popentiu Vladicescu, "UNESCO Chair in Information Technologies",
    University of Oradea, Romania
Dr. Celia Shahnaz, senior member of IEEE, fellow member of Institution of
    Engineers, Bangladesh (IEB)

Technical Program Committee Chair

Dr. J. K. Mondal, University of Kalyani, India
Dr. Paramartha Dutta, Visva-Bharati University, India
Abhishek Bhattacharya, Institute of Engineering and Management, India

Technical Program Committee Co-chair

Dr. Satyajit Chakrabarti, Institute of Engineering and Management, India
Dr. Subrata Saha, Institute of Engineering and Management, India
Dr. Kamakhya Prasad Ghatak, University of Engineering and Management, India
Dr. Asit Kumar Das, IIEST, Shibpur, India

Editorial Board

Dr. Ajith Abraham, Machine Intelligence Research Labs (MIR Labs), USA
Dr. J. K. Mondal, University of Kalyani, India
Dr. Paramartha Dutta, Visva-Bharati University, India
Abhishek Bhattacharya, Institute of Engineering and Management, India
Soumi Dutta, Institute of Engineering and Management, India

Advisory Committee

Dr. Mahmoud Shafik, University of Derby
Dr. MohdNazri Ismail, National Defence University of Malaysia
Dr. Bhaba R. Sarker, Louisiana State University
Dr. Tushar Kanti Bera, University of Arizona, USA
Dr. Shirley Devapriya Dewasurendra, University of Peradeniya, Sri Lanka

Dr. Goutam Chakraborty, Professor and Head of the Intelligent Informatics Lab, Iwate Prefectural University, Japan

Dr. Basabi Chakraborty, Iwate Prefectural University, Japan

Dr. Kalyanmoy Deb, Michigan State University, East Lansing, USA

Dr. Vincenzo Piuri, University of Milan, Italy

Dr. Biswajit Sarkar, Hanyang University, Korea

Dr. Raj Kumar Buyya, the University of Melbourne

Dr. Anurag Dasgupta, Valdosta State University, Georgia

Dr. Prasenjit Mitra, the Pennsylvania State University

Dr. Esteban Alfaro-Cortés, University of Castilla–La Mancha, Spain

Dr. Ilkyeong Moon, Seoul National University, South Korea

Dr. Izabela Nielsen, Aalborg University, Denmark

Dr. Prasanta K Jana, IEEE Senior Member, Indian Institute of Technology (ISM), Dhanbad

Dr. Gautam Paul, Indian Statistical Institute, Kolkata

Dr. Malay Bhattacharyya, IIEST, Shibpur

Dr. Sipra Das Bit, IIEST, Shibpur

Dr. Jaya Sil, IIEST, Shibpur

Dr. Asit Kumar Das, IIEST, Shibpur

Dr. Saptarshi Ghosh, IIEST, Shibpur, IIT KGP

Dr. Prof. Hafizur Rahman, IIEST, Shibpur

Dr. C. K. Chanda, IIEST, Shibpur

Dr. Asif Ekbal, Associate Dean, IIT Patna

Dr. Sitangshu Bhattacharya, IIIT Allahabad

Dr. Ujjwal Bhattacharya, CVPR Unit, Indian Statistical Institute

Dr. Prashant R.Nair, Amrita Vishwa Vidyapeetham (University), Coimbatore

Dr. Tanushyam Chattopadhyay, TCS Innovation Labs, Kolkata

Dr. A. K. Nayak, Fellow and Honorary Secretary CSI

Dr. B. K. Tripathy, VIT University

Dr. K. Srujan Raju, CMR Technical Campus

Dr. Dakshina Ranjan Kisku, National Institute of Technology Durgapur

Dr. A. K. Pujari, University of Hyderabad

Dr. Partha Pratim Sahu, Tezpur University

Dr. Anuradha Banerjee, Kalyani Government Engineering College

Dr. Amiya Kumar Rath, Veer Surendra Sai University of Technology

Dr. Kandarpa Kumar Sharma, Gauhati University

Dr. Amlan Chakrabarti, University of Calcutta

Dr. Sankhayan Choudhury, University of Calcutta

Dr. Anjana Kakoti Mahanta, Gauhati University

Dr. Subhankar Bandyopadhyay, Jadavpur University

Dr. Debabrata Ghosh, Calcutta University

Dr. Rajat Kr. Pal, University of Calcutta

Dr. Ujjwal Maulik, Jadavpur University

Dr. Himadri Dutta, Kalyani Government Engineering College, Kalyani

Dr. Brojo Kishore Mishra, C. V. Raman College of Engineering (Autonomous), Bhubaneswar

Dr. S. Vijayakumar Bharathi, Symbiosis Centre for Information Technology (SCIT)

Dr. Govinda K., VIT University, Vellore

Dr. Ajanta Das, University of Engineering and Management, India

Technical Committee

Dr. Vincenzo Piuri, University of Milan, Italy

Dr. Mahmoud Shafik, University of Derby

Dr. Bhaba Sarker, Louisiana State University

Dr. Mohd Nazri Ismail, University Pertahanan National Malaysia

Dr. Tushar Kanti Bera, Yonsei University, Seoul

Dr. Birjodh Tiwana, LinkedIn, San Francisco, California

Dr. Saptarshi Ghosh, IIEST, Shibpur, IIT KGP

Dr. Srimanta Bhattacharya, Indian Statistical Institute, Kolkata

Dr. Loingtam Surajkumar Singh, NIT Manipur

Dr. Sitangshu Bhattacharya, IIIT Allahabad

Dr. Sudhakar Tripathi, NIT Patna

Dr. Chandan K. Chanda, IIEST, Shibpur

Dr. Dakshina Ranjan Kisku, NIT Durgapur

Dr. Asif Ekbal, IIT Patna

Dr. Prasant Bharadwaj, NIT Agartala

Dr. Somnath Mukhopadhyay, Hijli College, Kharagpur, India, and Regional Student Coordinator, Region II, Computer Society of India

Dr. G. Suseendran, Vels University, Chennai, India

Dr. Sumanta Sarkar, Department of Computer Science, University of Calgary

Dr. Manik Sharma, Assistant Professor, DAV University, Jalandhar

Dr. Rita Choudhury, Gauhati University

Dr. Kuntala Patra, Gauhati University

Dr. Helen K. Saikia, Gauhati University

Dr. Debasish Bhattacharjee, Gauhati University

Dr. Somenath Sarkar, University of Calcutta

Dr. Sankhayan Choudhury, University of Calcutta

Dr. Debasish De, Maulana Abul Kalam Azad University of Technology

Dr. Buddha Deb Pradhan, National Institute of Technology Durgapur

Dr. Shankar Chakraborty, Jadavpur University

Dr. Durgesh Kumar Mishra, Sri Aurobindo Institute of Technology, Indore, Madhya Pradesh

Dr. Angsuman Sarkar, Secretary, IEEE EDS Kolkata Chapter, Kalyani Government Engineering College

Dr. A. M. Sudhakara, University of Mysore

Dr. Indrajit Saha, National Institute of Technical Teachers' Training and Research Kolkata

Dr. Bikash Santra, Indian Statistical Institute (ISI)
Dr. Ram Sarkar, Jadavpur University
Dr. Priya Ranjan Sinha Mahapatra, Kalyani University
Dr. Avishek Adhikari, University of Calcutta
Dr. Jyotsna Kumar Mandal, Kalyani University
Dr. Manas Kumar Sanyal, Kalyani University
Dr. Atanu Kundu, Chairman, IEEE EDS, Heritage Institute of Technology, Kolkata
Dr. Chintan Kumar Mandal, Jadavpur University
Dr. Kartick Chandra Mondal, Jadavpur University
Mr. Debraj Chatterjee, Manager, Capegemini
Mr. Gourav Dutta, Cognizant Technology Solutions
Dr. Soumya Sen, University of Calcutta
Dr. Soumen Kumar Pati, St. Thomas' College of Engineering and Technology
Mrs. Sunanda Das, Neotia Institute of Technology Management and Science, India
Mrs. Shampa Sengupta, MCKV Institute of Engineering, India
Dr. Brojo Kishore Mishra, C. V. Raman College of Engineering (Autonomous), Bhubaneswar
Dr. S. Vijayakumar Bharathi, Symbiosis Centre for Information Technology, Pune
Dr. Govinda K., Vellore Institute of technology
Dr. Prashant R. Nair, Amrita University
Dr. Hemanta Dey, IEEE Senior Member
Dr. Ajanta Das, University of Engineering and Management, India
Dr. Samir Malakar, MCKV Institute of Engineering, Howrah
Dr. Tanushyam Chattopadhyay, TCS Innovation Lab, Kolkata
Dr. A. K. Nayak Indian Institute of Business Management, Patna
Dr. B. K. Tripathy, Vellore Institute of Technology, Vellore
Dr. K. Srujan Raju, CMR Technical Campus, Hyderabad
Dr. Partha Pratim Sahu, Tezpur University
Dr. Anuradha Banerjee, Kalyani Government Engineering College
Dr. Amiya Kumar Rath, Veer Surendra Sai University of Technology, Odissa
Dr. S. D. Dewasurendra, University of Peradeniya, Sri Lanka
Dr. Arnab K. Laha, IIM Ahmedabad
Dr. Kandarpa Kumar Sarma, Gauhati University
Dr. Ambar Dutta, BIT Mesra and Treasurer—CSI Kolkata
Dr. Arindam Pal, TCS Innovation Labs, Kolkata
Dr. Himadri Dutta, Kalyani Government Engineering College, Kalyani
Dr. Tanupriya Choudhury, Amity University, Noida, India
Dr. Praveen Kumar, Amity University, Noida, India

Organizing Chairs

Krishnendu Rarhi, Institute of Engineering and Management, India
Sujata Ghatak, Institute of Engineering and Management, India
Dr. Apurba Sarkar, IIEST, Shibpur, India

Organizing Co-chairs

Sukalyan Goswami, University of Engineering and Management, India
Rupam Bhattacharya, Institute of Engineering and Management, India

Organizing Committee Convener

Dr. Sajal Dasgupta, Vice-Chancellor, University of Engineering and Management

Organizing Committee

Subrata Basak, Institute of Engineering and Management, India
Anshuman Ray, Institute of Engineering and Management, India
Rupam Bhattacharya, Institute of Engineering and Management, India
Abhijit Sarkar, Institute of Engineering and Management, India
Ankan Bhowmik, Institute of Engineering and Management, India
Manjima Saha, Institute of Engineering and Management, India
Biswajit Maity, Institute of Engineering and Management, India
Soumik Das, Institute of Engineering and Management, India
Sreelekha Biswas, Institute of Engineering and Management, India
Nayantara Mitra, Institute of Engineering and Management, India
Amitava Chatterjee, Institute of Engineering and Management, India
Ankita Mondal, Institute of Engineering and Management, India

Registration Chairs

Abhijit Sarkar, Institute of Engineering and Management, India
Ankan Bhowmik, Institute of Engineering and Management, India
Ankita Mondal, Institute of Engineering and Management, India
Ratna Mondol, Institute of Engineering and Management, India

Publication Chairs

Dr. Debashis De, Maulana Abul Kalam Azad University of Technology, India
Dr. Kuntala Patra, Gauhati University, India

Publicity and Sponsorship Chair

Dr. J. K. Mondal, University of Kalyani, India
Dr. Paramartha Dutta, Visva-Bharati University, India
Abhishek Bhattacharya, Institute of Engineering and Management, India
Biswajit Maity, Institute of Engineering and Management, India
Soumik Das, Institute of Engineering and Management, India

Treasurer and Conference Secretary

Dr. Subrata Saha, Institute of Engineering and Management, India
Rupam Bhattacharya, Institute of Engineering and Management, India
Krishnendu Rarhi, Institute of Engineering and Management, India

Hospitality and Transport Chair

Soumik Das, Institute of Engineering and Management, India
Nayantara Mitra, Institute of Engineering and Management, India
Manjima Saha, Institute of Engineering and Management, India
Sreelekha Biswas, Institute of Engineering and Management, India
Amitava Chatterjee, Institute of Engineering and Management, India

Web Chair

Samrat Goswami
Samrat Dey

About this Book

This book features research papers presented at the International Conference on Emerging Technologies in Data Mining and Information Security (IEMIS 2018) held at the University of Engineering and Management, Kolkata, India, from February 23 to 25, 2018.

Data mining is a currently a well-known topic in mirroring the exertion of finding learning from information. It gives the strategies that enable supervisors to acquire administrative data from their heritage frameworks. Its goal is to distinguish legitimate, novel, and possibly valuable and justifiable connection and examples in information. Information mining is made conceivable by the very nearness of the expansive databases.

Information security advancement is an essential part to ensure open and private figuring structures. Data mining-based intrusion-detection frameworks are incredibly profitable in discovering security breaks. This article will give a layout of the use of data mining procedures in the information security territory. Notwithstanding how strict the security techniques and parts are, more affiliations are getting the chance to be weak to a broad assortment of security breaks against their electronic resources. Network-intrusion area is a key protect part against security perils, which have been growing in rate generally.

This book comprises high-quality research work by academicians and industrial experts in the field of computing and communication, including full-length papers, research-in-progress papers, and case studies related to all the areas of data mining, machine learning, Internet of things (IoT), and information security.

Contents

About the Editors

Dr. Ajith Abraham received his Ph.D. from Monash University, Melbourne, Australia, and his Master of Science from Nanyang Technological University, Singapore. His research and development experience includes over 25 years in the industry and academia, spanning different continents such as Australia, America, Asia, and Europe. He works in a multidisciplinary environment involving computational intelligence, network security, sensor networks, e-commerce, Web intelligence, Web services, computational grids, and data mining, applied to various real-world problems. He has authored/co-authored over 350 refereed journal/conference papers and book chapters, and some of the papers have also won the best paper awards at international conferences and also received several citations. Some of the articles are available in the ScienceDirect Top 25 hottest articles → http://top25.sciencedirect.com/index.php?cat_id=6&subject_area_id=7.

He has given more than 20 plenary lectures and conference tutorials in these areas. He serves the editorial board of several reputed international journals and has also guest-edited 26 special issues on various topics. He is actively involved in the hybrid intelligent systems (HISs), intelligent systems design and applications (ISDAs), and information assurance and security (IAS) series of international conferences. He is General Co-Chair of the Tenth International Conference on Computer Modeling and Simulation (UKSIM'08), Cambridge, UK; Second Asia International Conference on Modeling and Simulation (AMS 2008), Malaysia; Eighth International Conference on Intelligent Systems Design and Applications (ISDA'08), Taiwan; Fourth International Symposium on Information Assurance and Security (IAS'07), Italy; Eighth International Conference on Hybrid Intelligent Systems (HIS'08), Spain; Fifth IEEE International Conference on Soft Computing as Transdisciplinary Science and Technology (CSTST'08), Cergy Pontoise, France; and Program Chair/Co-chair of Third International Conference on Digital

Information Management (ICDIM'08), UK, and Second European Conference on Data Mining (ECDM 2008), the Netherlands.

He is Senior Member of IEEE, IEEE Computer Society, IEE (UK), ACM, etc. More information is available at: http://www.softcomputing.net.

Dr. Paramartha Dutta was born in 1966. He completed his bachelor's and master's degrees in statistics from the Indian Statistical Institute, Calcutta, in the years 1988 and 1990, respectively. He afterward completed his Master of Technology in computer science from the same institute in the year 1993 and Doctor of Philosophy in Engineering from the Bengal Engineering and Science University, Shibpur, in 2005. He has served as a research personnel in various projects funded by the Government of India, which are done by Defence Research Development Organization, Council of Scientific and Industrial Research, Indian Statistical Institute, etc. He is now Professor in the Department of Computer and System Sciences of the Visva-Bharati University, West Bengal, India. Prior to this, he served Kalyani Government Engineering College and College of Engineering in West Bengal as Full-Time Faculty Member. He remained associated as Visiting/Guest Faculty of several universities/institutes such as West Bengal University of Technology, Kalyani University, Tripura University.

He has co-authored eight books and has also seven edited books to his credit. He has published more than two hundred technical papers in various peer-reviewed journals and conference proceedings, both international and national, and several chapters in edited volumes of reputed international publishing houses like Elsevier, Springer-Verlag, CRC Press, John Wiley. He has guided six scholars who had already been awarded their Ph.D. apart from one who has submitted her thesis. Presently, he is supervising six scholars for their Ph.D. program.

He is Co-Inventor of ten Indian patents and one international patent, which are all published apart from five international patents which are filed but not yet published.

He, as investigator, could implement successfully the projects funded by All India Council for Technical Education, Department of Science and Technology of the Government of India. He has served/serves in the capacity of external member of Board of Studies of relevant departments of various universities encompassing West Bengal University of Technology, Kalyani University, Tripura University, Assam University, Silchar. He had the opportunity to serve as the expert of several interview boards organized by the West Bengal Public Service Commission, Assam University, Silchar; National Institute of Technology, Arunachal Pradesh; Sambalpur University, etc.

He is Life Fellow of the Optical Society of India (FOSI), Institution of Electronics and Telecommunication Engineers (FIETE), Institute of Engineering (FIE), Life Member of Computer Society of India (LMCSI), Indian Science Congress Association (LMISCA), Indian Society for Technical Education (LMISTE), Indian Unit of Pattern Recognition and Artificial Intelligence (LMIUPRAI)—the Indian affiliate of the International Association for Pattern Recognition (IAPR), and Senior

Member of Associated Computing Machinery (SMACM), and Institution of Electronics and Electrical Engineers (SMIEEE), USA.

Dr. Jyotsna Kumar Mandal received his M.Tech. in computer science from the University of Calcutta in 1987. He was awarded Ph.D. in computer science and engineering by Jadavpur University in 2000. Presently, he is working as Professor of computer science and engineering and Former Dean, Faculty of Engineering, Technology and Management, Kalyani University, Kalyani, Nadia, West Bengal, for two consecutive terms since 2008. He is Ex-Director, IQAC, Kalyani University, and Chairman, CIRM, Kalyani University. He was appointed as Professor in Kalyani Government Engineering College through Public Service Commission under the Government of West Bengal. He started his career as Lecturer at NERIST, under MHRD, Government of India Arunachal Pradesh, in September 1988. He has teaching and research experience of 30 years. His areas of research are coding theory, data and network security, remote sensing and GIS-based applications, data compression, error correction, visual cryptography, and steganography. He has guided 21 Ph.D. scholars, 2 scholars have submitted their Ph.D. thesis, and 8 are pursuing. He has supervised 03 M.Phil. and more than 50 M.Tech. dissertations and more than 100 MCA. dissertations. He is Chief Editor of CSI Journal of Computing and Guest Editor of MST Journal (SCI indexed) of Springer Nature. He has published more than 400 research articles, out of which 154 articles are in various international journals. He has published five books from LAP Germany and one from IGI Global. He was awarded A. M Bose Memorial Silver Medal and Kali Prasanna Dasgupta Memorial Silver Medal in M.Sc. by Jadavpur University. India International Friendship Society (IIFS), New Delhi, conferred "Bharat Jyoti Award" for his meritorious service, outstanding performance, and remarkable role in the field of computer science and engineering on August 29, 2012. He received "Chief Patron" Award from CSI India in 2014. International Society for Science, Technology and Management conferred "Vidyasagar Award" in the Fifth International Conference on Computing, Communication and Sensor Network on December 25, 2016. ISDA conferred Rastriya Pratibha Award in 2017.

Abhishek Bhattacharya is Assistant Professor of computer application department at the Institute of Engineering and Management, India. He did his Masters in computer science from the Biju Patnaik University of Technology and completed his Master of Technology in computer science from BIT, Mesra. He remained associated as Visiting/Guest Faculty of several universities/institutes in India. He has three books to his credit. He has published twenty technical papers in various peer-reviewed journals and conference proceedings, both international and national, and chapters in edited volumes of the reputed international publishing houses. He has teaching and research experience of 13 years. His areas of research are data mining, network security, mobile computing, and distributed computing. He is the reviewer of a couple of journals of IGI Global, Inderscience Publications, and Journal of Information Science Theory and Practice, South Korea.

He is Member of International Association of Computer Science and Information Technology (IACSIT), Universal Association of Computer and Electronics Engineers (UACEE), International Association of Engineers (IAENG), Internet Society as a Global Member (ISOC), the Society of Digital Information and Wireless Communications (SDIWC) and International Computer Science and Engineering Society (ICSES); Technical Committee Member of CICBA 2017, 52nd Annual Convention of Computer Society of India (CSI 2017), International Conference on Futuristic Trends in Network and Communication Technologies (FTNCT-2018), ICIoTCT 2018, ICCIDS 2018, and Innovative Computing and Communication (ICICC-2018); Advisory Board Member of ISETIST 2017.

Soumi Dutta is Assistant Professor at the Institute of Engineering and Management, India. She is also pursuing her Ph.D. in the Department of Computer Science and Technology, Indian Institute of Engineering Science and Technology, Shibpur. She received her B.Tech. in information technology and her M.Tech. in computer science securing first position (gold medalist), both from Techno India Group, India. His research interests include social network analysis, data mining, and information retrieval. She is Member of several technical functional bodies such as the Michigan Association for Computer Users in Learning (MACUL), the Society of Digital Information and Wireless Communications (SDIWC), Internet Society as a Global Member (ISOC), International Computer Science and Engineering Society (ICSES). She has published several papers in reputed journals and conferences.

Part I
High Performance Computing

Proposed Memory Allocation Algorithm for NUMA-Based Soft Real-Time Operating System

Vatsalkumar Shah and Apurva Shah

Abstract Memory management algorithms for operating system have been explored broadly, but inadequate devotion has been concentrated on the real-time characteristic. Furthermost advanced algorithms are general-purpose and do not fulfill the requirements of real-time systems. Additionally, a few allocators combining with real-time systems do not focus well on multiprocessor architecture. The emergent needs for high-performance computational processing can be satisfied by NUMA architecture-based systems having multicore system. In this paper, memory allocation algorithm has been proposed for soft real-time operating system for NUMA-based architecture which will be useful to achieve constant execution time and less fragmentation.

Keywords Real-time operating system · NUMA · Memory allocator

1 Introduction

This algorithm is basically for memory management in NUMA-based soft real-time operating system. Real-time operating system (RTOS) is nothing but completing the task in given deadline. RTOS provides so many features like synchronization, interrupt-handling, memory management, task preemption. Diwase et al. and Shah and Shah [1, 2], but the memory management is fundamental part of each operating system. Memory management is essentially allocating or deallocating memory blocks to the specific process on demand [3–5].

Since last few years, so much development has been done and recent computing tools have been used for embedded devices like smart phones, tablets, palmtops. But

V. Shah · A. Shah (✉)
Computer Science and Engineering Department, M. S. University,
Vadodara 390001, Gujarat, India
e-mail: apurva.shah-cse@msubaroda.ac.in

V. Shah
e-mail: vatsal.shah-cse@msubaroda.ac.in

© Springer Nature Singapore Pte Ltd. 2019
A. Abraham et al. (eds.), *Emerging Technologies in Data Mining and Information Security*, Advances in Intelligent Systems and Computing 814,
https://doi.org/10.1007/978-981-13-1501-5_1

for improving any kind of services like high-performance computing tools such as parallel and cloud computing, systems with single processor are not capable to meet the essential request [6–8].

As "high performance" word is come into the picture, to overcome the single-processor systems cons, ccNUMA architecture is the optimal solution. ccNUMA architecture-based systems are nothing but the multiprocessor systems having shared memory in distributed nature. NUMA-based systems are more mountable and flexible compared to other multiprocessor architecture such as symmetric multiprocessor systems. These systems offer a distinct address space and are comprehensively cache-coherent in nature. Any application which can execute on ccNUMA-based system can easily execute on symmetric multiprocessor systems without any changes. This one is a key feature as application can execute any platform. Moreover, NUMA-based system affects the performance of application as it can allocate thread on closest node to take the advantage [9].

As the growth and need of real-time systems have been increasing, in nearer future it may happen that real-time application will share the resource with other real-time application, and to implement this, NUMA-based system will be more preferable though it has some complexity [10, 11, 5].

2   Related Works

There are various algorithms which are useful for memory allocation as well as deallocation for general-purpose as well as real-time operating system. These all algorithms are categorized as conventional and unconventional algorithms [1, 2, 9]. Conventional algorithms are (1) sequential fit algorithm, (2) buddy allocator algorithm, (3) indexed fit, (4) bitmapped fit. Unconventional algorithms are (1) Doug Lea (DLmalloc), (2) half-fit, (3) TLSF, (4) tcmalloc. In this section, all algorithms are briefly explained.

2.1   Conventional Algorithms [1, 2, 9]

(1)   Sequential Fit:

There are four different types of strategies named as best fit, worst fit, first fit, and next fit. In best-fit strategy, entire list of unallocated blocks will be searched and smallest memory block which is large enough to satisfy the request will be allocated, while in first fit the list of unallocated block will start searching from beginning and first memory block which is enough to satisfy the request will be allocated. In next fit, searching will be started from last seek point where the searching has been stopped and returned the first available block which is enough to satisfy the request, while in worst fit the first biggest block will be allocated.

(2) Buddy Allocator

In this allocator, an array of link lists of free memory blocks will be used—each and every list for acceptable memory block size. It means link list of 2 N memory block size like 200, 400, 800 kb. This allocator searches a smallest block which is large enough to satisfy the request from list of unallocated blocks. If no block is available in unallocated block list, then it will search a requested memory block from other link list which is larger than request, then select, and fragment the block [12, 13]. A block must be divided into a same size blocks; it means 800-kb block will be divided into two 400-kb blocks. Also for merging the blocks, it will use same strategy.

(3) Indexed Fit

This allocator uses structured index to implement desired fit strategy. As per allocation strategy, tree or hash table type of data structure will be used for searching unallocated blocks.

(4) Bitmapped Fit

This allocator uses a bitmap. This map is a simple vector of one-bit flag with one bit corresponding to each word of heap area [12, 13]. Every bit in the map has value either 1 or 0. 1 suggests in use else 0. The advantage of this allocator is memory overhead is lesser because search time is subject to bitmap size.

2.2   Unconventional Algorithms

(1) Doug Lea(DLmalloc):

This allocator has been designed by Doug Lea in 1996. And its extension has been designed by Gloger in 2006 [14] and by Free Software Foundation in 2012. This allocator works separately for small memory block and large memory block. The limit of small memory block is up to 256 byte, and the blocks whose size is greater than 26 byte will be considered as large block. For allocation of small memory block, it has large number of static size arrays called as small bins. For small block allocation, best-fit policy has been used (Fig. 1).

(2) Half-Fit

In 1995, this algorithm has been invented by Ogasawra [16]. Bitmapped fit strategy is used to achieve execution time as constant. Due to bitmap policy, it is getting slow. To maintain status of unallocated lists of memory block, bitmap has been used. Its time complexity is nothing but the O(1).

It uses single-level segregated list for unallocated memory block. In this list, all free memory blocks of various sizes are linked. It takes desired blocks of a required size from unallocated block list to satisfy the request. Figure 2 shows this example.

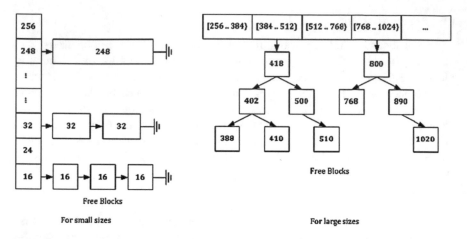

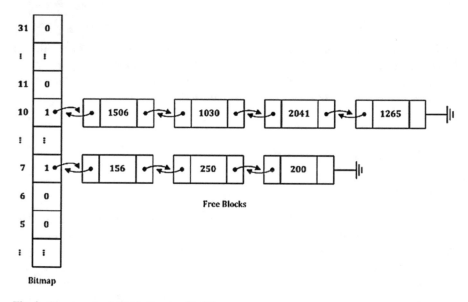

Fig. 1 Structure of DLmalloc allocator [9, 15]

Fig. 2 Structure for half-fit allocator [9, 15]

(3) TLSF

TLSF stands for two-level segregated fit. Unlike other algorithms, this allocator occupies two different levels of segregated list for storing unallocated memory blocks.

The first-level of list (FLI) splits unallocated memory blocks in parts which are apart by power of two like 2, 4, 8, 16 onward. The secondary known as second-level lists split each first-level list by a user-defined variable known as second-level index. TLSF structures are shown in Fig. 3.

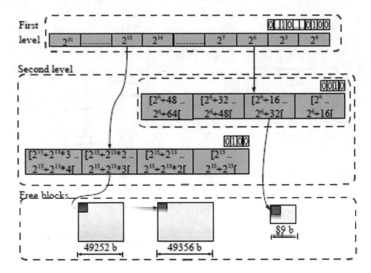

Fig. 3 Example of TLSF allocator [2, 9]

(4) tcmalloc

Sanjay Ghemawt has developed tcmalloc in 2010 [17]. It uses both global heap structure and thread private heap multiprocessor architecture. For small memory block allocation whose size ranges from 4 byte to 32 kb, it allocates private local heap to each thread.

For allocating large memory blocks whose size ranges from 32 kb to 1 Mb, it occupies a global heap structure which is collectively used by all available threads. To achieve mutual exclusion between thread, it employs spin-lock mechanism. For large memory block request greater than 1 Mb, it forwards the request to the existing operating system using system call or APIs. Its time complexity is O(1). Figure 4 shows the structure of tcmalloc allocator.

3 Objectives

As we are designing proposed allocator for NUMA-based real-time system, there are two main objectives of this algorithm: (1) minimum memory fragmentation and (2) constant execution time.

(1) Minimum Memory Fragmentation:

The span of real-time applications is typically longer than simple applications. It may be for 1 day or 1 month as well. Hence, during span of real-time application, it may free memory segment of arbitrary size, which can create holes generally known as unallocated memory block in the memory. It may happen in these holes; size is

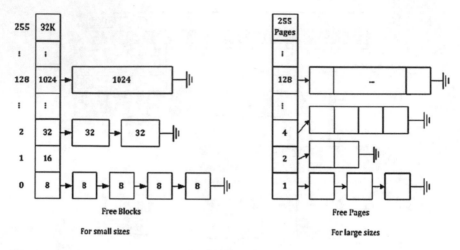

Fig. 4 Example of tcmalloc allocator [9, 15]

too small and cannot be used again. That's why decreasing memory fragmentation is a main objective.

(2) Constant Execution Time:

Applications of real-time systems should satisfy their timing conditions, and to meet its deadlines, the dynamic memory management allocator should be developed in such a way that application can be accomplished within specified limit.

4 Proposed Algorithm

We have proposed one memory allocator which can work on NUMA-based architecture for real-time operating system. Figure 5 shows schematic diagram of NUMA-based architecture for RTOS. As shown in figure, there are total four nodes where each node has two processors; each processor inside a node is connected with a bus, and all nodes are connected with shared memory. Also each processor has their local (private) memory.

Whenever any processor required any memory block, it will first check into its local memory; if the required memory block is available, then it will allocate it from the local memory, and if not then it will try to access from the shared memory; if the block is there, then it will allocate, but if it is not there, then it will ask for the another processor which is lightly loaded. Now what is lightly loaded processor? Let us check it.

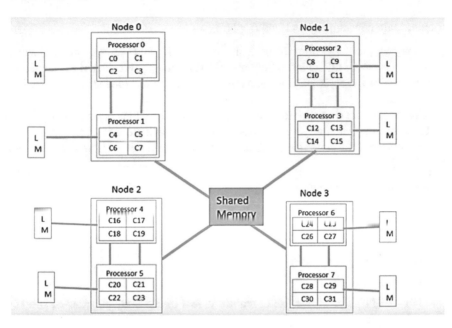

Fig. 5 Proposed algorithm structure

According to memory utilization, each processor will be categorized in four categories.

(1) Ideal
(2) Heavily loaded
(3) Normal loaded
(4) Lightly loaded

First step is to calculate the load average for memory utilization for all processors using following equation [18, 19].

$$\text{Mem}_{u\_avg} = \frac{\text{Mem}_{u1} + \text{Mem}_{u2} + \text{Mem}_{u3} + \cdots + \text{Mem}_{un}}{n} \tag{1}$$

Second step is to find the upper and lower threshold values for memory utilization using following equation.

$$T_U = H \times \text{Mem}_{u\_avg}$$
$$T_L = L \times \text{Mem}_{u\_avg}$$

where

T_U upper limit of threshold,
T_L lower limit of threshold,

U and L are constants. ($U > 1$ and $L < 1$)

In the proposed algorithm, U and L are set to be 1.3 and 0.7, respectively, which interpret if memory utilization is 30% above the Mem_{u_avg}, it is heavily loaded. And if memory utilization is 70% of the Mem_{u_avg}, it is lightly loaded; otherwise, it is normally loaded.

And then, select appropriate processor's memory for allocating memory.

Till now, which memory will be used to allocate the memory is decided.

Now, the turn for how to allocate block and for the same we have proposed one memory allocator.

How to allocate memory?

First of all memory will be divided into three categories as per requirement.

(1) Small blocks
(2) Normal blocks
(3) Large blocks

(1) Small Blocks
    For small blocks whose size of request is less than 512 bytes, exact fit strategy will be used.
(2) Normal Blocks
    For normal blocks whose size of request is between 512 bytes to 2 MB, good fit strategy will be used which is used in TLSF algorithm.
(3) Large Blocks
    For large blocks, worst fit will be used whose request exceeding the threshold or (some predefined size)

So, overall proposed allocator is taking advantage of all existing algorithms in order to achieve constant response time and less fragmentation.

5 Conclusion

In real scenario, different memory allocators are available for soft real-time systems, but we have proposed algorithm for soft real-time system for NUMA-based architecture which provides less fragmentation and constant response time with respect to algorithm mentioned in the literature review.

References

1. Diwase, D., Shah, S., Diwase, T., Rathod, P.: Survey report on memory allocation strategies for real time operating system in context with embedded devices. Int. J. Eng. Res. Appl. **2**(3), 1151–1156 (2012)

2. Shah, V., Shah, A.: An analysis and review on memory management algorithms for real time operating system. Int. J. Comput. Sci. Inf. Security (IJCSIS) **14**(5) (2016)
3. Budzinski, R.L., Davidson, E.S., A comparison of dynamic and static virtual memory allocation algorithms. IEEE Trans. Software Eng. **SE-7**(1) (1981)
4. Hanson, D.R.: Fast allocation and deallocation of memory based on object life times. Software-Practice and Experience **20**(1), 5–12 (1990)
5. Liu, J.W.S.: Real-time System, (published by Person Education, Page no. 20–40)
6. Ogasawara, T.: An algorithm with constant execution time for dynamic memory allocation. In: Proceedings of Second International Workshop on Real-Time Computing Systems and Applications, 25–27 Oct 1995, Pg No. 21–25
7. Puaut, Real-time performance of dynamic memory allocation algorithms. In: Proceedings of the 14th Euromicro Conference on Real-Time Systems (ECRTS'02), June 2002
8. Sun, X.H., Wang, J.L., Chan, X.: An improvement of TLSF algorithm
9. Kim, S.: Node-oriented dynamic memory management for real-time systems on ccNUMA architecture systems. Published in 2013
10. Shalan, M.A.: Dynamic memory management for embedded real-time multiprocessor system on a chip. In: A Thesis in Partial Fulfillment of the Requirements for the Degree of Doctor of Philosophy from School of Electrical and Computer Engineering, Georgia Institute of Technology, November 2003
11. Kim, R.M.J., Lee, W., Chung, Y.: Smart dynamic memory allocator for embedded systems. In: Proceedings of 23rd International Symposium on Computer and Information Sciences, ISCIS '08, 27–29 Oct. 2008
12. Wilson, P.R., Johnstone, M.S., Neely, M., Boles, D.: Dynamic memory allocation: a survey and critical review. In: Proceedings of the Memory Management International Workshop IWMM95, Sep 1995
13. Masmano, M., Ripoll, I., Crespo, A.: TLSF: a new dynamic memory allocator for real-time systems. In: Proceedings of 16th Euromicro Conference on Real-Time Systems, Catania, Italy, 79–88 July 2004
14. Lea, D.: A memory allocator. http://g.oswego.edu/dl/html/malloc.html. Unix/Mail December, 1996
15. Shah, V., Shah, A.: Critical analysis for memory management algorithm for NUMA based real time operating system. Published in IEEE Xplorer in December 2017
16. Ogasawara, T. (1995) An algorithm with constant execution time for dynamic storage allocation. In: RTCSA'95: Proceedings of the 2nd International Workshop on Real-Time Computing Systems and Applications, 21–25, Washington, DC, USA. IEEE Computer Society
17. Sanjay Ghemawat, P.M.: Tcmalloc: thread-caching malloc. Published in 2010
18. Werstein, P., Situ, H., Huang, Z.: Load balancing in a cluster computer. In: Seventh International Conference on Parallel and Distributed Computing, Applications and Technologies (PDCAT'06), 2006
19. Shah, V., Patel, K.: Load Balancing algorithm by Process Migration in Distributed Operating System. Int. J. Comput. Sci. Inf. Technol. Security (IJCSITS), ISSN: 2249-9555, **2**(6) (December 2012)

Mathematical Modeling of QoS-Aware Fog Computing Architecture for IoT Services

Prasenjit Maiti, Jaya Shukla, Bibhudatta Sahoo
and Ashok Kumar Turuk

Abstract The fog computing approach has come up as a distributed mechanism for capturing of data, its further processing, and allocation of resources associated with the Internet of things (IoT). The IoT services require several quality of service (QoS) parameters such as bandwidth utilization, resource provisioning, energy consumption, service delay. A new architecture for fog computing based on QoS parameters has been designed. A distributed solution for cloud-IoT has been presented where data is distributed optimally among several fog nodes/mini-clouds. The virtual machines (VMs) located in the edge devices are facilitated by these distributed fog nodes/mini-clouds to take care of IoT traffic. However, very little research has been done on designing any QoS-aware architecture for fog computing. The mathematical formulation for the presented model has also been proposed, and hence, the performance analysis of the system is shown in terms of the QoS metrics.

1 Introduction

The Internet of things (IoT) depicts a major change in data management. Real-time data management is associated with distributed objects and their associated smart sensors. Smart sensors data needs to be stored and retrieved efficiently on demand for IoT services. IoT devices are growing rapidly, and it is anticipated that about 50 billion devices will be deployed in 2020. Existing cloud solution provides services for a large amount of data. But in some scenarios, it can face limitations due to increased

P. Maiti (✉) · J. Shukla · B. Sahoo · A. K. Turuk
National Institute of Technology, Rourkela 769008, Odisha, India
e-mail: pmaiti1287@gmail.com

J. Shukla
e-mail: jayashukla2192@gmail.com

B. Sahoo
e-mail: bibhudatta.sahoo@gmail.com

A. K. Turuk
e-mail: akturuk@nitrkl.ac.in

© Springer Nature Singapore Pte Ltd. 2019
A. Abraham et al. (eds.), *Emerging Technologies in Data Mining and Information Security*, Advances in Intelligent Systems and Computing 814,
https://doi.org/10.1007/978-981-13-1501-5_2

traffic of the entire network and thus delay in processing the services. Different IoT services such as health care, face recognition, military, disaster management require real-time response with very low latency. To overcome this problem, a new architecture needs to be proposed, and thus, fog computing emerged to take care of these challenges. Fog computing is a concept that provides services at the network edge and involves smart gateways name fog smart gateways (FSG). Fog nodes are deployed in the network near the users to handle the services. In this architecture, the data is processed locally before sending it to the cloud. The major issues and challenges of architecture design for edge-centric IoT services are discovering fog nodes, data caching, partitioning, and offloading tasks using fog nodes publicly and securely. Due to constrained resources, resource discovery and resource allocation are the challenges of architecture design which meets the quality of service (QoS) and service-level agreement (SLA). Fog node controller plays a major role to identify the best candidate for placement a fog node and IoT services according to the resource availability. Sensor data virtualization is a requirement for IoT services that allows an application to retrieve and manipulate data with live data discovery and monitoring. Service allocation or service scheduling of all services to minimize the delay for each service allocation is done by considering both resource availability and the device condition. An efficient service can be provided with the help of service nodes placement, service nodes selection, service placement with a balanced and efficient pairing or matching strategy in a sensor-virtualization environment for edge nodes which is crucial for achieving Service Level Agreement (SLA) and Quality of Services (QoS). Containers handle processing of IoT traffic hosted by distributed fog nodes. Only a few literature are available for edge-centric architecture design for IoT services with above parameters. Therefore, we develop an economically feasible architecture that reduces service response time for overall network traffic.

The rest of the paper has been organized in the following manner. The related work has been mentioned in Sect. 2 that describes the prerequisites for the work. Section 3 describes the details of fog computing architecture and mathematical model. Section 4 presents QoS metrics of fog computing architecture. Section 5 gives the conclusion of the work presented in the paper.

2 Related Work

A sensor-fog architecture provides a platform to the users to easily provide services. Bonomi et al. [1] proposed fog computing as a distributed, highly virtualized platform that provides storage, computation and networking services. A fog network is a collection of fog nodes, and each fog node resides in any edge devices like router, switch, base station, access point, gateway, or smart phone. Three-layer architecture has been proposed by many researchers that includes devices as end points, fog nodes which are present at the network edge and cloud data centers. Depending on the definition, the fog nodes may be a router in the core network, a switch in the WAN, and even wireless access points(APs) and smart phones are included in the fog layer.

A smart gateway has been proposed as a fog node in [2, 3], the proposal of micro-data centers in [4], or proposing fog nodes that serve as caches in a information-centric networking in [5]. In [6, 7] fog nodes are represented as mini-clouds. Now, the key aspect is where fog nodes are located. Locating the fog nodes in various highly capable devices like smart gateways or routers has been proposed by some of the authors as in [2–4, 8]. Bonomi et al. [9–11] proposed intermediate compute nodes as fog nodes which has no dependency on specific devices. Tang et al. [7] recommended using three layers of fog computing to carry out big data analytics in the smart cities. In [12], a different application has been explained that shows the application of the fog computing in an industrial environment. It implements fog computing in Cisco edge routers as originally proposed by Cisco in [13]. Abdullahi et al. [5] and Skala et al. [6] proposed routers as a candidate for deployment of fog nodes. An interesting example has been given in [14] that describes sharing the computation resources present in a smart phone only in case when they are connected to the grid, but the approach is not sufficient enough for scenarios that are highly demanding. In this paper [15], author discussed that fogs are not limited anymore to either execute a task or forward it to the cloud, but also to communicate with other fogs to process the job request. This minimizes the overall network delay. Distributed cloud data centers with the facility of data replication have been developed by Kumar et al. [16] that are referred to as mini-clouds. A Fog-to-Cloud architecture was introduced by Masip-Bruin et al. [17] that consists of a layered structure for managing a hierarchical architecture of various heterogeneous layers of fog. Souza et al. [18] proposed a QoS-aware service distribution strategy in fog-to-cloud scenarios. A low delay service allocation has been aimed in the work by the use of service atomization that involves decomposition of services into separate sub-services known as atomic services that enables the parallel execution. The fog-to-cloud architecture comprises a control plane that distributes various atomic services to the edge nodes available. The process of service allocation aims at reduction of delay during service allocation, provision of load balancing and balancing of usage of energy among the fogs. Also, the position of deployment of fog node may lead to reduction in the network traffic. An optimal strategy for placement of mini-cloud has been proposed by Narendra et al. [19] that aims at minimization of latency during collection of data from the IoT devices and its migration onto the mini-clouds to address issues concerning storage capacity along with access latency minimization. The work described in [20] shows lower application latency as well as higher utilization of server though the strategy used for deployment by individual network operator and cities' geographic specifications decides the best approach. Most of the research issues in fog computing [21–24], the service latency, network traffic, and power consumption are reduced by fog computing architecture. Luan et al. [25] and Hong et al. [26] described the concept of fog computing as mobile fog and showed that mobile users can use fog computing to improve QoS, and reduce different shortcomings like usage bandwidth, energy consumption, end-to-end delay and network traffic. But, there is no consideration where to deploy or place the fog nodes. A fog node has capability to run multiple virtual machines (VMs) on its own physical machine. The VMs can be flexible placed in fog network, based on the traffic distribution and moving pattern of mobile users. Dynamic fog node placement in fog network systems incurs a significant cost on latency, energy, and bandwidth consumption of the network links.

3 Fog Computing Architecture and Mathematical Model

In our architecture (Fig. 1), the IoT service network consists of four layers. The task of processing and aggregation of data produced by IoT devices is performed by the networking elements present in the architecture.

(a) Tier 1: This is the ground-level layer encompassing all the smart sensor nodes (SSNs) that are assigned unique IPv6 addresses, suitably compressed according to the 6LoWPAN protocol, and form a mesh network. SSN is a collection of sensors and actuators. SSN senses environmental data and sends to the upper layer. There can be instructions from the upper layer to the actuator to perform an action. IoT devices or IoT nodes is a collection of SSN (mobile phones, smart vehicles, and smart meters etc.). SSN is distributed uniformly at random. A coordinate value is assigned to each SSN. A CD is known differently in different networks, namely cluster head(CH) in sensor networks, access point (AP) in WiFi networks, and reader in radio-frequency identification (RFID) network. S is a set of n number of static smart sensor nodes distributed uniformly at random in the area of $p \times q$. A SSN denoted by $\chi \in S$ is defined as a eleven-tuple.

$$\chi = < S_{id}, S_{st}, S_{\max\_v}, S_{\min\_v}, S_{ts}, S_e, S_l, S_h, S_{cid}, S_k, S_B >$$

S_{id} is representing the unique IP address of the sensor. A status of a SSN, S_{st}, is represented by a boolean value $S_{st} = \{0, 1\}$, which defines the sensor node is in

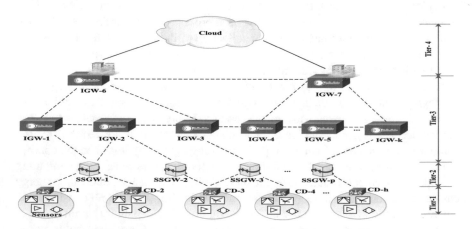

Fig. 1 Edge-centric architecture for IoT services

active state (value 0) or dead state (value 1). The maximum and minimum values that sensor can return (in the appropriate SI unit) are described by S_{max_v} and S_{min_v}. S_{ts} represents time in millisecond when the sensor value sends. One SSN can sense multiple environment events. S_e and S_l indicate the type of event and node location of a SSN. The tuple S_h express the specifications of a SSN which includes its hardware details. S_k represents the encryption key. S_B represents the battery level of a SSN. If the value of S_B is less than the threshold value, then SSN close its communication. S_{cid} indicates the cluster id of a SSN. SSNs send data to the CD. CD is a set of coordinating devices. Each CD have a certain transmission range r_{c_d}. The IoT network consists of different technologies (RFID, ZigBee, Bluetooth, BLE, 802.11 a/g, WiFi, etc.). TL is a set of technologies which are associated with CDs.

(b) Tier 2: CDs need to transmit their data to the Internet for efficient execution of their corresponding applications. This transmission of data is facilitated by device known as solution specific gateways ($SSGW$s) or IoT gateway (IGW). $SSGW$ is a set of solution specific gateway, and G is a set of IoT gateway (IGW). r_{sg} is a range of each $SSGW$. Wireless mesh network is as close as it can get to the IoT network with one fundamental difference. All gateway in a wireless mesh network support the same set of technologies, whereas $SSGW$ in IoT supports different sets of technologies. $SSGW$ is described by three-tuple.

$$SG =< SG_{id}, SG_{d[m]}, SG_{sp} >$$

Each $SSGW$ has unique IP address. SG_{id} represents ID of a $SSGW$. Each SSGW maintains a neighbor list that records the connected CDs. $SG_{d[m]}$ is non-empty 1D array of size m which stores the IDs of connected CDs. SG_{sp} dictates the hardware specification like processor, memory, wireless communication technologies involved such as RFID, ZigBee, Bluetooth, WiFi, etc.

(c) Tier 3: This layer consists of set of edge devices, such as gateway, router, switch etc. A fog node (FN) is placed within an IGW specific to a geographic location, and other IGW is served by the co-located FN. FNs are capable of load balancing and service orchestration. A fog device, F_d, is represented by the three-tuple. Each fog device id F_{id} is different from each other. The type (such as gateway, router, switch) of the fog computing device is represented by F_{type}. F_{sp} is the hardware specification of the device. An IoT service SE_{IoT} is defined as a four-tuple. The service is the main motivation of the IoT.

$$SE_{IoT} =< SE_{id}, SE_{type}, SE_{sp}, SE_{req} >$$

Each instance of the IoT service must have a unique identification. SE_{id} is the service ID. The purpose has to be specified by the IoT service, i.e., its functionality as well as its responsibilities. A set of operations that may be provided to the test of IoT services defines the functionality of the service. SE_{type} type denotes the purpose of use for the service (such as finance, medical, entertainment, utility, education, and gaming). The particulars of base framework required for running of application that includes the memory, processor, and operating system is managed by the SE_{sp}. SE_{req}

is the resource requirement (such as bandwidth, computation capability, storage) to run a service. An application A is represented as a three-tuple which is running at the end of a user.

$$A = < A_{id}, A_{sreq} >$$

A_{id} is the unique ID of an application. REQ is a set of requests from the users. A_{sreq} is an user request for a particular service.

(d) Tier 4: The cloud computing layer is top-most layer. Several physical servers are comprised within a data center, and there exists a high bandwidth link to the Internet from each physical server and an interconnection of high-speed LAN-network. Each IGW is connected to a cloud data center by a wired network. The cloud computing environment is with a number of heterogeneous physical hosts in a data center.

4 QoS Metrics

4.1 Service Latency

The service delay is the requested transmission delay and processing delay. We assume that the communication delay between SSNs is considered insignificant. Let Δ_{cd_sg} and Δ_{sg_igw}, Δ_{igw_sfg} be the delays in transmission of a data packet from a CD to the corresponding SSGW, from a SSGW to the corresponding IGW, and from IGW to a smart fog gateway, respectively. η_{sg}, η_{igw}, and η_{sfg} are the processing latency of SSGW, IGW, and smart fog gateway for a data packet. Thus, the mean transmission latency, σ_{sfg}, for the data packets of req_i request running within mc_i is given by

$$\sigma_{sfg} = \left(\Delta_{cd\_sg}\mu + \Delta_{sg\_igw}\theta + \Delta_{igw\_sfg}\tau \right) + \left(\eta_{sg}\mu + \eta_{igw}\theta + \eta_{sfg}\tau \right) \quad (1)$$

where μ, θ, and τ ($\mu > \theta > \tau$) are the total number of packets sent by CD, SSGW, and IGW.

4.2 Energy Consumption

Since the energy consumed by the data from CD to SSGW and from the SSGW to IGW is represented by λ_{cd_sg} and λ_{sg_igw}, respectively, λ_{igw_sfg} is the energy expansion from the IGW to the intelligent fog gateway for unit byte data transfers. The energy demand to process unit byte of data within the SSGW, gateway, and smart fog gateway is represented by ω_{sg}, ω_{igw}, and ω_{sfg}, respectively. Total energy consumption of a data packet is transmission energy and processing energy of a data packet. The rate of energy dissipation of a data packet is represented as

$$
\xi(t) = \left(\begin{array}{l} \left(\lambda_{cd\_sg} \sum_{i=1}^{h} \sum_{j=1}^{p} \gamma_{i,j} + \lambda_{sg\_igw} \sum_{i=1}^{p} \sum_{j=1}^{k} \alpha_{i,j} + \lambda_{igw\_sfg} \sum_{i=1}^{k} \sum_{j=1}^{t} \beta_{i,j} \right) + \\ \left(\omega_{sg} \sum_{i=1}^{h} \sum_{j=1}^{p} \gamma_{i,j} + \omega_{igw} \sum_{i=1}^{p} \sum_{j=1}^{k} \alpha_{i,j} + \omega_{sfg} \sum_{i=1}^{k} \sum_{j=1}^{t} \beta_{i,j} \right) \end{array} \right) \tag{2}
$$

where $\gamma_{i,j}$, $\alpha_{i,j}$ and $\beta_{i,j}$ $\left(\gamma_{i,j} > \alpha_{i,j} > \beta_{i,j} \right)$ are the total number of bytes being transmitted from cd_i to sg_j, sg_i to igw_j, and igw_i to $sf g_j$ at time t.

5   Conclusion and Future Work

With the rapid growth of IoT services, service management, QoS, and SLA are becoming a critical issues. Efficient, in time scheduling, resource management, minimum energy consumption and service latency not only allows fog nodes to perform according to the situations, but also helps customer satisfaction. In this paper, we have presented a model to reduce service latency and energy consumption through fog computing. Our future work involves designing an architecture based on efficient resource utilization, service composition and orchestration, and sensor virtualization environment.

Acknowledgements This research was supported by Media Lab Asia (Visvesvaraya Ph.D. Scheme for Electronics and IT, Project Code-CSVSE) under the department of MeitY, Government of India and carried out at Cloud Computing Research Laboratory, Department of CSE, National Institute of Technology Rourkela, India.

References

1. Bonomi, F., Milito, R., Zhu, J., Addepalli, S.: Fog computing and its role in the internet of things. In: Proceedings of the first edition of the MCC workshop on Mobile cloud computing, pp. 13–16, ACM, Helsinki, Finland (2012). https://doi.org/10.1145/2342509.2342513
2. Aazam, M., Huh, E.N.: Fog computing and smart gateway based communication for cloud of things. In: IEEE International Conference on Future Internet of Things and Cloud, pp. 464–470, IEEE, Barcelona (2014). https://doi.org/10.1109/FiCloud.2014.83
3. Gia, T.N., Jiang, M., Rahmani, A.M., Westerlund, T., Liljeberg, P., Tenhunen, H.: Fog computing in health-care internet of things: a case study on ECG feature extraction. In: IEEE International Conference on Computer and Information Technology; Ubiquitous Computing and Communications; Dependable, Autonomic and Secure Computing; Pervasive Intelligence and Computing, pp. 356–363, IEEE, Liverpool (2015). https://doi.org/10.1109/CIT/IUCC/DASC/PICOM.2015.51

4. Aazam, M., Huh, E.N.: Fog computing micro datacenter based dynamic resource estimation and pricing model for IoT. In: IEEE 29th International Conference on Advanced Information Networking and Applications, pp. 687–694 , IEEE, Gwangiu (2015). https://doi.org/10.1109/AINA.2015.254
5. Abdullahi, I., Arif, S., Hassan, S.: Ubiquitous shift with information centric network caching using fog computing. In: Computational Intelligence in Information Systems. Advances in Intelligent Systems and Computing, vol. 331, pp. 327–335 , Springer (2014). https://doi.org/10.1007/978-3-319-13153-5-32
6. Skala, K., Davidovic, D., Afghan, E., Sojat, Z.: Scalable distributed computing hierarchy: cloud, fog and dew computing. Open J. Cloud Comput. (OJCC) **2**(1), 16–24 (2015)
7. Tang, B., Chen, Z., Hefferman, G., Wei, T., He, H., Yang, Q.: A hierarchical distributed fog computing architecture for big data analysis in smart cities. In: Proceedings of the ASE BigData and Social Informatics, , ACM , Kaohsiung, Taiwan (2015). https://doi.org/10.1145/2818869.2818898
8. Rahmani, A., Thanigaivelan, N., Gia, T., Granados, J., Negash, B., Liljeberg, P., Tenhunen, H.: Smart eHealth gateway: bringing intelligence to internet-of-things based ubiquitous healthcare systems. In: 12th Annual IEEE Consumer Communications and Networking Conference (CCNC), pp. 826–834, IEEE, Las Vegas, USA (2015). https://doi.org/10.1109/CCNC.2015.7158084
9. Bonomi, F.: The smart and connected vehicle and the internet of things
10. Bonomi, F., Milito, R., Zhu, J., Addepalli, S.: Fog computing and its role in the internet of things. In: Proceedings of the first edition of the MCC workshop on Mobile cloud computing, pp. 13–16, ACM, Helsinki, Finland (2012). https://doi.org/10.1145/2342509.2342513
11. Bonomi, F., Milito, R., Natarajan, P., Zhu, J.: Fog computing: a platform for internet of things and analytics. In: Bessis N., Dobre C. (eds.) Big Data and Internet of Things: A Roadmap for Smart Environments, vol. 546 , pp. 169–186, Springer (2014)
12. Gazis, V., Leonardi, A., Mathioudakis, K., Sasloglou, K., Kirikas, P., Sudhaakar, R.: Components of fog computing in an industrial internet of things context. In: 12th Annual IEEE International Conference on Sensing, Communication, and Networking - Workshops (SECON Workshops), pp. 1–6, IEEE, Seattle, USA(2015). https://doi.org/10.1109/SECONW.2015.7328144
13. Cisco fog computing solutions: unleash the power of the internet of things http://www.cisco.com/c/dam/en_us/solutions/trends/iot/docs/computing-solutions.pdf
14. Busching, F., Schildt, S., Wolf, L.: DroidCluster: towards smartphone cluster computing- the streets are paved with potential computer clusters. In: 32nd International Conference on Distributed Computing Systems Workshops (ICDCSW), pp. 114-117, IEEE, Macau, China (2012). https://doi.org/10.1109/ICDCSW.2012.59
15. Masri, W., Ridhawi, I.A., Mostafa, N., Pourghomi, P.: Minimizing delay in IoT systems through collaborative fog-to-fog (F2F) communication. In: 9th IEEE International Conference on Ubiquitous and Future Networks (ICUFN), pp. 1005-1010, IEEE, Milan (2017). https://doi.org/10.1109/ICUFN.2017.7993950
16. Kumar, A., Narendra, N.C., Bellur, U.: Uploading and replicating internet of things (IoT) data on distributed cloud storage. In: IEEE 9th International Conference on Cloud Computing (CLOUD), pp. 670-677,IEEE, San Francisco, CA (2016). https://doi.org/10.1109/CLOUD.2016.0094
17. Masip-Bruin, X., Marn-Tordera, E., Tashakor, G., Jukan, A., Ren, G.J.: Foggy clouds and cloudy fogs: a real need for coordinated management of fog-to-cloud computing systems. In: IEEE Wireless Communications, vol. 23, no. 5, pp. 120–128 ,IEEE(2016). https://doi.org/10.1109/MWC.2016.7721750
18. Souza, V.B., Masip-Bruin, X., Marin-Tordera, E., Ramirez, W., Sanchez, S.: Towards distributed service allocation in fog-to-cloud (F2C) scenarios. In: IEEE Global Communications Conference (GLOBECOM), pp. 1–6, IEEE, Washington, DC (2016). https://doi.org/10.1109/GLOCOM.2016.7842341
19. Narendra, N.C., Koorapati, K., Ujja, V.: Towards cloud-based decentralized storage for internet of things data. In: IEEE International Conference on Cloud Computing in Emerging Markets (CCEM), pp. 160–168, IEEE, Bangalore (2015). https://doi.org/10.1109/CCEM.2015.9

20. Malandrino, F., Kirkpatrick, S., Chiasserini, C.F. : How close to the edge? delay/utilization trends in MEC. In: Proceedings of the ACM Workshop on Cloud-Assisted Networking, pp. 37–42, ACM, CA, USA (2016)
21. Hu, P., Ning, H., Qiu, T., Zhang, Y., Luo, X.: Fog computing based face identification and resolution scheme in internet of things. In: IEEE Transactions on Industrial Informatics, vol. 13, no. 4, pp. 1910–1920 , IEEE (2017). https://doi.org/10.1109/TII.2016.2607178
22. Chen, N., Chen, Y., You, Y., Ling, H., Liang, P., Zimmermann, R.: Dynamic urban surveil-lance video stream processing using fog computing. In: 2nd IEEE International Conference on Multimedia Big Data (BigMM), pp. 105–112, IEEE, Taipei (2016). https://doi.org/10.1109/BigMM.2016.53
23. Brogi, A., Stefano, F.: QoS-aware deployment of IoT applications through the Fog. In: IEEE Internet of Things Journal, vol. 4, no. 5, pp. 1185–1192, IEEE (2017). https://doi.org/10.1109/JIOT.2017.2701408
24. Sarkar, S., Misra, S.: Theoretical modelling of fog computing: a green computing paradigm to support IoT applications. In: IET Networks, vol. 5, no. 2, pp. 23–29, IEEE (2016). https://doi.org/10.1049/iet-net.2015.0034
25. Luan, T.H., Gao, L., Li, Z., Xiang, Y., Wei, G., Sun, L.: Fog computing: focusing on mobile users at the edge (2015). In: arXiv preprint arXiv:1502.01815
26. Hong, K., Lillethun, D., Ramachandran, U., Ottenwlder, B., Koldehofe, B.: Mobile fog: a programming model for large-scale applications on the internet of things. In: SIGCOMM workshop on Mobile cloud computing, pp. 15–20, ACM , Hong Kong, China (2013). https://doi.org/10.1145/2491266.2491270

Analysis of Attention Level of Human Body in Different Forms

Ankit Anand, Lala Shakti Swarup Ray, Ramesh K. Sahoo and Srinivas Sethi

Abstract The attention level of human beings is a very complicated property of the brain. It has demanded a lot of research work which is still being continued to understand how it works on different situations as the human brain shows a mixed level of attention. The attention level is generally derived from the alpha and beta waves processed in the human brain. With the help of an EEG device which is alternatively called as a brain–computer interface (BCI), it can retrieve the data from the human brain. The sole purpose of this study is to find out how the attention level of the human being gets affected when it turns the moods from normal to the stressed. How much variation does the two moods show in terms of attention. It can also be taken into consideration in finding out how much difference is there in the attention levels of boys and girls keeping for both the moods of the human beings. It has been studied how does the age group affect the brains' attention level when considered a young and an older group of students. It has been collected data from 30 students each of which were girls and boys and of age groups 20 and 22 years. It has been experimented on the data collected and found that with the normal mood the human showed a better level of concentration and attention level than that of the stressed mood. It has been also noticed that the girls irrespective of their moods showed greater value of attention than the boys. The last results showed that the elder group showed a better level of attention than the younger one concluding that with the increase in age, the attention level gets increased.

Keywords Brain–computer interface · MindWave mobile · Attention · Normal and stressed moods

A. Anand (✉) · L. S. S. Ray · R. K. Sahoo · S. Sethi
Department of CSEA, IGIT, Sarang, India
e-mail: infinitron23@gmail.com

L. S. S. Ray
e-mail: lalashakti96@gmail.com

R. K. Sahoo
e-mail: ramesh0986@gmail.com

S. Sethi
e-mail: srinivas_sethi@igitsarang.ac.in

1   Introduction

In the recent years of studies of computer science and neuroscience, the brain waves have been one of the most argued fields. Each individual possesses brain characteristics of human body. Humans have always focused on communication between machines and devices merely as a function of thoughts and emotions [1]. This is eventually achieved by the brain–computer interface (BCI) devices. The BCI acts as a channel for communication between the brain and the machines [2]. It captures the signal and processes those to determine some properties. The brain waves act as the function of electrical signals generated. These signals are in the form of waves and each wave has its own frequency. The waves are classified into infra-low, alpha, beta, gamma, and theta. The brainwave study helps us to determine the various characteristics of the brain. Attention and meditation are examples of such characteristics.

Attention of a human being as always been defined as the process by which a person concentrates on specific features of an environment regardless of the others. The attention level of human beings has shown varied mixed results. This is basically because of different nature of brains for different human beings. Due to this different human brain nature, it is very difficult to understand the attention as a whole. However, the attention can be broken down into smaller aspects whereby it helps us to recognize some of its basic characteristics. Attention can be easily observed using the EEG-based spectral–spatial pattern [3]. But several researchers have somehow neglected the fact that attention can be a great asset to understand consciousness [4].

With a change in mood, there might be a case that the average attention level of human beings gets deviated from the usual. It could get higher or lower when they are in a stressed mood. This could affect significant changes in the overall interaction of a human being while concentrating on some topic. Another thing that has been seen is the brains of girls and boys differ in a huge amount. This might lead to a variable attention and meditation level between the comparisons of the two. The age group can also play a vital role in determining if there is any variation in the level of attention with respect to its increase or decrease. Thus, it may happen that the humans act differently on the same problem according to their age.

This paper will deal with finding out how the attention level of human beings varies with respect to their moods such as normal and stress, their gender namely male and female, and their age.

2   Objectives and Hypotheses

Researchers around the globe have been consistently working out on finding a similar type of behavior in human brains. The attention property of the brain has given a lot of mixed behaviors as stated above. The attention level may vary from person to person and even from situation to situation. Thus, the attention level is needed to be analyzed as smaller function. The main objective of our study is to analyze the attention level of human, breaking it into some smaller aspects. It will be found out how the attention level of a human being can vary for its relaxed and stressed moods. Moreover, it will also be studied how it varies with their gender and their age.

So the hypotheses of our study will be

H1: The attention level of human varies with the change in their moods.
H2: Different genders, say male and female, show different attention levels for a particular mood.
H3: With the increase in age, the human being will show higher level of concentration or attention.

3   Literature Review

In [5], authors tried to evaluate a BCI and find its stability so that the attention and meditation level could be found optimally. Their research showed that using the well-recognized psychological examination like the Stroop test they were able to find the stability of the NeuroSky headset. Their successful study allowed them to test a new method namely the Tower of Hanoi.

Authors in [6] focused on the issue of two-dimensional cursor control by designing a new BCI and its implementation by combining the brain signals of Mu/Beta rhythm and P300 Potential

Authors in [7] conducted a research by which they investigated the use of EEG signals to enrich the attention recognition in human beings. They applied numerous algorithms and methods to identify a group of features so that they could actively implement those into the attention recognition process. The results showed that the classification rate achieved when the attention is divided into five classes is 51.9%, whereas when it is divided into three classes, the classification rate is 63.9%.

There has been a lot of work done on the brain waves of the human beings. Salle Dhali [1] focused on finding out how the brain waves can be seen and be used as a research setting. He focused on finding out how these waves differ for known and unknown objects and what is the consistency they show with respect to the relaxing and focusing states. His work gave an overall stable outcome for the meditation, but he is not able to achieve a stable outcome for the attention and hence received mixed outputs.

He et al. [8] focused on the problem of driving fatigue. They used the MindWave to collect the attention and meditation and found the relation between them when the subject is in the state of relaxation, fatigue, concentration, and sleep. After the study based on correlation coefficient of drivers' attention and meditation, new driving fatigue detection technique was given by them.

Authors in [9] worked on finding out the levels of attention and mediation for different color combinations. They tested this on a group of students without visual impairment. They found that the yellow background with black color font gave them the best average attention. They also found that the black and the red font gave equal slope but the red helped in more meditation, whereas black gave more attention. Adding to this, they concluded that black background and white font is more effective than white background and black font in terms of attention and meditation.

Chu1 and Chui [10] discussed on finding out how the mobile interaction for mobile games affected the attention and meditation level of players differs on different mobile inputs. They found that the attention level for a player is less on the hard key than on the soft key, whereas the opposite of this is seen in the meditation level. There is less meditation on the soft key, and the hard key showed more meditation level.

4 Methodology

Thirty students, 20 male, and 10 female took part in the successful conduct of the experiment. The students were of age groups 20 and 22 years. Ten boys and 10 girls were of the age group of 20 years, and 10 boys were of the age group of 22 years. The students were put into three groups according to the hypotheses for our study. Each group is then evaluated according to their brain waves for obtaining the result.

The NeuroSky's MindWave Mobile device [11] is used as the channel from which the brain waves were read. The device has Bluetooth enabled into it which acts as a communication channel between the computer and the device. The electrical signals generated in the mind are retrieved as pulses by the MindWave Mobile headset, and the data is transmitted to the computer. The data is then processed onto the computer. The attention level for all the students in two moods, namely normal and stressed, was recorded. The data consisted of the attention level for each student when he or she is in normal mood and when he or she is in stressed mood. The data is then used to process according to the aim of our study.

For the first hypothesis, the data of the students of 20 years is taken and the mean data for boys and girls is calculated for each mood. This is to get a general idea about how the human brain behaves with two moods, normal and stressed. The attention level on both the moods for the boys and girls is averaged, and the variation is recorded for each mood. The average result is then plotted against the two moods to see how the attention varies with the change in the moods of a human being.

For the second hypothesis keeping the age group same (20 years), the data of boys and girls is taken for evaluation. The average attention for boys and girls is calculated based upon the data collected from the first experiment. The data is so averaged on both the moods to check if there is any difference between the girls' attention and the boys' attention. The mean attention level of boys and girls for the two moods is plotted against each other.

The third hypothesis deals with the fact that the attention seen in human beings generally increases with the age. To test this hypothesis, 20 boy students with 10 as age groups of 20 years and 10 as age groups of 22 years were taken. This could have given us the knowledge about how well the attention level differs with respect to the increase or decrease in the age groups. The total attention level of both the

Fig. 1 Flow of process

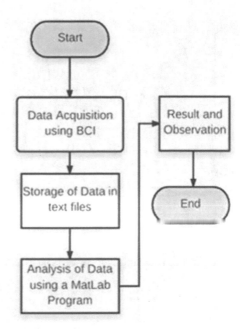

age groups is calculated as the mean of individual student, and the data collected is mapped in a graph to see the variation.

Our study however dealt with two moods in every aspect of the hypotheses whether it would be differentiating the attention levels in those two, the attention variation in boys and girls, or it be the variation of attention with respect of the age groups. The two moods selected for our study are the basic moods which could and will be able to optimally give the result on the above three hypotheses.

The whole process can be represented using the flow diagram as per Fig. 1.

5   Result and Discussion

The data generated by the NeuroSky device has been collected in a.txt (text) file, and the mean for each group has been found. After performing the experiments for the above-mentioned problems, the following result has been found and has been discussed.

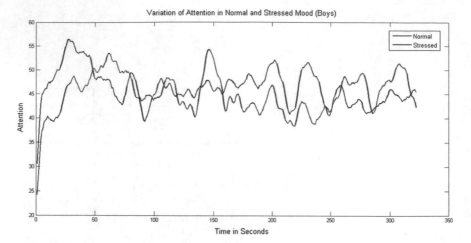

Fig. 2 Attention in normal and stressed moods of boys

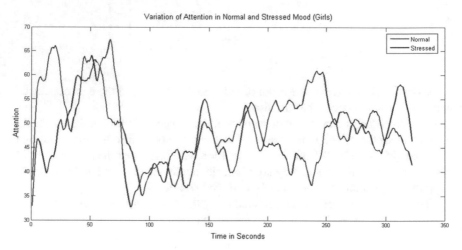

Fig. 3 Attention in normal and stressed moods (girls)

5.1 Hypothesis 1

The collected information of the attention in normal and stressed moods has been plotted for both the boys and girls. Figures 2 and 3 show the variation in boys and girls, respectively.

The red line in both the figures is the level of attention in the normal mood, whereas the blue line indicates the level of attention in the stressed mood. From the figure, it can be seen that in both girls and boys the level of attention becomes higher in the normal mood than in the stressed mood. In the stressed mood, the mind works upon several instructions at a particular instance. This causes the brain to less focus

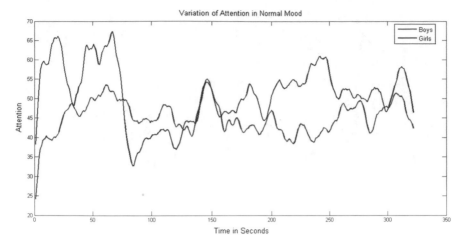

Fig. 4 Attention in normal mood of boys and girls

on a specific task which cannot be seen in the normal mood where the mind does not have to worry about several things. In the normal mood, the brain can act more efficiently and can have greater amount of focus on a particular task, thus helping the concentration level to be increased in the normal mood.

5.2 Hypothesis 2

In the study, it was seen that a human being can achieve greater amount of focus in the normal mood as compared to the stressed mood. The human brain for both the girls and boys differ, and hence, it may happen that the level of attention varies with the difference of gender. When this statement has been tested and the level of attention for boys and girls has been plotted as in Figs. 4 and 5.

The red line indicates the value of attention for the boys and the blue line for the girls. From the figures, it can be noticed that for both the moods whether the stressed or the normal, the gender shows significant changes in the amount of attention one can give for a particular task. The brains of girls and boys differ in a significant way. Moreover, it can also be noticed that the girls have a higher power of attention than the boys. The girls can possess a greater amount of concentration than the boys for the same tasks.

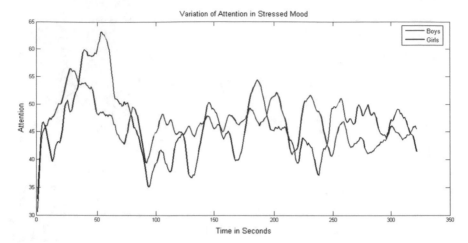

Fig. 5 Attention in stressed mood of boys and girls

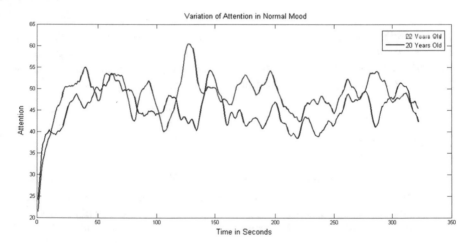

Fig. 6 Attention in normal mood between different age groups

5.3 *Hypothesis 3*

It is now known that attention has different characteristics with respect to moods and gender. Taking our study to a further extend we collected the data for different age groups in relaxed and stressed moods and plotted as in Figs. 6 and 7.

In the above two figures, the red line represents the attention level for 22 years old and the blue line indicated the level of attention for the 20 years old. It can be noticed that the 22 years old shows more level of attention than the 20 years old. The older we get the attention power of human being gets increased. We can say that with the increase in age, the level of maturity increases and our brain is able to recognize

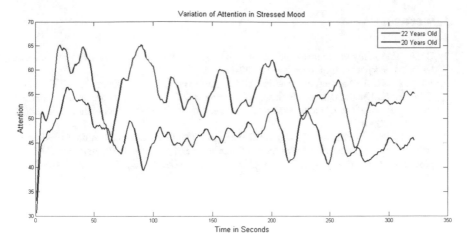

Fig. 7 Attention in stressed mood between different age groups

and solve more complex problems. Thus, the older we get the level of attention of the human beings increases.

6 Conclusion

In this paper, it has been tried to find how the brain shows the attention level in normal and stressed moods. It has been found out that in human beings the attention level is more in the normal mood than the stressed mood. This is because the mind is free in the normal mood than in stressed mood and hence can concentrate more efficiently on a particular topic. It has been also observed that the brains of girls and boys differ in a great amount; thus, their level of attention differs according to their gender. It has been found that the girls have a greater power of attention over boys in both the normal and the stressed moods. Moreover, it has been also focused on finding out the difference of attention levels with the age groups. It has been observed that with the level of maturity the brain enhances its level of attention. It has been noticed that the elder students were able to show greater level of attention than the younger ones. And hence, it is concluded that the elder gets the better attention level.

Acknowledgements All procedures performed in studies involving human participants were in accordance with the ethical standards of the institutional and/or national research committee and with the 1964 Helsinki declaration and its later amendments or comparable ethical standards. Informed consent was obtained from all individual participants included in the study.

References

1. Dhali, S.: A study of brainwave eSensing activity. 2015. Pdf - https://www.overleaf.com/artic les/bci/mcsvkjwhcffb.pdf. (Retrieved on 22-11-2017)
2. Wolpaw, J.R., Birbaumer, N., Heetderks, W.J., McFarland, D.J., Peckham, P.H., Schalk, G., Donchin, E., Quatrano, L.A., Robinson, C.J., Vaughan, T.M., et al.: Brain-computer interface technology: a review of the first international meeting. IEEE Trans. Rehabilitation Eng. 8(2), 164–173 (2000)
3. Hamadicharef, B., Zhang, H., Guan, C., Wang, C., Phua, K.S., et al.: Learning EEG-based spectral-spatial patterns for attention level measurement. In: 2009 IEEE International Symposium on Circuits and Systems (ISCAS2009), May 2009, Taipei, Taiwan. pp. 1465–1468, 2009
4. Taylor, J.G.: Paying attention to consciousness. Opin. Trends in Cognitive Sci. 6(5), (2002)
5. Crowley, K., Sliney, A., Pitt, I., Murphy, D.: Evaluating a brain-computer interface to categorise human emotional response. In: 10th IEEE International Conference on Advanced Learning Technologies 2010
6. Li, Y., Long, J., Yu, T., Yu, Z., Wang, C., Zhang, H., Guan, C.: An EEG-based BCI system for 2-D cursor control by combining mu/beta rhythm and P300 potential. IEEE Trans. Biomed Eng 57(10), (2010)
7. Xiaowei, L., Bin, H., Qunxi, D., Campbell, W., Moore, P., Hong, P.L EEG-based attention recognition. In: 6th International Conference on Pervasive Computing and Applications (ICPCA) (2011, 26-28 Oct. 2011)
8. He, J., Liu, D., Wan, Z., Hu, C.: A non-invasive real-time driving fatigue detection technology based on left prefrontal attention and meditation EEG. In: International Conference on the Multisensor Fusion and Information Integration for Intelligent Systems (MFI), (2014, 28-29 Sept. 2014)
9. Aydogan, H., Audogan, S.D.: EEG attention and meditation responses of students on different presentation slide colors. J. Educational and Instructional Studies in the World 6(1), Article-16. 2016
10. Chu1, K., Wong, C.Y.: Players attention and meditation level of input devices on mobile gaming. In: 3rd International Conference on User Science and Engineering (i-USEr) 2014
11. The introduction to MindWave Mobile device. https://store.neurosky.com/pages/mindwave. (Retrieved on 9-11-2017)

Analytical Survey on Standards of Internet of Things Framework and Platforms

Sumanta Kuila, Namrata Dhanda, Subhankar Joardar and Sarmistha Neogy

Abstract The Internet of things platforms, which are also sometimes called IoT middleware, are the software that activates different hardware components and/or devices of the Internet of things. It is the software which makes the bridge between data and device of the IoT platform. The goal of this research is that we study different standard IoT platforms and analyze the efficiency and suitability of these platforms. IoT platforms generate large amount of data that require acquisition, organization, transformation, processing, and analysis. This puts emphasis on cloud-based IoT platform with huge user base that makes the platform robust and applicable in our home, automobiles, smart city, smartphone, manufacturing equipments, etc. The goal of this paper is to study and analyze several available platforms and frameworks which are used to design and develop different IoT applications.

Keywords Internet of things (IoT) · Platform · Framework · Application programming interfaces (APIs)

S. Kuila (✉) · S. Joardar
Haldia Institute of Technology, Haldia, India
e-mail: sumanta.kuila@gmail.com

S. Joardar
e-mail: subhankarranchi@yahoo.co.in

N. Dhanda
Amity University, Noida, Uttar Pradesh, India
e-mail: ndhanda@lko.amity.edu

S. Neogy
Jadavpur University, Jadavpur, India
e-mail: sarmisthaneogy@gmail.com

© Springer Nature Singapore Pte Ltd. 2019   33
A. Abraham et al. (eds.), *Emerging Technologies in Data Mining and Information Security*, Advances in Intelligent Systems and Computing 814,
https://doi.org/10.1007/978-981-13-1501-5_4

1 Introduction

The Internet of things applications and its related development have boosted the Web world for more than 10 years. A recent survey made by Ericsson and Cisco reveals that more than 50 billion Internet-connected devices will be operational globally by 2020 [1]. So the device and connectivity is huge and obviously the raw data passed through IoT is also huge. So we need various types of IoT platforms to properly and efficiently deal with the data. In industrial landscape today Internet of things (IoT) has a major attraction and it is the most projected emerging technologies. IoT is in the "peak of inflated expectations," as revealed in the study of Gartner's Hype Cycle for Internet of Things, 2015 [2]. Various IoT platforms are in development phase also. The expansion of IoT platforms is determined by the requirement to make possible machine-to-machine (M2M) connectivity mapping which is budding at extraordinary rate. Today, the prediction is that M2M connectivity will rise up to 12 billion in 2020 compared to 2 billion in 2012. Cisco estimates that the value of the IoT market will be \$19 trillion by 2020 [3]. It is also revealed by the same that only 0.6% of the physical objects are now linked. The Internet of things describes the set of devices with hardware and middleware softwares that be linked real-world sensors and actuators to the Internet. This is a continuous list of request including wearable devices, Internet connection of cars, fitness and health-monitoring devices, smart objects, home automation systems, wireless sensor networks for monitoring weather, smartphones, smart meters, lighting controls, etc. [4]. In an IoT platform, an operating system is targeted at sensor technologies and an end-to-end system that joins sensors to the cloud and provides for data analytics [5]. In this paper, we deal with the issues by considering IoT reference architecture, which is based on several recently used IoT platforms. The orientation architecture is a conceptual idea to ensure extensive applicability. Many open architectures present a detailed and concert idea on IoT platforms. The additional details of each reference architecture adopt more new ideas as they use and analyze the characteristics of different IoT platforms individually [6] (Fig. 1).

2 Internet of Things Platform Components

The platforms of IoT are the middle section in the Internet of things which join the virtual and the real worlds and permit message between objects. About 40% of the whole activity which can not be done in some specific IoT platforms requires another platform to work. So, we can say that an IoT platform's activity is that it works as a connection between different Web objects. Device information is essential to monitor, run, and analyze IoT devices. Library API information is needed to be set up for each devices for proper integration among various types of sensors and implementation platforms. Advanced devices provide application programming interface (API) that allows for a uniform message interface to the platform.

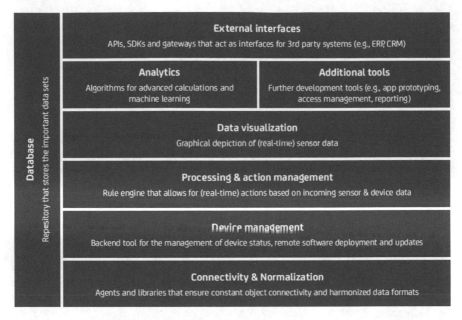

Fig. 1 Eight major building blocks of IoT platform [35]

2.1 Connectivity

Connectivity layer brings different data formats and protocols together to create a software interface. It confirms that all devices interact with each other correctly and read data correctly. Software agents are responsible for establishing stable and secure connection. Different software libraries activate individual hardware devices so that it can, in turn, activate appropriate sensor that sends analogue signals to other IoT components. Machine-to-Machine (M2M) paradigm is an important part as it ensures end-to-end communication which is executed automatically or on remote request. Networks that enable end-to-end communication act as core network backhaul which enables connectivity and related properties of session management, security, mobility management, etc. [7].

2.2 Device Management

The goal of device management is that it ensures connected objects are producing proper output and are updated from time to time. Remote configuration, device provisioning, troubleshooting, firmware/software updates are the tasks in this layer. The responsibility of device management is that it provides coordination among millions of various devices to become part of IoT facilitate solutions. Immense activity and mechanism are required to control expenditure and decrease manual effort [8].

2.3 Database

The collected data are stored in IoT platform which uses the concept of cloud-based database solution. It uses the concept of both big data and structured (SQL) and unstructured data (NoSQL) data.

2.4 Processing and Action Management

The data which are used in the connectivity and advanced module is activated in this module. Some protocol-based event action trigger is there that acts on specific sensor data for particular task. These features support various activities in smart home, smart city, healthcare sector, to name a few.

2.5 Analytics

The collected data from IoT go for complex data analysis which provides necessary action management to specific use cases. The analysis engine uses algorithms provided by IoT platform to learn about the characteristics of the specific events of that operation.

2.6 Data Visualization

It is sometimes treated as visual analytics, and different rule-based engines are used. Data visualization is mostly underrated as human brain and eye are much more superior than automated sensor-based visualization. Basically, the task of data visualization is to analyze the collected data pattern, its 2D, 3D models, different types of visualization diagrams, and mathematical models, etc. Here, the input from human visualization is important because it may work in parallel with the analytic and rule-based engine for IoT data visualization [9].

2.7 Extra Tools

IoT platform often requires different set of tools for its advanced setup to deliver the accurate IoT solution. Different advanced tools have different activities. For example, an access organization tool establishes the users who access which specific data at what time [10].

2.8 External Interfaces

IoT enabled system has business value which has huge impact on the IT industries with greater business outcome in the ERP system. IoT is an integral part of distributed system and by using agreed protocol any article can communicate with each other. All activities are performed by methods and technologies, like near-field communication (NFC), radio-frequency identification (RFID), infrared (IR) sensors. The IoT paradigm here uses three main goals: semantic-oriented (knowledge), things-oriented (sensors), and Internet-oriented (middleware). In 2013, machine-to-machine (M2 M) connection reaches 195 million worldwide and it maintains a growth rate of nearly 40% between 2010 and 2013 [11].

3 Frameworks for Internet of Things

3.1 Arrowhead

This is a light framework whose aim is to apply Internet of things for industrial purpose. Its goal is to improve interaction between different applications. Service-oriented architecture (SOA) is implemented here. The Arrowhead Framework holds general solutions for activities in the areas of Information communications, Information Assurance and systems management, and application services for process automation. Developer needs to know the process of deployment and execution of the overall system; i.e. the way systems, services, system of services are described and defined [12]. Configuration also contains documentation templates, design patterns, and guidelines that aim at helping systems, newly developed or legacy, to conform to Arrowhead Framework specifications. For high level of interoperability, this framework targets a common and unified approach for all of its users [13].

3.2   Lightweight Machine-to-Machine

3.2.1   Domoticz [14]

Domoticz is the home automation system which has pretty wide library of supported devices, ranging from smoke detectors to remote controls to weather forecasting, with a big number of additional third-party approach recognized on the project's Web site. It is available from both most modern smartphones as well as desktop browsers, and it is a lightweight application. Raspberry Pi is the low-power device and is the application of this lightweight machine. C/C++ is the basic language for Domoticz under GPLv3, and its source code can be browsed on GitHub.

3.2.2   Home Assistant [15]

Another open-source IoT platform is Home Assistant. It is designed to be easily deployed on systems which can run NSA device, Python 3, Raspberry Pi in addition to a Ducker container to make deploying on other systems. It adds with a number of open source as well as it makes huge industrial application also.

3.2.3   OpenMotics [16]

The advanced home automation system, OpenMotics, works with both software and hardware under open-source licenses, designed at providing a total system for controlling devices rather than working jointly many devices from different providers. OpenMotics focuses on a hardwired solution. GPLv2 provides the license for Open-Motics source code and is available in GitHub. Different home automation enthusiasts go with different solutions, or even decide to roll their own. Some other potential options to collaborate with OpenMotics include Smarthomatic, LinuxMCE, Mister-House, PiDemo, etc.

3.2.4   Calaos [17]

Another home automation platform Calaos defines a Web application, smartphone applications, server application, touchscreen interface, native mobile applications for Apple iOS and Android, and an advanced Linux operating system to run the application. Version 3 of the GPL gives the license of Calaos. The code is visible in the source of GitHub.

3.2.5   OpenHAB [18]

Open Home Automation Bus (OpenHAB) is one of the best familiar home automation systems among open-source availabilities, with huge market requirements, and its products capture a big section of the users. Java is the basic software for OpenHAB, and it is portable across most major operating systems and even runs on Raspberry Pi. OpenHAB is designed to be device agnostic while making it easier for developers to add their own devices or plugins to the system, and it also supports additional hundreds of similar devices. OpenHAB also ships Android apps and Apple's iOS for advanced device control, as well as relevant design tools so that user may create own user interfaces (UIs) for their own home system. OpenHub source code is available on GitHub under the Eclipse Public License.

3.2.6   AllJoyn Framework [19]

It is an open-source API framework designed for enabling interoperability for industrial lighting and house automation. Distributed software bus is implemented by AllJoyn framework, and the bus affords the "medium" which enables AllJoyn applications, which is the firmware on mobile device applications, conventional applications on PCs/servers, and it provides APIs for different microcontroller-based devices. Service and client are the two parts of an application, and a peer-to-peer system connection is there to communicate messages that map directly high-level programming languages' APIs. Two types of nodes are there in AllJoyn: bus architecture applications (App) and AllJoyn Router (R). App is only connected to R, App can surround R, R is connected to another R, and thus, AllJoyn framework works like mesh of star network.

3.2.7   IoTivity [20]

This framework is developed by open-source consortium whose initial target is implementation in smart homes and other advanced IoT applications. Constrained Application Protocol CoAP is the key component, and its basic building blocks are resource introspection (RI) layer and connectivity abstraction (CA) layer. Companies like Cisco, GE, Intel, MediaTek, and Samsung launched the Open Interconnect Consortium (OIC) to implement standards for connectivity. Interoperability of billions of IoT devices is ensured by OIC. IoTivity is an open-source software framework which implements OIC standards. It confirms flawless device-to-device connectivity to fulfill the rising needs of the IoT. IoTivity discovery is the activity of detection of resources and devices in remote and local areas, and its data control maintains data control and swap based on streaming and messaging model.

3.2.8   IPSO Alliance [21]

This is a forum that works worldwide and focuses on activating IoT devices to understand, communicate, and trust each other for their activity based on open standards. Some advanced radio manufacturing company implemented IEEE 802.15.4, where the wireless connection for low-power personal area networks (LoWPANs) 802.15.4 is extensively used in different embedded applications, like structural observation to bridge integrity and track building, weather monitoring to provide agricultural inputs, industrial control for low-cost productivity. These applications usually need low-cost nodes and cover multiple hops to communicate in a wide geographical region, mostly using modest batteries and unattended for years.

4   Platforms for Internet of Things

4.1   Platforms Connecting Devices

4.1.1   IzoT [22]

It is the application of a communication stack to create peer-to-peer connection. It works on different protocol services of higher level, which works on top of user datagram protocol (UDP). The seamlessly integrated wireless and wired solution decreases installation price where media is optimized for the application. Fully integrated BACnet/IP and LonTalk applications work here and reduce installation cost by integrating existing BACnet and LON devices. This platform supports lighting control, and the cost of integrating building and lighting controls is reduced using huge installed support of LON building controls. IzoT SDK, U60 module, Premium Edition, U60 DIN are the Echelon products used for building controllers and IzoT SDK Premium Edition, U60 module, U60 DIN, IzoT Router work with open systems integration. To build custom network and custom router, smart communication tool is developed using standard ARM/Linux processors. Today, the popularly used IoT platforms BeagleBone Black and Raspberry Pi toolkits use this technology.

4.1.2   Thingsquare [23]

It is a Swedish company founded in 2012, and the leading technology used by Thingsquare is an operating system named Contiki, suitable for memory-constrained, low-power connected machines. Contiki is a Thingsquare Mist, and under the BSD license, it affords training and services for open-source products like Fedora Linux. It gives emphasis to the use of smartphone apps as a physical product for connecting people. IAR Embedded Workbench is used for developing its Thingsquare Mist, and this connectivity platform supports building and home automation. Globally most broadly used tool chain is IAR Embedded Workbench that uses highly optimizing c/c++ compiler and it has independent integrated development environment (IDE) for building, developing, and debugging embedded applications. The IDE is "Thingsquare Code" whose online built applications are popular.

4.2 Cloud-Based Platforms

4.2.1 Cumulocity [24]

Cumulocity is a German-based company which has developed software on their own for IoT platform. It is acquired by Software AG on March 2017. It provides full of IoT platform which has inventory base that holds all master information. The devices and other assets maintain measurements, events, alarms, audit logs, and other database-related operations to maintain the repository. For device communication and application, Cumulocity system uses the same application programming interfaces (APIs), implying that everyone can access APIs with the same functions equally on devices and system applications. It offers an IoT platform for private cloud, public cloud, and on-premises operations. It affords functionality in main platform areas and checks possible integration with third-party systems if all requirements are matched with the built-in features [25].

4.2.2 Xively [26]

It is an advanced provider of end-to-end solutions that help to speed up creating of associated result. A central message bus is present in this platform that routes messages among devices with different network protocols. LogMeIn Gravity infrastructure is the base for Xively platform, and it uses worldwide brick and mortar data centers of LogMeIn. Study reveals that this platform connects 55 m + users and 255 + m devices in 240 countries. Rescue and Boldchat are the enterprise solutions of LogMeIn that give more support and safeguard for IoT solutions. Xively cloud services use platform-as-a-service (PaaS) and applications for IoT business setup, supervise and securing IoT devices. Customers can navigate technical and business problems using Xively's business and systems' incorporation facilities. The Xively predicts IoT ideas. It deals with pilot IoT projects to recognize and resolve issues. It also works for accelerating product development, accelerating time to market, and increasing business value [27].

4.2.3 ThingWorx [28]

Using model-driven implementation, this platform integrates applications, services, and sensors. Collected data sources are compressed and interconnected through virtual bus. ThingWorx supports cloud and any hybrid development. It operates flaw-lessly on-premise and in private or public clouds. Due to open-platform strategy, it gets the facility to use full force of device connectivity and existing cloud. Millions of connected devices are scaled, and it operates huge number (hundreds of thou-sands) transactions/second. The total connectivity includes file transfer, application tunneling to a device, and bi-directional messaging. The process, application, or solution is represented by "thing" model. The context into IoT data is provided by "things," and for application developers, it also operates the basic building blocks. It has relationship and structure to represent real global objects.

4.3   Commercially Leading IoT Platforms

IBM Watson [29] IoT platform uses the power of cognitive analytics engine in the IoT. It builds integrated specialized solutions to solve different business challenges and is secure, easy-to-use, and scalable. Using cognitive APIs tools and services analyze, manage and connect secure IoT data and devices.

Microsoft Azure [30] IoT platform facilitates connectivity to both new and exist-ing devices and provides new approach by harnessing power to capture and analyze untapped data for business transformation. For common IoT scenarios using precon-figured solutions, it can perform quickly connecting millions of devices and integrate the business system. Using preconfigured remote monitoring solution in real time, it can collect and analyze device data, which triggers automatic actions and alerts for specific business requirements.

Intel provides a wide spectrum of cross-ecosystem and interoperable Internet of things [31] solution which can build, optimize, and deliver trusted data. The workflow of Intel IoT follows connecting devices and accumulating data, raw data verification, storing data into insight, visualizing data, and end-to-end security.

Samsung ARTIK [32] is the IoT platform that implements an open data exchange environment that provides interoperable, secure, and intelligent IoT services and products. The key capabilities of this platform include quick deployment of gate-ways and edge devices, fast onboard devices to the cloud, easily activate devices to communicate with cloud devices or other services, protection to every app, device and give priority to secure data privacy.

HITACHI Lumada [33] is an open, secure, adaptable, verified IoT platform for implementing digital transformation. Operational technology (OT) and information technology (IT) worlds are connected by Lumada IoT platform, and it supplies anal-ysis tools and data which are necessary to implement different enterprise solutions. The open core platform Lumada creates a valuable legacy of innovation, which estab-lishes Hitachi to be a leading provider of "things" that are the base foundation of IoT.

Oracle Internet of things (IoT) [34] uses the concept of platform-as-a-service (PaaS) cloud-based service for business implementations. Connect, analyze, and inte-grate are the three pillars of its operation. Reliable and secure connection between the devices makes data security high which reflects the success of its business. Analyze performs operation with real-time big data from different domains. The role of integrate is that it ensures rich data is available for exact application at the exact time which implements Oracle's software-as-a-service (SaaS) and also platform-as-a-service (PaaS).

5   Conclusion

The goal of the survey is that to study some popular, commercially used Internet of things frameworks and platforms and their characteristics and overall acceptability In the commercial domain. Cloud based implementations are also discussed here. Users can select the framework and platform appropriate for their projects. A large variety of IoT-based applications are in the market which require proper presentation in terms of framework and platforms. The present work tries to portray a few of them so that all stakeholders of the IoT-based system are aware of the availability and adopt and/or adapt suitably.

References

1. Derhamy, H., Eliasson, J., Delsing, J, Priller, P.: A survey of commercial frameworks for the internet of things, IEEE Xplore, September 2015
2. Zdravkovi, M., Trajanovi, M., Sarraipa, R.: Survey of Internet-of-Things platform,https://hal. archives-ouvertes.fr/hal-01298141, Submitted on 5 Apr 2016
3. Nair, S.: Why is "internet of everything" a future growth driver for Cisco? http://finance.yaho o.com/news/why-interneteverything-future-growth-130020843.html
4. Acharya, C.: IOT platforms: a brief study. Int. J. Adv. Res. Comput. Sci. Technol. 4(2), 10–13 (2016)
5. Anderson, M.: Technical trade-offs of IoT platforms http://ThePTRGroup.com, mike@theptrgroup.com
6. Guth, J., Breitenbücher, U., Falkenthal, M.: Comparison of IoT platform architectures: a field study based on a reference architecture. Institute of Architecture of Application Systems, University of Stuttgart, Germany
7. Katole, B., Sivapala, M., Suresh, V.: Principle elements and framework of internet of things. Int. J. Eng. Sci. 3(5), 24–29 (2013)
8. Device Management in Internet of Things, Wind, an Intel company http://events.windriver.co m/wrcd01/wrcm/2016/08/WP-HDC-Device-Management.pdf
9. Gubbi, J., Buyya, R.: A Framework for IoT sensor data analytics and visualisation in cloud computing environments, http://www.cloudbus.org/students/Krishnakumar-IoT-Project2011.pdf
10. Tools for IoTSystem Developers, Delphin Systems, July 2016. http://delphiansystems.com/w p-content/uploads/2016/06/SR-Tools-for-Developers-071116.pdf
11. Chan, C.Y.: Internet of things business models. J. Service Sci. Manage. 552–568 (2015)
12. Delsing, J.: IoT automation value creation enabled by the Arrowhead Framework, www.arro whead.edu
13. http://www.arrowhead.eu/about/arrowhead-common-technology/arrowhead-framework/
14. http://www.domoticz.com/

15. https://home-assistant.io/
16. https://www.openmotics.com/solutions/home-automation/
17. https://calaos.fr/en/
18. https://www.openhab.org/
19. Spencer, B.: AllJoyn framework system overview brian, August, 2014, http://docs.huihoo.co m/alljoyn/AllJoyn-Framework-System-Overview.pdf
20. Subash, A.: IoTivity—connecting things in IoT, tizen developer summit, July 30–31, 2015, Bangaluru
21. http://www.ipso-alliance.org/wp-content/media/6lowpan.pdf
22. Buckland, M.: IZOT^{TM} platform overview May 2015, Echelon Corporation
23. http://mb.cision.com/Main/386/9311406/49601.pdf
24. Kalsi, J.: IoT service platform architecture, technology, communication and transport degree programme in information technology, Master's thesis June 2017
25. https://cumulocity.com/pdf/MachNation_Cumulocity_2016_AEP_ScoreCard_Snapshot.pdf
26. https://d15n4q3o4x3svq.cloudfront.net/assets/Xively-Overview-0e657c417cc50132d08339e b34f1e4ee.pdf
27. Xively Sets the bar for IoT scale and performance; Boston, Dec. 07, 2016, News Release, LogMeIn
28. https://www.liveworx.com/downloads/exclusive-content/The-ThingWorx-Platform-101.pdf
29. http://www.redbooks.ibm.com/redbooks/pdfs/sg248387.pdf
30. https://www.glasspaper.no/Documents/AzureIoTSuite.pdf
31. https://www.intel.com/content/dam/www/public/us/en/documents/solution-briefs/iot-platfor m-solution-brief.pdf
32. https://static.artik.io/files/Samsung_ARTIK_Overview.pdf
33. https://www.hitachivantara.com/en-us/pdf/solution-profile/hitachi-solution-profile-lumada-io t-core-platform.pdf
34. http://www.oracle.com/us/solutions/internetofthings/iot-cloud-service-ds-3209769.pdf
35. IoT Platforms, The central backbone for the Internet of Things. White paper, November 2015
36. Introduction to big data: infrastructure and networking considerations leveragingn hadoop-based big data architectures for a scalable, high-performance analytics platform, White Paper, Juniper Networks

GIS-Based Model for Optimal Collection and Transportation System for Solid Waste in Allahabad City

Shailendra Chaudhary, Chaitanya Nidhi and Nek Ram Rawal

Abstract Solid waste management (SWM) is one of the major problems faced by many countries in the world. The functional element of solid waste collection process includes not only the gathering of waste from the source, but also the transportation of these waste to common collection sites. It has been estimated that about 60–80% money spent for only the collection, and transportation of solid waste to the total amount is spent on the SWM system. An effective municipal solid waste management system is composed of optimized routes, systematic collection, transport, and transfer of wastes. The geographical information system (GIS) can be effectively applied for cost minimization and maximization of waste collection and transportation efficiencies in any area. The route optimization depends on many factors, including the location of waste bins, collection details, types of vehicle, travel impedances, and integrity of road network. In this study, to design better travel routes in city for enhance efficiency by GIS to develop model for collection and transportation of waste using Esri's ArcGIS 3.2, Network Analyst software. The ArcGIS is used to determine optimal zones and optimal routes for each zone. In addition, a simple optimal routing model is proposed to achieve the minimum transportation cost, distance, time, and number of vehicles. The developed model may be utilized by MSW management agencies for optimized waste collection and transportation activities.

S. Chaudhary (✉) · N. R. Rawal
Motilal Nehru National Institute of Technology, Allahabad 211004, India
e-mail: rce1504@mnnit.ac.in

N. R. Rawal
e-mail: nrrawal@mnnit.ac.in

C. Nidhi
Indian Institute of Technology, BHU, Varanasi 221005, India
e-mail: chaitanya.nidhi@gmail.com

© Springer Nature Singapore Pte Ltd. 2019   45
A. Abraham et al. (eds.), *Emerging Technologies in Data Mining and Information Security*, Advances in Intelligent Systems and Computing 814,
https://doi.org/10.1007/978-981-13-1501-5_5

1   Introduction

Solid waste management (SWM) is one of the major problems faced by many countries in the world. Huge quantities of solid waste are generated due to exponential increasing of population and development. After industrialization, the different types of waste such as solid, semisolid, liquid, and gases. All cities produce tonnes of waste everyday from the industries, household, hospitals, agricultural fields, etc. The management of this enormous amount of waste in terms of collection, handling, and disposal with conventional methods has become challenges for the engineers [1]. The simple solid waste management (SWM) consists of generation, collection, transfer, treatment, and final disposal. Municipal solid waste collection (MSWC) is the first step of the process of SWM. The functional element of solid waste collection process includes not only the gathering of waste from the source, but also the transportation of these waste to common collection sites. It has been estimated that about 60–80% money spent for only the collection, and transportation of solid waste to the total amount is spent on the SWM system [2]. Therefore, waste collection and transportation are an important issue in order to make efficient and economical of SWM system.

The geographical information system (GIS) technology has been used worldwide as an optimization tool. Now, this GIS technology used as optimization of collection and transportation process in SWM in order to reduced the cost of SWM. Several researchers [3–9] applied GIS for optimization of MSW collection routes. To optimize of vehicle route in Allahabad City, the vehicle routing problems (VRP) solver which is a component of Esri's ArcGIS 3.2a Network Analyst software extension introduced in version 9.3 used in this study. The VRP provides minimum operating cost by using optimum route for effective MSW management. The methodology proposed in this study uses optimization-based techniques for comprehensive management of MSW in Allahabad City. The aim of this study is to develop a model for optimal zone for collection of solid waste and the optimum route for collection and transportation of solid waste in SWM based on GIS technology.

2   Background and Context of Study Area

Allahabad, one of the ancient and holy cities of India, situated at the confluence of Ganga, Yamuna and Sarasvati rivers. It is located at $25°25'N$ latitude and $81°58'E$ longitude at the height of 98.0 m above the mean sea level. Allahabad stands at a strategic point both geographically and culturally. The Municipal Corporation of Allahabad (AMC) governs the proper solid waste management in Allahabad City. The entire operation of solid waste management system is performed under four heads, namely cleaning, collection, transportation, and disposal. The city of Allahabad encompasses an area of 70.05 km^2 approximately. The entire city is divided into 97 municipal wards within 20 sanitary wards. About 502.83 MT of solid waste is

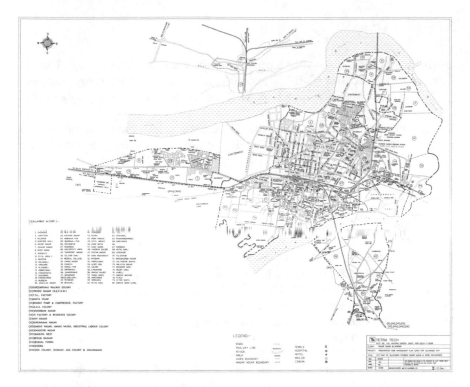

Fig. 1 Wardwise location map of Allahabad

generated every day and solid waste generate at the rate of about 440 gm/day/person. In order to demonstrate the usefulness of the optimization models in real field situation, data for the performance evaluation are taken according to the situation in Allahabad City shown in Fig. 1.

3 Data Collection and Sources

The required datasets such as location of wards, population, waste generation, details of waste collection bins, etc., are collected by conducting field visits of study area. Also, the governing agency Allahabad Development Authority (ADA) provided useful information regarding route and vehicles capacity, and geography of the area.

3.1 Bins Data

In Allahabad City, there are two types of dustbins used for MSW storage. They are
Compacter Placer Bins (CPB): These bins can be placed on the compacter placer
vehicles, and they are not transported to the dumping sites. The capacity of these
bins is 0.3 MT.
Dumper Placer Bins (DPB): This type of bins are generally placed on the truck and
carried to the disposal site for emptying it. The capacity of these bins is 0.3 MT.

3.2 Location of Landfill Sites

The three open landfill sites used in the city for disposal of waste. The location of
these landfill sites are as follows:
Kareli: It is located near Yamuna River. The bounding coordinates of this landfill
site are 81.81 N and 25.41 E. This landfill is largest in size and covers approximately
25-hectare area. The disposal site exists in ward no. 36.
Alopibagh: This disposal site has an area of 5 ha. It is located behind Alopibagh
Temple near the Ganga River. The disposal site exists in ward no. 53. The landfill
site is located on 81.87 N and 25.44 E coordinates.
Phaphamau: This landfill has an area of 3 hectare and is located near Lucknow
road. The disposal site exists in ward no. 21. The geographical location of this site
is 81.86 N and 25.52 E.

3.3 Location of Truck Depot

For this study, it is assumed that there are three truck depots (yards) in the city and
they lies in the following wards:

1. The truck depot is located in ward no. 33 (Chakia). The coordinates of this depot
   are 81.805379 and 25.437434.
2. The truck depot is located in ward no. 24 (Madhvapur). The coordinates of this
   depot are 81.865194 and 25.442276.
3. The truck depot is located in ward no. 27 (Myorabad). The coordinates of this
   depot are 81.85766 and 25.47912.

Table 1 GIS database

Theme	Fields	Descriptions
Road network	ID Name Length Time	Road id Name of the road Length of the road Time to pass the road
Landfill3	Order_Name2 Name Address	Doctor's name Name of hospital Address of hospital
Compacter placer	Order_Name1 Address Service time Demand	Name of the bin Address of the bin Time elapsed at the bin location for collecting waste Capacity of bins
Dumper placer	Order_Name1 Address Service time Demand	Name of the bin Address of the bin Time elapsed at the bin location for collecting waste Capacity of bins
Truck depot	Name Address	Name of the depot Address of depot location
Ward map	Name Population Total SW generation	Name of each ward Population of each ward Total solid waste generated in each ward

3.4 GIS Database

The GIS database is divided in seven themes. The details of these themes and associated information such as field information and respective description of each field is shown in Table 1.

3.5 Map Data and Attributes Data

For the study, a spatial geo-database is designed and implemented using GIS software (ESRI, Arc GIS 3.2a). The datasets such as route network, location of MSW collection bin, and geographical boundary are obtained from ADA and field visits. Further, additional information such as name and type of road, average speed of vehicles, details of bin, and time taken by vehicle for collection of MSW. Furthermore, spatial attributes of road network are registered and shown in Figs. 2, 3, 4, 5, 6, 7, 8, 9.

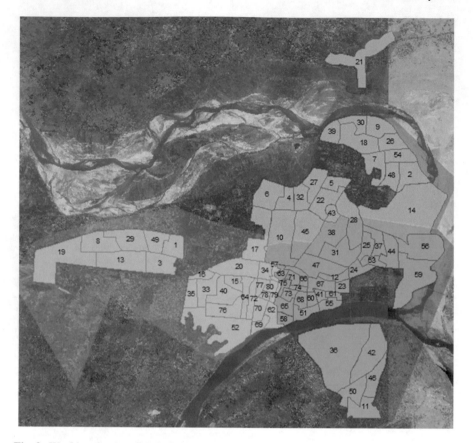

Fig. 2 Ward location in Allahabad

4 Methodologies

4.1 Spatial Database Development

In order to analyze the spatial data for the optimization of the waste collection process, a spatial database (SDB), within a GIS framework, is constructed. The data collected from field visits, toposheet of study area, and digitized map of study area are main sources of SDB.

Table

New_Shapefile(2)

FID	Shape *	Id	Ward_Name	Population	Total_SW_G
0	Polygon ZM	19	Mundera	8490	3.74
1	Polygon ZM	8	Neem Sarai	20340	8.95
2	Polygon ZM	49	Preetam Nagar	15619	6.87
3	Polygon ZM	1	Sulem Sarai	18130	7.98
4	Polygon ZM	13	Harwara	13337	5.87
5	Polygon ZM	29	Transport Nagar	12970	5.71
6	Polygon ZM	3	Jayantipur	14887	6.55
7	Polygon ZM	57	Sarai Gadi	5887	2.59
8	Polygon ZM	34	Minhajpur	15303	6.73
9	Polygon ZM	63	Shah Gunj	14943	6.57
10	Polygon ZM	66	Mohatsim Gunj	12919	5.68
11	Polygon ZM	71	Narain Singh Nagar	8615	3.79
12	Polygon ZM	73	Meel Gunj	11166	3.37
13	Polygon ZM	60	Mutthi Gunj Part -1	7371	3.24
14	Polygon ZM	77	Bakshi Bazar	13090	5.76
15	Polygon ZM	88	Daira Shah Ajmal	11600	5.11
16	Polygon ZM	78	Atala	14679	6.46
17	Polygon ZM	65	Dariabad – 1	17763	7.82
18	Polygon ZM	73	Malviya Nagar	11431	5.03
19	Polygon ZM	68	Mutthi Gunj Part –2	13535	5.98
20	Polygon ZM	74	Bahadur Gunj	10716	4.72
21	Polygon ZM	51	Katghar	12470	5.49
22	Polygon ZM	58	Dariabad – 2	11451	5.04
23	Polygon ZM	70	Tulsipur	11056	4.86
24	Polygon ZM	69	Sadiapur	12330	5.43
25	Polygon ZM	62	Meerapur	10824	4.76
26	Polygon ZM	20	Jhule lal Nagar	11541	5.08
27	Polygon ZM	15	Himmat Gunj	8662	3.81
28	Polygon ZM	72	Sultanpur Bhava	18664	8.21
29	Polygon ZM	64	Pura Manohar Das	11212	4.93
30	Polygon ZM	16	Chak Niratul	9593	4.22
31	Polygon ZM	33	Chakia	19415	8.54
32	Polygon ZM	35	Om Prakash Sabhasad Nagar	24898	10.96
33	Polygon ZM	40	Beni Gunj	19792	8.71
34	Polygon ZM	52	Karela Bagh	23080	10.16

|◄ ◄   1 ► ►| |▤ ▭| (0 out of 80 Selected)

New_Shapefile(2)

Fig. 3 Attribute data of ward

4.2 Routing Network Analysis

The optimized root creation using GIS requires preparation of data, configuration of network data, and analysis of route optimization. The vehicle routing problem (VRP) solver and closest facility provided by the ArcGIS Network Analyst is used to produce new collection routes and optimal zone. In this study, depots, routes, and route zones are considered to generate new collection routes and multiple settings for these input parameters are experimented.

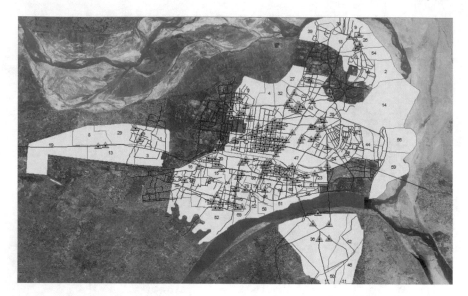

Fig. 4 CP bins point in Allahabad

4.3 Road Network Dataset for Route Analysis

The geo-database is utilized for creation of network datasets using features present in dataset. The road segments and attribute table are linked together. Further, during root optimization analysis, the attribute fields are utilized as individual parameter. The road attributes are updated: name, class, length, width, lanes, speed, and drive times. The files related to network datasets are stored in a separate feature database. The project file geo-database is utilized to store this information.

4.4 Closest Facility Solver for Optimal Zoning

Network Analyst closest facility solver is used to find the closest facility that can be served for the different types of bins present in the Allahabad municipality. In this, 247 bins locations inside the study area are added as incidents and three truck depot locations are added as facilities. The solver after solving finally analyzes 247 routes connected by different facilities.

FID	Shape *	Id	Demand	ServiceTim	Name
0	Point	0	4	15	DP1
1	Point	0	4	15	DP2
2	Point	0	4	15	DP3
3	Point	0	4	15	DP4
4	Point	0	4	15	DP5
5	Point	0	4	15	DP6
6	Point	0	4	15	DP7
7	Point	0	4	15	DP8
8	Point	0	4	15	DP9
9	Point	0	4	15	DP10
10	Point	0	4	15	DP11
11	Point	0	4	15	DP12
12	Point	0	4	15	DP13
13	Point	0	4	15	DP14
14	Point	0	4	15	DP15
15	Point	0	4	15	DP16
16	Point	0	4	15	DP17
17	Point	0	4	15	DP18
18	Point	0	4	15	DP19
19	Point	0	4	15	DP20
20	Point	0	4	15	DP21
21	Point	0	4	15	DP22
22	Point	0	4	15	DP23
23	Point	0	4	15	DP24
24	Point	0	4	15	DP25
25	Point	0	4	15	DP26
26	Point	0	4	15	DP27
27	Point	0	4	15	DP28
28	Point	0	4	15	DP29
29	Point	0	4	15	DP30
30	Point	0	4	15	DP31
31	Point	0	4	15	DP32
32	Point	0	4	15	DP33
33	Point	0	4	15	DP34
34	Point	0	4	15	DP35

I◄ ◄   0 ► ►I   ▢▢ (0 out of 75 Selected)

Fig. 5 Attribute data of CP bins

4.5   Route Optimization Using VRP

The new MSW collection routes are obtained through VRP. The VRP provides detailed explanation of workflow, properties, settings, and data used during route analysis and construction. Most of the settings and values are described from the initial analysis for the entire 247 points inside the study area. The workflow of VRP solver is illustrated in Fig. 10.

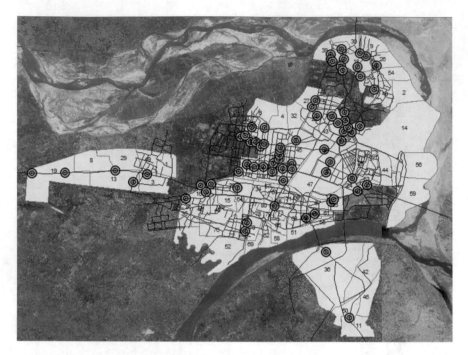

Fig. 6 DP bins point in Allahabad

5 Results and Discussions

Based on various criteria and restrictions results for optimal zones are obtained by configuring and running the Network Analyst closest facility solver. Also, the present study reviewed solution by configuring and running the Network Analyst vehicle routing problem solver for new routes.

5.1 Optimal Zoning from the Network Analyst Closest Facility Solver

The Network Analyst closest facility solver creates 247 routes for the 247 waste bin point locations in 80 wards of the Allahabad City which is illustrated in Fig. 11. Thus, each truck depot has a closest facility which serves 247 waste bins on the basis of shortest route and divides the whole wards into three zone groups Zone I, Zone II, and Zone III. Zone I is a group of 27 wards. The truck depot is located in ward no. 33 (Chakia), and the disposal site is located in ward no 36 (Kareli). Zone II consists of 33 wards. The truck depot is located in ward no. 24 (Madhvapur), and the disposal site is located in ward no 53 (Alopibagh). Zone III is a group of 22 wards. The truck

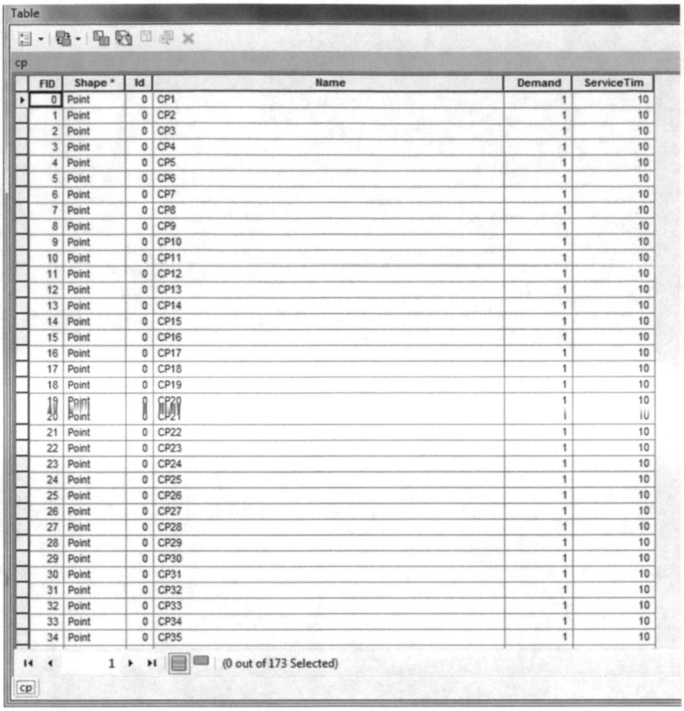

FID	Shape *	Id	Name	Demand	ServiceTim
0	Point	0	CP1	1	10
1	Point	0	CP2	1	10
2	Point	0	CP3	1	10
3	Point	0	CP4	1	10
4	Point	0	CP5	1	10
5	Point	0	CP6	1	10
6	Point	0	CP7	1	10
7	Point	0	CP8	1	10
8	Point	0	CP9	1	10
9	Point	0	CP10	1	10
10	Point	0	CP11	1	10
11	Point	0	CP12	1	10
12	Point	0	CP13	1	10
13	Point	0	CP14	1	10
14	Point	0	CP15	1	10
15	Point	0	CP16	1	10
16	Point	0	CP17	1	10
17	Point	0	CP18	1	10
18	Point	0	CP19	1	10
19	Point	0	CP20	1	10
20	Point	0	CP21	1	10
21	Point	0	CP22	1	10
22	Point	0	CP23	1	10
23	Point	0	CP24	1	10
24	Point	0	CP25	1	10
25	Point	0	CP26	1	10
26	Point	0	CP27	1	10
27	Point	0	CP28	1	10
28	Point	0	CP29	1	10
29	Point	0	CP30	1	10
30	Point	0	CP31	1	10
31	Point	0	CP32	1	10
32	Point	0	CP33	1	10
33	Point	0	CP34	1	10
34	Point	0	CP35	1	10

Fig. 7 Attribute data of DP bins

depot is located in ward no. 27 (Myorabad), and the disposal site was located in ward no 21 (Phaphamau). Details of wards in zones are shown in Figs. 11, 12, 13, 14.

5.2 *New Routes from the Network Analyst VRP Solver*

VRP solver has analyzed to obtain optimal routes in each zone. The best route calculated by the solver reflects minimum travel distance for collection and transportation of solid waste in each zone. Figures 15, 16, 17, 18, 19, 20 show optimize route from truck depot to landfill site in Zone I, Zone II, and Zone III for compacter placer bins and dumper placer bins. After optimization of route, a number of compacter placer bins and dumper placer bins are found to be 74, 42, 47 and 14, 21, 33 in Zone I, Zone II, and Zone III, respectively.

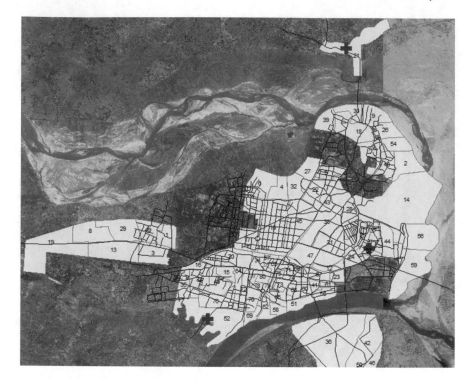

Fig. 8 Landfill point in Allahabad

Table 2 shows minimum total distance, minimum daily transportation cost, and minimum total time covered by different capacity vehicle. This table clearly illustrates how the parameters are changing with varying vehicle capacity and indicates that with increase in the vehicle capacity, and the number of vehicles for transfer the waste is reduced as well as total distance and total time also reduced.

Result based on VRP solver in Zone I for compacter placer bins, with vehicle capacity of 6, 8 and 10 ton, the decrease in the total distance (km) is found to be 27.46, 14.92, and 11.27%, respectively, when compared with 4-ton vehicle capacity. The decrease in total time (min.) elapsed is found to be 19.31, 14.85, and 8.78%, respectively. The decrease in daily transportation cost (INR) is found to be 14.73, 2.17, and 0.23%, respectively. In Zone II for compacter placer bins, with vehicle capacity of 6 ton, 8 ton, and 10 ton, the decrease in the total distance (km) is found to be 26.78, 14.78, and 14.86%, respectively, when compared with 4-ton vehicle capacity. The decrease in total time (min.) elapsed is found to be 19.89, 12.06, and 9.41%, respectively. The decrease in daily transportation cost (INR) is found to be

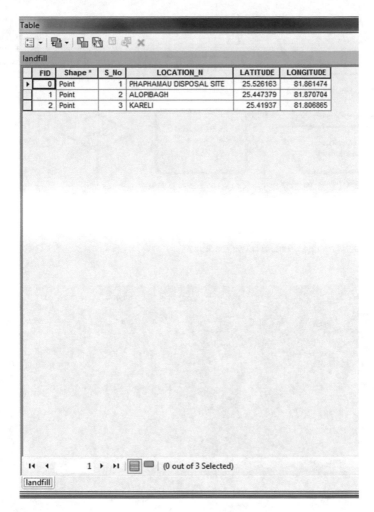

Fig. 9 Attribute data of landfill

13.89, 2, and 3.18%, respectively. In Zone III for compacter placer bins, with vehicle capacity of 6, 8, and 10 ton, the decrease in the total distance (km) is found to be 30, 20, and 16%, respectively, when compared with 4-ton vehicle capacity. The decrease in total time (min.) elapsed is found to be 21.72, 13.61, and 8.5%, respectively. The decrease in daily transportation cost (INR) is found to be 17.71, 8.09, and 5.11%, respectively.

Table 3 shows minimum total distance, minimum daily transportation cost, and minimum total time covered by 12-ton capacity vehicle. The optimized numbers of vehicles required for dumper placer bins for whole ward are 23 with vehicle capacity of 12 ton. The optimum daily transportation cost for the dumper placer bins covering the whole ward is Rs. 9491 for 12-ton vehicle.

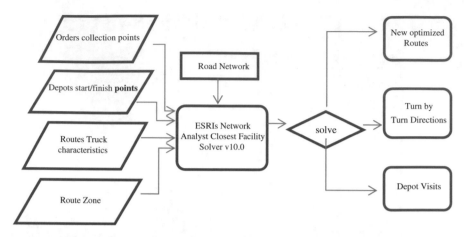

Fig. 10 Vehicle routing problem solver workflow

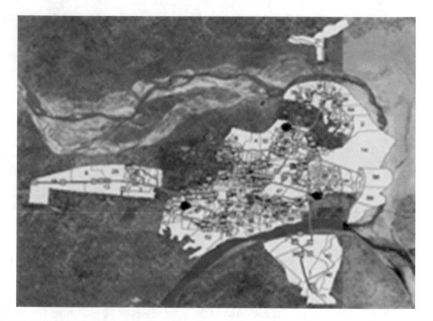

Fig. 11 247 routes from three truck depots

6 Conclusions

The GIS-based Esri's Network Analyst software extension and its optimization VRP solvers are successful demonstrate the developed model for optimal zoning of wards and optimization of route in each of the three zones. The model uses various geographical data (road network, number of bins, location of waste bins, etc.) in cooperation with advanced spatial analysis GIS tool, vehicle routing problem solver. The

Fig. 12 Details of wards in Zone I

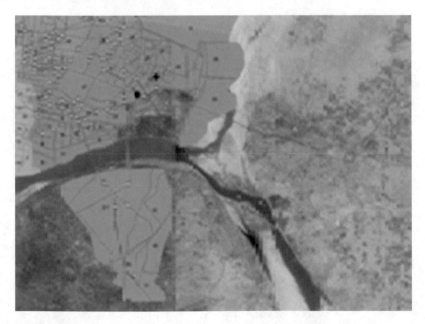

Fig. 13 Details of wards in Zone II

Fig. 14 Details of wards in Zone III

Fig. 15 Optimize route in Zone I for CPB

Fig. 16 Optimize route in Zone I for DPB

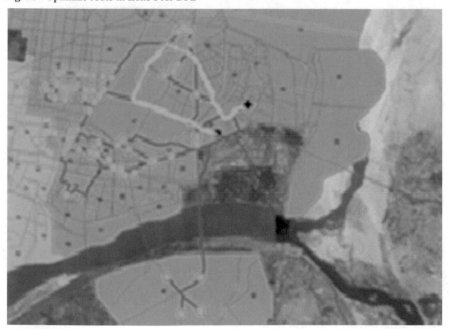

Fig. 17 Optimize route in Zone II for CPB

Fig. 18 Optimize route in Zone II for DPB

Fig. 19 Optimize route in Zone III for CPB

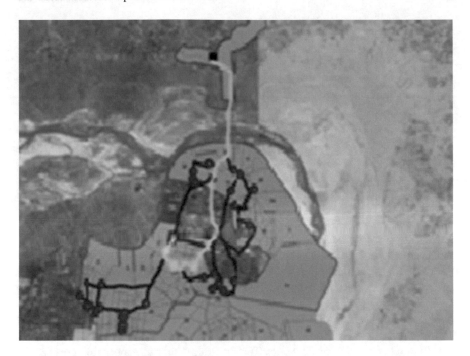

Fig. 20 Optimize route in Zone III for DPB

Table 2 Various parameters for vehicle of different capacity in Zone I, Zone II, and Zone III for compacter placer bins

Capacity (ton)	No. of vehicle	Total transportation cost (INR)	Total time (min.)	Total distance (km)
Zone I				
4	10	2110	1160	92.73
6	7	1799	936	67.26
8	5	1760	797	57.22
10	4	1764	727	50.77
Zone II				
4	19	3354	2111	147.44
6	13	2888	1691	107.96
8	11	2829	1487	92
10	8	2739	1347	78.85
Zone III				
4	12	3619	1427	159.07
6	8	2978	1117	111.31
8	6	2737	963	89.02
10	5	2597	883	74.73

Table 3 Various parameters for vehicle capacity of 12 in Zone I, Zone II, and Zone III for dumper placer bins

Capacity (ton)	No. of vehicle	Total transportation cost (INR)	Total time (min.)	Total distance (km)
Zone I				
12	5	2260	607	58.31
Zone II				
12	7	2231	828	57.58
Zone III				
12	11	5000	1351	129.04

model is applied to examine routing optimization of the existing scenario and its improvement through route optimization. An optimized route model is developed using GIS. The parameters such as population of area, waste generation capacity of area, road, routes for collection of MSW, and details of bins are considered during the analysis.

The results indicate that the three optimal zone are obtained from the closest facility solver, namely Zone I consisting of 27 wards, Zone II having 33 wards, and Zone III consisting of 22 wards. The optimum daily transportation cost for the compacter placer bins covering the whole ward is Rs. 9083 for 4-ton vehicle, Rs. 7665 for 6-ton vehicle, Rs. 7326 for 8-ton vehicle, and Rs. 7100 for 10-ton vehicle. This GIS based model can help to minimize transportation cost, number of vehicles, time, and distance for efficient collection and transportation of solid waste.

References

1. Tamilenthi, S., Chandra Mohan, K., Vijaya Lakshmi, P., Suja Rose, R.S.: The data base, land use and land cover and solid waste disposalsite—using remote sensing and GIS-A case study of Sakkottai –block, Sivagangai District, Tamil Nadu, India, Pelagia Research Library, Advances in Applied Science Research, ISSN: 0976- 8610, **2**(6), 379–392 (2011)
2. Karadimas, N.V., Doukas, N., Kolokathi, M., Defteraiou, G.: Routing optimization heuristics algorithms for urban solid waste transportation management. WSEAS Trans. Comput. **7**(12), 2022–2031 (2008)
3. Santos, L.A., Rodrigues, J.C.: Implementação em SIG de uma heurística para o estudo da recolha da resíduos sólidos urbanos. Research Report No. 6/2003. Instituto de Engenharia de Sistemas e Computadores (INESC) Coimbra, Portugal (2003)
4. Armstrong, J.M., Khan, A.M.: Modelling urban transportation emissions: role of GIS. Comput. Env. Urban Syst. **28**, 421–433 (2004)
5. Tarantilis, C.D., Diakoulaki, D., Kiranoudis, C.T.: Combination of geographical information system and efficient routing algorithms for real life distribution operations. Eur. J. Oper. Res. **152**, 437–453 (2004)
6. Viana, M.N.: A aplicação de heurísticas para geração de rotas em ambiente SIG. Dissertação de Mestrado em Investigação Operacional e Engenharia de Sistemas. Departamento de Engenharia Civil, Instituto Superior Técnico, Lisboa, Portugal (2006)

7. Tavares, G., Zsigraiová, Z., Semião, V., Carvalho, M.G.: A case study of fuel savings through optimisation of MSW transportation routes. Manage. Env. Qual.: Int. J. **19**, 444–454 (2008)
8. Tavares, G., Zsigraiová, Z., Semião, V., Carvalho, M.G.: Optimisation of MSW collection routes for minimum fuel consumption with 3D GIS modelling. Waste Manage. **29**, 1176–1185 (2009)
9. Tavares, G., Beijoco, A.F., Zsigraiova, Z., Semiao, V.: Minimisation of operation costs by vehicle fleet routes optimisation in waste collection activities. In: Proceedings Venice 2010, Third International Symposium on Energy from Biomass and Waste, Venice, Italy. Environmental Sanitary Engineering Centre, Italy (2010)

Higher Order Low-Pass Filter Using Single Current Differencing Buffered Amplifier

Mourina Ghosh, Bal Chand Nagar and Vishal Tiwari

Abstract This paper proposes a higher order low-pass filter using only one current differencing buffered amplifier (CDBA) as an active element. The proposed circuits realize third-order, fourth-order, and fifth-order low-pass filter responses. The PSpice simulation results using 0.5 μm CMOS technology agree well with the theoretical proposition. Power consumption and percentage of total harmonic distortion are very low for the proposed circuits which is useful for low-power VLSI design.

1 Introduction

These current mode (CM) circuits have received considerable attention because of their potential advantages [1, 2]. As a consequence, many CM building blocks have been developed, which are basically variants of current conveyors such as differential voltage current conveyor transconductance amplifier (DVCCTA) [3], differential voltage current controlled current conveyor transconductance amplifier (DVCCCTA) [4], current controlled differential difference current conveyor (CCDDCC) [5]. Most of these elements are unity gain amplifiers. There are many applications of these building blocks both in voltage mode and current mode operations.

Another active building block, which is called a current differencing buffered amplifier (CDBA), has been introduced in [6]. The input terminals of CDBA are internally grounded, thereby eliminating parasitic capacitances at the inputs. It can

M. Ghosh (✉) · V. Tiwari (✉)
Department of Electronics and Communication Engineering, Indian Institute of Information
Technology, Guwahati 781001, India
e-mail: mourina_06@rediffmail.com

V. Tiwari
e-mail: tiwarivishal9951@gmail.com

B. C. Nagar
Department of Electronics and Communication Engineering, National Institute of Technology,
Patna 800005, India
e-mail: balchandnagar@gmail.com

© Springer Nature Singapore Pte Ltd. 2019
A. Abraham et al. (eds.), *Emerging Technologies in Data Mining and Information
Security*, Advances in Intelligent Systems and Computing 814,
https://doi.org/10.1007/978-981-13-1501-5_6

also offer some advantageous features such as high slew rate, a wide bandwidth, and simple circuitry. Some analog filters using CDBA have been reported in [7–10] and references cited therein. The CDBA can also be realized using a commercially available current feedback operational amplifier (CFOA) IC AD844 [11].

Nowadays, higher order analog filters are getting more attention because they have the sharp cutoff frequency which is closer to the ideal response of the filter. To get a better quality of performance, they are used in audio and video signal processing, hard disk drive, CDMA, ultra-wideband wireless access technology, telephone, TV, and also for higher order filter design. High-order filters, such as third, fourth, and fifth order, are usually formed by cascading together single first-order and second-order filters or state variable method or signal flow graph. Although CDBA-based third-order filter is reported in the literature [12], they all used more than one CDBA. In this paper, the authors proposed a third-order, fourth-order, and fifth-order voltage mode (VM) low-pass filter (LPF) employing single CDBA as an active element.

2 Circuit Description

The symbol of CDBA is shown in Fig. 1a, and its port relationship is defined in matrix form as:

$$\begin{bmatrix} I_z \\ V_w \\ V_p \\ V_n \end{bmatrix} = \begin{bmatrix} 0 & 0 & 1 & -1 \\ 1 & 0 & 0 & 0 \\ 0 & 0 & 0 & 0 \\ 0 & 0 & 0 & 0 \end{bmatrix} \begin{bmatrix} V_z \\ I_w \\ I_p \\ I_n \end{bmatrix} \tag{1}$$

A high-performance CMOS-based CDBA shown in Fig. 1b is used in this work [7].

A generalized nth-order filter topology using single CDBA is shown in Fig. 2. After analyzing Fig. 2, transfer function can be found as

$$\text{Transfer Function} \,[T(F)] = \frac{V_{out}}{V_{in}} = \frac{Y_P - Y_N}{Y_Z - Y_W} \tag{2}$$

The proposed third-, fourth-, and fifth-order LPF is obtained from Fig. 2 by proper selection of admittance Y_p, Y_n, Y_w, and Y_z. The proposed third-order LPF is shown in Fig. 3a. The routine analysis gives output voltage as

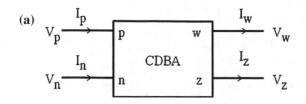

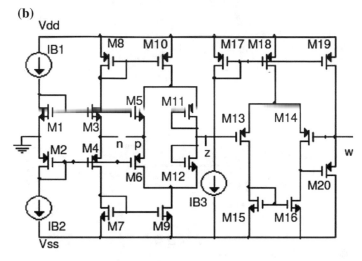

Fig. 1 CDBA: **a** symbol and **b** CMOS realization

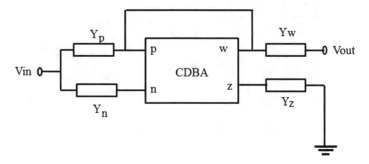

Fig. 2 Generalized nth-order filter

$$T(s) = \frac{V_{out}}{V_{in}} = \frac{R_Z(1 + sC_W R_W)}{R_{P1}(1 + sC_N R_N)} * \frac{\begin{aligned}&[1 + s^2 C_{P1} C_N (R_N R_{P2} + R_N R_{P1} - R_{P1} R_{P2}) \\ &+ s(C_P R_{P2} + C_N R_N + C_P R_{P1} - C_N R_{P1})]\end{aligned}}{s^3 + s^2 \left\{ \frac{C_Z R_Z C_W R_W + C_N R_N C_W R_W + C_N R_N C_Z R_W - C_N R_N C_W R_Z}{C_N R_N C_Z R_Z C_W R_W} \right\}}$$

$$+ s \left\{ \frac{C_N R_N + C_Z R_Z + C_W R_W - C_W R_Z}{C_N R_N C_Z R_Z C_W R_W} \right\} + \frac{1}{C_N R_N C_Z R_Z C_W R_W}$$

(3)

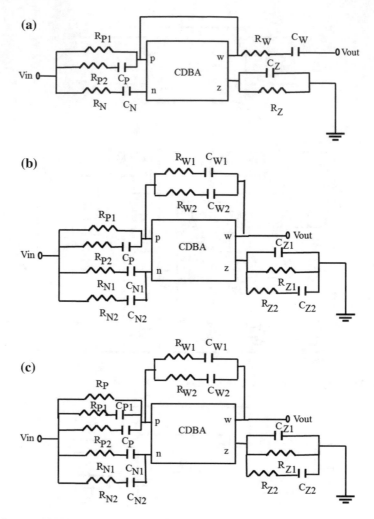

Fig. 3 Proposed LPF: **a** third-order, **b** fourth-order, **c** fifth-order

If $C_W R_W = C_N R_N$, $R_Z = R_{P1}$, $R_N R_{P2} + R_N R_{P1} = R_{P1} R_{P2}$, $C_P R_{P2} + C_N R_N + C_P R_{P1} = C_N R_{P1}$, then the transfer function (4) can be modified as

$$T(s) = \frac{V_{out}}{V_{in}} = \cfrac{1}{s^3 + s^2 \left\{ \frac{C_Z R_Z C_W R_W + C_N R_N C_W R_W + C_N R_N C_Z R_W - C_N R_N C_W R_Z}{C_N R_N C_Z R_Z C_W R_W} \right\} + s \left\{ \frac{C_N R_N + C_Z R_Z + C_W R_W - C_W R_Z}{C_N R_N C_Z R_Z C_W R_W} \right\} + \frac{1}{C_N R_N C_Z R_Z C_W R_W}} \tag{4}$$

Equation (4) shows that it is a third-order LPF, where cutoff frequency is given by:

$$\omega_0 = \frac{1}{\sqrt[3]{C_N R_N C_Z R_Z C_W R_W}} \tag{5}$$

Similarly, the fourth-order LPF is shown in Fig. 3b. The routine analysis gives output voltage as

$$
T(s) = \frac{V_{out}}{V_{in}} = \frac{R_Z(1+SC_{Z2}R_{Z2})(1+SC_{W1}R_{W1})(1+SC_{W2}R_{W2})R_{P1}(1+SC_P R_{P2})(1+SC_{N1}R_{N1})(1+SC_{N2}R_{N2})
\begin{Bmatrix} 1 + s^3 C_P C_{N1} C_{N2}\{R_{N1}R_{N2}(R_{P1}+R_{P2}) - R_{P1}R_{P2}(R_{N1}+R_{N2})\} \\ +s^2[C_{N1}R_{N1}C_{N2}R_{N2} + (C_P R_{P1}+C_P R_{P2})(C_{N1}R_{N1}+C_{N2}R_{N2}) \\ -\{C_{N1}C_{N2}R_{P1}(R_{N1}+R_{N2}) + C_P R_{P1}R_{P2}(C_{N1}+C_{N2})\}] \\ +s\{C_{N2}(R_{N2}-R_{P1}) + C_{N1}(R_{N1}-R_{P1}) + C_P R_{P1}\} \end{Bmatrix}}
{\begin{Bmatrix} s^4 + s^3 \left\{ \dfrac{C_{Z1}R_{Z1}C_{Z2}R_{Z2}(C_{W1}R_{W1}+C_{W2}R_{W2})+ C_{W1}R_{W1}C_{W2}R_{W2}(C_{Z1}R_{Z1}+C_{Z2}R_{Z2}+C_{Z2}R_{Z1})}{C_{Z1}R_{Z1}C_{Z2}R_{Z2}C_{W1}R_{W1}C_{W2}R_{W2}} \right\} \\ + s^2 \left\{ \dfrac{C_{Z1}R_{Z1}(C_{Z2}R_{Z2}+C_{W2}R_{W2}) + C_{W2}R_{W2}C_{Z2}(R_{Z1}+R_{Z2})+ C_{W1}R_{W1}(C_{Z1}R_{Z1}+C_{Z2}R_{Z2})(C_{Z2}R_{Z1}+C_{W2}R_{W2})}{C_{Z1}R_{Z1}C_{Z2}R_{Z2}C_{W1}R_{W1}C_{W2}R_{W2}} \right\} \\ + s\left\{ \dfrac{C_{Z1}R_{Z1}+C_{Z2}R_{Z2}+C_{Z2}R_{Z1}+C_{W2}R_{W2}}{C_{Z1}R_{Z1}C_{Z2}R_{Z2}C_{W1}R_{W1}C_{W2}R_{W2}} \right\} + \dfrac{1}{C_{Z1}R_{Z1}C_{Z2}R_{Z2}C_{W1}R_{W1}C_{W2}R_{W2}} \end{Bmatrix}} \tag{6}
$$

If $R_Z = R_{P1}$, $C_{Z2}R_{Z2} = C_P R_{P2}$, $C_{W1}R_{W1} = C_{N1}R_{N1}$, $C_{W2}R_{W2} = C_{N2}R_{N2}$,
$R_{N1}R_{N2}(R_{P1}+R_{P2}) = R_{P1}R_{P2}(R_{N1}+R_{N2})$,
$C_{N1}R_{N1}C_{N2}R_{N2} + (C_P R_{P1}+C_P R_{P2})(C_{N1}R_{N1}+C_{N2}R_{N2})$
$= \{C_{N1}C_{N2}R_{P1}(R_{N1}+R_{N2}) + C_P R_{P1}R_{P2}(C_{N1}+C_{N2})\}$,
$C_{N2}R_{N2} + C_{N1}R_{N1} + C_P R_{P1} = C_{N2}R_{P1} + C_{N1}R_{P1}$

Then, the transfer function (6) can be modified as

$$
T(s) = \frac{V_{out}}{V_{in}} = \frac{1}{\begin{Bmatrix} s^4 + s^3 \left\{ \dfrac{C_{Z1}R_{Z1}C_{Z2}R_{Z2}(C_{W1}R_{W1}+C_{W2}R_{W2})+ C_{W1}R_{W1}C_{W2}R_{W2}(C_{Z1}R_{Z1}+C_{Z2}R_{Z2}+C_{Z2}R_{Z1})}{C_{Z1}R_{Z1}C_{Z2}R_{Z2}C_{W1}R_{W1}C_{W2}R_{W2}} \right\} \\ + s^2 \left\{ \dfrac{C_{Z1}R_{Z1}(C_{Z2}R_{Z2}+C_{W2}R_{W2}) + C_{W2}R_{W2}C_{Z2}(R_{Z1}+R_{Z2})+ C_{W1}R_{W1}(C_{Z1}R_{Z1}+C_{Z2}R_{Z2})(C_{Z2}R_{Z1}+C_{W2}R_{W2})}{C_{Z1}R_{Z1}C_{Z2}R_{Z2}C_{W1}R_{W1}C_{W2}R_{W2}} \right\} \\ + s\left\{ \dfrac{C_{Z1}R_{Z1}+C_{Z2}R_{Z2}+C_{Z2}R_{Z1}+C_{W2}R_{W2}}{C_{Z1}R_{Z1}C_{Z2}R_{Z2}C_{W1}R_{W1}C_{W2}R_{W2}} \right\} + \dfrac{1}{C_{Z1}R_{Z1}C_{Z2}R_{Z2}C_{W1}R_{W1}C_{W2}R_{W2}} \end{Bmatrix}} \tag{7}
$$

Equation (7) shows that it is a fourth-order LPF, where cutoff frequency is given by:

$$\omega_0 = \frac{1}{\sqrt[4]{C_{Z1}R_{Z1}C_{Z2}R_{Z2}C_{W1}R_{W1}C_{W2}R_{W2}}} \tag{8}$$

Similarly, the fifth-order LPF is shown in Fig. 3c. The routine analysis gives output voltage as

$$T(F) = \frac{V_{out}}{V_{in}}$$

$$\cfrac{R_{Z1}(1 + SC_{Z2}R_{Z2})(1 + SC_{W1}R_{W1})(1 + SC_{W2}R_{W2})\left\{\begin{array}{l}\{1 + S^4 C_{N1}C_{N2}C_{P1}R_{P1}C_{P2}R_{P2}[R_{N1}R_{N2}\{R_{P1}R_{P2} \\ +R_{P1}R_P + R_{P2}R_P\} - R_P(R_{N1} + R_{N2})] \\ +S^3[C_{P1}C_{P2}(R_{P1}R_{P2} + R_{P1}R_P + R_{P2}R_P)(C_{N1}R_{N1} + C_{N2}R_{N2}) \\ -C_{N1}C_{N2}R_P(C_{P1}R_{P1} + C_{P2}R_{P2})(R_{N1} + R_{N2}) \\ -C_{P1}R_{P1}C_{P2}R_{P2}R_P(C_{N1} + C_{N2})] + S^2 \\ [C_{P1}C_{P2}(R_{P1}R_{P2} + R_{P1}R_P + R_{P2}R_P) \\ +(C_{P1}R_{P1} + C_{P2}R_{P2} + C_{P1}R_P + C_{P1}R_P) \\ (C_{N1}R_{N1} + C_{N2}R_{N2}) + C_{N1}R_{N1}C_{N2}R_{N2} - C_{N1}C_{N2}R_P(R_{N1} \\ +R_{N2}) - R_P(C_{P1}R_{P1} + C_{P2}R_{P2})(C_{N1} + C_{N2})] \\ +S\begin{bmatrix}C_{P2}(R_{P2} + R_P) + C_{P1}(R_{P1} + R_P) + C_{N1}R_{N1} \\ +C_{N2}R_{N2} - R_P(C_{N1} + C_{N2})\end{bmatrix}\end{array}\right.}{\begin{array}{l}R_P(1 + SC_{P1}R_{P1})(1 + SC_{P1}R_{P2})(1 + SC_{N1}R_{N1}) \\ (1 + SC_{N2}R_{N2})S^4 C_{W1}R_{W1}C_{W2}R_{W2}C_{Z1}R_{Z1}C_{Z2}R_{Z2} \\ +S^3\{C_{Z1}C_{Z2}R_{Z1}R_{Z2}(C_{W2}R_{W2} + C_{W1}R_{W1}) \\ +C_{W1}R_{W1}C_{W2}R_{W2}(C_{Z1}R_{Z1} + C_{Z2}R_{Z2} + C_{Z2}R_{Z1}) \\ -C_{Z2}R_{Z2}C_{W1}C_{W2}R_{Z1}(R_{W2} + R_{W1})\} + S^2\{C_{Z1}R_{Z1}C_{Z2}R_{Z2} \\ +(C_{Z1}R_{Z1} + C_{Z2}R_{Z2} + C_{Z2}R_{Z1})(C_{W2}R_{W2} + C_{W1}R_{W1}) \\ +C_{W1}R_{W1}C_{W2}R_{W2} - C_{W1}C_{W2}R_{Z1}(R_{W1} + R_{W2}) \\ -C_{Z2}R_{Z2}R_{Z1}(C_{W1} + C_{W2})\} + S\{C_{Z1}R_{Z1} + C_{Z2}R_{Z2} + C_{Z2}R_{Z1} \\ +C_{W2}R_{W2} + C_{W1}R_{W1} - R_{Z1}(C_{W1} + C_{W2})\} \\ +1\end{array}}$$

If the following condition is satisfied,

$$C_{Z2}R_{Z2} = C_{N2}R_{N2}, C_{W1}R_{W1} = C_{P1}R_{P1}, C_{W2}R_{W2} = C_{P2}R_{P2}$$
$$R_{N1}R_{N2}\{R_{P1}R_{P2} + R_{P1}R_P + R_{P2}R_P\} = R_P(R_{N1} + R_{N2})$$
$$C_{P1}C_{P2}(R_{P1}R_{P2} + R_{P1}R_P + R_{P2}R_P)(C_{N1}R_{N1} + C_{N2}R_{N2})$$
$$= C_{N1}C_{N2}R_P(C_{P1}R_{P1} + C_{P2}R_{P2})(R_{N1} + R_{N2})$$
$$+C_{P1}R_{P1}C_{P2}R_{P2}R_P(C_{N1} + C_{N2})C_{P1}C_{P2}(R_{P1}R_{P2} + R_{P1}R_P + R_{P2}R_P)$$
$$+(C_{P1}R_{P1} + C_{P2}R_{P2} + C_{P1}R_P + C_{P1}R_P)(C_{N1}R_{N1} + C_{N2}R_{N2})$$
$$+C_{N1}R_{N1}C_{N2}R_{N2} = C_{N1}C_{N2}R_P(R_{N1} + R_{N2}) + R_P(C_{P1}R_{P1} + C_{P2}R_{P2})(C_{N1} + C_{N2})$$
$$C_{P2}(R_{P2} + R_P) + C_{P1}(R_{P1} + R_P) + C_{N1}R_{N1} + C_{N2}R_{N2} = R_P(C_{N1} + C_{N2})$$

$$\tag{9}$$

The transfer function (9) can be modified as

$$\frac{V_{out}}{V_{in}} = \frac{R_{Z1}}{R_P} \frac{1}{\begin{bmatrix} S^5 C_{N1}R_{N1}C_{W1}R_{W1}C_{W2}R_{W2}C_{Z1}R_{Z1}C_{Z2}R_{Z2} \\[6pt] \begin{aligned} +S^4 &(C_{W1}R_{W1}C_{W2}R_{W2}C_{Z1}R_{Z1}C_{Z2}R_{Z2} \\ &+C_{N1}R_{N1}C_{Z1}C_{Z2}R_{Z1}R_{Z2}(C_{W2}R_{W2} + C_{W1}R_{W1}) \\ &+C_{W1}R_{W1}C_{W2}R_{W2}(C_{Z1}R_{Z1} + C_{Z2}R_{Z2} + C_{Z2}R_{Z1}) \\ &-C_{Z2}R_{Z2}C_{W1}C_{W2}R_{Z1}(R_{W2} + R_{W1})) \end{aligned} \\[6pt] +S^3 \left\{ \begin{aligned} &C_{Z1}C_{Z2}R_{Z1}R_{Z2}(C_{W2}R_{W2} + C_{W1}R_{W1}) + C_{W1}R_{W1}C_{W2}R_{W2}(C_{Z1}R_{Z1} \\ &+C_{Z2}R_{Z2} + C_{Z2}R_{Z1}) - C_{Z2}R_{Z2}C_{W1}C_{W2}R_{Z1}(R_{W2} + R_{W1}) \\ &+C_{N1}R_{N1}C_{Z1}R_{Z1}C_{Z2}R_{Z2} + (C_{Z1}R_{Z1} + C_{Z2}R_{Z2} \\ &+C_{Z2}R_{Z1})(C_{W2}R_{W2} + C_{W1}R_{W1}) + C_{W1}R_{W1}C_{W2}R_{W2} \\ &-C_{W1}C_{W2}R_{Z1}(R_{W1} + R_{W2}) - C_{Z2}R_{Z2}R_{Z1}(C_{W1} + C_{W2}) \end{aligned} \right\} \\[6pt] +S^2 \left\{ \begin{aligned} &C_{Z1}R_{Z1}C_{Z2}R_{Z2} + (C_{Z1}R_{Z1} + C_{Z2}R_{Z2} + C_{Z2}R_{Z1}) \\ &(C_{W2}R_{W2} + C_{W1}R_{W1}) + C_{W1}R_{W1}C_{W2}R_{W2} - C_{W1}C_{W2}R_{Z1}(R_{W1} + R_{W2}) \\ &-C_{Z2}R_{Z2}R_{Z1}(C_{W1} + C_{W2}) + C_{N1}R_{N1}C_{Z1}R_{Z1} + C_{Z2}R_{Z2} + C_{Z2}R_{Z1} \\ &+C_{W2}R_{W2} + C_{W1}R_{W1} - R_{Z1}(C_{W1} + C_{W2}) \end{aligned} \right\} \\[6pt] +S \left\{ \begin{aligned} &C_{Z1}R_{Z1} + C_{Z2}R_{Z2} + C_{Z2}R_{Z1} + C_{W2}R_{W2} + C_{W1}R_{W1} \\ &-R_{Z1}(C_{W1} + C_{W2}) + C_{N1}R_{N1} \end{aligned} \right\} \\[6pt] +1 \end{bmatrix}}$$

$$\tag{10}$$

Equation (10) shows that it is a fifth-order LPF, where cutoff frequency is given by:

$$\omega_0 = \frac{1}{\sqrt[5]{C_{N1}R_{N1}C_{W1}R_{W1}C_{W2}R_{W2}C_{Z1}R_{Z1}C_{Z2}R_{Z2}}} \tag{11}$$

The sensitivity of ω_o with respect to passive elements for third-, fourth-, and fifth-order LPF may be expressed as

Table 1 Aspect ratio of the transistors

Transistor	W(μm)/L(μm)
M_1–M_{10}	150/1
M_{11}, M_{12}	4/2
M_{13}, M_{14}, M_{17}, M_{18}	5/1
M_{15}, M_{16}	100/1
M_{19}	20/1
M_{20}	200/1

$$s_{C_N}^{\omega_0} = s_{R_N}^{\omega_0} = s_{C_Z}^{\omega_0} = s_{R_Z}^{\omega_0} = s_{C_W}^{\omega_0} = s_{R_W}^{\omega_0} = -\frac{1}{3}$$
$$s_{C_{Z1}}^{\omega_0} = s_{R_{Z1}}^{\omega_0} = s_{C_{Z2}}^{\omega_0} = s_{R_{Z2}}^{\omega_0} = s_{C_{W1}}^{\omega_0} = s_{R_{W1}}^{\omega_0} = s_{C_{W2}}^{\omega_0} = s_{R_{W2}}^{\omega_0} = -\frac{1}{4}$$
$$s_{C_{Z1}}^{\omega_0} = s_{R_{Z1}}^{\omega_0} = s_{C_{Z2}}^{\omega_0} = s_{R_{Z2}}^{\omega_0} = s_{C_{W1}}^{\omega_0} = s_{R_{W1}}^{\omega_0} = s_{C_{W2}}^{\omega_0} = s_{R_{W2}}^{\omega_0} = s_{C_{N1}}^{\omega_0} = s_{R_{N1}}^{\omega_0} - \frac{1}{5}$$

$$(12)$$

Equation (12) shows that the proposed circuit offers low sensitivity performance.

3 Simulation Results

To validate the theoretical prediction, the performance of the proposed LPF is evaluated using CMOS-based CDBA in PSpice. For the simulation, a CMOS-based CDBA using 0.5 μm MOSIS (AGILENT) CMOS technology is given in Fig. 1b [7] with supply voltage 2.5 V and bias current 30 μA. Aspect ratios used for different transistors are given in Table 1.

The designed values of resistances and capacitances to obtain third-, fourth-, and fifth-order low-pass filter output are given in Table 2. The simulated frequency responses of third-order LPF are shown in Fig. 4a for a cutoff frequency of $f_0 = 1$ MHz and Q = 1. The fourth- and fifth-order filter output is shown in Fig. 4b and c for a cutoff frequency $f_0 = 100$ kHz and $f_0 = 10$ kHz, respectively. To judge the quality of the output, total harmonic distortion (%THD) is obtained for third-, fourth-, and fifth-order LPF as shown in Fig. 5. It shows that the output distortion is very low and within 0.2% up to peak-to-peak input of 2 V for all the filter responses. It reveals that output of the filters is good of quality. Power consumption is very low for the proposed circuits which is almost 0.821 mW.

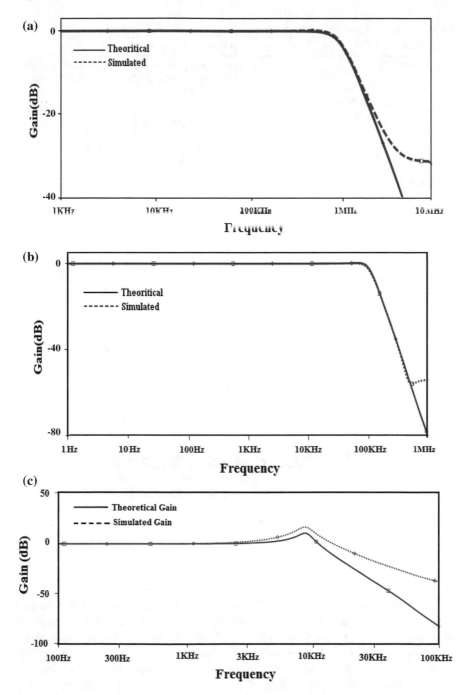

Fig. 4 Frequency response of LPF: **a** third order, **b** fourth order, **c** fifth order

Table 2 Designed values of passive components used for higher order filters

Type of filter	Component values
Third-order LPF	$R_P = R_{P1} = R_Z = 160$ kΩ, $C_P = 0.49$ pF, $R_N = 80$ kΩ, $C_N = C_Z = 1.99$ pF, $R_W = 53.33$ kΩ, $C_W = 1.49$ pF
Fourth-order LPF	$R_{P1} = R_{N2} = 480$ kΩ, $R_{P2} = R_{N1} = 160$ kΩ, $R_{W1} = 851.064$ kΩ, $R_{W2} = 13.979$ kΩ, $R_{Z1} = 479.904$ kΩ, $R_{Z2} = 35.35$ kΩ, $C_P = 4.97$ pF, $C_{N1} = 9.95$ pF, $C_{N2} = 1.10$ pF, $C_{W1} = 1.87$ pF, $C_{W2} = 37.9$ pF, $C_{Z1} = 19.9$ pF, $C_{Z2} = 22.5$ pF
Fifth-order LPF	$R_P = R_{P2} = 1.92$ MΩ, $R_{P2} = 0.32$ MΩ, $R_{N1} = R_{N2} = 0.48$ MΩ, $R_{Z1} = 1.77$ MΩ, $R_{Z2} = 5.72$ kΩ, $R_{W1} = 48.16$ kΩ, $R_{W2} = 4.16$ kΩ, $C_{P1} = 24.88$ pF, $C_{P1} = 2.07$ pF, $C_{N1} = 3.3$ pF, $C_{N2} = 11$ pF, $C_{Z1} = 199$ pF, $C_{Z2} = 927$ pF, $C_{W1} = 165$ pF, $C_{W2} = 956$ pF

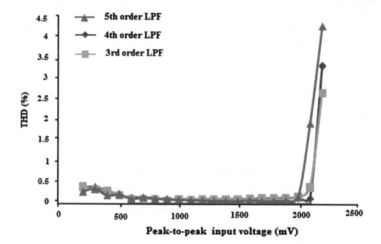

Fig. 5 %Total harmonic distortion for the propose filters

4 Conclusion

A CDBA-based higher order VM LPF is presented. In this work, some passive components matching criteria are required, but the proposed circuit deals with the following advantageous features: (i) only one CDBA is used as an active component, (ii) CDBA has low output impedance, so it appropriates for cascading, (iii) in this proposed work, all the capacitors are grounded, which is suitable for monolithic IC implementation, and (iv) for all kind of LPF, %THD is well within the acceptable limit which is 5%. All the filter structures are simulated using PSpice, and results are found to agree well with the theoretical values.

References

1. Toumazou, C., Lidgey, F.J., Haigh, D.G.: Analogue IC design: the current-mode approach. Peter Peregrinus Ltd. (1990)
2. Wilson, B.: Recent developments in current conveyors and current mode circuits. IEE Proc.G. **137**, 63–77 (1990)
3. Jantakun, A., Pisutthipong, N., Siripruchyanun, M.: A Synthesis of Temperature Insensitive/Electronically Controllable floating simulators based on DV-CCTAs. In: Proceedings of the 2009 6th international conference on electrical engineering/electronics, computer, telecommunication, and information technology (ECTI-CON 2009), pp. 560–563 (2009)
4. Jaikla, W., Siripruchyanun, M., Lahiri, A.: Resistorless dual mode quadrature sinusoidal oscillator using a single active building block. Microelectron. J. **42**, 135–140 (2010)
5. Prommee, P., Somdunyakanok, M.: CMOS-based current-controlled DDCC and its applications to capacitance multiplier and universal filter. Int. J. Electron. Commun. **65**, 1–8 (2011)
6. Acar, C., Ozoguz, S.: A new versatile building block: a current differencing buffered amplifier suitable for analog signal processing filters. Microelectron. J. **30**, 157–160 (1999)
7. Keskin, A.U., Hancioglu, E.: Current mode multifunction filter using two CDBAs. Int. J. Electron. Communi. **59**, 495–498 (2005)
8. Ozoguz, S., Toker, A., Acar, C.: Current mode continuous time fully integrated universal filter using CDBAs. Electron. Lett. **35**, 97–98 (1999)
9. Tangsrirst, W., Pukkalanun, T., Surakampontorn, W.: CDBA based universal biquad filter and quadrature oscillator. Act. Passive Electron. Compon. **2008**, 1–6 (2008)
10. Keskin, A.U.: Multifunction biquad using single CDBA. Electr. Eng. **88**, 353–356 (2006)
11. Ozoguz, S., Toker, A., Acar, C., Kuntman, H., Cicekoglu, O.: Single resistance controlled sinusoidal oscillators employing current differencing buffered amplifier. Microelectron. J. **31**, 169–174 (2000)
12. Acar, C., Sedef, H.: Realization of nth order current transfer function using current differencing buffered amplifiers. Int. J. Electron. **90**, 277–283 (2003)

Design of Voice-Controlled Smart Wheelchair for Physically Challenged Persons

**Khagendra Joshi, Rakesh Ranjan, Erukonda Sravya
and Mirza Nemath Ali Baig**

Abstract Physically challenged persons those who are suffering from different physical disabilities face many challenging problems in their day-to-day life for commutating from one place to another and even sometimes they need to have to be dependent on other people to move from one place to another. There have been many significant efforts over the past few years to develop smart Wheelchair platforms that could enable the person for its ease of operation without any ambiguity. The main aim of our paper is to develop the smart Wheelchair to make the life easier for physically challenged persons. This voice-controlled smart Wheelchair comes with enhanced features, like electric powered, voice control, line follower with the obstacle avoidance etc. The smart Wheelchair control unit consists of an integration of AVR microcontroller ATmega328 with Bluetooth module, GSM module SIM900, ultrasonic and infrared sensors, temperature sensor LM35 and motor driving circuit for controlling motor's speed.

Keywords Smart Wheelchair · Voice control · Arduino · Ultrasonic sensor
Bluetooth module HC-05 · Android phone · Line follower

K. Joshi (✉)
Department of ECE, Indraprastha Institute of Information Technology,
Delhi (IIIT-D), New Delhi 110020, India
e-mail: khagendra@iiitd.ac.in

R. Ranjan (✉) · E. Sravya (✉) · M. N. A. Baig
Department of ECE, CMR Engineering College,
Hyderabad 501401, Telangana, India
e-mail: rakeshranjan385@gmail.com

E. Sravya
e-mail: sravya.1724@gmail.com

M. N. A. Baig
e-mail: mirzanemath@gmail.com

1 Introduction

A smart Wheelchair is a Wheelchair which moves with the help of navigational controls and an electric motor instead of moving it using manpower. The navigational controls are usually controlled with the help of a small joystick placed near to the armrest, chin-operated joysticks, head switches, eye blinks, etc., that offers different operations to the Wheelchair [1–4, 8–10, 15].

Most of the physically disabled person uses traditional Wheelchairs. They are operated by hands or by a second person if the patient is unable to drive it. This is very difficult for that person if another person is not there for support. In that case, there is always a requirement of a second person [9]. Thus, the patients have to be dependent on another person.

What will happen if the Wheelchair starts moving with audio input like forward, backward, left, and right? The disabled person can move anywhere he wants without the help of the second person and independently. There will be no need to use hands for moving the Wheelchair. We are trying to implement this concept through our project 'smart Wheelchair' [2, 11–14]. The name itself indicates the meaning the Wheelchair which is intelligent. This Wheelchair takes commands from the user, and according to that, it moves in the required direction. The person who is unable to move chair by hands can move this Wheelchair just by giving the commands. This is the boon for paralyzed people. Hence using this chair, the patient can go anywhere independently. This is economical and fully automated. Hence, physically disabled people can use this Wheelchair easily and live their life happily.

In this paper, the design of a Smart Wheelchair is presented. The main motivation for this design came through the feedback from personal experiences of several colleagues as well as rehabilitation centers of the local hospitals that many elderly patients are unable to use the electronic Wheelchairs. This is primarily due to the fact that the types of illnesses common in this group of patients render damages to the motor system of the body affecting mostly arms and feet. Hence, using a joystick type control of an electronic Wheelchair is almost impossible. It was however noticed that the speech remains mostly unaltered for these patients, and hence, it is used in this work to be the main controlling agent for the motion of the Wheelchair [1, 3, 4, 8, 10]. ATmega328 microcontroller is the heart of the control circuit. The voice recognition is done by an android application, i.e., AMR voice. The Wheelchair moves according to voice commands given by the user.

The smart Wheelchair also includes obstacle detection with line follower [13]. As the condition of the patient/person with disability degrades, he/she might not be able to control the Wheelchair using joystick because of the shock from the situation or lack of energy. This loss of independence to move around freely using the Wheelchair would be a serious drawback. So we need a Wheelchair that is completely self-learning, and this particular smart Wheelchair owns it. It also provides a temperature sensor which continuously senses the temperature of the patient and sends a message to the doctor in the emergency.

2   Related Work

The researchers are showing a kin interest in the research area of smart controlling the Wheelchair. The Wheelchair is especially for those persons who are unable to move their body parts other than head and neck [2, 4, 11]. The movements are sensed using an in-mouth position sensor. Smart Wheelchairs are also under research. These Wheelchairs sense obstacles in the path of the user [13]. Sensors are mounted in the circuit to sense the objects. These Wheelchairs contain a voice recognition circuit and sensors.

Most of the systems have different options to perform tasks which are assumed to be difficult for a disabled person. In order to make the operation of the Wheelchair by the disabled persons much simpler and easier, we simplified and developed the control system with help of voice command system. This voice command system helps in moving the Wheelchair with short voice control commands. The number of physically disabled people in society have greatly aggravated in the past decade as a result of increasing war and ageing population. The demand for the improving the quality of life of such people has therefore been a major concern. The largest problem faced by all handicapped, paralyzed and disabled is immobility and the need for an artificial means of transport. The common and largely used solution in this regard is a Wheelchair, but a Wheelchair is associated with a number of disadvantages. The amount of user effort required for pulling and pushing the wheels, lack of security and stability prevents a physically weak user from operating a Wheelchair on his own and makes him depend on another person's help. With the busy lifestyles of people in today's society, this can be very in-affordable. Users who cannot use their arms have absolutely no way of using a traditional Wheelchair.

In the existing traditional Wheelchair where the person has to push the Wheelchair, though it has advantages like independence to move, exercise, and transportability in a cost-effective way but, it has the disadvantages, like lack of efficient assistance on inclines and irregular terrains, fatigue and repetitive stress injuries. These disadvantages are overcome by present-day power Wheelchairs. The power Wheelchairs with the joysticks placed near the armrest help in assisting without much effort. But most of the patients have injuries to their limbs, where they cannot move the joysticks. So, in the proposed systems, instead of moving the Wheelchair with a joystick, the Wheelchair is moved with the help of voice commands. Mounting of sensors, line detector, and obstacle detectors made the system avoid collisions or accidents. Equipping it with a temperature sensor intimates the family and concerned doctors about the patient temperature. This gives the patient independence to move freely without manual help and also helps the family to rest assure with a safe, collision avoidance system.

2.1 History of Wheelchair

Earlier a Wheelchair, used by the disabled people to move around while sitting in it, was propelled manually either by others or by the disabled person itself Nowadays, they are available by a little automated. Here is the brief explanation of the Wheelchair history used by the people with illness, injuries, or disabilities. The usage of the Wheelchair can be observed in the European continent from around the times of German Renaissance. A drawing dated 1595 of the Spain King, King Philips I1, shows him in a Wheelchair with foot and armrests. England recorded the use of Wheelchair from the 1670s. However, it was not able to be self-propelled. In 1783, Englishman John Dawson built the first Wheelchair that was self-propelled by pushing the wheels. With the invention of self-propulsion push rims, the modern Wheelchair has begun to take shape since 1881. In 1900, the wooden spiked wheels are replaced by the wire-spoked wheels. The first motorized Wheelchair was invented in 1918. The Wheelchair with voice activation had been used by a Norwegian law student where he used it for attending the classes without an attendant help [8–14].

2.2 Types of Disability

A report on disability by the World Health Organization (WHO) states that around 15% of the world population is living with some kind of disability [6]. Out of which about 2–4% had difficulties in functioning. United Nations Development Program (UNDP) estimated that around 80% of the disabled people live in developing countries. In India, the census 2011 which collected data for eight disabilities states that 20.5% of the disabilities lie in the movement [5]. The restriction in movement due to disability leads to low self-esteem, stress, isolation, fear of abandoning, etc. Arthritis patients and multiple sclerosis patients suffer from severe disabilities by which they cannot move the joystick mounted on Wheelchairs. The purpose of the proposed Wheelchair is to provide the multi-control-operated Wheelchair at a lower cost. Figure 1 depicts the statistics of the disabled population by the type of disability according to census 2011.

3 Proposed Wheelchair System

The 'smart Wheelchair' comprises of voice recognition, line follower circuit along with obstacle detection, temperature sensor along with GSM module. Figure 2 shows the block diagram of the proposed model.

The voice controller is clubbed with android smart phone at the side of Bluetooth module. The android smart phone supports BT voice control for Arduino [7, 8, 16, 17]. This application software uses android phone's internal voice recognition to forward voice commands to the microcontroller pairs with Bluetooth serial modules and sends in the recognized voice as a string. While the android device is paired with the microcontroller (ATmega328) via Bluetooth, we can give audio input, like forward, left, right, back, and line follower input to go to the specific section of the hospital. Microcontroller section also has Bluetooth device to receive the data string transmitted by android phone. Microcontroller executes these string inputs and gives output at PWM pins. This microcontroller has six PWM output pins. These PWM pins can be used to run DC motors. Here, we are using only four pins to drive DC motor driver circuits which will drive motor subsequently. We used Arduino IDE with embedded 'C' programming for the controller to implement the proposed task. The temperature sensor used here is LM35 which is interfaced with the microcontroller, measures the temperature of the patient in degree Celsius and thus together with the GSM module forms a patient monitoring system. As there are two motors so we used two separate motor drivers, one for each. Along with this, we have used 24 V battery

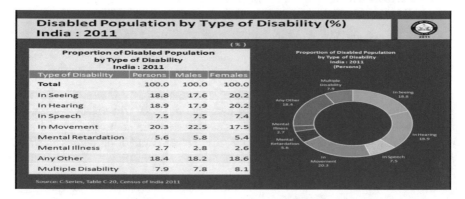

Fig. 1 Statistics of the disabled population by the type of disability according to census 2011 [5]

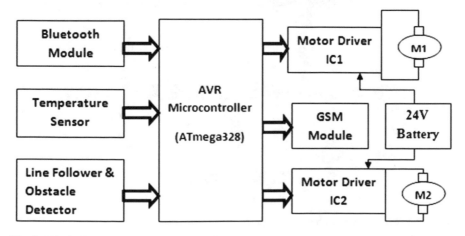

Fig. 2 Block diagram of the proposed model

Table 1 Voice control section

Audio command	Stands for	Motor A	Motor B	Wheelchair movement
F	Forward	Anticlockwise	Clockwise	Forward
L	Left	Anticlockwise	Anticlockwise	Left
R	Right	Clockwise	Clockwise	Right
B	Backward	Clockwise	Anticlockwise	Backward
S	Stop	Stop	Stop	Stop
K	Specific path by line follower			
M				
V				

as power source. Here, the motor drivers used are controlling the speed of the motors with the help of PWM. There are some specific audio commands given as input to the mobile via mic which are shown in Table 1.

3.1 Block Diagram Description

Microcontroller (AVR ATmega328) is the heart of this smart Wheelchair system. All the different types of module used in the smart system are controlled by the controller. ATmega328 is a high-performance, low-power Atmel 8-bit advanced RISC microcontroller.

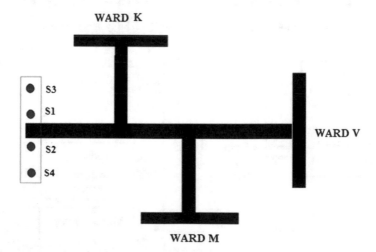

Fig. 3 Line follower path with respective position of IR sensors

Table 2 Wheelchair movement according to sensor status

Sensors				Wheelchair movement
S4	S2	S1	S3	
0	0	0	0	Stop
1	1	1	1	Forward
1	1	0	0	Left
1	1	0	1	Slight left
0	0	1	1	Right
				Slight right

a. Line Follower Section

Line follower section is used to reach the specific section of the hospital. This whole section also follows the same control procedure, i.e., the use of BT voice control application which is installed in an android mobile. The patient has to speak the name of the particular sections which is K, M, or V. Line follower section consists of four pairs of IR sensors and one pair of ultrasonic sensor. IR sensors are the main triggers of the whole line following action mechanism. IR sensor is basically a transceiver. It consists of one IR LED as a transmitter and one photodiode as a receiver. The transmitter section of IR sensor emits infrared light which will be received by receiver diode, and according to the intensity or pulse width of receiving light, the decision of the movement has been taken by the Wheelchair. There are two types of strips in line following the path which are black strip and white strip. There is no reflection from black strip, low or ('0') logic will reach to the microcontroller, and when there is white or reflective surface, there is high reflection and a high or ('1') reaches to the microcontroller. Figure 3 depicts line follower path with respective position of IR sensors.

When the Wheelchair is following a black line, the sensor will send the signal to microcontroller and the microcontroller executes these inputs and sends output to PWM pins in which motor driver is connected. In Table 2, the detailed Wheelchair movement according to sensor status has been shown.

b. Obstacle Detection

To avoid collision and obstacles, the ultrasonic sensor is used. Here, we have used HC-SR04 ultrasonic sensor. Whenever the Wheelchair is going on the desired path, the ultrasonic sensor transmits the ultrasonic waves without interruption. When an obstacle comes into the path, the sensor's wave is reflected by the object and the discontinuity in the reception of ultrasonic wave information is passed to the microcontroller. The distance of obstacle is continuously displayed on LCD. If any obstacle distant away less than 100 cm, Wheelchair will stop and will not move till the obstacle is passed or went away greater than 100 cm from Wheelchair.

Table 3 Motor driver control action

Dir(1)	Dir(2)	PWM(1)	PWM(2)	Wheelchair conditions
1	1	1	1	Forward
0	0	1	1	Backward
0	0	0	0	Stop
1	0	1	0	Left
0	1	1	0	Right

c. Temperature Monitoring Module with Emergency Messaging

Temperature sensor (LM35) is used to continuously monitor the temperature of the patient. The sensor is attached with the analog pin of the microcontroller. This temperature will display on LCD. GSM module is used to send a message to a specific mobile number. GSM module is interfaced with the controller to Tx/Rx pin. If the temperature of the patient gets higher than 37 °C, GSM will send a message to the doctor that the patient temperature is above the threshold so the doctor can reach to the patient and can give treatment to the patient.

d. Motor Driver Controller Circuit

Motor driver circuit MD10C is used to run motors. MD10C is PWM-enabled motor driver controller. The PWM output pins of the microcontroller are connected to the motor driver circuit. Motor driver controller has one PWM pin and one direction (DIR) pin. The Speed of motor is controlled by PWM, and direction pin is used to run the motor in clockwise or anticlockwise direction. Permanent magnet DC motors are connected to the motor drive circuit. In Table 3, motor driver control action is discussed.

e. Liquid Crystal Display (16 × 2)

LCD is used to monitor real-time temperature of the patient and display the distance of the obstacle from the Wheelchair. LCD is connected to microcontroller's digital pins to get digitized information of temperature from temperature sensor and distance of obstacle given by ultrasonic sensor.

4 Circuit Details and System Specifications

The smart Wheelchair control unit consists of an integration of AVR microcontroller ATmega328 with Bluetooth module, GSM module SIM900, ultrasonic and infrared sensors, temperature sensor LM35, and motor driving circuit for controlling motor's speed. Figures 4 and 5 presents the circuit diagram of smart Wheelchair system.

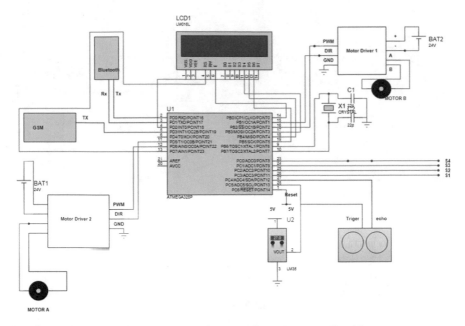

Fig. 4 Circuit diagram of smart Wheelchair system from microcontroller side

4.1 Features and System Specifications of Smart Wheelchair

The features of smart Wheelchair are the following:-

- Voice-controlled automatic Wheelchair.
- Smooth speed variations for patient comfort through PWM-controlled DC motor.
- Facility to convey message to the doctor in critical conditions and monitoring body temperature.
- Line follower to follow the definite path.
- Soft start and soft stop.
- Emergency STOP.
- Collision avoidance.

In Fig. 6, the designed smart wheel system is shown.
The system specifications are described in Table 4.

5 Software Design

We used embedded 'C' on Arduino IDE platform to code the controller. Voice control action is done by BT voice control android application. This application software uses android phone's internal voice recognition to forward voice commands to the

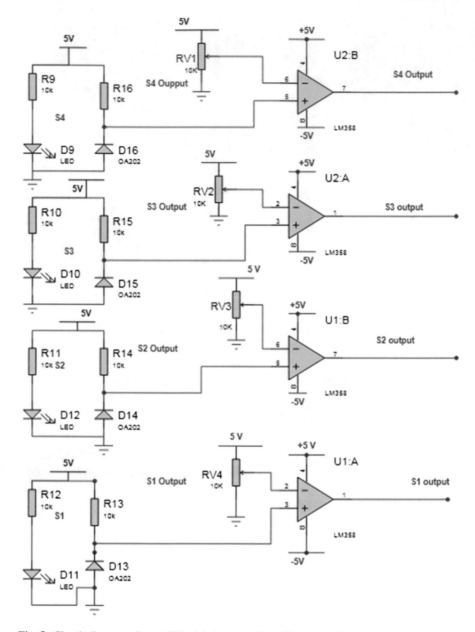

Fig. 5 Circuit diagram of smart Wheelchair system from IR sensors side

microcontroller pairs with Bluetooth serial modules and sends in the recognized voice as a string. While android device is paired with the microcontroller via Bluetooth, we can give audio input, like forward, left, right, back, and line follower input. In this

Fig. 6 Designed smart
Wheelchair

Table 4 Specifications of
smart Wheelchair

Parameters	Values
Height	4.1 ft.
Width	1.6 ft.
Weight	43 kg
Wheel diameter	36 cm
Speed	3.31 km/hr
Supply voltage	24 V, 5 V
Supply current	8 A
Load capacity	85 kg

application, if one says Chair the phone will return a sting *Chair# to your Bluetooth module. Here, '*' and '#' is for start and stop bits, respectively. Figure 7 shows the flowchart for voice control.

6   Conclusion

Our proposed smart Wheelchair provides a safe and reliable system with the presence of line follower and obstacle detector. It provides an easily accessible and a variety of functionalities. In this paper, we developed a Wheelchair system which includes ultrasonic and infrared sensors to automatically track the paths provided and also detects the obstacles in between the track along with a little intelligence of taking proper care to avoid the accidental mishaps, where we got the desired results. Thus,

Fig. 7 Flowchart for voice
control

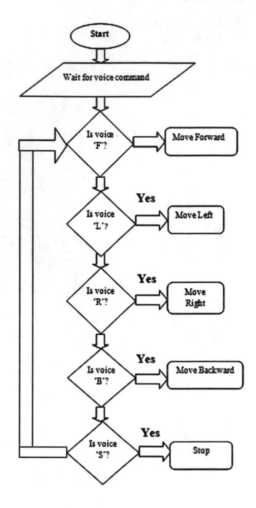

the disabled persons can be self-reliable, safe, and independent with the help of this easily controllable Wheelchair. Further improvement to the above-implemented system can be done by providing additional sensors which make the system more user-friendly and avoid accidents by self-learning. Also, security can be incorporated for accessing with the help of biometric authentication or including more control commands to pass through different types of doorways.

References

1. Kumaran, M.B., Renold, A.P.: Implementation of voice based wheelchair for differently abled. In: 4th IEEE International Conference on Computing, Communication and Networking Technologies, pp. 1–6, India (2013)

2. Wanluk, N., Visitsattapongse, S., Juhong, A., Pintavirooj, C.: Smart wheelchair based on eye tracking. In: 9th IEEE Biomedical Engineering International Conference (BMEiCON), 2016
3. Aruna, C., Dhivya, P., Malini, M., Gopu, G.: Voice recognition and touch screen control based wheelchair for paraplegic persons. In: IEEE International Conference on Green Computing Communication and Electrical Engineering, pp. 1–5, India (2014)
4. Chauhan, R., Jain, Y., Agarwal, H., Patil, A.: Study of Implementation of voice controlled wheelchair. In: 3rd International Conference on Advanced Computing and Communication Systems (ICACCS-2016), 2016
5. Census of India 2011 "Data on Disability"
6. World Health Organization "World report on Disability"
7. Yukesekkaya, B., et al.: A GSM. In: Internet and Speech Controlled Wireless Interactive Home Automation System. IEEE Transactions on Consumer Electronics, pp. 837–843. IEEE Press, New York (2006)
8. Nishimori, M., Saitoh, T., Konishi R.: Voice controlled intelligent wheelchair. IEEE conference, pp. 336–340, Japan (2007)
9. Asakawa, T., Nishihara, K., Yoshidome, T.: Experiment on operating methods of an electric wheelchair for a system of detecting position and direction. In: IEEE International Conference on Robotics and Biomimetics, pp. 1260–1265, China (2007)
10. Simpson, R.C., Levine, S.P.: Voice control of a power wheelchair. IEEE Trans. Neural Syst. Rehabilitation Eng. **10**(2), 122–125 (2002). IEEE Press, New York
11. Simpson, R.C.: Smart wheelchairs: a literature review. J. Rehabilitation Res. Dev. (2005)
12. Parikh, S.P., Grassi Jr., V., Kumar, V., Okamoto Jr., J.: Integrating human inputs with autonomous behaviors on an intelligent wheelchair platform. In: IEEE Computer Society, vol. 22(2), pp. 33–41, IEEE Press, New York (2007)
13. Murai, A., Mizuguchi, M., Nishimori, M., Saitoh, T., Osaki, T., Konishi, R.: Voice activated wheelchair with collision avoidance using sensor information. ICCAS-SICE IEEE Conference pp. 4232–4237, Japan (2009)
14. Ruzaij, M.F., Poonguzhali, S.: Design and Implementation of low cost intelligent wheelchair. IEEE international Conference on Recent Trends in Information Technology, pp. 468–471, India (2012)
15. Klabi, I., Masmoudi, M.S., Masmoudi, M.: Advanced user interfaces for intelligent wheelchair system. In: 1st IEEE Conference on Advanced Technologies for Signal and Image Processing, pp. 130–136, Tunisia (2014)
16. Kepuska, V.Z., Klein, T.B.: A novel wake-up word speech recognition system wake-up-word recognition task, technology and evaluation. Nonlinear Analysis 71, Science Direct, Elsevier, pp. 2772–2789, Elsevier (2009)
17. Alan, G.S.: Introduction to Arduino. Alan G. Smith Press (2011)

Distance-Based IoT Surveillance Alert System

Sivakumar Premkumar and T. S. Arthi

Abstract Technologies have now gone much far extent in the field of IoT. Internet of things mainly focuses on making an electronic device "live". It plays a vital role in the field of security with any system or environment that can be implemented more efficiently. The model is developed for the distance-based security system which can ensure low power usage with portability, mobility and installation facility. In this modern era, almost every device needs the above three features to be included as basic necessity. The minicomputer which is other name of the Raspberry Pi 3 model has been used to develop the system. This paper explains about the secured surveillance system which can be implemented in any smart environment where it is needed to monitor the particular object or scenario based on coverage distance. The system is built using the two modules—*distance-based motion capturing module* and *live message-based capturing module*. The first module deals with the activation of Pi camera to capture the images after nearby moving target comes within the range to alert the user through e-mail. The second module is concerned with capturing of current images through a messenger application lively. Hence, the proposed model is built using the Raspberry Pi 3 to control the above-mentioned modules. PIR motion sensor will always be actively waiting for any motion to be detected. The moment PIR detects, it will enable ultrasonic sensor to calculate the distance, and if the target is within the range, Pi camera will capture the image to alert the user through an e-mail. This process will continue as a loop, and if the user needs an immediate live image without any motion encountered, it can be obtained through a request message using TeleBot messenger application.

Keywords Raspberry pi 3 · Pi camera · PIR motion sensor · Ultrasonic sensor
TeleBot · Telegram messenger

S. Premkumar (✉)
Silver Oak College of Engineering and Technology, Ahmedabad 382481, India
e-mail: premambal@gmail.com

T. S. Arthi
Alpha College of Engineering and Technology, Pondicherry 607402, India
e-mail: arthitamilmani@gmail.com

© Springer Nature Singapore Pte Ltd. 2019

A. Abraham et al. (eds.), *Emerging Technologies in Data Mining and Information Security*, Advances in Intelligent Systems and Computing 814,
https://doi.org/10.1007/978-981-13-1501-5_8

1   Introduction

Internet of things in the today's technology has become a revitalizer of all electronic devices. The needs of Internet connecting all the devices have been increased in huge demand. The electronic machinery can be controlled by any aspects of technology which can be implemented with the help of IoT. It is recognized as a minicomputer, namely Raspberry Pi and Arduino, that is efficient in developing an IoT-based system. In the recent trends, industries and investors are much more involved in growing the scientific research of IoT incorporating product monitoring, customer monitoring and supply chain management [1]. The perspective of all the companies has started focusing on embedded technologies, such as sensors and communication devices, on their product itself.

IoT is really a very big innovative system for many large global companies which are going to raise their budget by 20% by 2018, $103 million as per the survey organized by TCS [1]. Many of the countries have already started adopting the IoT infrastructure to be an initiative on revenue, product and service customization and customer service tracking as shown in Fig. 1.

Across the four regions of the world, 47% companies leverage IoT technologies in the form of mobile applications to track resources. Smartphones are driving user's connections to the Internet of things as shown in Fig. 2 [1].

1.1   IoT Surveillance System

As per the survey shown in the graph in Fig. 2, the premise and product monitoring has more impact of integration in IoT. Home automation security involves communication among many sensors, controllers and home appliances controlled by a mobile application. There is much improvement and feasibility in the field of security through a Wi-fi-enabled system. The survey graph Fig. 3 shows the systems involved in security monitoring connected to Internet.

The majority of monitoring system developed using Raspberry Pi model is used in home automation, industrial automation and health care sectors to avoid any type of intrusion or to keep it more secured. The automatic alert system is been much developed, like if any intruder comes before the developed system, motion sensors detect the motion and capture images, and then using IMAP, it sends alert e-mails to the user or sends notification in mobile application [2]. The above whole process is handled by Raspberry Pi connected with Internet through Wi-fi having PIR motion sensor, Pi camera attached to it. The mentioned control will be in any of the iOS/Android-developed application to communicate with system [3]. Based on this manner, many revolutions have developed to surveil their environment by the users.

Fig. 1 Countries
implementing IoT

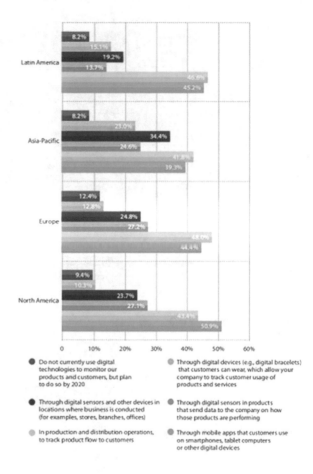

1.2 Motivation

The motion-based alert system is having a drawback of missing the target before capturing the image of it. The system depends on motion activation after the target moving here and there having chance of missing the picture of it. Also, it cannot be installed in such an environment where many necessary objects would be moving but not related to requirement. As mobile app is bridged to this system through the Internet, the administrator should possess the knowledge of smart app necessarily [4]. Keeping all these in notice, the technology on surveillance should be feasible in all different scenarios by monitoring the particular object or environment excluding unnecessary movements surrounding it. Instead of any smart application, messenger application can be implemented for the communication and control between the sensors.

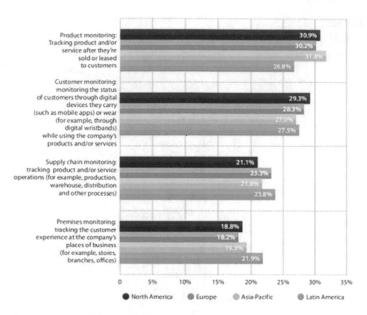

Fig. 2 Regions involved in IoT core field

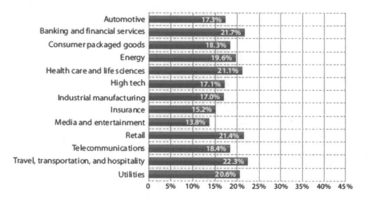

Fig. 3 Sectors involved in the security monitoring

1.3 Research Approach of This Paper

Approaches of this paper are as follows

- The secured architecture is been presented here which can be installed at any scenario or object to be monitored.
- System is defined in securing the object by differentiating intruder between many numbers of users present in the same environment.

- The hardware modules involved in it mainly focus on catching the target when it comes in some particular range and alerting through e-mail and messenger application called Telegram [5].

It is been tested with the aimed hardware model and present approaches.

The paper is organized as follows: Sect. 2 describes the preceding contribution by researchers in this field. Section 3 formulates the proper problem statement. Section 4 presents the framework with software. Here, the performance evaluation results are included scenario-wise in Sect. 5, and finally, Sect. 6 concludes our work with future scope.

2 Allied Scope of Work

The contributions in this domain using similar hardware modules:

Malche et al. [2] have proposed a home security control system, namely FLIP technology, monitoring the home environment. The user can avail the health risk by the alert. The system generates the information on air quality present at home with notification. In means of security also the user can surveillance the movements and trespass of windows and doors. It also sends e-mail notification received at user's registered account.

Chen and Li [6] have developed motion-triggered camera surveillance system through implementation of MF-IoT protocol on Atmel SAM R21 XPro with RIOT OS. Each board is used as a microcontroller and a PIR sensor as a relay. The system is connected with many motion sensors to detect motion and trigger the camera recording for surveillance, and the user can play back the old recordings at any time.

Joshi et al. [7] have proposed performance enhancement of IoT-based monitoring system using Raspberry Pi and Arduino inter-connection control through which the home appliances are monitored. The architecture is developed using application web server having alert alarm sensor with visualization of different protocols (MQTT, HTTP and CoAP). The analysis which is concluded is not appropriate.

Jyothi and Vardhan [4] have designed and implemented the surveillance system using Raspberry Pi. The system captures the image within the range and waits for interruption movements through which the camera snaps that image. The user can get the image through e-mail with SMS alerts automatically through GSM modem. Cloud server, that is YouTube, is used for live uploads of surveillance range. Local storage is adapted on unavailability of cloud and restored when the connection is reestablished. Any Internet-enabled device having Web browser can view the live video.

Gaikwadl et al. [8] have surveyed on smart home system which is compared and verified for various scenarios with different environments. Many key challenges have been put forward for the existing system that has to be considered for further same, like architecture implementation.

2.1 Aim

Subsequently, wrapping from the above survey of this concept of security in concern is given only through the movements of target before the camera and motion sensor, the distance parameters are not taken into consideration. There is a chance of moving target to move out of the camera range, and it can be implemented in the limited environments [9]. Here, the paper mainly focuses on installing the system in between mass of moving people and to detect the actual intruder. So the system can be used for protecting some object or particular range of space to be under surveillance. Therefore, this parameter is fulfilled by distance sensor. It also alerts through e-mail, messenger app (TeleBot) used here for unusual movements.

3 Problem Formulation

Alert system is the prime invention which should have prompt and precise information to detect any type of encounter or intervention. The setup should also be user friendly, adaptable for today's modern technology [10]. The user/client can easily habituate the working of technical controls without any scientific knowledge. System must not be vulnerable in means of Internet hacking or hardware rip-off. The Raspberry Pi used here connected to PIR motion sensor, ultrasonic sensor and Pi camera will increase the processor heat by all the sensors being active throughout the system running. Alert generated should be an instant one with significant information. In the present system, it has complication of missing the fast-moving target which must overtake through fulfilling the above parameters [11]. There is a need for rapid information transfer that also "live". The users nowadays have much exploration on using the messenger app which should be also infused in this alert system. So time, energy usage and feasibility of the security alert system are needed to be better chosen [8]. Next section defines about system working.

4 Framework and Components

Here, three scenarios are considered. First scenario is illustrated in Fig. 4a which shows the entire internal structure connected to each other working without trespass. Another scenario, as in Fig. 4b, explained the alert generated through ultrasonic sensor and e-mail sending after the intrusion detection. Third scenario, as in Fig. 4c, describes the current situation (i.e., instant image by taking snap) in front of the system derived from TeleBot—a messenger application [5].

The system proposed can be installed in the crowd environment, so here the working can be defined in the manner, like if administrator wants to secure the object even though having many people around the system alert can be generated

(a) Architecture without intrusion	(b) Architecture with intrusion detection

(c) Current scenario through
Telegram app

Fig. 4 **a** Architecture without intrusion, **b** architecture with intrusion detection and **c** current scenario through Telegram app

as soon as when someone tries to come near or touch the object in the presence of all [12]. The expected model can constitute all the requirements as shown in Fig. 5. Here, as shown in Fig. 5 the model will be having two range sensors working; once the target comes within the distance of sensor range, the alert module is triggered for the further procedure of sending data regarding interference.

Fig. 5 Proposed model installed in scenario

Table 1 Specifications of components

Component name	Description
Raspberry Pi	40-pin header, CPU: 4 × ARM Cortex-A53, 1.2 GHz GPU: RAM: 1 GB LPDDR2 (900 MHz) Networking: 10/100 Ethernet, 2.4 GHz 802.11n wireless Bluetooth: Bluetooth 4.1
Pi NoIR Camera V2	Infrared ray camera module for night vision
PIR sensor	HC-SR501 Pyroelectric Infrared Motion Sensor module
Ultrasonic sensor	5 voltage, range 3 cm to 4 m, dual transducers, 40 k Hz frequency

4.1 Hardware Model

Specifications of components used in this model is summarized in Table 1.

For the hardware implementation, the Raspberry Pi 3 Model B is used for controlling every process in this architecture. Pi camera is connected to Raspberry Pi for taking images after the trigger encountered by ultrasonic sensor. Then, the image is sent as an e-mail to the owner. The proposed hardware model is represented in Fig. 6.

The above model can ensure low power usage with portability, mobility and installation facility. The prototype for this system is shown in Fig. 7.

Fig. 6 Proposed hardware

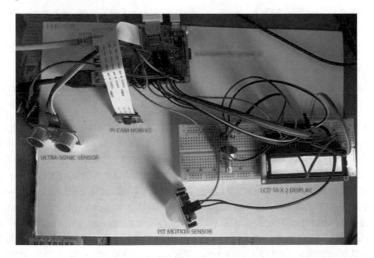

Fig. 7 Prototype of the proposed device model

4.2 Components Description

Raspberry Pi The Raspberry Pi is the heart of the system that handles other sensor devices connected via GPIO pins. It fetches the intrusion signals from PIR and ultrasonic sensors and then sends the signal to camera to take images, which is then sent via e-mail services.

PI Camera PI camera is used in the proposed model to snap the images of intervention of any encounter near the system. It stays idle and waits for the signal from Raspberry Pi; when distance sensor detects something, signal is sent to take images. It is also included with NOIR V2 that is night vision for more clarity.

PIR Sensor (Passive Infrared Sensor) PIR sensor connected to Raspberry Pi will be always active throughout. It is working on infrared light that is waiting for any type of movement detection. As soon as it detects any, it triggers the signal to activate the ultrasonic sensor for further procedure.

Ultrasonic Sensor Ultrasonic sensor is an integral part of the proposed system because it is the device which is helpful in consumption of low energy by the whole system. It waits for the signal from PIR sensor to be triggered; if yes, then it will activate the PI camera to take images of intrusion. This sensor activates only when some object comes within its range which is then converted into alert to the whole system.

LCD Screen (16 × 2): LCD screen is used to show the distance calculated by the device and the range of object distance.

4.3 Software Module

Telegram Application Telegram application is the most common chat application which does not need a technical knowledge to access. It is also having a feature of automatic synchronization with multiple communication devices, so the user can get the message easily at any manner and even reply. The Telegram is providing a third-party platform having Bot feature through which Bot API can be created to implement own procedure. Here, a personal TeleBot is generated according to the user. The user-defined TeleBot will be having a list of commands through which it will connect our system installed. So the procedure is like when no type of movement occurs still admin can get message from the proposed device. Message denotes the image or text data of current scenario before that system. It focuses on getting the information with or without any type of intervention.

Fig. 8 Proposed flow of the whole system

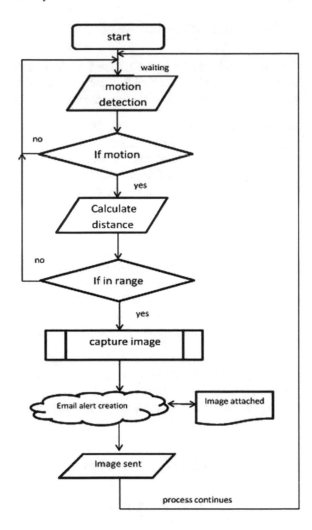

4.4 Flowchart

The flow of the whole structure is defined in Fig. 8. The input is the movement detection; as soon as the movement is detected, the distance is been calculated. If the distance calculated is within the range specified, then the snap of the target is triggered, and using IMAP e-mail service image taken is sent to the user as an alert. This process runs as a loop as shown in the flowchart till the system turns off.

5   Implementation and Results

The proposed hardware model is tested in the real-time environment. The sensors connected to the Wi-Fi-enabled Raspberry Pi are all kept active for testing the reliability inside the home. Here, two scenarios are considered:

Distance-Based Motion Capturing Module This scenario is where intruder movements are detected within the range and to create an alert for the user regarding the intrusion in the room. From this process, two scenes are created as shown in Fig. 9a. Movements will occur but device will not take snap, and the system will wait till target comes near as shown in Fig. 9b through which e-mail sent is shown in Fig. 9c.

Live Message-Based Capturing Module The scenario mainly focuses on live updates whenever it is necessary, and the user can avail through an instant message itself. This module is not concerned with or without intrusion in front of the proposed model; if the user wants to get the information of current environment at instant level, it can be availed easily just by a text message. Image and text information of the scene in front can be viewed at any time. It is been experimented with a 4G smart mobile phone as shown in Fig. 10. After the e-mail alert, user can use this module for conformation either for situation after the intrusion.

The modules defined above are tested with the live prototype with all specifications accordingly. Here, the waiting time for the PIR motion and ultrasonic sensor distance detection is taken approximately 60 s. It is stated that both the sensors will wait till the targets come within or move out of the range.

6   Conclusion and Future Scope

Smart city, in the context of Internet of things there are many revolution in monitoring and security. Here, only those issues fulfilling the consideration of low power usage with portability, mobility and Installation facility are considered. In this paper, we have proposed a portable device which constitutes Raspberry Pi with Wi-fi, PIR motion, ultrasonic sensor and Pi camera having an LCD display also attached. This proposed device can be installed in any scenario where monitoring of particular object or environment within the specific range of distance is demanded. Even though the system installed in the scenario has mass group or people surrounded, it can easily detect the intrusion which is proved by results. The device can be monitored by the user at any time once installed through a simple messenger application (Telegram). User can get the live remote status of device installed and synchronized with user smart mobile. By this new technique, the result of proposed approach has met our expectations with sensors working.

Once the proposed system starts without any type of admin input, it automatically e-mails to the admin as soon as intrusion is detected near the range and through

(a) Scene with Movements   (b) Scene with Movements
outside the range   inside the range

(c) Alert email is sent to the
Administrator

Fig. 9 **a** Scene with movements outside the range, **b** scene with movements inside the range and **c** alert e-mail is sent to the administrator

messenger application admin needs to give input in form of text message to receive the remote images or the status. It can be concluded that security of any environment can be achieved by this system within the required range. This paper also confirms the minimal storage because as soon as the images taken by device are sent through mail the next image replaces old one in the device storage. Raspberry Pi is used, and it fulfils low cost and mobility. As an add-on feature, the user can get the images in their messenger app itself which has removed the barrier of exclusive smart app for this architecture.

Preliminary results and analysis show encouraging outputs in NOOBS OS with Python code. In future, the device will be further modified to be implemented in

Fig. 10 Messages interacted
with TeleBot(Telegram app)

a clumsy environment like forest with more enhanced security. We will be trying in manner of bifurcating each sensor attached to this system in different locations connected wirelessly and then gathering the data from each to increase the security level and minimize the propagation delay.

References

1. TCS Global Trend Study 2015, Internet-of-things-the-complete-reimaginative-Force.pdf, http://sites.tcs.com/internet-of-things/
2. Malche, T., Maheshwary, P: Internet of Things (IoT) for building Smart Home System. In: 2017 International Conference on IEEE I-SMAC (IoT in Social, Mobile, Analytics and Cloud) (I-SMAC), 2017
3. Kesavan, G., Sanjeevi, P., Viswanathan, P.: A 24 hour IoT framework for monitoring and managing home automation. In: International Conference on 2017 Inventive Computation Technologies (ICICT)
4. Jyothi, S.N., Vardhan, K.V.: Design and implementation of real time security surveillance system using Iot. In: International Conference IEEE on 2017 Communication and Electronics

Systems (ICCES)
5. Telegram, Telegram-Bot, https://core.telegram.org/bots/api
6. Chen, J., Li, S.: Demo Abstract: Motion-Triggered Surveillance Camera using MF-IoT. In: 2017 IEEE/ACM Second International Conference on 2017 Internet-of-Things Design and Implementation (IoTDI)
7. Joshi, J., Rajapriya, V., Rahul, S.R., Kumar, P., Polepally, S., Samineni, R., Tej, "Performance Enhancement and IoT Based Monitoring for D.G.K.: Smart home. In: International Conference IEEE on 2017 Information Networking (ICOIN), 2017
8. Gaikwadl, P.P., Gabhane, J.P., Golait, S.S.: A survey based on smart homes system using internet-of-things. In: 2015 IEEE International Conference Computation of Power, Energy Information and Commuincation (ICCPEIC)
9. Tanwar, S., Patel, P., Patel, K., Tyagi, S., Kumar, N., Obaidat, M.S.: An Advanced Internet of Thing based Security Alert System for Smart Home. In: 2017 International Conference on IEEE 2017 Computer, Information and Telecommunication Systems (CITS)
10. Reddy, P.S.N., Reddy, K.T.K., Reddy, P.A.K., Kodanda, G.N.: An IoT based home automation using android application. In: International conference on Signal Processing, Communication, Power and Embedded System (SCOPES)-2016
11. Kumar, P., Pati, U.C.: IoT based monitoring and control of appliances for smart home. In: IEEE International Conference On Recent Trends In Electronics Information Communication Technology, May 20–21, 2016, India
12. Motlagh, N.H., Bagaa, M., Taleb, T.: UAV-Based IoT platform: a crowd surveillance use case. IEEE Communications Magazine, February 2017

A Predictive Resource Management Technique in Grid

Sukalyan Goswami, Ajanta Das and Kuntal Mukherjee

Abstract Computational grid is made up of virtual resources and differs from high-performance computing (HPC), as it is used in scientific and technological computation-intensive problem-solving. In computational grid, resources are classified based on their load factors. Utilization of these resources through co-ordination of various loads is always considered an optimization problem. In order to achieve this, approach of adaptive resource ranking is applied in this research. This paper proposes to improve the efficiency of earlier proposed NDFS algorithm by introducing historical or average load of each resource along with their current load for a defined period or interval. This adaptive methodology of resource co-ordination is necessary for balancing load in the computational grid. Jobs are scheduled for adaptively ranked resources, thus meeting the service quality agreement (SQA). The grid test bed experimental set-up is made by Globus Toolkit 5.2, and benchmark codes of matrix multiplication and fast Fourier transform are executed to demonstrate the results in this paper.

Keywords Grid computing · Adaptive resource ranking · Resource management
Load balancing · Nearest deadline first scheduled (NDFS) algorithm

1 Introduction

Grid computing [1] has opened new windows of research and development because of its co-ordination capabilities of large-scale resources over distributed environ-

S. Goswami (✉) · A. Das
University of Engineering and Management, Kolkata, India
e-mail: sukalyan.goswami@gmail.com

A. Das
e-mail: ajanta.desarker@gmail.com

K. Mukherjee
Birla Institute of Technology, Mesra, Lalpur Campus, Ranchi, India
e-mail: kmukherjee@bitmesra.ac.in

© Springer Nature Singapore Pte Ltd. 2019   109
A. Abraham et al. (eds.), *Emerging Technologies in Data Mining and Information Security*, Advances in Intelligent Systems and Computing 814,
https://doi.org/10.1007/978-981-13-1501-5_9

ment. Computational grid ensures more cost-effective usage of resources, thereby increasing the computational capability by a large extent. The resource broker in grid is entrusted with the responsibilities of optimized scheduling of clients' requests to appropriate resources. And to achieve this, a major area of research on load balancing has opened up for computational grid, where a lot of improvement could be achieved. Load balancing becomes even more demanding because grid is heterogeneous in nature and computation-data separation becomes prime complexity to be tackled.

This research work proposes a solution for the above-discussed problems by an improved nearest deadline first scheduled (NDFS) algorithm [2]. Average load of each resource has also been considered along with their current load for a defined period or interval. Co-ordination of resources by the broker has been achieved by this adaptive methodology, which finally ensures balancing of load in the computational grid. Jobs are scheduled for adaptively ranked resources, thus meeting the service quality agreement (SQA). The grid test bed experimental set-up is made by Globus Toolkit 5.2 [3], and benchmark codes [4] of multiple instances of matrix multiplication and fast Fourier transform (FFT) are executed to demonstrate workload balancing in computational grid.

The paper is organized as follows: Sect. 2 discusses relevant in grid. Section 3 discusses improvised resource management technique in NDFS algorithm. Results and discussion of benchmark codes are presented in Sect. 4. Conclusion of the paper is presented in Sect. 5.

2   Related Works

Resource scheduling, ensuring job deadline meets and achieving balanced distribution of workload are major challenges in grid. A few relevant approaches are discussed in this section.

The process of balancing the workload needs to consider dynamic load factors of the resources, and scheduling of jobs must be done accordingly [5]. Higher the number of deadlines met, better is the performance of the grid. An execution scheme, which is adaptive in nature, has been proposed by [6]. This approach achieves guaranteed performance based on the service level agreements. A resource management framework is used for implementing a set of agents for performance monitoring and optimized behaviour of the grid.

Michael Stal [7] had proposed a client–server broker architectural framework for distributed systems. But that framework is not suitable in computational grid, because the characteristics of grid are considerably different from a general distributed system's properties. Moreover, the model does not cater to the problem of load balancing. Comparison of different task scheduling approaches in grid is discussed in [8]. In addition, there has been an X-dimension binary tree data model for dynamic job scheduling and load balancing in grid environment proposed by [9].

Fault tolerant load balancing policy for grid has been proposed by [10]. But, the results have limited success in balancing the load across grid resources.

In this paper, an approach along with experimental results of benchmark code is presented to improve the makespan and reduce the number of job deadline misses by catering to the challenge of resource scheduling in the grid. Since resource scheduling is also a multiple criteria decision-making process, Saaty's AHP model [11], which supports MCDM, is used in this research work. Resource ranking is achieved by AHP model. In our research, broker schedules the submitted jobs in ranked resources based on job deadlines and requirements. This resource ranking ensures that categorization of grid resources based on load, i.e. *overloaded, loaded* and *underutilized*, becomes a theory of past, and practically load remains balanced across resources in grid.

Nearest deadline first scheduled (NDFS) [12] algorithm uses job deadline as the prioritized parameter for scheduling, and simulation results of NDFS compared to other existing algorithms in GridSim [13] prove that this approach is betterment over existing models. Girdlets increasing, makespan and resubmissions decrease when NDFS is used. This ensures maximum number of tasks meets their respective target times; overall makespan reduces resulting in enhancement of performance of grid. The current and average load scenarios of the resources are also considered while scheduling tasks and happen solely based on the capability and load factor of resources. Process of ranking of resources in NDFS is redefined in this research work by incorporating average load of the resource. This approach is advantageous, because the clients need to submit the jobs only once and then the broker will monitor the SQA, which was signed between client and broker prior to the submission of the job. The results obtained by executing benchmark codes ensure achievement of balanced load across the computational grid.

3 Proposed Resource Management Technique

In NDFS, job allocation vis-à-vis resource utilization is performed with the help of ranking of resources. In grid, resource ranking is dynamic due to the dynamic resources. Hence, adaptive approach of resource ranking is applied in this research. However, job scheduling in computational grid is an optimization problem.

Proposed Model:
The enhanced NDFS algorithm performs job allocation using assignment and scheduling problem of linear programming.
Let us explain the problem considering set of jobs and resources, J_a and R_b, respectively. In this paper, the proposed algorithm highlights the critical scenario where number of jobs submitted in the grid is more than number of available resources.

Inputs:
At first, clients submit the following parameter values of itself associated with particular job:

- # of processor cores
- CPU utilization (current)
- CPU clock frequency
- Utilization of network (current)
- RAM availability (current)

$$\text{Set of Jobs submitted to grid} : J_a, 0 \leq a \leq n \qquad (1)$$

$$\text{Set of Resources available in grid} : R_b, 0 \leq b \leq m \qquad (2)$$

$$\text{Assumptions} : J_a > R_b \qquad (3)$$

Processing Phases:
Execution of job is only initiated by the preparation of the job queue (based on Job$_{\text{weightage}}$) and adaptive ranking of resources (Resource$_{\text{weightage}}$). Hence, calculation of the two valuable parameters Job$_{\text{weightage}}$ and Resource$_{\text{weightage}}$ is explained according to the AHP model, developed by Thomas Saaty [11], in the subsequent steps. As final processing, signing of SQA between client and the broker is always mandatory.

Outputs:
Objective and outcome of the algorithm are to *allocate maximum number of jobs into minimum number of available resources* in grid such that SQA for each job is verified. To allocate job successfully to specified resource, Job$_{\text{weightage}}$ and Resource$_{\text{weightage}}$ need to be matched.

$$\text{Job Assignment} : \text{Max}(J_a) \Rightarrow \text{Min}(R_b), \qquad (4)$$

Such that, Job$_{\text{weightage}}$ of $J_a \leq$ Resource$_{\text{weightage}}$ of R_b
SQA verification,

$$J_a(C_t) \leq J_a(dl), \qquad (5)$$

where $J_a(C_t) =$ completion time of job J_a and
$J_a(dl) =$ deadline of job J_a.
Hence, successful allocation implies that all jobs (J_a) are allocated to specific resources (R_b) and SQAs for the job, J_a verified. The necessary step related to compliance of SQA is explained in later section.

Processing Phases:
After successful submission of the job by the clients in the computational grid, the broker needs to process three different phases:

(i)   Finding out the details of job and preparation of the job queue by Job$_{\text{weightage}}$
(ii) Adaptive ranking of the available resources by Resource$_{\text{weightage}}$
(iii) Signing of SQA

Preparation of Job Queue:
The broker calculates $\text{Job}_{\text{weightage}}$ using AHP [14].
In order to optimize weightage calculation procedure in NDFS algorithm, $\text{Job}_{\text{weightage}}$ is calculated by assigning weights to parameters submitted by client,
$x_1 = \#$ of processor cores
$x_2 =$ the total CPU processing power available $= (1 - \text{CPU utilization}/100)$
$x_3 =$ the total specific RAM availability
$x_4 =$ the total specific clock frequency
$x_5 =$ the total network bandwidth available $= (1 - \text{network utilization}/100)$

$$f\left(\text{Job}_{\text{weightage}}\right) = \sum w_i x_i \qquad (6)$$

where, $\sum w_i = 1$, $i = (1, 2, 3, 4, 5)$
Intensity of important values for above-mentioned factors is assigned in the following manner.

Number of cores is of highest priority, because higher number of cores ensures faster completion of job execution under equal load condition among resources. Availability of CPU and RAM percentage is given next level of importance. Network utilization parameter and clock frequency are least prioritized in comparison with the other three factors.

Resource ranking is achieved by using AHP decision matrix, as shown in [14].
Equation (6) is updated as

$$f\left(\text{Job}_{\text{weightage}}\right) = 0.464x_1 + 0.195x_2 + 0.195x_3 + 0.073x_4 + 0.073x_5 \qquad (7)$$

Adaptive Resource Ranking:
Load scenarios of the resources in grid keep on changing continuously. Hence, resource ranking has to be a dynamic procedure, to ensure its effectiveness. So, broker maintains records of average loads of all participating resources. Therefore, during job submission to the broker, resources submit the values of their specific system variables (as discussed earlier). In order to make it more dynamic and realistic, the CPU load utilization needs to be considered in two equal parts: current and average availability of CPU. The current availability means how much processing power is available at that moment, while average availability means how much processing power is available on an average for certain defined interval. So, the CPU processing power availability gets amended to

$$\text{CPU Availability}(x_2) = (0.5 * \text{current availability} + 0.5 * \text{average availability}) \qquad (8)$$

Hence, Eq. (7) is redefined as presented in Eq. (9):

$$f\left(\text{Resource}_{\text{weightage}}\right) = 0.464x_1 + 0.0975x_{21} + 0.0975x_{22} + 0.195x_3 + 0.073x_4 + 0.073x_5 \qquad (9)$$

Broker calculates Resource$_{weightage}$, the resource having highest Resource$_{weightage}$ value is assigned as Res$_{rank}$ = 1, next high Resource$_{weightage}$ valued resource is assigned as Res$_{rank}$ = 2, and so on. Broker then places the resources on a priority queue according to their ranks. Higher a loaded resource, lesser will be the Resource$_{weightage}$ value, lower will be the resource rank.

SQA Endorsement
Next, broker checks for matching the values of Resource$_{weightage}$ and Job$_{weightage}$. If higher value of Resource$_{weightage}$ is found compared to the Job$_{weightage}$ value of particular job, then a bipartite agreement, SQA, is signed between client and broker, regarding that particular job.

Outputs:

(i)   Assignment of jobs to appropriate resources, as depicted in Eq. (4)
(ii) Verification and compliance of SQA, as shown in Eq. (5)

After the SQA is signed, broker allocates job to the particular resource whose Resource$_{weightage}$ is immediate higher to Job$_{weightage}$ of submitted job. Broker sends job execution results to the client, after verification and validation of SQA.

The heterogeneous grid environment changes continuously, and resources' load factors are dynamic in nature. So, the probability of new job submission during execution of algorithm is high and hence requires dynamic ranking of resources with continuous scheduling of tasks in appropriate resources.

Implementation results of the algorithm are presented in next section, by executing several instances of benchmark codes in grid test bed.

4   Experimental Results

Globus Toolkit 5.2 20 is used to set up the real grid test bed. Java 6 is being installed on top of Globus. The System Information Gatherer And Reporter (SIGAR) [15] API is used to retrieve system parameters. The grid test bed is set up consisting of six clients, five resources and a grid broker.

Client nodes' specifications: Processor—dual core

RAM—2 GB
HDD—160 GB

Broker and resources' specifications: Processor—quad core

RAM—4 GB
HDD—500 GB

Execution of Multiple Jobs: To substantiate implementation of improved NDFS in the grid, heterogeneous benchmark codes of highly computation-intensive jobs of fast Fourier transform (FFT) [4] and matrix multiplication [4] are executed. In this research work, a critical situation is represented with two instances of FFT job and two instances of matrix multiplication job submitted from four clients and these jobs are executed in available five resources.

The operations are presented below:

1. Broker receives "FFT1.java" job details along with deadline specification from Client1 (192.168.30.3).
2. Then, broker fetches system parameters from Client1 and all available resources in grid. The broker calculates $Job_{weightage}$ and $Resource_{weightage}$. By this time, another job details arrive.
3. Now, broker receives "FFT2.java" job details along with deadline specification from Client3 (192.168.30.9).

   3.1 Broker then fetches system parameters from Client3 and all the resources. $Job_{weightage}$ and $Resource_{weightage}$ values are computed. They are represented in Table 1.

   3.2 Upon matching of values between $Job_{weightage}$ and $Resource_{weightage}$, broker signs SQAs with both Client1 and Client3.

   3.3 According to $Job_{weightage}$ and $Resource_{weightage}$ matching criteria of NDFS, FFT1.java is allotted to Resource1 and FFT2.java is allotted to Resource4 (presented in Table 1). After this, both the jobs are executed, and results are sent to broker, which subsequently are forwarded to the respective clients. The SQAs are met for both of these scenarios.

4. Next, broker receives "MatMul1.java" job details along with deadline specification from Client2 (192.168.30.8).

   4.1 Similar steps are followed for this job also; i.e. system parameters are fetched from Client2 and all the resources. Then, weightages are calculated, SQA is signed, and MatMul1.java is allocated to Resource5 [presented in Table 1]. Subsequently, this instance of the matrix multiplication job is executed, results are sent to the client through broker, and SQA is met.

5. Next, broker receives "MatMul2.java" job details along with deadline specification from Client4 (192.168.30.15).

   5.1 Similar steps are followed for this job also; i.e. system parameters are fetched from Client4 and all the resources. Then, weightages are calculated, SQA is signed, and MatMul2.java is allocated to Resource3 [presented in Table 1]. Subsequently, this instance of the matrix multiplication job is executed, results are sent to the client through broker, and SQA is met.

6. Finally, broker receives "FFT3.java" and "MatMul3.java" job details along with deadline specifications from Client5 (192.168.30.16) and Client6 (192.168.30.17), respectively.

Table 1 Job$_{weightage}$ and resource$_{weightage}$ values and job allocation

Weightages	FFT1.java (Client1)	FFT2.java (Client3)	MatMul1.java (Client2)	MatMul2.java (Client4)	FFT3.java (Client5)	MatMul3.java (Client6)
Job$_{Weightage}$	2.33	2.27	1.96	1.99	2.28	1.97
Resource$_{Weightages}$ (after respective job submissions)	Resource1: 2.42 Resource2: 1.85 Resource3: 1.96 Resource4: 2.09 Resource5: 2.13	Resource1: 1.73 Resource2: 1.89 Resource3: 1.97 Resource4: 2.36 Resource5: 2.17	Resource1: 1.58 Resource2: 1.88 Resource3: 1.93 Resource4: 1.87 Resource5: 1.99	Resource1: 1.91 Resource2: 1.93 Resource3: 2.01 Resource4: 1.92 Resource5: 1.71	Resource1: 2.30 Resource2: 2.06 Resource3: 1.58 Resource4: 2.13 Resource5: 1.82	Resource1: 1.68 Resource2: 2.01 Resource3: 1.84 Resource4: 2.28 Resource5: 1.93

Similar steps are followed for these jobs also; i.e. system parameters are fetched from Client5 and Client6 and all the resources. Then, weightages are calculated, SQAs are signed, and FFT3.java and MatMul3.java are allocated to Resource1 and Resource2, respectively (presented in Table 1). Subsequently, the jobs are executed, results are sent to the clients through broker, and SQAs are met.

5   Conclusion

Computational grid is constituted by geographically dispersed participating resources whose workloads vary by a large extent. Efficient scheduling of jobs among these resources results in workload balancing across the grid, and eventually number of deadline meets increased substantially. This research work achieves this target by improving the efficiency of an already proposed NDFS algorithm by inducting average load along with current load of resource. Resource allocation and management are achieved by this adaptive methodology, and SQAs are met by proper job scheduling in ranked resources. The grid test bed experimental set-up is made by Globus Toolkit 5.2, and benchmark codes of multiple instances of matrix multiplication and fast Fourier transform (FFT) are executed to demonstrate workload balancing in computational grid, thus justifying the proposition presented in the paper.

References

1. Foster, I., Kesselman, C., Tuccke, S.: The Anatomy of the grid. Int. J. Supercomputer Appl. (2001)
2. Goswami, S., Das, A.: Deadline stringency based job scheduling in computational grid environment. In: Proceedings of the 9th INDIACom; INDIACom-2015. 11th to 13th March, 2015
3. Globus Toolkit. http://toolkit.globus.org
4. http://introcs.cs.princeton.edu
5. Goswami, S., Das, A.: An adaptive resource allocation scheme in computational grid. Int. J. Control Theory Appl. ISSN: 0974–5572, vol. 9, Issue 41, pp: 721–736, December 2016
6. De Sarkar, A., Roy, S., Ghosh, D., Mukhopadhyay, R., Mukherjee, N.: An adaptive execution scheme for achieving guaranteed performance in computational grids. J. Grid Comput. (2010)
7. Stal, M.: The Broker Architectural Framework
8. Goswami, S., De Sarkar, A.: A Comparative study of load balancing algorithms in computational grid environment. In: Fifth International Conference on Computational Intelligence, Modelling and Simulation, pp. 99–104 (2013)
9. Abo Rizka, M., Rekaby, A.: Dynamic Job Scheduling and Load balancing algorithm in grid environment via X-dimension binary tree data model. Int. J. Intell. Computing and Inf. Sci. 12(2) (2012)
10. Balasangameshwara, J., Raju, N.: A hybrid policy for fault tolerant load balancing in grid computing environments. J. Network Comput. Appl. 35, 412–422 (2012)
11. Saaty, T.L.: Decision making with the analytic hierarchy process. Int. J. Services Sci. 1(1), 83–98 (2008)

12. Goswami, S., Das, A.: Handling resource failure towards load balancing in computational grid environment. In: Fourth International Conference on Emerging Applications of Information Technology (EAIT 2014) at Indian Statistical Institute, Kolkata during Dec 19–21, 2014
13. Buyya, R., Murshed, M.: GridSim: a toolkit for the modelling and simulation of distributed management and scheduling for Grid computing. J. Concurrency Comput.: Practice Experience **14**, 13–15 (2002)
14. Goswami, S., Das, A.: Optimisation of workload scheduling in computational grid. In: Proceedings of the FICTA-2016
15. https://github.com/hyperic/sigar
16. Di, S., Kondo, D., Cirne, W.: Google hostload prediction based on Bayesian model with optimized feature combination. J. Parallel Distributed Comput. **74**, 1820–1832 (2014)
17. Jaiswal, S., Mishra, A., Bhanodia, P.: Grid host load prediction using gridsim simulation and hidden markov model. Int. J. Emerging Technol. Adv. Eng. **4**(7), 775–781 (2014)
18. Kant Soni, V., Sharma, R., Kumar Mishra, M.: An analysis of various job scheduling strategies in grid computing. In: 2nd International Conference on Signal Processing Systems (ICSPS), 2010
19. Karthick Kumar, U.: A dynamic load balancing algorithm in computational grid using fair scheduling Int. J. Comput. Sci. 8, Issue 5, No 1, pp 123–129, September 2011
20. Keerthika, P., Kasthuri, N.: A hybrid scheduling algorithm with load balancing for computational grid. Int. J. Adv. Sci. Technol. **58**, 13–28 (2013)
21. Rajavel, R.: De-centralized load balancing for the computational grid environment. In: International Conference on Communication and Computational Intelligence, Tamil Nadu, India, 2010
22. Ray, S., De Sarkar, A.: Resource allocation scheme in cloud infrastructure. In: International Conference on Cloud and Ubiquitous Computing and Emerging Technologies, 2013
23. http://www.visual-paradigm.com

Ontology-Based Information Retrieval System for University: Methods and Reasoning

Mohammad Aman Ullah and Syed Akhter Hossain

Abstract The purpose of this paper is to search for a general framework for ontology development. This paper also implemented ontology on university domain, proposed a general framework for ontology searching and explained searching mechanism through university ontology. Also, it presents different ways of reasoning the ontology. In general, ontology classifies the variables in need for some computations and creates interrelationships between them. It is also an essential part of the semantic web. The introduction of semantic web poses the demands for creating ontology in many domains. This paper emphasized mostly on conceptualizing the university as a whole and was developed using standard tools protégé 4.3. This paper tries to fill the gap between existing works by including all the concepts and their related data and object properties. The reasoning of our created ontology was done through Fact++ and Hermit 1.3.8 reasoner.

Keywords Ontology · Semantic web · OWL · Protégé · Reasoning

1 Introduction

Ontology is an essential part of the semantic web. The objectives of semantic web are to permit more advanced knowledge engineering by organizing knowledge in classes with the use of automated system [1]. The introduction of the semantic web has motivated the creation of ontology in different domains, so that, the information could be represented and interpreted semantically and syntactically by the computer and make easily machine processable. Ontologies can also be used for modeling and generation of meta-data elements [2, 3, 4].

M. A. Ullah (✉)
International Islamic University Chittagong, Chittagong 4203, Bangladesh
e-mail: ullah047@yahoo.com

S. A. Hossain
Daffodil International University, Dhanmondi, Dhaka 1207, Bangladesh
e-mail: aktarhossain@daffodilvarsity.edu.bd

© Springer Nature Singapore Pte Ltd. 2019
A. Abraham et al. (eds.), *Emerging Technologies in Data Mining and Information Security*, Advances in Intelligent Systems and Computing 814,
https://doi.org/10.1007/978-981-13-1501-5_10

Selecting an appropriate ontology language and tool for characterizing the semantics is an important task in designing ontology application and knowledge engineering. The tools should be able to handle all the classes and objects that are in need to be addressed for the system to work. Ontologists use Web Ontology Language (OWL) for ontology development in any domain. Among many famous ontology editing tools, we have used Protégé 4.3.1, for its capability to handle OWL ontologies by OWL-DL [1]. The reasoning of our created ontology was done through Fact++ and Hermit 1.3.8 reasoner. Graph viz software was used to create the graphical view of the ontology such as Figs. 2 and 3.

University ontology consists of university's academic and administrative working details. Appropriate selection of classes, object properties, data properties, annotation properties, and applying suitable restrictions and axioms related to the different issues of the university helps build a complete ontology which is as effective as efficient in providing necessary information as quickly as possible with the appropriate reasoning for decision making in any issues of the university. From among many issues and concepts, this paper emphasizes mostly on departments of the university.

In this paper, we have built up the general framework for ontology development. This paper also presents the implementation procedure of the university ontology based on this framework, builds the general framework for ontology searching, and demonstrates this framework through university ontology using Description Logic (DL) query. Also, it presents different ways of reasoning the ontology.

The remaining work is organized as follows: Sect. 2 includes literature review, Sect. 3 represents detail methodology, whereas Sect. 4 includes reasoning and information retrieval and Sect. 5 includes conclusions.

2   Literature Review

Many ontologists have designed the university ontology. Among them, review of some of them is included here as related work. In [5], "knowledge-based university examination ontology" was developed for helping the examination system of the university to large extent with the use of protégé tools. They mainly divided the examination system into some classes and subclasses including object and data properties to some extend. An ontology on university was constructed in [4], where they focused on the methods of university ontology construction by the use of protégé 4.0 alpha tool.

Another ontology on university was developed by Sanjay et al. (2010), highlighting the creation of classes and instances for the concepts of ontology [6]. In [7], university ontology was created using protégé OWL tool, demonstrating the ways of creating superclasses, subclasses, instances, query process, and visualization system. The ontology was built only on Rajiv Gandhi Technical University, Bhopal, India. Abir et al. (2016) developed an ontology to help internship assignment process of the university with the use of OWL and proposed a semantic recommender system for the same process. For the improvements in query results, they have also proposed semantic matching system [8].

Zeng et al. [9] developed a model ontology for university courses using bottom-up approach of course ontology, which can illustrate one or more open courses and could pull out domain knowledge from open courses and help adult learners to find their favorite courses [9, 10] constructed an ontology-based information retrieval system with the use of fuzzy ontology framework in the domain of university scientific research management. As per them, their system resolves the problems with other fuzzy ontology models and allows semantic information retrieval through fuzzy concepts.

Gil et al. [11] dealt with a meta-model which is incrementally developed. They have proposed open-source tools for ontological learning experiments. They have also illustrated the methodology for improved ontological development using semi-supervised learning method. Richard (2010) applied the ontological learning methods on university relational database to retrieve data in semantics form for further use in any domain. They have also demonstrated the conversion process from relational data to semantics. Including the required concepts and information of the university, an ontology was developed in [12] to evaluate students achievements. They have also shown the ways data could be mapped to the data warehouse.

3 Methodology

Figure 1 shows the general framework for ontology development, and the detail descriptions of each step of this framework are given below.

The first step in any ontology creation ontology is to gather all the information regarding the working procedure of the targeted domain (university). Identifying and defining all the super- and subclasses of the targeted domain is the second step. The third step is the identification and definition of the properties (such as object, data, and annotation) that exist among defined classes. Identifying and applying appropriate restriction constraints to the classes in order to interrelate them to work together and help in reasoning are the fourth step. Bringing all the above steps together to have a single design view of the ontology in order to discover any design inconsistency is the fifth step. The sixth step is to practically implementing the designed ontology using Protégé 4.3.1. The next step is to check for consistency and reliability through reasoning (generating asserted and inferred model) by the use of Fact++ and Hermit 1.3.8 reasoner. If the reasoner result is not OK with the desired result, then we need to rethink of gathering the appropriate information for building the ontology again, then all the above steps should be repeated. The next step is the finding of our desire information. Last step is to save or export the created ontology in our desire place.

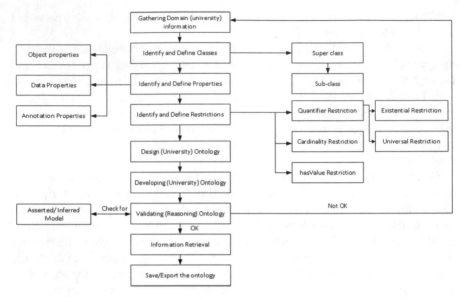

Fig. 1 General framework for ontology development

4 Implementation

University ontology development begins from creation of classes, defining properties, applying restrictions to practically implementing with the ontology using software tools.

4.1 Class Creation

Figures 2 and 3 represent asserted and inferred view of people class. Though our main concern is the department class of the university, due to space constraint we have shown only people class. All other superclasses of our ontology are departments, divisions, program, courses, clubs, events, publications. We have subclasses for almost all the superclasses. In Fig. 2, we could see the subclasses of people class such as student, administrator, and teacher. All these classes has been divided to subclasses as shown in Fig. 2. In Fig. 3, Inferred view of people class is shown, which was generated by applying the inference rules to the ontology and those rules were written in restriction section of the ontology creator software Protégé 4.3. This view is generated by applying the rules that we have defined in our restriction section.

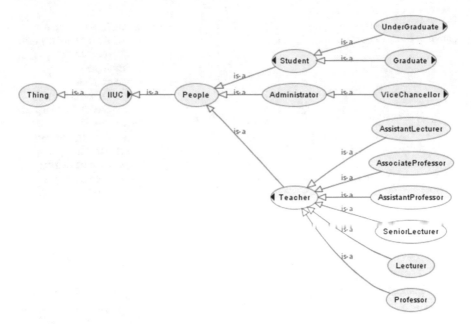

Fig. 2 Asseted view of people class

Fig. 3 Inferred view of people class

4.2 Defining Properties

Three types of properties exist in OWL such as object, data, and annotation properties. We have created three of them in our ontology. Figures 4 and 5 show the data and object properties, respectively. Object properties link the individuals of one class to the individuals of another class, whereas data properties link individual to data values [4]. Though the university ontology may contain large number of object and data properties, here we have shown only the related properties of students and departments.

4.3 Applying Restrictions

OWL restrictions can be the quantifier, cardinality, and value restrictions. Also, we could apply necessary and sufficient conditions, and closure axioms. We almost apply all the restrictions in our ontology, which is depicted in next section with Fig. 7.

Fig. 4 Data property of our ontology

Fig. 5 Object property of our ontology

5 Reasoning and Information Retrieval

For retrieving appropriate information, ontology plays a vital role, but it requires appropriate facts and rules to be defined and use for the interconnections of the classes of the ontology. Reasoning basically helps in deriving the facts or rules with classes to retrieve appropriate information. Figure 6 presents a general framework for searching from ontology. This searching framework is not domain restricted, rather could be used in any domain's ontology searching. Figure 7 represents the student class with reasoning; i.e., all form of relationship exists with other classes due to the matching of rules or facts are shown. Figures 8 and 9 represent different ways of DL query application in our ontology. From all the figures, it is clear that we could search for our required information individually or in a class, which will result us with Description Logic query of our ontology.

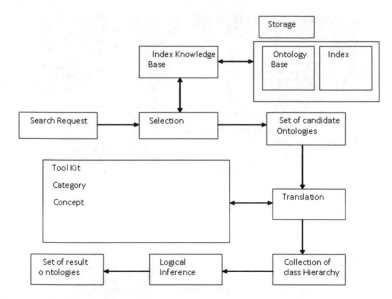

Fig. 6 General framework for search command processing

Fig. 7 Class view after reasoning

Fig. 8 DL query view of student in our ontology

Fig. 9 DL query view of
department class

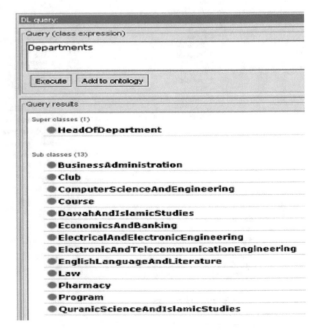

6 Comparison with Existing System

There are several related ontologies being developed so far but they have only considered very narrow domain of the university; i.e., some of them consider only examination system of the university [5], few of them considered only courses [9], some of them taken into account scientific research management [10], some have proposed part of data warehouse of the university [12], some has developed an ontology to guide internship process [8], some of them included few of subdomains [4, 7]. But in our ontology, we have brought together all the necessary concepts, their object and data properties to help university and their expert and inexpert staffs in all their related concerns.

7 Conclusions and Future Works

In this age of semantic web and ever-demanding ontology, this paper has introduced every aspect of the ontology from gathering information to creation and reasoning by providing the general framework and utilizing protégé 4.3.1 ontology development tool. We have made university ontology emphasizing departments of the university. The reliability and constancy of our ontology were verified utilizing FACT++ and Hermit 1.3.8 reasoner and were discovered working well. Different concepts of the university were mapped by different subclasses. We have clearly defined all the restrictions on the different super- and subclasses individually. The user will get all form of reasoning support using our ontology and could reuse this ontology by the alteration of the restrictions. Our general framework for searching will help efficient and effective searching and access to information through DL query. In future, we would like to expand our ontology including all the concepts and related restrictions of the university.

References

1. Grigoris, A., Frankvan, H.: Semantic Web Premier. MIT press Cambridge, Massachusetts, London (2008)
2. Bhaskar, K., Savite, S.: A comparative study ontology building tools for semantic web application. Int. J. web and Semantic Technol. (IJWest) **1**(3) (2010)
3. Staab, S., Studer R.: Handbook on ontologies. In: International Handbooks on Information System Springer (2004)
4. Ameen, K., Khan, R., Rani, B.P.: Construction of university ontology. In: Information and Communication Technologies (WICT), 2012 World Congress, pp. 39–44 (2012)
5. Venkataraman, D., Haritha, K.: Knowledge representation of university examination system ontology for semantic web. In: 2017 4th International Conference on Advanced Computing and Communication Systems (ICACCS) pp. 1–4 (2017)
6. Malik, S.K., Nupur, P., Rizvi, S.A.M.: Developing an University Ontology in Education Domain using Protégé for Semantic Web. Int. J. Eng. Sci. Technol. **2**(9) (2010)

7. Naveen, M., Nishchol, M., Santosh S.: Developing University Ontology using protégé OWL tool: process and reasoning. Int. J. Sci. Eng. Res. 2(9) (2011)
8. M'Baya, J., Laval, N., Moalla, Y., Ouzrout, A.: Ontology based system to guide internship assignment process. In: Signal-Image Technology & Internet-Based Systems (SITIS), 2016 12th International Conference on, pp. 589–596 (2016)
9. Zeng, L., Zhu, T., Ding, X.: Study on construction of university course ontology: content, method and process. In: International Conference 2009 Computational Intelligence and Software Engineering (CiSE), pp. 1–4 (2009)
10. Zhai, J., Liang, Y., Jiang, J., Yu, Y.: Ontology-based information retrieval for university scientific research management. In: 4th International Conference on Wireless Communications, Networking and Mobile Computing (WiCOM'08), pp. 1–4 (2008)
11. Gil, R., Borges, A.M., Contreras, L., Martin-Bautista, M.J.: Improving ontologies through ontology learning: a university case. In: 2009 WRI World Congress on Computer Science and Information Engineering, pp. 558–563, (2009)
12. Tanuska, et al.: The proposal of Ontology as a part of University Data warehouse. In: 2010 IEEE 2nd International Conferences on Education Technology and Computer (ICETE), pp. v3–21–24 (2010)

Methodology to Secure Agricultural Data in IoT

L. Vidyashree and B. M. Suresha

Abstract Internet of things has made significant changes in the real world, penetrates all aspects of human life, and controls on many things. Agriculture plays a considerable role in the development of the agricultural countries. Securing agricultural data is also a challenging issue in the agricultural field. In this paper, we focus on securing agriculture data from hackers. Especially in the agriculture field, due to lack of manpower, farmers are going to remote control devices. So in the future, we need to ensure the security of the devices and the data about the land and control of the precious devices. To strengthen the security of an existing system, in this paper extension of AES 128-bit encryption method has been proposed with additional methods like checksum creation, data segmentation, and shuffling of data to give more security. In this paper, we compare encryption and decryption time and processing time for both with and without using proposed encryption method for securing agriculture data. In this system, we convert the raw data into JSON format, split the data, and also jumbled it. We claim that our method is more secure, reliable, and efficient compared to the existing method.

Keywords IoT · Security · Agriculture · Authentication · Encryption
Decryption · Sensor data integration

1 Introduction

The Internet of things enables physical objects to see, hear, think, and perform tasks by having them "talk" together to coordinate decisions and to share information. Eventually, all aspects regarding physical, the social world will be interconnected and intelligent in the smart world [1]. Agricultural IoT is an era of farming agri-

---

L. Vidyashree (✉) · B. M. Suresha
DoS in Computer Science, University of Mysore, Mysore, India
e-mail: vid14.1987@gmail.com

B. M. Suresha
e-mail: sureshabm@yahoo.co.in

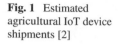

Fig. 1 Estimated
agricultural IoT device
shipments [2]

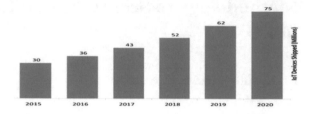

cultural activities using sensing media to give better services to farmers. Various sensor devices are deployed in agriculture land to collect essential information like temperature, humidity, light, carbon dioxide, soil moisture, acidity, raindrop, and correct information is indispensable to take the proper initiative during cultivation. Based on this type of monitoring, it will be very easy for the agricultural survey. The four areas that present a cyber-security threat in agriculture are access to services, personal privacy areas, proprietary information, and IP. There are various types of sensors to monitor any kind of variance in the atmosphere. The parameter what we choose according to that parameter particular sensor will be used for collecting the agriculture data. When doing smart agriculture, various smart technologies can be used. The data can be smartly created and protected by using IoT technology. We focus on some basic issues like encrypting the data and privacy issues.

The usage of IoT device for agriculture is gradually increasing and shipments of IoT device for agriculture are growing year by year. Figure 1 shows the estimation of IoT device shipments in agriculture by the end of 2020, and it is represented in x-coordinates in terms of year, and the percentage of IoT device shipments is represented in y-coordinates; the agriculture field will have 75% usage of IoT device [2]. The usage of IoT devices in agriculture is challenging.

The aim of this paper is to focus on securing agricultural data to protect from hackers. We have an algorithm to protect agricultural data from an unauthorized person. Basically, according to the present situation, the weather condition at a particular agricultural area was not considered while starting their farming process. To avoid this, we can make a weather bureau which consists of a different sensor that can get different parameters in a particular area and can get a real-time report from that data. In this weather bureau, we have to get the air quality where it should get different gas in the atmosphere that is harmful, temperature, and humidity have to be detected, dust particle in the air, light intensity, and noise, in particular, area all these parameters have to be considered. These are the different types of data we are going to secure. All these information can be stored in a cloud. Strangers or competitors may be doing the vulnerable activities if it comes in bigger market; at that time, we should be well evolved in security. An attacker is having a chance to access them and destroy them, so that it will affect the farmers.

The rest of paper is summarized as follows. The state of existing technologies and algorithm applied to agriculture is given in Sect. 2. The need of IoT in agriculture is provided in Sect. 3. The methodology of proposed encryption and decryption method for securing agriculture data and the experimental results of the proposed method is depicted in Sect. 4. The conclusion and the future steps in the development are discussed in Sect. 5.

2   Related Work

Verma Amit et al. proposed an improved AES, which will be more secure and good in performance as compared to existing AES. To encrypt the text or image data and to increase the speed and time, the data to be encrypted is pipelined. AES is combined with validation and segmentation. To make the data more secure, key expansion is done. The improved version of AES is compared with blowfish, but as compared to the improved AES result of the blowfish is fast and more secure encryption algorithm [3].

R. R. Agale et al. provide to get the information about farm-related data and proposed a IoT technology for monitoring soil data. To analyze data related to soil and crops is difficult for farmers. IoT technology can be used in the agricultural environment to collect and store data. Using IoT technology, soil monitoring is a major concept and small farmers can manage their farm from anywhere in the world [4].

Carlos Cambra et al. proposed a powerful irrigation tool that uses real-time data such as the variable rate irrigation and some parameters taken from the field. The field parameters, the index vegetation, and the irrigation events such as flow level, pressure level, or wind speed are periodically sampled. The developed multimedia platform can be controlled remotely by a mobile phone. Finally, bandwidth consumed measured when the system is sending different kinds of commands and data [5].

Pitchaiah et al. presented a 128-bit AES encryption and decryption and collaborated using virology code. The algorithm consists of three main parts, and they are a cipher, inverse cipher, and key expansion. The meaningless form data is transformed by a cipher is known as ciphertext. Cipher and inverse cipher process uses key schedule, and it is created by key expansion. Cipher and inverse cipher consist of a special number of rounds [6].

Verma O.P. provides two main features that differentiate one encryption algorithm from another and its potential to secured data opposed to attacks, its speed and also provides a performance differentiation between four of the most popular encryption algorithms: DES, 3DES, AES, and Blowfish [7].

3   Need of Internet of Things in Agriculture

Many industries are focusing on agriculture to meet the food requirement. It can achieve by involving the IoT into the agriculture and lack of workers in the field; it is not possible for mass production. So that we need to go for IoT to monitor the farms and also to maintain.

Fig. 2 Mass production farm [8]

3.1 Where We Need IoT Devices?

Nowadays, IoT devices are increasingly used at the very high rate for all applications. Figure 2 shows the mass production and highly sensitive plants need proper temperature and light in perfect timing to deliver the hundred percent yields. We need to maintain the environmental condition properly. For this work, we need to deploy many people to collect the data and need to control the temperature maintaining devices, irrigation system, and fertilizers. Instead of people, the IoT devices take place in order to collect all these types of data.

3.2 Why We Need to Be Secure?

Previously, data was maintained on a paper or a document, and a huge number of these data is not maintained securely; sometimes these data can be lost or stolen or theft. That time the data will be affected by some serious problems. This security can be maintained to protect a loss of information or accessing data from an unauthorized person. This will help to protect from an unauthorized person accessing a data. In future, many agriculture farms will use IoT devices and control the farms based on the sensor data. Trained employees will manage the system based on the sensor data. The whole controlling system is based on the sensor data, and vulnerabilities may cause because of hackers disruption of data. In order to avoid all these effects, the security can be given for agricultural data.

4   Experimental Result

To secure agricultural data, the following phases are carried out.

4.1   Agriculture Data Collection

Based on the some parameters like air quality, humidity, temperature, sound, lighting, and smoke, we are collected more than one thousand of the data from agriculture field. In this paper, to collect an agriculture data, the STM32F407VG microcontroller has been used. It is 32-bit ARM Cortex M4 microcontroller; it runs with clock frequency up to 168 MHz. Microcontroller reads the sensor data, and all sensors connected to the microcontroller's port pins. We are using hardware generated random data for the experiment. In Fig. 3, x-coordinates show the agriculture data collection for various parameters and data can be collected for each parameter individually and y-coordinates show the collected sensor data which are plotted based on the time per value.

4.2   Proposed Encryption Method for Data Security

AES 256-bit will be heavy for the embedded devices, to reduce the load AES 128-bit is used in the paper. AES is the majorly used encryption algorithm for encrypting and decrypting data. However, in the existing AES, there are some drawbacks such as low throughput and security issues. All these problems are resolved in the extension of AES. To give better security for the data, the proposed method has been depicted with three additional methods and they are checksum creation, segmentation of data, and shuffling of data segments which are used. This system uses both symmetric and asymmetric encryption with both private and public key methods encrypt the data. Here, this system having two types of keys one is private key and another is a public key. Private key means it would not be transmitted to the Internet. Actually, we can say this as a nonvolatile key. Sender and receiver should know the private key. The public key is a volatile key, and it will get updated over a time interval. The sender (sensor device) only creates the public key and shares this public key to a receiver (cloud server). If a receiver has both the keys, then only receiver can read the data.

Figure 4 shows encryption process between the sensor and cloud side. The sensor data is collected from agriculture field and stored securely in the cloud. Actually, cloud in the sense our Internet server storage space. The device formatting the sensor data and data get framed into JSON format. We are using two keys to encrypt the data, and they are private and public key. In this, public key will be a random number and frequently will get updated and further, the following additional methods are used in the proposed encryption to give more security.

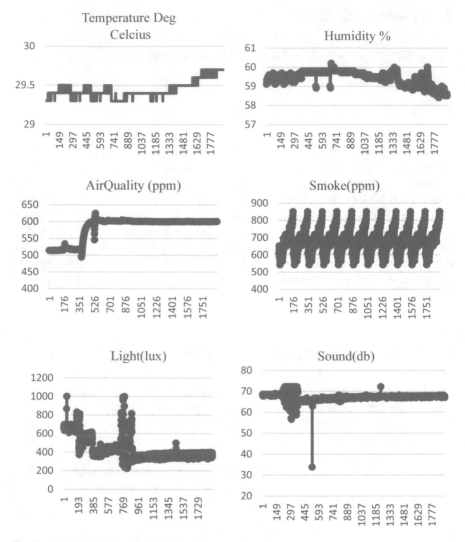

Fig. 3 Agriculture data collection based on various parameters

4.2.1 Checksum Creation

In this system, both symmetric and asymmetric and with private and public key methods are used to encrypt the data. After encrypting the formatted data, the CRC checksum will be added to data. We are using CBC method which is suitable for microcontroller processing. CBC is one of the AES128 encryption methods, there are so many mechanisms in AE128 like ECB, CTR, OCB, CFB.

The formula for CBC encryption is

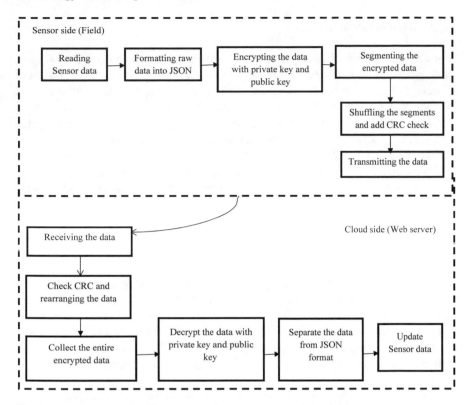

Fig. 4 Proposed encryption block diagram between sensor and cloud side

$$C_i = E_k(P_i X \text{ or } C_{i-1})$$
$$C_0 = IV \tag{1}$$

While the formula for CBC decryption is

$$Pi = (DK(C_i)) X \text{ or } C_{i-1}$$
$$C_0 = IV \tag{2}$$

i. **CRC Adding Process:**

Every block contains 16 bytes. We are going to add one more byte as a cyclic redundancy check (CRC) which helps to identify the received data is exactly transmitted data or corrupted data. With that encrypted data, we are adding some more protection so we are using the following CRC formula.

$$CRC = SUM(Byte1, Byte2 \ldots Byte16) \tag{3}$$

This CRC byte will be the last byte of the block. Now the block length is 17 byte.

4.2.2 Segmentation of Data

Segmentation is the process of slicing the data string in constant length. The segmentation process is needed for the shuffling process. Data is segmented into a number of packets, and every new packet having its own unique key and header will be created for each packet. Suppose 10 chunks of data taken, the private and public keys of the data to be different, and this will be encrypted securely. We cannot easily find which is an original packet or header because every time it keeps on changing.

Suppose for example: Data length = 512 byte.

We are going to shuffle the data with frame length = 16-byte length.

While segmenting the data 512/16 = 32 segment, data will be in separate locations. After this process, shuffling will be added.

4.2.3 Shuffling Process

After adding the CRC, the block size will be increased and these sequential data blocks will get shuffled and one more sequence ID byte gets added. This encrypted string is going to transmit over the internet. Even if an intruder tries to decode the data, it is impossible to decode the data without the key and this sequence number. In successful completion of all the encrypting phases, the data cannot able to read without a secret key and shuffling technique. This secured data only will send to the cloud.

In Table 1, Tx and Rx are the transmitter and receiver. Data shows the encryption, decryption, and packet framing process time taken by the controller in different length of data. According to Table 1, the encryption side data processing time is 10 microseconds per byte (10 μs/byte) and the decryption side data processing time is about 16 microseconds per byte (16 μs/byte).

Table 1 Process time with encryption

Tx side versus Rx side with encryption

Data length (byte)	16	32	48	64	128	256	512
Encryption side (μs)	240	360	510	700	1330	2200	2930
Decryption side (μs)	350	510	790	1080	2530	3850	4500
Encryption time per byte (μs)	15	11.25	10.62	10.93	10.39	8.59	5.7
Decryption time per byte (μs)	21.87	15.93	16.45	16.87	19.76	15.03	8.78

Fig. 5 Timing chart with encryption

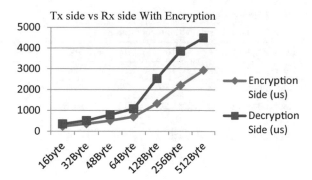

Fig. 6 Timing chart without encryption

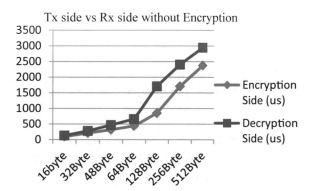

Figure 5 shows the data encryption time getting increased based on the data count. Almost encryption and decryption time are directly proportional to the data length. Figure 6 shows the data processing time getting increased based on the data count. Almost transmitter side and receiver side time are directly proportional to the data length.

The processing time per byte will be calculated by using the following formula.

$$\text{Process time per Byte} = \frac{\text{Total time taken for the string}}{\text{Total number of Bytes in the string}}$$

In Table 2, the encryption side data processing time is 6 microseconds per byte (6 μs/byte) and the decryption side data processing time is about 9 microseconds per byte (9 μs/byte) and based on Tables 1 and 2, the below calculation has been made.

TX side time per byte with encryption $= 10$ μs
TX side time per byte without encryption $= 6$ μs
TX side time increment by encryption process $= 4$ μs

Rx side time per byte with encryption $= 16$ μs
Rx side time per byte without encryption $= 9$ μs
Rx side time increment by encryption process $= 7$ μs

Table 2 Process time without encryption

Tx side versus Rx side without encryption

Data length (byte)	16	32	48	64	128	256	512
Encryption side (μs)	100	210	320	440	850	1710	2376
Decryption side (μs)	130	280	470	660	1710	2405	2945
Encryption time per byte (μs)	6.25	6.56	6.66	6.87	6.64	6.67	4.64
Decryption time per byte (μs)	8.12	8.75	9.79	10.31	13.35	9.39	5.75

According to the results of the analysis after including the encryption process into the system, 4 μs/byte getting increased. In receiver side, 7 μs/byte getting increased because of the decryption process.

Performance Metrics

According to our experiment, we need to consider some metrics to give proper security with less processing time.

$$\text{Process time TX side with encryption tTxEnc} = 10 \ \mu s$$
$$\text{Process time RX side with encryption tRxEnc} = 16 \ \mu s$$
$$\text{Maximum data process rate by controller} = 1 \ s/tTxEnc$$
$$= 1 \ s/10 \ \mu s$$
$$= 1,000,000/10$$
$$= 100 \ Kb$$

5 Conclusion

In this paper, an extension of AES 128-bit has been proposed to give better security, some additional methods checksum creation, segmentation, and shuffling are used in the algorithm. This system uses symmetric and asymmetric with both private and public key methods to encrypt the data. We have focused on the comparison of encryption and decryption time and processing time for both with and without using proposed encryption method for a securing an agriculture data. In order to perform all these things, proposed algorithm focused on mitigating the complexity of securing agricultural data. In this system, we are converting the raw data into JSON format, split the data, and also jumbled it. Once the data gets jumbled nobody can easily convert the data into the original format. So, we conclude our algorithm is more secure, reliable, and efficient as compared to the existing method. In the future development of this work, we will concentrate on the improvement of algorithm speed and decryption time reduction.

References

1. Nandyala, C.S., Kim, H.-K.(2016) Green IoT agriculture and healthcare application (GAHA). Int. J. Smart Home **10**(4), 289–300 (2016)
2. https://www.ibm.com/blogs/watson/2016/12/five-ways-agriculture-benefit-artificial-intelligence/
3. Verma, A., Kaur, S., Chhabra, B.: Improvement in the performance and security of advanced encryption standard using AES algorithm and comparison with blowfish (2016)
4. Agale, R.R., Gaikwad, D. P.: Web based intelligent irrigation and security system using internet of things. Int. J. Innovative Res. Comput. Commun. Eng. (An ISO 3297). 2007 Certified Organization) **5**(2), (February 2017)
5. Cambra, C., et al.: An IoT service-oriented system for agriculture monitoring. In: 2017 IEEE International Conference on Communications (ICC), IEEE, 2017
6. Philemon, P., Praveen, D.: Implementation of advanced encryption standard algorithm. Int. J. Sci. Eng. Res. (IJSER) **3**(3), ISSN: 2229-5518, 2012
7. Verma, Peformance analysis of data encryption algorithms. In: International Conference on Electrical and Computer Technologies (ICECT), 3rd IEEE, vol. 5, No. 7, pp. 399–403, 2011
8. https://www.permaculture.co.uk/articles/aquaponics-farming-fish

Docker Containers Versus Virtual Machine-Based Virtualization

Anuj Kumar Yadav, M. L. Garg and Ritika

Abstract Cloud computing is a paradigm based on IT that enables ubiquitous access to large pools of configurable resources which can be shared (such as computer networks, servers, storage, applications, and services) and rapidly provisioned with least effort. Cloud computing implementation in traditional ways is done using virtual machines, but nowadays a new concept of Docker containers is also gaining popularity due to its features. Containerization in some cases is treated as lightweight virtualization technique. Virtualization is used by cloud computing environments and data centres to disassociate the tools and applications from the underlying hardware. To validate this, hardware virtualization and OS-level virtualization is used. Apart from virtualization, a new technique is, containers, gaining popularity and many cloud-based deployments are using this technique. In this paper, both of these technologies are compared such that end-user can use these according to the requirement to get benefitted.

Keywords Container · Docker engine · Hypervisor · VMs

1 Introduction

Cloud computing provides many benefits to end-users on the basis of basic characteristics it poses. The characteristics include elasticity, availability, on-demand self-service, and scalability. Due to these, various users from different domains like industry, business, and other application hosting agencies are increasingly adopting cloud-based services.

A. K. Yadav (✉) · M. L. Garg · Ritika
DIT University, Dehradun 248001, India
e-mail: anujbit@gmail.com

M. L. Garg
e-mail: dr.ml.garg@dituniversity.edu.in

Ritika
e-mail: riti_79@rediffmail.com

© Springer Nature Singapore Pte Ltd. 2019
A. Abraham et al. (eds.), *Emerging Technologies in Data Mining and Information Security*, Advances in Intelligent Systems and Computing 814,
https://doi.org/10.1007/978-981-13-1501-5_12

Cloud implementation is totally based on the virtualization that is used to obtain elasticity of large pool of shared resources. There are many benefits of using virtualization like it provides flexibility of allocating the physically available resources to visualized applications and the entire task is done dynamically. In addition to this, virtualization also helps in multi-tenancy that allows more than single resource instances of virtualized applications to share the same physical resource. Multi-tenancy also helps the data centres to integrate applications into small server that reduces the operating cost. Virtualization also provides scalability and replication of applications [1].

Virtual machines (VMs) have been the central part of the cloud computing that acts as infrastructure layer which is used to provide operating system virtualization. With the help of virtual machines, various cloud services can be provided to end-users such as infrastructure, platform, and software. VMs basically create illusion that user is working on a machine that is dedicated to him/her. All the peripheral devices are also emulated using the virtualization. Virtual machine provides user a complete operating system to work upon where user can use various application software. We can treat virtualization as an additional layer between the operating system and user [2].

A Docker container is a similar concept like virtual machines but it is lightweight in terms of time consumption and resource needs. They have been suggested as a solution for more interoperable application packaging in the cloud. Containers and host share the same kernel; that is why they are called lightweights as compared to virtual machines (VMs) that have an additional layer of isolation. In a container, almost all the components (software and hardware) are shared between the applications and host OS. Isolation among the application is provided by the host operating system itself [3]. Container generally is a package that includes multiple applications as well as dependencies associated with these applications, and the package can be made shareable to the users. Along with the shareable feature, containers can be used in variety of cloud deployment models whether it is public cloud or private cloud. In addition to this, it helps in reducing the management overhead as a single OS needs to be managed. A limitation in some cases arise in terms of OS compatibility due to availability of shared kernel; in other words, we can say no Linux container can be used on Windows-based host OS and vice versa [4].

Both cloud implementation techniques are based on virtualization but handle different kind of problems and provide solution to them. The little difference is that virtual machines are more suitable for hardware allocation and hardware management, and containers are more suitable tools for software delivery or we can say prime focus is on PaaS.

So both of these are task-specific like for portable applications in the cloud computing we need a lightweight distribution of application package for development and deployment and virtual machines are preferred [1, 5].

2 Paper Organization

The rest of the paper is organized as follows. In Sect. 3, basics of virtual machines have been discussed. In Sect. 4, Docker containers' working and basics have been discussed. In Sect. 5, these two virtualization techniques are compared, and finally, the paper concludes in Sect. 6.

3 Virtual Machines

All the virtual machines are managed and controlled by virtual machine managers. These are generally termed as VMM or hypervisor, and their main focus is to provide abstraction to the underlying hardware. The system on which VMM is installed or run is termed as host machine, and all other virtual machines running inside the host machine are termed as guest. Both host and guest use almost same interface for using the different applications. Host machine and all the available guest machines running on host machine are independent of each other [6].

There are various hypervisors provided by different organizations. These hypervisors are capable of controlling the hardware and create a secure virtualized environment for users to work upon. These hypervisors can be divided into two categories: Type 1 and Type 2.

Type 1 hypervisors basically run on the top of system hardware, and due to this, they are named as native virtual machine. In other way, we can say this Type 1 hypervisor takes the place of OS and they can directly access the available hardware for their use. Type 1 hypervisors have one favour that if any virtual machine fails or does not respond for any reason then other guest OS does not get affected. Type 1 runs in kernel mode and because of which has exclusive physical CPUs. Examples of Type 1 hypervisors are Microsoft Hyper-V, VMware ESXi Server, Citrix/Xen Server, etc. On the other side, Type 2 hypervisors run within the OS installed on top of hypervisor or in other words we can say they are just like any other application software which runs under OS. Type 2 hypervisors are also known as hosted virtual machines. Examples of Type 2 hypervisors are Microsoft Virtual PC, VMware Workstation, and Oracle Virtual Box, etc. [7].

3.1 Virtual Machine Benefits

There are several benefits of using virtual machines; some of the benefits are as follows:

- As there are many virtual machines running on a single host machine, they have their own basic security zones which cannot be accessible via other virtual machines. In addition to this, there is one security layer of hypervisor as well, so we can say there are many security zones available to provide the security [8].

- In virtual machines, all the OS are isolated as well as their applications. This isolation provides better separation between the application and various activities of operating systems.
- Virtualization provides better resource utilization and improved performance as compared to traditional systems. All the underutilized resources can be used in much more efficient way by using virtualization.
- Virtualization provides better fault-tolerant environment as compared to traditional systems as if there is any server failure due to any reason, there are other servers that are available to work upon.
- Virtualization also reduces the many server requirements as it can be implemented using various virtual machines.
- Process migration is one of the major advantages provided by the virtual machines. Processes can be migrated at run-time from high-loaded virtual machines to less-loaded virtual machines. This process migration helps in saving of energy and balancing of loads. Another advantage is that due to this migration, no activity got disrupted.
- Traditional data centres generally have large pool of computing resources, and with the increasing demand, the capacity of various resources also needs to match. Due to this, large amount of energy is required to operate those systems as well as to keep them cool. With the use of virtual machines, the requirement can be minimized and resource utilization becomes efficient. And most importantly, this is also helpful for the environment.

3.2 Limitation of Virtualization

Apart from various benefits as discussed, there are several issues as well that needs to be taken into consideration while opting for virtual machine-based solution.

- Virtual machines generally share their data and interact with each other, so if the communication is not secure, then it can be exploited by the attacker and which in turn leads to security-based attack.
- Like other technologies, virtualization also has possibilities of attacks due to vulnerability like buffer overflow.
- There is a single point of failure in Type 1 hypervisor as there exists a single hypervisor. If hypervisor stops working, then the entire system gets affected.
- While using virtualization in cloud computing, one data needs to be stored far away from the local machine, and also data generally moves time-to-time from one tenant to another tenant. This movement leads to concern of leakage of data, and it can be a risk from security point of view [9].
- As Internet connection is the basic requirement of virtualization-based cloud computing, it can lead to various kinds of security risks.

- In cloud-based virtualization, user data always resides on cloud server, so user is always concerned about security of his/her data.

4 Docker Containers

Docker is an open-source project that is used for automation in a systematic way for fast deployment of applications running under a container. Docker engine is required to run the Docker containers like VM runs inside hypervisor.

Containers are a more lightweight virtualization concept, i.e. less resource- and time-consuming [10, 11]. They can be seen as more flexible tools for packaging, delivering, and orchestrating both software services and applications. Containers are built on recent advances in virtualization and therefore allows for better portability and interoperability while still utilizing operating systems' virtualization techniques [12]. Docker container is just like a directory. It contains all the things that are required for an application. In containers, isolation is done at kernel level. Docker is a platform that is used to design, deploy, and run various applications. With the help of Dockers, applications can be isolated from the available infrastructure and user can view the infrastructure as a managed application [13].

In recent times, OS-based virtualization gained popularity in terms of software to run predictably and transferring from one environment to other. By using the containers, all these isolated systems can be run on a single host operating system.

Containers lie on top layer over a server and its host operating system. Operating system can be Windows or Linux or any other operating systems. Every container not only shares the host OS but also the libraries and binary files as well that are required for application to run. All the shared components are generally read-only, and due to this, containers are lightweights, just some Mb in size. As the containers are less in size, they need few seconds to start up. By using the Dockers, jobs of application developers and system administrators become simple [14].

4.1 Different Container Models

There are different containers models or we can say delivery models according to the different operating systems. Few of them are listed as:

- Linux: Docker, LXC Linux containers, OpenVZ,
- Windows: Sandboxie,
- Cloud PaaS: Warden/Garden (in Cloud Foundry), LXC (in Openshift).

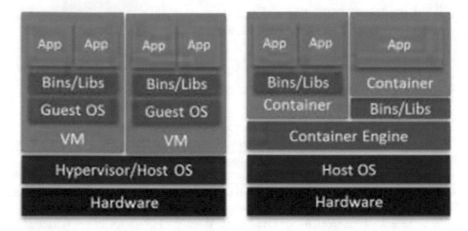

Fig. 1 Architecture comparison of virtual machines and hypervisors

4.2 Need for Containers

By the invention of virtual machines, various issues related to cloud computing like scheduling, packaging, resource management have been resolved. As the applications can be made isolated with the help of virtual machines, due to this security can be improved [15]. Cloud needs to answer the requirements of application management and packaging. Containers can give solution to these requirements in efficient way.

A container is a package that contains ready-to-deploy application parts, business logic, and middleware as shown in Fig. 1 [16].

Containers are highly scalable and safe to use. These are easy to deploy when we compare it with the virtual machines. So we can say Docker is an open-source platform that helps users and programmers to isolate application dependencies.

5 Containers Versus Virtual Machines

Both of these technologies generally provide an illusion that a single host machine can be used to run multiple machines. All of these machines running under the host machine need to be isolated from one another and also from the host machine. The difference comes in that how both of these technologies are able to achieve isolation between the different machines. A brief difference is shown in Fig. 2[17], according to which we can say that containers generally are executed on host OS and virtual machines run on hypervisor. A container engine is generally combined with the kernel of the host OS.

Further, both of these can be compared based on certain factors like:

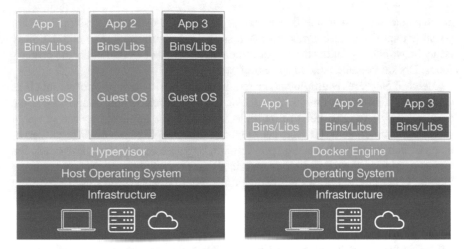

Fig. 2 Virtual machines versus hypervisors

Operating System Support: As per the architecture for both the virtual machine and Docker containers, the operating system support differs. A virtual machine contains a host OS which is able to run multiple guest OS inside different virtual machines, whereas containers need to be hosted on a single server that contains a shareable OS. The guest OS can be anything irrespective of the available host OS. On basis of this, we can say that both of these technologies can be used in different situations according to the requirement. If one wants to run many applications on a single OS kernel, then Dockers need to be preferred, and when user has many applications that need support of different operating systems, then virtual machines should be preferred. As the host OS is shared between the containers, it leads them to boot in very short span of time. So we can say maintenance overhead of containers is less than virtual machines [17].

Host/Guest Architecture: Virtual machine provides the facility to run the guest kernel that is different from the host kernel; that is not possible with containers as kernel needs to be shared.

Booting: Booting started as it starts in normal operating system, and the speed depends on the applications. Containers can start up rapidly when we compare it with virtual machines as they are less resource-centric.

Standardization: Virtual machines are generally like a complete standard operating system having all the features. On the other hand, containers are more application-specific [18].

Portability: Docker containers are the separate package which can run the needed application. As Dockers do not contain any separate operating system, so applications can be ported out easily across various platforms which is not possible in case of virtual machines. Containers can be switched on and off within seconds, much faster than that of virtual machines, because of their lightweight nature. Due to this feature, containers can be easily deployed on the servers. Virtual machines on the other side are separate server instance that is isolated with their operating system. It is not possible to port the virtual machines across various platforms due to compatibility issue. So we can say for the developers where application development is the primary focus, Docker containers need to be preferred.

Need of Servers: Multiple server requirements in Dockers are not as much as compared to virtual machines. As Dockers are lightweight and contain only the applications, there is no need of multiple servers. These applications can run on a single physical server. But if user needs to run multiple applications on different server instance and these servers need specific operating system, then the user needs the virtual machines. Virtual machines contain all the necessary library files, supporting files, and most importantly the entire operating system to work upon which is required by the particular application. So we can say lesser number of virtual machines can be accommodated to the same server if we compare it with the Dockers. As the number of virtual machines hosted on a single server is less than Dockers, we can say that the server density is lesser with virtual machines. Due to this feature, one can say that Dockers are cost-effective application hosting solution when we compare it with virtual machines.

Performance Evaluation: Both of the virtualization techniques have their specific purposes so comparison of performance evaluation is not fair.

But we can say as the containers are lightweight virtual architecture, they are less resource-intensive when we compare them with virtual machines. Due to this, start-up time of containers is very much less than that of virtual machines. Resource allocation in containers is not permanent as resource usage can vary with the load. Replication and elasticity are also much easier in containers in comparison with virtual machines as containers do not require a separate operating system.

Security: Security can be an overhead in the case of Dockers, as the host kernel is shareable among all the containers, so a single vulnerable point can lead to hacking of entire server. Due to this security concern, superuser access to the applications and also running them with root user privileges are not recommended. While in the case of virtual machines such applications are run, those need more security and privilege. Apart from this as we know that each virtual machine runs under separate or its own operating system, due to which they can use their own security features and kernel features.

Low Redundancy: Containers just need the applications to run on host operating system unlike virtual machines where entire operating system needs to install before proceeding. This results in lots of duplicity of various components. Thus on the basis of this, we can say that containers result in low redundancy when we compare them with the standard virtual machines.

Hardware Access: Applications that run under the containers have direct access to the hardware, which is not possible with virtual machines.

Resource Distribution: Containers generally require very less resources, only those which are required at that particular time, unlike virtual machines which require permanent resource allocation before start-up of virtual machines. So we can say resource distribution is optimal in case of containers [19].

Table 1 Virtual machines versus hypervisors

Comparison factor	Virtual machines	Containers
OS support	Requires guest OS	Host OS can be shared
Booting time	Require time to boot up just like traditional OS	Boots very fast
Standardization	Virtual machines are generally OS-specific	Containers are application-specific
Portability	Less portable	More portable
Need of servers	More servers required	Less number of servers required
Security	Each OS that runs in a virtual machine can provide security to user data. Security depends upon hypervisor as well	Security can be overhead as kernel is shared among the applications. So we can say it lacks security measures
Low redundancy	More redundant information as each virtual machine has its own OS	A single OS is shared so not much redundant information
Hardware access	No direct hardware access is possible	Direct hardware access is possible
Resource distribution	Large number of resources required	Fewer resources needed as compared to VM
Memory requirement	Large memory requirement as each guest has their own OS	Less memory is required as host OS is shared among applications
Files and library sharing	Sharing of file and library files is not possible	Files can be shared using Linux commands such as scp

Memory Usage: Virtual machines need complete operating system for each of the guest, due to which it requires large memory when we compare it with containers. Containers use less memory as it shares the host operating system.

Files and Library Sharing: Each virtual machine has its own OS, which contains large number of files and libraries. These files cannot be shared between different virtual machines. On the other side, containers run under host OS, no separate OS is needed by each application. So files and libraries can be shared using Linux commands.

All these differences are summarized in Table 1.

6 Conclusion and Future Scope

In this paper, virtual machines are compared with containers on the basis of various parameters. Both the techniques are based on virtualization and solve specific purpose; in some cases virtual machines can be used and in some cases containers can be preferred. If the requirement is to provide high availability and scalability, then containers are more suited. If the requirement is to create secure system, then virtual machines need to be preferred. While working in heterogeneous environment, Docker containers focus on applications and dependencies associated with them, whereas flexibility can be achieved using the virtual machines. So we can say both of these technologies are not to replace one another but these can be used simultaneously depending upon the requirement of the user. While adopting these two technologies, we can say VM provides better IaaS solution (machine portability, security and greater isolation) and Docker provides better SaaS solution (application/software portability) to end-users. If one can build hybrid architecture, then it will surely benefit variety of users.

References

1. Ranjan, R.: The cloud interoperability challenge. IEEE Cloud Comput. **1**(2), 20–24 (2014)
2. Goldberg, P.: Survey of virtual machine research. IEEE Comput. **7**(6), 34–45 (1974)
3. Soltesz et al.: Container-based operating system virtualization: A scalable, high-performance alternative to hypervisors. ACM, **41**, 275–287(2007)
4. https://www.taksatech.com/containers-vs-vms/
5. Di Martino, B.: Applications portability and services interoperability among multiple clouds. IEEE Cloud Comput. **1**(1), 74–77 (2014)
6. Silberschatz, P.B. Galvin, Gagne, G.: Operating System Concepts, 9th ed. ch. 8.5, 16, pp. 366–377, 711–740. ISBN: 978-1-118-06333-0, (2013)
7. http://www.golinuxhub.com/2014/07/comparison-type-1-vs-type-2-hypervisor.html
8. Sarna, D.E.Y.: Implementing and Developing Cloud Computing Applications. Taylor and Francis Group, LLC (2011)
9. Almond C.: A Practical Guide to Cloud Computing Security (2009)
10. Scheepers, M.J.: Virtualization and Containerization of Application Infrastructure: A Comparison. In: Presented at the 21st Twente Student Conference on IT, Twente The Netherlands June 23 (2014)
11. Pahl, C., Lee, B.: Containers and clusters for edge cloud architectures-a technology re View (2015)
12. Ranjan, R.: The cloud interoperability challenge. IEEE Cloud computer. **1**, 20–24 (2014)
13. https://docs.docker.com/introduction/understanding docker/
14. Merkel, D.: Lightweight linux containers for consistent development and deployment. Linux J. 239, 2 (2014)
15. Mao, M., Humphrey, M.: A performance study on the VM startup time in the cloud. In: 5th International Conference on Cloud Computing (CLOUD), IEEE, pp. 423–430. (2012)
16. Soltesz, S., Pötzl, H., Fiuczynski, M.E., Bavier, A., Peterson, L.: Container-based operating system virtualization: a scalable, highperformance alternative to hypervisors. ACM SIGOPS Operating Systems Rev. **41**(3), 275–287 (2007)
17. https://bobcares.com/blog/docker-vs-virtual-machines/
18. Bernstein, D.: Containers and cloud: from LXC to Docker to Kubernetes. IEEE Cloud Comput. **1**(3), 81–84 (2014)
19. http://www.channelfutures.com/technology/docker-vs-virtual-machines-understanding-performance-differences

Social IoT-Enabled Emergency Event Detection Framework Using Geo-Tagged Microblogs and Crowdsourced Photographs

Anbalagan Bhuvaneswari and Chinnaiah Valliyammai

Abstract Social Internet of things (SIoT) has obtained significant attention to address the computational intelligence for handling emergencies which can be sensed through the Internet of smart social things. In recent years, humans act as a social sensor to disseminate the information via microblogs in Online Social Networks (OSNs) such as Twitter, Weibo. At times, the reaction to certain microblogging prompts thousands of people to rethink and impulsively react on that which can handle humanity unrest situations during earthquakes, floods, Tsunamis, etc. In this paper, the streaming microblog tweets are investigated to detect the disaster events for a specified time and location. Firstly, the real-time Twitter streaming data is sensed via Apache Flume service agent. Secondly, the tweets are preprocessed and filtered tweets are clustered using the microblog-DBSCAN algorithm. The tweets belonging to various sliding window from diversified geographic event locations during disasters are aggregated. In addition, the crowdsourced photographs are added with geo-tags (location) during the period of analysis stood up as supplementary evidence to detect an event which acts as crisis recovery. The dynamic hive queries are executed to filter location-level tweets for the analysis. In contrast to conventional approaches which mainly focus the microblog textual content, we incorporate significant metadata features, namely photographs and its geo-tag, to precisely identify the events in real time.

Keywords Online social networks · DBSCAN clustering · Geo-tags
Event detection

A. Bhuvaneswari (✉) · C. Valliyammai
Department of Computer Technology MIT, Anna University, Chennai 600044, India
e-mail: bhuvana.cse14@gmail.com

C. Valliyammai
e-mail: cva@annauniv.edu

© Springer Nature Singapore Pte Ltd. 2019
A. Abraham et al. (eds.), *Emerging Technologies in Data Mining and Information Security*, Advances in Intelligent Systems and Computing 814,
https://doi.org/10.1007/978-981-13-1501-5_13

1 Introduction

The rapid advancements of Next-Generation Networks facilitate broad mainstream of network connections which are not only be accomplished among humans, rather among smart devices [1]. The typical notions, guidelines, interface methods, and changing aspects of social networking must inevitably be prolonged to the networks of smart things. The tremendous growth of ubiquity, communication speed, and cross-platform availability of social networking services has progressively deliberated as a major bridge for emergency alerting media during and after crisis situations such as disasters. In most urban zones, different categories of systems, such as cellular VoIP, WI-FI, fixed-line, and Mobile WiMAX, can deliver intersecting coverage [2] for Internet connection availability. Therefore, during the period of emergencies, when a certain type of telecommunication infrastructure is destroyed, individuals can still make use of best alternative means to communicate via OSNs. Meanwhile, Social Internet of things (SIoT) has revolutionized remarkable information production and mobile group handoff toward online user community [3]. During emergency situations, the Social IoT turns out to be virtual where each individual and device consumes an addressable geographic location, timestamp, and legible identity using Web of things (WoT) is shown in Fig. 1.

Meanwhile, the new evolution of socially connected things on the Internet is that they are able to communicate with other things in an independent and secured way with respect to the individuals [4]. The social and non-social sensor data [5] can be

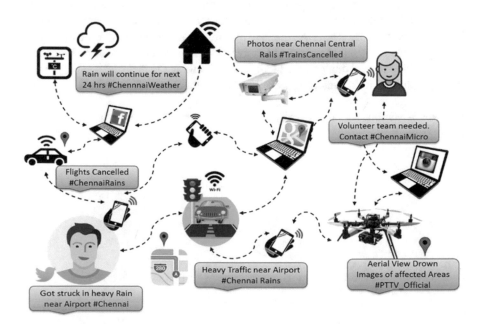

Fig. 1 Internet of social things connected during emergency situation

obtained from billions of IoT devices. In this paper, event detection is carried out by analyzing and clustering significant metadata features from microblogs. The main contributions of the proposed framework are (i) we obtain suitable strategies for the establishing and handling social connections concerning smart objects in such a way that the subsequent OSN communications are highly navigable geographically during emergency situations; (ii) we describe a possible exploratory data analytics technique for the SIoT that describes the functional prototype needed to incorporate all possible things into a social network. The rest of the paper is organized as follows. Section 2 covers the related work in the area of IoT and social stream analysis. The microblog-DBSCAN clustering approach is presented in Sect. 3. The event detection framework is proposed in Sect. 4. Our experimental settings and result analysis are discussed in Sect. 5. Finally, the paper concludes in Sect. 6 along with suggestions for future work.

2   Related Work

The data which is generated from SIoT is in structured, semi-structured, or unstructured format. Social media event monitoring [6] is growing day by day wherein analysis of social streams plays a vital role to understand human behavior particularly during disaster events. Social media trust [7] has empowered information sharing at an unprecedented scale; however, it has challenged small to large organizations for extracting relevant information at a rapid scale. Within the humanitarian domain, SDN-enabled big data analytics [8] often provides sophisticated tools and techniques to improve the distributed work practices by mining novel data sources across the globe.

The process of mining behaviors from social media crowds [9] as well as intelligence from Web sources for real-world event traces presents an opportunity to assimilate relevant knowledge. The multiple services and procedures with complex humanitarian organizations during emergency situations during landslide detection [10] are resolved using social networks. The significance of fragmented accessibility leads to failure in cross-platform architecture. It will still prevent implementing an authentic SIoT environment [11] on top of complex applications where it can be deployed. The Web of things (WoT) was the commonly referred for implementing Web-based protocols with specific gateways and proxies of devices. Mostly, the existing applications use Representational State Transfer (RESTful) APIs and Device Profile for Web Services (DPWS). It is also provisioning APIs for thing-specific versions, such as Thing-REST. The WoT paradigm has certain restrictions, caused by the complications in accessing, determining, retrieving, and exploiting the devices and its facilities [12].

Our proposed work is quite different from various pioneering research works to accomplish dynamic event databases for the acceleration of event detection by indexing tweets [13] and make use of multiple tweets from all users. Our work addressed these challenges using multi-pass clustering and the semantic similarity distance to competently detect events in large streaming social networks. The major drawback of event detection is the volume of tweets with maximum noisy tweets. A piece of visual exploration [14] of passenger flows work which tried to explore the necessity of physical sensors data and social media like Twitter, Weibo, YouTube, Facebook information has a noteworthy motivation to our work.

3   Microblog-DBSCAN Clustering Approach

The DBSCAN [10] is density-based spatial clustering of applications with noise which is used for cluster formation. The microblog-DBSCAN approach which is proposed to use typically discovers clusters using arbitrary silhouettes and size for a large volume of twitter data. The clusters are uncovered by examining the tweets that are densely or loosely grouped together. The more similar tweets in the geographic space that are often divided by regions of low-density tweets have been loosely connected. Each tweet is considered as a data point in a given geographic region of radius R (in meters) in our experimentation. The clusters are created when the quantity of neighbors is larger than or equivalent to MinPts which is shown in Fig. 2, where the circle radius is given by ε and minimum number of neighbor points (MinPts) is set to a positive value. The data point is moved to a specific cluster with its neighborhood at a distance ε, at least a minimum number of data points (MinPts) with two constraints as follows.

$$\text{DistPred}(A, A') \equiv |A - A'| \leq \varepsilon \tag{1}$$

$$\text{MinWeight}(A) \equiv \{|A \in D||A - A'| \leq \varepsilon\} \geq \text{MinPts} \tag{2}$$

The initial point is grouped with adjacent neighbors to a cluster. Later, the initial tweet data point is marked as visited. The algorithm recursively evaluates the method for its entire neighbors to create a cluster of similar tweets that represents an event.

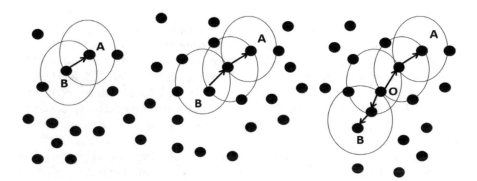

Fig. 2 Directly density-reachable and density connectivity in data space

As a result, all these density-connected data points in space are the clusters with similar tweets which represent an event. If the total number of neighborhood points is fewer than MinPts, the data point is noted as noise and can be ignored from further iterations. According to the above definitions of Eqs. (1) and (2), the prerequisites obtain the maximal density-connected data point spaces to the data points containing cluster in each feature space. The neighborhood tweet can be predicted in the entire course of event detection within the same geographic location where the event occurs. Particularly, microblog–DBSCAN uses positively valued parameters as the threshold for tweet, tweet-time ε_1, tweet-location ε_2 and photograph tags, photograph-time ε_3, photograph-location ε_4. The neighborhood predicate for a tweet T in microblog–DBSCAN is expressed as follows.

$$\text{DistPred}(T \ T') = |A_1 \quad A_1'| \leq \varepsilon_1, |A_2 \quad A_2'| \leq \varepsilon_2, |A_3 - A_3| \leq \varepsilon_3, |A_4 - A_4'| \leq \varepsilon_4 \tag{3}$$

where $|A_i - A_i'|$ represents the distance functions for the dimensions, namely text, text-time, text-location, photograph-time and, photograph-locations. The cosine similarity distance technique is used to measure cohesion distance between tweets within clusters which are defined by the cosine distance. The similarity distance can be given as follows.

$$\text{Sim}(d, c) = \frac{WF^m . WF^n}{\|WF^m\| \cdot \|WF^n\|} \tag{4}$$

where WF is the word frequency for mth cluster and nth tweet. In particular, the cosine distance is used for the temporal and spatial dimensions of both tweet texts and photographs' geo-tags given that latitude and longitude coordinates of the event occurred. In addition, Term Frequency–Inverse Document Frequency (TF-IDF) is a statistical weighting method which allocates each term within a tweet a weight that reveals the term's significance inside the document. The weight of a tweet is the cumulative sum of all individual term weights within the tweet. The significance of a term is determined as follows.

$$TF_{IDF} = tf_{ij} * \log_2 \frac{N}{df_j} \tag{5}$$

The TF-IDF value contains two parts. The term frequency component (tf) allocates extra weight to words that take place repeatedly within a document. The inverse document frequency component (IDF) represents certain common stop words which are frequent.

4 Proposed Event Detection Framework

The proposed work identifies the disaster-related event based on the tweet. The twitter tweets are collected using Apache Flume via Twitter Streaming API. The tweets are accessed iteratively to retrieve the most recently collected tweets. The tweets were collected and parsed for effective text analysis. The tweets are aggregated, preprocessed, and prepared for further analysis. The event detection framework is shown in Fig. 3.

The actual message containing the text portion of the tweet and photographs collected through crowdsourcing is considered. The date, time, geographic location embedded in the tweets and photographs are taken as input to the framework. The dataset undergoes microblog-DBSCAN clustering algorithm to cluster the similar tweets which represent event detection in the online social stream.

4.1 Data Collection and Preprocessing

The Hadoop framework is proficient enough to build up applications competent to run on clusters of commodity hardware, and they could execute required statistical analysis for vast amounts of data. Apache Hadoop and Hive support open-source software framework for distributed processing and storage of massive online data on clusters. Apache Flume collects high volume of streaming data and online logs in real time from multiple sources. Flume belongs to Hadoop ecosystem component which is used to gather, aggregate, and shift a large amount of logbook data from diverse sources to a central data repository. It starts with the client (Web Server) which transfers the event to a source functioning in the agent. When spikes in client-side activity are produced quicker than what the provisioned capability on the destination terminus can handle, the size of the channel will be increased automatically. The Twitter Streaming API source obtains the tweets related to an event and will be delivered to collection Channels is shown in Fig. 4.

Fig. 3 Event detection framework

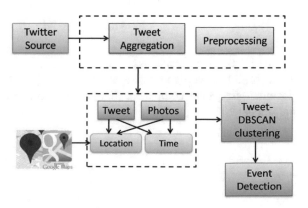

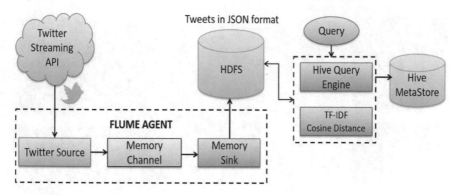

Fig. 4 Data collection and workflow overview

In preprocessing stage, tweet IDs are generated by Twitter to identify each tweet which is removed for privacy. Stop words are also removed; tweet Identity (ID) is the number produced by Twitter to recognize each tweet. In the content investigation, stemming process is carried out to reduce abnormal assemblies. For example, the words 'now' and 'past' indicate the timeline of the event.

4.2 Microblog—Cluster-Level Features

The microblog-DBSCAN is used to cluster a stream of Twitter messages in real time. It is mandatory to choose a clustering algorithm that is highly scalable and that does not require a priori knowledge of the number of clusters as follows.

Algorithm : Proposed Microblog -DBSCAN algorithm

Input: Tweets $T_1, T_2, \ldots T_n$ and global threshold $\varepsilon_1, \varepsilon_2, \varepsilon_3, \varepsilon_4$
Output: Cluster of Tweets
begin
  1: Select a point A initially
  2: **for all** data points
  3:   Compute cosine distance with respect to $\varepsilon_1, \varepsilon_2, \varepsilon_3, \varepsilon_4$ and MinPts
  4:     **if** (A = border point) **then**
  5:         no other data points are density-reachable from A
  6:             **else if** (A = core point)
  7:         Discover all points density-reachable from A
  8:     **end if**
  9:         Continue until all the tweet data points are processed
  10: **end for**
end

Fig. 5 Crowdsourced tweets and photograph geo-tags in Chennai

Twitter tweet generation and photograph posts are persistently evolving, and fresh events will be added to the flume agent stream over time. Based on these observations, the microblog-DBSCAN clustering is used with a threshold parameter that is tuned empirically during a training phase.

5   Experiments and Results

For implementation, a dataset of around 1000 tweets captured in the year 2015 is used. The tweets are collected using Twitter Streaming API using Flume agent to identify the corpus or word that bursts out to detect a disaster event. Out of 1000 tweets, only 670 valid tweets are filtered in JSON format and moved to HDFS. The potential metadata features, namely time, location, tweet text, and photograph geo-tags, are extracted. In turn, the file is stored in indexed table format, wherein the location, time, and text used in the dataset are queried using Hive. The dataset containing tweets is clustered using microblog-DBSCAN, and the performance of the clustering algorithm is evaluated. The sample microblog tweet text is associated with photograph geo-tags. It plots the exemplary photographs into a Google Map as shown in Fig. 5.

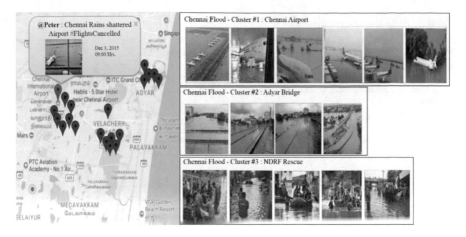

Fig. 6 Visualization of photographs and geo-tags in clusters

Its metadata describes from where the photograph is taken, time of the photograph and posted along with geo-tags. The multiple metadata features are fused from both tweet text and photographs; the mean and variance of the special features such as geo-tags, photograph-time which are most likely not in similar range, in turn, lead to sub-optimal clustering. In order to overcome the uneven feature availability, the tweets and photographs are cleaned and filtered from certain sources; we simply assign weights for each significant feature multiple times. Nearly 100 photographs are manually crawled and stored into a csv file with photograph identity and metadata descriptor for the same. In the first step, stop words are removed and content-specific terms are identified such as *'Chennai,'* *'Airport,'* *'Railway Junction.'* Furthermore, from each photograph's geo-tagging, the precise location and the corresponding place name, for instance, *'Adyar, Chennai,'* are identified using geo-names and it is mapped to Google Map. Using photograph metadata features, the pairwise text similarity is calculated between the photographs related to the events. Our experiment has been crowdsourced with a team of four members to whom the task is to verify the related photographs for clustering. Being native habitat to the location 'Chennai,' the crowdsourcing task is successful which yields a more accurate result to identify the event. It is performed by varying the thresholds for microblog time, location, tweet text, and photograph geo-tags. For good measure, we mapped the tweets from the cluster with the second-most tweets with the word 'Chennai' in it as well in Fig. 6.

The algorithm returns the list of clusters that have tweeted with 'Chennai, Airport, Rains' in them; the one with the most hits is cluster 3.04 which has 600 tweets in it. We can use Google Maps API to visualize these points on the map to provide some spatial context. From Table 1, the precision, recall, and accuracy are set as performance evaluation metric for event detection by setting 0.00781 as ε_75 for obtaining a maximum number of clusters.

The subset of 100 random tweets is visualized from the cluster for processing speed estimation. In the default configuration, the Twitter tweets are collected

Table 1 Performance evaluation

Features configuration	Tweet text		Photographs tags		# of clusters	Precision	Recall	Accuracy
	ε_1	ε_2	ε_3	ε_4				
Without photographs	12	50	–	–	31	0.51	0.71	0.62
Without text	–	–	25	81	55	0.64	0.73	0.67
With text + photograph tags	20	73	29	90	60	0.52	0.79	0.79
With text + tags + Date	25	81	35	10	36	0.65	0.81	0.81
Proposed features	30	150	45	120	47	0.71	0.86	0.90

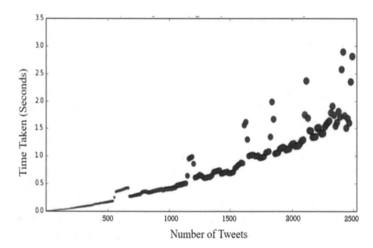

Fig. 7 DBSCAN processing time at ε_75

and evaluated. The ground truth is that the Flood in Chennai 2015. The various threshold settings are used on tweet-time ε_1, tweet-location ε_2 and photograph tags, photograph-time ε_3, photograph-location ε_4. Later, the tweets are clustered based on neighborhood distance and minimum tweet points. The computation of clustering is iteratively carried out to find ε_75 that will yield a maximum number of clusters. The approximation of 100 km per degree is estimated precisely as $\varepsilon = $ km/100. The outcomes show that the implementation, with the value of ε, does scale-up reasonably well with varying sample size. It processes 10,000 samples within an acceptable range of 10–20 s. The implementation is carried out using Sklearn (Python) machine learning package, and its processing time for analyzing tweets is shown in Fig. 7.

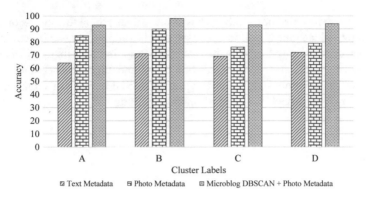

Fig. 8 Event detection accuracy

For all clustering experiments, the threshold settings for all photographs with geo-located tags are used and externally retrieved event days by considering the ground truth and herein the result only slightly drops which is reasonably acceptable. Further, the location and time window of the events to 30 min and 150 km does help moderately to our experiments. The results suggest that the exact time, date, photographs, and associated geo-tags are more descriptive to identify the event. After clustering the tweets up to 75 clusters, the manual annotation is carried out to verify the correctness of cluster precision, recall, and accuracy. The human annotators have chosen four clusters that are highly significant to detect the event and labeled it as A, B, C, and D for which the accuracy for text metadata, photograph metadata, text and photograph metadata is shown in Fig. 8. The proposed work significantly improved the accuracy which uses photograph metadata when compared to the baseline approach which uses only text metadata in a tweet.

6 Conclusion and Future Work

The integration of social networking notion into the Internet of things comprises more critical technical challenges in exploratory data analysis. A semi-clustering approach for event detection problem in location-based social networks is proposed in this paper. The basis is a pairwise similarity calculation between metadata features in microblogs. Microblog-DBSCAN clustering is validated to be a very effective technique to extract event clusters of the identical spatial entities in diversified contexts (text, photographs, tags, etc.) with an accuracy of 90%. The mining on the geo-tagged text of the clusters along with crowdsourced photograph metadata is proposed to mine similar tweets per cluster. It is observed that the semi-supervised clustering technique on photograph metadata is far more accurate than on text metadata. The relevance accuracy is verified by crowdsourcing the photographs with the mined event clusters. Our observed evaluation confirmed some previous empirical results and also revealed new findings based on various global thresholds of the sliding window and location check-ins. As a part of the future work, we intend to extend our framework for supervised deep learning approach for multi-linguistic users. Various imminent researchers focus toward on the stimulating demand of exchange in information built upon social notions, typically by real humans in social networks, with a vision toward deploying Social Internet of things.

References

1. Jara, A.J., Parra, M.C., Skarmeta, A.F.: Participative marketing: extending social media marketing through the identification and interaction capabilities from the Internet of things. Pers. Ubiquit. Comput. **18**, 997–1011 (2013)
2. Guo, B., Yu, Z., Zhou, X., Zhang, D.: Opportunistic IoT: Exploring the social side of the internet of things. IEEE 16th International Conference on CSCWD, 925–929. (2012, May)
3. Huang, C.M., Shao, C.H., Xu, S.Z., Zhou, H.: The social internet of thing (S-IOT)-based mobile group handoff architecture and schemes for proximity service. In: IEEE Transactions on Emerging Topics in Computing. (2017)
4. Li, S., Da Xu, L.: Securing the Internet of Things. Syngress (2017)
5. Guo, B., Zhang, D., Yu, Z., Liang, Y., Wang, Z., Zhou, X.: From the internet of things to embedded intelligence. WWW **16**(4), 399–420 (2013)
6. Bello-Orgaz, G., Jung, J.J., Camacho, D.: Social big data: recent achievements and new challenges. Inf. Fusion **28**, 45–59 (2016)
7. Lin, Z., Dong, L.: Clarifying trust in social internet of things. (2017). arXiv preprint arXiv:17 04.03554
8. Hu, H., Wen, Y., Gao, Y., Chua, T.S., Li, X.: Toward an SDN-enabled big data platform for social TV analytics. IEEE Network **29**(5), 43–49 (2015)
9. Loke, S.W.: Social links for crowds and things. In: Crowd-Powered Mobile Computing and Smart Things, pp. 55–63. Springer International Publishing, Berlin (2017)
10. Musaev, A., Wang, D., Pu, C.: LITMUS: a multi-service composition system for landslide detection. IEEE Trans. Serv. Comput. **8**(5), 715–726 (2015)
11. Kranz, M., Roalter, L., Michahelles, F.: Things that twitter: social networks and the internet of things. In: What can the Internet of Things do for the Citizen (CIoT) Workshop at 8th International Conference on Pervasive Computing, 1–10 (2010)
12. Atzori, L., Iera, A., Morabito, G.: Siot: Giving a social structure to the internet of things. IEEE Commun. Lett. **15**(11), 1193–1195 (2011)
13. Cai, H., Huang, Z., Srivastava, D., Zhang, Q.: Indexing evolving events from tweet streams. IEEE Trans. Knowl. Data Eng. **27**(11), 3001–3015 (2015)
14. Itoh, M., Yokoyama, D., Toyoda, M., Tomita, Y., Kawamura, S., Kitsuregawa, M.: Visual exploration of changes in passenger flows and tweets on mega-city metro network. IEEE Transactions on Big Data **2**(1), 85–99 (2016)

Part II
Image Encryption

Cryptanalysis of a Chaotic Key-Based Image Encryption Scheme

Pratyusa Mukherjee, Krishnendu Rarhi, Abhishek Mishra and Abhishek Bhattacharya

Abstract Security of multimedia data is a major concern due to its widespread transmission over various communication channels. Hence, design and study of good image encryption schemes has become a major research topic. During the last few decades, there has been an increase in chaos-based cryptography. This paper proposes an attack on a recently proposed chaos-based image encryption scheme. The cryptosystem under study proceeded by first shuffling the original image to disturb the arrangement of pixels by applying a chaotic map several times. Second, a key stream is generated using Chen's chaotic system to mix it with the shuffled pixels to finally obtain the cipher image. A chosen-ciphertext attack can be done to recover the system without any knowledge of the key. It simply demands two pairs of plaintext–ciphertext to completely break the cryptosystem.

Keywords Cryptanalysis · Chaos · Key stream · Shuffle · Chosen-ciphertext attack · Image encryption

1 Introduction

With rapid dissemination of digital multimedia data over the Internet, security of digital information is a major concern. Recently, due to advancements in the theory and application of chaos, many researchers are focusing on the chaotic cryptography.

P. Mukherjee (✉)
Indian Institute of Engineering Science and Technology, Shibpur, India
e-mail: pratyusa.mukherjee@gmail.com

K. Rarhi · A. Mishra · A. Bhattacharya
Institute of Engineering & Management, Kolkata, India
e-mail: rarhik@gmail.com

A. Mishra
e-mail: amkolkata13@gmail.com

A. Bhattacharya
e-mail: abhishek.bhattacharya@iemcal.com

© Springer Nature Singapore Pte Ltd. 2019
A. Abraham et al. (eds.), *Emerging Technologies in Data Mining and Information Security*, Advances in Intelligent Systems and Computing 814,
https://doi.org/10.1007/978-981-13-1501-5_14

Chaotic maps have attracted the attention of cryptographers because of the following fundamental properties [1]:

- Chaotic maps are deterministic, meaning that their behavior is predetermined by mathematical calculations.
- Chaotic maps are unpredictable and nonlinear because they are sensitive to initial conditions. Even a very slight change in the starting point can lead to a significantly different outcome.
- Chaotic maps appear to be random and disorderly but, in fact, they are not; beneath the random behavior is an order and pattern.

These properties can be connected with the "confusion" and "diffusion" property in cryptography. Hence, chaos are used to enrich the design of new ciphers. We are mainly interested on those schemes dedicated to image encryption. Image encryption is a bit different from text encryption due to certain characteristics of an image such as redundancy of data, a strong correlation among adjacent pixels, and a minute change in the attribute of any pixel that does not drastically degrade the quality of the image. Some of the encryption schemes have been cryptanalyzed and have been found to be insecure. In this paper, we propose a break for the image encryption algorithm proposed in [1].

Chaos-based encryption schemes can be broadly divided into two phases: "permutation" and "substitution." Permutation is used to move the image pixels from one position to another, whereas substitution is used to make the statistics of the cipher independent on the plaintext. Similarly, the algorithm proposed in [1] also constitutes two parts: first shuffling the original image to disturb the arrangement of pixels by applying Arnold cat map [1] several times and second, a key stream is generated using Chen's chaotic system [2, 3] to mix it with the shuffled pixels to finally obtain the cipher image.

Initially, this paper introduces the cryptosystem, as a basis of the whole letter in Sect. 2. In Sect. 3, we describe the method proposed to totally break the system. We also provide an illustrative example followed by the simulation results in Sect. 4. The paper concludes with final remarks in Sect. 5.

2 Description of the Cryptosystem

2.1 Arnold Cat Map

Assume that we have an $N \times N$ image P with pixel coordinates $I = \{(x, y)|x, y = 0, 1, 2 \ldots N - 1\}$.

Arnold cat map is given by

$$\begin{bmatrix} x' \\ y' \end{bmatrix} = A \begin{bmatrix} x \\ y \end{bmatrix} = \begin{bmatrix} 1 & p \\ q & pq + 1 \end{bmatrix} \begin{bmatrix} x \\ y \end{bmatrix} \bmod N \qquad (1)$$

where p, q are positive integers and x', y' are the coordinate values of the shuffled pixel. After iterating the map "n" times, we have

$$\begin{bmatrix} x' \\ y' \end{bmatrix} = A^n \begin{bmatrix} x \\ y \end{bmatrix} \bmod N = M \begin{bmatrix} x \\ y \end{bmatrix} \bmod N \qquad (2)$$

where

$$M = \begin{bmatrix} m1 & m2 \\ m3 & m4 \end{bmatrix} = A^n \bmod N \qquad (3)$$

The shuffled image S is related to the original image P as $S(x', y') = P(x, y)$ where $0 \leq x, y \leq N-1$.

2.2 Chen's Chaotic System [2]

Chen's chaotic system is a set of differential equations given as

$$\dot{x} = a(y - x).$$
$$\dot{y} = (c - a)x - xz + cy.$$
$$\dot{z} = xy - bz. \qquad (4)$$

where $a = 35$, $b = 3$, and $c \in [20, 28.4]$ for Chen's system to be chaotic.

2.3 Encryption Algorithm

Secret keys of the algorithm are parameters p, q, n of Arnold cat map and initial values x_0, y_0, z_0 of Chen's chaotic system [4].

- Shuffle the image P using Arnold cat map and obtain the shuffled image S. Scan the image S row by row and arrange its pixels as sequence $S = \{s_1, s_2 \ldots s_{N \times N}\}$
- Iterate Chen's chaotic system and obtain the real values x_i, y_i, z_i, where $1 \leq i \leq N_0$ and $N_0 = (N \times N)/3$.
- Obtain the key space $K = \{k_1, k_2 \ldots k_{N \times N}\}$ from the key generator.
- Obtain the encrypted system as
  $C = \{c_1, c_2 \ldots c_{N \times N}\}$ as $c_i = s_i \oplus k_i$
  where $\oplus$ represents bitwise OR operation.

3 Cryptanalysis of the Existing Cryptosystem

Here the intruder intercepts a cipher image and has to retrieve the original plain image from it. He has other information available. It is assumed that he has access to the communication channel and the sender or recipient end to gather valuable information which he can utilize later to decrypt the intercepted cipher image.

So there are two tasks: one, to retrieve the key [5] and the other to obtain the original image from shuffled image [6].

3.1 Retrieving the Key

According to the encryption algorithm, the key is constructed by iterating the Chen's equation. Since it is not possible to guess the initial parameters, we cannot calculate the particular solution of Chen's equations and obtain the key.

Hence, a method that does not depend on reconstructing the key has to be designed.

As per the definition of chosen-ciphertext attack, the intruder can choose arbitrary ciphertext and have access to plaintext decrypted from it [7].

Let us assume that we have transient admittance to the encrypting system through Lunch Break Attack.

The task of encryption machine is to take input from user and XOR it with the key and produce the output. The key is kept hidden from user.

If a user sends "0" as input, then the following steps take place:

$$
\begin{aligned}
\text{Output} &= \text{Input} \oplus \text{Key} \\
&= 0 \oplus \text{Key} \\
&= \text{Key}
\end{aligned}
$$

Hence, unknowingly, the encryption machine actually provides the key as output. This key can be used to decrypt other messages. We will use this idea in the following section.

3.2 Obtaining the Original Image from Shuffled Image

Assume that we have an $N \times N$ cipher image $C = C_1 \, C_2 \, C_3 \, \ldots \, C_{N \times N}$, to decrypt without knowing the secret parameters. Let us again assume that a Lunch Break Attack has been performed. The steps leading to the recovery of the plain image P from the intruded cipher image C are described below:

Step 1   We construct an image $P1$ of size 4×4, whose all the elements are 0. For example, let

$$P1 = \begin{pmatrix} 0\,0\,0\,0 \\ 0\,0\,0\,0 \\ 0\,0\,0\,0 \\ 0\,0\,0\,0 \end{pmatrix}$$

Since all elements of image $P1$ is 0, all elements of the corresponding shuffled image $S1$ will also be 0.

Step 2   We now request the cipher image of $P1$ from the encryption machinery. Let this cipher image be $C1$

According to the algorithm

$C1 = $ Key $\oplus$ Shuffled Image $= $ Key $\oplus\, 0 = $ Key

Hence, we have obtained the key without even knowing the secret parameters or method of its construction.

Step 3   Now we have already obtained the key so perform the following steps:

Actual shuffled image for original plain image say $S = C \oplus$ Key

Where C is the intercepted cipher image

Hence, we have obtained our actual shuffled image. Now we simply have to get the original image corresponding to this shuffled image.

Step 4   We next construct another image $P2$ of size 4×4, whose elements are 1, 2, 3 ... 4×4.

$$P2 = \begin{pmatrix} 1 & 2 & 3 & 4 \\ 5 & 6 & 7 & 8 \\ 9 & 10 & 11 & 12 \\ 13 & 14 & 15 & 16 \end{pmatrix}$$

Step 5   We now request the cipher image of $C2$ of $P2$ from the encryption machinery.

Step 6   Since we have already obtained the key, perform the following steps:

Shuffled image for $P2 = S2 = C2 \oplus$ Key

Step 7   Now we feed the intercepted cipher image C, the obtained actual shuffled image S, image $P2$, and its corresponding shuffled image $S2$ to the analyzer machine. The analyzer machine converts the images into their corresponding matrices and does the following jobs:

- Compare the matrices $P2$ and $S2$ to understand the shuffling algorithm. Analyzer machine searches for the new position of $P2$'s first element in $S2$ and similarly repeat the same process for other elements to find their new positions and understand the shuffling.
- Use the relation to move the elements of S into another matrix P.
- This P is the original image we are looking for.

Table 1 gives a detailed description and example of the proposed method. Figure 1 gives simulation results on a 100×100 image of a sunflower.

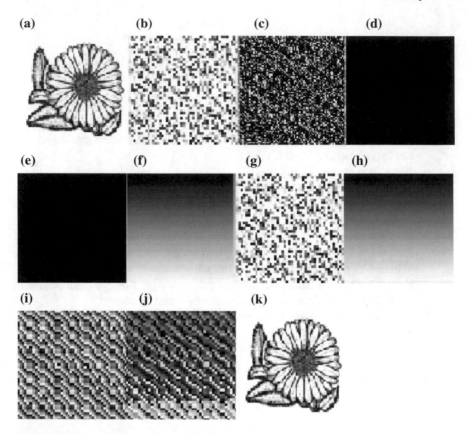

Fig. 1 **a** Original plain image P; **b** shuffled image S; **c** cipher image C (intercepted by intruder); **d** first constructed image $P1$; **e** corresponding shuffled image $S1$; **f** corresponding cipher image $C1$; **g** actual shuffled image calculated; **h** second constructed image $P2$; **i** corresponding shuffled image $S2$; **j** corresponding cipher image $C2$; **k** recovered image M

Table 1

Steps to recover the plain image P of size 4×4 from the intercepted cipher image C using chosen-ciphertext attack:

$$
\begin{array}{ccc}
\text{PLAIN IMAGE } (P) & \text{KEY } (K) & \text{CIPHER IMAGE } (C) \\
\begin{pmatrix} 23 & 45 & 64 & 32 \\ 179 & 180 & 26 & 58 \\ 67 & 136 & 139 & 20 \\ 17 & 99 & 220 & 100 \end{pmatrix} &
\begin{pmatrix} 186 & 24 & 39 & 72 \\ 23 & 87 & 47 & 13 \\ 221 & 49 & 50 & 2 \\ 44 & 32 & 65 & 110 \end{pmatrix} &
\begin{pmatrix} 174 & 123 & 7 & 252 \\ 6 & 23 & 156 & 134 \\ 240 & 11 & 186 & 102 \\ 54 & 99 & 157 & 121 \end{pmatrix}
\end{array}
$$

Step 1   We construct an image $P1$ of size 4×4, whose all elements are 0.

$$P1 = \begin{pmatrix} 0\,0\,0\,0 \\ 0\,0\,0\,0 \\ 0\,0\,0\,0 \\ 0\,0\,0\,0 \end{pmatrix}$$

Step 2 We now request the cipher image of $C1$ of image $P1$ from the encryption machinery.

According to the encryption algorithm, the machinery will perform the following operation using the key, i.e., hidden from user.

$C1 =$ Shuffled Image $\oplus$ Key

$$C1 = \begin{pmatrix} 186 & 24 & 39 & 72 \\ 23 & 87 & 47 & 13 \\ 221 & 49 & 50 & 2 \\ 44 & 32 & 65 & 110 \end{pmatrix} \oplus \begin{pmatrix} 0\,0\,0\,0 \\ 0\,0\,0\,0 \\ 0\,0\,0\,0 \\ 0\,0\,0\,0 \end{pmatrix} = \begin{pmatrix} 186 & 24 & 39 & 72 \\ 23 & 87 & 47 & 13 \\ 221 & 49 & 50 & 2 \\ 44 & 32 & 65 & 110 \end{pmatrix}$$

Hence, we obtained the key without even knowing the secret parameter or method of its construction.

Step 3 Now since we have already obtained the key, we can obtain the original shuffle image.

Actual Shuffle Image $=$ Intercepted Cipher Image $\oplus$ Key

$$\begin{pmatrix} 174 & 123 & 7 & 252 \\ 6 & 23 & 156 & 134 \\ 240 & 11 & 186 & 102 \\ 54 & 99 & 157 & 121 \end{pmatrix} \oplus \begin{pmatrix} 186 & 24 & 39 & 72 \\ 23 & 87 & 47 & 13 \\ 221 & 49 & 50 & 2 \\ 44 & 32 & 65 & 110 \end{pmatrix} = \begin{pmatrix} 20 & 99 & 32 & 180 \\ 17 & 64 & 179 & 139 \\ 45 & 58 & 136 & 100 \\ 26 & 67 & 220 & 23 \end{pmatrix}$$

Step 4 We next construct another image $P2$ of size 4×4, whose all elements are 1, 2, 3, … 16.

$$P2 = \begin{pmatrix} 1 & 2 & 3 & 4 \\ 5 & 6 & 7 & 8 \\ 9 & 10 & 11 & 12 \\ 13 & 14 & 15 & 16 \end{pmatrix}$$

Step 5 Now let us request the cipher image of $C2$ of $P2$ from encryption machinery

$$C2 = \begin{pmatrix} 182 & 22 & 35 & 78 \\ 26 & 84 & 42 & 6 \\ 223 & 57 & 56 & 18 \\ 43 & 41 & 78 & 111 \end{pmatrix}$$

Step 6   Perform the XOR operation of ciphertext with the key to obtain shuffled image of plain image.

$$
\begin{pmatrix} 182 & 22 & 35 & 78 \\ 26 & 84 & 42 & 6 \\ 223 & 57 & 56 & 18 \\ 43 & 41 & 78 & 111 \end{pmatrix} \oplus \begin{pmatrix} 186 & 24 & 39 & 72 \\ 23 & 87 & 47 & 13 \\ 221 & 49 & 50 & 2 \\ 44 & 32 & 65 & 110 \end{pmatrix} = \begin{pmatrix} 12 & 14 & 4 & 6 \\ 13 & 3 & 5 & 11 \\ 2 & 8 & 10 & 16 \\ 7 & 9 & 15 & 1 \end{pmatrix}
$$

Step 7   Now we need to feed the intercepted image C, the obtained actual shuffled image S, image $P2$, and its corresponding shuffled image $S2$ to analyzer machine. The analyzer machine converts the image into their corresponding matrices and does the following job

- Compare the matrices $P2$ and $S2$ to understand the shuffling algorithm. Analyzer machine searches for the new position of $P2$'s $(1, 1)$ element in $S2$. From our example, $P2$'s $(1, 1)$ element has moved to $(4, 4)$ in $S2$. Thus, move $(4, 4)$ element of S to $(1, 1)$ position in another matrix M.
- Similarly, repeat the process for all other elements to find their new position in the resultant matrix M.
- This M is the original image we are looking for.

$$
P2 = \begin{pmatrix} 1 & 2 & 3 & 4 \\ 5 & 6 & 7 & 8 \\ 9 & 10 & 11 & 12 \\ 13 & 14 & 15 & 16 \end{pmatrix} \qquad S2 = \begin{pmatrix} 12 & 14 & 4 & 6 \\ 13 & 3 & 5 & 11 \\ 2 & 8 & 10 & 16 \\ 7 & 9 & 15 & 1 \end{pmatrix}
$$

$$
M = \begin{pmatrix} 23 & 45 & 64 & 32 \\ 179 & 180 & 26 & 58 \\ 67 & 136 & 139 & 20 \\ 17 & 99 & 220 & 100 \end{pmatrix} \qquad S = \begin{pmatrix} 20 & 99 & 32 & 180 \\ 17 & 64 & 179 & 139 \\ 45 & 58 & 136 & 100 \\ 26 & 67 & 220 & 23 \end{pmatrix}
$$

4   Results

See Fig. 1.

5 Conclusion

In this paper, we have cryptanalyzed a recently proposed image encryption scheme by chosen-ciphertext attack. It is seen that the reuse of the key stream more than once makes it weak against chosen-ciphertext attack [8]. The generated key stream is totally independent of the plaintext as well as ciphertext, which makes it unchangeable in every encryption process. The process of shuffling is also predictable, and we can extract the original image from the shuffled image. Two couples of plaintext/ciphertext are sufficient to break the system in a chosen-ciphertext attack scenario. We can make the cryptosystem more secure and harmless to described attack by changing the key stream for every encryption procedure. It can be achieved by using one-time pads as either keys or generate the key stream dependent to the plaintext or ciphertext. One-time pad [9] is a kind of stream cipher that never reuses its key, but in this case distribution of the key is a problem and hence one-time pad is impractical in a real secure communication. The solution of making the generation of key stream dependent to the plaintext or the ciphertext is more adequate, secure, and practical than one-time pads, and our future works concern the same.

References

1. Al-Maadeed, S., Al-Ali, A., Abdalla, T.: A new chaos-based image-encryption and compression algorithm. J. Electr. Comput. Eng. **2012**, Article ID 179693
2. Chen, G., Ueta, T.: Int. J. Bifur. Chaos **9**(7), 1465 (1999)
3. Alvarez, G., Li, S.J.: Int. J. Bifurcat. Chaos **16**, 2129 (2006)
4. Çokal, C., Solak, E.: Cryptanalysis of a chaos-based image encryption algorithm. Phys. Lett. A **373**, 1357–1360 (2009)
5. http://pages.physics.cornell.edu/~sethna/teaching/562_S03/HW/pset02_dir/catmap.pdf
6. Maniccam, S.S., Bourbakis, N.G.: Pattern Recogn. **37**, 725 (2004)
7. Fridrich, J.: Int. J. Bifur. Chaos **8**(6), 1259 (1998)
8. Rhouma, R.: Cryptanalysis of a spatiotemporal chaotic cryptosystem. Chaos Solitons Fractals **41**, 1718–1722, (Aug 2009). https://doi.org/10.1016/j.chaos.2008.07.016
9. Scharinger, J.: J. Electron. Imaging **7**, 318 (1998)
10. Huang, C.K., Nien, H.H.: Opt. Commun. **282**, 2123 (2009)
11. Hu, J.K., Han, F.L.: J. Netw. Comput. Appl. **32**, 788 (2009)
12. Gao, T.G., Chen, Z.Q.: Chaos Solitons Fractals **38**, 213 (2008)
13. Gao, T.G., Chen, Z.Q.: Phys. Lett. A **372**, 394 (2008)
14. Mazloom, S., Eftekhari-Moghadam, A.M.: Chaos, Solitons Fractals **42**, 1745 (2009)
15. Arrifin, M.R.K., Noorani, M.S.M.: Phys. Lett. A **372**, 5427 (2008)

Image Steganography Using Edge Detection by Kirsch Operator and Flexible Replacement Technique

Barnali Gupta Banik, Manish Kumar Poddar and Samir Kumar Bandyopadhyay

Abstract This paper proposes a novel data hiding technique in image using Kirsch operator, which has unique feature to find maximum edge strength in different orientations. Depending on a threshold value for the Kirsch operator and the intensity value of each pixel of cover image, a scale with 3 ranges would be created. This scale is the basis for choosing flexible, i.e., 2, 3, or 4 LSB replacements for steganographic encoding. The threshold value is sharable to intended receiver as a key and if appropriate reverse approach is taken, secret image can be successfully retrieved without any data loss. This work has been carried on different grayscale images, maximum payload has been calculated, PSNR and SSIM values are measured to get satisfactory result of quality. To ensure un-detectability, Bhattacharyya distance has been calculated and for robustness test, RS steganalysis attack has been performed.

Keywords Data privacy · Information security · Image edge detection
Image quality · Robustness

1 Introduction

Steganography is an age-old technique of data hiding. Digital steganography is a technique of hiding data onto digital objects like text, audio, video, image, and network protocols [1]. In digital steganography, the object where the secret data is embed-

B. Gupta Banik (✉)
Department of Computer Science & Engineering, St. Thomas' College of Engineering & Technology, Kolkata 700023, India
e-mail: barnali.guptabanik@stcet.ac.in

M. K. Poddar
St. Thomas' College of Engineering & Technology, Kolkata 700023, India
e-mail: stcetmanish@gmail.com

S. K. Bandyopadhyay
University of Calcutta, Kolkata 700098, India
e-mail: skb1@ieee.org

© Springer Nature Singapore Pte Ltd. 2019
A. Abraham et al. (eds.), *Emerging Technologies in Data Mining and Information Security*, Advances in Intelligent Systems and Computing 814,
https://doi.org/10.1007/978-981-13-1501-5_15

ded known as cover object. In this research article, an innovative secured method for image steganography has been proposed using Kirsch operator and flexible LSB replacement technique.

2 Literature Survey

The LSB substitution algorithm exploits the fact that human eye cannot perceive small changes [2]. This algorithm states least significant bit of every byte of an image is substituted with secret message bit. In this process, a secret key can be used as stego key which is shared between sender and receiver—this is used during data encoding and decoding [3]. In [4], author has proposed a technique where not only LSB is used to hide data but 2, 3, and 4-bit LSB can also be used to hide data. In [5], author has shown that by 5-bit LSB substitution, secret message starts revealing its existence.

2.1 Edge Detection

Edges of an image can be defined as sharp brightness change or discontinuities in intensity [6]. To detect edges in an image, three approaches can be taken:

(1) Gradient
(2) Second derivative operator
(3) Gaussian

Gradient or Hamilton operator is a vector operator. This operator is denoted by ∇ named as delta. Gradient has different meaning and utilities in mathematics. In simplest form, it means 'slope'. In image processing, this is used for two-dimensional vector like $f(x, y)$ and gradient (gr) can be defined by Eq. (1).

$$\text{gr} = \nabla f(x, y) = \begin{bmatrix} \frac{\partial f}{\partial x} \\ \frac{\partial f}{\partial y} \end{bmatrix} \tag{1}$$

If $f_x = \frac{\partial f}{\partial x}$ and $f_y = \frac{\partial f}{\partial y}$ then the magnitude and direction of gr can be defined by Eqs. (2) and (3), respectively.

$$\|\text{gr}\| = \sqrt{f_x^2 + f_y^2} \tag{2}$$

$$\angle \text{gr} = \tan^{-1}\left(\frac{f_x}{f_y}\right) \tag{3}$$

To detect the edges by Gradient method, it is required to compute maximum and minimum in the first derivative of the image. There are some popular techniques in this Gradient method like Robert, Sobel, and Prewitt operator.

In the second derivative operator, it searches for zero crossing in second derivative to find edges. There are two methods in this approach—Laplacian and second directional derivative.

In Gaussian method, edge detection takes the advantage of Gaussian smoothing filter to detect edge in the direction of steepest change [7, 8].

In [9], author has proposed a method where an edge image is created using hybrid edge detector composed of canny edge operator and fuzzy edge detector. In this paper, edge image is divided into pixel blocks and variable LSB technique applied to different pixel blocks. Another approach is discussed in [10] where region of data embedding is selected using pixel differencing and steganography is done using LSBM, i.e., LSB Matching. There is a difference between LSB and LSBM approach, where if the bit of secret message does not match with LSB then +1 or −1 randomly done over the pixel value. In [11], one more approach has proposed where canny edge detector is used to detect edges and pseudorandom number is used to generate secret key to embed data in edge pixel and non-edge pixel using XOR operation. In [12] LSBM is used in accordance to edge image created by Sobel Operator.

2.2 Kirsch Operator

Kirsch operator is a nonlinear edge detector operator and named after scientist Russell A. Kirsch. This operator takes a mask and rotates it in 45° to N, NW, W, SW, S, SE, E, and NE directions [13]. Any type of mask can be chosen for Kirsch operator; the only constraint is that sum of the coefficients of each mask must be zero [14]. This is because edge is actually high-frequency regions of the image. Since sum of all coefficients is zero, they eliminate all low-frequency components and returns maximum edges. Kirsch operator is very useful for detecting weak edges as well.

In this research article, to detect edge on the cover image, the following Kirsch operators have been considered for eight directions:

$$H_0^k = \begin{bmatrix} -1 & -1 & -1 \\ 0 & 0 & 0 \\ 1 & 1 & 1 \end{bmatrix} \quad H_1^k = \begin{bmatrix} -1 & -1 & 0 \\ -1 & 0 & 1 \\ 0 & 1 & 1 \end{bmatrix} \quad H_2^k = \begin{bmatrix} -1 & 0 & 1 \\ -1 & 0 & 1 \\ -1 & 0 & 1 \end{bmatrix}$$

$$H_3^k = \begin{bmatrix} 0 & 1 & 1 \\ -1 & 0 & 1 \\ -1 & -1 & 0 \end{bmatrix} \quad H_4^k = \begin{bmatrix} 1 & 1 & 1 \\ 0 & 0 & 0 \\ -1 & -1 & -1 \end{bmatrix} \quad H_5^k = \begin{bmatrix} 1 & 1 & 0 \\ 1 & 0 & -1 \\ 0 & -1 & -1 \end{bmatrix}$$

$$H_6^k = \begin{bmatrix} 1 & 0 & -1 \\ 1 & 0 & -1 \\ 1 & 0 & -1 \end{bmatrix} \quad H_7^k = \begin{bmatrix} 0 & -1 & -1 \\ 1 & 0 & -1 \\ 1 & 1 & 0 \end{bmatrix}$$

By using Kirsch operator, the edge magnitude is calculated as maximum magnitude in all eight directions using the kernel as shown in Eq. (4)

$$\|c_{m,n}\| = \max_{i=1,2,\ldots 8}(C_i) \tag{4}$$

where $\|c_{m,n}\|$ is the edge magnitude, i is the Kirsch direction,
   C_i is the response of the kernel at the pixel position i, which is defined by Eq. (5)

$$C_i = \sum_{p=-1}^{1} \sum_{q=-1}^{1} H_i^k f_{m+p,n+q} \tag{5}$$

3   Proposed Method

3.1   Embedding Procedure

Input:
(a)   Cover image: Any grayscale image
(b)   Threshold value: range between 0 and 255
(c)   Secret image: Any grayscale image
Output: Stego grayscale image

At first a grayscale image is taken as a cover image. To use Kirsch operator, a threshold value 'T' is taken and compared with pixel gradient 'P' as shown below

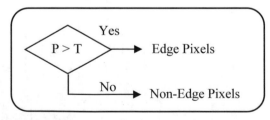

After getting the pixel values, Kirsch operator is used to detect edge image from the given image. This edge image would be used to embed secret data.
   Each pixel of a grayscale image composed of a byte, i.e., 8 bits. So, a pixel can have values between (2^0-1) and (2^8-1), i.e., 0–255. If the threshold value is 'T', then a scale of three ranges can be created by Eq. (6).

Table 1 Mapping between regions and LSB method

Pixels in region	Flexible LSB
$R1$	2-bit LSB
$R2$	3-bit LSB
$R3$	4-bit LSB

$$R = (255 - T)/3 \tag{6}$$

Leaving first T values, a scale can be created as shown in Eqs. (7), (8) and (9).

$$R1 = (T \text{ to } T + R) \tag{7}$$

$$R2 = (T + R + 1 \text{ to } T + 2R + 1) \tag{8}$$

$$R3 = (T + 2R + 2 \text{ to } T + 3R + 2) \tag{9}$$

The flexible type of LSB method would be chosen by Table 1.

Now a calculation shall be done to check what would be the safe length to encode using this cover image. If the number of edges in the image is 'E' and each edge having average 'I' pixels in $R1$, 'J' pixels in $R2$, and 'K' pixels in $R3$, then the encoded safe length ('SLen') can be calculated by Eq. (10).

$$\text{SLen} = E * (I * 2 + J * 3 + K * 4) \tag{10}$$

Now the length of the secret text is denoted by 'SecLen'. if SecLen > SLen, then "steganography cannot be performed"; otherwise "steganography can be performed".

3.2 Extraction Procedure

During the process of secret data retrieval, it has been assumed that the threshold value is shared with receiver end. So, receiver will apply analogous approach to detect edges and find where 2, 3, or 4 bits of LSB substitution have been done. Depending on those information, secret image bytes can be rebuilt and thereafter complete secret image can be extracted.

The following flowcharts shown in Figs. 1 and 2 explain the steps for embedding and extraction technique, respectively.

4 Results and Quality Analysis

This proposed technique has been tested on several grayscale images. However, due to the space limit, here the result has been discussed for only three popular images

Fig. 1 Flowchart showing
embedding procedure

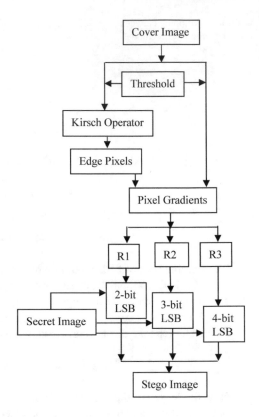

like Sergio De Arrola from RAW Cycling Magazine, The Californian Girl Haim
and Dennis Hopper HD Desktop wallpaper for thresholds 10, 100, and 200 with
different secret images taken from DECSAI image database. The result is analyzed by
three image quality metrics—peak signal-to-noise ratio (PSNR), structural similarity
(SSIM), and Bhattacharyya distance (D_B).

PSNR is the ratio between the original and the noisy signal in an image, given in
decibels. It is calculated in terms of mean squared error (MSE) which is the average
squared difference between a reference image and noisy image [15] and given by
Eq. (11).

$$\text{MSE} = \frac{1}{mn} \sum_{0}^{m-1} \sum_{0}^{n-1} [f(i, j) - g(i, j)]^2 \tag{11}$$

where f and g represent the matrix data of original and degraded images; m and n
represent the numbers of rows and columns of pixels of the image; i and j represent
the index of row and the index of column. PSNR can be calculated by Eq. (12).

$$\text{PSNR} = 20 \log_{10}\left(\frac{\text{Max}_f}{\sqrt{\text{MSE}}}\right) \tag{12}$$

Fig. 2 Flowchart showing
extraction procedure

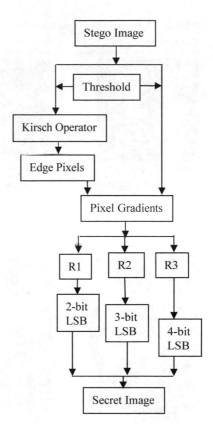

where Max_f is maximum signal value of original image.

SSIM is the metric to analyze quality of the image in accordance to structural content of the image. It compares images with respect to luminance, contrast, and structure [15]. If the structural content of two images is exactly same, then the value turned to be 1 otherwise it would be <1. SSIM can be represented as Eq. (13).

$$S(h, t) = \frac{(2\mu_h\mu_t + c_1)(2\sigma_{ht} + c_2)}{\left(\mu_h^2 + \mu_t^2 + c_1\right)\left(\sigma_h^2 + \sigma_t^2 + c_2\right)} \tag{13}$$

where μ_h and μ_t are the respective mean value of image h and t; σ_h and σ_t are the standard deviations of image h and t, respectively; σ_{ht} is cross-correlation of h and t after removing their means.

Bhattacharyya distance (D_B) can be used as a measure of detectability [16], which defines how much the stego image reveals the existence of secret message. The D_B is measured between two discrete or continuous probability distributions Pd and Pc over the space Ω as shown in Eq. (14).

$$D_B(\text{Pd}, \text{Pc}) = -\ln \ \rho B(\text{Pd}, \text{Pc}) \tag{14}$$

Table 2 Creating stego images with different thresholds

Cover image	Edge image by Kirsch operator	Stego image
Dennis Hopper	Threshold = 10	2-bit LSB
Sergio De Arrola	Threshold = 100	3-bit LSB
Californian Girl	Threshold = 200	4-bit LSB

where $\rho B(\text{Pd}, \text{Pc}) = \int_{\Omega}^{0} \sqrt{\text{Pd}(\omega)\text{Pc}(\omega)} d\omega$.

After the calculation of D_B using Eq. (14), it needs to be normalized between 0 and 1 where 0 means full detectability and 1 means zero detectability.

Table 2 illustrates the stego images which have been generated by applying different LSB methods using diversified threshold values. These stego images are visually similar to the cover image; hence, they do not reveal the existence of secret embedded image anyway.

Table 3 demonstrates the results of quality metric for original secret versus extracted secret.

Table 4 shows the quality of result analysis for one of the cover images. The outcome of PSNR and SSIM values demonstrate that the stego image is perceptually equivalent to the original image. The values of D_B are nearer to 1 signifies less detectability of this proposed method.

Table 3 Quality analysis of the original versus extracted secret images

Original secret	Extracted secret	Quality results
		PSNR: 65.0172 SSIM: 0.9989
		PSNR: 73.8014 SSIM: 0.9849
		PSNR: 76.7042 SSIM: 0.9726

Table 4 Quality result analysis for image Dennis Hopper

Threshold (T)	Safe length (SLen)	Secret length (SecLen)	PSNR	SSIM	D_B
10	74,622	50	74.4487	1.000	0.9946
10	74,622	20,000	58.5316	0.9994	0.9970
10	74,622	74,622	39.5316	0.9729	0.9966
100	23,423	50	69.3848	1.000	0.9970
100	23,423	2000	53.9571	0.9995	0.9988
100	23,423	23,423	42.3665	0.9944	0.9946
200	12,089	50	67.9071	1.000	0.9966
200	12,089	5000	47.2873	0.9986	0.9970
200	12,089	12,089	42.9289	0.9964	0.9946

Table 5 Result of proposed method on Sergio De Arrola image

Threshold	Embedding capacity (in bit)	PSNR
153	126,616	42.3422
20	525,612	39.4166
8	600,256	39.0417

Table 6 Result of related work [17] on Sergio De Arrola image

Variable embedding ratio	Embedding capacity (in bit)	PSNR
4:0	126,616	38.21
4:1	525,612	37.11
4:2	600,256	36.47

4.1   Comparison with Existing Method

The related work of this method found in [17] where authors have used variable embedding ratio (VER) which specifies the number of bits embedded in edge and non-edge pixels. The proposed method of this article is novel from this approach in three ways:

- The authors in [17] have used canny operator whereas here Kirsch operator is used for edge detection which has lot of advantages over canny. Most of the edge detectors including canny compute an average value for each side and then compute the earth mover's distance whereas Kirsch operator allows to exist several values to each side.
- The authors of [17] have used both of edge and non-edge pixels to embed data whereas in this current approach only edge pixels are used for embedding data. This enhances the quality of data hiding. Also, the comparison shown in the Tables 5 and 6 proves that the capacity of related work [17] is less than the capacity of proposed method.
- The projected approach used threshold values to flexibly choose 2-, 3-, or 4-bit LSB replacement method which has not been done in the related work in [17]—there was no scope to choose variability in their method.

Tables 5 and 6 compare the result obtained from both techniques on Arrola 512 × 512 image which clearly depicts proposed method generates higher value of PSNR when measuring with same embedding capacity (max. bit to hide).

4.2   Robustness Test by RS Steganalysis Attack

As LSB substitution steganography is prone to RS steganalysis attack, hence to measure the robustness of the proposed method, regular singular (RS) steganalysis

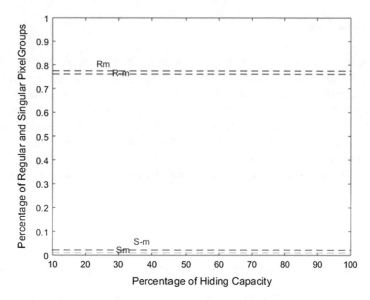

Fig. 3 Outcome of RS steganalysis on 'Sergio De Arrola' image

has been applied on the stego image which splits that into n groups of adjacent pixels [18]. The smoothness of each group is calculated by function shown in Eq. (15).

$$f(I) = \sum_{i=1}^{n-1} |I_{i+1} - I_i| \tag{15}$$

where I is the group of n image pixel. There is also a concept of masking 'm' which can have combination of -1, 0, and 1. There are three flipping functions:

T_0: refers to no change in pixel value; T_1: refers to $2t \leftrightarrow 2t + 1$; T_{-1}: refers to $2t-1 \leftrightarrow 2t$;

The above functions are applied on each group of pixels to identify whether it belongs to regular group (R_m) where $f(t_m(I)) > f(I)$; singular group (S_m) where $f(t_m(I)) < f(I)$, or unusable group (U_m) where $f(t_m(I)) = f(I)$. When no message is hidden, then the relationship among these groups can be represented in Eq. (16).

$$|R_m| = |R_{-m}| \text{ and } |S_m| = |S_{-m}| \tag{16}$$

The mask $-m$ can be achieved by negating all the elements of m. The graphical plot in Fig. 3 shows the result of RS steganalysis attack on Sergio image.

5 Conclusion

Steganography is an ancient method of secret communication. In this paper, a new method of image steganography has been proposed using edge detection by Kirsch operator along with flexible LSB replacement. This technique shows highly acceptable results with respect to the values obtained from several metrices, namely PSNR, SSIM, and Bhattacharyya distance. Also, robustness of this method has been measured against RS steganalysis attack. This proposed technique can be further improved by using LSB matching revisited (LSBMR) algorithm, which is the latest form of LSB with the advantage of hiding RGB secret data inside a RGB image.

References

1. Mishra, R., Bhanodiya, P.: A review on steganography and cryptography. In: Publisher IEEE, pp. 119–122. https://doi.org/10.1109/icacea.2015.7164679
2. Johnson, N., Jajodia, S.: Exploring steganography: seeing the unseen. Computer **31**(2), 26–34. https://doi.org/10.1109/mc.1998.4655281
3. Sutaone, M., Khandare, M.: Image based steganography using LSB insertion technique. In: IET International Conference on Wireless, Mobile and Multimedia Networks. ISBN: 978-0-86341-887-7 (2008)
4. Deshpande, N., Kamalapur, S., Jacobs, D.: Implementation of LSB steganography and its evaluation for various bits. In: Publisher IEEE, pp. 173–178. https://doi.org/10.1109/icdim.2007.369349
5. Gupta Banik, B., Bandyopadhyay, S.K.: An image steganography method on edge detection using multiple LSB modification technique. J. Basic Appl. Res. Int. **9**(2), 75–80. International Knowledge Press
6. Zhang, R., Zhao, G., Su, L.: A new edge detection method in image processing. In: Publisher IEEE, pp. 430–433. https://doi.org/10.1109/iscit.2005.1566889
7. Cui, F., Zou, L., Song, B.: Edge feature extraction based on digital image processing techniques. In: Publisher IEEE, pp. 2320–2324. https://doi.org/10.1109/ical.2008.4636554
8. Zheng, Y., Rao, J., Wu, L.: Edge detection methods in digital image processing. In: Publisher IEEE, pp. 471–473. https://doi.org/10.1109/iccse.2010.5593576
9. Chen, W., Chang, C., Le, T.H.N.: High payload steganography mechanism using hybrid edge detector. Expert Syst. Appl. **37**(4), 3292–3301. Elsevier. https://doi.org/10.1016/j.eswa.2009.09.050
10. Zhenhao, Z., Tao, Z., Baoji, W.: A special detector for the edge adaptive image steganography based on LSB matching revisited. In: Publisher IEEE, pp. 1363–1366. https://doi.org/10.1109/icca.2013.6564938
11. Alam, S., Vipin, K., Waseem, A.S., Ahmad, M.: Key dependent image steganography using edge detection. In: Publisher IEEE, pp. 85–88. https://doi.org/10.1109/acct.2014.72
12. Fouroozesh, Z., Jaam, J.: Image steganography based on LSBMR using sobel edge detection. In: Publisher IEEE, pp. 141–145. https://doi.org/10.1109/icend.2014.6991368
13. Kirsch, R.A.: Computer determination of the constituent structure of biological images. Comput. Biomed. Res. **4**(3), 315–328 (1971). https://doi.org/10.1016/0010-4809(71)90034-6
14. Sharma, S.: Fundamentals of Digital Image Processing. Publisher S. K. Kataria and Sons. ISBN: 978-93-5014-504-3
15. Hore, A., Ziou, D.: Is there a relationship between peak-signal-to-noise ratio and structural similarity index measure? IET Image Proc. **7**(1), 12–24 (2013). https://doi.org/10.1049/iet-ipr.2012.0489

16. Korzhik, V., et al.: On the use of bhattacharyya distance as a measure of the detectability of steganographic systems. Trans. Data Hiding Multimedia Secur. III **4920**, 23–32 (2008). https://doi.org/10.1007/978-3-540-69019-1_2. Springer, Berlin
17. Geetha, C.R, Basavaraju, S., Puttamadappa, C.: Variable load image steganography using multiple edge detection and minimum error replacement method. In: Publisher IEEE, pp. 53–58. https://doi.org/10.1109/cict.2013.6558061
18. Manoharan, S.: An empirical analysis of RS steganalysis. In: Publisher IEEE, pp. 172–177. https://doi.org/10.1109/icimp.2008.15

Blind RGB Image Steganography Using Discrete Cosine Transformation

Emlon Ghosh, Diptasree Debnath and Barnali Gupta Banik

Abstract In this paper, a new method for RGB image steganography is being proposed which hides RGB secret image in a RGB cover image by dividing both the
secret and cover image in three color planes (red, green, blue). Then among those
color planes, the actual color value of an image pixel has been hidden within the same
color plane of the cover image using discrete cosine transformation. To hide the secret
bit, here the desired cover bit can be chosen from a set of 8 bits, which provides better
security. This proposed technique has been tested using different color images, and
experimental results have been analyzed by various quality metrics which assures
better imperceptibility.

Keywords Data privacy · Information security · Discrete cosine transformation
Image quality

1 Introduction

The word "steganography" originated from two Greek words, "steganos" which
means "covered" and "graphein" which means "writing." Steganography is a process
of encrypting messages or information within another unclassified digital media.
Image steganography is the technique of encoding a secret message within an image
such that its presence remains hidden. In this case, the secret can be a text message
or a stream of bits or even an image, but the cover will have to be an image. The
objective of image steganography is to send a secret through a common transmission

E. Ghosh (✉) · D. Debnath · B. Gupta Banik
Department of Computer Science & Engineering, St. Thomas' College
of Engineering & Technology, Kolkata 700023, India
e-mail: emlonghosh@gmail.com

D. Debnath
e-mail: diptasree.debnath@gmail.com

B. Gupta Banik
e-mail: barnali.guptabanik@stcet.ac.in

© Springer Nature Singapore Pte Ltd. 2019
A. Abraham et al. (eds.), *Emerging Technologies in Data Mining and Information
Security*, Advances in Intelligent Systems and Computing 814,
https://doi.org/10.1007/978-981-13-1501-5_16

channel without letting the network administrator know the presence of the secret and preventing it from being extracted by any intruder except the desired recipient of the secret.

Images can be of different types like RGB, grayscale, binary. A binary image only consists of pixel values of 0 and 1 which results in black and white image. On the other hand, grayscale image has a range of 0–255 for pixel value which gives different shades of gray. Here each pixel value has 8 bits, whereas a RGB image has 24 bits representation of pixel value where the image is divided in three color planes (red, green, and blue). If there is even very small amount of change in RGB image, the colors get distorted. That is why RGB image steganography is usually avoided.

Direct modification of the pixel values of the cover image to hide the secret can extensively expose secret data. Therefore, converting it to frequency domains can be a good option. This transformation can be done using different techniques like discrete wavelet transform (DWT), discrete cosine transform (DCT), discrete Fourier transform (DFT). Conversion in frequency domain is implemented to reduce the visible impact on the cover image.

2   Literature Survey with Related Knowledge

Covariance-based steganography using DCT-based algorithm can hide a grayscale image in another grayscale image efficiently [1]. This kind of process can be applied to RGB images considering each of the red, blue, and green color plane individually as grayscale image.

While hiding a plain text in a RGB image using least significant bit (LSB) algorithm, the stego image produces a good PSNR value but this method does not explain how images can be treated as a secret to hide [2]. Using pixel value differencing method, message bits stream can be hidden under a RGB cover image but to hide an image using that method, transformation of the secret image into streams of message bits will be required [3]. Now using LSB and improved pixel indicator method together, secret data can be hidden in RGB cover image but only in binary form. Therefore, if the image needs to be hidden through this method, it must be a binary image [4]. Using DCT, text message can be hidden in a RGB image and the extraction method would be blind [5]. Merging the ideas of random pixel manipulation method and LSB method, data can be hidden in an RGB image. In this method, least two significant bits of any one of the color channel indicate existence of data in other two-color channels. However, the extraction method requires a secret key [6].

Using DWT and integer wavelet transform (IWT), grayscale image can be hidden in RGB image but RGB image cannot be hidden [7]. An improved LSB method can be used to hide grayscale image into RGB image to better the human visual features and enhanced privacy. It starts with the improvement of the randomness of the LSB embedding position, and encryption of the message which controls embedded position, so the hidden information cannot be extracted without the corresponding private key [8]. Same can be done using DWT and Arnold transformation and in that case, the extraction method is blind [9]. But again, none of these methods explain how a RGB image can be embedded in another RGB image.

Hiding a RGB secret in a RGB cover image, using DWT and alpha-blending technique gives quite low values of PSNR (best is around 29 but mostly below 20). This method is also not blind, i.e., it needs the cover image to extract the secret from the stego image [10]. Hence, there is a scope for betterment to innovate a blind data hiding technique with enhanced privacy and improved visual presentation, which has been proposed and implemented here.

2.1 Discrete Cosine Transform (DCT)

In today's world, where image processing, image compression, and steganography techniques are rapidly developing, DCT holds a key to the success of all these technologies. One of the secret behind changing a picture but still making it not detectable is that these changes can be made in the basic frequency component of the pixel value of an image. Though DCT and DFT are quite related; however, when DCT is applied to an image, a set of numbers called coefficients are given as an output.

By the help of these DCT coefficients, the image can be changed without the change being easily detected. When the DCT coefficients vary a lot, any change in those coefficients can be easily detected but when changes are made in those coefficients are minimal, chance of getting detected is quite low, and these changes are not visually prominent.

A DCT block can be subdivided into three categories—high-band coefficients, mid-band coefficients, and low-band coefficients. Changes made in the mid-band are most likely to be the safest as that is least detectable. Equation for DCT is shown in Eq. (1)

$$\text{DCT}(i, j) = \frac{1}{\sqrt{2N}} C(i)C(j) \sum_{x=0}^{N-1} \sum_{y=0}^{N-1} \text{pixel}(x, y) \cos\left[\frac{(2x+1)i\pi}{2N}\right] \cos\left[\frac{(2y+1)j\pi}{2N}\right]$$

$$(1)$$

where $C(x) = \frac{1}{\sqrt{2}}$ if x is 0 else 1 if $x > 0$.

2.2 RGB Image

An RGB image is a tricolor image formed by the combination of the pixel values of three colors—red, green, and blue. It is generally represented as $m * n * 3$ where $m * n$ represents the general image size as height and width and has 3 as the third dimension as to get a RGB image, it is necessary to know the pixel values of all the three (red, green, and blue) pixel components of the image. All these individual

Fig. 1 Original RGB image

Fig. 2 Red Component

components are grayscale image and by manipulating these components (also known as image planes), the image can be modified. A RGB image is stored as 24 bits in graphics file formats with 8 bits to each of the three components. By this way, 16 million combinations of colors can be formed.

As shown, Fig. 1 is the original RGB image having a dimension of 512 * 512 * 3. After separating the three components of the RGB image, three image components have been received—out of which Fig. 2 is shown as the image represents the red component, Fig. 3 corresponds to the green component, and Fig. 4 denotes the blue component, and it can be also observed that all the three images are grayscale images and visually different from each other.

Fig. 3 Green component

Fig. 4 Blue component

3 Proposed Method

Here the objective is to embed a RGB secret into a RGB cover image. To do this, the RGB cover image is divided in three color channels, then divided the matrices into cells of required size. Now DCT is applied on each block of cell which transforms the image into the frequency domain. Now from each block of cell, a DCT coefficient is chosen within the mid-band region to replace with the value of a pixel of the secret image. Choosing more number of coefficients will increase the capacity of hiding. The following sections contain detailed discussion of this proposed method and its experimental results to validate the efficiency.

3.1   Embedding Procedure

As both cover and secret are RGB images, so to attain a more stable and effective embedding system, the respective image plane of the secret can be embedded on to the respective cover plane (i.e., a red component of the secret is embedded in the red component of the cover only). Initially, cover components are divided into several cells depending upon the size of the secret. For example, if the cover size is 512 * 512 and the secret size is 64 * 64, then the cover is needed to be divided into 8 * 8 cells so that there are exactly 64 * 64 number of block of cells and in each block, one value can be embedded from the 64 * 64 sized secret. Post this step, DCT has been applied to the individual cells for all three components. Then an element of the 8 * 8 block is selected and its value is manipulated according to the pixel value of the respective component of the secret and the alpha factor. This is done for all the 64 * 64 cell of each of the three components of the RGB cover image, and it is saved in three separate matrices. The main purpose of using alpha factor is to keep the DCT coefficient of the stego image (i.e., post-embedding of the secret image into cover's DCT coefficients) almost in the same range of cover's DCT coefficients, such that increase in DCT values due to embedding is not visible. Value of this alpha factor should be set according to the pixel values of the secret image. Lastly, IDCT is applied to the individual cells to get the pixel values back from frequency components. Then, these cells of individual components are converted back to matrix form to get three separate $m * n$ matrices. These three matrices are the embedded components of the final stego RGB image. Finally, these components are stored in a new $m * n * 3$ RGB image with $m * n * 1$ component having the values of the matrix corresponding to the red component of the cover. Similarly, $m * n * 2$ is for green and $m * n * 3$ is for blue. Thus, this final $m * n * 3$ matrix is representing stego image.

3.2   Embedding Algorithm

STEP-1: Input the cover image and secret image (both RGB)
STEP-2: Find out cover image size and accordingly decide maximum secret message size that can be embedded in cover based on block size.
STEP-3: Divide the original RGB cover image into the 3 separate Red, Green, and Blue components of each individual pixel values.
STEP-4: Similarly divide the secret RGB image in the 3 separate Red, Green, and Blue components of each individual pixel values.

STEP-5: Divide these individual cover image components into 8*8 block of cells (this size has been chosen based on the size of the secret to be embedded)
STEP-6: Set the alpha factor
STEP-7: For each of the 8*8 block cell of the cover image apply DCT
STEP-8: For each of these cells replace the least detectable positional element with each element of the respective components of the secret image multiplied by the alpha factor
STEP-9: Now for each of these 8*8 cells apply IDCT, i.e., just the inverse of DCT.
STEP-10: Now combine these modified components of the cover image to give the Stego image (which is thus also a RGB image)

Flowcharts for embedding and extraction process have been shown in Figs. 5 and 6, respectively.

3.3 Extraction Procedure

One of the advantages of this method is that the extraction process is blind, i.e., the cover image is not required to get the secret back from the stego. Here value of the alpha factor can be used as key, which would be needed both during embedding as well as extraction process and can be shared to the intended receiver.

For extraction of secret image, initially the RGB image planes needed to be separated from the RGB stego and put them into three different $m * n$ matrices. Thereafter those matrices are divided into the number of cells depending upon the size of the secret to be retrieved. Then, DCT is applied on these cells to get the respective frequency components and the DCT coefficient values are divided by the alpha factor to get the original pixel values of the secret image. Finally, once all such calculated image pixel values are retrieved, those would be combined into $m * n * 3$ matrix to get the secret image without any data loss.

3.4 Extraction Algorithm

STEP 1: Read the Stego RGB image
STEP 2: Divide the stego image in respective Red, Green, and Blue components.
STEP 3: Divide each of these components into 8*8 block of cells

STEP 4: Refer the same alpha factor which has been used in embedding
STEP 5: Now apply DCT on each of these 8*8 blocks
STEP 6: Now the least detectable position which was determined during embedding has been saved here and

Fig. 5 Flowchart for embedding process

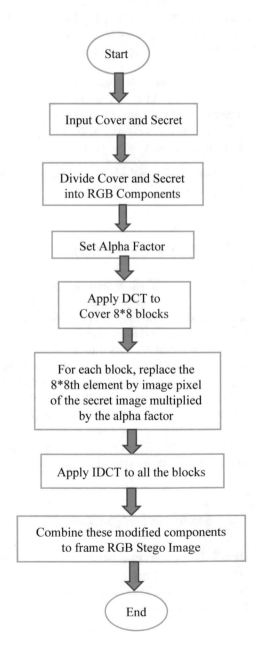

value of each cell should be divided by the alpha factor and consider as an element of the retrieved image.

STEP 7: Now combine the components retrieved to get the retrieved secret image

Fig. 6 Flowchart for extraction process

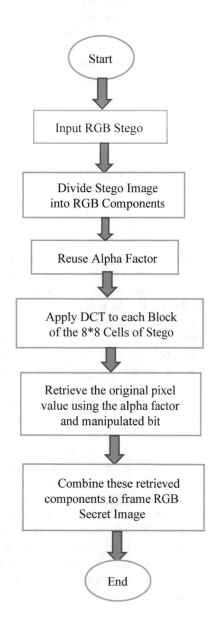

4   Results and Quality Analysis

Here peak signal-to-noise ratio (PSNR) and structural similarity index (SSIM) metrics have been used for quality analysis as these are most popular for measuring perceptual transparency.

PSNR is most commonly used to measure the quality of the modified signal as compared to the original one. Thus, it can be said that if the noise is more, then the value of PSNR will be low as PSNR is inversely proportional to the noise value. For undetectable steganography process, it is necessary to have a high PSNR value between cover and stego image as well as between embedded and extracted secret image. The expression that is used to calculate the PSNR is shown in Eq. (2)

$$PSNR = 20 \log_{10}\left(\frac{\text{Max}_I}{\sqrt{\text{MSE}}}\right) \tag{2}$$

where MSE is the mean squared error and is represented in Eq. (3).

$$\text{MSE} = \frac{1}{mn}\sum_{i=0}^{m-1}\sum_{j=0}^{n-1}[I(i,j) - K(i,j)]^2 \tag{3}$$

$I(i, j)$ is pixel of cover image; $K(i, j)$ pixel of stego image, and Max_I is the maximum pixel value of cover image.

SSIM is another useful method to measure the quality of the modified image. For SSIM test, two images are required—first one considered as ideal and other one as reference image, which is compared with the ideal one. For optimum steganography method, similarity index should be higher as the changes will be least detectable. SSIM is given in range between 0 and 1, whereas 1 indicates exact same image and 0 being the completely different image. SSIM is calculated as per Eq. (4)

$$\text{SSIM}(x, y) = \frac{\left(2\,\mu_x\,\mu_y + C_1\right)\left(2\sigma_{xy} + C_2\right)}{\left(\mu_x^2 + \mu_y^2 + C_1\right)\left(\sigma_x^2 + \sigma_x^2 + C_2\right)} \tag{4}$$

where, $\mu_{x,}$, $\mu_{y,}$, σ_x, σ_y and σ_{xy} are the local mean, standard deviation, and cross-covariance of images x and y respectively.

In Table 1, quality analysis of cover versus stego images has been performed.

In Table 2, result analysis of original secret vs. extracted secret has been shown.

The proposed method has performed better as compared to other existing methods discussed in Sect. 2 with respect to two key points—(a) secret extraction process is blind, which provides better security as there is no need to have the original cover image during extraction, (b) PSNR outcome is consistently above 30 between embedded versus extracted secret image and around 40 between cover versus stego images.

Table 1 Quality comparison between cover and stego image

Cover image	Secret image embedded	Stego image	PSNR	SSIM
			39.2374	0.9497
			42.5654	0.9928
			42.2765	0.9724
			41.5177	0.9981
			39.8333	0.9934

5 Conclusion

This method successfully embeds RGB secret image into RGB cover image. Unlike most steganography algorithms where secret image must be converted into grayscale before embedding, here that is not required as RGB secret image can be embedded and extracted directly without any data loss. Moreover, as RGB image has three image planes, hence pixel value of secret image can be inserted into all these three planes which would enhance the embedding capacity as well as that would ensure

Table 2 Quality analysis of embedded vs. extracted secret image

Cover image	Secret image embedded	Secret image extracted	PSNR	SSIM
			30.7177	0.9895
			34.1347	0.9772
			32.3891	0.9454
			31.5704	0.9699
			32.8391	0.9904

better picture quality of secret image upon retrieval. This proposed method can be further improvised to apply in video steganography, as there are relatively very few blind video steganography algorithms exist which can embed RGB secret and later retrieved that from stego without any data/color loss.

References

1. Sathisha, N., Suresh Babu, K., Raja, K.B., Venugopal, K.R.: Patnaik, L.M.: Covariance based steganography using DCT. In: International Conference on Advances in Computing and Communications, in Communications in Computer and Information Science, Springer. https://doi.org/10.1007/978-3-642-22714-1_66
2. Lenka, S.K., Swain, G.: A novel approach to RGB channel based image steganography technique. Int. Arab J. e-Technol. 2(4), 181–186
3. Prasad, S., Pal A.K.: An RGB colour image steganography scheme using overlapping block-based pixel-value differencing. R. Soc. Open Sci. 4(4) 161066. https://doi.org/10.1098/rsos.161066
4. Amirtharajan, R., Behera, S.K., Swarup, M.A., Mohamed Ashfaaq, K., Rayappan, J.B.B.: Colour guided colour image steganography. Univ. J. Comput. Sci. Eng. Technol. 1(1), 16–23
5. Abdullatif, F.A., Shukur, W.A.: Blind colour image steganography in spatial domain. IBN Al-Haitham J. Pure Appl. Sci. 24(1)
6. Gutub, A., et al.: Pixel indicator high capacity technique for RGB image based steganography. In: 5th International Workshop on Signal Processing and its Applications (WoSPA 2008), At Keyin Print Media, Editorial Department of China
7. Hemalatha, S., Acharya, U.D., Renuka, A., Kamath, P.R.: A secure colour image steganography in transform domain. Int. J. Crypt. Inf. Secur. 3(1). https://doi.org/10.5121/ijcis.2013.3103
8. Gong, W., Fu, W.L., Zhou, X., Jin, L.J.: An improved method for LSB based colour image steganography combined with cryptography. In: 15th International Conference on Computer and Information Science, Publisher IEEE. https://doi.org/10.1109/icis.2016.7550955
9. Pourarian, M.R., Hanani, A.: Blind steganography in color images by double wavelet transform and improved Arnold transform. Indonesian J. Electr. Eng. Comput. Sci. 3(3), 586–600. https://doi.org/10.11591/ijeecs.v3.i3
10. Roy, A.B., Dey, S., Dey, N.: A novel approach of color image hiding using RGB color planes and DWT. In: Int. J. Comput. Appl. 36(5). https://doi.org/10.5120/4487-6316

Logistic Map-Based Image Steganography Scheme Using Combined LSB and PVD for Security Enhancement

Shiv Prasad and Arup Kumar Pal

Abstract In recent, the confidentiality of important data is indispensable for Internet users. Although both the encryption and steganography schemes are well suited to protect the confidentiality of data, the combination of encryption and steganography approaches enhances the level of security. In this paper, an image steganography scheme is implemented where the security has been improved without incorporating any encryption process with steganography. The secret information is embedded into cover images directly either by LSB or PVD approaches. The selection of data embedding approach is decided based on a secret key, where the logistic map-based secret key sequence is used. This approach directly enhances the security without considering any conventional cryptography scheme. The experimental results ensure the acceptable results in terms of decent quality of stego-images. In addition, the proposed scheme provides a high level of security since the embedding process is realized based on the secret key sequence.

Keywords Image security · Logistic map · LSB substitution
Pixel value differencing (PVD)

1 Introduction

In the current century, we are the witness of the exponential growth of digital data due to the advancement of information technology. The availability of the high-speed Internet has not only made the digital data transmission popular among users, but also brought several security threats during transmission. One of the common security threats may be misused of data by the adversary.

S. Prasad (✉) · A. K. Pal
Department of Computer Science and Engineering, Indian Institute of Technology (Indian School of Mines), Dhanbad, Jharkhand 826004, India
e-mail: psad.shiv@gmail.co

A. K. Pal
e-mail: arupkrpal@gmail.co

© Springer Nature Singapore Pte Ltd. 2019
A. Abraham et al. (eds.), *Emerging Technologies in Data Mining and Information Security*, Advances in Intelligent Systems and Computing 814,
https://doi.org/10.1007/978-981-13-1501-5_17

Since the Internet itself is not a protected or secured communication medium, so the transmitted data may be effortlessly intercepted by the dishonest users. In general, the confidential digital data are preprocessed before their transmission over the public channel. So the security is an important part during the transmission of the secret message. Different security mechanisms, like confidentiality, integrity, and authentication [1], have been adopted to ensure the security of vital data during their transmission. The protection of confidentiality is achieved through the transformation of the meaningful messages into unreadable content. Only the authorized persons are allowed to retrieve the original content from the transformed message. The confidentiality and property are mainly protected by either cryptography [1, 2] or steganography [3] approach. The main concern of both cryptography and steganography is to transmit data securely over the public channel or the Internet but their working principle is different. Cryptography is a secret writing mechanism for converting the original message, i.e., known as plaintext into the corresponding worthless format, i.e., known as ciphertext. The steganography hides secret data into the digital cover media, like image, text, video, after hiding secret data to outline a meaningful message that is known as a stego-media, which store the secret content in a visually imperceptible manner. Both approaches ensure excellent security in a different way, but the steganography is a superior substitute from the cryptography.

The fundamental steganography scheme is known as an LSB substitution [4]. In this mechanism, the secret message bits are directly embedded into the cover media after replacing some insignificant information from the LSBs position of each pixel. The LSB substitution is mostly used up to three bits for hiding the secret data into the stego-image because more than 3LSB modification causes visual artifacts in the stego-image [4]. In the literature, LSB-based several steganography variants [4–6] are found with different intentions, like to improve the hiding capacity and to retain good visual quality of the stego-image. Later, Wu and Tsai [7] suggested a different kind of steganography approach which is widely popular as pixel value differencing (PVD). In this mechanism, the secret data are embedded by comparing the differences involving the intensity values of two successive non-overlapping pixels. Several researchers have proposed PVD-based variants steganography schemes [7–10]. Both the LSB and PVD approaches hide the secret data directly into the cover image sequentially. In these particular approaches, if the attacker someway is able to identify any image as stego-image, then the sequentially extracted information may reveal the original content. So, several researchers [11–13] have clubbed both the cryptography and steganography for enhancing the security during data transmission. In their approaches, firstly, the secret information is encrypted by suitable encryption algorithms, like 3DES, AES, IDEA. Thereafter, the encrypted content is hidden into the cover image either by LSB or PVD approaches. This combined approach is computationally intensive since the processing overhead in encryption process is high. In this paper, we have enhanced the security only considering the steganography approaches. We have used both the LSB and PVD during data hiding process into the cover image. The secret data are embedded based on a secret key sequence which is generated by logistic map [14]. The proposed scheme is suitable to retain a high level of security without using any conventional encryption algorithms.

The remainder of the paper is organized as follows. Preliminaries behind this work are presented in Sect. 2. The proposed scheme, including the secret data embedding and extraction process, is discussed in Sect. 3. The experimental results of the proposed scheme are demonstrated in Sect. 4. Finally, the conclusion related to the proposed scheme is drawn in Sect. 5.

2 Preliminaries

The proposed steganography scheme is implemented using a secret binary sequence, which is obtained by logistic map. Later, the data hiding techniques, like LSB and PVD, are discussed in the following subsections.

2.1 Logistic Map-Based Chaotic Sequence

In the proposed scheme, a secret binary sequence is required which is derived by the logistic map. It effectively generates the random sequence using Eqs. 1 and 2, respectively. The initial parameters like β and α_0 are considered as seed values. The subsequently generated sequences are as follows:

$$\alpha_1 = \beta \times \alpha_0 \times (1 - \alpha_0) \tag{1}$$

$$\alpha_i = \beta \times \alpha_{(i-1)} \times (1 - \alpha_{(i-1)}) \tag{2}$$

The sender and the receiver are able to produce the same sequence if they are using the same seed values, i.e., β and α_0.

2.2 LSB-Based Steganography

The LSB scheme conceals the secret bit stream into the cover image directly, where numbers of the least significant bit are replaced by the secret bits sequentially into the cover image. So, after secret bits are replaced into the cover image, then the modified cover image or the stego-image are transmitted to the receiver. The secret bits are embedding procedure is done by Eq. 3.

$$P' = P - \mod(P, 2^k) + SM \tag{3}$$

The extraction procedure is given by Eq. 4.

$$SM = \mod(P', 2^k) \tag{4}$$

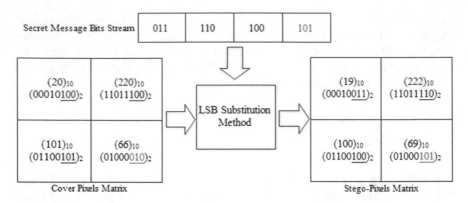

Fig. 1 3-LSB substitution method

where P is a cover pixel, k is number of bits, *SM* is secret message in decimal value, and P' is a modified stego-pixel.

It observed that up to three least significant bits (LSB) replacement is suitable to maintain the satisfactory visual quality of the stego-image. If more than 3LSB modification of the stego-image is done, then the high visual artifacts and the distortion are easily detectable by eavesdropper's eyes. The 3LSB substitution procedure is shown in Fig. 1

2.3 PVD-Based Steganography

The PVD is another type of data hiding approach. Initially, the cover image is decomposed into non-overlapping block of size 1×2. Subsequently, the secret message bits are concealed into each block based on the nature of pixels Q_i and Q_{i+1} of each block. The difference value, *diff$_i$*, between two pixels is computed by $\text{diff}_i = |Q_i - Q_{i+1}|$. The *diff$_i$* value is further quantized into several regions as shown in Fig. 2. Each region is identified by the lower and upper bound values, i.e., [lower$_i$, upper$_i$]. The number of embedding bit sequence (k) in each block depends on the quantization range table, and it is determined as $k = \lfloor \log_2(\text{upper}_i - \text{lower}_i + 1) \rfloor$. The obtained bit sequence is converted into decimal value k_d. The new difference value (diff'_i) is obtained as $\text{diff}'_i = \text{diff}_i + \text{lower}_i$.

The modified pixel values are computed based on Eq. 5.

R_i	R_1	R_2	R_3	R_4	R_5	R_6
Range	[0 - 7]	[8 - 15]	[16 - 31]	[32 - 63]	[64 – 127]	[128 – 255]
Width	8	8	16	32	64	128
Payload/Capacity (In bits)	3	3	4	5	6	7

Fig. 2 Quantization range

$$(Q'_i, Q'_{i+1}) = \begin{cases} (Q_i + \lceil l/2 \rceil, Q_{i+1} - \lfloor l/2 \rfloor), \text{ if } Q_i \geq Q_{i+1} \text{ and diff}'_i > \text{diff}_i \\ (Q_i - \lfloor l/2 \rfloor, Q_{i+1} + \lceil l/2 \rceil), \text{ if } Q_i < Q_{i+1} \text{ and diff}'_i > \text{diff}_i \\ (Q_i - \lceil l/2 \rceil, Q_{i+1} + \lfloor l/2 \rfloor), \text{ if } Q_i \geq Q_{i+1} \text{ and diff}'_i \leq \text{diff}_i \\ (Q_i + \lceil l/2 \rceil, Q_{i+1} - \lfloor l/2 \rfloor), \text{ if } Q_i < Q_{i+1} \text{ and diff}'_i \leq \text{diff}_i \end{cases} \quad \cdots$$

$$(5)$$

where $l = \left| \text{diff}'_i - \text{diff}_i \right|$.

Later, pixels Q_i and Q_{i+1} of each block are replaced by the stego pixels Q'_i and Q'_{i+1}. The receiver will find the difference of ith block diff$_i$ = $|Q'_i - Q'_{i+1}|$. The difference diff$'_i$ is used to search the number of conceal bit streams in ith block using the quantization range given in Fig. 2. Finally, the secret bit streams are obtained after converting the decimal value of (diff$'_i$ − lower$_i$) into binary form.

3 Proposed Scheme

In this paper, we have proposed a grayscale image steganography scheme where the level of security has been improved without incorporating any encryption process with steganography. The secret messages bits are embedded into the cover image either by LSB or PVD techniques. The choice of data hiding approach is decided based on a secret binary sequence generated by logistic map. The logistic map generates the chaotic sequence values using Eqs. 1 and 2. The sequence values are found in the range from 0 to 1. Later, this sequence is converted into binary sequence based on Eqs. 6 and 7.

$$B_i = \text{round}(\alpha_i \times 255) \qquad (6)$$

$$A_i = \text{mod}(B_i, 2) \qquad (7)$$

where ith chaotic sequence is α_i, ith sequence in round is B_i, and ith binary bit is denoted by A_i.

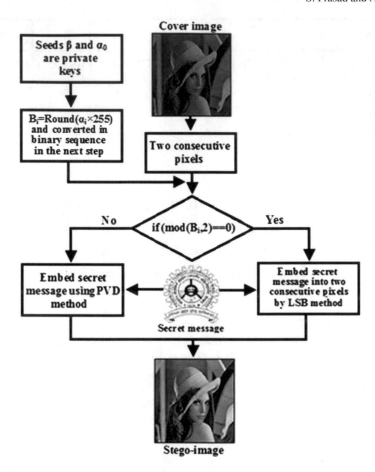

Fig. 3 Block diagram for the secret message embedding procedure

The data embedding and extraction process of the proposed image steganography scheme are given in Algorithm 1 and Algorithm 2, respectively. The schematic diagram is further given in Figs. 3 and 4.

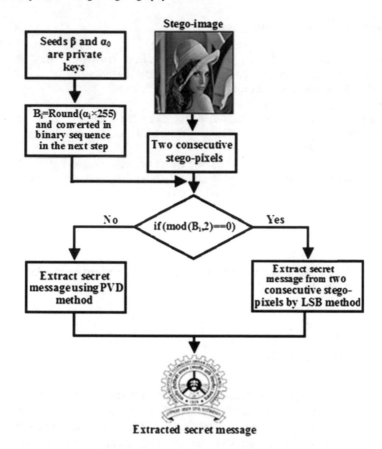

Fig. 4 Block diagram for the secret message extraction procedure

Algorithm 1: Embedding Procedure

Input: A grayscale cover image size of M×N, secret binary sequence generated by Logistic-map and the secret message bit streams.
Output: Stego-image size of M×N.
  Step:
  1 Read two consecutive non overlapping pixels Q_i and Q_{i+1} size of 1×2 block from gray-scale cover image.
  2 If $(mod(B_i, 2) == 0)$
     a. Then, 3 LSB substitution method applied into two consecutive non overlapping pixels block
     Else
     a. Apply pixel value differencing (PVD) into two consecutive non overlapping pixels blocks
     b. Obtain stego-pixels are Qs_i and Qs_{i+1}
  3 Process rest of the gray-scale pixels using step 1-2

Algorithm 2: Extraction Procedure

`Input:` A gray-scale stego-image size of M×N and secret binary sequence generated by Logistic-map.
`Output:` The secret message bits streams.
`Step:`
   1   Read two consecutive non overlapping stego pixels size of 1×2 block from gray-scale stego-image.
   2   If $(\text{mod}(B_i, 2) == 0)$
       a. Then, extract 3 bits secret message by LSB substitution method from two consecutive non overlapping stego-pixels blocks
       Else
       a. Extract k bits secret message by pixel value differencing(PVD) from two consecutive non overlapping stego-pixels blocks
   3   Process rest of the gray-scale stego-pixels using step 1 to 2.

4   Experiment Results

The image steganography scheme is tested on several grayscale images, but in this scheme we have produced the results for only four standard grayscale images, i.e., 'Lena', 'Jet', 'Peppers', and 'Baboon' with size of 512×512 and each has 256 intensity levels (as shown in Fig. 5a–d). The histograms of the corresponding cover images are given in Fig. 6a–d. The obtained stego-images and their corresponding histograms are also presented in Figs. 7a–d and Fig. 8a–d, respectively. The presented stego-images retain high visual quality and closely similar to the original cover images. Accordingly, these similarities are further revealed based on presenting histograms for both the cover images and the stego-images. The disparity between histograms of corresponding categories of images is less. This dissimilarity can be noticed in Fig. 9a–d. In addition, the security is estimated when the opponent does not know the secret binary sequence. For that, we have considered Fig. 10a as cover image and correspondingly the secret image as shown in Fig. 10b is embedded. Similarly, Fig. 10c is considered as cover images and corresponding secret image is Fig. 10d. Firstly, we present the corresponding stego-images in Figs. 11a and 12a, respectively. Later, we have extracted secret images based on 3 LSB and PVD. We have received meaningless secret images as shown in Figs. 11b, c and 12b, c, respectively. The correct secret images are shown in Figs. 11d and 12d when the concerned person knows the correct binary sequence. So, it is difficult to reveal the actual secret content without knowing the secret binary sequence.

In addition, the obtained stego-images retain high visual quality even after embedding large volume of secret message. Table 1 shows comparative results in terms of

(a)   **(b)**   **(c)**   **(d)**

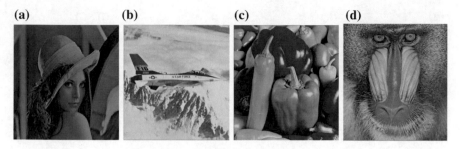

Fig. 5 Cover images are **a** lena, **b** Jet, **c** peppers, and **d** baboon

(a)   **(b)**   **(c)**   **(d)**

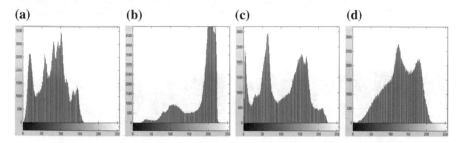

Fig. 6 Cover image histograms are **a** lena, **b** Jet, **c** peppers, and **d** baboon

(a)   **(b)**   **(c)**   **(d)**

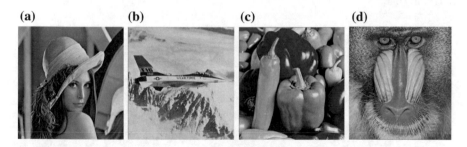

Fig. 7 Stego-images are **a** lena, **b** Jet, **c** peppers, and **d** baboon

(a)   **(b)**   **(c)**   **(d)**

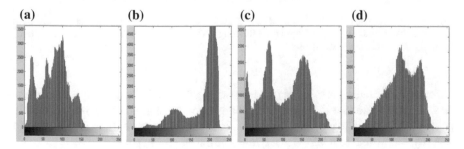

Fig. 8 Stego-image histograms are **a** lena, **b** Jet, **c** peppers, and **d** baboon

(a)   (b)   (c)   (d)

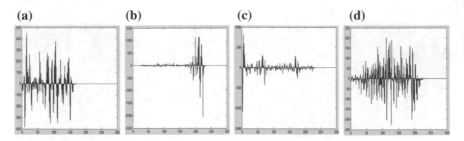

Fig. 9 Difference images histograms are **a** lena, **b** Jet, **c** peppers, and **d** baboon

(a)   (b)   (c)   (d)

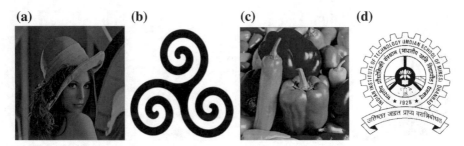

Fig. 10 **a** Lena cover image, **b** secret message (logo), **c** peppers cover image, and **d** secret message (IIT(ISM) Logo)

(a)   (b)   (c)   (d)

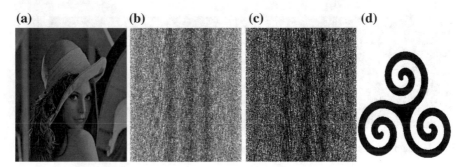

Fig. 11 **a** Stego-image Lena, **b** extracted secret message (logo) from Fig. 11a using 3LSB, **c** extracted secret message (logo) from Fig. 11a using PVD, and **d** extracted secret message (logo) from Fig. 11a using proposed scheme

different data hiding process, payload capacity, and PSNR values. The experimental results indicate that the stego-images have good visual quality with reasonable payload/capacity. The proposed technique is suitable to achieve high level of image security and is able to retain visual quality of the stego-image.

(a)   (b)   (c)   (d)

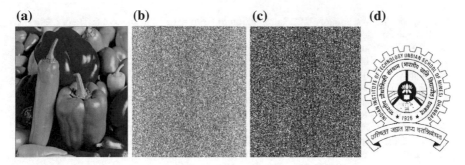

Fig. 12 **a** Stego-image peppers, **b** extracted secret message (IIT(ISM) logo) from Fig. 12a using 3LSB, **c** extracted secret message (IIT(ISM) logo) from Fig. 12a using PVD, and **d** extracted secret message (IIT(ISM) logo) from Fig. 12a using proposed scheme

Table 1 Comparative results

Image name (512 × 512)	Proposed method		PVD method		3LSB method		PVD and LSB [8]	
	Payload (bits)	PSNR (dB)	Payload (bits)	PSNR (dB)	Payload (bits)	PSNR (dB)	Payload (bits)	PSNR (dB)
Lena	593118	39.56	404080	42.14	786432	37.92	528512	38.80
Jet	595634	39.12	409187	40.80	786432	37.94	530048	37.63
Peppers	595004	39.11	407643	40.46	786432	37.93	528256	37.50
Baboon	620200	37.38	457105	36.92	786432	37.93	544056	33.33

5 Conclusion

In this paper, the scheme has embedded the secret message into grayscale cover image using combined of 3LSB and PVD. The security scheme is enhanced using logistic map-based generated secret binary sequence. The experimental results indicate that the proposed method has attained the appropriate embedding capacity/payload and excellent visual quality of stego-image compared to other relevant approaches. In addition, this scheme has achieved high security where it is not feasible to extract the secret message without knowing the secret binary sequence. The presented simulation results confirm the validity of the proposed scheme.

References

1. Stalling, W.: Cryptography and Network Security: Principles and Practices, 4th edn. Pearson Education India (2007)

2. Trapppe, W., Washington, L.C.: Introduction to Cryptography with Coding Theory, 2nd edn. Pearson Prentice Hall (2011)
3. Lu, C.-S.: Multimedia Security: Steganography and Digital Watermarking Techniques for Protection of Intellectual Property. Idea Group Publishing, Hershey (2005)
4. Chan, C.K., Cheng, L.M.: Hiding data in images by simple LSB substitution. Pattern Recogn. **37**(3), 469–474 (2003)
5. Wang, R.Z., Lin, C.F., Lin, J.C.: Image hiding by optimal LSB substitution and genetic algorithm. Pattern Recogn. **34**(3), 671–683 (2001)
6. Pal, A.K., Pramanik, T.: Design of an edge detection based image steganography with high embedding capacity. LNICST **115**, 794–800 (2013)
7. Wu, D.-C., Tsai, W.-H.: A steganographic method for images by pixel value differencing. Pattern Recogn. Lett. **24**(9–10), 1613–1626 (2003)
8. Wu, H.C., Wu, N.I., Tsai, C.S., Hwang, M.S.: Image steganographic scheme based on pixel-value differencing and LSB replacement methods. IEE Proc. Vis. Images Sig. Process **152**(5), 611–615 (2004)
9. Wang, C.-M., Wu, N.-I., Tsai, C.-S., Hwang, M.-S.: A high quality steganographic method with pixel-value differencing and modulus function. J. Syst. Softw. **81**(1), 150–158 (2007)
10. Prasad, S., Pal, A.K.: An RGB colour image steganography scheme using overlapping block-based pixel-value differencing. R. Soc. Open Sci. **4**(4), 161066 (2017)
11. Abikoye, O.C., Adewole, K.S., Oladipupo, A.J.: Efficient data hiding system using cryptography and steganography. Int. J. Appl. Inf. Syst. **4**(11), 6–11 (2012)
12. Qian, Z., Zhang, X., Feng, G.: Reversible data hiding in encrypted images based on progressive recovery. IEEE Sig. Process. Lett. **23**(11), 1672–1676 (2016)
13. Liśkiewicz, M., Reischuk, R., Wölfel, U.: Security levels in steganography-insecurity does not imply detectability. Theor. Comput. Sci. **692**, 25–45 (2017)
14. Trivedy, S., Pal, A.K.: A logistic map-based fragile watermarking scheme of digital images with tamper detection. Iranian J. Sci. Technol. Trans. Electr. Eng. **41**(2), 103–113 (2017)

Blind Digital Image Watermarking for Copyright Protection Based on Hadamard Transform

Sanjida Sharmin, Md. Khaliluzzaman, Md. Mahiuddin and Abdullahil Kafi

Abstract The reason behind the extension of digitized media is a speedy augmentation of multimedia data transaction over the Internet. For this reason, there arises the need for copyright fortification for multimedia data. On this, time digital watermarking can be the best solution to this problem. Following this, the basis on Hadamard transform for the image authentication a blind watermarking process is proposed in this succeeding paper. To implant the watermark, the proposed method used breadth-first search algorithm for finding the efficient embedding point. Additionally, PSNR and NCC have been taken into account for finding the performance of watermark. The result of the experiment showed significant sturdiness against enhancement and noise addition attacks such as compression by JPEG, cropping, sharpening, and filtering. In comparison with other available watermarking techniques, the result of regarding experiment has shown better inaudibility and vigor.

1 Introduction

The method of hiding the information that is related to the digital signal i.e., an image, audio, and video in the signal itself presents the technique that is used to verify its authenticity of its owners is known as digital watermarking [1, 2]. To use the original image in the process of watermark extraction, the process can be classified as either blind or non-blind. In the blind process, the original image is required to

S. Sharmin · Md. Khaliluzzaman (✉) · Md. Mahiuddin · A. Kafi
Department of Computer Science and Engineering, International Islamic
University Chittagong (IIUC), Chittagong 4318, Bangladesh
e-mail: khalilcse021@gmail.com

S. Sharmin
e-mail: ssharmin114@gmail.com

Md. Mahiuddin
e-mail: mmuict@gmail.com

A. Kafi
e-mail: abkafi@gmail.com

© Springer Nature Singapore Pte Ltd. 2019
A. Abraham et al. (eds.), *Emerging Technologies in Data Mining and Information
Security*, Advances in Intelligent Systems and Computing 814,
https://doi.org/10.1007/978-981-13-1501-5_18

215

detect and verify the watermark, where in the non-blind process original image is not used in the phase of watermark detection and verification [3]. Our proposed method follows the blind watermarking technique. In terms of the utilization, the watermarking techniques are divided according to the fragile, semi-fragile, as well as robust. The sensitivity of the fragile is very high. This watermarking technique is altered with slight modification at the watermark signal, where semi-fragile watermarking depends on the specific threshold. If the process exceeded that specific threshold, the watermarking process is also broken. On the other hand, robust watermarking depends on the various threads, i.e., filtering, scaling, and cropping, as well as compression [4–7].

This topic has gained importance for the purpose of automatic authentication of the security systems. As far as security is concerned, researchers have found many methods to improve the accuracy of the present watermarking methods. For example, in [8], a blind image watermarking method is proposed. This method is based on Hadamard transform. In this method, best-first search algorithm is used to find the increasing sequence of each block to embed watermark. In [9], a DWT-DCT method is utilized to propose a watermarking algorithm. By this algorithm, various watermark attacks can be obstructed. In this work, authors utilized the DWT instead of embedding the mid-frequency of the DCT to the watermark. Finally, for updating the security system the watermark is scrambled and embedded in a spread spectrum pattern.

A blind dual-color image watermarking method is proposed in [10]. In this method, singular value of decomposition (SVD) is used. Earlier, hybrid image watermarking system is presented in [11]. Here, the authors employed the DWT and SVD methods to improve the performance, where watermarks are not combined implicitly on the information of singular values of the sub-bands of cover image's DWT.

In [12], a new method based on the blind watermarking is presented in [12], where gray-level watermarking system is utilized. In this system, the input image is divided into some non-overlapping block that is 4 × 4. A joint DWT-DCT method is used for digital image watermarking in [13]. This method is used for the process of copyright protection which is imperceptible [13, 14].

This paper proposed a method on Hadamard transform. Breadth-first search algorithm is used to select the positive connected coefficient to embed the watermark. The major task is to select the best embedding point to avoid distortion. The embedding point is selected in the first column because the rate of deformation is less in this column.

The paper is presented as follows. In Sect. 2, 2D-Hadamard transform is described. In Sect. 3, the proposed method of watermarking system is described. In the next section, the experimental results are explained. Finally, the paper is concluded in Sect. 5.

2 2D-Hadamard Transform

The Hadamard transform (HT) is also represented as the Walsh–Hadamard transform, Walsh transform, Hadamard–Rademacher–Walsh transform, or Walsh–Fourier transform. The 2D-Hadamard transform (2D-HT) which is well-known transform system is used extensively in the field of image processing and image compression [15]. The 2D-HT can be represented by Eq. (1), where U is the original image and V is the transformed image [16].

$$[V] = \frac{H_n[U]H_n}{N} \tag{1}$$

where an $N \times N$ Hadamard matrix is represented by H_n and the value of N is $2n$, $n = 1, 2, 3\ldots$ with element values either $+1$ or -1.

The main advantage of the HT is its transform matrix H_n which is simple. The values of this matrix are binary as well as real. The orientation of the H_n is orthogonal. The inverse 2D-HT is represented by (2).

$$[U] = H_n^{-1}[V]H_n^* = \frac{H_n[V]H_n}{N} \tag{2}$$

In this work, the original image is subdivided into 8×8 blocks. The matrix of Hadamard transform that is used in this work is H_3 as shown in (3).

$$
H =
\begin{bmatrix}
1 & 1 & 1 & 1 & 1 & 1 & 1 & 1 \\
1 & -1 & 1 & -1 & 1 & -1 & 1 & -1 \\
1 & 1 & -1 & -1 & 1 & 1 & -1 & -1 \\
1 & -1 & -1 & 1 & 1 & -1 & -1 & 1 \\
1 & 1 & 1 & 1 & -1 & -1 & -1 & -1 \\
1 & 1 & 1 & -1 & -1 & 1 & -1 & 1 \\
1 & 1 & -1 & -1 & -1 & -1 & 1 & 1 \\
1 & -1 & -1 & 1 & 1- & 1 & 1 & -1
\end{bmatrix}
\tag{3}
$$

3 Proposed Method

The proposed method is divided into two major parts, i.e., the embedding process of watermark and the extraction process of the watermark.

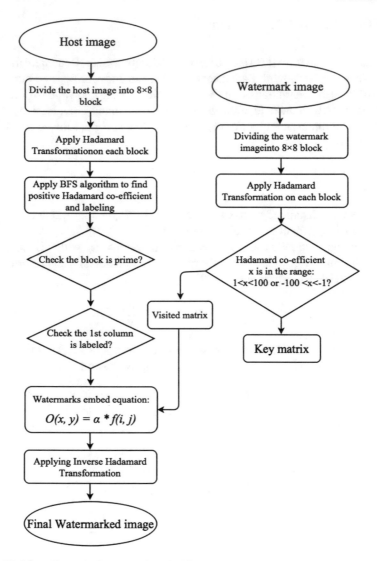

Fig. 1 Workflow diagram of watermark embedding process

3.1 Watermark Embedding Process

In this section, the watermark embedding process is described. The workflow diagram of the proposed embedding method is shown in Fig. 1. The steps of the proposed embedding method are described as follows.

Step 1: *Divide input image*

The input image is divided into some blocks that are 8 × 8.

Step 2: *Applying Hadamard transform*

Hadamard transform is employed on every block to get Hadamard transform coefficient for the blocks. These Hadamard transform coefficients are needed for next selection. Each block consists of a DC value which is avoided for any embedding. Each 8*8 block has 64 coefficients. Here, one coefficient is avoided and remaining 63 coefficients are used for the embedding. The remaining 63 coefficients are called AC value of the block.

Step 3: *Applying breadth-first search algorithm*

Now, BFS algorithm is used on the Hadamard coefficient and finds the positive connecting component. A spanning tree is generated from the Hadamard coefficient. An order for breadth-first searching is given. Here, C is center point and 1, 2, 3, 4, 5, 6, 7, 8 is the order of searching.

Step 4: *Prime number system*

In this stage, the prime numbering system is used for selecting the blocks. Non-prime blocks are selected for embedding. After selecting the Non-prime block refers to the number of blocks which are not prime like 2, 4, 6, 8 etc. Then, select the labeling that positions in the first column. In the next, those points are selected to embed watermarking. In this way, mark the watermark point. According to the embedded process, only first column is selected. This is because the rate of distortion of the watermarked image is very small in that column.

Step 5: *Divide watermarked image and apply Hadamard transform*

This section takes the input watermarked image. The input watermarked image is resized as 64×64 pixels. The watermarked image is divided into 8×8 blocks and applies HT on every block. After applying Hadamard transform, extract the points that are in the range of $1 < x < 100$ or $-100 < x < -1$. These points are preserved in a visited matrix. The components that are not extracted from the image are preserved in another storage matrix, i.e., key visited matrix. This key visited matrix is used in the phase of the decoding process.

Step 6: *Applying embedding equation*

In this phase, use Cox's equation for embedding watermark Hadamard coefficient into host Hadamard coefficient.

$$O(x, y) = \alpha * f(i, j) \qquad (4)$$

where $O(x, y) =$ host Hadamard coefficient, $\alpha =$ scaling factor, and $f(i, j) =$ Hadamard coefficient of watermark.

Step 7: *Applying inverse Hadamard transform*

After completing the embedding process again apply the inverse Hadamard transform to produce a watermarked image.

Fig. 2 Workflow diagram of
watermark extraction process

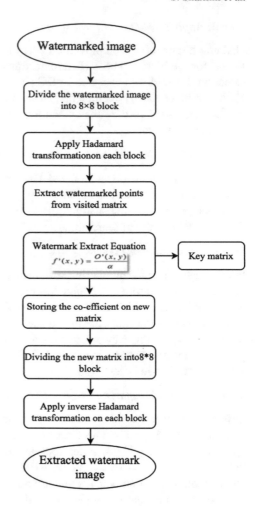

3.2 Watermark Extraction Process

This section describes the process for extracting the security watermarked image from Hadamard transform watermarked image. Figure 2 presents the working diagram of the proposed extraction procedure. The procedures are explained in the following steps.

Step 1: *Dividing the original image*

The input image is divided into some blocks that are 8×8.

Step 2: *Applying Hadamard transform*

After dividing the input image, the Hadamard transform is employed on every block of the image.

Step 3: *Applying extraction equation*

The visited matrix is used for watermark extraction. This watermark extraction process is used by the inverse embedded transform as shown in (5). Here, the inverse embed Eq. (5) is used to extract watermarked image.

$$f'(x, y) = \frac{O'(x, y)}{\alpha} \tag{5}$$

Here, $f'(x, y)$ is watermark extract coefficient as well as $O'(x, y)$ is the Hadamard coefficient of watermark.

Step 4: *Applying inverse Hadamard transform*

After extracting Hadamard coefficient, the inverse Hadamard transform is applied to extract Hadamard coefficient to get the extracted watermarked image.

4 Experimental Result and Analysis

In this paper, simulation is carried out for grayscale images. The simulation was performed using MATLAB environment.

The mean square error (MSE) and peak signal-to-noise ratio (PSNR) are typically used to measure the quality of the receiving image with respect to the input image. The MSE and PSNR are measured using (6) and (7).

$$\text{MSE} = \frac{1}{MN} \sum_{i=0}^{M-1} \sum_{j=0}^{N-1} \| I(i, j) - K(i, j) \|^2 \tag{6}$$

$$\text{PSNR} = 10 \log_{10}^{\frac{\| I(i,j) - R(i,j) \|^2}{\text{MSE}}} \tag{7}$$

The N and M define the width and height of the image. The input image and watermarked image are presented as $I(i, j)$ and $K(i, j)$, respectively, where (i, j) is the coordinates.

The normalized correlation coefficient (NCC) is evaluated after extracting the watermarked image. For that, input and extractive watermark are used to examine the presence of watermark. It is also used to estimate the faultlessness of the extractive watermark. The NCC is defined as (8).

$$\text{NCC} = \frac{1}{m \times n} \sum_{i=1}^{m} \sum_{j=1}^{n} w(i, j) \times w'(i, j) \tag{8}$$

Here, watermark height and width are defined as m and n, respectively. $w(i, j)$ is the input image, and $w'(i, j)$ is the extractive watermarked image value at the coordinate of (i, j), respectively.

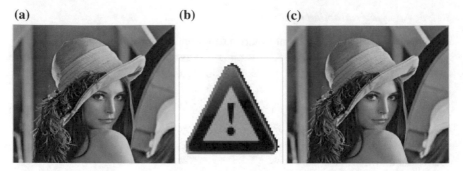

Fig. 3 Experiment of (lenna): **a** original image, **b** watermarked image, and **c** watermarked image

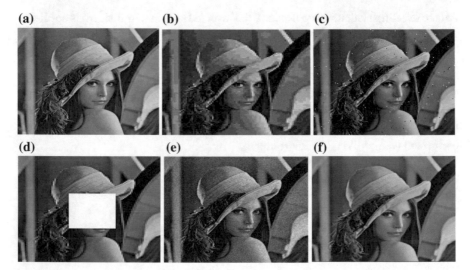

Fig. 4 Different attacks applied to watermarked image: **a** original image, **b** JPEG compression, **c** salt and pepper, **d** cropping, **e** Gaussian noise, **f** changing aspect ratio

Table 1 PSNR and NCC for without alteration of image

Name of image	PSNR	NCC
Lenna	39.21	1

The original image and watermarked image are shown in Fig. 3. The result of after applying the different attacks on the watermarked image is shown in Fig. 4. The PSNR and NCC value for without alteration of Lena image is shown in Table 1.

To measure the quality and robustness of the image, PSNR and NCC are very important. These values are measured from the extracting watermarked image with respect to host watermarked image. The measured NCC and PSNR values for the proposed method and Sarker's method [8] with the different attacks are shown in

Table 2 Performance comparison for NCC

SI. no	Attack	Sarker's method [8]	Proposed method
1	JPEG compression (QF=30)	0.9842	0.9867
2	JPEG compression (QF=60)	0.9909	0.9919
3	JPEG compression (QF=90)	1	1
4	Median filtering [3 × 3]	0.9873	0.9786
5	Weiner filtering	0.9711	0.9867
15	Rotation angle = 5	0.8875	0.8976
7	Rotation angle = 10	0.9915	0.9908
8	Rotation angle = 20	0.9956	0.9963
9	Gaussian noise (average = 0, density = .002)	0.9873	0.9878
10	Speckle noise (density =. 01)	0.9887	0.9890
11	Salt and pepper noise (strength = 01)	0.9856	0.9867
12	Cropping [32 × 32]	0.9982	0.9989
13	Cropping [64 × 64]	0.9894	0.9987
14	Cropping [128 × 128]	0.9595	0.9689

Tables 2 and 3, respectively. From these values of PSNR and NCC, it is seen that the proposed method shows the better performance than that of Sarker's method.

In [8], Hadamard transform is proposed where the embedding point is selected by best-first search algorithm. Applying best-first search, some increasing sequences are found. Longest sequence from the extracted sequence is selected. For embedding process, the highest value from the longest sequence is used.

On the other hand, in the proposed method breadth-first search algorithm is used to select the embedding point. By using this algorithm, positive connecting component is selected. A spanning tree is formed from the Hadamard coefficient. Next, the prime numbering system is used for selecting the blocks. In the proposed method's embedded process, select the embedding point only in the first column. Since the rate of distortion of the watermarked image is very small in that column, the proposed method has shown the values of PSNR and NCC for different attacks are better than the method proposed in [8] (Table 4).

Table 3 Performance comparison for PSNR

SI. no	Attack	Sarker's method [8]	Proposed method
1	JPEG compression (QF = 30)	3 3.7029	34.4321
2	JPEG compression (QF = 60)	36.3765	36.7654
3	JPEG compression (QF = 90)	38.4525	38.342
4	Median filtering [3 × 3]	34.4961	35.316
5	Weiner filtering	36.2677	37.578
6	Rotation angle = 5	24.1601	24.876
7	Rotation angle = 10	22.5655	23.232
8	Rotation angle = 2D	21.4640	2 2.43 6
9	Gaussian noise (average = 0, density = .002)	26.8266	27.91
10	Speckle noise (density = .01)	26.5200	26.99
11	Salt and pepper noise (strength = .01)	24.7564	25.87
12	Cropping [32 × 32]	30.1313	32.476
13	Cropping [64 × 64]	25.4391	26.786
14	Cropping [128 × 128]	16.3047	17.345

Table 4 Performance comparison of NCC between Anthony's method [15] and proposed method

Different attack	Anthony's method [1]	Proposed method
Sharpening 3 × 3	0.9573	0.9933
1 row and 1 column removed	0.9866	0.9883
Frequency mode Laplacian removal	0.9580	0.9754
Scaling 0.75 (75%)	0.9354	0.9554
JPEG compression of factor 30	0.8688	0.9042
Changing aspect ratio	0.8199	0.8558

5 Conclusion

This paper proposes an improved blind digital image watermarking scheme based on the Hadamard transform which provides an effective method to embed and extract the watermark information. In watermark extraction process, neither host nor the original watermark is required which made the proposed method as a blind watermarking technique. This method of embedding persists the visual robustness of the image. The comparative results show that proposed method has better invisibility and robustness against various types of common signal processing attacks. Analyzing the performance of the proposed method, it is seen that the proposed technique gives the better result than the other existing methods.

References

1. Husain, A.: A survey of digital watermarking techniques for multimedia data. MIT Int. J. Electron. Commun. Eng. **2**(1), 37–43 (2012)
2. Lee, S.J., Jung, S. H.: A survey of watermarking techniques applied to multimedia. In: IEEE International Symposium on Industrial Electronics, vol. 1, pp. 272–277 (2001)
3. Su, Q., Niu, Y., Zou, H., Liu, X.: A blind dual color images watermarking based on singular value decomposition. Appl. Math. Comput. **219**, 8455–8466 (2013)
4. Deb, K., Al-Seraj, M.S., Hoque, M.M., Sarker, M.I.H.: Combined DWT DCT based digital image watermarking technique for copyright protection. In: IEEE International Conference on Electrical and Computer Engineering (ICECE) (2012)
5. Sathik, M., Sujatha, S.S.: An improved invisible watermarking technique for image authentication. In: Int. J. Adv. Sci. Technol. (IJAST), **24** (2010)
6. Nikolaidis, A., Pitas, I.: Asymptotically optimal detection for additive watermarking in the DCT and DWT Domains. IEEE Trans. Image Process. **12**(5), 563 (2003)
7. Wu, X., Sun, W.: Robust copyright protection scheme for digital images using overlapping DCT and SVD. Appl. Soft Comput. **13**, 1170–1182 (2013)
8. Sarker, M.I.H., Khan, M.I.: An improved blind watermarking method in frequency domain for image authentication. In: 2013 International Conference on Informatics Electronics and Vision (ICIEV). IEEE (2013)
9. Feng, L.P., Zheng, L.B., Cao, P.: A DWT-DCT based blind watermarking algorithm for copyright protection. In: 3rd IEEE International Conference on Computer Science and Information Technology (ICCSIT). IEEE (2010)
10. Su, Q., Niu, Y., Zou, H., Liu, X.: A blind dual color images watermarking based on singular value decomposition. Appl. Math. Comput. **219**, 8455–8466 (2013)
11. Lai, C.C., Tsai, C.C.: Digital image watermarking using discrete wavelet transform and singular value decomposition. IEEE Trans. Instrum. Measur. **59**(11), 3060 (2010)
12. Saryazdi, S., Nezamabadi-pour, H.: A blind digital watermark in Hadamard domain. In: Proceedings of World Academy of Science, Engineering and Technology, vol. 3 (2005)
13. Al-Haj, A.: Combined DWT-DCT digital image watermarking. J. Comput. Sci. **3**(9), 740–746 (2007)
14. Zhao, R., Hua, L., Pang, H., Hu, B.: A watermarking algorithm by modifying AC coefficies in DCT domain. In: IEEE International Symposium on Information Science and Engineering (ISISE) (2008)
15. Ho, A.T.S., Shen, J., Chow, A.K., Woon, J.: Robust digital image-in-image watermarking algorithm using the fast hadamard transform. In: IEEE International Symposium on Circuit and system (ISCAS'03), vol. 3, pp. 826–829. IEEE (2003)
16. Kountchev, R., Rubin, S., Milanova, M., Todorov, V.: Resistant image watermarking in the phases of the complex Hadamard transform coefficients. In: IEEE International Conference on Information Reuse and Integration (IRI), pp. 159–164 (2010)

Part III
Image Processing

An Analysis of Various Techniques for Leaf Disease Prediction

P. Niveditha, H. L. Gururaj and V. Janhavi

Abstract The detection of plant diseases at early stage can be the best precaution taken by any farmer to avoid great loss. If that work of detecting the plant is made automatic, then it would be the essential topic for discovery. Mainly, plant diseases are caused by fungi, bacteria, and virus. All three affect the plant in different way which can be identified. This feature helps in detecting the particular disease. When it comes to fungi, they can be classified with their morphology that is based on their reproductive structures. Bacteria have the unique property of increasing their number in short duration, the single cell dividing into two and hence multiplying in number, but compared to fungi, they have simple life cycle. Viruses are the smallest particle found. They are made up of proteins and genetic materials. The method for detecting involves five stages, in the first stage, the image is selected through the inputs, the second stage involves processing of image, the third involves dividing the image, the fourth involves finding unique attributes, and final step involves analysis.

Keywords Plant leaf diseases · Morphology · Segmentation · Symptoms
Resampling · Automatic identification

1 Introduction

In developing countries like India, there are many who are depending on agriculture for their shelter and food. According to the recent survey, about 60–70% of the total population is dependent on agriculture. This holds an important position in the Indian economy. There are many products which are depending on the agriculture for

P. Niveditha · H. L. Gururaj (✉) · V. Janhavi
Vidyavardhaka College of Engineering, Mysuru 570002, India
e-mail: gururaj1711@vvce.ac.in

P. Niveditha
e-mail: spendlikal@gmail.com

V. Janhavi
e-mail: janhavi@gmail.com

© Springer Nature Singapore Pte Ltd. 2019   229
A. Abraham et al. (eds.), *Emerging Technologies in Data Mining and Information Security*, Advances in Intelligent Systems and Computing 814,
https://doi.org/10.1007/978-981-13-1501-5_19

their production such as they may be used as raw materials in some industries, may be small-scale or large-scale industries. To give example, sugar factory depends on sugarcane. All the productions need best quality raw materials; this proves that even industrial sector also depends on agriculture. So this becomes vital topic in research field, agriculture, and its productivity.

Usually, farmers find out all the diseases by keen observation using naked eyes. But there are few diseases which are microscopic and cannot be noticed using naked eyes. Symptoms may be microscopic, but there are chances where farmers consider it as healthy which may not be. When it comes to large farm when they use classical method is applied it may be difficult to get the proper and accurate result.

The other scenario is when the farmers detect the disease, it will be too late. They have to travel long distance to reach the experts; this may take long time and may be expensive. Every time it is difficult to consult experts for the identification. To overcome all these problems, automatic detection of plant disease is used; it is more effective even for the large areas. It senses the change in the plant using image processing techniques and shows the appropriate results. This is a way of disease management in plants. Diseases itself mean disasters, which cause crucial problems to the farmers as mentioned above. Diseases are an impairment of health or a condition of abnormal functionality. Plant diseases are caused by fungi, virus, and bacteria. It may be due to changes in the environment. They should be diagnosed regularly and immediately after the identifying the diseases which prevents heavy loss. Diseases in plants can be identified in many parts such as fruits, seeds, flowers, stem, and root; hence, the symptoms are different for every part; it is important to diagnose at right time [1].

Structuring of remaining paper is as follows: Sect. 2 provides related work, Sect. 3 focuses on suggested methodology, Sect. 4 focuses on comparative analysis and shows the similar products in market, and Sect. 5 concludes the paper.

2 Related Work

There are various techniques used in the identification of plant disease. A survey on some related papers shows the following methods namely,

The paper uses the technique of vision image enhancement. Here, there is the use of grayscale image, color conversion, and histogram equalization to increase quality visibility and contrast of the image. This is used in detection of plant leaf diseases [2].

This paper uses the technique of neutral network in the methodology. The precreated database is compared with input image which undergoes k-means clustering, and classification is based on gray-level co-occurrence matrix (GLCM) and finally uses a rich platform of parallel processing called neural network. This application uses back-propagation algorithm [3].

This paper uses the technique of machine learning for automated decision support system approach to detect the plant disease. This approach includes preprocessing, subtracting the unwanted background data, analysis of texture using opposite color local binary pattern (LBP) [4].

This paper makes use of combination of two techniques, namely k-means clustering for image segmentation and fuzzy logic for classification the image. This method may be difficult to understand the structure of algorithm and to determine optimal parameters [5].

This paper uses soft computing techniques to detect the plant diseases in leaf, as healthy plants have more green pigments. Segmentation is done by masking the green pixels, threshold-based segmentation using intensity or gray levels. This helps in identifying the diseased portion and is compared with preprocessed database to lead to a conclusion [6].

This paper gives a homogeneous pixel counting technique for cotton disease detection (HPCCDD). Segmentation is done by using canny and Sobel edge detection homogeneous techniques, and for analysis and classification, HPCCDD algorithm is applied [7].

This paper approached in identifying the diseases caused in sugarcane leaf. This system is applicable for only three verified diseases, namely rust spot, ring spot, and yellow spot. Considering 30 testing data, among which 9 are of rust diseases, 7 of ring disease, remaining of yellow spot disease. Feature extraction is done using SVM classifier. It has mainly four kernels, namely linear, quadratic, radial, and polynomial which are tested, but linear shows the efficient result [8].

This paper deals with the diseases which can be detected in abiotic conditions such as winter. There are new spectral indices (NSI) introduced which are sensitive to abiotic conditions. The efficiency of new NSIs will improve with the image of hyperspectral data [9].

This paper concentrates on a particular plant "chilli" and uses MATLAB software to obtain results by keeping the standard image characteristics. In the feature extraction, color clusters are used [10].

This paper along with the detection of diseases also helps in detecting fruit grading. Here, the word "smart farming" is used to include the technology for better productivity. Image segmentation here is done with the back-propagation technique [11].

In this paper, there is a proposed approach for plant disease detection using SVM classifier and SIFT algorithm. The experimental analysis of 120 images of soya bean proves that algorithm correctly recognizes the plant species based on leaf shapes. The average accuracy is as high as 93.79% [12].

This paper deals with the diseases in the fruit plants. Feature extraction includes blob analysis, pattern matching. Blob analysis deals with color brightness, whereas pattern matching compares the tokens with preexisting patterns. ANN is applied for pattern matching [13].

3 Plant Disease Identification

The basic procedure of the proposed observe based technique is in this paper. Initially, the images of various leaves are collected using a digital camera. In Fig. 1, there is a picture of healthy leaf of chilly plant, and Fig. 2 gives the picture of affected leaf of chilly plant. Then using image processing techniques to the input images to acquire useful features that are necessary for further analysis [14]. Figure 3 shows the flow of the different stages in the current methodology.

3.1 Image Processing

In this stage, in image processing, it can be defined as acquiring the input images from different sources. To acquire the images, usually used resources are cameras, scanners. Depending on the type of input source, processing is done.

Fig. 1 Healthy leaf of chilly plant

Fig. 2 Affected leaf of chilly plant

Fig. 3 Flow of the plant
disease identification

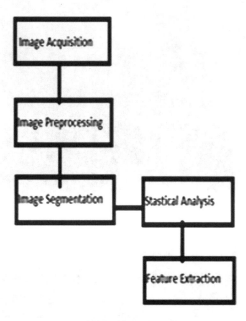

3.2 Image Preprocessing

Image processing is also known as image construction. Image reconstruction is the lowest level of abstraction which aims in improving the image data, and it suppresses the unwanted distortions and increases the quality of image [15].

It includes the resampling of the image, increasing the intensity of image, and includes noise removal. Resampling means changing the dimensions of the pixel according to the need. Enhancement includes the brightness and quality of the image. Noise removal is correcting the errors in the pixel level. Figure 4 depicts the resampling of image.

3.3 Image Segmentation

In this part of image processing, it includes the division and integration of the parts of the same image. An image is divided into segments and then segments into pixels which are given unique value. This is done to identify the region where it is infected with diseases. To do this, there are various techniques followed among them; the frequently used technique is k-means clustering. Figure 5 shows the image of k-means clustering.

Consider collection of data points $X = \{x_1, x_2 \dots x_n\}$. Each point in the set is called as potential cluster center.

Fig. 4 Resampling of image

Fig. 5 Gray scaling for image segmentation

$$P_n = P_n \sum_{i=1}^{n} e^{\frac{-4x_n - x_i^2}{r_a^2}}$$

where r is hypersphere cluster radius.

Find the potential of each data points, and select data point having maximum potential as first cluster center. Calculate the revise potential for each data point recursively and find the highest potential as next cluster. Continue until there are a sufficient number of clusters.

Consider an image with resolution of $Y \times Z$, and the image has to be converted into k number of clusters.

Let P(Y, Z) be input pixels and xK be element centers.
Algorithm:

1. Initialize the value of k and center.
2. For each pixel in an image, calculate the distance d from center to pixel using

$$d = P(Y, Z) - xK.$$

3. Based on the distance, assign all the pixels to their nearest clusters.
4. After all pixels are assigned, calculate new center,

$$xK = \sum P(Y, Z) YxK, ZxK$$

5. Repeat it until the sufficient clusters are obtained.
6. Consider the clusters form an image.

3.4 Feature Extraction

This is the main step where we identify that it is infected with the disease based on some features. As in this methodology there is usage of image processing, the feature includes color, texture, shape. These are the common visual features used.

Color

The input image obtained which is in RGB format should be converted into HSV color representation. In RGB, spectral components such as red, green, and blue are based on Cartesian coordinate system. This format matches with perspective of human eye to the primary colors. To avoid this problem, it is converted to HSV. The hue and saturation components lead to further analysis. These are carried out on each pixel. Spectral components of an image are shown in Fig. 6.

RGB values range from [0, 1], whereas HSV values have different ranges such as h = [0, 360], s = [0, 1], v = [0, 1].

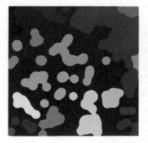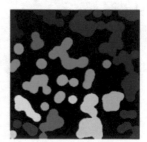

Fig. 6 Spectral components of image

Algorithm:

1. min=min(r,g,b)
2. max=max(r,g,b)
3. delta=max-min
4. if(r==max)
   h=(g-b)/delta
5. if (g==max)
   h=2+(b-r)/delta
6. Else
   h=4+(r-g)/delta
7. if (h<0)
   h+=360

Texture

This feature shows the uniqueness which is easy to identify and is the mostly used feature of image processing. Current trending researchers are targeting texture as the important feature to classify plants [15].

Shape

Shape in this context means the boundary or region of the infected area. This can be visually observed, and hence, it is visual feature. This also helps in classifying the plants.

Image Classification

This is the main phase of the methodology. This phase is a challenging technique as it involves the analysis of many attributes. The main purpose to classify is to predict the correct value based on the observation made and comparing it with previous statistics to lead to a conclusion. In this, classification is based on whether the plant is affected or not. There are various methods used to classify [1]. SVM classifier is efficiently used. The flow of classifier is shown in Fig. 7.

An image can be classified based on subspaces.

There are three layers: input, hidden, and output.

If there are k subspaces, there are k called classification results such as cl_ss1, cl_ss2….cl_ssk

$$CL = 1/k \sum cl\_ss_i$$

where i ranges from 1 to k.

Fig. 7 Flow of SVM
classifier

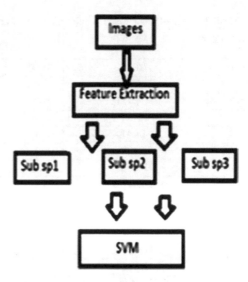

4 Results and Discussion

Table 1 has 4 attributes with 20 samples. There are three primary colors considered, namely yellow, cyan, and green. The values of their intensity are recorded. The fourth column indicates whether the plant is healthy or affected.

In Table 1, 20 samples of chilly leaves are considered. Using MATLAB software and image processing techniques, the intensity of primary colors is recorded for various samples. Ten leaves of affected leaves and other ten of non-affected leaves are considered. The variation in the intensity of the colors concludes if the plant is affected or not which is depicted in Table 1.

5 Conclusion

The above methodology proposed uses visual features to detect the plant diseases. This early detection is necessary as mentioned above. Firstly, RGB is converted into HSV format because HSV gives better color descriptor. Image segmentation is based on k-means clustering and feature extraction using visual attributes and using classifier drawing a conclusion. This work will be progressed with different algorithms enhancing the classifier. Further am intended to develop a tool which works on android using the above mentioned methodology named as "PHOTOsynthesis".

Table 1 Table of comparison of color intensity of chilly plant to detect affected or not

Sample	Cyan	Yellow	Green	Affected/not affected
1.	0.000	0.000	4.259	Not affected
2.	180519	0.190	0.000	Not affected
3.	24.711	0.000	0.050	Not affected
4.	1.454	0.040	0.793	Not affected
5.	14.474	0.413	0.000	Not affected
6.	0.000	0.521	31.742	Not affected
7.	0.060	0.828	20.417	Not affected
8.	0.443	0.226	7.926	Not affected
9.	19.670	0.000	5.455	Not affected
10.	0.462	0.000	6.128	Not affected
11.	3.234	4.022	0.639	Affected
12.	0.001	1.825	17.16	Affected
13.	0.000	2.656	2.595	Affected
14.	0.000	4.117	8.305	Affected
15.	0.000	3.774	12.555	Affected
16.	0.000	4.066	4.595	Affected
17.	0.000	2.539	3.789	Affected
18.	0.002	9.809	7.415	Affected
19.	0.000	6.074	4.008	Affected
20.	0.000	12.981	28.157	Affected

References

1. Bharwad, V.S., Dangarwala, K.J.: Recent research trends of plant disease detection. Int. J. Sci. Res. ISSN (online): 2319–7064 (2015)
2. Thangadurai, K., Padmavathi, K.: Computer vision image enhancement for plant leaves diseases detection. In: 2014 World Congress on Computing and Communication Technologies (2014)
3. Dhakate, M., Ingole, A.B.: Diagnosis of pomegranate plant disease using neural network. In: 2015 IEEE. 978-1-4673-8564 (2015)
4. Waghmare, H., Kokare, R., Dandawate, Y.: Detection and classification of Diseases of grape plant using opposite colour local binary pattern feature and machine learning for automated decision support system. In: 2016 3rd International Conference on Signal Processing and Integrated Networks (2016)
5. Pathy, M.J.B., Kumar, D.D., Manish, L., Choudhry, L.: Leaf disease detection using k-means clustering and fuzzy logic classifier. Int. J. Eng. Tech. Approach. ISSN-2395-0900 (2016)
6. Francis, J., Anto Sahaya Dhas D., Anoop B.K.: Identification of Leaf Diseases in Pepper Plants using Soft Computing Techniques
7. Revathi, P., Hemalatha, M.: Classification of cotton leaf spot diseases using image processing edge detection techniques. In: IEEE International Conference on Emerging Trends in Science, Engineering and Technology (INCOSET), Tiruchirappalli, pp 169–173 (2012)
8. Ratnasari, E.K., Mentari, M., Kartika Dewi, R., Hari Ginardi, R.V.: Sugarcane leaf disease detection and severity estimation based on segmented spots image. In: IEEE International Con-

ference on Information, Communication Technology and System (ICTS), Surabaya, pp. 93–98 (2014)

9. Huang, W., Guan, Q., Luo, J., Zhang, J., Zhao, J., Liang, D., Huang, L., Zhang D.: New optimized spectral indices for identifying and monitoring winter wheat diseases. IEEE J. Select. Top. Appl. Earth Observ. Remote Sens. **7**(6) (2014)

10. Husin, Z.B., Shakaff, A.Y.B.M., Aziz, A.H.B.A., Farook, R.B.S.M.: Feasibility study on plant chilli disease detection using image processing techniques. In: 2012 Third International Conference on Intelligent Systems and Modelling and Simulation (2012)

11. Jhuria, M., Kumar, A., Borse, R.: Image processing for smart farming: detection of disease and fruit grading. In: 2013 IEEE Second International Conference on Image Processing (2013)

12. Dandawate, Y., Kokare R.: An automated approach for classification of plant disease towards development of furistic decision support system in Indian perspective. In: IEEE 2015. 978-1-9299-8792-4 (2015)

13. Awate, A., Deshmankar, D., Amrutkar, G., Bagul, U., Sonavane, S.: Fruit Disease detection using color, texture analysis and ANN. In: 2015 International Conference on Green Computing and Internet of Things (ICGCIOT) (2015)

14. Dhaygude, S.B., Kumbhar, N.P.: Agricultural plant leaf disease detection using image processing. Int. J. Adv. Res. Electr. Electron. Instrum. Eng. (2013)

15. Kiran, R.G., Ujwalla, G., Kamal, O.H.: Unhealthy region of citrus leaf detection using image processing techniques. In: IEEE International Conference on Convergence of Technology (12CT), Pune, 1 June 2014

Image Compression Using VQ for Lossy Compression

Rishav Chatterjee, Alenrex Maity and Rajdeep Chatterjee

Abstract The process to minimize the total number of bits required to depict an image is known as image compression. The main goal of image compression is to minimize the transmission cost and to reduce the storage space. Vector quantization is a most popular technique for lossy compression due to its high compression rate and simple decoding algorithm. The key technique of VQ is the codebook design. In this paper, we have compared and found out the compression ratios of jpg and tiff images using VQ for lossy compression and k-means clustering.

Keywords Encoding · Decoding · Lossy · K-means · VQ

1 Introduction

Image compression is the process, where we decrease the entire total number of bits required to display an image. The process of identifying and matching of pixel identity vectors into binary vectors resulting in a lesser number of possible reproductions is known as vector quantization, and it is one of the major image compression algorithms. An image compression method will remove the irrelevant and duplicate information and will encode the part which remains. Image compression is considered as one of the important aspects of quality transmission and of proper storage of images [1].

Vector quantization is one of the most used techniques which is used for image compression. Vector quantization (VQ) has following ways—(i) vector formation, (ii) training set selection, (iii) codebook generation, and (iv) quantization. The first and foremost among them is vector formation, and it involves in set of multiple vectors by dividing the input image. The vector formed is in bigger form, and the

R. Chatterjee (✉) · A. Maity · R. Chatterjee
School of Computer Engineering, KIIT University, Bhubaneswar, India
e-mail: rishavpiku@gmail.com

A. Maity
e-mail: alenrex8@gmail.com

© Springer Nature Singapore Pte Ltd. 2019
A. Abraham et al. (eds.), *Emerging Technologies in Data Mining and Information Security*, Advances in Intelligent Systems and Computing 814,
https://doi.org/10.1007/978-981-13-1501-5_20

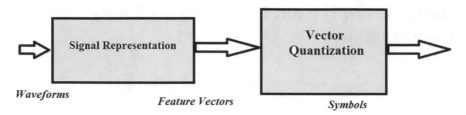

Fig. 1 Acoustic modeling flow

subsets of this vector are used as the training sequence. The codebook is calculated by a randomized continuous clustering algorithm. Lastly, in quantizing a vector among the others, which has been taken as input, the most closer words of the codebook is found, and their corresponding tag line of this found code word is radiated. Data compression could be possible as the address could be transmitted, and at the same time, it requires lesser number of bits than transmitting vector itself. The process of data quantization has been extended from scalar to vector data of any flexible dimension. Vector quantization employs a set of representation vectors and matrices as output. The set of vectors is defined as "codebook," and input of those are defined as "code words." Vector quantization has been found to be an effective coding technique due to its ability to exploit the high correlation between the pixels which are present around it [2]. The quality of depiction requires exhibition of a distortion measure. There are many algorithms in form of papers, which have been [1, 3–7] published on how to generate a codebook. VQ algorithm which is commonly used is the one presented by Linde, Buzo, and Gray, (LBG) [1, 8] and generalized by Lloyd [1, 9] 1982. VQ puts the input data into cluster of "vectors" based on the distortion measure. Before adding as input, vectors must be well defined. Nowadays, VQ is widely used basically in the encoding/compression; in many applications, VQ is widely used like face detection, pattern recognition, neural network design, face detection [1, 10, 11] (Fig. 1).

2 Lossy Compression

Lossy image compression is one of the popular methods which is used for image compression due to the abrupt increase of the chunks of data, which is being employed for Internet and other applications. We have to make sure that it is of great importance just to ensure that the compression part is not degrading the quality of the image in a lossy compression. The working principle of a lossy compression algorithm is evaluated, as far as conflicting aspects are concerned. PSNR value is used to determine compression ratio and image quality [12] (Fig. 2).

(a)   Input tif Image   **(b)**   Reconstructed tif Image

Fig. 2   **a** Normal input image, **b** reconstructed image

3 Vector Quantization and K-means Clustering Algorithms

We will describe how both the algorithms work together—the one being acoustic modeling technique, which involves both VQ and k-means. The signal representation produces feature vector sequence. Now, there are few ways to process the multidimensional sequence—firstly, the methods which can directly model continuous space or secondly, the methods which involve quantizing and modeling of discrete symbols. Clustering comes under unsupervised learning. Here, we are not aware of the number and forms of classes, and the data is not present in a structured format. Clustering algorithm plays the important role in this.

4 Proposed Work

The proposed method is as follows.

1. Convert the image into grayscale image tif.
2. Then, I have to read the image.
3. Fix the block size; then, each block does VQ (image, parameter bits) using VQ encoder.
4. Then, in decoder part, take input as encoder output, then decode it, and output will come VQ (original image and reconstructed image).
5. Apply k-means algorithm in this lossy compression.
6. Then, in final by varying parameters and bits, i can plot rate and distortion.
7. Change the bits and for each block do for rest and plot it.

We will elucidate the method which we have proposed. We have taken input image in code, and the purpose is to compare the size of the grayscale uncompressed tiff image with the length of the encoded sequence that you have stored, which is "encodedSeq.bin," and it represents your compressed version of the image. This

(a)   **Input Image**   (b)   **Reconstructed Image**

Fig. 3 **a** Input tif image, **b** reconstructed tif image

(a)   **Input jpg Image**   (b)   **Reconstructed jpg Image**

Fig. 4 **a** Input jpg image, **b** reconstructed jpg image

quantity is what should be less than 1 for your code being effective. We have seen that the input image size and reconstructed image size are same, and it is coming out that the compression ratio is less than 1. The encoder takes an image in input and gives in output a binary sequence which is the compressed file, and the encoder function must save at the end the encoded sequence as a binary file which is the encode sequence (Fig. 3).

Secondly, the decoder takes the encoded sequence (.bin file) and reconstructs the image. Then, all the information that the decoder needs for the decoding must be written in the encoded sequence (.bin file): Then, you should have the decode function calling (Fig. 4).

We need to keep in mind that the function decoder takes as an input, among the other arguments, the image Img1 itself. In real applications, the decoder would know the original image; then, it could throw away all the other information and has a perfect reconstruction. We have specified that the compression ratio you get is: CR = (num. bit for encoding without compression)/(number bit of the encoded sequence). (Clearly, num. bit for encoding without compression = size (image)*—in

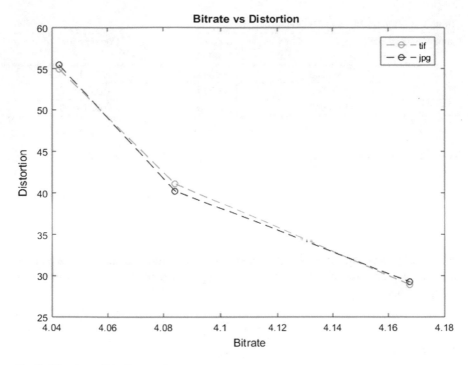

Fig. 5 Bit rate vs distortion graph

the script distortion. ask to the user' enter the no. of iteration(n):' it does not make sense. For the parameters L and K, I suggest to fix L (the smallest one, 4) and iterate on K. Actually, you should iterate on the bit rate and not on K. Since VQ is very computationally heavy, you can consider only very low values for the bit rate. (1, 2, 3, maybe no more that this value)

After the encoding and the decoding functions have been called, our task is converting the reconstructed tif image into unsigned 8-bit (1-byte) integers of class uint8 along with its size in the memory. Then, we calculate the compression ratios after displaying it. File size * 8 = no. of bit in the binary file; file size is in byte, so we are converting it into bit. Same goes for the jpg image (Fig. 5).

5 Results

The compression ratios of the two types of images could be compared, and we plot a graph between bit rate and distortion after we have calculated the Euclidean distance and the bit rate formula for the given inputs (Table 1).

Table 1 Output for JPG and TIF images

Input memory size		Memory size compressed	Compression ratio
1.	196608	99348	5.053101e−01
2.	196608	100372	5.105184e−01
3.	196608	102420	5.209351e−01

6 Conclusion

Image compression algorithms and effective methods play an important role in the field of pattern recognition, data mining, computer vision, and human interaction. Image compression is an effective approach where we can code the digital image in order to minimize the number of bits required in depicting it. Numerous image compression techniques are already present in the literature, but they do have certain limitations. This paper could propose a novel approach using VQ for effective image compression.

References

1. Chatterjee, R.: Image compression and resizing using vector quantization and other efficient algorithms. IJESC **7**(6), 13243
2. Salomon, D.: Data Compression: The Complete Reference, 3rd edn. Springer, New York (2004)
3. LockerGnome: Real World Application of Image Compression (2011)
4. Gautam, B.: A Thesis on Image Compression using Discrete Cosine Transform and Discrete Wavelet Transform. National Institute of Rourkela, Rourkela
5. Shelley, P., Li, X., Han, B.: A Hybrid Quantization Scheme for Image Compression. University of Alberta (2003)
6. Quispe-Ayala, M.R., Asalde-Alvarez, K., Roman-Gonzalez, A.: Image Classification Using Data Compression Techniques. Author Manuscript, 26th Convention of Electrical and Electronics Engineers, IEEE, Israel (2010)
7. Pantaesaena, N., Sangworaisl, M., Nantajiwakornchai, C., Phanpraist, T.: Image Compression Using Vector Quantization. ReCCIT, Thailand (2005)
8. Pragada, S., Sivaswamy, J.: Image denoising using matched biorthogonal wavelets. In: Sixth Indian Conference on Computer Vision, Graphics Image Processing, pp. 25–32 (2008)
9. Thresholds for wavelet 1-D using Birgé-Massart strategy—MATLAB wdcbm. www.mathworks.com (2017)
10. Ergen, B.: Signal and Image Denoising Using Wavelet Transform. InTech., London (2012)
11. How to get SNR for 2 images—MATLAB answers—MATLAB central. www.mathworks.com (2017)
12. Ahamed, M.U., Chikkannan, E., Ramakrishan, K.: Lossy image compression based on prediction error and vector quantization. EURASIP J. Image Video Process. **1**, 35 (2017)

Oil Spill Detection Using Image Processing Technique: An Occupational Safety Perspective of a Steel Plant

Anima Pramanik, Sobhan Sarkar and J. Maiti

Abstract Oil spill at workplace is one of the potential hazards in industry. Though it has not attracted more importance from research point of view, it can lead to economic loss for the industry through the occurrence of accident phenomena like slipping, firing, or pollution to the environment. Hence, oil spill detection should be considered as an essential research issue. In order to address this, the present study endeavors to use image processing technique for oil spill detection using the image data retrieved from an integrated steel plant in India. Results reveal that the technique adopted for oil spill detection is an effective and efficient way. This method, though used in steel plant, can be used in any other industry like construction, manufacturing.

Keywords Occupational safety · Oil spill detection · Image processing · Steel industry

1 Introduction

Occupational safety is a serious concern of every industry. If it is not ensured and accidents are taken place, the consequences directly affect human life, industry, society, and the nation, as a whole by causing considerable damage to physical, financial, and environmental resources. In addition to these direct effects, they also impart indirect consequences such as psychological trauma. Every worker in his/her working place is subject to occupational hazards during the working hours. There exist a number of hazardous situations which may lead to occupational accidents or incidents. The phenomena like slip/trip/fall (STF), falling from height, accidents

A. Pramanik · S. Sarkar (✉) · J. Maiti
Department of Industrial & Systems Engineering, IIT Kharagpur, Kharagpur, India
e-mail: sobhan.sarkar@gmail.com

A. Pramanik
e-mail: apramanik17@gmail.com

J. Maiti
e-mail: jhareswar.maiti@gmail.com

© Springer Nature Singapore Pte Ltd. 2019
A. Abraham et al. (eds.), *Emerging Technologies in Data Mining and Information Security*, Advances in Intelligent Systems and Computing 814,
https://doi.org/10.1007/978-981-13-1501-5_21

during material handling, collisions of objects, accidents during the use of toxic chemicals are found to be the prominent causes or factors behind the occupational accidents. Of them, STF is a big issue for the industry. According to statistics from the Health and Safety Executive (HSE), slips and trips are the single most common cause of injuries at work and account for over a third of all major work injuries. They cost employers more than £512 m a year in lost production and other costs. The main causes behind this are (i) uneven floor area, (ii) wet floors, (iii) change in levels, (iv) poor lighting, etc. Out of them, the condition of wet surface of floors is playing an important role behind the occurrence of accidents. Due to the spillage of water or oil or any other liquid substances during transportation within or outside the working place, the level of risk of slipping hazards raises to a higher extent. In particular, due to the presence of oil, which is an highly lubricant material, the surface becomes extremely slippery, which may lead to the occurrence of accidents.

Oil spill, often being a term of lower importance, is, in fact, acting as a significant hazardous substance in the workplace since importance is only given to phenomena of large-scale oil spills, such as the Deepwater Horizon incident [1]. However, even minor spills in the workplace result in a dangerous situation for the workers. Indeed, oil spill facilitates a number of conditions of potential risks at workplaces, for example, (i) risk of physical injury to the workers due to the slipping on a patch of oil spilled on the pathway; (ii) risk of fire while there is a proximity of open flame or heat source to the oil spillage; and finally (iii) risk of harmful runoff, if it gets in contact with nearby water sources, which consequently pollutes the water or environment. Therefore, a proper perspective should be undertaken by the industry for the prevention of accidents caused by oil spillage. To handle the industrial issues, conventional data mining or machine learning techniques have recently been used by many researchers using classification, clustering, or rule mining algorithms [2–11]. However, use of image processing technique in the solution of industrial issues is rather new. In recent research works, the focus of oil spill detection has been concentrated in analysing satellite images, either after oil tank explosion [12] or after oil pollution in the oceans [13]. Synthetic-aperture radar (SAR) images, in this respect, have largely been used for the analysis of two- or three-dimensional images of objects like oil spill. For example, Robla et al. analyzed SAR images using image equalization, binarization, and morphological operations for segmentation of the image retrieved. Once the segmentation is done, active contour generated around oil spill surface helps to find out the shape and deformation of oil spill area over time [12]. Using video surveillance, Qu et al. proposed a system for detecting oil leak on petroleum platform [14]. In 2014, Mera et al. developed an automatic decision support system based on SAR data for detection of hydrocarbon spillage on the ocean surface [15]. Artificial neural network has been used to focus on edge detection. Shape, contextual, and physical information of oil spill area are considered as feature vectors, which are later used for classification. Akkartal and Sunar developed an oil spill monitoring and detection system using radar images [16]. In a recent work by Song et al. in 2017, classification of ocean oil spillage has been shown using RADARSAT-2 based on an optimized wavelet neural network [17].

Based on the previous studies on analyses of oil spill problems mentioned above (though not illustrated in details due to space constraint), it is found out that researches have been carried out broadly on oil platform accidents or oil spill detection on ocean for ship tracking or other purposes. But, till now, oil spill detection, being an important issue for the industry perspective as mentioned above, has not been carried out. Hence, there exists a need for research. If the oil spill on the floor area in plant is detected early, occurrence of incidents or accidents could be prevented. Realizing the need of such a system that helps the industry to detect the oil spill early, our present study endeavors to contribute in early oil spill detection using image processing technique. The reason behind adopting this approach is that the proposed image processing-based technique can remove noises easily, it can correct image density and contrast for clarity in visualization, and it can easily be stored in and retrieved from computers for analysis. Thus, using the image data of oil spill retrieved from an integrated steel plant in India, the aim of the present study is to propose a methodology for early detection system of oil spillage based on image processing technique. However, the the scope of the present study is limited to steel plant only.

The remainder of the paper is organized as follows. Section 2 presents the proposed methodology for this analysis. In Sect. 3, results are discussed. Finally, in Sect. 4, conclusions with future scopes are given.

2 Methodology

In this study, digital image processing technique has been used. The proposed methodology used in this study is depicted in the flowchart (refer to Fig. 1). The entire process consisting of three main tasks, i.e., (i) data acquisition, (ii) data pre-processing, and (iii) object (here, oil spill) detection, is described below in short.

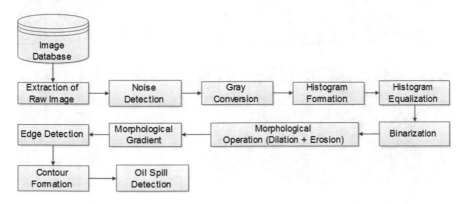

Fig. 1 Proposed methodological flowchart

2.1 Data Acquisition System

All the images used in this study were acquired from the video surveillance system of an integrated steel plant. A database comprising ten images of oil spillage at workplace in steel plant is used to develop this detection system.

2.2 Data Preprocessing

Once the raw data in the form of images were collected from the video surveillance system, they were then preprocessed since they contain a large amount of noises. In Fig. 2a, one of the raw images used in this study is shown. To reduce the noises from the raw image, data preprocessing is done in two steps: (i) smoothing operation and (ii) histogram equalization, which are explained below.

2.2.1 Smoothing Operation

To reduce noise and to prepare raw image for further processing, smoothing operation is done. The simplest smoothing algorithm is moving of average filter throughout the original raw image. The output of smoothing by linear filter is simply the average of

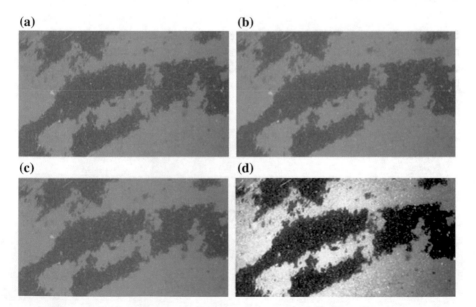

Fig. 2 **a** Raw image; **b** blurred image; **c** gray image; and **d** histogram equalized image

the pixel contained in the neighborhood of the filter mask [18]. In this experiment, moving of average filter is used as filtering technique which resulted in blurred image (refer to Fig. 2b).

2.2.2 Histogram Equalization

In histogram equalization, there are three steps involved: (a) generation of gray scale image, (b) formation of histogram, and (c) generation of histogram equalized image. They are explained below in steps.

2.2.3 Generation of Grayscale Image

Raw images acquired from video surveillance system are originally color images. There are three color space R, G, and B present in each pixel. In RGB color space, the components R, G, and B represent not only color information but also luminance which may vary due to the different level of lighting conditions. Therefore, if further processing is done on color image, then it may result in an unreliable identification. To get only one intensity value at each pixel, blurred image is transformed to grayscale image, which is shown in Fig. 2c.

In color image, there are three attributes, namely lightness, chroma, and hue, which represent R, G, and B, respectively. The color images are represented by 24 bits. Each R, G, and B contains 8 bits, so the combination of RGB contains 24 bits, which supports $2^{24} = 1,67,77,216$ different colors. Grayscale image is represented by luminance using 8 bits. The intensity of a pixel value of grayscale image ranges from 0 to 255 [19]. If the grayscale image is used for further processing, then simple scalar operation can be done, but in case of color image to obtain some better results, vector calculation is required, which is computationally more complex than scalar operation. However, vector calculation cannot provide guaranteed better result compared to scalar operation.

Initially, Y, U, and V components are calculated from R, G, and B components. Here, Y determines the brightness of the color and both U and V determine the color itself. Equations (1–3) are used to calculate *luminance* and *chrominance* values of the source blurred color image. In Eqs. (4–6), the R, G, and B values are approximated using R, G, and B components, respectively.

$$Y = 0.299 \times R + 0.587 \times G + 0.114 \times B \tag{1}$$

$$U = (B - Y) \times 0.565 \tag{2}$$

$$V = (R - Y) \times 0.713 \tag{3}$$

$$R_a = \frac{R \times 0.299 + R \times 0.587 + R \times 0.114}{3} \tag{4}$$

$$G_a = \frac{G \times 0.299 + G \times 0.587 + G \times 0.114}{3} \tag{5}$$

$$B_a = \frac{B \times 0.299 + B \times 0.587 + B \times 0.114}{3} \tag{6}$$

It is to be noted that intensity value of grayscale image for each pixel is the average of UV, R_a, G_a, and B_a, where $UV = \text{sum}(U, V)$.

2.2.4 Formation of Histogram

The histogram of a grayscale image with intensity levels in the range (0 to 255) is a discrete function $h(r_k) = n_k$ where r_k represents kth intensity value and n_k is the number of pixels in the image with intensity r_k. The normalized histogram is denoted by $P(r_k) = \frac{n_k}{MN}$ for $k = 0, 1, \ldots, 255$, and M and N are the row and column dimensions of the image. Histogram can be viewed graphically simply as plots of $h(r_k) = n_k$ versus r_k or $P(r_k) = \frac{n_k}{MN}$ versus r_k. For the same image with different levels of illumination, we can get the different histogram. For dark and light images, histogram is concentrated on low level and high level of intensity values, respectively. Moreover, low contrast and high contrast images have a narrow and wider histogram located at the middle of intensity value, respectively [18]. Once the desired histogram is formed, it is equalized, which is described below.

2.2.5 Generation of Histogram Equalized Image

To nullify the effect of illumination, histogram equalization has been performed. It is a process of transformation of low range value to high dynamic range. An image, whose pixels tend to occupy the entire range of possible intensity level and distributed uniformly, will have an appearance of high contrast and will exhibit a large variety of gray tones [18]. Basically, this process follows three steps which are given below:
   Step 1: Compute intensity mapping of the gray image.

$$S_k = \frac{L - 1}{MN} \sum_{j=0}^{k} n_j \tag{7}$$

where $L = 2^p$, p is the number of bits representing grayscale image, and $k = 0, 1, 2, \ldots, (L - 1)$. Compute S_k using Eq. (7). Round-off the resulting values for S_k to the integer range $(0, 1, 2, \ldots, (L - 1))$. r_k represents kth intensity value and S_k represents intensity mapping of r_k.

Step 2: Compute transformation of specified histogram

$P_z(Z_i)$ is the ith value of the specified histogram. All values of the transformed functions G are calculated using Eq. (8). The values of G are rounded off to the integer range from 0 to $(L - 1)$.

$$G(Z_q) = (L - 1) \sum_{i=0}^{q} P_z(Z_i) \qquad (8)$$

where $q = 0, 1, 2, \ldots, (L - 1)$

Step 3: Compute intensity mapping of histogram specified image to histogram equalized image

For every value of S_k, different values of G are used to find corresponding value of Z_q, so that $G(Z_q)$ that becomes closest to S_k. Hence, the mapping from S to Z is stored. If more than one value of Z_q satisfies S_k, then the smallest value is chosen. Gray image and histogram equalized image are shown in Fig. 2c, d.

2.3 Object Detection

Once the histogram equalization is performed, object, i.e., oil spillage, is detected following three important steps:

2.3.1 Step 1: Generation of binary image

In this step, histogram equalized image is transformed to binary image maintaining an optimal threshold value determined from the histogram of equalized image. Threshold value, in this study, is considered as the valley point of the histogram. Here, pixel intensity is kept as either zero if Z_q is less than threshold value, or 255, otherwise.

2.3.2 Step 2: Morphological Operation (Dilution and Erosion)

Morphological operations usually include a broad set of image processing operations that process images based on shapes. Due to the presence of noises in the original oil spill image, binary image may contain holes. To reduce the size of holes, the basic morphological operation, i.e., dilation, has been used. Another morphological operation is erosion. Dilation adds pixels to the boundaries of objects in an image, while erosion removes pixels on object boundaries. The area of foreground pixels shrinks in size and holes within those areas becomes larger. For detailed understanding of these basic operations, readers may refer to [14, 20, 21]. Once dilation and erosion are completed, morphological gradient, which is the difference between dilation and erosion of the given image, is calculated, which can be expressed as $G(f) = f \oplus b - f \ominus b$, where the notations $\oplus$ and $\ominus$ denote dilation and erosion, respectively, f is the original binary image, and b is the grayscale structuring element [12]. To separate the dark zone due to the presence of oil in the workplace from the other dark areas due to the presence of other factors, morphological gradient is used. In this process, not only oil spill areas are detected, but non-oil surfaces are also identified.

2.3.3 Step 3: Labeling of Contour

To search the coincident area between binary and gradient image, contour extraction has been done. To obtain the contours of the gradient image, *Canny edge detection* algorithm was applied on both binary image and gradient image [22]. The contours of gradient image, which are coincident with labeled areas of possible oil spillage in binary image, have been defined. To get the labeled contours of gradient image, the contours have been multiplied with labeled image (n values). To detect oil spill area, the degree of coincidence of the labeled contour of gradient image with the labeled binary image has been computed. For each label, the number of existing pixel has been counted for both images. The labels with high percentage are considered as area with high probability of oil spills. All the images, i.e., from binary image to oil spill detected image, which have been obtained from the above-mentioned sequential operations, are depicted in Fig. 3a–f.

3   Results and Discussions

The algorithm of each step of flowchart is used over the total imagery. Raw image in the form of RGB color space is used for oil spill detection which is shown in Fig. 2a. To minimize the level of noise from raw image, mean filtering is used, which produces blurred resultant image as depicted in Fig. 2b. To get the intensity value of each pixel, color image is converted to gray scale image, which is shown in Fig. 2c. For contrast adjustment, histogram equalization is performed on it to get uniform distribution of intensities. This step creates an equalized image as shown in Fig. 2d. To find the region of interest, gray image is converted into binary image (refer to Fig. 3a). Due to the presence of noise, binary image contains some holes. For further noise reduction, morphological operation is done. To enhance variations of pixel intensity in a given neighborhood, morphological gradient is used, the result of which is shown in Fig. 3d. To get the actual position of oil spill area, edge detection and contour formation are done. The results obtained from the entire analyses are shown in Fig. 3a–f. It can be observed that the textures are different in each image. Nevertheless, the flowchart results in a good approximation to the localization of oil spill; the segmentation method produces some false positive results; yet oil spill detection is achieved approximately. However, segmentation can be improved by

(a) (b)

(c) (d)

(e) (f)

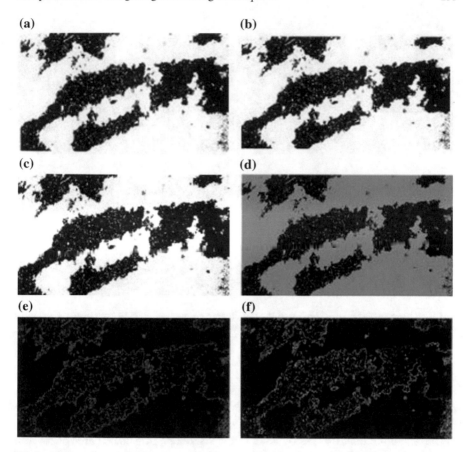

Fig. 3 **a** Binary image; **b** dilated image; **c** eroded image; **d** morphological gradient image; **e** edge detected image; and **f** oil spill area detected image

improving threshold value. The images of same location at different time are shown in Fig. 4a, d, where it shows that oil spill detected on workplace with label contour from raw image (refer to Fig. 4a, b) gets deformed over time (refer to Fig. 4c, d).

4 Conclusions

The proposed method attempts to detect the oil spill at workplace. The advantage of this technique is not only to detect the oil spill over time but also it provides the information about the evolution of residues over time. Though the segmentation produces some false positive output, oil spill was detected successfully. Active contour is done to labeled and localized oil spill area in a given image. In addition, it can also be detected how deformation is taken place on oil spill over the span of time. Image

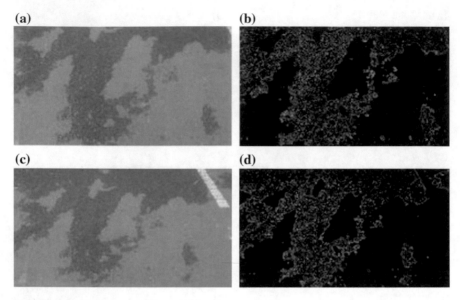

Fig. 4 **a** Raw image; **b** level contour of the raw image; **c** image of oil spill with different shape at workplace in different time; and **d** level contour of the raw image (**c**)

of oil spill at workplace at different time is shown here. Segmentation results can be improved by determining proper threshold value. This method gives acceptable result, but the result can be further improved by combining some information like the altimeter data and, of course, some other video surveillance images in order to discriminant the false oil spill by image analysis. Shape of detected area of oil spill may also be defined. This is for further investigation purpose. This may give more prominent results than before. If several oil spill are unified in an image, the active contour of the following image can be identified which results in unified contours in resultant image. In addition, image registration may be used before going for object identification technique which may lead to better results. Hence, this kind of study can be done in other domain also such as construction, manufacturing, aviation.

References

1. Kujawinski, E.B., Kido Soule, M.C., Valentine, D.L., Boysen, A.K., Longnecker, K., Redmond, M.C.: Fate of dispersants associated with the deepwater horizon oil spill. Environ. Sci. Technol. **45**(4), 1298–1306 (2011)
2. Sarkar, S., Vinay, S., Maiti, J.: Text mining based safety risk assessment and prediction of occupational accidents in a steel plant. In: 2016 International Conference on Computational Techniques in Information and Communication Technologies (ICCTICT), pp. 439–444. IEEE (2016)

3. Krishna, O.B., Maiti, J., Ray, P.K., Samanta, B., Mandal, S., Sarkar, S.: Measurement and modeling of job stress of electric overhead traveling crane operators. Saf. Health Work **6**(4), 279–288 (2015)
4. Gautam, S., Maiti, J., Syamsundar, A., Sarkar, S.: Segmented point process models for work system safety analysis. Saf. Sci. **95**, 15–27 (2017)
5. Sarkar, S., Patel, A., Madaan, S., Maiti, J.: Prediction of occupational accidents using decision tree approach. In: India Conference (INDICON), 2016 IEEE Annual. pp. 1–6. IEEE (2016)
6. Sarkar, S., Vinay, S., Pateshwari, V., Maiti, J.: Study of optimized svm for incident prediction of a steel plant in india. In: India Conference (INDICON), 2016 IEEE Annual. pp. 1–6. IEEE (2016)
7. Sarkar, S., Lohani, A., Maiti, J.: Genetic algorithm-based association rule mining approach towards rule generation of occupational accidents. In: International Conference on Computational Intelligence, Communications, and Business Analytics, pp. 517–530. Springer (2017)
8. Sarkar, S., Verma, A., Maiti, J.: Prediction of occupational incidents using proactive and reactive data: A data mining approach. In: Industrial Safety Management, pp. 65–79. Springer (2018)
9. Verma, A., Chatterjee, S., Sarkar, S., Maiti, J.: Data driven mapping between proactive and reactive measures of occupational safety performance. In: Industrial Safety Management, pp. 53–63. Springer (2018)
10. Sarkar, S., Pateswari, V., Maiti, J.: Predictive model for incident occurrences in steel plant in india. In: 8th ICCCNT, pp. 1–5. IEEE (2017)
11. Sarkar, S., Vinay, S., Raj, R., Maiti, J., Mitra, P.: Application of optimized machine learning techniques for prediction of occupational accidents. Comput. Oper. Res. (2018)
12. Robla, S., Llata, J., Torre, C., Sarabia, E.: An approach for tracking oil slicks by using active contours on satellite images. In: OCEANS 2009-EUROPE, pp. 1–8. IEEE (2009)
13. Brekke, C., Solberg, A.H.: Oil spill detection by satellite remote sensing. Remote Sens. Environ. **95**(1), 1–13 (2005)
14. Qu, L., Wang, J., Xin, S., Qin, M., Dong, J.: A system for detecting sea oil leak based on video surveillance. In: 2011 Third Pacific-Asia Conference on Circuits, Communications and System (PACCS). pp. 1–3. IEEE (2011)
15. Mera, D., Cotos, J.M., Varela-Pet, J., Rodríguez, P.G., Caro, A.: Automatic decision support system based on sar data for oil spill detection. Comput. Geosci. **72**, 184–191 (2014)
16. Akkartal, A., Sunar, F.: The usage of radar images in oil spill detection. Int. Arch. Photogram. Remote Sens. Spat. Inf. Sci. 37(Part B8), 271–276 (2008)
17. Song, D., Ding, Y., Li, X., Zhang, B., Xu, M.: Ocean oil spill classification with radarsat-2 sar based on an optimized wavelet neural network. Remote Sens. **9**(8), 799 (2017)
18. Gonzalez, R.C., Woods, R.E., Eddins, S.L.: Processing (1987)
19. Saravanan, C.: Color image to grayscale image conversion. In: 2010 Second International Conference on Computer Engineering and Applications (ICCEA), vol. 2, pp. 196–199. IEEE (2010)
20. Haralick, R.M., Sternberg, S.R., Zhuang, X.: Image analysis using mathematical morphology. IEEE Trans. Pattern Anal. Mach. Intell. **4**, 532–550 (1987)
21. De Natale, F.G., Boato, G.: Detecting morphological filtering of binary images. IEEE Trans. Inf. Forensics Secur. **12**(5), 1207–1217 (2017)
22. Goulart, J.T., Bassani, R.A., Bassani, J.W.M.: Application based on the canny edge detection algorithm for recording contractions of isolated cardiac myocytes. Comput. Biol. Med. **81**, 106–110 (2017)

Blood Vessel Extraction from Fundus Image

Abheek Ray, Ayantika Chakraborty, Dipankar Roy, Barun Sengupta
and Mainak Biswas

Abstract Almost 6.5 million people are affected due to glaucoma, diabetics every year in the USA (CDC, Diabetic retinopathy. National for chronic disease prevention and health promotion [Online]. Atlanta, GA [1]). The paper presents a computer-aided blood vessel detection from fundus imaging. This is mainly necessary for premature baby and diabetic patient (Karperien et al. Clin Ophthalmol 2(1):109 [2]). By proper screening of the vessel of patients, the medicine can be modified accordingly. For this case, DRIVE database (Staal et al. IEEE Trans Med Imaging 23(4):501–509 [3]) is chosen. And this detection scheme is fully automated and initially based on top-hat filtering on the input image. Iteratively, region growing and median filtering are applied on the initial segmented output and result in 94.02% accuracy when compared with the ground truth. Diabetic retinopathy analysis and retinal vessel tortuosity in premature infants can be determined using the proposed approach in this paper.

Keywords Diabetic retinopathy · Top-hat filter · Median filter · Vessel tortuosity · Image processing

A. Ray (✉) · A. Chakraborty · D. Roy · B. Sengupta · M. Biswas
Department of Electrical Engineering, Techno India College of Technology,
Block – DG Action Area 1 New Town, Kolkata 700156, India
e-mail: sanu.ray13@gmail.com

A. Chakraborty
e-mail: ayantikach123@gmail.com

D. Roy
e-mail: dipankarroyee@gmail.com

B. Sengupta
e-mail: bsg10.messigerrard@gmail.com

M. Biswas
e-mail: mainakbiswas041@gmail.com

© Springer Nature Singapore Pte Ltd. 2019   259
A. Abraham et al. (eds.), *Emerging Technologies in Data Mining and Information Security*, Advances in Intelligent Systems and Computing 814,
https://doi.org/10.1007/978-981-13-1501-5_22

1 Introduction

Diabetic retinopathy and retinal vessel tortuosity in premature infants harm retinal blood vessels adversely causing eye problems [2]. The above-mentioned disease is becoming common day by day, so their diagnosis is one of the important researches going on in the whole modern world. There are several indications of retinal vessel problems like hemorrhages, dilated retinal veins, cotton wool spots [4]. Ignorance of such problems can lead to some other serious problems such as serious diabetes, stroke, and various cardiovascular diseases. So early investigation of retinal vessel is very important so that the patient can take appropriate steps while the sickness is yet in its initial stage. So automatic segmentation of retinal vessel image is a hot topic in today's world because this can help in diagnoses of the eye problems.

Visual deficiency is one of the causes due to diabetic retinopathy which generally harms the retinal blood vessels. The structure and appearance of the blood vessels in the retinal image can be used for diagnoses of a defected eye. In this paper, we used an approach which is less computational, more simplified, and fully automatic procedure for detection of retinal vessels which yield promising results. Top-hat filters [5] have been used for enhancement and median filters to remove the unwanted noise from the output image. The output image was then compared with the ground truth data that has been accurately marked by the experts. This paper is the one-step extended version of Roychowdhury et al. [6]. The main advantage in the paper is that this process is faster than reported other benchmark procedures.

2 Procedure

The concept of masking is also known as spatial filtering. In this concept, the filtering operation is performed directly on the given image. The process of filtering is also named as convolving a mask with the given image. Filters are applied on image mainly due to blurring the image, reducing noise of the given image and are also used for edge detection. Mainly, masking is done to separate an image from its background either to cause the image to stand out on its own or to place the given image over another background.

Diagnosing the given image is one of the important aspects of image processing. It just does not enhance the image quality but also increase compression efficiency and also improve the robustness of the subsequent algorithms such as pattern classification or objection detection. To enhance the image, it is important that we should work in green channel [6]. The enhancement process conducted in green channel is appropriate to enhance the image quality.

For every image processing, the primary step is the preprocessing which is done to remove the noisy pixels whose color is distorted than normal from background and to enhance the given image. In the first step of preprocessing, green channel is taken because green channel shows high intensity and approximation as compared to blue and red channels. In green channel, most of the minute details of image can be viewed. Using the red channel, only boundaries can be made visible and in blue channel, the image mainly shows much noise. For all those reasons, green channel is preferred over all other channels.

2.1 Top-Hat Filter

In digital image processing and mathematical morphology, top-hat transform [5] is an operation that mainly extracts minute details and elements from the given image. Mainly, there are two types of top-hat transform:

(1) The white top-hat transform mainly used as the difference between the input image and its opening by another structuring element.
(2) The black top-hat transform mainly used as the difference between the closing and the input image.

For various image processing tasks, top-hat transform can be used. Top-hat transform can be used for background equalization, feature extraction, enhancement of the image, and many more.

The main difference between white top-hat transformation (WTHT) and black top-hat transformation (BTHT) is that, in case of WTHT, the output image contains the stuffs which are not only smaller than elements but also the surrounding is brighter than input image. For black, the output image which contains the components which are not only reduced in size than input elements but also the neighboring is obscurer than input image.

Top-hat transforms are images hold only positive values [6, 7].

By differentiating the linear ramp function having a width of ¥, this transformation is formed, while the limiting value of the width can be determined by δ function.

The solution can be stated from the Eqs. 1 and 2.

where $\partial : S \rightarrow R$ is an input grayscale 2D image, which can be mapped from points of Euclidean space S (such as J_2 or O_2) into the real line.

Then, WTHT may be noted as:

$$T_{\omega}(\partial) = \partial - \partial \circ q \qquad (1)$$

where o glosses the opening operation.

The BTHT or bottom-hat transform of δ is termed by:

$$T_q(\partial) = \partial \bullet q - \partial \qquad (2)$$

where • is morphological closing.

2.2 *Median Filter*

Median filter [8] is one of the important nonlinear digital filters generally used to remove noise from an image. Noise reduction is one of the main preprocessing steps in image processing to improve the results of later processing. In digital image processing, median filter is widely used as it mainly preserves edges while removing noise [9] from the image. The main function of median filter is to run through the image step by step, replacing each entry with the median of neighboring entries. The neighbors' pattern is called the window which slides step by step over the entire image. Mainly, median filters are used to reduce impulsive or salt–pepper noise. Impulsive and salt–pepper [10] noises can mainly occur due to a random bit error in a communication channel. Like low-pass filters, median filters smoothen the image and thus indirectly reduce noise.

2.3 *Proposed Segmentation Scheme*

The proposed scheme is elaborated where in the first stage, the mask is created then the green channel is extracted, and thus, it is enhanced to some extent for major blood vessels' segmentation. Then the ground truth is taken for comparison purpose. The result of the comparison is then noted. The details of flowchart of extraction are given below in Fig. 1.

The details of flowchart of training are given below in Fig. 2, where main part belongs to Roychowdhury et al. [6].

Fig. 1 Extraction flowchart

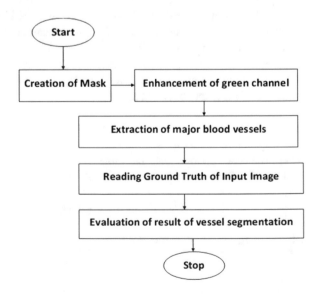

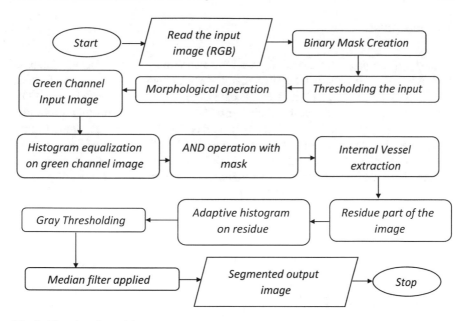

Fig. 2 Flowchart for training

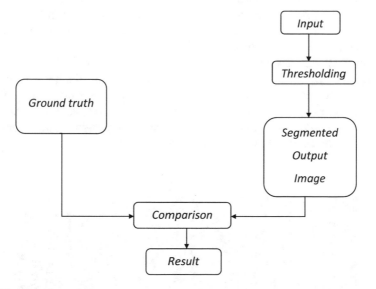

Fig. 3 Flowchart for testing

The details of flowchart of testing the input images with the provided ground truth images are given below in Fig. 3.

3   Database

To perform the proposed modified vessel segmentation method, DRIVE database is chosen [3]. The dataset DRIVE contains forty images with 45 degree field of view (FOV). Each image in the DRIVE dataset is of 768 × 584 pixels. The FOV of every images is circular with a diameter of 540 pixels approximately. From that database, one is the training set which is named as DRIVE Train containing twenty images and the other one is a test set named as DRIVE Test also containing twenty images. Two manual segmentations are available for the test cases. All the images contained in the DRIVE database were actually used for making clinical diagnoses. The performance of the technique used for vessel segmentation is evaluated on the images from DRIVE Test dataset with respect to the ground truths.

Figure 4 shows the input image and the desired segmented output image, called ground truth.

4   Experimental Result

This proposed method is implemented in MATLAB using (Core i3 6th gen CPU, 2.40 GHz, 8 GB RAM). To make the experiment harder, the algorithm is run several times, and the mean detection rate along with standard deviation is recorded. The list for required formulas is shown below (Fig. 5).

The segmentation results are displayed here step by step in the Fig. 6. First image is the input image and second is the created binary mask of the given input. Third figure shows the green channel of that, while fourth shows histogram equalized image. Next image is the output after applying top-hat filter, and sixth image is the ANDing with mask. Seventh image is the initial segmentation result, eighth image is the final output, and last image is the desired ground truth.

(a)   **(b)**

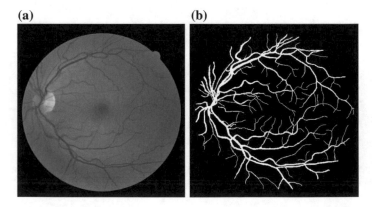

Fig. 4 **a** Input image, **b** ground truth

Sensitivity or true positive rate (TPR) = True Positive / (True Positive + False Negative)

Specificity (SPC) or true negative rate = True Negative / (True Negative + False Positive)

Precision or positive predictive value (PPV) = True Positive / (True Positive + False Positive)

Negative predictive value (NPV) = True Negative / (True Negative + False Negative)

True positive fraction = (total positive – false negative)/total positive

False positive fraction = 1- (total negative – true negative)/total negative

Fall-out or false positive rate (FPR) = False Positive / (False Positive + True Negative)
= 1- Specificity

False negative rate (FNR) = False Negative / (True Positive + False Negative)
= 1- True Positive Rate

False discovery rate (FDR) = False Positive / (True Positive + False Positive)
= 1- Positive Predictive Value

Accuracy (ACC) = (True Positive + True Negative) / (True Positive + False Positive + False Negative + True Negative)

Fig. 5 List of formula

Table 1 Comparison of results of proposed model with existing works on the DRIVE datasets with respect to ground truth

Methods	Accuracy	Specificity	Sensitivity
Jiang and Mojon [11]	0.89	0.90	0.83
Nguyen et al. [12]	0.94	–	–
Perez et al. [13]	0.92	0.967	0.644
Proposed	**0.94**	**0.9901**	**0.7708**

Tables 1 and 2 show the comparative study with some well-known and benchmark procedures.

5 Future Work

Huge progress has been accomplished in the area of fundus image analysis in recent years. Currently, many researches are going on worldwide in retinal blood vessels segmentation of smaller vessels, differentiating retinal arteries from the veins, determining eye vessel tortuosity, and measuring vessel diameter, and last but not least vessel tree analysis.

Huge work is also going on retinal lesions for mainly detecting irregularities in shapes of hemorrhages. Detection of rare pathology is also going on, and research has increased extensively. Detection of retinal lesion distribution pattern is also going on side by side.

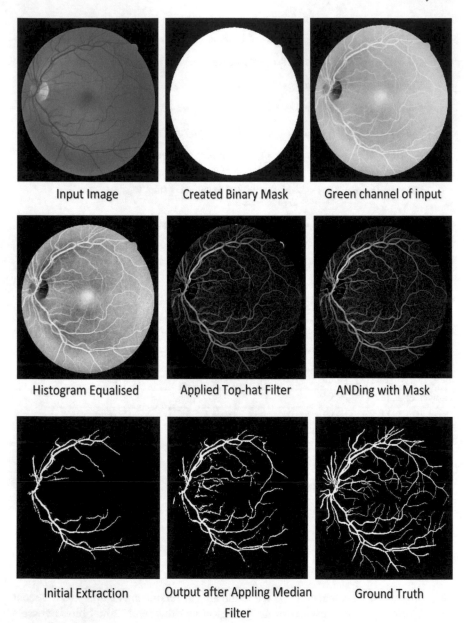

Fig. 6 Segmentation result

Table 2 Comparison of time taken for the proposed model with existing works on the DRIVE datasets with respect to ground truth

Methods	Time taken*	System configuration
Staal et al. [3]	15 min	1.0 GHz, 1-GB RAM
Soares et al. [14]	~3 min	2.17 GHz, 1-GB RAM
Marin et al. [15]	~90 s	2.13 GHz, 2-GB RAM
Fraz et al. [16]	~100 s	2.27 GHz, 4-GB RAM
Mendonca and Campilho [17]	2.5 min	3.2 GHz, 980-MB RAM
Al-Diri et al. [18]	11 min	1.2 GHz
Lam et al. [19]	13 min	1.83 GHz, 2-GB RAM
Perez et al. [13]	~2 min	Parallel Cluster
Miri et al. [20]	~50 s	3 GHz, 1-GB RAM
Roychowdhury et al. [21]	3.11 s	2.6 GHz, 2-GB RAM
Roychowdhury et al. [6]	2.45 s	2.6 GHz, 2-GB RAM
Jiang and Mojon [11]	8-36 s	600 MHz PC
Nguyen et al. [12]	2.5 s	2.4 GHz, 2-GB RAM
Budai et al. [23]	11 s	2.0 GHz, 2-GB SDRAM
Budai et al. [22]	~5 s	2.3 GHz, 4-GB RAM
Proposed	**1.3 s**	**i3 6th gen, 8 GB RAM**

*Segmentation time is recorded for one image only

6 Conclusion

The paper proposes an unsupervised algorithm to extract blood vessel using fundus images and compared with ground truth provided by the experts in datasets of DRIVE. The algorithm extracts vessel from a provided morphological image with enhanced blood vessel by using green channel, masking, and filtering (by top-hat and median filters) to extract major parts of blood vessels. After running the algorithm, 94.02% accuracy is achieved, which may be used for further future works.

References

1. CDC: Diabetic Retinopathy. National for Chronic Disease Prevention and Health Promotion [Online], Atlanta, GA. http://www.cdc.gov/visionhealth/pdf/factsheet.pdf (2011, Mar 23)
2. Karperien, A., Jelinek, H.F., Leandro, J.J., Soares, J.V., Cesar Jr, R.M., Luckie, A.: Automated detection of proliferative retinopathy in clinical practice. Clinical Ophthalmol. (Auckland, NZ) **2**(1), 109 (2008)
3. Staal, J., Abramoff, M.D., Niemeijer, M., Viergever, M.A., Van Ginneken, B.: Ridge-based vessel segmentation in color images of the retina. IEEE Trans. Med. Imaging **23**(4), 501–509 (2004)

4. Lakshmana, M. et al.: Central Retinal Vein Occlusion. https://emedicine.medscape.com/articl e/1223746-overview (2016)
5. Jackway, P.T.: Improved morphological top-hat. Electron. Lett. **36**(14), 1194–1195 (2000)
6. Roychowdhury, S., Koozekanani, D.D., Parhi, K.K.: Iterative vessel segmentation of fundus images. IEEE Trans. Biomed. Eng. **62**(7), 173–1749 (2015)
7. Serra, J.: Image Analysis and Mathematical Morphology, vol. 1. Academic Press, New York (1982)
8. Wang, Z., Zhang, D.: Progressive switching median filter for the removal of impulse noise from highly corrupted images. IEEE Trans. Circ. Syst. II: Analog Digital Sig. Process. **46**(1), 78–80 (1999)
9. Sun, T., Neuvo, Y.: Detail-preserving median based filters in image processing. Pattern Recogn. Lett. **15**(4), 341–347 (1994)
10. Chan, R.H., Ho, C.W., Nikolova, M.: Salt-and-pepper noise removal by median-type noise detectors and detail-preserving regularization. IEEE Trans. Image Process. **14**(10), 1479–1485 (2005)
11. Jiang, X., Mojon, D.: Adaptive local thresholding by verification-based multithreshold probing with application to vessel detection in retinal images. IEEE Trans. Pattern Anal. Mach. Intell. **25**(1), 131–137 (2003)
12. Nguyen, U.T., Bhuiyan, A., Park, L.A., Ramamohanarao, K.: An effective retinal blood vessel segmentation method using multi-scale line detection. Pattern Recogn. **46**(3), 703–715 (2013)
13. Palomera-Perez, M.A., Martinez-Perez, M.E., Benitez-Perez, H., Ortega-Arjona, J.L.: Parallel multiscale feature extraction and region growing: application in retinal blood vessel detection. IEEE Trans. Inf. Technol. Biomed. **14**(2), 500–506 (2010)
14. Soares, J.V., Leandro, J.J., Cesar, R.M., Jelinek, H.F., Cree, M.J.: Retinal vessel segmentation using the 2-D Gabor wavelet and supervised classification. IEEE Trans. Med. Imaging **25**(9), 1214–1222 (2006)
15. Marin, D., Aquino, A., Gegundez-Arias, M.E., Bravo, J.M.: A new supervised method for blood vessel segmentation in retinal images by using gray-level and moment invariants-based features. IEEE Trans. Med. Imaging **30**(1), 146–158 (2011)
16. Fraz, M.M., Remagnino, P., Hoppe, A., Uyyanonvara, B., Rudnicka, A.R., Owen, C.G., Barman, S.A.: An ensemble classification-based approach applied to retinal blood vessel segmentation. IEEE Trans. Biomed. Eng. **59**(9), 2538–2548 (2012)
17. Mendonca, A.M., Campilho, A.: Segmentation of retinal blood vessels by combining the detection of centerlines and morphological reconstruction. IEEE Trans. Med. Imaging **25**(9), 1200–1213 (2006)
18. Al-Diri, B., Hunter, A., Steel, D.: An active contour model for segmenting and measuring retinal vessels. IEEE Trans. Med. Imaging **28**(9), 1488–1497 (2009)
19. Lam, B.S., Gao, Y., Liew, A.W.C.: General retinal vessel segmentation using regularization-based multiconcavity modeling. IEEE Trans. Med. Imaging **29**(7), 1369–1381 (2010)
20. Miri, M.S., Mahloojifar, A.: Retinal image analysis using curvelet transform and multistructure elements morphology by reconstruction. IEEE Trans. Biomed. Eng. **58**(5), 118–1192 (2011)
21. Roychowdhury, S., Koozekanani, D.D., Parhi, K.K.: Blood vessel segmentation of fundus images by major vessel extraction and subimage classification. IEEE J. Biomed. Health Inform. **19**(3), 1118–1128 (2005)
22. Budai, A., Bock, R., Maier, A., Hornegger, J., Michelson, G.: Robust vessel segmentation in fundus images. Int. J. Biomed. Imaging (2013)
23. Budai, A., Michelson, G., Hornegger, J.: Multiscale blood vessel segmentation in retinal fundus images. In: Bildverarbeitung für die Medizin, pp. 261–265 (2010)

Wavelet and Pyramid Histogram Features for Image-Based Leaf Detection

Al Amin Neaz Ahmed, H. M. Fazlul Haque, Abdur Rahman,
Md. Susam Ashraf and Swakkhar Shatabda

Abstract Image recognition-based methods have been widely used in leaf detection. Leaf detection can play an important role in agriculture in detecting potential plant diseases, phenotyping and taxonomy. In this paper, we present an analysis on the wavelet and pyramid histogram-based features for leaf detection. We have used Haar wavelet transform and pyramid histograms from original image as feature extraction method. The experiments were done on a standard dataset of leaf images of 36 different species of plants. We have tested the effectiveness of different types of features over a large variety of supervised machine learning algorithms. We propose the use of Random Forest as the best performing classifier on this dataset using the selected features. Our proposed method achieved significantly better accuracy in comparison to the previous methods.

Keywords Supervised learning · Leaf detection · Wavelet transform · Feature selection

A. A. N. Ahmed · H. M. F. Haque · A. Rahman · M. S. Ashraf · S. Shatabda (✉)
Department of Computer Science and Engineering, United International University,
Madani Avenue, Satarkul, Badda, Dhaka 1212, Bangladesh
e-mail: swakkhar@cse.uiu.ac.bd

A. A. N. Ahmed
e-mail: neazahmedneaz@gmail.com

H. M. F. Haque
e-mail: fazlulhaquue.jony@gmail.com

A. Rahman
e-mail: fuhad1987@gmail.com

Md. S. Ashraf
e-mail: susam.ashraf.s4@gmail.com

© Springer Nature Singapore Pte Ltd. 2019
A. Abraham et al. (eds.), *Emerging Technologies in Data Mining and Information Security*, Advances in Intelligent Systems and Computing 814,
https://doi.org/10.1007/978-981-13-1501-5_23

269

1 Introduction

Plant species detection plays a very important role in agriculture and crop industry. Many of the plant species detection methods are developed using image-based
sensors and thus uses pattern recognition and image processing techniques [8]. The
accurate prediction and detection of leafs can lead to many application like disease detection, growth monitoring, phenotype monitoring, population tacking and
medicinal plant research [11].

The success of the detection algorithms often depends on the feature extraction
methods. Many features have been so far used in the literature of plant species detection via image-based features. Among them are shape, margin, texture [11], Gabor
filters [9], histogram [13], etc. There are mainly three types of feature: pixel-based,
filter-based and computationally generated features. Computationally generated features depend on sophisticated algorithms that often perform segmentation and other
algorithms to detect various features like shape, margins. On the other hand, pixel-
based features like histograms are simple to calculate and use but depends on the
classification algorithm's capability. Classification algorithms that incorporate structure representation such as artificial neural networks (ANN) works well with pixels.
However, filter-based methods often transform the original image and make it easy
to work for simple feature extraction methods.

In this paper, we propose the application of wavelet and pyramid histogram-
based features taken from leaf images for prediction of plant species. We use the
original image and its wavelet image for extracting histogram and pyramid histogram
features. The extracted are tested on benchmark leaf dataset using a large variety of
classification algorithms. The best set of features and classifier was selected after
analysis. Our proposed method is able to produce significantly improved prediction
over previous state-of-the-art method on the same benchmark.

2 Related Work

Detection of plant species or various diseases are possible by identification of plant
leaves. Many researchers have tried to develop automatic prediction algorithms using
various machine learning algorithms. The success of these algorithms often depends
on feature extraction techniques. In this section, we present a brief overview of the
related work in this area. Several review papers exist in the literature that provide
a survey on the prediction methods for leaf detection and disease detection via leaf
images [5, 10, 16]. Silva et al. [17] extracted and used a set of image-based features
for leaf species detection. In their method, they have used various techniques for
feature reduction and analysed the performance of different features. Pape et al.
[14] used top views of leafs in a proposed detection method where they took three-
dimensional histograms as features and used Euclidean distance-based methods. In
another work, Arivazhagan et al. [1] used a dataset with 500 plant leaves to detect
diseases using texture-based features.

Different machine learning algorithms are used in detection of plant leaf and disease detection. Among them are artificial neural network (ANN) [9], support vector machines (SVM) [8], self-organizing maps (SOM) [15], etc. Kulkarni et al. [9] used segmentation and Gabor filters for feature extraction in their methods. They employed artificial neural networks as their classification algorithm. Support vector machines were used in [8] to detect rice blast disease. Another work of rice plant diseases was done in [15] by Phadikar et al. where they used self-organizing maps.

Wavelet-based histograms have been previously used in the literature of object detection [4]. Pyramid histograms have been used previously with gradient images to detect pedestrians [2], protein structures [7], etc. Histogram-based features were previously used in detection of leaves of plant species in [13].

3   Our Method

In this section, we are going to describe about our methodology. The dataset that we used consists of RGB images. We took the RGB images from the dataset and extracted four types of features for all the images belonging to different classes or species. We have used fivefold cross-validation to sample the dataset and each time use a different partition of the dataset as test. The experiments were done on different combinations of features using different classification algorithms. After all the experiments, the best feature group combination with the highest percentage of correctly classified instances were selected as best features with the best performing classification algorithm. The block diagram of the methodology used in this paper is given in Fig. 1.

3.1   Dataset

The dataset used in this paper is taken from UCI machine learning repository and provided by Silva et al. [17]. There are total 40 different plant species. There are total 443 RGB leaf images in the data set belonging to different plant species. Each leaf specimen was photographed over a coloured background using an Apple iPad 2 device. The 24-bit RGB (Red–Green–Blue) images recorded have a resolution of 720×920 pixels.

3.2   Feature Extraction

In this section, we will discuss about the different types of features extracted in this study. As our features are various histograms, the intensity range is between 0 and

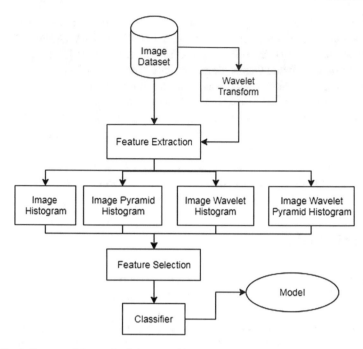

Fig. 1 Block diagram of the methodology used in our paper

255 and there are total 256 bins in each histogram. From each histogram, we are getting 256 attributes.

Image Histogram: Histograms are simple count of values that fall into a particular bin. Here in case of images, it counts the number of pixels that fall into a particular bin that corresponds to range denoted by the intensity level of colour. In this paper, we have used four colours for taking colour histograms: red, blue, green and grey. To calculate grey histogram, the RGB image has to be converted into a greyscale image first. We have used 252 bins for each colour. As there are four colour histograms, we would get total 1024 features for each image instance.

Image Pyramid Histogram: Pyramid histograms are taken level-wise, and in each level the original image is divided into two parts and histograms are calculated for each parts. We have used only second level of the pyramid, and thus, only the halves of the original image are considered. There are two types of pyramids: Gaussian pyramid and Laplacian pyramid. Gaussian pyramid downsamples the images and Laplacian pyramid upsamples an image lower in the pyramid. For our paper, we have upsampled and downsampled the dataset by twofold. From both of the converted image pyramids, we collected red, green, blue and grey histograms. As there are total eight histograms, the total number of the feature group would be 2048 features.

Wavelet Histogram: For generating wavelet histograms, first the RGB image is wavelet transformed. Haar wavelet transform [18] is used for the transformation of the images. This transformation is based on lifting scheme. Lifting scheme was

Table 1 Feature groups

Identifier	Feature group name	Number of features
A	Blue-green-red-grey histogram	1024
B	Image pyramid histogram	2048
C	Wavelet histogram	1024
D	Wavelet pyramid histogram	2048

developed by Wim Sweldens [19]. The original image is converted into three two-dimensional matrices for storing blue, green and red values of each pixel in the original RGB image. The three matrices have the same number of rows and columns as the rows and columns as of the RGB image. The three matrices are then Haar wavelet transformed . Blue, green, red and grey histograms are collected from the wavelet transformed image as mentioned earlier in this section. The total attribute of each image is 1024.

Wavelet Pyramid Histogram: To generate this feature group , the RGB image is transformed into a wavelet transformed image described in the previous section. After the transformation, the pyramid histogram features are extracted like in the previous section. There are total eight histograms so this feature group would have total 2048 attributes.

Summary of all the feature groups used in this paper is given in Table 1.

3.3   Feature Selection

We have used feature group-based forward selection. Four groups were used for classification. And the best performing feature group was selected. In the next phase, other three feature groups were added with this feature group one group at a time. The best combination was once again chosen. This process was iteratively continued as long as there were significance improvement in the performance of the classifiers.

3.4   Classifiers

In this section, we discuss about the classifier algorithms used with this classifiers. We have used five different classifiers: K-nearest neighbour (KNN), Naive Bayesian classifier, support vector machines (SVM), Adaptive Boosting (AdaBoost) and Random Forest. Brief description of each of these classifiers is given in this section.

K-Nearest Neighbour (KNN): K-nearest neighbour algorithm (KNN) [12] is a similarity-based lazy classification technique. In this method, the whole dataset has to be stored and for each instance, the K-nearest neighbours are calculated using some distance metrics. The label of the test instance is decided based on the labels of the nearest neighbours. KNN works poorly for high-dimensional data. Various similarity measures like Euclidean distance, Hamming distance, Manhattan distance, Minkowski distance, Tanimoto distance and Jaccard distance are used.

Naive Bayesian Classifier: Naive Bayesian classifier [12] is based on probabilistic inference from samples observed where the decision variable and the features form a very naive structure of Bayesian network. Naive Bayesian classifiers work best for image recognition and text mining. From the observed samples in the training dataset, likelihood is calculated. For a test instance, posteriori probability of the decision variable is calculated and decided based on the maximum likelihood.

Support Vector Machine (SVM): Support Vector Machines (SVM) [12] works by creating a hyper-plane in the dataset for which samples in different classes are separated by maximum width. SVM classifiers often work by transforming the sample feature space by using sophisticated kernel functions like polynomial, gaussian, sigmoid.

Adaptive Boosting (AdaBoost): AdaBoost classifier [12] is a meta-classifier that uses weak classifiers. Weak classifiers are those whose performance is marginally better than the random classification. In each iteration of the AdaBoost classifier, samples that are incorrectly classified by the ensemble meta-classifier are penalized using weights, and a new classifier is learned and added to the ensemble meta-classifier.

Random Forest: Random Forest is an ensemble classifier that creates decision tree in each iteration with randomly taken features and samples selected using bootstrap aggregating [12]. Random Forest works effectively for high-dimensional data.

3.5 Performance Evaluation

To ensure the robustness of the methods, most researchers use either separate independent test set or cross-fold sampling method. In performance estimation of the classifiers, the k-fold cross-validation technique is preferred as it overcomes the problem of over-fitting. This method does not use the entire dataset to build the model; it splits the data into k partitions. Each partition is then used as test set at each iteration where the rest of the data is used as training data.

In this paper, we have used accuracy as the performance metric to compare between learned models. Accuracy is calculated as the percentage of correctly classified instances to the total number of instances. Note that ours is a multi-class classification problem and thus the traditional evaluation metrics for binary classification does not work here. However, it is possible to create a binary classification problem for each of the labels in the dataset and measure class-wise binary classification performance.

4 Experiments Results

In this section, we are going to describe the experiments carried out in this study. All the experiments were carried out in a Dell Inspiron 15 Laptop Computer of 3000 series with 4 GB RAM and 240 GB SSD hard drive. All the programs were implemented in Java language using the Eclipse IDE and Java 8 standard edition. The OpenCV software library [3] was used to generate the features. The classification algorithms used in this paper were as implemented in Weka tool [6]. The wavelet transform code was taken from an available implementation.[1]

4.1 Parameters Used for Classifiers

For the classifiers, different set of parameters were used. A linear searching was used with no distance weighting for KNN. In case of the Naive Bayesian classifier, SVM, a polynomial kernel was used with $c = 1.0$ and $\varepsilon = 1.0w^{-2}$. Data was normalized before feeding to the classifier. J48 decision tree classifier was used in AdaBoost classifier as the weak base classifier. For Random Forest classifier, the number of iterations was set to 100.

4.2 Analysis of Features

In this section, we are going to discuss about the analysis of the features. Each of the four feature groups was selected one by one and every classifier are run to get the percentage of correctly classified instances. Results in terms of average accuracy in fivefold cross-validation is reported in Table 2. After running the experiments for our four feature groups A, B, C and D, we can see that feature group D has the best percentage of correctly classified instances with every classifier. Among them, Random Forest performed the best. This feature group D with the highest percentage of correctly classified instances is selected as the base for the next step of adding other feature groups and selecting the next best feature group combination. We combine the other three classifiers now with D. So there are three two-group combinations AD, BD and CD. Among these three groups, we could notice that the best performing group is AD. AD has the best accuracy for all the classifiers except in the case of Naive Bayesian. Here too, the best classification algorithm was Random Forest. Next, we move to combinations with three feature groups. We add other two feature groups at a time with AD. So the two options are: ABD and ACD. Analysing the results of ACD and ABD, we could note that the best performing feature group combination is ACD and Random Forest is again the best performing classifier. Since the accuracy is increasing, at the last step, we add the last feature group to combine all of them.

---

[1] http://www.jeejava.com/haar-wavelet-transform-using-java/.

Table 2 Average accuracy of different classifiers achieved using different combinations of features

Features	Classifiers				
	KNN	Naive Bayesian	SVM	AdaBoost	Random Forest
A	69.0745	64.1084	70.8804	64.5598	78.5553
B	66.14	61.851	69.3002	64.7856	78.781
C	81.7156	71.1061	86.456	81.2641	86.456
D	84.6501	81.0384	86.456	79.0068	88.7133
AD	85.7788	67.9458	86.9074	81.9413	90.7449
BD	79.2325	63.4312	82.8442	78.5553	87.8104
CD	68.623	**69.9774**	75.1693	72.6862	83.9729
ABD	63.4312	63.4312	73.1377	70.4289	82.8442
ACD	**88.0361**	67.9458	**90.0677**	**82.8442**	**91.1964**
ABCD	58.6907	63.8826	–	73.1377	80.3612

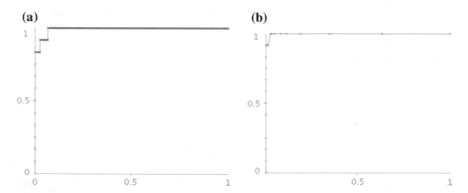

Fig. 2 Receiver operating characteristic (ROC) curves for **a** AdaBoost classifier and **b** Random Forest classifier using feature groups ACD for dataset label 1

We note that the feature group combination ABCD fails to improve over the previous results achieved by ACD. Thus, we conclude that the best performing feature group combination is ACD, and the best classifier is Random Forest classifier.

The bold faced values of the table indicate the highest percentage of correctly classified instances achieved for each of the classifiers. Note that, our method could not learn fivefold model for SVM due to shortage of memory. We also show Receiver Operating Characteristic (ROC) curve for the class 1 of the dataset in Fig. 2. ROC curves plot true-positive rate against the false-positive rate for a binary classifier with probabilistic output. As we know from the ROC curve that the closer the curve is to the border of the left then the top border of the ROC space, the more accurate the test. We can see from our two ROC curves of class one of the feature group combination ACD that the curves are more closer to the left border. The ROC curves for other classes are also similar.

Table 3 Comparison of the method proposed features in this paper with those by Silva et al. [17]

Features	Feature analysis table				
	KNN	Naive Bayesian	SVM	AdaBoost	Random Forest
Silva et al. [17]	62.05	74.11	50.88	73.23	77.94
This paper	88.03	67.94	90.06	82.84	91.19

4.3 Comparison with Other Methods

We compare the performance of our proposed method with that of a state-of-the-art algorithm proposed in [17]. In their method, the authors used fourteen features: eccentricity, aspect ratio, elongation, solidity, stochastic convexity, isoperimetric factor, maximal indentation depth, lobedness, average intensity, average contrast, smoothness, third moment, uniformity, and entropy. The first seven features are related to the shape and the rest with texture. We conducted experiments with different classifiers using the same parameters as we did for feature analysis with the feature groups. The results are given in Table 3. We could notice that except in the case of Naive Bayesian classifier, our features were able to outperform the features in Silva et al. [17].

4.4 Discussion

We have demonstrated the superiority of the features used in this paper over the shape- and texture-based methods. Note that all these shape- and texture-based methods are dependent on specialized algorithms to calculate these features. On the other hand, we mainly used histograms of original image and wavelet images. Histograms are very easy to calculate. However, with increase in the level of pyramid histograms, the number of features generated drastically increases and thus have a adverse effect on the performance of the classifier.

5 Conclusion

In this paper, we show the effectiveness of wavelet histogram and pyramid histogram of leaf images in detecting plant species. With the advent of deep learning techniques, nowadays feature extraction is being done from pixels of the input image based on convolution and many fascinating applications are possible. In future, we wish to explore the application of the features explored in this paper for disease detection of plants from leaf images. We also wish to explore other domains such as protein structure similarity search and human face emotion detection.

References

1. Arivazhagan, S., Shebiah, R.N., Ananthi, S., Varthini, S.V.: Detection of unhealthy region of plant leaves and classification of plant leaf diseases using texture features. Agricultural Engineering International: CIGR J. **15**(1), 211–217 (2013)
2. Bai, Y., Guo, L., Jin, L., Huang, Q.: A novel feature extraction method using pyramid histogram of orientation gradients for smile recognition. In: 2009 16th IEEE International Conference on Image Processing (ICIP). pp. 3305–3308. IEEE (2009)
3. Bradski, G.: The OpenCV Library. Dr. Dobb's J. Softw, Tools (2000)
4. Dalal, N., Triggs, B.: Histograms of oriented gradients for human detection. In: IEEE Computer Society Conference on Computer Vision and Pattern Recognition. CVPR 2005. vol. 1, pp. 886–893. IEEE (2005)
5. Ghaiwat, S.N., Arora, P.: Detection and classification of plant leaf diseases using image processing techniques: a review. Int. J. Recent Adv. Eng. Technol. (IJRAET) 2347–2812 (2014). ISSN (Online)
6. Hall, M., Frank, E., Holmes, G., Pfahringer, B., Reutemann, P., Witten, I.H.: The WEKA data mining software: an update. ACM SIGKDD Explor. Newslett. **11**(1), 10–18 (2009)
7. Karim, R., Al Aziz, M.M., Shatabda, S., Rahman, M.S., Mia, M.A.K., Zaman, F., Rakin, S.: CoMOGrad and PHOG: from computer vision to fast and accurate protein tertiary structure retrieval. Scien. Rep. **5**, 13275 (2015)
8. Kaundal, R., Kapoor, A.S., Raghava, G.P.: Machine learning techniques in disease forecasting: a case study on rice blast prediction. BMC Bioinform. **7**(1), 485 (2006)
9. Kulkarni, A.H., Patil, R.A.: Applying image processing technique to detect plant diseases. Int. J. Mod. Eng. Res. (IJMER) **2**(5), 3661–3664 (2012)
10. Li, L., Zhang, Q., Huang, D.: A review of imaging techniques for plant phenotyping. Sensors **14**(11), 20078–20111 (2014)
11. Mallah, C., Cope, J., Orwell, J.: Plant leaf classification using probabilistic integration of shape, texture and margin features. Sig. Process. Pattern Recogn. Appl. **5**(1) (2013)
12. Mohri, M., Rostamizadeh, A., Talwalkar, A.: Foundations of machine learning. MIT Press, Cambridge (2012)
13. Naikwadi, S., Amoda, N.: Advances in image processing for detection of plant diseases. Int. J. Appl. Innov. Eng. & Management 2(11) (2013)
14. Pape, J.M., Klukas, C.: 3-d histogram-based segmentation and leaf detection for rosette plants. ECCV Workshops **4**, 61–74 (2014)
15. Phadikar, S., Sil, J.: Rice disease identification using pattern recognition techniques. In: 11th International Conference on Computer and Information Technology, ICCIT 2008. pp. 420–423. IEEE (2008)
16. Sankaran, S., Mishra, A., Ehsani, R., Davis, C.: A review of advanced techniques for detecting plant diseases. Comput. Electron. Agric. **72**(1), 1–13 (2010)
17. Silva, P.F., Marcal, A.R., da Silva, R.M.A.: Evaluation of features for leaf discrimination. In: International Conference Image Analysis and Recognition, pp. 197–204. Springer (2013)
18. Stanković, R.S., Falkowski, B.J.: The haar wavelet transform: its st0atus and achievements. Comput. Electr. Eng. **29**(1), 25–44 (2003)
19. Sweldens, W.: The lifting scheme: a construction of second generation wavelets. SIAM J. Math. Anal. **29**(2), 511–546 (1998)

A Comparative Analysis on Currency Recognition Systems for iCu₹e and an Improved Grab-Cut for Android Devices

Vishwas Raval and Apurva Shah

Abstract Every country has its own currency in terms of coins and paper notes. Each of the currency of Individual County has its unique features, colors, denominations, and international value. In the twenty-first century, various issues like terror funding, smuggling have lead to the printing of fake currencies. Due to this, many times a person would never be able to know that the currency which one is holding is genuine or fake. This can only be decided if one knows all the features of the currency. However, for a common man, it is not possible to remember the features of the currency. Though the denomination can easily be recognized for a currency, it becomes difficult to identify a counterfeit currency from the real one. Especially for the blind people, it is not at all possible to do this check. This paper carries out an in-depth comparison of all the currency recognition systems and algorithms developed so far in order to propose a new system for Indian currency recognition iCu₹e, especially, for blind people.

1 History of Currency

From the moment, barter systems ended, currencies came into existence for day-to-day life. It dates back 2000 BC where it used to be in the form of coins. Later, in between 618 AD and 907 AD, in pre-modern China, during Tang dynasty, use of paper-based currencies started. In the medieval Islamic world, during seventh–twelfth century, the paper-based currency was introduced that became base for a stable-high-valued currency dinar. Sweden was the first country to introduce paper-based currency in Europe in 1661. Each country in this world has its own currency and each currency has a specific denomination that indicates its monetary value. US

V. Raval (✉) · A. Shah
CSE Department, Faculty of Technology & Engineering, The M S University of Baroda, Vadodara 390001, India
e-mail: vishwas.raval@gmail.com

A. Shah
e-mail: apurva.shah-cse@msubaroda.ac.in

© Springer Nature Singapore Pte Ltd. 2019
A. Abraham et al. (eds.), *Emerging Technologies in Data Mining and Information Security*, Advances in Intelligent Systems and Computing 814,
https://doi.org/10.1007/978-981-13-1501-5_24

279

Table 1 List of currencies according to their market share

Rank	Currency	ISO 4217 code (symbol)
1	US dollar	USD ($)
2	Euro	EUR (€)
3	Japanese yen	JPY (¥)
4	Pound sterling	GBP (£)
5	Australian dollar	AUD (A$)
6	Canadian dollar	CAD (C$)
7	Swiss franc	CHF (Fr)
8	Renminbi	CNY (元)
9	Swedish krona	SEK (kr)
10	New Zealand dollar	NZD (NZ$)
11	Mexican peso	MXN ($)
12	Singapore dollar	SGD (S$)
13	Hong Kong dollar	HKD (HK$)
14	Norwegian krone	NOK (kr)
15	South Korean won	KRW (₩)
16	Turkish lira	TRY (□)
17	Russian ruble	RUB (□)
18	Indian rupee	INR (₹)
19	Brazilian real	BRL (R$)
20	South African rand	ZAR (R)

dollars, British pound, Japanese yen, Euro are the examples of currencies of different countries. Following is the list of prominent currencies of countries across the world according to their market share (Table 1).

In India, the Reserve Bank of India is the chief controlling authority for the issuance of the currency. The symbol for Indian rupee is ₹ was designed by D. Udaya Kumar which was conferred an official sign for Indian rupee by Government of India on July 15, 2010.

In earlier days, India currency was from 1 aana to 100 rupees. The currency had denomination like 1, 5, 10, 20, 25, and 50 in terms of paisa. However, except ₹1 and ₹2 coins, the other coins have been discontinued and new coins of ₹5, ₹10, ₹20, ₹50, ₹100, and ₹500 have been introduced. The currency is available in a denomination value of ₹1, ₹2, ₹5, ₹10, ₹20, ₹50, ₹100, ₹500, and ₹2000.

On November 8, 2016, Honorable Prime Minister of India declared cancellation of existing currency of ₹500 and ₹1000. The reasons that Prime Minister cited in his address were to curb the black money, corruption menace, to stop terror funding and Hawala business.

2 Motivation for iCuₑe, Existing Tools, Applications, and Related Work

After being cheated with a counterfeit currency of ₹500, one of the authors got ignited a thought in his mind. It has been observed that, especially for Indian currency, be it a coin or note, currency identification, is really a herculean task for the blind people. This laid down the foundation for the development of iCuₑe.

In USA, to make it easier to identify the denominations of all currencies being same size, the Government provides some way to help them to tell apart the different money denominations, whereas the countries like Australia and Malaysia have every denomination of distinct width and length making the identification easier for the blind people. In Canada, Dollars have provision of Braille dots to indicate a specific denomination. Blind people can easily touch and read that dots to know the amount denomination they are holding. In India, the RBI has introduced an embossed pattern for every currency note but as the currency gets older, the embossed spot gets faded.

Apart from these basic features of different currencies, there are many applications which are available on Internet. For example, Smart Saudi Currency Recognizer (SSCR) [1] is a currency recognizer for Saudi Arabian currency. LookTel another App [2] which supports the US dollar, Australian dollar, Bahraini dinar, Brazilian real, Belarusian ruble, British pound, Canadian dollar, Euro, Hungarian forint, Israeli shekel, Indian rupee, Japanese yen, Kuwaiti dinar, Mexican peso, New Zealand dollar, Polish zloty, Russian ruble, Saudi Arabian riyal, Singapore dollar, and United Arab Emirates dirham. EyeNote [3] and MoneySpeaker [4] are the other applications used for currency recognition.

Apart from these applications, many research-oriented attempts, in the form of algorithms, have been made to provide a solution for currency recognition across the world since 1992. For all the work that has been carried out, the preprocessing and feature extraction are the common techniques which have been used in the initial phase. The following is the list of the classification approaches that have been used in the individual work in order to recognize the currency accurately (Table 2).

In India, the real implementations of currency recognition techniques are available in ATMs and Banks which everyone cannot afford and not handy as well. So, there is a real need for a system that can help the blind people in India to recognize currency properly, especially in the situation like demonetization when real currency also becomes unuseful. The next section discusses the proposed work and initial work that has been completed.

Table 2 List of currency recognition algorithms/work carried out for various currencies

Sr.#	Currency recognition approach	Currency	Accuracy	Year
1	Neural network [7]	USD	98.08	1993
2	Neural network, optimized mask, and genetic algorithm [8]	USD	>95%	1995
3	Hybrid neural network [9]	USD and Japanese	92	1996
4	Multilayered perceptrons in NN [10]	USD	High	1996
5	Neural NN with Gaussian distribution [11]	USD, CA$, AU$, Krone, Franc, GBP, Mark, Pesetas	High	1998
6	Neural NN and axis-symmetrical mask [12]	Euro	97	2000
7	Neural NN and principal component analysis [13]	USD	95	2002
8	Back propagation NN [14]	Chinese renminbi	96.6	2003
9	Speeded-up robust feature (SURF) [15]	USD	High	2007
10	Markov models [16]	USD, Euro, dirham, rial	95	2007
11	Ada-Boost classification [17]	USD	High	2008
12	Artificial neural network [18]	SL rupee	High	2008
13	Neural network [19]	Malaysian ringit	High	2008
14	Data acquisition [20]	Chinese renminbi	100	2008
15	Bio-inspired image processing [21]	Euro	100	2009
16	Ensemble neural network with negative correlation learning [22]	Bangladeshi taka	98	2010
17	Local binary patterns [23]	Chinese renminbi	100	2010
18	Wavelet transform [24]	Rials	81	2010
19	Intersection change [25]	Chinese renminbi	97.5	2010
20	Image processing and neural network [26]	Indian rupee	High	2010
21	Support vector machine [27]	Chinese renminbi	87.097	2011
22	Speeded-up robust feature (SURF) [28]	USD	100	2012
23	Wavelet transform and neural network [29]	Dirham	99.12	2012
24	Local binary patterns and RGB space [30]	Mexican	97.5	2012
25	Quaternion wavelet transform and generalized Gaussian density [31]	USD, renminbi, and Euro	99.68	2013

(continued)

Table 2 (continued)

Sr.#	Currency recognition approach	Currency	Accuracy	Year
26	Basic feature extraction using Euler numbers [32]	Pakistani rupee	High	2013
27	Number recognition [33]	Chinese renminbi	95.92	2014
28	Instance retrieval and indexing [34]	Indian rupee	96.7	2014
29	Radial basis Kernel function [35]\	Dirham	91.51	2015
30	Segmentation, feature extraction [36]	Bangladeshi taka	High	2015
31	Region of interest (ROI), discrete wavelet transform, linear regression and SVM [37]	Indian rupee	High	2015

3 The Proposed Solution for Indian Currency Recognition—iCu₹e and Improved Grab-Cut Algorithm for Android Devices: cGrab-Cut

Looking toward the need for having an affordable application that can help the visually challenged people in currency identification, the authors are developing a computationally lighter image processing algorithm and the iCu₹e [5] an Android App for Indian currency recognition in Indian vernacular languages. Figure 1 shows the overall flow of the process.

Currently, it is in its infant stage and is being developed. The preprocessing phase has been completed, and the care has been taken to remove any outliers in the image

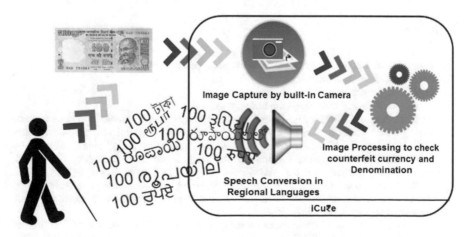

Fig. 1 iCu₹e process

in the form of background noise or any other background images. For this, Grab-Cut [6] has been implemented and improved in terms of execution time. The native implementation of Grab-Cut is faster and quite time-efficient but Java-based implementation using OpenCV is time-consuming. This has been improved in terms of time. Though it has been achieved with a small trade-off, yet by losing a little accuracy in terms of cuts, the time is reduced by 33%. It has been tested on Android devices of different configurations and has been observed that irrespective of RAM or processor speed, the improved Android implementation of Grab-Cut takes one-third time. Currently, image classification is under development. The improved Grab-Cut implementation has been named as cGrab-Cut, a Compromised Grab-Cut.

4 Conclusion

This paper carries out the comparative analysis of all the work done in the direction of currency recognition for the various currencies across the world. It shows that there are systems available for foreign currencies, but for Indian currencies, for blind people, there is no such product or application available to identify the denomination and check if the currency is real or fake. It also discusses a brief about a system being developed, named, iCu₹e for Indian currency recognition with initial implementation and a small implementation about a compromised Grab-Cut to improve the time efficiency.

Acknowledgements We are thankful to the omnipotent God for making us able to do something for the society. We are thankful to our parents for bringing us on this beautiful planet. We are grateful to our department and University for providing support and resources for this work. Finally, we acknowledge the authors and researchers whose papers helped us to move ahead for this work.

References

1. https://play.google.com/store/apps/details?id=sscr.imagemanipulations
2. http://www.looktel.com/moneyreader
3. http://www.eyenote.gov/
4. https://www.androidpit.com/app/com.hmi.moneyspeaker
5. Raval, V., Shah, A.: iCu₹e—An IoT application for indian currency recognition in vernacular languages for visually challenged people. In: Confluence-2017, 7th International Conference on Cloud Computing, Data Science and Engineering, Amity University, Noida (2017)
6. Rother, Carsten, Kolmogorov, Vladimir, Blake, Andrew: Grabcut: Interactive foreground extraction using iterated graph cuts. ACM Trans. Graph. (TOG) **23**(3), 309–314 (2004). ACM
7. Takeda, F., Omatu, S., Onami, S.: Recognition system of US dollars using a neural network with random masks. In: IJCNN'93-Nagoya. Proceedings of 1993 IEEE International Joint Conference on Neural Networks (1993)
8. Takeda, F., Omatu, S.: A neuro-paper currency recognition method using optimized masks by genetic algorithm. In: IEEE International Conference on Intelligent Systems, Man and Cybernetics for the 21st Century, vol. 5, pp. 4367–4371 (1995)

9. Takeda, F., Sigeru, O.: High speed paper currency recognition by neural networks. IEEE Trans. Neural Netw. **6**, 73–77 (1995)
10. Frosini, A., Gori, M., Priami, P.: A neural network-based model for paper currency recognition and verification. IEEE Trans. Neural Netw. **6**, 1482–1490 (1996)
11. Masahiro, T., Takeda F., Ohkouchi K., Michiyuki Y.: Recognition of paper currencies by hybrid neural network (1998)
12. Takeda, F., Toshihiro N.: Multiple kinds of paper currency recognition using neural network and application for Euro currency. In: Proceedings of the IEEE International Joint Conference on Neural Networks, vol. 2, pp. 143–147 (2000)
13. Ahmadi, A., Sigeru O., Michifumi Y.: Implementing a reliable neuro-classifier for paper currency using PCA algorithm. In: Proceedings of the 41st SICE Annual Conference, vol. 4, pp. 2466–2468 (2002)
14. Zhang, E.-H., Bo, J., Jing-Hong, D., Zheng-Zhong, B.: Research on paper currency recognition by neural networks. In: International Conference on Machine Learning and Cybernetics, vol. 4, pp. 2193–2197 (2003)
15. Chen, W., Yingen, X., Jiang, G., Natasha, G., Radek, G.: Efficient extraction of robust image features on mobile devices. In: Proceedings of the 2007 6th IEEE and ACM International Symposium on Mixed and Augmented Reality, pp. 1–2 (2007)
16. Hassanpour, H., Yaseri, A., Ardeshiri, G.: Feature extraction for paper currency recognition. In: 9th International Symposium on Signal Processing and Its Applications, pp. 1–4 (2007)
17. Liu, X.: A camera phone based currency reader for the visually impaired. In: Proceedings of the 10th International ACM SIGACCESS Conference on Computers and Accessibility, pp. 305–306 (2008)
18. Gunaratna, D.A.K.S., Kodikara, N.D., Premaratne, H.L.: ANN based currency recognition system using compressed gray scale and application for Sri Lankan currency notes-slcrec. In: Proceedings of World Academy of Science, Engineering and Technology, pp. 235–240 (2008)
19. Nurlaila, H.: Currency recognition and converter system. Ph.D. Disserion, Universiti Malaysia Pahang (2008)
20. He, J., Zhigang, H., Pengcheng, X., Ou, J., Minfang, P.: The design and implementation of an embedded paper currency characteristic data acquisition system. In: International Conference on Information and Automation, pp. 1021–1024 (2008)
21. Parlouar, R., Florian, D., Marc, M., Christophe, J.: Assistive device for the blind based on object recognition: an application to identify currency bills. In: Proceedings of the 11th International ACM SIGACCESS Conference on Computers and Accessibility, pp. 227–228 (2009)
22. Debnath, K., Sultan, U., Shahjahan, Md, Kazuyuki, Mu: A paper currency recognition system using negatively correlated neural network ensemble. J. Multimedia **5**(6), 560–567 (2010)
23. Guo, J., Yanyun, Z., Anni, C.: A reliable method for paper currency recognition based on LBP. In: 2nd IEEE International Conference on Network Infrastructure and Digtal Content, pp. 359–363 (2010)
24. Daraee, F., Saeed, M.: Eroded money notes recognition using wavelet transform. In: 6th Iranian Conference on Machine Vision and Image Processing, pp. 1–5 (2010)
25. Shao, K., Yang, G., Na, W., Hong-Yan, Z., Fei, L., Wen-Cheng, L.: Paper money number recognition based on intersection change. In: Third International Workshop on Advanced Computational Intelligence (IWACI), pp. 533–536 (2010)
26. Gopal, K.: Image processing based feature extraction of indian currency notes. M. Tech thesis, Thapar University (2010)
27. Yeh, C., Wen-Pin, S., Shie-Jue, L.: Employing multiple-kernel support vector machines for counterfeit banknote recognition. Appl. Soft Comput. **11**(1), 1439–1447 (2011)
28. Hasanuzzaman, F., Xiaodong, Y., Yingli, T.: Robust and effective component-based banknote recognition for the blind. IEEE Trans. Syst. Man Cybern. (Appl. Rev.) **42**(6), 1021–1030 (2012)
29. Ahangaryan, F., Mohammadpour, T., Kianisarkaleh, A.: Persian banknote recognition using wavelet and neural network. In: International Conference on Computer Science and Electronics Engineering (ICCSEE). pp. 679–684 (2012)

30. García-Lamont, F., Jair, C., Asdrúbal, L.: Recognition of Mexican banknotes via their color and texture features. Expert Syst. Appl. **39**(10), 9651–9660 (2012)
31. Gai, S., Guowei, Y., Minghua, W.: Employing quaternion wavelet transform for banknote classification. Neurocomputing **118**, 171–178 (2013)
32. Ali, A., Mirfa, M.: Recognition system for Pakistani paper currency. World Appl. Sci. J. **28**(12), 2069–2075 (2013)
33. Yu, H., Yingyong, Z.: Study on money number recognition arithmetic. Int. J. Multimedia Ubiquitous Eng. **9**(11), 189–196 (2014)
34. Suriya, S., Shushman, C., Vishal, K., Jawahar, C.: Currency recognition on mobile phones. In: Proceedings of IEEE 22nd International Conference on Pattern Recognition, pp. 2661–2666 (2014)
35. Sarfraz, M.: An intelligent paper currency recognition system. Proc. Comput. Sci. **65**, 538–545 (2015)
36. Saifullah, S., Rahman, M., Hossain, Md: Currency recognition using image processing. Am. J. Eng. Res. **4**(11), 26–32 (2015)
37. Pham, T., Danh, Y., Seung, Y., Dat, T., Husan, V., Kang, R., Dae S., Sungsoo, Y.: Recognizing banknote fitness with a visible light one dimensional line image sensor. **9** 21016–21032 (2015)

A Memory-Efficient Image Compression Method Using DWT Applied to Histogram-Based Block Optimization

Amiya Halder, Aritra Kundu, Apurba Sarkar and Kanik Palodhi

Abstract Image compression is an essential task for storing images in digital format. In this communication, an improved and hugely memory-efficient block optimization technique is presented that incorporates byte compression and discrete wavelet transform (*DWT*). Instead of the common method of nulling insignificant *DWT* coefficients, all the *DWT* coefficients are stored. The only lossy part comes from block optimization without noticeable degradation in the decompressed images. The method shows huge improvement in compression and reduces image storage space. The results obtained from this technique are compared to JPEG and JPEG2000 standard which shows this can be a fast alternative to other compression methods.

Keywords Image compression · Histogram · *PSNR* · *DWT*

1 Introduction

Images stored in digital format are extensively used for natural and scientific purposes, captured in a variety of cameras with different resolutions. Scientific applications like biomedical imaging, space explorations and forensics are routinely using images of different resolutions. Therefore, storage of digital images, particularly, "true colour" multimedia images, and huge digital databases, is a great challenge, since

A. Halder (✉) · A. Kundu
STCET, 4-D.H. Road, Kolkata 700023, India
e-mail: amiya.halder77@gmail.com

A. Kundu
e-mail: aritrakundu99@gmail.com

A. Sarkar
IIEST Shibpuer, Howrah 711103, India
e-mail: as.besu@gmail.com

K. Palodhi
University of Calcutta, Kolkata 700009, India
e-mail: kanikpalodhi@gmail.com

© Springer Nature Singapore Pte Ltd. 2019
A. Abraham et al. (eds.), *Emerging Technologies in Data Mining and Information Security*, Advances in Intelligent Systems and Computing 814,
https://doi.org/10.1007/978-981-13-1501-5_25

they are shared on Internet platforms and are transmitted on a regular basis. Fast and accurate digital compression techniques have come into the forefront of the computer science research [1–3]. A digital image is, essentially, a sampled version of an analog scene, and typical images are mostly in three colour format with 8-bit (24-bit colour/3 bytes). Apart from special cases, the neighbouring pixels have high correlation, which can be utilized for compressing the images, noting that image compression techniques are sensitive to the content of the image [4]. Broadly, image compression methods can be classified as lossless and lossy compressions. Lossless compression techniques retrieve the exact data of the original images upon decompression but compression ratio achieved in these cases is low, at the most of the order of five [5–8]. On the other hand, lossy compression, as the name suggests, leads to loss of original data and degradation of the quality of images but compression achieved in this case can be quite high. This technique has its usage in many cases, like transferring images through Internet, and this low file size leads to fast transfer of data. If the technique is applied sensibly, the loss may not be apparent between the original image and the retrieved (decompressed) image. There are different ways of achieving lossy compression, and in most of the cases, it is compared to JPEG and JPEG2000 standard [9–19]. In the recent past, particularly, *DCT* and *DWT* are extensively used for image compression with variable quantization where flexible compression ratio can be exercised [16–20]. In these cases, after computing *DCT* and *DWT* coefficients, some of the coefficients with insignificant contributions are nulled to achieve image compressions. In [21], a lossy compression technique with an efficient algorithm based on block optimization and byte compression are discussed. In this communication, a novel and much improved technique is presented where each pixel on the block, after block optimization, is replaced by the pixel with maximum value of the histogram. Finally, compression is achieved after bytes pressing with *DWT* without changing the coefficients, but, simply, by storing them, a novel feature not applied before. The method is fast, and a huge compression of the order of 1000 is obtained without any appreciable loss to the image. This efficient method is termed as histogram-based *DWT* compression method, and in first four sections, the steps of the techniques are explained with the next section compares the experimental results applied to the sample images with detailed comparison with JPEG and JPEG2000 standard.

2 Block Optimization with Histogram

Neighbouring pixels in an image generally have close correlation since image scene does not contain rapidly varying brightness which is utilized in block optimization [10, 11]. In this process, a basic building block of 4×4 is considered and an image of size $M \times M$ is separated according to the building blocks. For each basic block, histogram is computed and all the pixels of the basic blocks are replaced by the peak value of the histogram. After that, an augmented image is created where these basic

Fig. 1 An example of basic block of dimension 4 × 4

21	23	30	23
22	23	27	23
23	25	24	25
31	23	30	22

blocks are replaced by pixels with highest histogram values. If the initial image is represented by A, then it can be written as a

$$A = \sum_{p=1}^{M} \sum_{q=1}^{N} a_{pq} \tag{1}$$

where, $a_{pq} = \{R_{pq}, G_{pq}, B_{pq}\}$ provides the brightness values in red, green and blue and the range varies between [0–255] for 8-bit images with maximum values of row and columns given by M and N. After separating the image in terms of basic building blocks and replacing the each basic building with the peak of the intensity from the histogram of the basic block, the new augmented image A_{au} obtained from A can be written as,

$$A_{au} = \sum_{p=1}^{M/4} \sum_{q=1}^{N/4} h_{pq} \tag{2}$$

where $h_{pq} = max(\sum_{p=1}^{4} \sum_{q=1}^{4} a_{pq}^{4 \times 4})$ is the peak of the intensities (red, green and blue) from the histogram of the basic block.

For example, let us consider a basic block as shown in Fig. 1. For this block, clearly, the peak of the histogram is at the value 23 as shown in Fig. 2. Therefore, the entire block will be replaced by single pixel with value 23. This leads to a compression of the order of four and also loss of resolution, but for images with higher sizes, the effect of loss is much less.

3 Byte Compression

In the byte compression technique, 3-byte RGB (each byte for each colour) image is reduced to 2-byte data, resulting in a compression of 33%. Here, the first byte that belongs to red as shown in Fig. 3 is divided by 8 which is binary equivalent of left shift of 3 places (since $log_2^8 = 3$), and other 5-bits remain from red. The operation is repeated for the next byte that belongs to green (left shift of 3-bits). Finally, the last byte that belongs to blue is divided by 4 leading to a reduction of 2-bits resulting in

Fig. 2 Histogram of the block

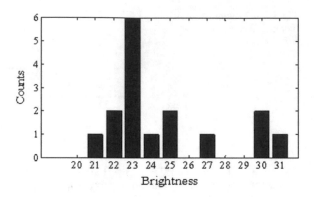

Fig. 3 Illustration of byte compression technique

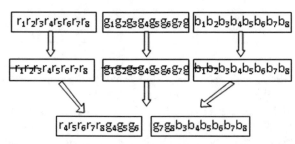

an overall reduction of 8-bits/1-byte. Obviously, this is a lossy compression but the information loss in the augmented image is not at all appreciable with no significant change in *PSNR* [8]. The byte compression technique of the RGB component is shown in Fig. 3. This can be represented in terms of another augmented image A_{au}^b which has same spatial dimension as A_{au} but with reduced size.

$$A_{au}^b = \sum_{p=1}^{M/4} \sum_{q=1}^{N/4} h_{pq}^b \tag{3}$$

where h_{pq}^b is the peak of the intensities with reduced byte.

4 Discrete Wavelet Transform

A wavelet is a particular irregular form of wave of limited duration with an average value of zero and represents signals that are local in time and scale. Different types of wavelets are used to represent signals with appropriate shift and scale which generates a set of coefficients [5, 8]. Therefore, a two-dimensional dataset or an image in this case can be reduced to a set of components wavelets, a property ideally suited for image compression. In this case, after the byte optimization and byte compression, *DWT* is applied to the existing augmented image. This results in a greatly reduced dataset, and only application of *DWT* without block optimization and byte compression provides a compression of the order of 1000. Together with these steps, *DWT* achieves a compression of the order of 5000. Theoretically, this can be written as in terms of scaled and translated basis functions $\varphi_{j,m,n}$ and $\psi_{j,m,n}$ which are given by,

$$\varphi_{j,m,n}(p, q) = 2^{\frac{j}{2}}\varphi(2^j p - m, 2^j q - n) \tag{4}$$

$$\psi_{j,m,n}^k(p, q) = 2^{\frac{j}{2}}\psi^i(2^j p - m, 2^j q - n), k = \{H, V, D\} \tag{5}$$

where H, V and D represent horizontal, vertical and diagonal directions. The byte-compressed augmented image A_{au}^b can, therefore, be transformed in terms of *DWT* coefficients (in Eqs. 4 and 5) as,

$$W_\varphi(j_o, m, n) = \frac{4}{\sqrt{MN}}\sum_{p=1}^{M/4}\sum_{q=1}^{N/4}h_{pq}^b\varphi_{j_o,m,n}(p, q) \tag{6}$$

$$W_\psi^k(j, m, n) = \frac{4}{\sqrt{MN}}\sum_{p=1}^{M/4}\sum_{q=1}^{N/4}h_{pq}^b\psi_{j,m,n}^k(p, q) \tag{7}$$

where j_o is the arbitrary staring scale and $W_\varphi(j_o, m, n)$ coefficients are approximations of h_{pq}^b at scale j_o. The $k = H, V, D$ details for scales $j \geq j_o$ are added by $W_\psi^k(j, m, n)$.

5 Decompression

The decompression or decoding of the images are done by first applying inverse *DWT* to obtain the byte-compressed augmented image elements h_{pq}^b from the stored *DWT* coefficients, according to Eqs. 6 and 7 [5, 8]. The byte-compressed 2-byte augmented image A_{au}^b an image with reduced size ($\frac{M}{4} \times \frac{N}{4}$) is obtained from image elements h_{pq}^b (Eq. 3). Each elements of A_{au}^b, i.e., h_{pq}^b are then subjected to byte release by multiplying 8 (equivalent of right shift of 3 places) to blocks of first of the two 5-bits and then multiplying 4 to final block of 6-bits (equivalent of right shift of 2 places) to obtain the peak of the intensities of a 4 × 4 block, h_{pq} (Eq. 2). Essentially, with these set of steps, a 3-byte augmented image A_{au} with *RGB* components and reduced size ($\frac{M}{4} \times \frac{N}{4}$), is obtained from 2-byte A_{au}^b. Finally, size is expanded to get original image A where each pixel is replaced with 4 × 4 basic building block with the peak value of the histogram h_{pq} (Eq. 1). Though the original brightness values cannot be retrieved due to the final step where original brightness values of 4 × 4 block are replaced only by the peak value, still the degradation is not apparent for most of the images. The process, obviously, works much better if applied to high-definition image with huge image size. The experimental results presented in the Sect. 6 will clarify our claim.

6 Experimental Results

The proposed algorithm is developed in *MATLAB*, and it is applied on images of different sizes ranging from 60.7 KB to 25.7 MB, and nine of them are presented in comparison tables (Tables 1 and 2). Apart from these images, the algorithm is tried on numerous other images where it has produced excellent results. The comparison of proposed method with other different techniques such as block optimization, JPEG and JPEG2000 are shown below in Table 1. It is based on compression in terms of bits per pixel (bpp) which presents compression in RGB model for each pixel, defined as,

$$bpp = \frac{24 \times compressed\ file\ size}{Actual\ file\ size} \tag{8}$$

Clearly, the proposed method, histogram based *DWT*, shows far superior performance in terms of compression ability with *bpp* about 100–1000 times less compared to JPEG standards. As previously argued, for high-definition images, the proposed method histogram based *DWT* compression works best, which can simply be verified from analysis of the result for images *Test*2 and *Test*3.

In Table 2, the previous images are compared with respect to the *PSNR* which are kept same for each of the original image using the same compression methods as in Table 1. Here, *PSNR*, which measures image quality, is defined, as follows.

$$PSNR = 20log_{10}\frac{255}{RMSE} \tag{9}$$

Table 1 Compression results of different compression techniques in bits per pixel (bpp)

Name of images	Original image size	Block optimization	JPEG compression	JPEG2000 compression	Proposed algorithm
Apocalypse	5.49 MB	1.003	0.653	6.95	0.00089
Boat	1.29 MB	1.019	0.608	8.866	0.00332
Building	9.61 MB	1.502	0.921	6.193	0.00041
Car	1.37 MB	1.019	0.410	2.754	0.00281
Parrot	750 KB	1.033	0.537	7.424	0.00443
Test2	25.7 MB	0.999	0.236	1.316	0.00019
Test3	10.8 MB	1.009	0.221	1.590	0.00059
Zelda	1.29 MB	1.019	0.426	8.711	0.00302
Waterfall	60.7 KB	1.383	0.525	9.054	0.03398

Table 2 Compression results (in size) for different compression of different images and *PSNR* of the proposed method

Name of images	Original image size	Block optimization	JPEG compression (KB)	JPEG2000 compression	Proposed algorithm (Byte)	PSNR
Apocalypse	5.49 MB	235 KB	153	1.59 MB	214	30
Boat	1.29 MB	56.1 KB	33.5	488 KB	187	37
Building	9.61 MB	616 KB	378	2.48 MB	171	49
Car	1.37 MB	59.6 KB	24	161 KB	168	35
Parrot	750 KB	32.3 KB	16.8	232 KB	142	31
Test2	25.7 MB	1.07 MB	259	1.41 MB	213	52
Test3	10.8 MB	465 KB	102	733 KB	279	51
Zelda	1.29 MB	56.1 KB	28.9	591 KB	170	46
Waterfall	60.7 KB	3.5 KB	1.33	22.9 KB	88	72

where the root mean square error, $RMSE = (\frac{1}{MN}\sum_{p=1}^{M}\sum_{q=1}^{N} |f_s(p,q) - \hat{f}_s(p,q)|^2)^{\frac{1}{2}}$ with original image denoted as $f_s(p,q)$ and decompressed image $\hat{f}_s(p,q)$ and $s = R, G, B$; i.e., it includes red, green and blue component in RGB colour model. Histogram based *DWT* shows compression ratio of the order of 1000 or more in terms of storage and for the image with highest size it about 106 times (image Test2 in Tables 1 and 2). JPEG and JPEG2000 at same *PSNR*, ideally, provide maximum compression with equal image quality, yet the proposed method gives better results than that. Another important observation is that even for images with smaller sizes, the proposed method may produce compressed images with bigger size, e.g., Test2 with original size 25.7 MB and Test 3 with original size 10.8 MB produce compressed images by proposed method of size 213 and 279 byte, respectively. This is due to the fact that neighbouring pixels have different correlations for each image that changes for each basic 4×4 block during block optimization. In Figs. 4 and 5, four of the images are reproduced with original and decompressed images shown together which demonstrates that the image degradation is not significant in most of the cases. It is also possible to enhance the speed of the method by increasing the block size for high-definition images with bigger sizes, though image resolution will be compromised.

Fig. 4 Different original images

Fig. 5 Decompressed images of the proposed method

7 Conclusion

In this work, an improved version of compression technique is presented where together with block optimization and byte compression, *DWT* is applied for image compression. The method is far superior to the JPEG standard and other similar techniques, in terms of compression ratio measured in bits per pixel, as pointed out in Table 1 with typical compression ratios of the order of 1000. It does not degrade the image quality significantly, and if compared with other techniques such as JPEG and JPEG200 at the same *PSNR*, it simply shows far superior compression. The method, histogram-based *DWT*, is quite fast and can be applied to a wide variety of image files. In case of file transfers through communication links, this can make immense contribution since the method works particularly well for image files that need bigger storage space. Therefore, a fast and highly memory efficient method is presented in this communication, here, that is far superior to JPEG and JPEG2000 in terms of compression ratio with huge utilities.

References

1. Khuri, S., Hsu, H.C.: Interactive packages for learning image compression algorithms lists. ITiCSE 2000. Helsinki, Finland
2. Bhattacharjee, S., Das, S., Choudhury, D., R., Chouduri, P. Pal.: A pipelined architecture algorithm for image compression. In: Proceedings of the Data Compression Conference. Saltlake City, USA, March 1997
3. Ritter, J., Molitor, P.: A pipelined architecture for partitioned DWT based lossy image compression using FPGA. International Symposium on FPGA, pp. 201–206 (2001)
4. Halder, A., Kole, D. K., Bhattacharjee, S.: On-line colour image compression based on pipelined architecture. ICCEE 2009. Dubai, UAE (2009)
5. Pratt, W.K.: Digital Image Processing. PIKS Scientific Inside (2007)
6. Wallace, G.K.: The JPEG still picture compression standard. Commun. ACM 31–44 (1991)
7. Pennebaker, W.B., Mitchell, J.L.: JPEG: Still Image Data Compression Standard. Van Nostrand Reinhold, New York (1993)
8. Gonzalez, R.C., Woods, R.E.: Digital Image Processing. Pearson Education, Harlow (2002)
9. Acharya, T., Tsai, P.S.: JPEG2000 Standard for Image Compression
10. Delp, E.J., Mitchell, O.R.: Image compression using block truncation coding. IEEE Trans. Commun. 1335–1342 (1979)

11. Kuo, C.H., Chen, C.F., Hsia, W.: A comression algorithm based on classified interpolative block truncation coding and vector quantization. J. Inf. Sci. Eng. 1–9 (1999)
12. Amerijckx, C., Legaty, J.D., Verleysenz, M.: Image compression using self-organizing maps. Syst. Anal. Model. Simul. **43**(11) (2003)
13. Namphol, A., Chin, S., Arozullah, M.: Image compression with a hierarchical neural network. IEEE Trans. Aerosp. Electron. Syst. 32(1) (1996)
14. Sentiono, R., Lu, G.: Image compression using a feedforward neural network. In: International Conference on Neural Networks (1994)
15. Jiang, J.: Image compression with neural networks—A survey. Image Communication, vol. 14, no. 9. Elsevier (1999)
16. Telagarapu, P., Naveen, V.J., Lakshmi Prasanthi, A., Santhi, G.: Vijaya.: Image compression using DCT and Wavelet transformations. Int. J. Sig. Process. Image Process. Pattern Recogn. **4**(3), 61–74 (2011)
17. Aullinas, F.C.: General embedded quantization for wavelet-based lossy image coding. IEEE Trans. Signal Process. **61**(6), 1561–1574 (2012)
18. Douak, F., Benzid, R., Benoudjit, N.: Color image compression algorithm based on the DCT transform combined to an adaptive block scanning. Int. J. Electron. Commun. 16–26 (2011)
19. Zhang X.: Lossy compression and iterative reconstruction for encrypted image. IEEE Trans. Inf. Forensics Secur. **6**(1) (2011)
20. Lee, S.: Compressed image reproduction based on block decomposition. IET Image Process. **3**(4), 188–199 (2009)
21. Halder, A., Dey, S., Mukherjee S., Banerjee, A.: An efficient image compression algorithm bades on block optimization and byte compression. ICISA-2010 (2010)

Forensic Image Reconstruction Based on Efficient Morphological Operational Model

Sajib Bhawal and Mehnaz Tabassum

Abstract Fragmented forensic image recovering from unallocated space plays an important role in computer forensics and investigation. For clear-cut investigation, it is necessary to reconstruct the recovered forensic images. Morphological operations can reconstruct the recovered images. Existing paper applies different methods to a damaged image based on structuring element, but cannot recuperate damaged image properly. This paper proposed a morphological image processing model that applies some operations like H-maxima transform, bottom-hat transformation, erosion, opening, dilation, closing operations based on structuring elements to sharp the final output image. This model can fill all holes of the damaged image properly compared to reference model. Structural Similarity Index (SSIM), Peak Signal-to-Noise Ratio (PSNR) of the proposed model is higher than the reference model. This paper applies security mechanism such as watermarking with Discrete Cosine Transform (DCT) to hide the image in the forensic workstation.

Keywords Forensics · Carving · Morphological processing · Top-hat transformation · H-maxima transform · Bottom-hat transformation SSIM, PSNR · DCT

1 Introduction

Nowadays, digital images can be used as evidence in computer forensics and investigation. Images are stored in the digital media. If it is related to crime can be used as evidence in investigation. Criminals can delete image from their computer and USB device. The deleted image can be fragmented in the unallocated space of the

S. Bhawal · M. Tabassum (✉)
Department of Compute Science and Engineering, Jagannath University,
Dhaka, Bangladesh
e-mail: mtabassum2013@gmail.com

S. Bhawal
e-mail: sajibbhawal@gmail.com

© Springer Nature Singapore Pte Ltd. 2019
A. Abraham et al. (eds.), *Emerging Technologies in Data Mining and Information Security*, Advances in Intelligent Systems and Computing 814,
https://doi.org/10.1007/978-981-13-1501-5_26

suspect drive. Carving of deleted image plays an important role in a digital forensic and investigation. After recovering the fragmented image, it can be used in the investigation. If the recovered image from the suspect device is noisy and damaged, it is necessary to reconstruct the image. Many researchers tried to recover images and files from digital media. In this paper, two forensic software are used to recover fragmented image from unallocated space of the suspect disk. This paper tries to design a proposed morphological operational model to reconstruct the damaged image using the above morphological operations. The major part of this paper has three steps: carving image, morphological image processing, and image security using watermarking with Discrete Cosine Transform.

2 Literature Review

Effective morphological image processing techniques and image reconstruction [1] are described by applying morphological operations like opening, closing, erosion, and dilation on image with a structuring element to reconstruct image and a morphological filter known as top-hat transformation applied to enhance the image. Another research paper titled an application of morphological image processing to forensics [2] shows the effect of single morphological operation on forensic fingerprint image. Research paper on image restoration based on morphological operations [3] applied only erosion and dilation on image. In this paper, no sharpening filter is used for noise removal. Research paper [4] describes morphological grayscale reconstruction of image. H-maxima transform [5] is used to select the useful segments from image and also to smooth noisy images. Research paper about medical image enhancement techniques by bottom-hat and median filtering [6] presents the effect of bottom-hat filtering to enhance image. Medical image enhancement using morphological transformation [7] applied Top-hat transformation on medical image for enhancement. Research papers [8, 9] describe Structural Similarity Index (SSIM) to compare reconstructed image with input image. Paper [10, 11] applied image and file carving technique from unallocated storage disk. Research on image watermarking algorithm based on DCT [12] is described for image watermarking using Discrete Cosine Transform (DCT). Research paper about a novel robust color image digital watermarking algorithm based on Discrete Cosine Transform [13] described the process of color image watermarking using Discrete Cosine Transform. A DCT-based imperceptible color image watermarking scheme [14] proposed a scheme for color image watermarking based on DCT.

3 Proposed Methodology

This paper proposed three steps to perform its task. First step is the recovering images from unallocated space of the suspect drive. Second step is the morphological image

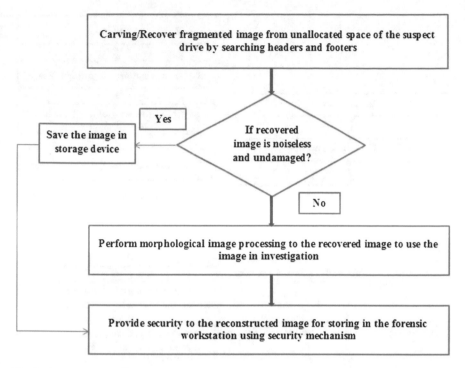

Fig. 1 Steps of the working procedure

processing to the damaged or noisy images to reconstruct images. Third step is used to provide security for storing the reconstructed image. The working procedure is depicted as follows in Fig. 1.

Here, fragmented image recovering from unallocated space of the disk using image carving method. For image hiding in the forensic workstations' storage device, watermarking using Discrete Cosine Transform is used in this paper. In this paper, we compare our proposed model with a reference model based on [1], depicted as below in Fig. 2.

To overcome the hole filling problem of the reference model, a morphological model is proposed to reconstruct damaged image. The proposed model of morphological operations can fill all holes properly.

The security model is used to hide the image in the forensic workstations' storage device (Fig. 3).

This security model takes color image as input image. Then a watermark cover image is applied to the input image. Discrete Cosine Transform is performed to embed input image and cover image to hide the input image. Then de-watermarking is performed to extract watermark from the watermarked image. The output image is same as the input image.

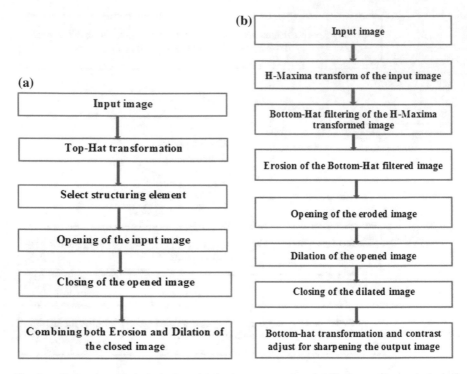

Fig. 2 **a** Reference morphological model for image reconstruction. **b** Proposed morphological model for image reconstruction

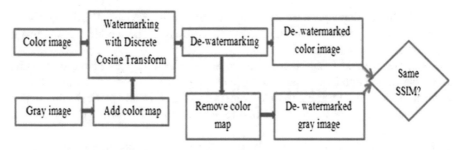

Fig. 3 Image security model using watermarking using DCT

3.1 *Watermarking with Discrete Cosine Transform (DCT)*

This section describes about watermarking with Discrete Cosine Transform (DCT) to hide the image in forensic workstation to prevent unauthorized access. This paper applies DCT on a color image and a gray image. Working procedure of watermarking with DCT is depicted below (Fig. 4).

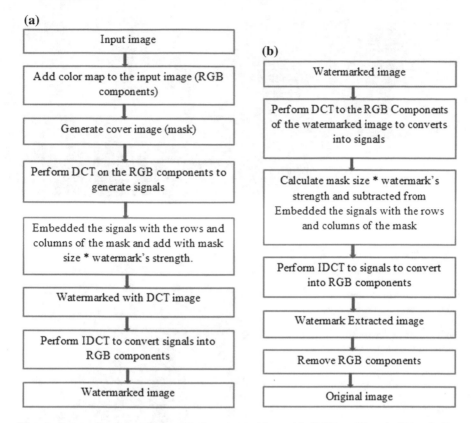

Fig. 4 Steps of watermarking and de-watermarking with DCT. **a** Watermarking. **b** De-watermarking

3.2 Result Analysis

A fingerprint database from FVC2002 [15] is used in this paper. This database is downloaded from FVC2002 Web site and stored in a partition of the working hard disk. Then the reference of this partition is deleted. Then many fingerprint images may be fragmented with other files in the unallocated space of the disk. Image carving method can recover this fragmented image using forensic tools such as AccessData FTK Imager and Hex Workshop Hex Editor. Simulation result of image recovering with carving method using AccessData FTK Imager is given below (Figs. 5 and 6).

Then recover this fragmented image using another tool Hex Workshop Hex Editor. Here two images are found using two different tools. The following figure shows the difference between two recovered images (Fig. 7).

Simulation result of the morphological image reconstruction by reference model (Fig. 8).

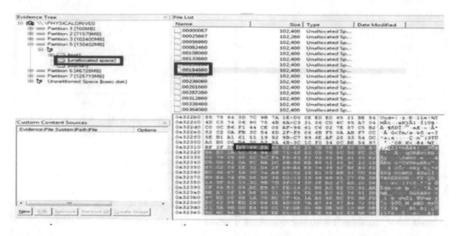

Fig. 5 Image carving using AccessData FTK Imager. The recovered image using AccessData FTK Imager is

Fig. 6 Recovered image using AccessData FTK Imager (Image courtesy to the FVC2002 Fingerprint Verification Competition.)

The noisy and holes containing input image and some holes containing output image using MATLAB zoom-in operations are given below.

From the above comparison, it is clear that the reference morphological model cannot fill all holes properly. Simulation result of the proposed model is as follows.

From above simulation results, it is shown that figure shows the bottom-hat transformation of the closed image to sharp the final output image. The proposed morphological operational model can reconstruct the damage image and can fill all holes properly. The noisy and holes containing input image and all holes filled output image using MATLAB zoom-in operations are given below.

Figure a shows the noisy and damaged input image and b shows the reconstructed image which has no holes and noise. This proposed morphological model can clear all noise from the background by using H-maxima transform and sharpen the image

Fig. 7 Comparison result of the two recovered images and simulation result of the morphological image reconstruction by reference model

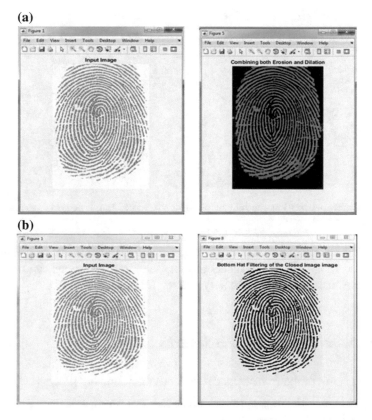

Fig. 8 Simulation results of (tiff image). **a** Reference morphological model. **b** Proposed morphological model

Fig. 9 Difference between input and output image for **a** reference morphological model, **b** proposed morphological model

applying bottom-hat transformation. All holes are filled properly using opening operation by this model. Proposed model of morphological operations can reconstruct the recovered damaged image properly and can fill all holes (Fig. 9).

3.3   Simulation Result of the Security Model

Simulation result of the security model includes the following process

– Watermarking with Discrete Cosine Transform (DCT) for gray image
– Watermarking with Discrete Cosine Transform (DCT) for color image.

Simulation result of watermarking using DCT for gray image. DCT and watermark embedding of RGB components: Watermarked with DCT image, de-watermarked image, and original color map extracted image (Fig. 10).

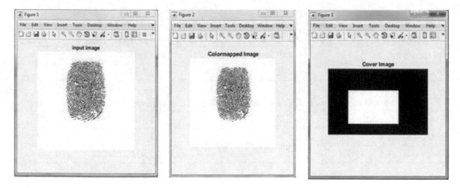

Fig. 10 Simulation results of watermarking using DCT for gray image

4 Performance Analysis

In this thesis, two performance metrics are used to analyze the performance of the proposed methodology. Name of two performance metrics are:

- Peak Signal-to-Noise Ratio (PSNR)
- Structural Similarity Index (SSIM) (Fig. 11).

4.1 Comparison of PSNR Value for .tif Format Fingerprint Image of Reference Model and Proposed Model

This section compares the performance of reference and proposed morphological models. Here performance comparison is done by comparing PSNR and SSIM value. Performance comparisons are shown here in table and graphical format. Following

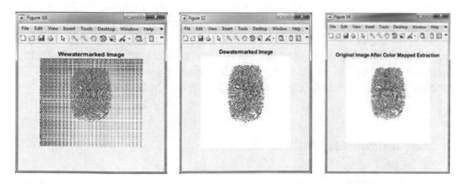

Fig. 11 Simulation results of watermarking using DCT for color image

Table 1 SSIM and PSNR value of a tiff format image for morphological models

Morphological model	SSIM	PSNR
Reference model	0.449	12.033
Proposed model	0.791	13.572

Table 2 SSIM and PSNR value of color and gray image for watermarking using DCT

Watermarking using DCT	SSIM	PSNR
For color image	1	130.206
For gray image	1	130.027

table shows the SSIM and PSNR values for the reference model and proposed morphological model for a tiff format fingerprint image (Table 1).

The SSIM values of a tiff format fingerprint image for reference and proposed morphological models are 0.449 and 0.791, respectively. The PSNR values are 12.033 and 13.572, respectively.

Following table shows the comparison of SSIM and PSNR values for color and gray images of watermarking using DCT technique (Table 2).

From the above table, it is seen that SSIM value for color image and gray image using watermarking with DCT is 1. So, this technique can watermark and de-watermark color image and gray image perfectly. The PSNR value for color and gray images is almost same.

5 Conclusions

This paper shows recovering fragmented images from unallocated space of the disk and USB device using image carving method. Some recovered images may be noisy and damaged. Morphological image reconstruction model is used to reconstruct noisy and damaged image. Here a reference morphological model and a proposed morphological model are used to reconstruct the damaged image. Performance of the proposed morphological model for image reconstruction is better than reference model. This paper used watermarking with Discrete Cosine Transform (DCT) to hide image in the forensic workstation. Here watermarking using DCT technique for color images is applied on gray images. SSIM value of the watermarking with DCT for color and gray images is equal to 1, obtained from simulation.

6 Limitations and Future Works

Watermarking using DCT can hide images. This technique is not secure enough. This technique cannot be used for secure image transmission. In future work, secure cryptography algorithm will be used for image transmission. Advanced Encryption Standard (AES) and secure Hashing Algorithm like SHA-512 can be used for secure image transmission.

References

1. Priya, M., St Annes, F., Nawaz, G.K.: Effective morphological image processing techniques and image reconstruction
2. Dar, S.A., Lone, S.A.: An application of morphological image processing to forensics. Int. J. Comput. Eng. Technol. **6**(8) (2015)
3. Raid, A., Khedr, W., El-Dosuky, M., Aoud, M.: Image restoration based on morphological operations. Int. J. Comput. Sci. Eng. Inf. Technol. **4**(3) 2014
4. Vincent, L.: Morphological grayscale reconstruction in image analysis: applications and efficient algorithms. IEEE Trans. image Process. **2**(2), 176–201 (1993)
5. Ghazali, K., Xiao, R., Ma, J.: Road lane detection using h-maxima and improved hough transform. In: 2012 Fourth International Conference on Computational Intelligence, Modelling and Simulation (CIMSiM), pp. 205–208. IEEE (2012)
6. Das, S.S.: Medical image enhancement techniques by bottom hat and median filtering.'
7. Firoz, R., Ali, M.S., Khan, M.N.U., Hossain, M.K., Islam, M.K., Shahinuzzaman, M.: Medical image enhancement using morphological transformation. J. Data Anal. Inf. Process. **4**(01), 1 (2016)
8. Wang, Z., Bovik, A.C., Sheikh, H.R., Simoncelli, E.P.: Image quality assessment: from error visibility to structural similarity. IEEE Trans. Image Process. **13**(4), 600–612 (2004)
9. Premaratne, P., Premaratne, M.: New structural similarity measure for image comparison. In: ICIC, vol. 3, pp. 292–297. Springer (2012)
10. Ravi, A., Kumar, T.R., Mathew, A.R.: A method for carving fragmented document and image files. In: 2016 International Conference on Advances in Human Machine Interaction (HMI), pp. 1–6. IEEE (2016)
11. Xu, M., Dong, S.: Reassembling the fragmented jpeg images based on sequential pixel prediction. In: International Symposium on Computer Network and Multimedia Technology, CNMT 2009, pp. 1–6. IEEE (2009)
12. Xu, Z., Wang, Z., Lu, Q.: Research on image watermarking algorithm based on dct. Procedia Environmental Sciences **10**, 1129–1135 (2011)
13. Wang, T.-Y., Li, H.-W.: A novel robust color image digital watermarking algorithm based on discrete cosine transform. JCP **8**(10), 2507–2511 (2013)
14. Abraham, J., Paul, V.: A dct based imperceptible color image watermarking scheme. Int. J. Sig. Process. Image Process. Pattern Recogn. **9**(7), 137–146 (2016)
15. Fingerprint verification competition. http://bias.csr.unibo.it/fvc2000. Accessed 10 November 2017

Classification of Geographical Features from Satellite Imagery

Md. Haidar Sharif, Sahin Uyaver, Md. Haris Uddin Sharif,
Ibrahim Furkan Ince and Zaid Zerdo

Abstract It is a challenging task to classify heterogeneous geographical features from satellite imagery. This paper addresses 31 straightforward classification algorithms based on predominantly pixels to classify miscellaneous geographical features from satellite imagery. The addressed algorithms can extract and process the features of a large dataset with high-resolution images expeditiously. A total of 606 red-green-blue satellite images of the Bosnian city of Banja Luka are exercised to comprehend their performances for classifying cemeteries, fields, houses, industries, rivers, and trees. The recorded experimental results demonstrate that the best average performance can come into possession of 87%.

1 Introduction

To classify miscellaneous geographical features from the satellite imagery is not new. For example, given a large set of satellite images of some cities on the Earth, how can we classify geographical features, e.g., cemeteries, fields, houses, industries, rivers,

Md. H. Sharif (✉)
International Balkan University, 1000 Skopje, Republic of Macedonia
e-mail: msharif@ibu.edu.mk

S. Uyaver
Turkish-German University, 34820 Beykoz, Istanbul, Turkey
e-mail: uyaver@tau.edu.tr

Md. H. U. Sharif
University of the Cumberlands, Williamsburg, KY 40769, USA
e-mail: msharif6291@ucumberlands.edu

I. F. Ince
Kyungsung University, Busan 608736, South Korea
e-mail: furkanince@ks.ac.kr

Z. Zerdo
IUS, Hrasnicka cesta 15, 71210 Sarajevo, Bosnia and Herzegovina
e-mail: zerdo.zaid@gmail.com

© Springer Nature Singapore Pte Ltd. 2019
A. Abraham et al. (eds.), *Emerging Technologies in Data Mining and Information
Security*, Advances in Intelligent Systems and Computing 814,
https://doi.org/10.1007/978-981-13-1501-5_27

309

streets, forests, and mountains from those images? Cemeteries may contain lots of small and white patches or rectangles, while the industrial class would be quite varied and much more difficult to detect. Houses in the housing class may have red roofs, which would make the classification process a bit easy after extracting necessary features properly. But forests and rivers are quite difficult to differentiate due to their similarity. Many efforts have been evolved to solve the existing problem of classifying geographical features from the satellite images of the Earth with great accuracy [1–4]. For example, Risojevic [1] showed how the images can be classified using convolutional neural networks. The interesting part of his approach is that he used convnets to extract features. Since such a neural network needs a lot of data to produce decent results from SAT-4 and SAT-6 datasets [5], which contain about 330,000 images taken in the USA. The results are quite impressive, reaching a result of about 99% accuracy considering a neural network trained with a total of 400,000 images and tested on 10,000 images. Risojevic and Babic [2] analyzed the importance of texture in remote image classification. As there were correlations between the coefficients at various scales, the Gabor wavelet coefficients played a vital role in their method. They also used support vector machines. In the end, their method produced 85% correct classification rate. Haralick et al. [3] showed how texture plays a vital role in image classification of satellite imagery, photographic imagery, and photomicrograph of 1:20,000 scale. They used two kinds of decision rules, one for convex polyhedra and the other for rectangular parallelepipeds. The results for the classification of satellite imagery were recorded up to 83%. Cleve et al. [4] compared pixel and made object-based classifications considering high-resolution aerial photographic images. Their object-based classification scored more than 40% of accuracy. Zhang [6] described a method which attempted to detect urban buildings in two stages. The first stage was a conventional multispectral classification. In the second stage, the classification of buildings was improved by means of their spatial information through a modified co-occurrence matrix-based filtering. The direction dependence of the co-occurrence matrix is utilized in the filtering process. The method was tested by using TM and SPOT Pan merged data for the whole area of the city of Shanghai, China. Lu and Weng [7] performed a survey of image classification methods and techniques for improving classification performance. Hester et al. [8] suggested that conventional spectral-based classification methods can be used to generate highly accurate maps of urban landscapes using high spatial resolution imagery.

We are more interested in pixel-based classification rather than object-based classification to classify geographical features from the satellite images on Earth. To conduct experiments, we have used 606 red-green-blue satellite images of the Bosnian city of Banja Luka to classify cemeteries, fields, houses, industry, rivers, and trees. The Banja Luka dataset [9] consists of 28 cemetery images, 178 field images, 143 houses images, 75 industry images, 77 river images, and 105 tree images. Each image has a size of 128 × 128 pixels. Figure 1 shows sample images of cemetery, fields, houses, industry, river, and trees from Banja Luka dataset [9]. In our implementation, a series of Java programs are used to run on the dataset to extract the necessary

(a) Cemetery   (b) Fields   (c) Houses

(d) Industry   (e) River   (f) Trees

Fig. 1 Sample images of Banja Luka dataset [9]. How can we classify miscellaneous geographical features automatically?

features and then WEKA[1] (Waikato Environment for Knowledge Analysis) is used to train neural networks to classify them. WEKA is a suite of machine learning software written in Java. It was developed at the University of Waikato, New Zealand. It provides a series of tools for preprocessing, classification, visualization, and so on. We take into account the quantity of images. For example, cemeteries in Banja Luka dataset have only 28 images, which is about 4.62% of the total sum of images. A neural network will have problems learning patterns from such a small sample of images. To solve this problem, we try to find unique values to that set of images. Basically, our proposed 31 disparate algorithms are straightforward and they are based on mainly pixels. Yet the experimental results demonstrate an acceptable range of their applicability and performance.

The rest of the paper is organized as follows: Sect. 2 illustrates necessary implementation steps. Section 3 discusses the experimental results followed by a few hints of future works. Section 4 concludes the paper.

---

[1] https://www.cs.waikato.ac.nz/~ml/weka/.

2 Implementation Steps

In this section, we have explained all the necessary steps needed to classify satellite images. We have addressed five key algorithms, namely x_1, x_2, x_3, x_4, and x_5, followed by their combination algorithms, i.e., a total of 31 algorithms.

2.1 Algorithm x_1 : Sum of Red-Green-Blue (RGB) Pixels

In computer graphics, the RGB color model is an additive color model in which red, green, and blue light are added together in various ways to reproduce an ample array of colors. Nowadays, this model is used for the sensing, representation, and display of images in electronic systems. Before the electronic epoch, the model was theoretically explained by human perception of colors. However, one of the simplest ways to classify various geographical features from satellite images is the sum of all RGB pixels for each image. The values can be viewed separately; i.e., three features can be put independently. The reason behind the inclusion of RGB features is that they can sometimes distinguish between certain images quite easily. For example, cemeteries usually have large amounts of white pixels due to the gravestones, which could be seen from a high count of all RGB features. Rivers do not contain many red pixels, as much as housing images possess. Normally, housing images have many red roofs. Fields and trees have a lot of green pixels. Nevertheless, they have much less red and blue pixels. Putting together these common cases, an effective and efficient classification can be made.

2.2 Algorithm x_2 : White-Green-Red (WGR) Pixels Count

Naturally, summing up all RGB pixels is not enough. Henceforth, we can go further to count exactly how many WGR pixels exist in a given image. The difference between WGR and RGB features is that a white pixel contains a lot of all three colors; i.e., a white pixel cannot be considered as apparently red even though it contains a high level of red. As a result, we have counted how many WGR pixels exist according to the following threshold conditions:

1. A pixel is regarded as white if it contains red, green, and blue levels higher than 200. And then the average of the three channels is taken.
2. A pixel is regarded as green if it contains a level smaller than a 100 for its red value, smaller than 150 for its blue channel, and higher than 100 for green.
3. A pixel is regarded as red if it contains a red value higher than 190, while both green and blue are less than 200.

The aforementioned threshold values (i.e., 200, 100, 150, 100, 190, and 200) are experimentally defined; henceforth, there would not exist any theoretical background.

The values for a given condition can be taken after observing what each value needs to distinguish. For instance, red pixels need to make it easier for the machine to identify housing districts with high levels of red.

2.3 Algorithm x_3 : Texture (Tex) Differences

The meaning of texture varies based on the application areas of music, arts, science, and technology. For example, in music a texture would mean an overall sound created by the interaction of aspects of a piece of music. A texture in visual arts would point to an element of design and its application in art. In science and technology, the texture would mean the smoothness or roughness of the surface of an object. However, there exists no formal mathematical definition of texture. One of the first quantitative or physiological texture descriptions can be found in [10], where six basic textural features, namely coarseness, contrast, directionality, line-likeness, regularity, and roughness, had been approximated in computational. Bharati et al. [11] discussed various methods for image texture analysis. An image texture can be a series of metrics computed in image computation intended to measure the sensed texture of an image. Image texture provides knowledge about the spatial organization of color or image intensities or exclusive domain of an image. We have used texture differences in certain colors and images.

2.4 Algorithm x_4 : White Rectangle Count (WRC)

Due to the count of how many white shapes are existing on a given image, the cemeteries are much easier to classify by employing WRC feature. We have used a flood-fill algorithm to go from one white pixel to another and then count the number of white pixels. We have implemented the following steps for x_4.

1. Convert the image to a gray one using a simple averaging method.
2. Convert the gray image into a binary one considering a simple thresholding technique that distinguishes pixels between extremely white, e.g., higher than 230 in its gray value, and non-white pixels.
3. Flood every fifth pixel with respect to width and height. The flooding is recursive with the stopping criteria being the edges of image, not going into another indexed flood fill and if it hits a non-white pixel.
4. A map is made that demos how many pixels should contain each index; i.e., how many pixels should each flood-fill make.
5. Indexes that have a moderate pixel count, in our case between 5 and 30, are considered as white rectangles, ideally gravestones.

2.5   Algorithm x_5 : River Pixel Count (RPC)

The easiest way to target river images is to look at the individual pixels present in the image. The existing pixels on images of rivers are quite unique. The final status of a pixel whether it will be considered as a river pixel or not depends on the following few sessions of analysis:

1. The absolute difference between green and blue values is negligible, e.g.,

$$|Green - Blue| < 10. \tag{1}$$

2. The absolute difference between green and red is not negligible, e.g.,

$$|Green - Red| > 50. \tag{2}$$

3. The green value is not negligible, e.g.,

$$140 < Green < 200. \tag{3}$$

The aforementioned stipulations result in a feature that almost intelligibly recognize river images from other images.

2.6   Combination of x_1, x_2, x_3, x_4, and x_5 Algorithms

By the way of addition, we have proposed all possible combinations of x_1, x_2, x_3, x_4, and x_5 algorithms. Explicitly, a total of

$$\binom{5}{1} + \binom{5}{2} + \binom{5}{3} + \binom{5}{4} + \binom{5}{5} = 5 + 10 + 10 + 5 + 1 = 31 \tag{4}$$

distinct algorithms have been put forward. A concatenated algorithm of $x_1|x_2$ suggests that the algorithm brings together the sum of red-green-blue pixels and white-green-red pixels count. Correspondingly, a concatenated algorithm of $x_1|x_2|x_3|x_4|x_5$ addresses that the algorithm consists of the sum of red-green-blue pixels, white-green-red pixels count, texture differences, white rectangle count, and river pixel count.

2.7   Training of Neural Networks

Following classification methods of WEKA to train neural networks can be used.

1. Multilayer Perceptron (MulPer): It gives evidence of a neural network with feed-forward capabilities and backpropagation that results in a better learning outcome. The number of hidden layers can be calculated as:

$$a = \frac{\text{Number of attributes} + \text{Number of classes}}{2}. \tag{5}$$

   If a number of attributes and classes are given as 4 and 2, respectively, then the number of hidden layers should be $a = 3$.
2. Logistic Regression (LogReg): It shows one of the possible regression models in which dependent variables are grouped but not continuous. In the case of a binary dependent variable, the output can take only two values 0 and 1 (e.g., pass/fail, win/lose, alive/dead, or healthy/sick). Cases where the dependent variable has more than two outcome categories may be analyzed in multinomial logistic regression.
3. Sequential Minimal Optimization (SeqOpt): It makes clear and visible that one of the solutions to quadratic programming problems which present in the training of support vector machines. SeqOpt is an iterative algorithm for solving the optimization problem. It breaks the underlying problem into a series of smallest possible subproblems, which are then solved analytically.
4. Random Forest (RanFor): It gives evidence of an algorithm in which a series of random decision trees are constructed and then giving statistical information such as the mode of the classes (classification) or mean prediction (regression) of the individual trees.

3 Experimental Results

3.1 Experimental Setup

Predominantly, all experiments were conducted on a computer of 8-Core CPUs at 3.50 GHz with 16 GB RAM. WEKA was used to train neural networks. Satellite images of Banja Luka dataset [9] were used. The analysis of algorithmic outputs became easy and fast as the ground truths are given. The number of hidden layers in the MulPer was considered as 3.

3.2 Results of Geographical Features Classification

We have performed a quantitative study with the given ground truths. Table 1 depicts the recognition results of our algorithms to classify various geographical features from Banja Luka dataset without using the classification algorithms of WEKA. It

Table 1 Performance of our algorithm to classify various geographical features from Banja Luka dataset without using the classification algorithms of WEKA

Algorithms	Classification performance without WEKA						
	Cemeteries (%)	Fields (%)	Houses (%)	Industries (%)	Rivers (%)	Trees (%)	Mean value (%)
x_1	69	72	77	80	73	86	76.17
x_2	65	68	71	60	62	67	65.50
x_3	58	53	55	47	58	64	55.83
x_4	70	64	65	64	69	60	65.33
x_5	78	94	95	84	79	90	86.66
$x_1 \vert x_2$	62	65	62	64	45	47	57.50
$x_1 \vert x_3$	52	50	52	62	70	48	55.67
$x_1 \vert x_4$	42	50	42	62	45	49	48.33
$x_1 \vert x_5$	46	53	62	52	72	54	56.50
$x_2 \vert x_3$	49	60	61	55	55	58	56.33
$x_2 \vert x_4$	45	52	57	42	58	53	51.17
$x_2 \vert x_5$	62	60	62	61	65	58	61.33
$x_3 \vert x_4$	43	66	51	72	38	68	56.33
$x_3 \vert x_5$	62	60	46	72	66	67	62.17
$x_4 \vert x_5$	62	70	47	52	55	59	57.50
$x_1 \vert x_2 \vert x_3$	52	45	42	52	55	49	49.17
$x_1 \vert x_2 \vert x_4$	56	51	50	57	65	53	55.33
$x_1 \vert x_2 \vert x_5$	72	70	72	74	65	66	69.83
$x_1 \vert x_3 \vert x_4$	39	60	62	64	58	56	56.50
$x_1 \vert x_3 \vert x_5$	55	67	62	72	74	83	68.83
$x_1 \vert x_4 \vert x_5$	63	60	62	70	64	57	62.67
$x_2 \vert x_3 \vert x_4$	47	70	62	63	69	65	62.67
$x_2 \vert x_3 \vert x_5$	52	60	72	74	75	68	66.83
$x_2 \vert x_4 \vert x_5$	55	51	47	57	64	67	56.83
$x_3 \vert x_4 \vert x_5$	53	58	72	71	70	68	65.33
$x_1 \vert x_2 \vert x_3 \vert x_4$	82	90	92	83	75	89	85.17
$x_1 \vert x_2 \vert x_3 \vert x_5$	83	92	91	85	77	88	86.00
$x_1 \vert x_2 \vert x_4 \vert x_5$	80	86	85	83	81	87	83.67
$x_1 \vert x_3 \vert x_4 \vert x_5$	84	91	93	80	76	84	84.67
$x_2 \vert x_3 \vert x_4 \vert x_5$	82	87	91	84	78	89	85.17
$x_1 \vert x_2 \vert x_3 \vert x_4 \vert x_5$	84	91	92	83	81	90	86.83

is perceptible that x_5 algorithm outperformed the congregate algorithms of x_1, x_2, x_3, x_4. This is widely due to the features of x_5 algorithm highly fits the conditions at Eqs. 1, 2, and 3 with images of fields, houses, industries, rivers, and trees. But those conditions did not suit very well for the case of cemetery images. As a result, x_5 algorithm was limited with 78% success rate in that case. It is noticeable that the performance does not increase proportionally if more features of an algorithm are added into another existing algorithm. The average highest performance was

Table 2 Performance of our algorithm to classify various geographical features from Banja Luka dataset by using the classification algorithms of WEKA

Algorithms	Normalized performance of miscellaneous algorithms With WEKA								Mean
	SeqOpt		MulPer		LogReg		RanFor		
	Success (%)	Gain (%)	Success (%)	Gain (%)	Success (%)	Gain (%)	Success (%)	Gain (%)	Success (%)
x_1	62.62	0	73.78	0	72.81	0	70.87	0	70.02
x_2	61.71	−1.45	73.78	0	72.81	0	70.87	0	70.02
x_3	60.51	−3.37	73.78	0	72.81	0	70.87	0	70.02
x_4	61.19	−2.28	73.78	0	72.81	0	70.87	0	70.02
x_5	70.86	13.16	73.78	0	72.81	0	70.87	0	70.02
$x_1|x_2$	71.84	14.72	83.49	13.16	78.15	7.33	83.00	17.12	79.12
$x_1|x_3$	71.84	14.72	83.49	13.16	78.15	7.33	83.00	17.12	79.12
$x_1|x_4$	71.84	14.72	83.49	13.16	78.15	7.33	83.00	17.12	79.12
$x_1|x_5$	71.84	14.72	83.49	13.16	78.15	7.33	83.00	17.12	79.12
$x_2|x_3$	71.84	14.72	83.49	13.16	78.15	7.33	83.00	17.12	79.12
$x_2|x_4$	71.84	14.72	83.49	13.16	78.15	7.33	83.00	17.12	79.12
$x_2|x_5$	71.84	14.72	83.49	13.16	78.15	7.33	83.00	17.12	79.12
$x_3|x_4$	71.84	14.72	83.49	13.16	78.15	7.33	83.00	17.12	79.12
$x_3|x_5$	71.84	14.72	83.49	13.16	78.15	7.33	83.00	17.12	79.12
$x_4|x_5$	71.84	14.72	83.49	13.16	78.15	7.33	83.00	17.12	79.12
$x_1|x_2|x_3$	75.24	20.15	84.32	14.29	82.83	13.83	85.31	20.38	81.93
$x_1|x_2|x_4$	75.24	20.15	84.32	14.29	82.83	13.83	85.31	20.38	81.93
$x_1|x_2|x_5$	75.24	20.15	84.32	14.29	82.83	13.83	85.31	20.38	81.93
$x_1|x_3|x_4$	75.24	20.15	84.32	14.29	82.83	13.83	85.31	20.38	81.93
$x_1|x_3|x_5$	75.24	20.15	84.32	14.29	82.83	13.83	85.31	20.38	81.93
$x_1|x_4|x_5$	75.24	20.15	84.32	14.29	82.83	13.83	85.31	20.38	81.93
$x_2|x_3|x_4$	75.24	20.15	84.32	14.29	82.83	13.83	85.31	20.38	81.93
$x_2|x_3|x_5$	75.24	20.15	84.32	14.29	82.83	13.83	85.31	20.38	81.93
$x_2|x_4|x_5$	75.24	20.15	84.32	14.29	82.83	13.83	85.31	20.38	81.93
$x_3|x_4|x_5$	75.24	20.15	84.32	14.29	82.83	13.83	85.31	20.38	81.93
$x_1|x_2|x_3|x_4$	79.70	27.28	86.93	17.82	85.48	17.40	87.46	23.41	84.89
$x_1|x_2|x_3|x_5$	79.70	27.28	86.93	17.82	85.48	17.40	87.46	23.41	84.89
$x_1|x_2|x_4|x_5$	79.70	27.28	86.93	17.82	85.48	17.40	87.46	23.41	84.89
$x_1|x_3|x_4|x_5$	79.70	27.28	86.93	17.82	85.48	17.40	87.46	23.41	84.89
$x_2|x_3|x_4|x_5$	79.70	27.28	86.93	17.82	85.48	17.40	87.46	23.41	84.89
$x_1|x_2|x_3|x_4|x_5$	81.52	30.18	89.60	21.44	88.45	21.48	89.77	26.67	87.34

recorded as 86.83% upon executing the concatenated algorithm of $x_1|x_2|x_3|x_4|x_5$, which consists of all features from the algorithms of x_1, x_2, x_3, x_4, and x_5. Table 2 presents the performance of our algorithms to classify various geographical features from Banja Luka dataset using the classification algorithms of WEKA. The outputs of some randomly selected algorithms are depicted in Fig. 2. The average highest performance was obtained as 87.34% on executing the concatenated algorithm of $x_1|x_2|x_3|x_4|x_5$, which includes all features of corresponding algorithms. The gain

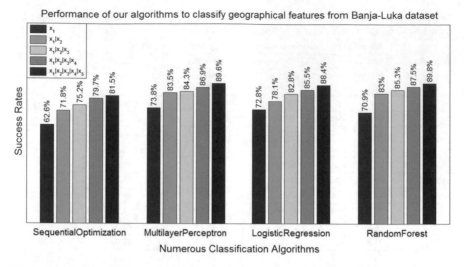

Fig. 2 Performance of some randomly selected algorithms (e.g., x_1, $x_1|x_2$, $x_1|x_2|x_3$, $x_1|x_2|x_3|x_4$, and $x_1|x_2|x_3|x_4|x_5$) using data from Table 2. The algorithm $x_1|x_2|x_3|x_4|x_5$ demonstrates the best performance

with respect to x_1 algorithm in Table 2 indicates the performance increment when a new feature is added into the existing features. The worst result, average normalized performance approximately 70.02%, was resulted from the cases when the algorithms of x_1, x_2, x_3, x_4, and x_5 were employed separately. They were expected to be the worst cases, but they represented somewhat modest results. The reason behind this proposition is that cemeteries have large levels of all three colors, due to the color white being present a lot. Housing usually has a lot more red, due to the red colored roofs. Rivers are abundant in the color blue, although some of them have a lot of green as well, making it more difficult for the machine to distinguish them from tree images. Notwithstanding, by observing the results of both Tables 1 and 2, it is easy to affirm that with a view to classifying various geographical features from Banja Luka dataset the average performances of our proposed algorithms with WEKA were slightly better than those of without WEKA.

In concluding remark, our simple classification algorithms are based on standard features extraction methods but their overall performances play some sort of important role in the machine learning applications including the classification of various geographical features from satellite imagery.

3.3 Comparison with State-of-the-Art Method

The method of Risojevic and Babic [2] produced 85% correct classification results from Banja Luka dataset. The mean performances of our some selected algorithms

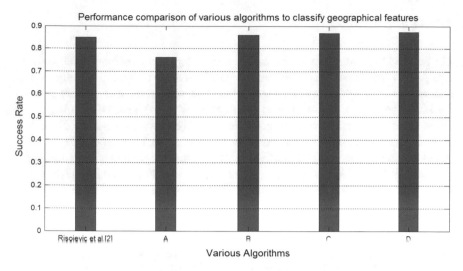

Fig. 3 Performance comparison for classifying various geographical features from Banja Luka dataset. A, B, and C represent the algorithms of x_1, $x_1|x_2|x_3|x_5$ and $x_1|x_2|x_3|x_4|x_5$, respectively, from Table 1. D demonstrates the algorithm of $x_1|x_2|x_3|x_4|x_5$ from Table 2

with the algorithmic performance of Risojevic and Babic [2] are depicted in Fig. 3. It appears to exist that our simple algorithms having up to 87% correct classification outputs function more or less in parallel with the state-of-the-art method (e.g., Risojevic and Babic [2]) without any serious concern of efficacy.

3.4 Our Observation

The most striking thing to us was the fact that simple features extracted can play a vital role in the machine learning step. A percentage of 70% performance for a simple x_1 algorithm that extracts RGB features only for a dataset that contains six classes is not negligible. It may not be ideal or good, but the neural network managed to classify many correctly just by the sum of their pixels. Naturally, that does not take into account any shapes in the image; i.e., no spatial data are extracted. That represents the unexpected finding, no spatial data are needed for decent classification of satellite imagery. High performance is the result of that. All features need a short amount of time to be extracted. This could be quite amazing in cases where extremely large datasets are needed to be analyzed and classified quickly. Edge cases could be executed on more sophisticated algorithms, while high-resolution images could be resized or only a portion of the whole image is taken into consideration.

3.5 Future Works

The proposed algorithms would be tested against a large series of other datasets. More data signify more possible observations to make in order to enhance the algorithms. Neural networks would have an easier time to learn patterns and identify possible classification criteria. As mentioned in Haralick et al. [3], textural features can be relied upon for satellite imagery classification. We have not performed much of texture features and object extraction. For the aerial classification of images, more advanced features can be extracted, especially features regarding texture and object recognition. Houses in the housing images could be easily distinguished from other images by counting the amount of rectangular shapes that contain a lot of red, which is quite unique for those types of images. The distinction between trees and fields was one of the challenging tasks for us. Usually, fields are lighter in green than forests, but some of them have almost no difference. The only possible way, we can imagine, is to look at the texture. In general, trees have darker green neighboring lighter green, due to sunlight hitting the branches of trees. Fields usually have lines passing through various angles that can be used to easily distinguish them. Over time, more and more features would be extracted and would result in the problem being too complex and have a dimensionality problem. We would have to apply a dimensionality reduction or principal component analysis to find exactly which features do not affect the solution as much as we need. One more thing is that we did not take into account the machine learning part. We only used machine learning algorithms that stood out with their results. No explanation is provided on the reasons of our choices, besides them being algorithms we are accustomed to. To improve the results even more, the algorithms would need to be analyzed and their functionalities learnt to see which model fits the data best. Multiple comparisons with statistical tests [12] and areas under the receiver operating characteristic curves [13] would be performed by considering the proposed algorithms and state-of-the-art algorithms. Time complexity [14] of each algorithm with cache memory effect [15] would be computed and compared.

4 Conclusion

We proposed a group of pixel-based classification algorithms to classify geographical features from the satellite images of the Earth. The proposed algorithms are unsophisticated and easy to implement. A key advantage of them includes that the features of a large dataset with high-resolution images can be extracted and processed in a short amount of time. The experimental results on 606 red-green-blue satellite images demonstrated that the best average performance could be reached up to 87% to classify cemeteries, fields, houses, industries, rivers, and trees. Future study would include further increment of their performance by taking into account object recognition and spatial texture extraction.

References

1. Risojevic, V.: Analysis of learned features for remote sensing image classification. In: Symposium on Neural Networks and Applications (NEUREL), pp. 1–6 (November 2016)
2. Risojevic, V., Babic, Z.: Orientation difference descriptor for aerial image classification. In: International Conference on Systems, Signals and Image Processing (IWSSIP), pp. 150–153 (April 2012)
3. Haralick, R.M., Shanmugam, K.S., Dinstein, I.: Textural features for image classification. IEEE Trans. Syst. Man Cybern. 3(6), 610–621 (1973)
4. Cleve, C., Kelly, M., Kearns, F.R., Moritz, M.: Classification of the wildland-urban interface: a comparison of pixel-and object-based classifications using high-resolution aerial photography. Comput. Environ. Urban Syst. 32(4), 317–326 (2008)
5. Basu, S., Ganguly, S., Mukhopadhyay, S., DiBiano, R., Karki, M., Nemani, R.R.: Deepsat: a learning framework for satellite imagery. In: Proceedings of the 23rd SIGSPATIAL International Conference on Advances in Geographic Information Systems, Bellevue, WA, USA November 3–6, 2015 pp. 37:1–37:10 (2015)
6. Zhang, Y.: Optimisation of building detection in satellite images by combining multispectral classification and texture filtering. ISPRS J. Photogrammetry Remote Sens. 54(1), 50–60 (1999)
7. Lu, D., Weng, Q.: A survey of image classification methods and techniques for improving classification performance. Int. J. Remote Sens. 28(5), 823–870 (2007)
8. Hester, D.B., Cakir, H.I., Nelson, S.A., Khorram, S.: Per-pixel classification of high spatial resolution satellite imagery for urban land-cover mapping. Photogrammetric Eng. Remote Sens. 74(4), 463–471 (2008)
9. Risojevic, V.: Aerial image classification : Database of RGB+NIR aerial images (2017). http://dsp.etfbl.net/aerial
10. Tamura, H., Mori, S., Yamawaki, T.: Textural features corresponding to visual perception. IEEE Trans. Syst. Man Cybern. 8(6), 460–473 (1978)
11. Bharati, M.H., Liu, J., MacGregor, J.F.: Image texture analysis: methods and comparisons. Chemometr. Intell. Lab. Syst. 72(1), 57–71 (2004)
12. Kusetogullari, H., Sharif, M.H., Leeson, M.S., Celik, T.: A reduced uncertainty-based hybrid evolutionary algorithm for solving dynamic shortest-path routing problem. J. Circuits Syst. Comput. 24(5) (2015)
13. Sharif, M.H.: An eigenvalue approach to detect flows and events in crowd videos. J. Circuits Syst. Comput. 26(7), 1–50 (2017)
14. Sharif, M.H.: A numerical approach for tracking unknown number of individual targets in videos. Digit. Signal Proc. 57, 106–127 (2016)
15. Sharif, M.H.: High-performance mathematical functions for single-core architectures. J. Circuits Syst. Comput. 23(4) (2014)

Automated Detection of Glaucoma Using Image Processing Techniques

Mishkin Khunger, Tanupriya Choudhury, Suresh Chandra Satapathy
and Kuo-Chang Ting

Abstract Glaucoma is one of the most dreaded eye diseases and is a chronically progressive and ischaemic optical neuropathy leading to deterioration of vision generally caused due to increased pressure caused by increasing aqueous humour inside the eye. This is caused either due to reduced drainage or sometimes due to increased secretion. It causes damage of ischaemic to the optic nerve which results in nerve fibre layer damage and permanent loss of vision. Two kinds of primary Glaucoma are there, namely wide-angle glaucoma and narrow-angle glaucoma which have diverse mechanism of lessening watery surge and are in charge of increment in intraocular pressure. In the beginning of glaucoma, no detectable side effects show up. As the ailment advances, vision exacerbates and harm to visual field occurs. If undetected and untreated, it results in complete vision loss. Manual investigation of ophthalmic images is tedious, and accuracy relies upon the skill of the experts. Programmed examination of retinal pictures is turning out to be of great importance these days. It helps in detecting, diagnosing and anticipating of dangers related with glaucoma. Fundus pictures acquired from fundus camera have been utilized for the investigation. The systems specified in the present survey have certain positive and negative points. In view of this investigation, one can undoubtedly figure out which strategy gives the ideal outcome.

M. Khunger (✉)
Amity University, Noida, India
e-mail: mishkink@yahoo.com

T. Choudhury
UPES, Dehradun, India
e-mail: tanupriya1986@gmail.com

S. C. Satapathy
Department of CSE, PVP Siddhartha Institute of Technology, Vijayawada, Andhra Pradesh, India
e-mail: sureshsatapathy@gmail.com

K.-C. Ting
Minghsin University of Science and Technology, Xinfeng Hsinchu 30401, Taiwan
e-mail: ting@must.edu.tw

© Springer Nature Singapore Pte Ltd. 2019
A. Abraham et al. (eds.), *Emerging Technologies in Data Mining and Information Security*, Advances in Intelligent Systems and Computing 814,
https://doi.org/10.1007/978-981-13-1501-5_28

1 Introduction

The disease here which is glaucoma is a structural and functional disorder in eyes which damages optic nerve of the affected patient and causes deterioration of vision over time. It is an ailment that slaughters retinal ganglion cells. It is basically connected with the accumulation of high-pressure inside the eyeball. Glaucoma has a tendency to be acquired and may not appear till late in life. High measure of intraocular pressure (IOP) is one of the significant threats in this ailment. Present-day medicament focuses on decreasing IOP inside eyes to avert basic harm. The increased IOP can cause harm to the optic nerve, which is in charge of transmitting pictures to the cerebrum. Without proper and timely treatment, glaucoma can cause irreversible visual impairment within a couple of years.

The two primary kinds of glaucoma are as follows.

1.1 Wide-Angle Glaucoma

The name wide-angled glaucoma originates from the fact that in such type of glaucoma there is an open and wide edge between the cornea and the iris. Ninety percentage of the glaucoma-affected patients are wide angled. This occurs due to the obstructing of the drainage trenches in the eye. It is more commonly known by the name open-angle glaucoma.

1.2 Narrow-Angle Glaucoma

Additionally called angle closure glaucoma, this sort of glaucoma is less common in the West than in Asia. Poor drainage is caused on the grounds that the edge between the iris and the cornea is excessively thin and is physically hindered by the iris. This condition prompts a sudden development of IOP in the eye.

Valuation of retinal nerve fibre layer (RNFL), IOP and visual field examination is vital for the diagnosis. An assortment of different potential outcomes conceding mechanical and vessel structures has been used for neurotic procedure of glaucoma disease. It is one of the main sources of visual deficiency in individuals beyond 40 years old. Loss of fringe vision is the primary indication. If not treated on time, the field of vision will keep on deteriorating. In case of early diagnosis, further vision loss can be stopped. Enhancement of strategies for screening and treatment is necessary due to existing high costs. The clinical evaluation of optic nerve head (ONH) is carried out using the ISNT rule. It is suggested to observe the ONH on the inferior side as it shows the degree of damage done to the optic nerve.

Optic plate cup is a tough task because of the intensity variations within the optic disc and its vicinity. The optic disc is the shining portion in the normal images of fundus. A vital feature of interest in the optic cup segmentation is that the vessel curves at the boundary of the cup. Besides CDR, the NRR's thickness is equally important structural attribute for distinguishing between a glaucomatous and a non-glaucomatous eye. The inferior, superior, nasal, temporal (ISNT) rule should be followed by a normal eye. The temporal rim is the thinnest one. The inferior rim thickness is greater than that of the superior one in a healthy eye, which in turn is greater than that of the nasal rim. In a glaucomatous eye, ISNT rule will not be an entire assessment. OCT is used in the structural changes' detection in the layers of retina which is due to the glaucoma progression. Because of the glaucoma progression, the layers of retina, RNFL and macular layer (ML) thicknesses minimize because of the destruction of ganglion cells. Thus, RNFL and ML thicknesses are the main structural properties for distinguishing between a glaucomatous and a non-glaucomatous eye.

2 Automated Glaucoma Detection

This paper proposes two techniques of image processing for automated detection of glaucoma [1]. The retinal image used is the fundus image obtained by funduscopy. Preprocessing of retinal picture prompts change of picture information by smoothening undesirable bends and by upgrading some picture highlights for additional handling. The feature extraction from the image of retina integrates the algorithms, thus designed and then evolved in order to identify different desired features (shapes) from the images of the retina. The image of the Retina which is classified to analyze the different image characteristics that are extracted and then arranges the data into the categories as follows: Glaucoma likely or Glaucoma unlikely.

2.1 Automated Glaucoma Detection Using Histogram Features

This technique is suggested for the early recognition of the glaucoma disease with the aid of two highlights to be specific phase and magnitude highlights from the fundus pictures. Local binary patterns (LBPs) are used for extracting the features. Daugman's algorithm is used to carry out the calculations of the magnitude part. The processing of histogram highlights for the phase, and magnitude parts are carried out. The proposed framework comes about 96% accurate for sensitivity and specificity.

It incorporates selection of region of interest (ROI), Gabor separation, LBP ventures to remove the highlights and Daugman's calculation. Every subject is portrayed in the further areas in detail. Figure 1 illustrates the methodology of glaucoma detection using histogram features.

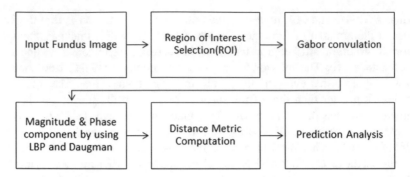

Fig. 1 Methodology of automated detection using histogram features

2.1.1 Selection of Points of Interest (POI)

POI is a chosen region inside the picture distinguished specifically for investigation. The strategy here is that the optic plate is distinguished by searching the POI of the specified glaucoma picture. It should be kept in mind that the end aims to gauge the POI, and the scientific morphological methods like enlargement and disintegration are done. The minor gaps in the picture get occupied smoothly after the execution of the morphological capacities.

The physical boundary for extricating the optic plate incorporates the evacuation of the veins in the retinal pictures. The widening and disintegration of A and B are characterized as

$$A \ominus B = \{ k \in S, B_k \subseteq A \} \tag{1}$$

Disintegration is performed to differentiate the boundaries of the picture. The output is a smooth picture with no veins. As a rule, a picture pixel is assigned value 1 and the surrounding pixel is assigned value 0. The minimum pixel is used for the extraction of optic plate.

2.1.2 Gabor Convolution

Gabor convolution is a linear filtering technique for image processing. It is used for analysing whether there is greater frequency of a particular component near the POI. Its main application is extracting the edges of an optical plane. The 2D Gabor filter has a real part and an imaginary part. Extraction can be carried out using the following equation

$$f(m, n, w, \alpha, \sigma_1, \sigma_2) = 1/2 \quad \pi\sigma_1\sigma_2\left\{-\frac{1}{2}\left[\left(\frac{m}{\sigma^1}\right)^2 + \left(\frac{n}{\sigma^2}\right)^2\right]\right| + jw(m\cos\alpha + n\sin\alpha)\right\} \tag{2}$$

where σ denotes diffusion and ω denotes the accuracy. The extraction of the real and imaginary parts is discussed below.

2.1.3 Local Binary Patterns (LBPs)

LBP is a kind of value that shows the spatial structure of an image. At a predefined pixel position (x_n, y_n), it is featured as a sequenced grouping of double correlations of pixels between the pixel which is set inside and the encompassing pixel. The computation of LBP denoted by L is carried out as follows:

$$L(x_n, y_n) = \sum_{e=0}^{7} b(k_e - k)2 \tag{3}$$

where k_e denotes grey scale of the centred pixels (x_p, y_p) and values of the eight corresponding pixels; value of $c(x)$ is given by:

$$c(x) = \begin{cases} 1 & \text{when } x > 0 \\ 0 & \text{when } x < 0 \end{cases} \tag{4}$$

2.1.4 Daugman's Calculation

Daugman's calculation is carried out for iris recognition. Iris is the part of eye which is responsible for controlling the quantity of light entering into the retina by monitoring the pupil's size. The manipulation is continued by performing integration over Gaussian surface through the following equation

$$\max(g, a_o, b_o)\left|G_\sigma(g) \times \frac{\partial}{\partial r_{g,a_o,b_o}} \oint \frac{I(a, b)}{2\pi r} ds\right| \tag{5}$$

Here, $G_o(g)$ denotes Gaussian function and I denotes iris picture. If the value is less than the predecided threshold value, then it is considered to be iris pixel, and if the value is greater, then it denotes the surrounding pixels (Fig. 2).

True positive rate (TPR) = TP/P; True negative rate (TNP) = TN/N

where TP is true positive, P is the sum of no. of true positives and false negatives, TN is true negative and N is the summation of true negative and false positive. Table 1

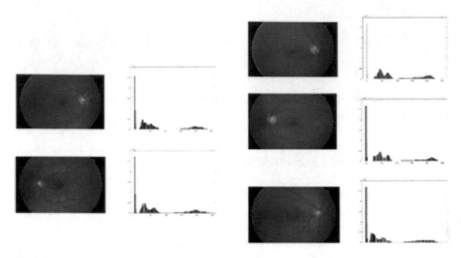

Fig. 2 Histogram outputs

Table 1 TPR calculations

Parameters	Glaucoma affected	No glaucoma	Total
Positive test	True positive: 24	False positive: 1	25
Negative test	False negative: 1	True negative: 24	25
Total	25	25	
	Sensitivity = (24/24 + 1)*100 = 96	Specificity = (24/1 + 24) * 100 = 96	

illustrates the calculation of TPR using histogram features on a data set of 50 eye fundus images.

2.2 Automated Glaucoma Detection Using Structural Features and Non-structural Features

Algorithm takes a fundus picture using fundus camera as the input and extracts optic plate and disc for the evaluation of CDR. For the testing of the classifier, intensity features and textural features are extracted. Result from both is combined to classify the image as high suspect, traits of glaucoma or no glaucoma (Fig. 3).

2.2.1 Detection of Optic Cup

Optic cup is the focal vibrant yellowish round structure present inside the plate. The shading force of optic cup and disc is indistinguishable; thus, it is hard to identify

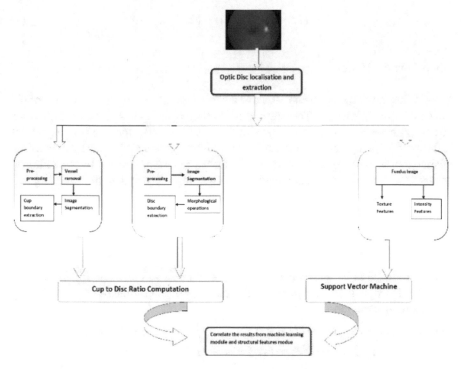

Fig. 3 Methodology

optic cup. Additionally, cup area contains the veins, as all the veins begin from the second pair of cranial nerve, which is situated on the focal point of the optic cup. The veins conceal the cup area, and only a very minute part is visible.

Preliminary Processing

Preliminary processing of the image is done to make a significant contrast in the intensities of cup and disc part. Mapping of the intensity value is done on a one-dimensional scale. One-dimensional mapping requires a greatest and a least value to be mapped on the scale. These operations are called point operations and are given by the following equation

$$= (a - a_2)b_1 - b_2a_1 - a_2 + b_2b = (a - a_2)b_1 - b_2a_1 - a_2 + b_2 \qquad (6)$$

Point operations are applied separately on the three coloured regions of the RGB plane. Inverted values are used to obtain negatives of the pictures. Negative can be obtained as follows

$$Y = (2L - 1) = (xy) = (2L - 1) - x \tag{7}$$

Abolishing the Vessel

Many veins cover the optic cup. To have a clear view of the cup, the intensity of the image of veins is lowered. Therefore, dim levelled opening is carried out utilizing a circular non-level organizing component (S) of size 40*40. Opening is a mathematical morphology operation characterized by disintegration took after by expansion. Disintegration will expel the undesirable items, i.e. veins, trailed by widening to ensure that alternate articles are unaffected. Opening will expel all the veins impression bringing about an even and regular container locale.

Extraction of the White Depressed Area Inside the Optic Nerve Head

Image produced in the precedent step is divided into sections using intensity mean measure (IMM) using a seed point. Average for the section S can be computed using the following equation

$$\mu r, q = 1(A \times B) \sum q = -B/2q = B/2 \sum r = -A/2r = A/2n(r, q)\mu r, q = 1(A \times B)$$
$$\sum q = -B/2q = B/2 \sum r = -A/2r = A/2n(r, q) \tag{8}$$

Segmentation

Opthalmic plate is the pale round section situated at the position where opthalmic nerve head goes away from the eye. Opthalmic nerve is in charge of correspondence amongst the eye and cerebrum, and therefore the opthalmic circle area is a delicate district which is also known as blind side because the poles and cones are being absent in this locale. Opthalmic plate discovery denotes a key and testing venture in numerous visual mechanized malady symptomatic frameworks in the light of the fact that numerous ophthalmic sicknesses are caused because of the basic changes related to optic circle area and optic circle locale is hindered by many veins. A calculation is proposed in this segment for identification of optic circle.

2.3 Feature Extraction

Characterized texture is represented as a particular course of action of forces inside a picture. Texture highlights can be analysed as measurable texture highlights and auxiliary texture highlights. In factual highlights. Factual highlights can be initial request, subsequent request or requests relying on quantity of pixels engaged with computations. LBP highlights are measurable highlights. Inside basic highlights,

Fig. 4 Features

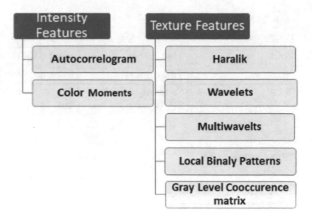

course of action power esteems is registered. Wavelets, multi-wavelets are auxiliary textural highlights. The grouping of highlights is appeared in Fig. 4.

2.3.1 Magnitude-Based Attributes

Highlights based on magnitude are utilized to break down the interrelationship of magnitudes. Shading or magnitude-based highlights are the minimal magnitude highlights. In pictures obtained by funduscopy, as glaucoma spreads it results in enlarged container estimate, bringing about a difference in intensity data of picture because of the way that glass district is grey or certain coloured. Magnitude-dependent highlights may likewise utilize for separating the disease affected and sound fundus pictures. Shaded and monochrome intensity filled highlights are excluded from the fundus picture in order to investigate the intensity connection in between the picture's pixels and to form a group as solid.

Colour Distribution

1. These are utilized to differentiate the picture on the grounds of magnitude. These represent a computed value of magnitudes amongst. In this procedure, the contingency of a particular shade inside a picture may be summarized as a functional discrete variable whose integral over any of the interval is the probability that the variate mentioned by it will stay within that interval. The three colour shades being utilized for examining shade resemblance of a picture are average and SD. Coloured picture can disintegrate three different shades of RGB.

Table 2 Sensitivity calculations

Parameters	Glaucoma affected	No glaucoma	Total
Positive test	True positive: 20	False positive: 20	25
Negative	True negative: 5	False negative: 20	25
Total	Sensitivity = (20/20 + 5)*100 = 80	Specificity = (20/5 + 20) = 80	50

Auto Correlogram

Correlograms are a decent portrayal of colour data of a picture utilizing 3-D maps. Autocorrelogram is an adjustment of colour correlograms which is responsible for catching data and is a connection between magnitudes with indistinguishable force esteems. Dissimilar to nearby and worldwide parameters of a picture, correlograms experience worldwide impacts of the dispersion. These are only substantial look changes. Along these lines, these possess great colour highlights for force-dependent characterization and segregation. Texture-dependent highlights expressed earlier may be extricated through pictures for figuring out the component value. Detached and crossover highlights (blend of surface and power highlights) are utilized to prepare and test utilizing double folds. Fifty pictures are utilized for preparing and rest 50 are utilized to testify the other way around. Highlight choice is finished utilizing PCA for decreasing measurement of highlight vector by choosing the critical element for arrangement. SVM with straight bit is utilized for preparing. SVM searches for the best plane isolating the subclasses with the most extreme separation (Fig. 5).

Table 2 illustrates calculations for detection of glaucoma by the use of structural and non-structural characteristics on a data set of 50 eye images.

3 Comparison Table

By comparing the automated detection using histogram features (ADHF) and automated detection using structural and non-structural features (ADSN), we infer that using a fundus image which we obtained using a fundus camera we can determine whether the patient is suffering from glaucoma or not. These detections are completely independent of the ophthalmologist's instinct as no analysis is required. ADHF involves Gabor convolution in which the magnitude and the phase components are extracted by the use of LBP and Dougman's algorithm while ADSN involves three types of extractions—cup extraction, disc extraction and feature extraction; cup and disc extractions are used for the evaluation of CDR. The results computed using MATLAB on 25 glaucoma-affected fundus images show that ADHF provides 96% accuracy while ADSN provides only 80% accuracy which make ADHF more suitable for use (Table 3).

Table 3 Comparison of the two techniques

Automated detection of glaucoma using histogram features	Automated detection of glaucoma using structural and non-structural features
• The process used here is Gabor convolution	• The process used here is extraction
• Gabor convolution is divided into two parts (a) Magnitude component(LBP) (b) Phase component (Daugman's calculation)	• Extraction is divided into three parts (a) Cup extraction (b) Disc extraction (c) Feature extraction
• Accuracy is 96% (results obtained in Table 1)	• Accuracy is 80% (results obtained in Table 2)
• Histogram pattern is 	• Histogram pattern is
• Due to higher efficiency and easy implementation, this technique is considered to be better	• This technique has easy implementation, but its accuracy is lesser as compared to automated detection using histogram features

Fig. 5 Output using
structural and non-structural
features

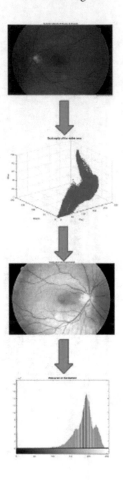

4 Conclusion

There is an irreversible damage done by glaucoma. The only solution is prior detection and glaucoma treatment as soon as possible. The two techniques discussed in this paper have easy implementation, but due to better efficiency automated detection using histogram features is considered to be better.

References

1. Glaucoma Research Foundation, Glaucoma Research Foundation (2013). [Online]. Available: http://www.glaucoma.org/glaucoma/typesofglaucoma.php
2. Quigley, H.A., Broman, A.T.: The number of people with glaucoma worldwide in 2010 and 2020. Brit. J. Ophthalmol. 90(3), 262 267 (2006)
3. Kumar, B., Naveen, R.P., Chauhan, Dahiya, N.: Detection of Gaucoma using image processing techniques; A review. In: 2016 International Conference on Microelectronics Computing and Communications (MicroCom) 2016
4. Salam, A.A., Khalil, T., Akram, M.U., Jameel, A., Basit, I.: Automated Detection of Glaucoma Using Structural and non Structural Features. Spingerplus
5. Liu, Y.Y., Chen, M., Ishikawa, H., Wollstein, G., Schuman, J.S., Rehg, J.M.: Automated macular pathology diagnosis in retinal OCT images using multi-scale spatial pyramid and local binary patterns in texture and shape encoding. Medical Image Analysis (2011)
6. Sun, X., Wang, J., Chen, R., Kong, L., She, M.F.: Directional Gaussian filter-based LBP descriptor for textural image classification. Procedia Eng.

DWT and DCT-Based Compressed Image Transmission Over AWGN Using C-QAM

Md. Khaliluzzaman, Shahela Pervin, Umme Moon Ima
and Deepak Kumar Chy

Abstract Image transmission in noisy channel demands integrity of image, less transmission time, and faster processing. These can be achieved when the transmitted image is compressed. Nowadays, various types of compression methods are used for image compression. In this paper, DCT and DWT are applied for compression of image. Performance analysis of QAM for DCT- and DWT-based compressed image transmissions over AWGN channel is shown in this work. In addition, for better transmission efficiency and achieving the required bandwidth 32 cross-constellations QAM (C-QAM) modulation scheme is used. Furthermore, for measuring performance of image quality, PSNR and MSE values are taken into account. The simulation results reveal that with lower signal-to-noise ratio, raised cosine filter exhibits better performance than employing no filter. Inter-symbol interference (ISI), which plays an important role in image transmission, degrades the image quality, which may be eliminated by employing filter.

Md. Khaliluzzaman (✉) · S. Pervin · U. M. Ima
Department of Computer Science and Engineering, International Islamic University
Chittagong (IIUC), Chittagong 4318, Bangladesh
e-mail: khalilcse021@gmail.com

S. Pervin
e-mail: shahelapervinrumpa@gmail.com

U. M. Ima
e-mail: ummemoonima@gmail.com

D. K. Chy
Department of Electrical and Electronics Engineering, Port City International University,
Chittagong, Bangladesh
e-mail: dk_chy53@yahoo.de

© Springer Nature Singapore Pte Ltd. 2019 337
A. Abraham et al. (eds.), *Emerging Technologies in Data Mining and Information
Security*, Advances in Intelligent Systems and Computing 814,
https://doi.org/10.1007/978-981-13-1501-5_29

1   Introduction

In multimedia application, the image data is described to be broadcasted. In the modern communication system, channel bandwidth is a vital issue for smooth data transmission. For efficient and lossless transmission, the image is compressed before transmitting through the channel. This method is also effective to the perfect utilization of the channel bandwidth. The compression is performed in such a way that the image quality is maintained in an unacceptable limit. The transmitted image is compressed to maintain the storage memory channel and transmitting time. Discrete cosine transform (DCT) and discrete wavelet transform (DWT) are most widely methods for the image compression in the last few decades. Both DCT and DWT have the capacity to show the image at various explorations. In addition, ISI degrades the image quality over a wireless channel and/or wired channel due to the band-limited characteristic of the transmitted signal. An appropriate filter technique may improve the image quality that is transmitted through the noisy channel.

This section explains some previous works on image compression and transmission and their limitations, such as [1] demonstrated the minimum MSE filter based on DCT coefficients of the statistical model. A compressed method based on the DCT is described in [2]. This method obtained the better performance for PSNR. Khaliluzzaman et al. [3] proposed raise cosine filter and C-QAM for high-quality image transmission. However, the proposed method with filter showed low PSNR values for higher E_b/N_0.

On the contrary, [4] suggested hierarchical quadrature amplitude modulation (HQAM) [5] for better protection of high-priority data during image transmission. Nevertheless, no channel model such as AWGN is considered in this work.

Here, only salt and pepper noise is taken into account. A raised cosine filter is used in the AWGN channel introduced in [6]. Here, authors evaluate the performance of the communication channel through the bit error rate (BER). In [7], M array QAM is used in AWGN channel for evaluating the polar coding technique. In this method, the characteristic of the band-limited and channels' inter-symbol interference are not considered.

In this paper, a raised cosine filter is proposed before transmission of compressed image through the noisy communication channel. The typical communication model is tested by transmitting the still image using cross-constellation 32-QAM modulation schemes. DCT and DWT compression methods are compared for measuring the performance of QAM. The simulation results proved that the proposed communication model for image transmission yields high-quality image when the channel is more vulnerable to noise and also proved that the DCT gives better results for QAM than DWT.

The rest of the paper is organized as follows. In Sect. 2, theoretical background of DCT and DWT compression, encoding, 32 C-QAM modulation, raised cosine filter, additive white Gaussian noise (AWGN) channel, bit error rate (BER), and signal-to-noise ratio (SNR) have been demonstrated. In Sect. 3, the proposed method is described and the experimental results are explained.

2   Theoretical Background

2.1   Discrete Cosine Transform (DCT) and Discrete Wavelet Transform (DWT)

DCT which is similar to the discrete Fourier transform, however, only uses the real numbers. In DCT, the complete image demonstrates as a block such as $I = f(u, v)$, where u and v are the range of $\{0 < u \leq M - 1\}$ and $\{0 < v \leq N - 1\}$. The image dimension is $M \times N$, and $f(u, v)$ is the gray-level value at the pixel of (u, v). The DCT coefficients are defined by (1) [8].

$$f(i, j) = \frac{2c(i)c(j)}{\sqrt{MN}} \sum_{u=0}^{M-1} \sum_{v=0}^{N-1} f(u, v) \cos\left[\frac{(2u + 1)i\pi}{2M}\right] \cos\left[\frac{(2v + 1)j\pi}{2N}\right] \quad (1)$$

where $c(:)$ is defined as:

$$c(i) = \begin{cases} \frac{1}{\sqrt{2}}, & \text{if } i = 0 \\ 1, & \text{if } i = 0 \end{cases}$$

Respectively, the inverse DCT coefficients are defined by (2).

$$f(u, v) = \sum_{u=0}^{M-1} \sum_{v=0}^{N-1} \frac{2c(i)c(j)}{\sqrt{MN}} f(i, j) \cos\left[\frac{(2u + 1)i\pi}{2M}\right] \cos\left[\frac{(2v + 1)j\pi}{2N}\right] \quad (2)$$

The square image is considered in this section, i.e., $(M = N)$.

In the image processing, the discrete wavelet transform sub-banded the input images from where the approximate values of the image are extracted. It is analyzed both in the numerical and function for wavelets sampling discretely. By this technique, the image pixel is transformed into wavelets. These pixels are then utilized for coding and wavelet-based compression. The definition of DWT is shown in (3) and (4).

$$W_\varphi(j_0, k) = \frac{1}{\sqrt{M}} \sum_z f(x)\varphi_{j_0,k}(x) \quad (3)$$

DWT and DCT Based Compressed Image Transmission over AWGN using C-QAM

$$W_\psi(j, k) = \frac{1}{\sqrt{M}} \sum_k f(x)\psi_{j,k}(x) \quad (4)$$

For $j \geq j_0$ and the inverse DWT (IDWT) is defined in (5).

$$f(x) = \frac{1}{\sqrt{M}} \sum_k W_\varphi(j_0, k)\varphi_{j_0,k}(x) + \sum_{j=j_0}^{\infty} \frac{1}{\sqrt{M}} \sum_k W_\psi(j, k)\psi_{j,k}(x) \qquad (5)$$

where $f(x)$ is the function of the discrete variable $x = 0, 1, 2, \ldots, M-1$. Normally, let $j_0 = 0$ and select M to be a power of 2 (i.e., $M = 2^J$) so that the summations in Eqs. (3), (4) and (5) are performed over $x = 0, 1, 2, \ldots, M-1, j = 0, 1, 2, \ldots, J-1$, and $k = 0, 1, 2, \ldots, 2^j - 1$.

2.2 Quadrature Amplitude Modulation (QAM)

M-QAM is defined as the cross-shape constellation. This constellation for the n bit number is defined as $Q = 2^n$. The constellation shape has a cross, so why it is called M cross constellation QAM, here the number of bits is n which is the size of the one symbol [9]. The rate of the symbol is T, which is also called the baud rate. In the transmission, the signal bandwidth relays on the T. The QAM of M cross-constellation carries $m = \log 2(M)$ bit information. So, per symbol M-QAM cross-constellation carries m bits; i.e., for $M = 32$, it is 5 bits. For M-QAM cross-constellation, the average energy is calculated that is repeated translation of square constellation of M-QAM. Here, the corner points are deleted as these points consumed more power compared to the rest of the points in M-QAM. The energy is not changed if the M-QAM is rotated. So, the total energy of M-QAM is same as it is firstly estimated and the four corner points' energy is evaluated. At last, M-QAM energy is subtracted from four corner points' energy. The final estimated energy of M-QAM is divided by M. This provides the average energy of M cross-constellation (Fig. 1).

Fig. 1 Square 16 QAM expanded to 32 cross-constellation QAM ($n = 5$)

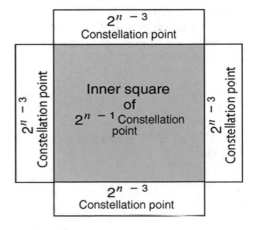

2.3 Raised Cosign Filter, AWGN Channel, and Signal-to-Noise Ratio

A pair of raised cosine filter is used in both transmitter and receiver sites, which act as square root-raised cosine filter. In practice, ISI cannot be zero when both filters are implemented due to some numerical decision error in both the design phase and implementation phase.

For ensuring no inter-symbol interference, the following conditions should meet when a pulse-shaping filter is employed in [6].

1. The pulse shape shows a zero crossing at the sampling point of all pulse intervals other than its own, that is minimized ISI.
2. The shape of the pulses is such that the magnitude decays readily outside of the pulse interval resulting in high stop band attenuation.

In practice, according to Shannon's capacity theorem, every channel is corrupted by noise. All transmitted signals are influenced by noise and are unpredictable in nature. Additive white Gaussian noise (AWGN) is common to every communication channel, which is the statistically random radio noise characterized by a wide frequency range with regard to a signal in the communication channel [10].

Measuring signal power in comparison with noise power is known as signal-to-noise ratio. Considering AWGN channel model for the complex and unpredictable signal as well as channels, signal-to-noise ratio (SNR) can be estimated as (6):

$$SNR = (Signal\,Power/Noise\,Power) \qquad (6)$$

3 Proposed Method

This section describes the proposed method in detail. The basic steps involved in the proposed method are: (1) converting RGB image into grayscale image, (2) applying DCT and DWT in grayscale image to be compressed and quantized, (3) encoding the quantized image and finally finding the compressed image, (4) utilizing modulation technique on encoded values to obtain the symbols, (5) before transmitting symbol value through AWGN channel, filtering the symbol data values with raised cosine filter and applying the inverse fast Fourier transform to convert the value from frequency domain to time domain, (6) passing the noisy values through the inverse raised cosine filter at the receiver site and then converting to time domain and frequency domain by applying Fourier transform, (7) demodulating these transformed values with 32-QAM, (8) decoding the demodulated values accordingly, (9) applying the IDCT and IDWT to get the original grayscale values, and (10) retrieving the original image. The workflow of the proposed framework is shown in Fig. 2.

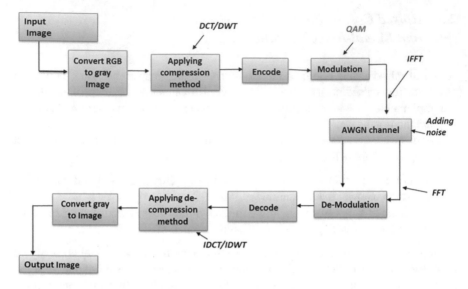

Fig. 2 Workflow of the proposed framework

3.1 RGB Image into Grayscale Image

In this section, initially the input RGB image is converted into grayscale image, which contains only one channel value, that is, intensity. The RGB image is converted to grayscale image by (7).

$$\text{Image} = (\text{red} * 0.299) + (\text{green} * 0.587) + (\text{blue} * 0.114) \tag{7}$$

3.2 Applying DCT and DWT to Compress Grayscale Image

In this work, the image that will be transmitted through the channel is compressed using the discrete cosine transform (DCT) and discrete wavelet transform (DWT).

DCT is the efficient method for image compression which is utilized in the algorithm of the JPEG image compression. In the DCT input, given image is blocked into 8×8 or 16×16. From each block, the 2D DCT coefficients are computed. After that, the quantized values are extracted from the DCT coefficients. In the quantized value, the zigzag method is applied to extract the nonzero quantized values. These quantized values are encoded in the next section to get the binary values. The source image is digitized through the DWT to get a signal. The signal is the string of the numbers.

3.3 Encode

In this section, the compressed quantization values are encoded to convert into binary code streams. Huffman and arithmetic encoders are most commonly used entropy encoders. For fast execution of the application, in this work, simple run-length encoding (RLE) method is used. This method is very effective for fast execution.

3.4 Applying Root-Raised Cosine Filter

This section uses the root-raised cosine filter to eliminate inter-symbol interference, which degrades the image quality severely at low signal-to-noise ratio. Single raised cosine filter at transmitter side cannot reduce the inter-symbol interference. Therefore, a pair of raised cosine filters is used in this paper, which acts as root-raised cosine filter. In addition, the original image block is in the frequency domain, but raised cosine filter and AWGN model follow real-time operation. For this reason, in this proposed method inverse fast Fourier transform (IFFT) is introduced before using raised cosine filter.

3.5 M-QAM Modulation Scheme

In this work, $M = 32$, i.e., 32-QAM is used, which has five I and Q values. As a result, the signal will have 32 possible states. As $M = 32$ is used, here, each symbol contains five bit. So, $32 = 2^5$, five bits per symbol can be sent at a time.

3.6 Additive White Gaussian Noise (AWGN)

AWGN channel embeds white noise to the signal that has been passed through it. All transmitted signals are corrupted by noise, and noise is unpredictable in nature. Precise mathematical operations cannot be done on noise. It represents adequately many real-world channels. And (fortunately) its treatment is straightforward. Hence, AWGN channel is chosen as channel model for simulation.

3.7 Retrieving Original Image

In this section, retrieving the original image at the receiver side is demonstrated. First, the noisy image data is passed through the inverse raised cosine filter to obtain

the modulated image data in time domain. Then, demodulation process is applied
followed by fast Fourier transform to obtain the encoded data in frequency domain.
After this, the decoding technique is implemented to obtain quantized image value.
Furthermore, inverse discrete cosine transform (IDCT) and inverse discrete wavelet
transform (IDWT) operations are performed to obtain the grayscale value. Finally, the
original RGB image is retrieved from the grayscale image by using the appropriate
method.

4   Simulation and Experimental Results

In this paper, simulation is carried out for grayscale images. The simulation was
performed using MATLAB software. The mean square error (MSE) and peak signal-
to-noise ratio (PSNR) are typically used to measure the quality of the receiving image
with respect to the transmitting image. The MSE and PSNR are measured using (8)
and (9) respectively.

$$\text{MSE} = \frac{1}{MN} \sum_{i=0}^{M-1} \sum_{j=0}^{N-1} \|I(i, j) - R(i, j)\|^2 \tag{8}$$

$$\text{PSNR} = 10 \log_{10} \frac{\|I(i, j) - R(i, j)\|^2}{\text{MSE}} \tag{9}$$

From the experimental results, it is seen that the received image quality improved
with the increase of E_b/N_0 (dB) values. Below figures show the processing example
of retrieving images with E_b/N_0 (dB) effect. Two images are taken into consideration
for the experiment, i.e., Lena and cameraman.

Figure 3a, b shows the original, gray, and compressed images for DCT for the
Lena and cameraman images, respectively. Figure 4a, b shows the DWT compressing
both for the Lean and cameraman images.

The processing examples for output image that passed the AWGN channel without
using raised cosine filter for 1 to 10 E_b/N_0 are shown in Fig. 5, where Fig. 4a shows
the Lena image and Fig. 4b shows the cameraman image. Same processing example
for the DWT is shown in Fig. 5.

The MSE and PSNR values with respect to E_b/N_0 (dB) with channel noise without
filter are shown in Table 1 for DCT and DWT.

The MSE and PSNR values are almost constant which is 37.1239 and 2.6376
when E_b/N_0 is six and more. On the contrary, DWT gives at clear images when
E_b/N_0 is nine. And the MSE and PSNR values are 3.1005 and 2.2007, respectively.
The value for DCT is larger than DWT at low SNR value. From both results, it is
clearly noticed that DCT gives clear images for 32-QAM.

Figures 6a, b and 7a, b show the processing example of retrieving images for Lena
and cameraman image with 1–10 E_b/N_0 (dB) effect when raised cosine filter is used.
From the experimental results, it is seen that the received image quality has improved

(a)   (b)

(c)   (d)

Fig. 3 Processing example of quantized image: original image, gray image, and compressed image using DCT and DWT: **a** Lena with DCT, **b** cameraman with DCT, **c** Lena with DWT, **d** cameraman with DWT

(a)

(b)

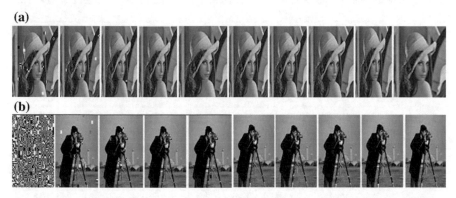

Fig. 4 Processing example of output images with different E_b/N_0 (dB) values without using raised cosine filter and using DCT: **a** Lena image and **b** cameraman image

with the increase of E_b/N_0 (dB) values. Images are noise-free where E_b/N_0 is two in case of the compression method of DCT.

The MSE and PSNR values are constant which is 13.8668 and 7.1364 when E_b/N_0 is 2 and further increased. However, for DWT, MSE and PSNR values are constant that is 2.2007 and 0.0441 and E_b/N_0 value is three and more. After using raised cosine filter, also DCT gives the better result. The PSNR value for DCT is larger with respect to DWT (Table 2).

It is seen that filter gives better results with respect to without filter. When the filter is not used, images are clear where E_b/N_0 is six for DCT and nine for DWT. On the contrary, after using filter images are clear at low SNR values that is where E_b/N_0 is two for DCT and three for DWT. Furthermore, the PSNR values are comparatively better after using filter. That is, the PSNR values are 2.6376, 0.0441 and 7.1364, 0.0441 without using filter and using filter, respectively.

(a)

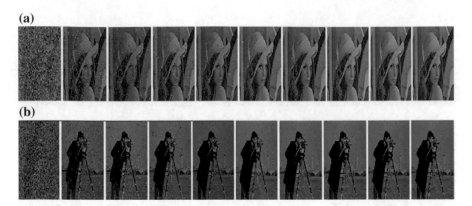

(b)

Fig. 5 Processing example of output images with different E_b/N_0 (dB) values without using raised cosine filter and using DWT: **a** Lena image, **b** cameraman image

Table 1 Experimental results of MSE and PSNR with channel noise without filter using DCT and DWT

E_b/N_0	DCT				DWT			
	Lena		Cameraman		Lena		Cameraman	
	MSE	PSNR	MSE	PSNR	MSE	PSNR	MSE	PSNR
1	NaN	NaN	NaN	NaN	5.51	0.02	9.92	0.01
2	NaN	NaN	Inf	0	4.49	0.02	6.33	0.02
3	Inf	0	Inf	0	5.15	0.02	5.05	0.02
4	Inf	0	Inf	0	3.62	0.03	5.59	0.02
5	974.90	0.10	31.59	3.10	3.74	0.03	6.95	0.01
6	37.12	2.63	31.59	3.10	2.60	0.04	4.77	0.02
7	37.12	2.63	31.59	3.10	2.87	0.03	4.77	0.02
8	37.12	2.63	31.58	3.10	2.20	0.04	4.75	0.02
9	37.12	2.63	31.58	3.10	2.20	0.04	4.75	0.02
10	37.12	2.63	31.59	3.10	2.20	0.04	4.75	0.02

Figures 8a, b and 9a, b show the original, gray, noise, and compressed images for DCT and DWT, respectively, for the two images, i.e., Lena and cameraman.

Figures 10a, b and 11a, b show the processing example of retrieving images with E_b/N_0 (dB) effect where 10% salt and pepper noise is added to the input image. In this case, DCT shows the better result than DWT. Noise eliminates in larger values of SNR that is the MSE and PSNR values are comparatively better than DWT (Table 3).

For DCT, the MSE and PSNR values are constant where E_b/N_0 is six and for DWT values are constant at nine. So, it can clearly reveal that the DCT compression method works better for QAM.

However, for external noise, there is no effect of the filter. That is, the filter does not eliminate external noise. The cause is that the compression is done after adding external noise. So, the noise is compressed and the filter has no effect in

(a)

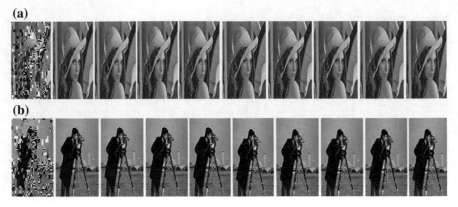

(b)

Fig. 6 Processing example of output images with different E_b/N_0 (dB) values using raised cosine filter and using DCT: **a** Lena image and **b** cameraman image

(a)

(b)

Fig. 7 Processing example of output images with different E_b/N_0 (dB) values using raised cosine filter and using DWT: **a** Lena image and **b** cameraman image

case of external noise. The processing example for external noise is shown only for comparison with channel noise.

In [7], PSNR value of 45 dB for $E_b/N_0 = 10$ dB is obtained. In addition, in [3] with filter shows a value of 4.1420 dB at the same value of E_b/N_0, whereas the proposed method with filter shows a value of 7.1364 for DCT and 0.0441 dB for DWT at the same value of E_b/N_0.

Table 2 Experimental result of MSE and PSNR using DCT and DWT with filter

E_b/N_0	DCT				DWT			
	Lena		Cameraman		Lena		Cameraman	
	MSE	PSNR	MSE	PSNR	MSE	PSNR	MSE	PSNR
1	NaN	NaN	NaN	NaN	5.58	0.017	8.13	0.012
2	13.87	7.14	37.12	2.64	2.20	0.04	4.75	0.02
3	13.87	7.14	37.12	2.64	2.20	0.04	4.75	0.02
4	13.87	7.14	37.12	2.64	2.20	0.04	4.75	0.02
5	13.87	7.14	37.12	2.64	2.20	0.04	4.75	0.02
6	13.87	7.14	37.12	2.64	2.20	0.04	4.75	0.02
7	13.87	7.14	37.12	2.64	2.20	0.04	4.75	0.02
8	13.87	7.14	37.12	2.64	2.20	0.04	4.75	0.02
9	13.87	7.14	37.12	2.64	2.20	0.04	4.75	0.02
10	13.87	7.14	37.12	2.64	2.20	0.04	4.75	0.02

Fig. 8 Processing example of quantized image: original image, gray image, 10% noisy image, and compressed image using DCT: **a** Lena image and **b** cameraman image

(a)

(b)

Fig. 9 Processing example of quantized image: original image, gray image, 10% noisy image, and compressed image using DWT: **a** Lena and **b** cameraman

(a)

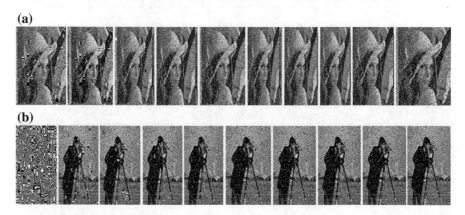

(b)

Fig. 10 Processing example of output images with different E_b/N_0 (dB) values using 10% external noise and DCT: **a** Lena and **b** cameraman image

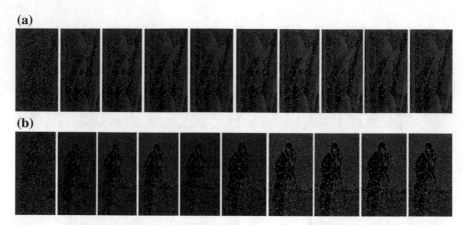

Fig. 11 Processing example of output images with different E_b/N_0 (dB) values using 10% external noise and DWT: **a** Lena image and **b** cameraman image

Table 3 Experimental results of MSE and PSNR with 10% external noise with DCT and DWT

E_b/N_0	DCT				DWT			
	Lena		Cameraman		Lena		Cameraman	
	MSE	PSNR	MSE	PSNR	MSE	PSNR	MSE	PSNR
1	NaN	NaN	NaN	NaN	8.31	0.01	1.14	0.008
2	NaN	NaN	NaN	NaN	8.24	0.01	1.08	0.01
3	NaN	NaN	NaN	NaN	8.208	0.01	0.99	0.01
4	Inf	0	1.67	0.0000	8.288	0.01	0.94	0.01
5	306.19	0.32	0.00	0.312	8.208	0.01	1.07	0.01
6	305.63	0.32	0.00	0.32	7.278	0.01	0.99	0.01
7	305.63	0.32	0.00	0.32	6.478	0.02	0.76	0.01
8	305.63	0.32	0.00	0.32	6.48	0.02	0.76	0.01
9	305.63	0.32	0.00	0.32	6.47	0.02	0.76	0.01
10	305.63	0.32	0.00	0.32	6.47	0.02	0.76	0.01

5 Conclusion

There are obvious advantages of employing the filter in the proposed method. For low SNR, when the compressed image is corrupted by noisy channel, it is hard to retrieve the original image. For that reason, for the lower value of E_b/N_0 the images are blurred. However, this problem is mitigated by introducing raised cosine filter in the proposed method. For higher SNR, the conventional communication model with 32-QAM works better. Furthermore, compression method DCT shows the better result than DWT.

References

1. Khan, S., Casseau, E., Menard, D.: High-performance discrete cosine transform operator using multimedia oriented subword parallelism. Ad. Comput. Eng. **2015**, 10 (2015)
2. Raid, A., Khedr, W., El-dosuky, M., Ahmed, W.: Jpeg image compression using discrete cosine transform-A survey. arXiv preprint arXiv:1405.6147, vol. 5, p. 47 (2014)
3. Khaliluzzaman, M., Chy, D.K., Deb, K.: Analyzing image transmission quality using filter and C-QAM. In: Computer, Communication and Electrical Technology: Proceedings of the International Conference on Advancement of Computer Communication and Electrical Technology (ACCET 2016), p. 131. CRC Press, Boca Raton, FL (2017)
4. Kader, M.A., Ghani, F., Badlishah, R.: Development and performance evaluation of hierarchical quadrature amplitude modulation (HQAM) for image transmission over wireless channels. In: 2011 3th International Conference on Computational Intelligence, Modelling and Simulation (CIMSiM), pp. 227–232 (2011)
5. Kader, M.A., Ghani, F., Ahmad, R.B.: Image transmission over noisy wireless channels using HQAM and median filter. In: The 8th International Conference on Robotic, Vision, Signal Processing & Power Applications, pp. 145–152. Springer, Singapore (2014)
6. Chy, D.K., Khaliluzzaman, M.: Comparative performance of BER in the simulation of digital communication systems using raised cosine filter. In: Third International Conference on Advances in Computing, Electronics and Electrical Technology-CEET, pp. 29–33 (2015)
7. Mishra, A., Sharma, K., De, A.: Quality image transmission through AWGN channel using polar codes. Int. J. Comput. Sci. Telecommun. **5**(1) (2014)
8. Telagarapu, P., Naveen, V.J., Prasanthi, A.L., Santhi, G.V.: Image compression using DCT and wavelet transformations. Int. J. Sign. Process., Image Process. Pattern Recogn. **4**, 61–74 (2011)
9. Hanzo, L., Webb, W., Keller, T.: Basic QAM Techniques in single and Multi Carrier Quadrature Amplitude Modulation: Principles and Applications for Personal Communications. Wiley, New York (2000)
10. Ghosh, A.: Comparative BER performance of M-ary QAM-OFDM system in AWGN & multipath fading channel. Int. J. Comput. Sci. Eng. (IJCSE), **4**(06) (2012)

Detection and Identification of Downy Mildew Diseased Leaf of Cucurbita Pepo (Pumpkin) Using Digital Image Processing

Prabira Kumar Sethy, Ansuman Bisoi, Gyana Ranjana Panigrahi and Santi Kumari Behera

Abstract Since India is a largest international producer of pumpkins, downy mildew disease of pumpkin diminishes the foremost fiscal progression toward the field of agriculture. Your after year from the very beginning to till the date, it has been noticed that Cucurbita pepo harvesting for our eatable purposes and remains indispensable harvest plant forever. Due to its edible demand, it is highly produced and available on a large scale for its commercial purposes. These plants have many diseases and infections but downy mildew is one of the most and common appearing disease among all diseases. This paper herewith proposed an original approach toward segmentation algorithm, i.e., K-means clustering for automatic detection and principal component analysis (PCA) for identifying the diseased leaves which help to indorse the early disease detection. The trial significances infer that the planned scheme might help to identify and distinguish the mentioned disease satisfactorily and successfully from the leaves of pumpkin.

Keywords K-means clustering · PCA

1 Introduction

Cucurbita pepo is one of the oldest domestic plants [1]. This species has a phenomenal diverse fiscal standing for its grown and undergrown raw green. It can be utilized in manifold area like vegetal, drug, and lasix. And moreover, it can be used as blood

P. K. Sethy (✉) · A. Bisoi · G. R. Panigrahi
Department of Electronics, Sambalpur University, Sambalpur 768019, Odisha, India
e-mail: prabirasethy@suniv.ac.in

A. Bisoi
e-mail: ansuman87637@gmail.com

G. R. Panigrahi
e-mail: gyan7420@gmail.com

S. K. Behera
Department of Computer Science and Engineering, VSSUT, Sambalpur 768018, Odisha, India
e-mail: b.santibehera@gmail.com

© Springer Nature Singapore Pte Ltd. 2019
A. Abraham et al. (eds.), *Emerging Technologies in Data Mining and Information Security*, Advances in Intelligent Systems and Computing 814,
https://doi.org/10.1007/978-981-13-1501-5_30

Fig. 1 Drone captured
image for identifying and
detecting downy mildew
diseased leaves of pumpkins

cleansing agent. Main remedial composition in ayurvedic drugs as for the cooling
agent, for dehydration and tiredness and provides relief from contraction of body
tissues and canals [2]. The green foliage of pumpkin can be taken as preventive
ingredient for painkiller and nausea and lift up the rate of hemoglobin contained in
the blood. Best medicine toward the treatment of chronic respiratory diseases and
pyrexia. In this technique, camera captures the leaves from the crops and analyses
the color of leaves then starts detecting for the infected leaves. Our aim is to design
a key for providing an authentic value for the farmers for reducing harvest input
contributions, enhancing crops, and increasing revenues. We provide an agricultural
intelligence services methodology where the drone equipped with camera fly over
the pumpkin field and capture the images. The captured image of pumpkin leaf sends
to a processor which processes the image, identifies, and detects the downy mildew
diseased leaf of pumpkin as in Fig. 1. Here also a second approach where the image
is captured by any smartphone camera and send to the server where the image is
processed and identify & detect the downy mildew diseased leaf as in Fig. 2.

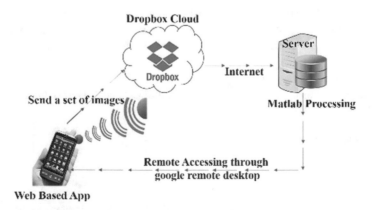

Fig. 2 System overview, sample image collected by smartphone for identifying and detecting
downy mildew diseased leaves of pumpkins

1.1 Downy Mildew Disease of Pumpkin

The leaf-infecting disease such as downy mildew on the leaf of pumpkin is very common. Downy mildew can be quite severe on pumpkin leaves and appear angular brown lesions on the leaves as in Fig. 3. Downy mildew first appears as brown patches. These patches gradually spread and a pumpkin that is severely affected may have a reduced yield with shorter lifetime and fruits with small size. Figures 3 and 4 show the downy leaf diseased and healthy leaf of pumpkin.

1.2 Literature Survey

A number of scholars shown and presented their state of interest toward the findings for the contaminated edible leaves using various ideas. Out of which below are some of the findings and they are

Fig. 3 Downy mildew diseased leaf of pumpkin

Fig. 4 Healthy leaf of pumpkin

i. Sethy et al. [3] developed Automatic Leaf Disease Detection by using K-means segmentation method. It has used the image segmentation approaches of K-means and Euclidean distance clustering methods for the calculation of diseased leaves.

ii. Sanjay et al. [4] discussed K-means clustering algorithm and CBIR-based system for detection of leaf disease. The degree of infection can be calculated using K-means clustering by calculating the leaf area of infection.

iii. Dubey et al. [5] denoted that K-means clustering algorithm can be taken for the identification of shortcoming segmentation of diseased fruits. And even experimentally shown that more the image clusters better the result is.

iv. Naikwadi and Amoda [6] projected that the extraction and image segmentation can be done using K-means clustering and their features like gray-level unison matrix and color feel touching for the classification and identification purposes.

v. Bashish et al. [7] presented that by using squared Euclidean distances of K-means clustering which moreover works efficiently for the detection of diseased area presented on a leaf. Furthermore, taking the benefits and theory of backpropagation under the neural network, disease sorting can also be possible.

vi. Wang et al. [8] suggested that by use of principal component analysis (PCA) and neural networks' plant diseases of wheat and fruits are recognized. In this approach, PCA is used for reducing dimensions in feature data processing and then neural networks including backpropagation (BP) networks, radial basis function (RBF) neural networks, generalized regression networks (GRNN), and probabilistic neural networks(PNN) were used as classifier to identify wheat and grape diseases. Strip and leaf erosion are the two results of optimum recognition based on PCA and BP network whose prediction accuracy is 100% appropriate. Again, it also results for two varieties of fruit diseases (downy mildew and powdery mildew) by use of GRNN and PNN with PCA as dimension reduction, and the prediction and fitting accuracy are 94.29 and 100%, respectively.

vii. Wang et al. [9] proposed key methods of threshold segmentation based on hue, iteration binarization, image morphological operation, and contour extraction were adopted for image processing and image segmentation; thereafter, the texture, color, and shape features were extracted systematically. Genetic algorithm especially used for getting approximation features. Finally, Fisher discrimination analysis was applied for the recognition of main maize leaf diseases. In this research, 28 characters including energy, informationization measure, fractal dimension, hue, cb, color moment, disease spot area, rotundity, figure factor, and others were extracted, and four approximate features were selected from 28 primeval features. The results indicated that the precision of the three kinds of maize disease recognition was higher than 90%.

viii. Khole et al. [10] presented a novel method for detection of downy mildew disease existing in the grape leaves based on fuzzy importance factor. The proposed method uses the operations of digital image processing and the concept of fuzzy set theory. They have experimented on thirty-one diseased and non-diseased images and got the result in the percentage scale of 87.09% success.

2   Materials and Methods

A.   Defect Area Segmentation Using K-means Clustering

Here from Fig. 5, it represents the basic flow steps diagram followed for detection of mildew diseased leaf.

(1)   **Image Acquisition**. The Pumpkin Leaf is captured in the form of primary color pattern as RGB. Then, the RGB leaf image is converted to L*a*b color space.

(2)   **Preprocessing**. In this process, the query image converted to respective color space where the set of rules might have functioned. By using several image preprocessing techniques like image resizing, image filtering, etc., we can extract the essential data competently using the image accurately. Hence using filters, we can able to remove glitches from the image and get the exact portion as per our need.

(3)   **K-means-Based Image Segmentation**. In this paper, we have focused on K-means algorithm, which is used for image segmentation purpose. Since, more the clusters better the quality it yields, hence these high-quality clusters can only be achieved by the collection of similar groups of $m \times n$ group of pixels in the order of high dpi rate which is faster and accurate in processing. Accurate measurements and assessments can only be achieved by seed points and data points along with some other parameters using the technique.

Following steps are for the used algorithm representation:

In the first step, we have to choose the value of K of our points as a partition center randomly. Next, the set point of remoteness with its center computed successfully, thereafter by saving its information for further use.

In final step calculations, we have to assign each point to the nearest cluster centers. So that we get the minimum distance calculated for index each point. Then, we add that the point of the specific partition set.

The distance between the cluster center and the pixels can be minimized by assigning individual dotted particles into its respective collection as mention in equation Eq. (1).

$$C^{(i)} = \operatorname{argmin}_{j \| x(i) - \mu j \|} \tag{1}$$

Fig. 5 Flow step diagrams of proposed methodology

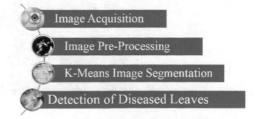

The cluster centers are computed by averaging all of the pixels present in the cluster according to the equation Eq. (2).

$$\mu_i = \frac{\lim_{i \to m} 1\{c(i) = j\}x(i)}{\lim_{i \to m} 1\{c(i) = j\}} \tag{2}$$

(4) **Defected Area Calculation**. For successful image examination determination, this algorithm will be highly effective result oriented among all. By using this algorithm, it is able to get the interest of the region. K-means Clustering is exploratory data analysis technique. Object juxtaposing can be done by the help of a non-graded method. The goal is to segregate the background and body by obscuring the border of the given image.

Step 1. First, read the input image of infected leaves.
Step 2. Then, the input image should have to convert from RGB to L*a*b color space. L*a*b color space consists of two chromaticity layers in *a and *b channels and a luminosity layer in L*, channel. All of the color information is present in the *a and *b layers only, so by using L*a*b color space computational speed can increase. In MATLAB software, the RGB image can convert to L*a*b color space by using makecform and applycform command directly.
Step 3. Colors are classified by using K-means clustering in *a*b space, and simultaneously, the difference between two colors can be evaluated with the help of Euclidean distance metric which is considered here.
Step 4. Each pixel of the image will be labeled with its assigned cluster.
Step 5. The pixels present in the input image have been separated by color using pixel labels which will result in the different image based on the number of clusters. The flowchart has been depicted in Fig. 6.

Figure 7 shows the unique and cluster indexed Image. Figure 8 shows the segmentation results of the pumpkin leaf using K-means clustering technique. The input image has been segmented into three clusters and it is very clear that the second

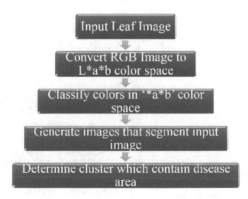

Fig. 6 Flowchart of detection of diseased leaf

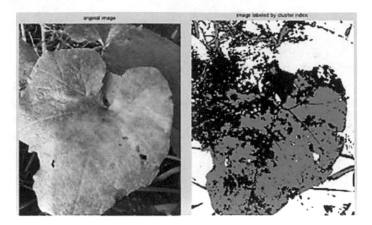

Fig. 7 Original image and cluster-indexed image

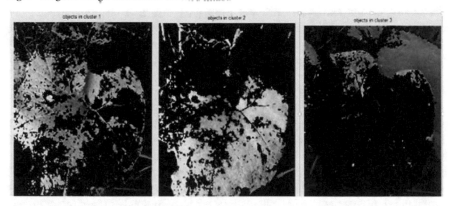

Fig. 8 Cluster1, Cluster2, and Cluster3 images

cluster correctly segments the infected portion of the image and contains diseased affected area. From the above observations, it is found that the 3-Means clustering technique gives good segmentation results. So, in this experiment, the input image is partition into three clusters as per the requirements. By using 3-Means clustering, the outputs are visualized more accurately. The investigational consequences indicate that the suggested technique is for the cluster-based segmentation which is a robust procedure because the degree of segregation is quite accurate for the healthy, infected, and background portion of the infected one.

B. Identification of Downy Mildew Diseased Leaf of Pumpkin using PCA

The use of principle component analysis (PCA) is to determine the most discriminating features between images. The function Eigenface Core gets a 2D matrix, containing all training image vectors, and returns three outputs which are extracted from training database. The three-extracted feature are- m- ($M*N \times 1$) Mean of the training database.

Eigenleaf—($M*N \times (P-1)$) Eigenvectors of the covariance matrix of the training database.

A—($M*N \times P$) Matrix of centered image vectors.

Our aim is to generate the basis vectors of same dimensions from a training set of leaf images, which can be stated as eigenleaf as like the original pumpkin leaf images. By inserting one new image into a topological space, the eigenleaf can be grouped by comparing its leaf position using the reference to the actual one. This can be expressed as, by considering a group of M eigenleaf images, they are i_1, i_2, … i_M having N × N size of similar images where the mean set of value can be stated by

$$\tau = \frac{1}{M} \sum_{j=1}^{M} ij \tag{3}$$

Vectorially, all the leaves are different from one another. Hence eigenvector can be estimate, by calculating the mean value of the product of the deviations of two variates from their respective means matrix C: As, A is $N_2 \times M$ dimensions. Hence, C covariance matrix

$$C = AA^T, \quad \text{Where } A = [\Phi_1, \Phi_2, \ldots \Phi_M] \tag{4}$$

having $N_2 \times N_2$ scopes, where for N_2 eigenvectors each will be $N_2 \times 1$ measurement. Therefore, there may be chance of system retard and low memory when the value of N is very large for the given value of N eigenvector at the time of real computation. Hence, the mean value of the product of the deviations of two variates from their respective means matrix with reduced planeness can be utilised which may not only curtail the significant volume of calculations but also diminishes eigenvector noises. Hence, mathematically this can be expressed as given below

$$C = A^T A, \quad \text{where } A = [\Phi_1, \Phi_2, \ldots \Phi_M] \tag{5}$$

Therefore, undoubtedly it may reduce to $M \times M$ scopes where for the M eigenvectors each will be $M \times 1$ measurement which confirms that $M \ll N_2$. Thus, it is apparent that K eigenvectors can be found in lower planeness in comparison to the higher planeness where the chosen value should accurate. From lower to higher planeness by the help of eigenvectors using $u_i = Av_i$ having, u_i for the ith eigenvector of original space and v_i is the ith eigenvectors of lower planeness. Out of these eigenleafs, $K(<M)$ eigenleaf are nominated and agree to the K maximum eigenvalues. The set of eigenleaf images, $\{i\}$, is reformed into its eigenleaf elements by the given equation

$$\omega_{nk} = b_k(i_k - \bar{\tau}) \tag{6}$$

where $n = 1, 2 \ldots, M$ and $k = 1, 2 \ldots, k$

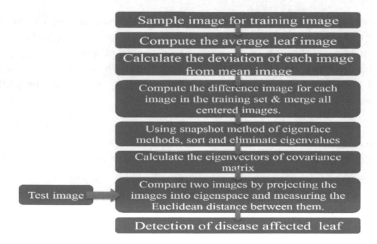

Fig. 9 Flowchart for Identification of downy mildew diseased leaf

The vector weight defines the role of each eigenleaf on behalf of the inputted image, representing the eigen-leaves are the source and root of all leaf images.

$$\tau_n = [\omega_{n1}, \omega_{n2}, \ldots \omega_{nk}] \tag{7}$$

When the image of the leaf they will be framed into the eigen space, the resemblance between any couple of printed leaf images may be reckoned by the result from $\|y_1 - y_2\|$, where, y_1 and y_2 assumed to be the Euclidean distance among the respective temporal vectors. Hence, it is very much apparent and clears that lesser the distance of temporal vectors, more the similarities of eigen-leaves. Thus, the score of similarities can be stated as $S(y_1, y_2)$ depending on the inverse Euclidean distance:

$$S(y_1, y_2) = \frac{1}{1 + |(y1 - y2)|} \in [0, 1] \tag{8}$$

Leaf recognition can be achieved by ensuring the score of similarities which further can be calculated between an input to the training leaf images. Finally, the matched leaf image which obtained that is having the score of highest similarities (where the rate of unit representing a precise match). Figure 9 shows the flowchart of proposed methodology.

3 Result and Discussion

An effective execution of algorithm using MATLAB yields the proposed methodology using PCA identifies the downy mildew diseased leaf successfully. In our experiment, as far as concerned with the experimental calculation 10 numbers of diseased and healthy leaves taken for training purposes and 15 numbers of diseased and 10 numbers of healthy leaves are used for testing purposes. As a result, 87% of accuracy is being obtained for identification. Again, for the detection and calculation of the affected area K-means clustering technique can be used precisely and satisfactorily.

4 Conclusion

A context which is being framed, proposed, and evaluated with the aid of principle component analysis for the detection and identification of downy mildew diseased leaves of pumpkin using group-based dissection by means of "K". The group suggested that the concerned schemes' ability is to detect and identify the area of diseased leaf under defect is more successful and efficient.

5 Future Scope

Further, the work may extend to identify different diseases of pumpkin and helps toward the calculation of the affected area.

Acknowledgements Our immense pleasure to convey heartiest thanks to Sh. Srikanta Sethy from village Kapundi, District Keojhar, Odisha, India, who help and allow us to collect sample pumpkin leaf images from his productive field.

References

1. Pitrat, M.: Vegetable crops in the Mediterranean Basin with an overview of virus resistance. Adv. Virus Res. **84**, 1–29 (2012)
2. Perez Gutierrez, R.M.: Review of Cucurbita pepo (pumpkin) its phytochemistry and pharmacology. Medicinal Chemistry **6**(1) (2016) ISSN: 2161-0444
3. Sethy, P.K., Negi, B., Bhoi, N.: Detection of healthy and defected diseased leaf of rice crop using K-means clustering technique. Int. J. Comput. Appl. **157**(1), 24–27 (2017)
4. Dudhane, T.M., Patil, S.B.: Leaf disease detection by the help of K-Means segmentation method and CBIR system. Int. Res. J. Eng. Technol. **2**(9) (2015)
5. Dubey, S.R., Dixit, P., Singh, N., Gupta, J.P.: Infected fruit part detection using K-means clustering segmentation technique. Int. J. Interact. Multimedia Artif. Intell. **2**, 65–72 (2013)
6. Naikwadi, S., Amoda, N.: Advances in image processing for detection of plant diseases. Int. J. Appl. Innov. Eng. Manage. **2**(11), 168–175 (2013)
7. Braik, M., Al Bashish, D., Bani-Ahmad, S.: Detection and classification of leaf diseases using K-means-based segmentation and neural-networks-based classification. Inf. Technol. J. **10**(2), 267–275 (2011)

8. Wang, H., Li, G., Ma, Z., Li, X.: Image recognition of plant diseases based on principal component analysis and neural networks. In: 8th International Conference on Natural Computation (2012)

9. Wang, N., Wang, K.R., Xie, R.Z., Lai, J.C., Ming, B., Li, S.K.: Maize leaf disease identification based on Fisher discrimination analysis. Sci. Agric. Sin. **42**(11), 3836–3842 (2009)

10. Kole, D.K., Ghosh, A., Mitra, S.: Detection of downy mildew disease present in the grape leaves based on fuzzy set theory. In: Kundu, M.K., Mohapatra, D., Konar, A., Chakraborty, A. (eds.) Advanced computing, networking and informatics—vol. 1. Smart Innovation, Systems and Technologies, vol. 27. Springer, Cham (2014)

A Survey on Rice Plant Disease Identification Using Image Processing and Data Mining Techniques

Tanmoy Bera, Ankur Das, Jaya Sil and Asit K. Das

Abstract Rice is cultivated almost all over the world. Various diseases and pests cause loss of rice productivity in spite of using several pesticides. Though use of pesticides increases the production, it reduces the quality of the crop. Thus, disease identification from rice images becomes one of the important research works in the field of Computer Science and Agriculture Department. In the paper, different rice plant diseases are described and identified using various image processing and data mining techniques. The image processing techniques are applied for collecting the disease portion of the image and extracting relevant disease-related features. On the other hand, data mining techniques are applied for extracting relevant hidden information useful for disease detection based on the extracted features. The techniques are very useful for the farmers to take the corrective actions in a very short period of time to save their crops from diseases. Here, we have studied several papers related to rice plant disease identification to explore the current scenario about the agricultural decision support system.

Keywords Image processing · Segmentation · Feature extraction
Disease identification · Data mining · Rule based classification

T. Bera (✉)
Department of Information Technology, Murshidabad College of Engineering & Technology, Murshidabad 742102, West Bengal, India
e-mail: tanmoybera06@gmail.com

A. Das
Department of Computer Science and Engineering, Calcutta Institute of Engineering and Management, Kolkata 700040, West Bengal, India
e-mail: ankurdas8017@gmail.com

J. Sil · A. K. Das
Department of Computer Science and Technology, Indian Institute of Engineering Science and Technology, Shibpur, Howrah 711103, West Bengal, India
e-mail: js@cs.iiests.ac.in

A. K. Das
e-mail: akdas@cs.iiests.ac.in

© Springer Nature Singapore Pte Ltd. 2019
A. Abraham et al. (eds.), *Emerging Technologies in Data Mining and Information Security*, Advances in Intelligent Systems and Computing 814,
https://doi.org/10.1007/978-981-13-1501-5_31

1   Introduction

Rice (scientific name Oryza sativa) is one of the most important food crops in the world on which over half of the world population relies. India is the second highest rice producer and the highest rice exporter in the world since 2015. Thus, Indian economy is very much dependable on rice production. But, farmers face a loss of production every year due to different diseases attack the crops. Sometimes, farmers cannot make attention to the diseases in time or face difficulty to identify the disease which cause too late to recover leading loss of production. One of the important aspects to reduce the ill-effect of the diseases is to identify, detect, and take timely solution. To identify the diseases, farmers had to depend on guidebooks or used their own experiences which lead to a very time-consuming process to select correct pesticides in time. An autonomous system can be developed for classification of rice disease through the latest development of image processing [1, 2] and data mining [3, 4]. The periodic images of the different parts of the rice plant are captured and then processed to recognize the diseases using the sequence of steps such as image collection, preprocessing of image, image segmentation, feature (e.g., texture, color, shape, position) extraction and classification (decision tree, neural network, support vector machine, naïve-based classifier, etc.) of diseases. Based on the accuracy level of image processing [1, 2] and data mining techniques [3, 4], efficiency of the system is measured. The performance of the system degrades due to several reasons, such as lack of a large number of objects in the dataset, wrong feature selection, bad selection of classifier. To mitigate the problem, researchers have developed different models for both image processing and data mining techniques. In this paper, we have made a survey for disease detection based on different techniques and approaches on different rice plant diseases. A comparative study of these approaches is also given in this paper.

The rest of the paper is organized as follows: Sect. 2 presents different types of rice plant diseases. Section 3 describes different image processing tasks applied for preparation of disease dataset. Data mining task is discussed in Sect. 4 for classification of diseases. Section 5 presents a comparative study among several related research works in rice disease detection, and finally, the paper is concluded in Sect. 6.

2   Rice Plant Diseases

This section gives the details of rice plant diseases that occur at different parts of the plant. In this paper, we have restricted our work within the common rice diseases in India. This section will provide some idea about which image processing techniques are suitable and what kind of features is relevant to detect disease accurately. To simplify the disease detection system, we have categorized the diseases into four classes based on the location of the infection in the plant body. Before going into details of the diseases, we must be familiar with the location of the infected parts of the mature rice plant for more accurate detection. The rice plant consists of grains/panicles, leaf, sheath, stem (including node, neck), and root. Initially, we want to separate four generic types of diseases occur in respective parts of the plant. Finally, for each type of generic disease, specific diseases are identified based on their exterior symptoms discussed below:

A. **Grains/Panicle—False Smut**: The infected plant produces velvety spores from rice grain. Growth of velvety spores encloses floral parts. Smooth and little flattened yellowish spores indicate that the spores are underdeveloped which is covered by membrane. Growth of spores results in broken membrane. Mature spores are orange and turn yellowish-green or greenish-black in color.

B. **Grains/Panicle—Grain Discoloration**: It occurs at the time of panicle initiation stage. At this time, normal grain color changes to brownish-white color.

C. **Leaf and Sheath—Blast (Leaf, Collar)**: The infected leaf with green-gray spots surrounded with deep green borders is the starting symptoms. At matured stage, the shape of the spots or lesions becomes elliptical or spindle along with reddish-brown borders. Some of the lesions become diamond shaped. They can grow together by enlarging themselves to completely destroy the leaf.

D. **Leaf and Sheath—Brown Spot**: Tiny circular lesions with brown or yellow-brown may affect coleoptiles at seedling stage. Lesions on the leaves can be noticed at tiller stage. Lesions at the early stage become tiny circular along with dark brown color. Matured lesions are gray centered encircled by reddish-brown margin.

E. **Leaf and Sheath—Narrow Brown Spot (Affects Leaf, Sheath, and Panicles)**: The infected leaf and sheath have lesions with brown color along with linear progression. They are usually 2–10 mm long and 1–1.5 mm wide. The lesions are getting enlarged to connect each other and make a region. The lesions on glumes are wider compared to leaf. Sometimes, pedicles have brown lesions. Netlike pattern on the discolored leaves is also noticed at leaf sheath.

F. **Leaf and Sheath—Tungro Disease**: Tungro-infected leaf becomes orange-yellow to yellow in color. The staining spreads from leaf tip to the lower part of the leaf. Stripy appearances and spot with rust-colored are noticed on the infected leaf. Leaf may contain planthoppers.

G. **Stem—Blast (Node, Neck)**: Node infection occurs in banded pattern. Due to node infection, plants' part that holds the panicle is broken. Grayish-brown lesions are noticed at node. Grayish-brown lesions at the neck cause panicle to collapse. Infection at neck at milky stage and later causes no grains and poor grains, respectively. Whiteheads are caused by these types of blast as well as stem borers. Stem borers can split the stem at the boring point. The stems are not split by the blast.

H. **Stem—Stem Rot**: Numerous tiny white and black sclerotia and mycelium are visible inside the infected culms. Infected culms lodges and caused unfilled panicles and chalky grain. Leaf sheath has asymmetrical and tiny black lesions close to water level. Lesions expand as the disease advances. Severe infection causes tiller death.

I. **Stem—Yellow Dwarf**: The leaves of the infected plants are pale green in color, and the plat becomes undersized. There are excessive tillering and leaves which became soft and droop slightly. Root growth is also reduced significantly. Sometimes, chlorosis is noticed which may scatter to sheath of the leaf. Leaf streaks are parallel in shape toward the veins.

J. **Stem—Rice Grassy Stunt**: The infected plants are severely stunted, excessively tiller with very upright growth, grassy and rosette appearance. They have yellowish-green leaves that are shorter and narrower than normal leaves that remain yellow even after application of sufficient nitrogen fertilizers and contain numerous small rusty spots or patches on leaves, which merge into blotches leaves with a mottled appearance that fail to produce panicles.

K. **Root—Bakanae/Foot Rot**: Infected seedlings look unhealthy, have lesions on roots, and can die before transplanting or immediately after. Infected plants are abnormal and are several inches taller and thinner than normal plants. The color of the leaves is pale and yellowish-green. The root or the lower part of the plant has white chalky growth. Roots are grown from aboveground nodes on the stem. The images of the above-mentioned diseases are shown in Fig. 1a–k, respectively.

3 Image Processing on Diseased Images

To prepare image database, image acquisition is required from live farm. The rice plant-diseased images are collected by different researchers from live firm using high-resolution digital camera like Nikon D80, Nikon Coolpix P4, Canon 450D, Canon Powershot G2. However, the original image is degraded due to different parameters like noises, shadow. This sort of degradation leads to the extraction of wrong pixel information as well as features from noisy and shadowed image. So, the initial work is to improve the quality of the images for further processing. For this reason, diseased plant image has to be separated from non-diseased image. Images are resized and cropped to reduce the processing time. The image is degraded because of dust, dewdrops, insect's excrements present on the plant; these things are considered as image noise. Furthermore, captured images may have distortion of some water drops and shadow effect, which could create problems in the segmentation and feature extraction stages. Image restoration by using different noise removal filters is done to remove distortion. There may be shadow and low contrasting captured images which results in improper feature extraction. The image can be improved by removing blurring effect with the help of Weiner filter [5] and increasing contrast using various contrast enhancement algorithms. In case of the images captured using high-definition cameras, the size of the pictures might be very large, so image resizing and cropping are essential for image processing. Also, image reduction helps in reducing the computing memory power. Image preprocessing can significantly increase the reliability of an optical inspection. The following preprocessing techniques are commonly used to enhance the image quality before image database preparation:

Fig. 1 Image of different rice diseases

A. ***Noise Removal***: The acquired image may contain different noises from environment, image capturing devices, etc., which are removed by various noise removal techniques [6, 7]. The average filter method is used to remove Gaussian noise. The median filtering is a nonlinear digital filtering technique, used to remove salt and pepper noise and to sharpen the edges. For noise removal and smoothing of images by edge preservation, bilateral filter is used.

B. ***Shadow Removal***: Shadow removal [8] is an essential step for feature extraction. Shadow often appears in the image due to improper alignment of camera with the direction of light source. The aim of the task is to isolate the shadow and background pixels from the image in order to identify the plant region with the help of different vegetative indices like normalized difference greenness index, redness index. A few pixels of the shadow region may appear in the plant region, which are difficult to locate using vegetative indices. Shadow regions do not contain any texture, whereas textures are obtained from the strips of a rice plant. For strip detection and shadow pixel removal, Canny edge detection algorithm is applied on the image.

C. **Image Enhancement**: The quality of the image is improved by different image enhancement methods [9, 10] including contrast stretching, histogram equalization, gamma correction, log transform. Contrast improvement is done by adaptive histogram technique. After equalization, the edges become more prominent compared to the original image.

D. **Image Segmentation**: It is used to differentiate infected regions from non-infected regions. Many segmentation methods are available in the literature [11, 12] for different types of applications. Some of segmentation techniques used for rice plant disease is fermi energy level segmentation, K-means clustering segmentation, Otsu's method, region segmentation, boundary segmentation, etc. Generally, there are two ways to carry out the image segmentation: (i) based on discontinuities and (ii) based on similarities. In the first way, an image is partitioned based on sudden changes in intensity values, e.g., edge detection, while in the second way images are partitioned based on the specific predefined criteria; e.g., threshold is done using Otsu's method. Phadikar et al. in [13] stated segmentation algorithms based on image luminance amplitude and color components. In [14], YCbCr color space has been used mainly for segmentation of rice diseases because of following reasons: (i) In YCbCr plane infected part can be easily detected, (ii) the color difference of human discrimination can be directly uttered by Euclidean distance in the YCbCr color space, (iii) the intensity and chromatic components can be used separately, and (iv) plant infected spots form small clusters in Cr space. Amar et al. [15] achieved the best segmentation result compared to RGB and YCbCr. They have applied Otsu's method on H-component of HVS color space for image segmentation. Zainon et al. [16] used K-mean clustering technique for segmentation, and the obtained segmented image is passed through pretrained neural network.

E. **Feature Extraction**: The visual features which seem to be better based on certain visual properties of an image, either globally for the entire image or locally for a small group of pixels, may be extracted from the images. Normally, healthy leaves of rice are green in color but when rice leaves are infected by any disease, color of the leaves is changed and the different colored spots are created on leaves. These spots are of different shapes. Hence, color and shape of the infected part are selected as features to identify the diseases. Some shape-based features are area, axis, and angle. Amar et al. [15] select feature from the image.

4 Data Mining for Disease Identification

The term data mining refers to the non-trivial extraction of implicit, previously unknown and potentially useful information from data in databases. It is the process which provides a concept to attract attention of users due to high availability of huge amount of data and need to convert such data into useful information. Data mining is also stated as essential process where intelligent methods are applied in order to extract the data patterns. In case of rice disease identification, different classification techniques are used by the researchers after feature extraction and feature selection from the diseased images. Amrita et al. [14] proposed an automated system for identifying and classifying different diseases like bacterial blight, rice blast, rice sheath rot, and brown spot. Identification of the diseases is the key to prevent qualitative and quantitative loss of agricultural yields. They used two classifiers, k-NN and MDC, and the accuracies achieved with these classifiers are 87.02 and 89.23%, respectively. Automated disease classification of leaf brown spot and leaf blast was proposed in [17] by Phadikar et al. The accuracies of Bayes' and SVM classifiers had been measured on 1000 diseased rice leaf images. Anitha [18] proposed a predictive modelling framework to forecast the crop yield, and identification of ideal condition for getting high yield of paddy crop was carried out scientifically by a data mining approach. Decision tree method was used for classification. Mohammed et al. in [19] analyzed different tree-based classification algorithms along with their performance on different diseased datasets of Egyptian rice. They proposed a methodology using C4.5 decision tree algorithm to discover classification rules for Egyptian rice diseases. They had considered only seven agronomic attributes for experimental purpose. Entropy and information gain were used as attribute selection measures. The performance of the neural network and C4.5 decision tree algorithm was observed as 96.4 and 97.25%, respectively.

5 Experimental Results and Performance Evaluation

This section presents the experimental results obtained by nine survey papers [13, 14, 20–26] on rice disease detection through image processing and data mining techniques. Table 1 shows the analysis of different image processing techniques applied on diseased rice images in detail so that further research can be made on this area to improve the overall performance of the rice disease detection system. The following notations are used in Table 1: leaf blast (LB), leaf brown spot (LBS), rice planthopper (RPH), bacterial leaf blight (BLB), sheath blast (SB), brown spot (BS), rice bacterial blight (RBB), rice blast (RB), rice brown spot (RBS), rice sheath rot (RSR), sheath rot (SR), bacterial blight (BB), tungro (TR), rice leaf blast (RLB), vegetation indices (VI), red, green, blue (RGB), hue, saturation, value (HSV). All the papers collected the diseased images from paddy field by high-resolution digital camera. Since the captured images are very large in size, some of the authors [13, 14, 23, 24, 26] reduce the size of the images by cropping. As images are collected directly from paddy field in different lightning conditions, it may contain different noises, contrast problem, outdoor illumination effect. The disease detection is basically divided into two steps: (i) First step consists of image collection, preprocessing, segmentation, feature extraction, and (ii) second step consists of classification and identification of diseases, as described in Tables 1 and 2, respectively.

Table 1 Comparative study of image processing techniques applied on rice images

Authors and years	Goals	Rice disease and pest	Image collection	Image preprocessing	Features extracted
Suresha M et al. (2017)	Identification of paddy leaf disease using geometrical features	Fungal disease LB, LBS	330 images captured with 18.1 mp camera in daylight with black background	Otsu's segmentation, color model: HSV	Geometrical features: area, minor and major axis length, perimeter
Tsung-Han Tsai et al. (2017)	Region of interest (ROI) method to detect rice planthopper s(RPHs) clearly before its growth	RPH	50 images about RPHs on rice stem are captured at the evening	Region of interest (ROI) to reduce the image area to close to the RPHs. Transformation from RGB to HSV	Color feature
Sofianita Mutalib et al. (2016)	Weed detection and recognition of leaf diseases before time	BLB LB SB	1207 leaves with diseases at different stages and 239 healthy leaves	1. Cropping and resized 2. RGB to gray conversion 3. Canny edge detection 4. Smoothing by using dilation and erosion	
Xiaochun Mai et al. (2016)	Segment lesion automatically from rice leaf blast disease images	RLB	71 images captured in rice fields so that each image contains at least one lesion region	1. Scaling to reduce size 2. Color adjustment 3. Simple linear iterative clustering algorithm is used for segmentation	Features extracted are regional, color, shape, and texture
Santanu Phadikar et al. (2016)	Automatic classification of rice plant disease by using VI-based segmentation	BS Blast	100 samples of both the blast and brown spot diseases are collected from the fields	1. Cropping to reduce size 2. Noise removal using median filter 3. Computation of VI indices 4. Segmentation: Otsu's method	Texture feature

(continued)

Table 1 (continued)

Authors and years	Goals	Rice disease and pest	Image collection	Image preprocessing	Features extracted
Amrita A. Joshi et al. (2016)	Diagnose and classify rice diseases	RBB RB RBS RSR	115 rice leaf images of four diseases	1. Scaling to reduce size 2. Normalization to remove outdoor illumination 3. YCbCr plane space used for segmentation	1. Color feature 2. Zone-wise shape feature
John W. O. et al. (2014)	Identification of rice plant disease using backpropagation artificial neural network	BLB BS RB	134 images (55 images for bacterial leaf blight, 37 images for brown spot, and 42 images for rice blast)	1. Noise reduction, contrast adjustment, conversion of RGB to HSV and intensity adjustment 2. Thresholding and masking for segmentation 3. LAB color space and A color plane	1. Arithmetic mean values for the R, G, B color component 2. Standard deviation of R, G, B color component 3. Mean values of H, S, and V
Santanu Phadikar et al. (2013)	To classify different types of rice diseases by using feature selection and rule generation	LBS RB SR BB	500 infected rice plant images consist of four diseases	Fermi energy-based segmentation	Total features (color, shape, position): 25(15,9,1)
K. Majid et al. (2013)	To develop a paddy plant disease identification system using probabilistic neural network	BS LB TR BLB	Images were taken from paddy fields in West Java, Indonesia	1. Cropping 2. Converting to gray scale 3. Enhancement using Laplacian filter	Fuzzy entropy-based feature extraction

Table 2 Comparative study of data mining techniques applied to detect rice disease

Authors and years	Classification techniques applied	Parameters of classification	Accuracy
Suresha M et al. (2017)	k-Nearest neighbor (k-NN) classifier	Geometrical features like area, major axis, minor axis, and perimeter	76.59%
Tsung-Han Tsai et al. (2017)	Decision tree algorithm to classify analytic data	Relative magnitude between color elements and deviation	100%
Sofianita Mutalib et al. (2016)	Backpropagation neural network	Three layers: input layer, hidden layer, and output layer	70–80%
Xiaochun Mai et al. (2016)	Random forest classifier	Color, shape and texture features of each superpixel	Mean dataset1: 97% dataset2: 80%
Santanu Phadikar et al. (2016)	Vegetation index-based method	Texture values calculated from VI index	84%
Amrita A. Joshi et al. (2016)	Minimum distance classifier (MDC) and k-nearest neighbor (k-NN) classifier	Color and zone-wise shape features	MDC: 89.23% k-NN: 87.02%
John W. O. et al. (2014)	Backpropagation neural network	Stopping criteria: 1. Maximum iteration/gradient are reached 2. Maximum validation checks	100%
Santanu Phadikar et al. (2013)	Rule generation IF-THEN classifier	1. (feature, value) pair 2. minimal feature subset	More than 90%
K. Majid et al. (2013)	Probabilistic neural network	Four layers: input layer, pattern layer, summation layer, and output layer, radial basis function for pattern layer	1. Brown spot: 100% 2. Leaf blast: 100% 3. Tungro: 76% 4. Bact. leaf blight: 96%

6   Conclusion

Every year a huge amount of damage of crops occurs due to rice plant diseases. Rice disease detection is an important research work in the field of Computer Science and Agriculture Department. In this study, nine papers are reviewed and their results are summarized on the basis of image processing and data mining techniques used in disease identification. We have made a comparative study of the related research papers based on feature selection, segmentation, and classification accuracy. Although different algorithms are already running to detect diseases, these algorithms can be improved further to provide more accuracy in rice plant disease detection. This study encourages us to propose an efficient method for plant disease identification, which is the future work of this paper. Adequate number of references has been given for further detailed understanding. We apologize to researchers whose important contributions are not given in the paper due to page limitation.

References

1. Fakhri, A., Nasir, A., Nordin, M., Rahman, A., Mamat, A.R.: A study of image processing in agriculture application under high performance computing environment. Int. J. Comput. Sci. Telecommun. **3**, 16–24 (2012)
2. Russ, J.C.: Image Processing Handbook, 3rd edn. CRC/IEEE Press, USA (1999)
3. Han, J., Kamber, M.: Data Mining: Concepts and Techniques. Morgan Kaufmann, San Francisco (2001)
4. Camargo, A., Smith, J.S.: Image pattern classification for the identification of disease causing agents in plants. Comput. Electron. Agric. **66**(2), 121–125 (2009)
5. Choi, H., Baraniuk, R.G.: Analysis of wavelet domain wiener filters. In: IEEE International Symposium Time-Frequency and Time-Scale Analysis, (Pittsburgh), pp. 613–616. IEEE (1998)
6. Sankur, B., Sezgin, M.: Survey over image thresholding techniques and quantitative performance evaluation. J. Electron Imaging **13**, 146–165 (2004)
7. Trier, O.D., Jain, A.K.: Goal-directed evaluation of binarization methods. IEEE Trans. PAMI **17**, 1191–1201 (1995)
8. Guo, R., Dai, Q., Hoiem, D.: Single-image shadow detection and removal using paired regions. In: CVPR. 175 (2011)
9. Gonzalez, R.C., Woods, R.E.: Digital Image Processing. Pearson Education, New Delhi, India (2007)
10. Li, H., Wang, Y., Liu, K., Freedman, M.T.: Computerized radiographic mass detection C Part I: lesion site selection by morphological enhancement and contextual segmentation. IEEE Trans. Med. Imaging **20**(4), 289–301 (2001)
11. Mardia, K.V., Hainsworth, T.J.: A spatial thresholding method for image segmentation. IEEE Trans. Pattern Anal. Mach. Intell. **10**, 919–927 (1988)
12. Pal, N.R., Pal, S.K.: A review on image segmentation techniques. Pattern Recogn. Soc. **26**(9), 1277–1294 (1993)
13. Phadikar, S., Goswami, J.: Vegetation indices based segmentation for automatic classification of brown spot and blast diseases of rice. In: International Conference on Recent Advances in Information Technology, pp. 284–289. IEEE (2016)
14. Amrita, A.J., Jadhav, B.D.: Monitoring and controlling rice diseases using image processing techniques. In: International Conference on Computing, Analytics and Security Trends (CAST), pp. 471–476. IEEE (2016)
15. Dey, A.K., Sharma, M., Meshram, M.R.: Image processing based leaf rot disease, detection of betel vine (Piper BetleL.). In: International Conference on Computational Modeling and Security, Procedia Computer Science, vol. 85, pp. 748–754 (2016)
16. Zainon, R.B.: Paddy Disease Detection System Using Image Processing. http://umpir.ump.edu.my/6994/1/CD7686.pdf (2017)
17. Phadikar, S., Sil, J., Das, A.K.: Classification of rice leaf diseases based on morphological changes. Int. J. Inf. Electron. Eng. **2**(3) (2012)
18. Anitha, A.: A predictive modeling approach for improving paddy crop productivity using data mining techniques. Turk. J. Electr. Eng. Comput. Sci. **25**, 4777–4787 (2017)

19. Mohammed, E., Mahmoud, W.: An empirical comparison of tree-based learning algorithms: an egyptian rice diseases classification case study. Int. J. Adv. Res. Artif. Intell. **5**(1), 22–26 (2016)
20. Phadikar, S., Sil, J., Das, A.K.: Rice diseases classification using feature selection and rule generation techniques. Comput. Electron. Agric. **90**, 76–85 (2013). Elsevier
21. Suresha, M., Shreekanth, K.N., Thirumalesh, B.V.: Recognition of diseases in paddy leaves using k-NN classifier. In: International Conference for Convergence in Technology, pp. 663–666. IEEE (2017)
22. Tsai, T.H., Lee, T., Chen, P.: The ROI of rice planthopper by image processing. In: Prior, Lam (eds.) Proceedings of the IEEE International Conference on Applied System Innovation, pp. 126–129 (2017)
23. Mutalib, S., Abdullah, M.H, Rahman, S.A., Aziz, Z.A.: A brief study on paddy applications with image processing and proposed architecture. In: IEEE Conference on Systems, Process and Control (ICSPC), pp. 124–129 (2016)
24. Xiaochun, M., Max, Q.: Automatic lesion segmentation from rice leaf blast field images based on random forest. In: International Conference on Real-time Computing and Robotics (RCAR), pp. 255–259. IEEE (2016)
25. William, J.O., Cruz, J.D., Agapito, L., Paul, J.S., Valenzuela, I.: Identification of diseases in rice plant (oryza sativa) using back propagation artificial neural network. In: International Conference on HNICEM, pp. 1–6. IEEE (2014)
26. Majid, K., Herdiyeni, Y., Rauf, A.M.: I-PEDIA: mobile application for paddy disease identification using fuzzy entropy and probabilistic neural network. In: International Conference on Advanced Computer Science and Information Systems, pp. 403–406. IEEE (2013)

A Survey on Comparison Analysis Between EEG Signal and MRI for Brain Stroke Detection

Snehashish Bhattacharjee, Sujata Ghatak, Soumi Dutta, Biswajoy Chatterjee and Mousumi Gupta

Abstract Encephalogram (EEG) provides the recordings of the brain and is used for detecting the brain diseases. In this paper, a detailed study has been carried out for a few applications in detecting brain diseases by EEG and MRI. In addition, a detail comparison study is made between EEG and MRI. This paper has been arranged in two phases out of which, in the first phase, a detailed study has been carried out for EEG processing. The next phase consists of a comparison study of the detection of brain diseases by both EEG and MRI.

Keywords Encephalogram · MRI · EEG

1 Introduction

The activity of human brain starts from the early stage of prenatal growth. From this beginning stage to whole life, the electrical signals produced by the human brain signify the condition of the entire body. Encephalogram, also known as EEG signal, is a measurement of brain activity, which records the electrical activity generated

S. Bhattacharjee (✉) · B. Chatterjee
University of Engineering & Management, Computer Science, Saltlake, Kolkata, India
e-mail: snehashishbhattacharjee@gmail.com

B. Chatterjee
e-mail: biswajoy.chatterjee@iemcal.com

S. Ghatak (✉)
Institute of Engineering & Management, Computer Science, Saltlake, Kolkata, India
e-mail: Sujata.ghatak@iemcal.com

S. Dutta
Indian Institute of Engineering Science and Technology, Shibpur, India
e-mail: Soumi.it@gmail.com

M. Gupta
Sikkim Manipal Institute of Technology, Computer Applications, Majitar, Sikkim, India
e-mail: mousumi.g@smit.smu.edu.in

© Springer Nature Singapore Pte Ltd. 2019
A. Abraham et al. (eds.), *Emerging Technologies in Data Mining and Information Security*, Advances in Intelligent Systems and Computing 814,
https://doi.org/10.1007/978-981-13-1501-5_32

from scalp. The fluctuations occur in voltage when the ionic current generated in the neurons that runs within the brain is measured by EEG. The frequency of EEG is classified into different ranges. The frequency range of alpha waves lies between 8 and 13 Hz and is rhythmic in nature. The amplitude of alpha wave is low. The characteristics of alpha rhythm are present in various regions of the brain, but particularly it is recorded from parietal and occipital region. The frequency of beta waves lies greater than that of alpha waves (greater than 13 Hz) and is irregular in nature. Beta wave is generally recorded from frontal lobe and temporal from brain. The frequency range of delta waves lies lower that of alpha and beta waves (4–7 Hz) and is rhythmic in nature. The lower frequency of delta waves provides higher amplitude. Theta waves are slow in nature, and its frequency range lies in less than 3.5 Hz. It can be found from adult in normal sleep rhythm. Gamma waves are the fastest brainwave frequency, and its frequency range lies between 31 and 100, and its amplitude is the smallest among the other frequency bands [1, 2]. Alpha, beta, theta, and delta are used to detect ischemic stroke by using relative power ratio technique [3]. Ischemic stroke is a kind of brain stroke that causes large number of people's death and permanent disability. Ischemic stroke occurs due to the blockage of an artery to the brain. The brain depends on the arteries to bring fresh blood from heart and lungs [4].

Computer tomography (CT) scans and magnetic resonance imaging (MRI) are commonly used examination methods for stroke detection because they can provide the image of the brain as well as blood vessel [5]. The magnetic field and resonance are used by MRI to capture the image inside the body. MRI is used to detect and identify a stroke lesion in patients that show stroke symptoms. When the patient is positioned to strong magnetic field, the hydrogen nuclei become excited and absorb the radio frequency and release it until it relaxes completely [6]; MRI costs very high and high duration for scanning are the two basic limitations of MRI in emergency situations. MRI is available for only multispecialty hospital in India [7]. Post-MRI biomedical image analysis by image processing is very important because doctors are unable to understand the abnormal growth by simple biomedical image. Pixel-based and morphological method are two existing types of image segmentation which are commonly used in brain stroke detection [8].

Now a days, huge number of research have been carried out on image segmentation for brain stroke detection [9, 10]. Global thresholding followed by morphological operations is used to detect stoke and image segmentation [9]. Convolution neural network introduced in 1989 is a popular technique for segmentation of MRI image and classification. Brain tumor is one of the reasons for causing stroke and for segmentation of tumor clustering; region-based segmentation techniques have been used in recent literatures. Fuzzy c-means algorithm is also mostly used technique to detect stroke and segmentation [8, 10–13].

Existing techniques available for EEG data analysis:

The voltage on the scalp is recorded using EEG, and EEG can be used to diagnose some patients with cerebrovascular disease who have additional clinical problem or complications such as coma or seizures [14]. Conventional EEG is able to detect acute ischemic stroke (AIS). However, the use of quantitative EEG (QEEG) may constitute an alternative examination where the MRI and CT are not available. QEEG has better capability for detection of brain ischemia [3, 15].

Evaluation of AIS can be done through suitable signal processing algorithm. Based on the application of discriminant function analysis (DFA) of stroke brainwaves, ischemic stroke detection has been selected as an input to DFA. Brainwaves classification are carried out using relations power ratio cluster analysis. A flow diagram based on the literature [9] is proposed to analyze EEG signal for ischemic stroke.

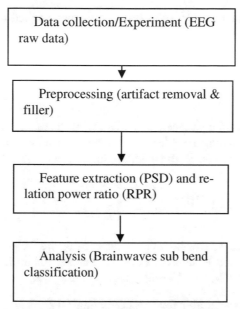

The standard approaches for detecting brain abnormalities through signal analysis are:

1. Extracting the spectral feature of EEG signal by fourier transform.
2. For multiscale basis, wavelet transform can be applied to the signal $f(t)$.
3. Detection of stroke by using various methods, like brain symmetric index.

For step 1, the Fourier transform can be applied to extract the spectral feature of EEG signal.

$$X(f) = \sum_{t \to -\infty}^{t \to \infty} x(t)e^{-ift} \tag{1}$$

where t denotes time sequence and f is the frequency sequence.

The STFT can be used to find both the spatial and temporal features of EEG signal. The frequency content can be determined by applying short-time Fourier transform and is given as

$$X(m, f) = \sum_{t \to -\infty}^{\infty} x(t) f(t - m) e^{-i\pi f} \tag{2}$$

The frequency content can be determined by applying short-time Fourier transform and is given as

$$x(m, f) = \sum_{t \to -\alpha}^{\alpha} x(t) f(n - m) e^{-j\pi f} \tag{3}$$

For step 2, continuous wavelet transform can also be applied to $f(t)$ which is the alternative to Fourier transform

$$W(a, b) = \frac{1}{\sqrt{a}} \int_{-\infty}^{\infty} f(t)'' \psi^* \left(\frac{t - b}{a} \right) dt \tag{4}$$

where a and b are scaling parameters and $\psi_{a,b}(t)$ is the mother wavelet [14].

For step 3, brain symmetry index (BSI) is another useful measure to monitor the changes of brain function and hence finding the abnormalities of the brain

$$\text{BSI}(t) = \frac{1}{M} \sum_{j=1}^{M} \left\| \sum_{i=1}^{N} \frac{R'_{ij} - L_{ij}}{R_{ij} + L_{ij}} \right\| \tag{5}$$

where R_{ij} and L_{ij} are the power spectral density using the Welch method of the right and left hemisphere and M and N are the total number of Fourier coefficients $j = 1$, $2, \ldots, M$ and total number of electrode pairs $i = 1, 2, \ldots, N$.

Existing techniques available for MRI data analysis:

For MRI, although there are different algorithms have already been put forwarded by many recent literatures, still there lies scope to research and improve the efficiency of the existing algorithms or proposing new algorithms which removes the limitations of that of the existing ones. Fuzzy based c-means clustering algorithm is an well known existing technique for MRI that converts the image into meaningful features so that the analysis becomes easier.

The fuzzy matrix (clustering object) denoted by μ with n rows and c columns in which 'n' denotes the number of data objects and 'c' denotes the number of clusters [11]. μ_{ij} denotes the element of the ith row and jth col which signifies the degree of association present in the ith object of the jth cluster. The equations of μ are mentioned below:

$$\mu_{ij}[0, 1], \quad i = 1, 2, \ldots, n. \quad j = 1, 2, \ldots, c \tag{6}$$

$$\Sigma \mu_{ij} = 1, \quad i = 1, 2, \ldots, n \tag{7}$$

$$0 < \Sigma \mu_{ij} < n, \quad j = 1, 2 \ldots, c \tag{8}$$

The position of data defines its membership toward the particular cluster center and the summation of membership of each data point should be equal to one. The main objective of fuzzy c-means algorithm is to minimize the objective function.

$$J(U, V) = \sum_{i=1}^{n} \sum_{j=1}^{c} \mu_{ij}^{m} \|x_i - v_j\| \ldots \tag{9}$$

where m ($m > 1$) denotes the scalar termed the weighting exponent and controls the fuzziness of the resulting clusters and '$\|x_i - v_j\|$' denotes the Euclidean distance between jth cluster center and ith data. Recently, in many research works fuzzy-based c-means clustering technique which exhibits its fuzzy nature is used for segmentation of MRI image. The existing fuzzy c-means clustering method is not up to the mark to segment MRI data efficiently. To overcome this issue, an improved algorithm based on FCM clustering is proposed for brain MRI segmentation. The modified conventional FCM algorithm allows the intensity inhomogeneity by introducing the regularization of the neighborhood influence and bias field [13].

2 Conclusion

Both EEG and MRI are established an area in modern science to detect brain stroke in current medical applications. Although several techniques have already been put forward in many literatures, still there lies a lot of future directions of research in EEG and MRI. The nonlinear nature of EEG attracts many researchers to study its parameters. On the other hand, detecting stroke and processing of MRI images are still a developing area. MRI costs very high and is available in only multispecialty hospital in India. In addition, high duration for scanning makes the basic limitations of MRI in emergency situations.

References

1. Sanei, S., Chambers, J.A.: EEG signal processing. Wiley, Centre of Digital Signal Processing Cardiff University, UK
2. Kalaivani, M., Kalaivani, V., Devi, V.A.: Analysis of EEG signal for the detection of brain abnormalities. In: International Conference on Simulations in Computing Nexus, ICSCN (2014)
3. Omar, W.R.W., Taib, M.N., Jailani, M., Fuad, N., Isa, R.M., Jahidin, A.H., et al.: Acute ischemic stroke brainwave classification using relative power ratio cluster analysis. Procedia Soc. Behav. Sci. **97**, 546–552 (2013)
4. http://www.strokecenter.org/patients/aboutstroke/ischemic-stroke/
5. Rahma, O.N., Wijaya, S.K., Prawito, Badri, C.: Electroencephalogram analysis with wavelet transform and neural network as a tool for acute ischemic stroke identification. In: Proceedings IASTEM International Conference, Bali, Indonesia, 9 Jan 2016, ISBN: 978-81-925751-9-3
6. Baird, A., Warach, S.: Magnetic resonance imaging of acute stroke. J. Cereb. Blood Flow Metab. **18**(6), 583–609 (1998)

7. Jacobs, M., Ibrahim, T., Ouwerkerk, R.: MR imaging: brief overview and emerging applications. Radiographics **27**(4), 1213–1229 (2007)
8. Nabizadeh, N.: Automated Brain Lesion Detection and Segmentation Using Magnetic Resonance Images. Electronic Theses and Dissertations (2015)
9. Alhawaimil, A.: Segmentation of brain stroke image. Int. J. Adv. Res. Comput. Commun. Eng. **4**(9) (2015), ISSN (Online) 2278-1021, ISSN (Print) 2319-5940
10. Gupta, S., Agrawal, M., Sharma, S.K.: Analysis of different brain tumor detection and segmentation techniques in MRI and other medical images. Int. J. Comput. Sci. Technol. IJCST **7**(1) (2016), ISSN : 0976-8491 (Online) I ISSN : 2229-4333 (Print)
11. Zhang, D.-Q., Chen, S.-C.: A novel kernelized fuzzy C-means algorithm with application in medical image segmentation. Artif. Intell. Med. **32**(1), 37–50 (2004)
12. Mangukiya, M., Lohiya, R.: A review for segmentation for brain MRI images using fuzzy logic. **2**(3) (2016) www.ijariie.com, IJARIIE-ISSN(O)-2395-4396
13. Li, M., Zhang, L., Xiang, Z., Castillo, E., Guerrero, T.: An improved fuzzy c-means algorithm for brain MRI image segmentation. In: Proceedings of 2016 International Conference on Progress in Informatics and Computing (PIC), pp. 23–25 (2016)
14. Vijaya, S.K., Badri, C., Misbach, J., Soemadri, T.P., Sutarno, V.: Electroencephalography (EEG) for detecting acute ischemic stroke. In: Proceeding of 4th International Conference on Instrumentation, Communications, Information Technology, and Biomedical Engineering (ICICI-BME) Bandung, 2–3 Nov 2015
15. Omar, W.R.W., Taib, M.N., Jailani, M., Mahamadad, A., Ahiddin, A.H.J., Sharif, Z.: Application of discriminant function analysis in ischemic stroke group level discrimination. In: 2014 IEEE 10th International Colloquium on Signal Processing and its Application (CSPA), pp. 229–232 (2014)

Part IV
Natural Language Processing

Enhancing Call Center Operations Through Semantic Voice Analysis

Charitha Samaranayaka and Saminda Premaratne

Abstract A call center is a central place where customer care agents handle queries of customers over the telephone to fulfill and satisfy their needs. Information searching delay is an increasingly and important problem in a call center environment because of multiple and not similar subsystems in operation. Developed application is running stand-alone and no need to install, and it uses .NET framework and NAudio open-source library for enhancing signals and converting enhanced voice signal into text. Application is consisted with a prediction algorithm and a categorization functionality which is capable for correcting falsely recognized text and eliminates the need of following highly time and human memory consuming file menus or hyperlinks. This will increase the accuracy of the recognized sentences, lead to better information searching functionality, and propose the most probable locations of the required information. Application was tested with several users and achieved 90% accuracy for information categorization.

Keywords Call center · NAudio · Voice to text · Information searching
Information categorization · Markov chain · Bayes Theorem

1 Introduction

In a call center, several metrics will be used to measure the performance such as service level, quality monitoring scores, customer satisfaction, adherence, occupancy, average handle time, number of calls offered, first call resolution rate [1, 2]. Average handle time (AHT) can be interpreted as the sum of average amount of time an agent spends talking and in post-call work in relation to a call. This is a metric which is directly related to agent skill and his/her performance. Average handle time

C. Samaranayaka · S. Premaratne (✉)
Faculty of Information Technology, University of Moratuwa, Moratuwa 10400, Sri Lanka
e-mail: samindap@uom.lk

C. Samaranayaka
e-mail: Charitha.Samaranayaka@gmail.com

© Springer Nature Singapore Pte Ltd. 2019
A. Abraham et al. (eds.), *Emerging Technologies in Data Mining and Information Security*, Advances in Intelligent Systems and Computing 814,
https://doi.org/10.1007/978-981-13-1501-5_33

will increase due to low skill level of the agent, inability to clearly understand the customer's query, delay of accessing relevant information, not up-to-date details of the systems, less searchability of the systems, less user-friendliness of the systems, etc. [3].

Danilo Garcia and others have conducted a research on managing time and customer satisfaction at call centers [4]. This research based on the study of relationship between customer satisfaction and waiting time for the service. Organization with intelligent information regarding customers uses them as an asset to their success [3]. Waiting time is an important factor in the service industry because increased waiting time will lead to customer dissatisfaction [2]. This will lead to work strategy of the organization. This research suggests managers and decision-makers might need to know which variables might influence recollections of satisfaction with waiting time in the queue. Also, research presented that people are more willing to wait in line (queue) depending on how much they value the service provided by the organization.

Customer handling time can be reduced by training the call center agents and employing proper tools and technologies to query customer requests efficiently [5]. Major media in a call center is voice calls, although other access methods are available to customers such as Web, text, SMS, mail [6–8]. It is a mistake to go with automated low-cost self-services by ignoring the human side of the customer [7]. Unplanned or unsuitable technology enhancements will lead to additional stress on call center staff [7]. It is important to improve the processes related to human answering of the customer's telephone calls by reducing the delay. We are using voice to text technology [9–11] to reduce the typing delay and understanding time of the customer query by using sentiment analyzing [12].

In this research, we are focusing on reducing the delay of accessing information which leads to reducing average handle time via voice to text conversion.

2 Principle of Voice to Text

2.1 Sampling

We are reading PCM 16-bit samples via .NET libraries and NAudio library. Product of the time signal and an impulse train gives the sampled signal Xs and the sampling function (also called a comb function). Below equation will explain the time domain sampling mathematically.

$$x_s(t) \sum = x(t).comb(t) = x(t) . \sum_{m=-\infty}^{\infty} \delta(t - mt_0) = \sum_{m=-\infty}^{\infty} x[m] . \delta(t - mt_0) \quad (1)$$

2.2 IIR Filter

In this research project, we have used a low pass time domain filter to remove background noise of the sound input. Below equation is describing the FIR filtering [13] mathematically.

$$y(n) = h(n) \otimes x(n) = \sum_{k=0}^{N-1} h(k)x(n - k) = \sum_{k=0}^{N-1} a_k x(n - k) \qquad (2)$$

where N is the size of the filter and $x(n - k) = 0$ if $(n - k) < 0$.

While FIR filter is using forward filter only, IIR filter will use both forward filter and feedback filter parts. In this project, we are using an IIR filter for filtering the sound input by using NAudio library. Below equation is describing the IIR filtering mathematically

$$y(n) = h(n) \otimes x(n) = \sum_{k=0}^{N-1} a_k x(n - k) - \sum_{k=1}^{M} b_k y(n - k) \qquad (3)$$

where N is the size of the forward filter, M is the size of the feedback filter, and $x(n - k) = 0$ if $n - k < 0$.

2.3 Markov Chain

In this research project, we are converting the filtered sound into texts, reject less accurate recognitions, and correct falsely identified words by predicting upon the previous recognitions. Below example will explain a two-state Markov chain.

$$P = \begin{pmatrix} 1 - \alpha & \alpha \\ \beta & 1 - \beta \end{pmatrix} \qquad (4)$$

(α) is the probability of changing from state 1 to 2. ($1-\alpha$) is the probability of staying at the same state. (β) is the probability of changing from state 2 to 1. Finally, ($1-\beta$) is the probability of staying at the same state.

As above example explained, in our project we have calculated the probabilities of occurring next word after a given word by providing training set initially. We have compared the recognized word from the recognition engine with the highest probable words and select the most matching word from the highest probable word list.

2.4 Levenshtein Distance

In this research project, to compare two words (strings), we have used the Levenshtein distance between two words. Mathematically, the Levenshtein distance between two strings a, b (of length $|a|$ and $|b|$, respectively) can be calculated by lev a, b ($|a|$, $|b|$) function where

$$\text{lev}_{a,b}(i, j) = \begin{cases} \max(i, j) & if\, \min(i, j) = 0, \\ \min \begin{cases} \text{lev}_{a,b}(i - 1, j) + 1 \\ \text{lev}_{a,b}(i, j - 1) \\ \text{lev}_{a,b}(i - 1, j - 1) + 1_{(ai \neq bj)} \end{cases} & \text{otherwise,} \end{cases} \tag{5}$$

where 1 $(ai \neq bj)$ is the indicator function equal to 0 when $ai = bj$ and equal to 1; otherwise, lev a, b (i, j) is the distance between the first i characters of a and the first j characters of b.

As above function explained, we have calculated the distance between the recognized words from the recognition engine with the highest probable words and select the lowest distance word from the probable words list.

2.5 Bayes Theorem

In this research project, for clustering, we have used naïve Bayes theory [12]. Equation of the naïve Bayes theory is explained below.

Bayes Theorem to predict the probability.

$$P(h/D) = [P(D/h)P(h)]/P(D) \tag{6}$$

where

$P(h)$	Prior probability of hypothesis
H	Predicted value
$P(D)$	Prior probability of training data D (value given)
$P(h/D)$	Probability of h given D (expected o/p).

Above basic equation is able to cluster word list into two categories only. We have enhanced the naïve Bayes theory to cluster a word list into several categories.

3 Methodology

The solution was designed as a stand-alone system with HTTP client embedded to that for the communication between application and the backend.

Application was built in .NET4 framework, and application has three modules, namely enhancing the voice signal and converting enhanced signal to text module, categorizing texts, and displaying searched results on Web browser module which will be described separately.

3.1 Input to the System

Customers will choose voice calls as a medium to contact the organization. Call center agents use headsets as the input device and a computer connected to Intranet to answer customer queries. This voice input via sound card will be taken as the input to the software. Voice input will be preprocessed by using an audio library before it sends to the voice to text converting algorithm which is coded inside windows library. Converted texts are presented to user in the interface of the software to manipulate manually by using key inputs. Search command will be received by the operator of the software, and output will be displayed in a Web browser.

3.2 Output of the System

Though the main output of the system is searched information with respect to a customer query, clustering of the input and predicting the level of customer query are also outputs of the system. Searched information will be presented on a Web browser. Searching criteria is displayed on the software as an intermediate output to the user.

3.3 Process of the System

This research project software contains three major processes, namely convert speech to text, mining text, and launch Web browser with results. First process contains reading voice signal from the input, filtering the samples, enhancing the samples, converting enhanced voice signal to text, and improving the outcome texts sub-processes. Second process contains categorization of text which was the outcome of the first process and saving output for reporting purposes sub-processes. Third process contains displaying results on Web browser and focusing the results sub-processes.

Enhancing the Voice Signal and Converting Enhanced Signal to Text This
module will read inputs of the system, enhance input, filter input signal, and convert
enhanced voice signal to text. In this section, we are explaining the implementation
of the module step by step. First step is reading the voice signal from the sound
card of the computer. We are using a headset with an audio jack to implement and
test the new solution. Input signal will be read as a PCM16 signal which means it
has 16 bits to represent a signal sample. We are using NAudio library to read inputs
from the sound card as explained. This library supports both native C++ and C#
.NET technologies, and we are using the C# .NET support functions of the library
to implement our solution. Below code segment will explain the preprocessing of
the retrieved audio signal which involves sampling and IIR filtering.

```
/**
* Preprocessing initialization
*/
waveSource = new WaveIn();
waveSource.DataAvailable            += new EventHandler <
WaveInEventArgs > (waveIn_DataAvailable);
waveSource.RecordingStopped += new EventHandler
        (waveSource_RecordingStopped);
filter = BiQuadFilter.LowPassFilter
     (waveSource.WaveFormat.SampleRate, 10000, 1);

/**
* Low pass filtering for retrieved PCM16 signal
        and creating intermediate output wav signal.
*/
for (int i = 0; i < floatBuffer.Count; i ++)
{
waveFile.WriteSample(filter.Transform(floatBuffer[i]));
waveFile.Flush();
}

/**
* Time domain low pass filtering algorithm
*/
   var result   =   a0 * inSample   +   a1 * x1   +
a2 * x2 - a3 * y1 - a4 * y2;
    // shift x1 to x2, sample to x1
    x2 = x1;
    x1 = inSample;
    // shift y1 to y2, result to y1
    y2 = y1;
    y1 = (float)result;
```

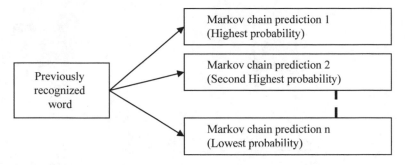

Fig. 1 Markov chain generation with respect to the previously identified word

For recognizing recorded sounds, first we need to initialize the recognizing engine which is implemented with .NET library. We are using English language for this experiment. So we need to select language parameter as en-US. We need to implement two important events of the recognizing engine as SpeechRecognized and SpeechRecognitionRejected. SpeechRecognized event will be triggered at a successful recognition of a sound signal. We are checking the confidence of the recognized word prior to accept it. This will improve the quality of the resultant texts.

The input for the Markov chain is a text file which contains a text paragraph. Logic will generate the Markov chain for the prediction logic. Each time a word was recognized, and it will be compared with the Markov chain predicted words and match it to the closest word by using Levenshtein distance algorithm.

As Fig. 1 shows, Markov chain will be generated with respect to the previous success output. Then these predictions will be matched top to bottom against the recognizing engine's output to correct the grammar and avoid false recognitions by using a threshold Levenshtein distance.

Categorizing Texts This module is responsible for categorizing the words which are converted from voice in the previous module. Theory behind this implementation is naïve Bayes theory which was explained in previously. Although naïve Bayes theory categorizes a word set into two groups, namely positive and negative, enhanced implementation is capable of calculating the probability of more than two groups. Firstly, we need to feed metadata into the calculation logic. This will be done via a text file.

For an example below, metadata will create the data structure explained in Fig. 2.

A:ADSL:router, adsl, internet, web
A:PEO:router, peo, tv
B:router:ZTE, DLINK.

When the categorize command issued, naïve Bayes theory will calculate the matching probability for each end node. Then it will select the parent node of the highest probable end node as the category for the converted texts. For an example,

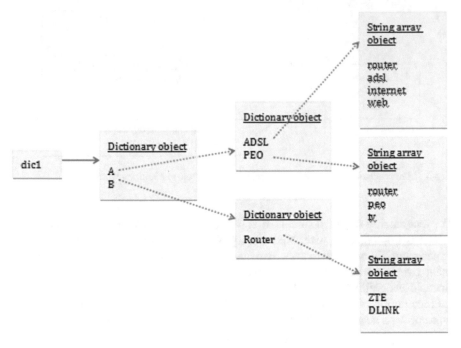

Fig. 2 Sample data structure of cluster metadata

if router and ZTE words are presented in the converted text, then they will fall into two categories, namely ADSL and router as A and B category.

Displaying Searched Results on Web Browser This module is responsible for launching Web browser with the results which are searched by the agents. Logic will take the resultant text which is derived from voice to text conversion and the users input and construct a URL with a predefined prefix and launch the Web browser with that URL.

Second important functionality of this module is handling predefined keystrokes such as SHIFT PLUS ONE. For this purpose, we have to use two system library methods, namely RegisterHotKey and UnregisterHotKey. SHIFT is the constraint, and the key is ONE. Both fields hold an integer value and uses Boolean operators to combine them. Then this combined value will register to receive triggers when user presses the above mention key combination.

3.4 Overall Features

In connection with the input, output, users, and process, the overall features of the developed system include the following characteristics.

- Multiple inputs (voice input and agents input)
- User-friendly
- Research platform for call centers (analysis voice signals).

4 Result and Discussion

Below test cases were conducted to measure the accuracy of the research application by using multiple users. Below sentences were spelled by multiple users, and the results were presented in "Table 1."

- internet,is,slow
- internet,browsing,is,slow
- new,connections,request
- mail,box,is,full
- order,a,television,channel
- bill,is,not,receiving.

Table 1 Comments of the performance test

Spelled sentences	Comments (speech to text)	Cluster	Information searched correctly
internet,is,slow	Internet word sometimes	○ ADSL	○ Yes
internet,browsing,is,slow	not recognized	○ ADSL	○ Yes
new,connections,request	Bill word did not identify	○ PSTN	○ Yes
mail,box,is,full		○ EMAIL	○ Yes
order,a,television,channel		○ PEO	○ Yes
bill,is,not,receiving		○ BILL	○ Yes
internet,is,slow	Internet word sometimes	○ ADSL	○ Yes
internet,browsing,is,slow	not recognized	○ ADSL	○ Yes
new,connections,request	Bill word did not identify	○ PSTN	○ Yes
mail,box,is,full		○ EMAIL	○ Yes
order,a,television,channel		○ PEO	○ Yes
bill,is,not,receiving		○ BILL	○ Yes
internet,is,slow	Internet word sometimes	○ ADSL	○ Yes
internet,browsing,is,slow	not recognized	○ ADSL	○ Yes
new,connections,request	Bill word did not identify	○ PSTN	○ Yes
mail,box,is,full		○ EMAIL	○ Yes
order,a,television,channel		○ PEO	○ Yes
bill,is,not,receiving		○ BILL	○ Yes
internet,is,slow	Internet word sometimes	○ ADSL	○ Yes
internet,browsing,is,slow	not recognized	○ ADSL	○ Yes
new,connections,request	Bill word did not identify	○ PSTN	○ Yes
mail,box,is,full		○ EMAIL	○ Yes
order,a,television,channel		○ PEO	○ Yes
bill,is,not,receiving		○ BILL	○ Yes

Below paragraph shows the training data set for the Markov model.

- Internet is not working. Internet browsing is slow. Internet usage limits. Need to reset password. Need to create portal account. Television program is not available. No television screen display. Order a television channel. New connections request. Bill is not receiving. Mailbox is full. Mail is not receiving. Unwanted mails.

Below paragraph shows the starting words for the training data set for the Markov model.

- Internet, need, no, order, new, bill, mail, unwanted, television

Below paragraph shows the cluster metadata for information clustering.

- A:ADSL:internet, adsl, web, slow
- A:PEO:internet, tv, television, channel
- A:PSTN:new, connections, telephone
- A:EMAIL:mail, box, receiving
- A:BILL:bill, receiving.

According to the conducted test cases, application meets the requirements of the proposed solution.

5 Conclusion

Our research was success with the accuracy of information categorization 90 and 100% of information querying from content management system via HTTP communication and displays it on the Web browser (Table 1). Test metadata and results were described in the results and discussion section. Although the voice to text conversion accuracy is varying with the sound cards performance, our prediction algorithm and signal enhancing algorithm increased the accuracy of the voice to text conversion than the normal voice to text conversion by using a voice to text engine. We concluded that proposed application can reduce the average handle time (AHT) in a call center due to enhanced capability of searching text across the multiple systems simultaneously.

References

1. Gans, N., Koole, G., Mandelbaum, A.: Telephone call centers: tutorial, review, and research prospects. Manufact. Serv. Oper. Manag. 5(2), 79–141 (2003)
2. McGuire, K.A., Kimes, S.E., Lynn, M., Pullman, M.E., Lloyd, R.C.: A framework for evaluating the customer wait experience. J. Serv. Manag. 21(3), 269–290 (2010)
3. Evgeniou, T., Cartwright, P.: Barriers to information management. Eur. Manag. J. 23(3), 293–299 (2005)

4. Garcia, D., Archer, T., Moradi, S., Ghiabi, B.: Waiting in vain: managing time and customer satisfaction at call centers. Psychology 3(2), 213–216 (2012)
5. Rasooli, P., Albadvi, A.: Knowledge Management in Call Centres. Electron. J. Knowl. Manag. 5(3), 323–332 (2007)
6. Kumar, A., Telang, R.: Does the web reduce customer service cost? empirical evidence from a call center. Inf. Syst. Res. 23(3-NaN-1), 721–737 (2012)
7. Oodith, D., Parumasur, S.B.: Technology in a call center: an asset to managing customers and their needs? Probl. Perspect. Manag. 12(1) (2014)
8. Fitzsimmons, J.: Service Management: Operations, Strategy, Information Technology with Student CD, 5th edn. McGraw-Hill/Irwin, Boston (2005)
9. Lee, L., Oun-Young, M.: Voice and text messaging-a concept to integrate the services of telephone and data networks. In: IEEE International Conference on Communications, ICC'88. Digital Technology-Spanning the Universe. Conference Record, pp. 408–412 (1988)
10. Tuerk, C., Monaco, P., Robinson, T.: The development of a connectionist multiple-voice text-to-speech system. In: 1991 International Conference on Acoustics, Speech, and Signal Processing, ICASSP-91, pp. 749–752 (1991)
11. DobriSek, S., Gros, J., Mihelic, F., Pavesic, N.: HOMER: a voice-driven text-to-speech system for the blind. In: Proceedings of the IEEE International Symposium on Industrial Electronics. ISIE'99, vol. 1, pp. 205–208 (1999)
12. Medhat, W., Hassan, A., Korashy, H.: Sentiment analysis algorithms and applications: a survey. Ain Shams Eng. J. 5(4), 1093–1113 (2014)
13. Shome, S.K., Vadali, S.R.K., Datta, U., Sen, S., Mukherjee, A.: Performance evaluation of different averaging based filter designs using digital signal processor and its synthesis on FPGA. Int. J. Signal Process. Image Process. Pattern Recognit. 5(3) (2012)

Integration of Wit API with Python Coded Terminal Bot

Sanyam Jain, Shivani Sharma and Ravi Tomar

Abstract This research paper consists of the development of a python chatbot integrated with specific packages and dependencies having capabilities to perform various real-life applications which can also be packaged with some integrated systems as a personal assistant. Various Web services are used using their APIs so as to make it more powerful and less dependent on python packages. It operates completely on Linux (Ubuntu) terminal. Python command with the __main__.py file has been used to run it. Terminal then becomes a live Web-based chatbot service with STT and TTS services. Natural language processing is further done by Wit.ai API which token key is set in the code. Each NLP is done by the API and then processed, and crunched text is sent from the wit engine direct to display terminal.

Keywords NLP · Chatbot · AI · Supervised learning · Python

1 Introduction

What do you think the first time you come across chatbot or personal assistant. It is been very first time in 1997 Microsoft came with clippy.js (Office assistant) in their Microsoft Office products though it is discontinued due to internal reasons. Nowadays, we have various engines which do all crunching work for us with just an integration of their API to our code. Even Google, Amazon, Apple provide their packages to import skills into our personal assistant. Google cloud platform, Amazon Alexa, Apple Siri all are popular engines to provide this kind of services. The chatbot

S. Jain (✉) · S. Sharma · R. Tomar
UPES, 248007 Dehradun, India
e-mail: sanyamjaincs@gmail.com

S. Sharma
e-mail: 1998shivisharma@gmail.com

R. Tomar
e-mail: ravitomar7@gmail.com

© Springer Nature Singapore Pte Ltd. 2019
A. Abraham et al. (eds.), *Emerging Technologies in Data Mining and Information Security*, Advances in Intelligent Systems and Computing 814,
https://doi.org/10.1007/978-981-13-1501-5_34

provides user both types of interfaces whether you can do speech recognition or writing commands.

Using terminal as chatbot screen, using text mode it is easy to just type whole sentence with a specific keyword pointing to the dictionary and giving output. For example, keyword "rain" having dictionary pointing to similar keywords like umbrella, weather, forecast, and other bot synonyms which all redirect to the statement "its rainy outside, good to take umbrella outside." This is chat based where your text-based input is processed by the code.

Using wit integration, it is operated using NLP having high success rate. The natural language given by the user is handled by code and output is given as "You said this ___" Now this output is used as an input for the text-based Web chatbot. This type of process requires time as the audio is sent to the wit server first, and then it is reverted to the specific session of that server (using API).

This research paper is mainly focused to the integration of text-based Web chatbot and wit API so as to make it text-to-speech and speech-to-text enabled in Python and performing those speech recognition tasks. The chatbot is already undergone many black box environments to yield 78% correct result [1]. Flow of this paper is as: (1) System architecture, (2) Software specifications, (3) Open-source approach, (4) System implementation, (5) Results, (6) Conclusion.

2 Proposed Work

The code is exclusively editable, and the libraries used can be changed anytime. The modularity of the software allows to extract each function separately, and the function can be modified according to user. Advanced users can embed their own APIs for more functionalities to the code. Wit engine is free-API-driven platform (python here), and anyone can generate its own API and start implementing NLP. Wit also provides the support for analyzing the processed natural language. This can be done on wit dashboard online.

The Agile development supports the bot to adapt the in-between errors and circumstances. The development model followed is incremental model, and each skill is developed step by step. The open-source development allows the user to add upcoming technologies to the existing package preserving the algorithms used.

3 Related Work

Earlier attempts have been made for the same. An intelligent Web-based voice chatbot was proposed in 2009. The working of the bot is fully dependent on Java. Compared to 2009 today Python is the most popular language in 2017. Popularity graph is still increasing for python comparison with other languages. The related work counts to making JavaScript chatbots which are limited to only several tasking like chat responses, using native system calls rather than exploiting new trend API. One major drawback till now is none of them is open source. This is completely open source so as to meet new trends and so as to integrate it. Also, all the processing is done on server so as to reduce system consumption.

4 Implementation

After fulfilling complete dependencies in the Linux or Mac system (including pip installations) this chatbot works on the terminal screen (*sudo pip install requirements.txt*). Requirements include various python packages with some specific versions for which the bot is coded compatible. For example, colorama $==0.3.7$, instantmusic $==2.2$, PySocks $==1.6.6$, pyttsx $==1.1$, aimlgTTS $==1.1.8$, etc.

System described in the bot consists of two possibilities:

CASE 1: Text-based Web chatbot
CASE 2: Wit integration for the NLP

Wit is API-based NLP processor, and it can also be integrated to any application than chatbot [1]. The processed text from wit server is the output for the text-based Web chatbot. The system architecture for the part (1) is described below, and the main functionalities of the bot are described as:

(1) Audio handler: It uses gTTS package and Google API to generate voice output [2].
(2) Api.chucknorris.io [3] provides the bot to grab random jokes from the Internet. This API memorizes the user inputs to the server and customize the future outputs of random jokes according to interests.
(3) Maps [2] module follows the directions API to find the optimal path and distance between two coordinates (mapps.directions (toCity, fromCity)).
(4) Evaluator or calculator is also embedded in the packages as a module, and all basic arithmetic operations can be performed using "evaluate some_number (operation) some_number."
(5) Empty __init__.py file must be attached so as to make the package folder treated as actual package by Python [4].
(6) Current Weather [2]: requires modules (WeatherIn, Maps, umbrella): getLocation() function is called firstly. As soon the (lat, long) are parsed from API, coordinates are then sent to API endpoint to further get weather result. The alternative function is also supplied if the (lat, long) are unable to fetch, traceIp() function fetches the exact location of the client and mapps.getLocation()[city] gets city. And finally, weather() function gets weather from openWeatherMap server (Fig. 3).
(7) News [2]: The very unique news API provides you the freedom to grab news from different sources according to the user interest. The bot is designed such as to memorize the old interest topics so as to ask next time about to continue

```
class News:

    def __init__(self, source="google-news", apiKey="7488ba8ff8dc43459d36f06e7141c9e5"):
        self.apiKey = apiKey
        self.source = source
        self.url = "https://newsapi.org/v1/articles?source=google-news&sortBy=top&apiKey=7488ba8ff8dc43459d36f06e7141c9e5"
        self.m = Memory()
```

Fig. 1 News fetching from Google News using newsApi.org

as the same user or to refresh the interested topics. You can select to see news from (BBC, BuzzFeed, Google, Reddit, or TechCrunch). You can also select a default source (Fig. 1).

(8)   Shutdown: Other system commands can also be performed using this os.system (**command here**).

(9)   The same mapps.getLocation() function gives the exact online date and time according to the coordinates of the traced ip (done by API) [4].

10)   ToDo module is same as the reminder function in usual chatbots. This feature is integrated with Twilio (api.twilio.com) [5] (Fig. 2). Python repository for the same is provided with the API set a new way to reminder system. Sample code is shown in this feature which sends direct text message to the user who set the reminder. The same can be performed in various ways as:

> Remind me through mail for keynote at cross street amphitheater tomorrow morning

The most important query done by user to the bot is "How's the weather like?", this feature comes under basic functionalities of a chatbot. The first and most asked query to Google and Apple Siri is about weather and stocks. An average Siri user ask minimum of three times about weather and five times about stocks. Weather now has become the most relevant and necessary query asked to a chatbot. The rain teller mechanism from api.openweathermap.com follows the following algorithm shown in.

(11)   Miscellaneous: The bot can do many more functionalities than described in the paper. Any new function can be modularized to the package, thus making it always updated. AIML also makes it more human-friendly responding to normal conversations for example, "How are you?", "What are you?" Sample code looks like shown in Figs. 4 and 7.

(12)   Keyword base: It includes all the keywords which can be added in the user input sentence. Table 1 shows the list of keywords that must be included in any sentence given by the user.

5   Software Engine Specifications

(1)   The bot described in this paper is text-based Web chatbot and integrated with wit engine to accept voice input and gives best applications of various APIs.

```
from twilio.rest import Client

# Your Account SID from twilio.com/console
account_sid = "ACXXXXXXXXXXXXXXXXXXXXXXXXXXXXXXXXXXX"
# Your Auth Token from twilio.com/console
auth_token  = "your_auth_token"

client = Client(account_sid, auth_token)

message = client.messages.create(
     to="+15558675309",
     from_="+15017250604",
     body="Hello from Python!")

print(message.sid)
```

Fig. 2 Twilio API working in Python (*Source* www.twilio.com/docs/libraries/python)

Table 1 Keywords

"check": ("ram", "weather", "time")	"evaluate"	"news"	"display": ("pics",)
"ask"	"hotspot": ("start", "stop")	"os"	"reboot"
"chuck"	"increase": ("volume",)	"remind"	"todo"
"enable": ("sound",)	"open": ("camera",)	"screen": ("off",)	"umbrella"
"update": ("location", "system"), "weather"			

(2) Python being most popular language, make the chatbot more accessible widely, and could do seamlessly many operations.

(3) This is the rise of APIs. Making complex software with less code and integrating more powerful APIs to serve user with most of the chatbot.

(4) The brain is set offline, and it performs the memory and parsing work. The actual dictionary-based keywords are parsed from the natural language given by user by brain and are processed by matching algorithm to generate required output. The brain also remembers the data to the PyCache to make it more personalized (Fig. 3).

(5) The self-learning engine of the bot enables it to memorize the addresses of the resources acquired during instantiation and hence prevents in deadlock.

(6) All the text-based conversations are performed locally in the system to provide fast output, and the emotion and chatting part are coded under Artificial Intel-

ligence Markup Language while the thinking part is done using Web services (Fig. 4).

(7) User have two types to follow the conversation, the former is text-based input which is comparatively faster than the latter voice-based conversation.

6   System Implementation and Application

The main language used throughout is the python. The software is embedded in Web using Django framework. First time user will never know about the backend of the bot. Developing programs for the Web have become viral since the increase of "Web 2.0" that focuses on user-generated content on Web sites.

It is endlessly been come-at-able to use Python for creating Web sites, but it had been a rather tedious task. Most HTTP servers are written in old languages, in order that are unable to run/compile Py code natively—API is required between the server and the program. These interfaces/APIs, outline, however, programs move with the server [6, 7]. Django could be a framework consisting of many tightly coupled parts (Fig. 5) that were written from scratch and work along okay.

The template engine is text-based and is intended to be usable for page designers who cannot write Python. It supports template inheritance and filters (which work like UNIX operating system pipes). Voice recognition is accomplished exploitation an open supply library known as wit. Speech recognition is often broken into two categories with five subsections in total, speakers and speech designs. Speakers compose single speaking and speaking severally wherever speech designs embody isolated word detection, connected word detection, and continuous word interpretation. Isolated word detection needs long breaks between all words to with success resultant words (Fig. 6).

Continuous word recognition additionally needs this, however, significantly shorter break. The last part of speech modeling is continuous speech detection whenever one will speak fluently or not need stopping or breaking between words. This has issues decoding similar vowel within the starting of words. An important a part of the speech system is contained among a configuration file. Using API and con-

```
If(Rain.value() >=300 && rain.value()<=500){
        print "take umbrella";}
If(Rain.value()>700)  {
            print "no umbrella";
        }else{
            print "take umbrella";
        }
```

Fig. 3 Rain strengths are decided by the values. The end result is judged by the values fetched from the API. The value is then manipulated locally

```
<aiml version="1.0.1" encoding="UTF-8">
<!-- basic_chat.aiml -->

    <category>
        <pattern>HELLO</pattern>
        <template>
            Well, hello!
        </template>
    </category>

    <category>
        <pattern>WHAT ARE YOU</pattern>
        <template>
            I'm a bot, silly!
        </template>
    </category>

    <category>
        <pattern>HOW DID I *</pattern>
        <template>
        <random>
            <li>I'm not sure I wasn't paying a bit of attention</li>
            <li>Well you see</li>
            <li>I do not recall</li>
        </random>
        </template>
    </category>
```

Fig. 4 Simple AIML code for chatting

figuration file words are matched and decoded with wordbook. AIML files or the chatting components of the bot are forming the acoustic model placed within the API (Fig. 7).

The speech recognition engine and the text bot require the basic configuration of fulfilled dependencies on the offline system, The packages, repositories, and modules all must be satisfied before running the application. User calls the run.py on the terminal and starts recorder. The voice is being recorded to the file which is going to be uploaded to the wit server. It responds with text to the terminal screen. Before displaying the actual query processed by wit, the text is treated as the text input from user. This gives the illusion to the user that the natural language is actually being processed locally to the system rather giving an API scenario.

The input sentence is split into words using regex (Fig. 6) rather than using string functions.

- Now each word is matched with the skill_set dictionary (using string matching algorithm with O(n2) complexity), simple recursion algorithm continues to check each word.
- Meaning association, determine the resultant meaning of the sentence.

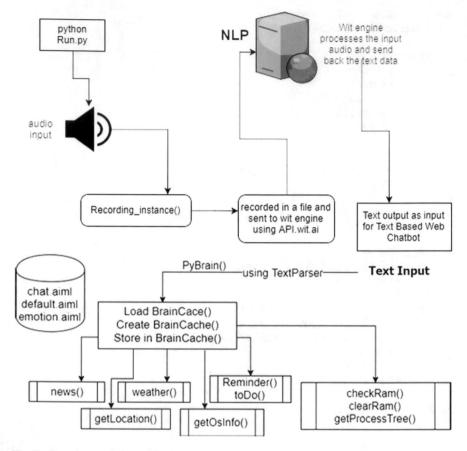

Fig. 5 Complete workflow of the bot

```
rgx = re.compile("(\w[\w']*\w|\w)")
s = "How's The weather like today? Should I take umbrella before leaving home?"
rgx.findall(s)

["How's", 'the', 'weather', 'like', 'today', 'should', "I", 'take', 'umbrella', 'before', 'leaving', 'home']
```

Fig. 6 Final emotion of the user query is judged by parsing keywords and matching them to the command interpreter to get desired output

- On matching word from dictionary call that particular function executing the required task. Else reply "Currently out of my capabilities."

The bot is targeted on supervised human guide for a learning method. This helps to highlight correct answers because the system memorizes from conversations with users and updates the AIML files from earlier mistakes. AIML files are structured to fit XML data format, with context descriptions among successive code part. A training set consists of queries and their responses as pattern and template. Thus, the system is restricted to the already skilled set of conversation. Worst case can be when

```
<category>
<pattern>ONE TIME I *</pattern>
<template>
    <random>
        <li>Go on.</li>
        <li>How old are you?</li>
        <li>Be more specific.</li>
        <li>I did not know that.</li>
        <li>Are you telling the truth?</li>
        <li>I don't know what that means.</li>
        <li>Try to tell me that another way.</li>
        <li>Are you talking about an animal, vegetable or mineral?</li>
        <li>What is it?</li>
    </random>
</template>
<category>
```

Fig. 7 Simple AIML code for chatting with the bot

there is no match to the query, and the bot can be trained to give special responses that time. System can generate a default response that time. This can be thought about a trick for inadequate knowledge, however, could be a commencement for any chatbot.

7 Results

The combination of voice and text-based input–output allows a very user-friendly environment. The simple architecture allows any user to integrate this bot to their more complex software machine. Bot can perform as many functions as any other bot can do. There are endless possibilities that can be integrated to the chatbot. One can also thin on very small scale to the embedded systems. This chatbot can be embedded to the average Linux-based systems having low configurations, portable systems integrated with Raspberry Pi, etc. The applications include roadside kiosks, navigation systems, story tellers, TODO reminder system, news retrievers, even on the personal computers. The bot results in all of its operations precisely and as expected. Each module is tested periodically (black box). This paper work results in an ideology or a cheat sheet for the future developing chatbots to add intelligent work with more advancements. The bot is capable of doing all major tasks that a virtual assistant posses with small system operations including shutdown and rebooting the system, osinfo(), etc. The aim of the paper is to apply a chatbot or intelligent online assistant with text and voice inputs integrated with smart and accurate responses. The Modularised nature of any system gives opportunities to add more updates and latest technologies to the engine. Open-source technologies also make developers feasible to use the bot as more personalized.

After having thorough knowledge of the subject, there are following conclusions:

- The bot fulfills all the basic functionalities to a great extent and more accurately.
- The location parsing algorithm follows tracing Internet IP and hence determines latitudes and longitudes further resulting to time zone.
- The weather API also responses about rain by giving rain.value() as output and judges whether to take umbrella or not.
- Any system command can be operated using os.system ("command_here").
- All the latest popular technologies about AI gives their API in python, thus making bot to sustain future advancements and/or updations.
- The core functioning is done by PyBrain and text parsing system. The short conversations handled by AIML file working.

8   Conclusion

Thus, it is concluded that this modularized development is successfully conducting all major tasks that are must have to be performed by any chatbot. The bot described here is able to perform all major tasks by the user voice or text input. The keywords are successfully parsed from the query and matched from the "cmdinterpreter.py", and it collects all the possible user inputs.

There are future possibilities to add more functionalities to the chatbot. Due to its modularizing nature and python development, it successfully accepts all new features. For example, face recognition system can also be added to the current functionalities to make results more user specific.

References

1. Wit-Docs Home, Wit engines. https://wit.ai/docs
2. Google Cloud Computing, Hosting Services and APIs | Google Cloud Platform, Google. https://cloud.google.com/
3. Chuck Norris Jokes Api, Chuck Norris Inc. https://api.chucknorris.io
4. TimeZoneDB- API access for developers. https://timezonedb.com/api
5. The Twilio Python Helper Library, Twilio. https://www.twilio.com/docs/libraries/python
6. Python Documentation. https://docs.python.org/
7. StackOverflow. https://stackoverflow.com/questions/tagged/django
8. EUROCON 2009. http://ieeexplore.ieee.org/abstract/document/5167660/

An Approach of Splitting Upsarg and Pratyay of Sanskrit Word Using Paninian Framework of Sanskrit Grammar

Bhavin Panchal, Abhijitsinh Parmar and Sadik Dholitar

Abstract Computational linguistic began more than sixty years ago. The designing of a system splits upsarg and pratyay of Sanskrit word using Maharishi Panini Sanskrit grammar framework and is a more challenging task. In Sanskrit, sentences are combinations of more than one words and word is derived from fix set of धातु. It means by adding upsarg and pratyay before and after of any word we can derive a new word धातु is a basic building block of Sanskrit grammar, from that we can derive thousands of new Sanskrit words by adding one or more धातु. Upsarg and pratyay are performing the major role to derive a new word, which is changing the meaning of the whole word, and it also eases the pronunciation of a word. In one word whether single or conjoined is broken to form two words. Splitting upsarg and pratyay of any word is one subtask for a complete analysis of input data. The work mainly focuses on the Maharishi Panini Sanskrit grammar framework and splits upsarg and pratyay of the input word and goes to the base of the word which is called धातु.

Keywords Natural language processing · Artificial intelligence
Panini Sanskrit grammar · धातु

1 Introduction

Sanskrit computational linguistics has gained impulse in recent time. Many efforts have been considered in the computational system for bringing Sanskrit texts, from these efforts some of the tools consider sandhi splitting and semantic analysis and

B. Panchal · A. Parmar (✉) · S. Dholitar (✉)
Pacific School of Engineering, Surat 394305, Gujarat, India
e-mail: abhijit.parmar@outlook.com

S. Dholitar
e-mail: dholitarsadik@gmail.com

B. Panchal
e-mail: bpanchal083@gmail.com

© Springer Nature Singapore Pte Ltd. 2019
A. Abraham et al. (eds.), *Emerging Technologies in Data Mining and Information Security*, Advances in Intelligent Systems and Computing 814,
https://doi.org/10.1007/978-981-13-1501-5_35

remaining tools consider parsing, morphological analysis, and part-of-speech (POS) tagging. However, there have been almost insignificant efforts in direction of Sanskrit compounds, after separation.

In the ancient world, Sanskrit was a primary and very rich language in compound formation. The compound forming being prolific, design, and an open-set, and it is also difficult to enlist all the compounds in a single dictionary. The compound form involves an irremissible sandhi or sandhi-vicheda but simply sandhi splitting does not help a reader to identify the meaning of a compound or upsarg and pratyay. In Sanskrit, a word can be the combination of more than one upsarg or pratyay. So that developing a parser splits upsarg or pratyay of input word to identify धातु.

Compare to any other languages, Sanskrit has superlative vocabularies and it has 102 arab 78 crores 50 lakh words. If we will use these words in our computer technology, then the usage of these words will increase in future. Compare to other languages, Sanskrit reduces sentence length without changing its meaning. America has the university dedicated to Sanskrit, and NASA too has a department in it to research on Sanskrit manuscript. Sanskrit is a highly regularized and word order free language. The only unambiguous spoken language on the planet is Sanskrit NASA declared this and NASA scientists believe that Sanskrit is helpful in speech therapy. An alphabet used in Sanskrit language and their correct pronunciation improves the tone of speech.

The paper majorly focuses on Panini Sanskrit grammar framework for the semantic analysis of natural language and approach for splitting upsarg or pratyay of Sanskrit words. Splitting of upsarg and pratyay is using धातु as an elementary component of the language. Chapter 2 discusses Maharishi Panini Ashtadhyayi Sanskrit grammar. In Chap. 3, we have discussed concepts of upsarg and pratyay in detail. In Chap. 4, we have discussed the approach of splitting upsarg and pratyay. Finally, in Chap. 5, we narrated conclusion of the work and future plan.

2 Panini Sanskrit Grammar: Ashtadhyayi Sutrapatha

2.1 A Representation of Ashtadhyayi

Following are three basic steps of Ashtadhyayi's grammatical process:

- **Prescription** of the basic elements that form the language.
- **Characterization** of these basic elements by adding some of the attributes.
- **Specification** of the process depends on basic elements and their attributes.

2.2 Component of Ashtadhyayi

It has four unique components/elements:

- *Ashtadhyayi* has nearly around 4000 grammatical rules.
- *SivasuTras* the inventory of phonological segments, partitioned by markers (anUbaNdhas) to allow classes of segments (pratyaharas) to be designated by a set of special conventions.
- *dhAtupaths* a list of about 2000 verbal roots, with subclassification and diacritic markers encoding their morphological and syntactic properties.
- *ganapAtha* a list of 261 lists of lexical items which are idiosyncratically subject to certain rules. Some of the lists are open-ended and

2.3   Sanskrit Grammar Compounds

The Sanskrit word "samaasah" whose compound is "samasanam" which means combining more than one word into a word conveys the same meaning. Whenever we combine the components resulted in a compound, omit some operations like loss of upsarg or pratyay, loss of accent. A Sanskrit compound has following features:

1. It is a single word *ekapAda.*
2. It has a single case pratyay *ekAvibhaktiKam* with an exception of aluk compounds such as yuDhisTirah, where there is no deletion of case pratyay of the first component.
3. The only single accent is their *ekAsvaRah.*
4. In a compound, component occurrence is fix.
5. A compound did not accept insertion of words.
6. The compound formation is binary with an exception of *dvAndva* and *bAhuPada bAhuvRuhi.*
7. Euphonic change (sandhi) is a must in a compound formation. Constituents of a compound may require special gender or number different from their default gender and number. For example, *PaniPaDam, paciKabhaRyah,* etc. [1].

Though compounds of two or three words are more frequent, compounds involve more than three constituent words with some compounds.

3   Sanskrit: Upsurge and Pratyay

Grammar of every language is built upon the unique fundamental concept that in any language word represents or symbolized object or entity. If we concern general architectural of any language, it is completely harmless. The computational representation of these concepts leads us to the need of prior knowledge about the word.

As the part of grammatical framework of Panini, his book Ashtadhyayi contains 2012 धातु and nearly 4000 rules from Sanskrit grammar. From the part of Sanskrit grammar, one of these is upsarg and pratyay. Upsarg, a word that join before of any word and change the meaning of that word is called upsarg and pratyay, a word that

Table 1 Words deriving using upsarg

उपसर्ग	अर्थ	शब्द् रुप
अति	अधिक, उपर	अतिरिक्त
उप	निकट, गौण	उपचार
नि	अभाव, विशेष	निबन्ध्
परि	सब और	परिजन
प्रति	सामना, उलता	प्रीतफुला

Table 2 Words deriving using pratyay

प्रत्यय	शब्द् रुप
क	पाठक
आ	भुखा प्यासा
वान	देयावन् गुणवान
एरा	चचेरा
ईक	धार्मिक

join at the end of any and create a new word and also change the meaning of word is called pratyay [2] (Table 1).

For the understanding, अति is upsarg which means far to. अधिक words derive from अति, that means far too, very likewise derive word—उपर, अधिक, etc. By adding upsarg and pratyay help to derive many other word from single as well as multiple धातु. Similarly, take upsarg उप, which means up, above. So, we derive word—उपचार, उपर by adding pratyay रुष् on that word, derive words like उपारम, ऊर्ण by using single or multiple धातु and adding upsarg or pratyay on that word we can derive thousands of new Sanskrit words. Likewise use upsarg नि, परि, प्रति and pratyay are क, आ, वान, एरा, and ईक (Table 2).

4　Approach of Splitting Upsarg and Pratyay of Sanskrit Word

Most of the languages are built upon the principle where words represent objects and entities, but this principle is responsible for much inefficiency in the processing of input languages. Sanskrit follows different principle where word represents properties of objects or entities.

The धातु is most foundational part of the Sanskrit language. They can be considered as the basic building blocks of the Sanskrit grammar. This foundation of the language is responsible that being an ancient language the structure of Sanskrit is not changed since its commencement. There are 2012 धातु presented in the Sanskrit grammar. Panini stated nearly 4000 rules for formation of new words in his book Ashtadhyayi.

As a fundamental block of language, each धातु contains its own meaning. From one धातु, we can form more than one word. By combining more than one धातु, we can design and form thousands of new words. This is the biggest advantage of Sanskrit language.

The grammatical formation of Sanskrit is very strong. The grammar is able to form new word by its own. A single धातु holds its own meaning, and multiple धातु together generate thousands of new words, as already discussed. In addition to this, there are 22 upsarg and various pratyay like ति, भी are available. For example, word आगच्छति means to coming back. It is derived from धातु गच्छ. The result of splitting the word is shown below.

<आ><गच्छ>< ति>

Here, आ is one of the upsarg, गच्छ is धातु and ति is a pratyay. We have discussed various steps and necessary database table. The system working with following table:

- **Dictionary table** (contain Sanskrit words and its meaning in English language)
- **Upsarg table** (contain Sanskrit upsarg)
- **Pratyay table** (contain Sanskrit pratyay)
- **dhAtu table** (contain Sanskrit dhAtu-roop)

Step I: Initially, we have taken input Sanskrit corpus (Fig. 1).

Step II: Further, we split the entire sentence into a number of possible words. (Fig. 2).

Step III: The system is initially not aware of all the words from the Sanskrit sentences. Here, we have taken upsarg, pratyay along with the धातु. If system finds an unknown word or a word not available in the dictionary table, it will split the word and derive the meaning using fundamental rules of Sanskrit grammar frameworks. Designing a system which takes input as Sanskrit sentences and applies reverser engineering on that and goes to Sanskrit धातु. (Figs. 3, 4, 5 and 6).

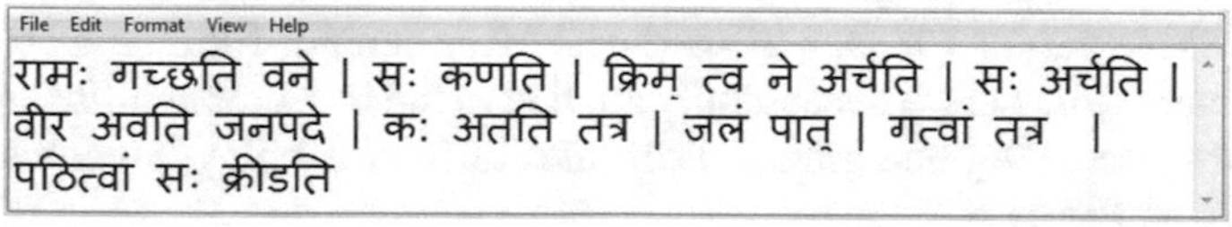

Fig. 1 Take input Sanskrit sentence

```
['\ufeffसः', 'कणति', 'रामः', 'गच्छति', 'वने', 'क्रिम्', 'त्वं', 'ने', 'अर्चति', 'सः', 'अ
र्चति', 'वीर', 'अवति', 'जनपदे', 'कः', 'अतति', 'जलं', 'पातु', 'गत्वां', 'तत्र', 'पठित्वां',
'सः', 'क्रीडति']
```

Fig. 2 Split sentence into word

```
सः
He is
Split उपसर्ग and प्रत्यय of word : कणति
धातु   0 ===> क
धातु   1 ===> कण
cry
प्रत्यय 11 ===> त
प्रत्यय 22 ===> ति
ing
रामः
The ram
Split उपसर्ग and प्रत्यय of word : गच्छति
धातु   0 ===> ग
going
वने
in forest
क्रिम्
Why
त्वं
you
ने
are
```

Fig. 3 Split upsarg and pratyay from word-I

Step IV: After apply splitting on each of the input word, the generated English translation of corpus in as below (Fig. 7).

5   Conclusion

From the study of Sanskrit grammar and initial understanding of Paninian rules, we found Sanskrit grammar as one of the finest NL grammar for natural processing. The grammar framework is successfully applied to the word for splitting upsarg and pratyay. The use of धातु and grammatical rules of Sanskrit plays significant role to split upsarg and pratyay.

Future Work

Future research in this area is to use the complete set of Panini Sanskrit grammar rules for all types of sandhi, upsarg, or pratyay splitting. The parser should be capable to parse the input Sanskrit sentences with higher accuracy in consideration of विभक्ति, सन्धि, समास् and लकार.

```
सः
He is
Split उपसर्ग and प्रत्यय of word : अर्चति
धातु  0 ===> अ
धातु  1 ===> अच
धातु  2 ===> अर्च
cry
प्रत्यय 11 ===> त
प्रत्यय 22 ===> ति
ing
वीर
Brave men
Split उपसर्ग and प्रत्यय of word : अवति
धातु  0 ===> अ
धातु  1 ===> अव
protect
प्रत्यय 11 ===> त
प्रत्यय 22 ===> ति
ing
```

Fig. 4 Split upsarg and pratyay from word-II

```
जनपदे
country
कः
Who is
Split उपसर्ग and प्रत्यय of word : अतति
धातु  0 ===> अ
धातु  1 ===> अत
continuously rodu
प्रत्यय 11 ===> त
प्रत्यय 22 ===> ति
ing
जलं
Water
Split उपसर्ग and प्रत्यय of word : पातु
धातु  0 ===> प
धातु  1 ===> पा
drinking
```

Fig. 5 Split upsarg and pratyay from word-III

```
drinking
प्रत्यय 11 ===> त
प्रत्यय 22 ===> तु
for
Split उपसर्ग and प्रत्यय of word : गत्वां
धातु 0   ===> ग
going
तत्र
There
Split उपसर्ग and प्रत्यय of word : पठित्वां
धातु 0   ===> प
धातु 1   ===> पठ
धातु 2   ===> पठि
study
प्रत्यय 11 ==> त
प्रत्यय 22 ==> त्व
प्रत्यय 33 ==> त्वां
after
सः
Split उपसर्ग and प्रत्यय of word : क्रीडति
धातु 0   ===> क
धातु 1   ===> क
धातु 2   ===> क्रीड
play
प्रत्यय 11 ===> त
प्रत्यय 22 ===> ति
ing
```

Fig. 6 Split upsarg and pratyay from word-IV

```
File  Edit  Format  View  Help
He is crying. The ram going in forest.
Why you are praying. Brave men protecting
country. Who is continuously roduing.
Water for drinking. There are going. He
is playing after study.
```

Fig. 7 Translated sentences into English

Acknowledgements We would like to express our gratitude to Mr. Vishavajit Bakrola and Mr. Dipak Dabhi for valuable advice and direction. We would like to thank the department of computer engineering for their support.

We also express our sincere and heartfelt thanks to Mr. Madhav Gopal, JNU, Delhi, for providing support and knowledge about Sanskrit language.

Finally, we would like to thank Mr. Ishank Sharma, Research Assistant, IIT Delhi, for giving us good knowledge about natural language processing tool.

References

1. Kumar, A., Mittal, V., Kulkarni, A.: Sanskrit compound processor. Sanskrit Comput. Linguist. **6465**, 57–69 (2010)
2. Kiparsky, P.: On the Architecture of Panini's Grammar. Sanskrit Computational Linguistics, pp. 33–94. Springer, Berlin (2009)
3. Gupta, P., Goyal, V.: Implementation of rule based algorithm for Sandhi-Vicheda of compound hindi words. arXiv preprint arXiv:0909.2379 (2009)
4. Deshmukh, R., Bhojane, V.: New Panvel PIIT.: Sandhi splitting techniques for different indian languages. Int. J. Eng. Technol. Manage. Appl. Sci. (ijetmas) **2**(7) (2014)
5. Teja, D., Kothuru, S.: Sanskrit in natural language processing. ijarcsse, March-2015, 2277 128x
6. Goyal, P., Arora, V., Behera, L.: Analysis of Sanskrit Text: Parsing and Semantic Relations. Sanskrit Computational Linguistics, pp. 200–218. Springer, Berlin (2009)
7. Kulkarni, A., Pokar, S., Shukl, D.: Designing a Constraint Based Parser for Sanskrit. Sanskrit Computational Linguistics, pp. 70–90. Springer, Berlin (2010)
8. Mishra, A.: Simulating the Paninian System of Sanskrit Grammar. Sanskrit Computational Linguistics, pp. 127–138. Springer, Berlin (2009)
9. Kak, S.C.: The Paninian approach to natural language processing. Int. J. Approximate Reasoning **1**(1), 117–130 (1987)
10. Saxena, H., Agrawal, R.: Sanskrit as a programming language and natural language processing. Global J. Manage. Bus. Stud. **3**(10), 1135–1142 (2013)

Collaborative Ranking-Based Text Summarization Using a Metaheuristic Approach

Pradeepika Verma and Hari Om

Abstract In the present work, a novel approach for improvement in automatic text summarization has been proposed. We introduce a different model for summarization problem by exploiting the strengths of different techniques like metaheuristic approaches and collaborative ranking. First, the sentences of document are scored via two methods. Method one assigns the weight to each text feature using a new metaheuristic approach 'Jaya' and scores the sentences by linearly combining these feature scores with their optimal weights. Method two scores the sentences by simple averaging the scores of each text feature. Moreover, the sentences are then ranked according to these scores which generates two sets of ranks for the documents. To calculate the final ranking of sentences, the concept of collaborative ranking has been adopted. The implementation of the proposed approach has been done in Python 3.5 in Anaconda environment. The experiments are performed on the DUC 2002 dataset using the co-selection-based performance parameter. We show empirically that the proposed method is viable and effective for extractive text summarization.

1 Introduction

With the exponential growth of electronically connected information, it becomes increasingly desirable to access core information from the document in abbreviated form. Consequently, automatic text summarization has achieved serious attention to handle the enormous amount of data. The automatic summaries can be generated in extractive and abstractive form. Extractive summarization aims at extracting most relevant information from the given text and the abstractive summarization rephrases the relevant information or generates new sentences from a group of relevant concepts in the text document. Most of the previous research work has been devoted to gen-

P. Verma (✉) · H. Om
IIT (ISM), Dhanbad 826004, India
e-mail: pradeepikav.verma093@gmail.com

H. Om
e-mail: hariom4india@gmail.com

© Springer Nature Singapore Pte Ltd. 2019
A. Abraham et al. (eds.), *Emerging Technologies in Data Mining and Information Security*, Advances in Intelligent Systems and Computing 814,
https://doi.org/10.1007/978-981-13-1501-5_36

erate extractive summaries which are based on human-engineered sentence scoring methods. These includes term frequency [11], title words, sentence position, stigma words [6], proper nouns [15]. Sentences are simply scored using these features to find the relevancy. Several core methods have been used to generate extractive summary like graph-based method [7, 17], latent semantic analysis [22], neural network-based algorithm [5, 14]. In recent works, sentences are scored by assigning optimal weights to sentence scoring methods using metaheuristic approaches like genetic algorithm [10, 12], particle swarm optimization [1–4], harmony search algorithm [21], cat swarm optimization [19], and cuckoo search [20] which produces effective results.

In this work, we have proposed a novel collaborative ranking-based single document summarization method which takes into account several text features to find the importance of sentences. A new metaheuristic approach [18] has been used to adjust the weights of text features. The rationale for using the Jaya is that it is very simple and has the least involvement of controlling parameters, which is absent in other existing approaches to the best of our knowledge. The contributions of this paper may are summarized as follows:

a. The adopted metaheuristic approach 'Jaya' has not been applied in the field of text summarization to our best knowledge.
b. The concept of collaborative ranking has been used to rank the sentences.
c. One of the significant sentence scoring methods, i.e., sentence position, is derived using the heuristic that the first few and last few sentences of a document are more relevant than any other sentences.

2   Overview

In this section, we formally define the problem of text summarization. Let a document D has a set of sentences S and set of words W. We aim to obtain a subset of sentences which summarizes D. In the following, we focus on scoring sentences with two different methods: $M1$ and $M2$. The sentences are then ranked ($R = [r_{kl}]_{M \times N}$, where R represents rank of a sentence, $k = 1, 2, ..., M$ represents number of methods, and $l = 1, 2, ..., N$ represents number of sentences) according to scores obtained using $M1$ and $M2$. It is noteworthy that the ranks of sentence given by two methods may be very different, so simple averaging of two ranks of a sentence may do not give accurate ranking. Suppose two sentences get the ranks $\{2, 3\}$ and $\{1, 4\}$ by $M1$ and $M2$. Then, by simple averaging both sentences get same ranking. But, there is noticeable difference in the ranking of second sentence as compared to the first which shows that both sentences should not get same ranking. Therefore, final ranking to each sentence is given by collaborative raking, which involves the Pearson correlation coefficient (PCC) for finding the similarities between the ranking of two methods.

3 Framework for Proposed Summarization Method

3.1 Preprocessing

Preprocessing is a preliminary process of our proposed algorithm that covers the areas of stop-word removal, stemming, normalization, and sentence separation. The stop words removal eliminates the most frequent words occurring in a document that does not have their stand-alone significant meaning like articles, prepositions, conjunctions, interrogations, helping verbs. Stemming is a process of converting a term into its morpheme. The sentence separation identifies each sentence in a document. We accomplish these processes by employing Natural language toolkit (NLTK). Here, we use the Porter stemmer for English text, provided for Python 3.5. The output of preprocessing is a document in the form of keywords.

3.2 Sentence Scoring Methods

A sentence consists of several text features. The relevant sentences can be recognized by scoring them using the following text features. Oliveira et al. [16] suggest such scoring techniques which are modified and briefly presented below.

1. **Sentence length**: This feature is used to filter out very short and very long sentences. The average length of a sentence can be calculated by Eq. 1.

$$AL(S) = \frac{\min(\text{sentence\_length}) + \max(\text{sentence\_length})}{2} \tag{1}$$

where $AL(S)$ is average length of a sentence in a summary. On the basis of average length, we calculate the sentence length score SL by Eq. 2.

$$\text{Score} = 1 - \frac{|AL(S) - SL|}{\max(\text{sentence\_length})} \tag{2}$$

2. **Sentence position**: It is one of the major features for sentence extraction. Generally, in a document, first few sentences and last few sentences consist of more significant sentences than others. So, we take into account this factor and calculate the score by Eq. 3.

$$\text{Score} = \frac{|n/2 - i|}{n/2} \tag{3}$$

where n is number of sentences, and i is index of sentence in document.

3. **Named entities**: This technique considers that the sentences containing named entities have higher probability to be included in the summary and it is calculated by Eq. 4.

$$\text{Score} = \frac{\text{Number of named entities in } S_i}{\text{Maximum number of named entities in a sentence}} \quad (4)$$

4. **Numerical data**: The sentences with numerical data generally indicate important information which is likely to be included in the summary and it is calculated by Eq. 5.

$$\text{Score} = \frac{\sum_{\text{ND} \in S} \text{Count}(\text{gram}_n)}{\sum_{w \in S} \text{Count}(\text{gram}_n)} \quad (5)$$

where ND denotes numerical data.

5. **Sentence centrality**: The sentence centrality refers to the degree of word overlap of a sentence with respect to other sentences. This method considers that the central sentences describe the significant information of a document more appropriately. It can be computed by Eq. 6.

$$\text{Cent}(S_i) = \frac{W_{S_i} \cap W_{S_j}}{W_{S_i} \cup W_{S_j}} \quad (6)$$

6. **Sentence with title words**: The title words in sentences show highly relevant sentences of a document. This score is calculated by Eq. 7.

$$\text{Score} = \frac{(\sum_{w \in \text{tw}} \sum_{\text{gram}_n \in S} \text{Count}_{\text{match}}(\text{gram}_n))^2}{\sum_{w \in \text{tw}} \text{Count}(\text{gram}_n) \sum_{w \in S} \text{Count}(\text{gram}_n)} \quad (7)$$

where tw denotes title word, w denotes a word, $\text{Count}_{\text{match}}(\text{gram}_n)$ is number of matching words, and $\text{Count}(\text{gram}_n)$ is number of words.

7. **Frequent words**: A sentence with the frequent words can also represent title of the document and it conveys significant information. For normalizing it, we divide the total count of frequent words by the total number of words. It is calculated by Eq. 8.

$$\text{Score} = \frac{\sum_{\text{fw} \in S} \text{Count}(\text{gram}_n)}{\sum_{w \in S} \text{Count}(\text{gram}_n)} \quad (8)$$

where fw denotes frequent words.

3.3 Sentence Scoring Using M1

In this method, sentences of each document have been scored by assigning optimal weights to text features using Jaya approach. The optimal weights are generated as follows and also illustrated in Fig. 1.

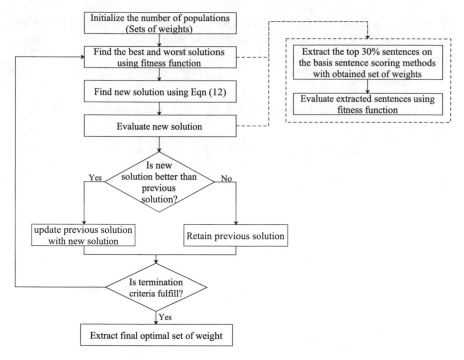

Fig. 1 Optimal weights using Jaya optimization algorithm

3.3.1 Initialization

The initial weights of text feature are generated randomly with population size as NP and dimension as ND in the range (0, 1). Here, we have considered seven text features. So, the dimension of every individual of the population is seven.

3.3.2 Fitness Function

The three major aspects, i.e., cohesion, non-redundancy, and readability, are important to generate a good summary. The sentences of a summary should have high conceptual relation between them. The contents of the summary should be non-redundant and readable at the word level as well as sentence level. These three aspects are defined by Eqs. 9, 10 and 11.

$$\text{Cohesion}(\text{Sum}_s) = \frac{\log(\text{avg}_{S_x \in \{\text{Sum}_s\}}(\text{Sim}(S_x)) \times 9 + 1)}{\log(\max_{S_x \in \text{Sum}_s}(\text{Sim}(S_x)) \times 9 + 1)} \tag{9}$$

where $\text{Avg}_{S_x \in \{\text{Sum}_s\}}$ is average of similarities of all sentences belonging to system summary and $\max_{S_x \in \text{Sum}_s}(\text{Sim}(S_x))$ is maximum of similarities in system summary.

$$\text{Non\_redundancy}(\text{Sum}_s) = 1 - \max_{y \in \text{Sum}_s}(\text{Sim}(S_x, S_y)) \tag{10}$$

where Sum_s denotes system summary, and $\text{Sim}(S_x, S_y)$ denotes similarity between S_x and S_y

$$\text{Readability}(\text{Sum}_s) = \frac{\sum_{0 \leq x < \text{Sum}_s} \text{Sim}(S_x, S_x + 1)}{\max \text{Sim}(S_x)} \tag{11}$$

Finally, the fitness function is formulated by the linear combination of these aspects with their weights $w1$, $w2$, and $w3$ by Eq. 12.

$$F(\text{Sum}_s) = \alpha \times \text{Cohesion}(\text{Sum}_s) + \beta \times \text{Non-redundancy}(\text{Sum}_s)$$
$$+ \gamma \times \text{Readability}(\text{Sum}_s) \tag{12}$$

Here, we use the values of α, β, and γ as 0.4, 0.3, and 0.3, respectively.

3.3.3 Solution Modification

Solution of populations are modified on the basis of best solution and worst solution. The best and worst solutions are identified using above-mentioned fitness function. The solution having largest value is considered as best solution and solution having lowest value is considered as worst solution. After obtaining best and worst solution, the solutions are modified for next iteration based on Eq. 13.

$$X_{j,i,k+1} = X_{j,i,k} + r_1(X_{j,\text{best},k} - |X_{j,i,k}|) - r_2(X_{j,\text{worst},k} - |X_{j,i,k}|) \tag{13}$$

If $F(X_{j,i,k+1}) > F(X_{j,i,k})$, then solution will be modified in $(i+1)$th iteration else solution remains same in $(i+1)$th iteration as in ith iteration.

3.3.4 Termination of Algorithm

If the termination condition is reached, then the best weights of text features which have the highest fitness value are selected as final weights of the scoring methods. If the termination condition is not reached, then the new weights are calculated to update the weights. Finally, the sentences are scored using $M1$ by Eq. 14.

$$\text{Score}(M1) = \frac{(w_1 \times \text{FS}_1 + w_2 \times \text{FS}_2 + \cdots + w_7 \times \text{FS}_7)}{7} \tag{14}$$

where FS_1 denotes first text feature score, and w_1 denotes weight of first feature score.

3.4 Sentence Scoring Using M2

According to this method, sentences are scored by simple averaging the scores of each text feature with respect to each sentence. The rationale behind simple averaging is to score sentences by giving equal weights to each text feature since it may be the case that some unimportant sentences are extracted due to text feature with higher weight. So, the sentences are scored using $M2$ by Eq. 15.

$$\text{Score}(M2) = \frac{w(\text{FS}_1 + \text{FS}_2 + \cdots + \text{FS}_7)}{7} \tag{15}$$

4 Sentence Ranking and Summary Generation

The sentences are ranked according to scores obtained by $M1$ and $M2$. In this way, we obtain two sets of ranks $\{r_{11}, r_{12}, \ldots, r_{1N}\}$ and $\{r_{21}, r_{22}, \ldots, r_{2N}\}$ for each sentence. These ranks are further used in finding final ranking using collaborative ranking. Collaborative filtering [24], which is different from simple averaging, takes into account similarity factor between two ranks. We have ranked each sentence using memory-based collaborative ranking. It predicts the ranking of each sentence S as an aggregation between two sets of ranking of sentences as shown in Eq. 16.

$$R_{S_i} = \text{aggre}_{m \in M, m' \in \{M-m\}} r_{m,S_i} r_{m',S_i} \tag{16}$$

where $M = \{M1, M2\}$ and m is any method in the set and m' is another method except m in the set. By taking into account the similarity factor between two methods, we can formally expressed the final ranking by Eq. 17.

$$R_{S_i} = \frac{k}{2} \sum_{m \in M} \sum_{m' \in \{M-m\}} \text{Sim}(m, m') r_{m,S_i} r_{m',S_i} \tag{17}$$

where k is the normalizing factor. $\text{Sim}(m, m')$ denotes similarity between two methods m and m'. r_{m,S_i} is the rank given by method m to sentence i. In collaborative filtering, Pearson correlation coefficient (PCC) has been widely used for similarity measurement. So, we have used PCC for measuring similarity between two methods by Eq. 18.

$$\text{Sim}(m, m') = \frac{\sum_{W \in S}(r_{m,W} - \bar{r}_m)(r_{m',W} - \bar{r}_{m'})}{\sqrt{\sum_{W \in S}(r_{m,W} - \bar{r}_m)^2}\sqrt{\sum_{W \in S}(r_{m',W} - \bar{r}_{m'})^2}} \tag{18}$$

where W denotes words of sentence S. $\bar{r}_m$ and $\bar{r}_{m'}$ are the average ranking values given by methods m and m', respectively. $r_{m,W}$ and $r_{m',W}$ denotes the ranking of words given by their tf-idf scores using method m and m', respectively.

Finally, the summaries are generated by extracting the top-ranking sentences and ordered them as in original document. Here, we have extracted 30% top-ranked sentences for summary generation.

5 Experimental Setup

The proposed method has been evaluated on 100 documents of DUC 2002 dataset. The Document Understanding Conference (DUC) provides a standard data collection which is widely used for evaluation of summarization system. DUC 2002 dataset contains test data together with reference summaries. The test data has 30 sets of documents and 10 reference summaries per document. The experiments of the proposed method are conducted on Linux operating system using the ROUGE tool implemented in Python 3.5. The ROUGE evaluates the proposed system on the basis of $n -$ gram, skip-bigram plus unigram, skip-bigram, longest common sequence, and weighted longest common sequence. It evaluates the performance in terms of three metrics: precision, recall, and F-score. Here, we have done the evaluation on the basis of ROUGE-1 and ROUGE-2 methods.

The comparison has been done with several methods like a benchmark summarizer (Msword), participating systems in DUC 2002 (System19 [8] and System31 [25]), hybrid genetic and mimetic algorithm-based summarizer (MA-SingleDocSum) [13], graph-based summarizer (UniRank) [23], and manifold ranking-based summarizer (MRank) [9].

6 Results and Discussions

Tables 1 and 2 show the comparison of proposed method with other six methods evaluated on the basis of recall, precision, and F_1 scores using ROUGE-1 and ROUGE-2, respectively, at 95% confidence interval. From the observation of the results shown in both tables, we can say that proposed method outperforms other methods with

Table 1 Comparison of proposed model with other methods using ROUGE-1

Method	Recall	Precision	F_1 score
Msword	0.40171	0.45156	0.42518
System19	0.40762	0.47585	0.43910
System31	0.07604	0.50595	0.13221
MA-SingleDocSum	0.43216	0.54755	0.48306
UniRank	0.43592	0.54967	0.48623
MRank	0.39210	0.45634	0.42179
Proposed	0.43226	0.58760	0.49810

Table 2 Comparison of proposed model with other methods using ROUGE-1

Method	Recall	Precision	F_1 score
Msword	0.17216	0.19373	0.18231
System19	0.18921	0.21203	0.19997
System31	0.03518	0.30842	0.06314
MA-SingleDocSum	0.19232	0.21774	0.20424
UniRank	0.18710	0.22915	0.20600
MRank	0.17822	0.20192	0.18933
Proposed	0.19061	0.23007	0.20849

$\approx$59% precision value and $\approx$50% F_1 value using ROUGE-1 and $\approx$23% precision value and $\approx$21% F_1 value. UniRank and MA-SingleDocSum outperform for recall value using ROUGE-1 and ROUGE-2, respectively. It can also be observed that MA-SingleDocSum performs good in generalization which is also based on metaheuristic approaches. So, we can conclude from this point is that the selection of appropriate text features with assignment of optimal weights using metaheuristic approaches can give better results. Besides these, proposed method focuses on improving the ranking of sentence by taking into account difference error in ranking given by different methods. So, better ranking of sentences also gives better results.

7 Conclusion

In this paper, we have discussed a collaborative ranking-based single document summarizer with metaheuristic approach to create a generic extractive summary. The purpose of employing metaheuristic approach is to get optimal weights of text features to score the sentences. It is quite possible that optimal results may not produce better results; therefore, we have also considered to score the sentences by assigning equal weights to each feature. Next, collaborative ranking is used to rank the sentences using ranks obtained from scores of sentences. The performance of proposed method is better than the other methods. For future work, we propose to improve the results of proposed summarization method by adding the module of semantic role labeling which can be accomplished by employing word2vec and WordNet.

References

1. Abbasi-ghalehtaki, R., Khotanlou, H., Esmaeilpour, M.: Fuzzy evolutionary cellular learning automata model for text summarization. Swarm Evol. Comput. **30**, 11–26 (2016)
2. Alguliev, R.M., Aliguliyev, R.M., Hajirahimova, M.S., Mehdiyev, C.A.: Mcmr: maximum coverage and minimum redundant text summarization model. Expert Syst. Appl. **38**(12), 14514–14522 (2011)

3. Asgari, H., Masoumi, B., Sheijani, O.S.: Automatic text summarization based on multi-agent particle swarm optimization. In: 2014 Iranian Conference on Intelligent Systems (ICIS), pp. 1–5. IEEE, New York (2014)
4. Binwahlan, M.S., Salim, N., Suanmali, L.: Fuzzy swarm diversity hybrid model for text summarization. Inf. Process. Manage. **46**(5), 571–588 (2010)
5. Cheng, J., Lapata, M.: Neural summarization by extracting sentences and words. arXiv preprint arXiv:1603.07252 (2016)
6. Edmundson, H.P.: New methods in automatic extracting. J. ACM (JACM) **16**(2), 264–285 (1969)
7. Erkan, G., Radev, D.R.: Lexrank: graph-based lexical centrality as salience in text summarization. J. Artif. Intell. Res. **22**, 457–479 (2004)
8. Harabagiu, S.M., Lacatusu, F.: Generating single and multi-document summaries with GIS-TEXTER. In: Document Understanding Conferences, pp. 11–12 (2002)
9. He, R., Qin, B., Liu, T.: A novel approach to update summarization using evolutionary manifold-ranking and spectral clustering. Expert Syst. Appl. **39**(3), 2375–2384 (2012)
10. Khan, A., Salim, N., Kumar, Y.J.: A framework for multi-document abstractive summarization based on semantic role labelling. Appl. Soft Comput. **30**, 737–747 (2015)
11. Luhn, H.P.: The automatic creation of literature abstracts. IBM J. Res. Dev. **2**(2), 159–165 (1958)
12. Meena, Y.K., Gopalani, D.: Evolutionary algorithms for extractive automatic text summarization. Procedia Comput. Sci. **48**, 244–249 (2015)
13. Mendoza, M., Bonilla, S., Noguera, C., Cobos, C., Leon, E.: Extractive single-document summarization based on genetic operators and guided local search. Expert Syst. Appl. **41**(9), 4158–4169 (2014)
14. Nallapati, R., Zhai, F., Zhou, B.: Summarunner: A recurrent neural network based sequence model for extractive summarization of documents. In: IAAA, pp. 3075–3081 (2017)
15. Nenkova, A., Vanderwende, L., McKeown, K.: A compositional context sensitive multi-document summarizer: exploring the factors that influence summarization. In: Proceedings of the 29th annual international ACM SIGIR conference on Research and development in information retrieval, pp. 573–580. ACM, New York (2006)
16. Oliveira, H., Ferreira, R., Lima, R., Lins, R.D., Freitas, F., Riss, M., Simske, S.J.: Assessing shallow sentence scoring techniques and combinations for single and multi-document summarization. Expert Syst. Appl. **65**, 68–86 (2016)
17. Parveen, D., Ramsl, H.M., Strube, M.: Topical coherence for graph-based extractive summarization (2015)
18. Rao, R.: Jaya: a simple and new optimization algorithm for solving constrained and unconstrained optimization problems. Int. J. Ind. Eng. Comput. **7**(1), 19–34 (2016)
19. Rautray, R., Balabantaray, R.C.: Cat swarm optimization based evolutionary framework for multi document summarization. Phys. A **477**, 174–186 (2017)
20. Rautray, R., Balabantaray, R.C.: An evolutionary framework for multi document summarization using cuckoo search approach: MDSCSA. Appl. Comput. Inf. (2017)
21. Shareghi, E., Hassanabadi, L.S.: Text summarization with harmony search algorithm-based sentence extraction. In: Proceedings of the 5th international conference on Soft computing as transdisciplinary science and technology, pp. 226–231. ACM, New York (2008)
22. Steinberger, J., Jezek, K.: Using latent semantic analysis in text summarization and summary evaluation. In: Proceedings ISIM'04, pp. 93–100 (2004)
23. Wan, X.: Towards a unified approach to simultaneous single-document and multi-document summarizations. In: Proceedings of the 23rd international conference on computational linguistics, pp. 1137–1145. Association for Computational Linguistics (2010)
24. Yang, X., Guo, Y., Liu, Y., Steck, H.: A survey of collaborative filtering based social recommender systems. Comput. Commun. **41**, 1–10 (2014)
25. Zajic, D., Dorr, B., Schwartz, R.: Automatic headline generation for newspaper stories. In: Workshop on Text Summarization (ACL 2002 and DUC 2002 meeting on Text Summarization), p. 65. Philadelphia (2002)

Chemical Reaction Optimization Algorithm for Word Detection Using Pictorial Structure

C. M. Khaled Saifullah, Md. Rafiqul Islam and Md. Riaz Mahmud

Abstract Word detection and recognition from natural scenes is very challenging and vastly popular research topic under the domain of computer vision. The problem comprises of various subproblems: character detection, optimal word formation from detected character, word recognition, etc. The paper focuses on optimal word formation from detected words only. Previously, pictorial structure model was used for optimal word formation where dynamic programming was applied. In this paper, chemical reaction-based meta-heuristics algorithm named as Chemical Reaction Optimization (CRO) has been used for word formation from detected character for larger instances. We have designed the solution generation, reaction operators and scoring function for the specific problem. Comparison with dynamic programming (DP) shows that DP generates optimal solution for small instances but for large instance it stuck due to memory limit exceed. CRO on the other hand has good optimality properties and better execution time than DP.

Keywords Word detection · Pictorial structure · Word formation · Text area detection · Chemical Reaction Optimization

1 Introduction

Scene text problem is reading text from natural images or scenes. For reading text, we need to detect the text area in the image first and then we need to recognize the characters inside the text area. After detecting the characters, we need to formulate

C. M. Khaled Saifullah (✉) · Md. R. Islam (✉) · Md. R. Mahmud (✉)
Computer Science and Engineering Discipline,
Khulna University, Khulna 9208, Bangladesh
e-mail: khaledkucse@gmail.com

M. R. Islam
e-mail: dmri1978@gmail.com

M. R. Mahmud
e-mail: riazmahmudcse@gmail.com

© Springer Nature Singapore Pte Ltd. 2019   427
A. Abraham et al. (eds.), *Emerging Technologies in Data Mining and Information Security*, Advances in Intelligent Systems and Computing 814,
https://doi.org/10.1007/978-981-13-1501-5_37

words optimally from those detected characters and finally, we can read the text in the images. The problem is a complex and has application in searching image according to the text [1], reading the text for robots, uneducated and blind persons in groceries [2]. A second application can be illustrated as: say a robot or a blind person goes to a super shop where the shop can be navigated using different types of direction board. Now, if scene text recognition system is installed in robot or spectacles of blind person, then they can explore the shop and can buy their desired products. Figure 1 shows some natural images captured by mobile phone or camera and the pictures are collected from [3–6]. It shows that text area in the images is detected. Then it recognizes the characters in the texts. From the detected characters, optimal words are formed and then the texts can be readable.

Scene text problem is decomposed into different subproblems such as text area detection, character recognition, optimal word formation, full-text recognition [7]. In this paper, we focus on optimal word formation problem. Optimal word formation problem states as given a set of characters detected in the text area of the image, our task is to form words optimally from those detected characters.

Fig. 1 Pictures are taken from [3–6] showing the text area detection in natural images

Word formation problem can be solved using pictorial structure model introduced by Fischler and Elschlager [8]. Pictorial structure model is modified for word formation problem where taking the position of each character and pairwise cost of neighbouring characters it formulates the cost of formation of a word. So, if $x = (c_1, c_2, \ldots, c_m)$ is a word having m characters and $p = (p_1, p_2, \ldots, p_m)$ be the position of each character in x, then $s(p_i, c_i)$ will be the score of the detection of ith character. The score of detection can be calculated by using Eq. 1 which is derived from [7].

$$s(p_i, c_i) = \log \left(\frac{p(c_{i|f})}{p(c_{bg}|f)} \right) \tag{1}$$

Here f is the feature vector and c_{bg} is the background character. Now, if $d(p_i, p_j)$ be the pairwise cost of the ith and jth character, then the cost of formation of a word L^* can be derived using the objective function as follows:

$$L^* = \underset{\forall i, p_i \in p}{\text{argmin}} \left(\sum_{i=1}^{m} -s(p_i, c_i) + \sum_{i=1}^{m-1} d(p_i, p_{i+1}) \right) \tag{2}$$

Recent studies show that word formation can efficiently be solved using dynamic programming by Wang et al. [7]. Besides, the pictorial structure is used for object recognition by Felzenszwalb [8]. Scene text recognition problem is solved using AdaBoost machine learning algorithm by Chen and Yuille [9], Stroke width transformation by Epshtein et al. [10].

In this paper, we have proposed a well-known nature-inspired metaheuristic algorithm named Chemical Reaction Optimization (CRO) for word formation problem using pictorial structure model. Due to its convergence nature using both local and global searches, CRO has been applied to varieties of NP-hard problem like 0-1 knapsack problem [11], Shortest Common Supersequence [12], quadratic assignment problem [13], multiobjective optimization problem [14]. We have designed the solution generation method, four basic reaction operators by which local and global searches are implemented and a scoring function for checking the validity of the word. Besides parameter settings are done by trial-and-error-based method and the outcomes of CRO algorithm are compared with dynamic programming proposed by Wang et al. [7].

2 Related Works

Different approaches are proposed and implemented to solve scene text problem. The challenge of getting better accuracy lies on how well the methods handle the noisy and blurry images. Epshtein et al. [10] firstly transform the image into greyscale space, and the Canny detector is used then to find the edges. To evaluate the stroke width for each pixel pairs of the parallel edge are used. Lastly, pixels having similar strokes are categorized into characters. It works fine with noisy and blurry images but it generates single segmentation for each character and that is not sufficient for OCR module [1].

Chen and Yuille [9] collect natural images by blind and normally sighted people. Using those images, they manually label and extract text regions. Then using statistical analysis of the text regions, features of images are extracted. Using the joint probability of features, the authors built a weak classifier. Later the weak classifier is fed into AdaBoost algorithm to train a strong classifier.

Wang et al. [7] introduce the methods where each character is detected via the sliding window. Due to a large number of the category, the Random Fern method is used. Then the detected character along with their scores is used for optimal word formation using pictorial structure model. Finally, the words are re-scored and the features are extracted and passed to SVM classifier for final training. Cross validation is used for testing. Since dynamic programming is used for word formation step, it cannot take many words as data set [1].

3 Chemical Reaction Optimization

The chemical reaction is an optimized natural phenomenon. Every unstable molecule in the world wants to get stabilized by colliding with the surface or with another molecule [12]. The phenomenon can be viewed in sense of optimization as it is a stepwise searching for an optimal solution as well as it will ensure optimal or near optimal solution with minimum fitness functional value [15]. Besides exchange of energy is very key characteristics of a chemical reaction. Keeping it in mind two laws of thermodynamics are considered for designing the algorithm. Firstly, energy in a system will remain constant that means the summation of potential energy, kinetic energy and the energy of the surroundings will be constant for the whole time. During algorithm design, potential energy is considered as the objective function whereas kinetic energy is referred as the value to accept the worst value. And the energy in the surroundings is coined as a buffer. Let potential energy and kinetic energy of a molecule be PE_i and KE_i and the buffer is termed as Buffer then first law of thermodynamic can be stated as:

$$\sum_{i=1}^{n}(PE_i + KE_i) + Buffer = C \tag{3}$$

Here C referred as constant in a system where n number of molecules is involved in the reaction.

Second law states as the entropy of a system have the tendency to flourish. And the entropy means the rate of having the disorder. When a molecule having high kinetic energy wants to move faster and thus create disorder and collide which results in a decrease in potential energy, thus it can react to an optimal situation. Mimicking these properties Lam and Li in 2010 [13] introduces Chemical Reaction Optimization (CRO) algorithm.

The algorithm is variable population-based metaheuristic algorithm that consists of four elementary reaction operators under two categories: unimolecular and intermolecular. Unimolecular collisions are the reaction where only one molecule is involved. Such reactions are on-wall ineffective collision and decomposition. On-wall ineffective collision involves a single molecule(solution) that collides with the wall of the container (small change) to produce a new molecule. Decomposition reaction creates two new molecules from one after the collision with the wall (big change). Intermolecular collisions are the reactions where more than one molecule is involved. An example of such reactions is intermolecular ineffective collision where two or more molecules collide to form the same number of product and synthesis reaction where two or more molecules collide to form a single molecule.

Like other evolutionary algorithms, CRO algorithm involves three stages: initialization, iteration, and finalization. In initialization stage population including their properties like potential and kinetic energy, molecular structure, initial hit number are assigned. Iteration stage includes one reaction out of four basic reactions each time. And in the finalization stage, the best solution along with the objective function value is returned.

4 Proposed Methods

Proposed method focuses only forming near optimal word from detected characters. Preprocessing includes characters detection which was done using random fern procedure described in [7]. The output of character detection was the location of a character and its score for that location. The score of the character had been calculated using Eq. 1. Then our proposed CRO algorithm uses those detected characters with their respective scores and location to form near optimal words.

4.1 Solution Generation

Taking a list of character along with the score as input, random solutions having different size are generated. First of all, from the detected characters we choose character randomly. Then we just measure the cost of the word, and for CRO, cost of the word is calculated considering the size of the word and difference of indexes of the characters. The cost function of a word for CRO is shown in Eq. 4.

$$Cost(w) = \sum_{i=1}^{n} u_i - \sum_{j=1}^{m} (l_{i+1} - l_i) \tag{4}$$

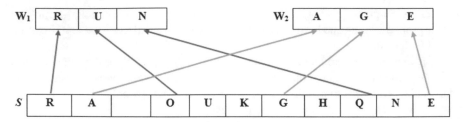

Fig. 2 Solution generation

Here u_i represents the score of ith detected character and l_i means the location of the ith character in the word formed by the solution generation method. Next, if the cost seems to be higher than a threshold value, we swap the last character with the previous one and measure again. This procedure continues until the cost is under the threshold value. Figure 2 shows an example of solution generation where two words $W1$ and $W2$ are formed from a set of ten detected characters S.

4.2 Reaction Operators

Four elementary reaction operators are considered for the problem. Reaction operators are discussed as follows.

4.2.1 On-Wall Ineffective Collision

The on-wall ineffective collision is where a single molecule collides with the wall of the container and produces a new solution. Very slight change occurs in the molecule structure of the molecule and thus local search is applied in the system. The two-exchange operator is used for on-wall ineffective collision. Here, two indexes of the solution are selected randomly and then characters of both indices are exchanged with each other. The scoring function is called after the reaction. Figure 3 shows the exchange operator.

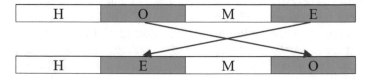

Fig. 3 Two-exchange operator

4.2.2 Decomposition

Here two molecules are formed from a single molecule. Massive change in the structure happens and change in the population is also occurring. Global search is implemented using decomposition reaction. For word formation problem, half-total exchange operator is used. Reactant molecule is divided in middle to form two new molecules. So one-half of the product is filled with reactants structure whereas other half is produced randomly from the set of detected characters. After the reaction scoring function operates on new solutions. Figure 4 represents the half-total exchange operator.

4.2.3 Intermolecular Ineffective Collision

Two or more molecules collide with each other to form two or more molecules. Molecularity does not change by this reaction. Changes occur in molecular structure in a small amount, and thus, local search is implemented. For our problem, we use two-point crossover operator between two molecules which is used for intermolecular ineffective collision reaction. Here two points are chosen randomly, and the odd parts of molecule 1 and the even part of molecule 2 form one new molecule and even part molecule 1 and odd parts of molecule 2 form another new molecule. The scoring function is operated for each molecule. The operator is shown in Fig. 5.

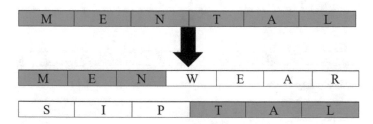

Fig. 4 Half-total exchange operator

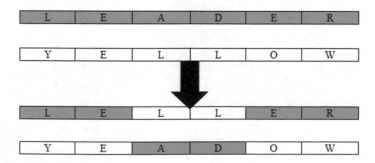

Fig. 5 Two-point crossover operator

4.2.4   Synthesis

Synthesis reaction produces single molecule by reacting with two or more molecules. It works opposite of decomposition and causes global search in the search space. We use the probabilistic select operator for synthesis reaction. Here a random number, is drawn and on the basis of that random number, we chose character randomly. The scoring function is called after the reaction. Figure 6 shows the probabilistic select operator.

4.3   Scoring Function

Costing function scores the newly produced solution. The cost is calculated on the basis of scores of the detected character and their location in the newly formed words. The mathematical equation of the cost function is described in Eq. 4. Then if the value exceeds the threshold value or becomes negative, it deletes the character for which deviation occurs. This process terminates whenever the score came less than the threshold. Our proposed method is represented in Fig. 7 and detailed in Fig. 8.

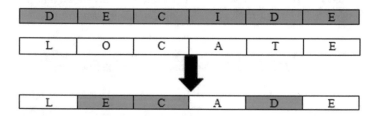

Fig. 6   Probabilistic select operator

Fig. 7   Steps of proposed system

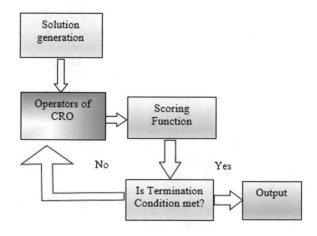

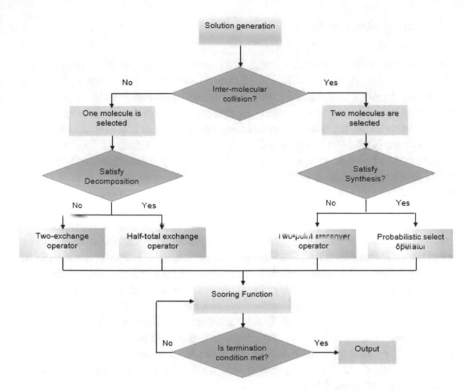

Fig. 8 Detail steps of proposed method

5 Experimental Analysis

We have experimented our proposed CRO algorithm and dynamic programming algorithm in a Toshiba personal laptop model number Satellite A305-S6905 having Core 2 Duo processor, 2.5 GB RAM. Both the algorithms are implemented in Java SE Development Kit 7 and Eclipse IDE. We generate a random set of detected character and their respective score. The score of each character is a random number between 1 and 100. The random data set consists of eight set of characters of length 10, 50, 100, 500, 1000, 1500, 2000, 3000, respectively. Experiments are done in two phases. Firstly, for different parameters of CRO, we did a parameter analysis experiment. Here different values of parameters especially decomposition threshold and synthesis threshold are tried, and converge speed is observed. In the second phase, CRO algorithm is compared with dynamic programming algorithm described in [7].

5.1 Parameter Analysis

Chemical Reaction Optimization algorithm has seven parameters: popsize, PE, KELossRate, MoleColl, InitialKE, α (decomposition threshold) and β (synthesis threshold). Besides we use score threshold named parameter used in population generation method and scoring function. Except for decomposition threshold and synthesis threshold, all other parameters are set. For decomposition and synthesis threshold, we have used the trial-and-error method. The values of parameters are: popsize = 20, PE = Cost function (C), KELossRate = 0.4, MoleColl = 0.3, InitialKE = 100 and scoreThreshold = Random [5–20]. Table 1 represents the comparative results of CRO algorithm for different values of decomposition threshold. The same experiment is run for 20 times, and the results show the average of the 20 runs. Table 1 shows that for the higher value of decomposition threshold, it gives a better result where it takes more computational time. Especially, the difference between $rand[50, 100]$ and $rand[100, 500]$ in the case of detection is minor whereas in the case of experimental time former one is much faster. So for rest of our experiment, we accept the value of $\alpha = rand[50, 100]$. Table 2 shows the comparative results CRO algorithm for different values of synthesis threshold. A very much similar picture is returned after this experiment too. Results are improved if we increase the threshold value. But computational time is also a big concern for the higher value of synthesis threshold. For synthesis threshold too, we consider $\beta = rand[50, 100]$ for rest of the experiments since it seems to have best if we consider both numbers of detected words and computational time.

Table 1 CRO results for different value of α

Number of word	α = rand[1, 10]		α = rand[10, 50]		α = rand[50, 100]		α = rand[100, 500]	
	Word detected	Time (s)	Word detected	Time (s)	Word detected	Time (s)	Word detected	Time (s)
10	7.1	410-5	7.6	710-4	8.3	0.002	8.5	0.01
100	84.4	0.003	86.7	0.009	89.2	0.04	91.2	0.8
1000	782.8	3.45	804.2	5.43	842.4	7.3	856.1	17.4

Table 2 CRO results for different value of β

Number of word	β = rand[1, 10]		β = rand[10, 50]		β = rand[50, 100]		β = rand[100, 500]	
	Word detected	Time (s)	Word detected	Time (s)	Word detected	Time (s)	Word detected	Time (s)
10	6.9	310-5	7.4	110-3	8.1	0.001	8.3	0.009
100	82.5	0.001	84.9	0.011	88.5	0.03	90.9	0.7
1000	778.4	3.25	801.5	6.14	841.2	6.9	852.4	16.1

5.2 Comparison with Dynamic Programming

In this phase, we compare our proposed CRO algorithm with dynamic programming algorithm for word formation problem. Dynamic programming (DP) is implemented using the procedure described in [7]. The number of words detected (N), objective functional value (f) and execution time (t) are the measures that we compared. Besides statistical measures like optimality percentage and speed rate in percentage of CRO with respect to DP is also shown. CRO algorithm is run for 20 times for each data set, and the average value is considered as the preferred output. The standard deviation for CRO is also given indicating the scattering properties of the output. Table 3 represents the simulation results. Optimality percentage (OP%) can be obtained using Eq. 5.

$$\text{Optimality Percentage (OP\%)} = \frac{N_{CRO}}{N_{DP}} \times 100\% \tag{5}$$

Here N_{CRO} means a number of words detected by CRO algorithm, whereas N_{DP} referred as the number of words detected by DP algorithm. Similarly, Speed Rate in percentage (SR%) can be demonstrated using Eq. 6.

$$\text{Speed Rate (SR\%)} = \frac{t_{DP} - t_{CRO}}{t_{DP}} \times 100\% \tag{6}$$

t_{CRO} is the time taken by CRO algorithm, and t_{DP} is the time taken by DP algorithm. Standard deviation (λ) for CRO algorithm is calculated using Eq. 7.

$$\text{Standard Deviation } (\lambda) = \sum_{i=1}^{n} (f' - f_i) \tag{7}$$

Here f' means the mean value of objective functional value for all run, whereas f_i is the functional value of the ith run and n is the number of runs. Simulation result of Table 3 shows that dynamic programming always formed words more than our proposed CRO algorithm but DP stuck with memory whenever a number of detected characters exceed 1500. Thats why we cant generate output for large instance with DP algorithm. Besides through CRO algorithm does not generate optimal result but the solutions of CRO have optimality of 79–82% in average which is very appealing. Most significant fact of results in Table 3 is execution time. CRO algorithm outperformed DP algorithm by 99.99 % in all cases by the execution time. Even for larger instances like a number of detected characters 3000, CRO algorithm able to generate solution within 3–4 min, whereas DP algorithm fails to generate a solution (Figs. 9, 10 and 11).

Table 3 Simulation results

Number of character	Dynamic programming			CRO					
	N_{DP}	f_{DP}	t_{DP}	N_{CRO}	OP%	f_{CRO}	λ	t_{CRO}	SR%
10	4	27	11.3	3.1	77.5	31.3	2.3	0.001	99.99
50	27	83	132.62	24.5	90.74	101.3	3.1	0.009	99.99
100	41	194	763.45	33.6	81.95	232.4	4.2	0.03	99.99
500	275	1048	2601.4	217.4	79.05	1365.2	5.8	1.7	99.99
1000	417	2174	87620.5	324.3	78.9	3376.4	8.4	6.9	99.99
1500	571	2592	128884112	452.7	77.4	3965.8	9.3	21.7	99.99
2000	–	–		593.3	–	4582.4	11.7	45.8	–
3000	–	–		801.7	–	6749.5	14.1	195.4	–

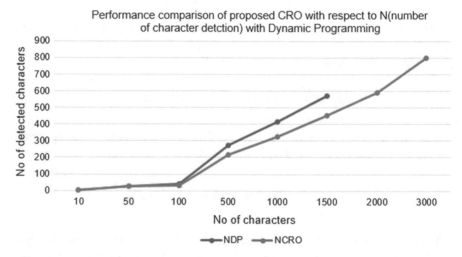

Fig. 9 Performance comparison of proposed CRO with respect to N (number of character detection) with dynamic programming

6 Conclusion

The paper concerned with a very practical and challenging problem in the domain of computer vision named as scene text problem. Scene text problem is subdivided into some subproblem, and one of them is word formation problem from a set of detected characters. We have reviewed some applications and related works and their limitation. A very well-known nature-inspired metaheuristic algorithm named as Chemical Reaction Optimization algorithm is proposed to solve the word formation problem. Experiment with dynamic programming suggests that CRO has a very

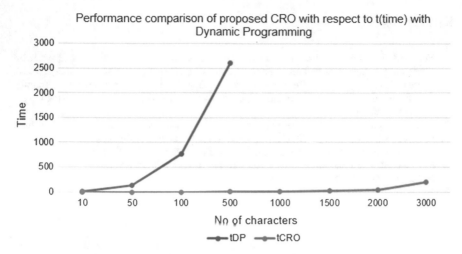

Fig. 10 Performance comparison of proposed CRO with respect to t (time) with dynamic programming

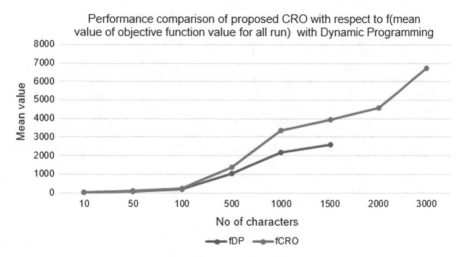

Fig. 11 Performance comparison of proposed CRO with respect to f (mean value of objective function value for all run) with dynamic programming

good optimality record with the much faster procedure. In future, a classifier will be designed using CRO as word formation methods. Besides detail study of parameters of CRO might improve the results of the problem.

References

1. Neumann, L., Matas, J.: Real-time scene text localization and recognition. In: 2012 IEEE Conference on Computer Vision and Pattern Recognition (CVPR), pp. 3538–3545. IEEE (2012)
2. Merler, M., Galleguillos, C., Belongie, S.: Recognizing groceries in situ using in vitro training data. In: IEEE Conference on Computer Vision and Pattern Recognition, 2007. CVPR'07, pp. 1–8. IEEE (2007)
3. Mare's computer vision study.: 07.06.15 http://study.marearts.com/2015_06_07_archive.html
4. Index of /mediawiki/images (2016). http://iapr-tc11.org/mediawiki/images
5. Tao wang stanford university (2016). http://ai.stanford.edu/~twangcat/
6. Modena, F.C.M.: Fbk—it—tev—semantic image labelling: Textinscene (2016). https://tev-static.fbk.eu/TeV/Technologies/TextInScene_files/I0.html
7. Wang, K., Babenko, B., Belongie, S.: End-to-end scene text recognition. In: 2011 IEEE International Conference on Computer Vision (ICCV), pp. 1457–1464. IEEE (2011)
8. Fischler, M.A., Elschlager, R.A.: The representation and matching of pictorial structures. IEEE Trans. Comput. **100**(1), 67–92 (1973)
9. Chen, X., Yuille, A.L.: Detecting and reading text in natural scenes. In: Proceedings of the 2004 IEEE Computer Society Conference on Computer Vision and Pattern Recognition, 2004. CVPR 2004. vol. 2, pp. II–II. IEEE (2004)
10. Epshtein, B., Ofek, E., Wexler, Y.: Detecting text in natural scenes with stroke width transform. In: 2010 IEEE Conference on Computer Vision and Pattern Recognition (CVPR), pp. 2963–2970. IEEE (2010)
11. Truong, T.K., Li, K., Xu, Y.: Chemical reaction optimization with greedy strategy for the 0–1 knapsack problem. Appl. Soft Comput. **13**(4), 1774–1780 (2013)
12. Saifullah, C.K., Islam, M.R.: Chemical reaction optimization for solving shortest common supersequence problem. Comput. Biol. Chem. **64**, 82–93 (2016)
13. Lam, A.Y., Li, V.O.: Chemical-reaction-inspired metaheuristic for optimization. IEEE Trans. Evol. Comput. **14**(3), 381–399 (2010)
14. Li, H., Wang, L., Hei, X.: Decomposition-based chemical reaction optimization (cro) and an extended cro algorithms for multiobjective optimization. J. Comput, Sci (2015)
15. Lam, A.Y., Li, V.O.: Chemical reaction optimization: a tutorial. Memetic Comput. **4**(1), 3–17 (2012)

Sentiment Prediction Based on Lexical Analysis Using Deep Learning

S. M. Mazharul Hoque Chowdhury, Sheikh Abujar, Mohd. Saifuzzaman, Priyanka Ghosh and Syed Akhter Hossain

Abstract This era of computing made everything computerized. So people started to interact in the virtual life more than real life. From shopping to talking, everything is controlled by computer. Because of that it has become more and more important to analyze and extract sentiment. Sentiment can be extracted from different type of data format like audio, text, image. In this paper, we proposed some methods to analyze text data in the paragraph level. Those methods can be implemented using bag of words, and priority was given to the lexical-based analysis.

1 Introduction

Nowadays, technologies became the most available than any other time. People are now dependent to the technologies more then they think. This huge use of technology creating huge amount of text data in different platforms and different format. It has become harder to analyze such amount of data using general methods and techniques. So it has become essential to create new method to make the analysis process reliable and smarter than before.

S. M. M. H. Chowdhury (✉) · S. Abujar · Mohd. Saifuzzaman
P. Ghosh · S. A. Hossain
Department of Computer Science and Engineering, Daffodil International University,
Dhanmondi, Dhaka 1205, Bangladesh
e-mail: mazharul2213@diu.edu.bd

S. Abujar
e-mail: sheikh.cse@diu.edu.bd

Mohd. Saifuzzaman
e-mail: saifuzzaman.cse@diu.edu.bd

P. Ghosh
e-mail: priyanka2378@diu.edu.bd

S. A. Hossain
e-mail: aktarhossain@daffodilvarsity.edu.bd

© Springer Nature Singapore Pte Ltd. 2019
A. Abraham et al. (eds.), *Emerging Technologies in Data Mining and Information Security*, Advances in Intelligent Systems and Computing 814,
https://doi.org/10.1007/978-981-13-1501-5_38

Machine learning and data mining were introduced to solve the problem about this huge amount of data. Sentiment analysis is a small part of data science which integrates natural language processing to solve its problems. Sentiment analysis focuses on only organized approach, methods, and techniques [1]. Many different methods have been applied and many are under research to increase the accuracy of the analysis [2].

In this research paper, some sentiment analysis methods were proposed based on lexical analysis and bag of words. Sentence-level techniques were used for the paragraph-level analysis. Full paragraph to single word, every aspect was taken under consideration. In this paper, we tried to introduce an adaptive model for the unknown types of words who do not have any sentiment value. So that proper sentiment can be extracted from text.

2 Literature Review

All over the world, thousands of scientists are working to improve sentiment analysis methods and their accuracy. In 2009, Paul Ferguson and his team worked on paragraph-level analysis to increase accuracy of document-level sentiment analysis [3]. Recently, two researchers named Pamungkas and Putri (2017) worked on word-sense disambiguation for the lexical-based sentiment analysis [4]. Till now, a lot of work has been done on domain-based analysis using word library. For example, Cruz Laura and his team (2017) wrote a chapter about applying lexical library for fixed domain [5].

Devina Ekawati and Masayu Leylia Khodra worked on (2017) aspect-based review analysis [6]. Sentence-based sentiment analysis was done by Alexandre Trilla and Francesc Alías (2013). They tried to implement this to get better accuracy in their text to speech program [7]. Parinya Sanguansat (2016) tried to implement paragraph to vector for business data analysis from social media [8]. Huy Nguyen and Minh-Le Nguyen (2017) worked on sentence-level sentiment analysis on different social medias like twitter, and their focus was on the improvement of the accuracy of the sentence-level analysis [9]. They have proposed a new method for the improvement of their result.

Researcher Mike Thelwall (2016) worked on the sentiment strength detection program named SentiStrength that was developed during the Cyber Emotions project [10]. It was developed to detect the strength of sentiments expressed in social Web texts. In his work, he described how SentiStrength works using lexical approach and using its own rules and terms. Researcher Soumi Dutta and his team (2015) worked on sentiment analysis of online content using WordNet [11]. In their research, they proposed a method using WordNet to detect sentiment from different social media.

Therefore, it can be said that a lot of research has been done and still in process to increase the accuracy, develop new solutions, create new tools. But very few researches can fulfill all the necessary requirements and very few limitations. So continuous work is required to manipulate the continuous growth of data.

All those researchers are continuously trying to improve the accuracy of the result than existing methods. They tried to find out different way of finding better accuracy and methods to handle the data. Main difference between those existing and this proposed method is the consideration of non-relevant words existing to the document. We tried to classify non-relevant words into positive or negative words according to the sentence-level polarity, and those unknown words library will be adaptive, so that it will be able to move its position in positive or negative side. Therefore, we proposed a model that can bring better accuracy in the calculation for sentiment analysis.

3 Proposed Model

From Fig. 1, it can be said that new data will be sent to the system then algorithm will extract data with the help of bag of words. If the data is matched with bag of words, then algorithm goes toward sentiment analysis process, if not then it goes toward unknown words storage which will make another further decision that "Is the data Exists on Unknown Dictionary?" If yes, then sentiment analysis process will start and if negative result occurs, the data will be sent to the temporary dictionary for unknown words which consist of negative/positive words based on the sentiment analysis (SA) report. For the next analysis, the unknown words storage will be matched with temporary dictionary for unknown words. When the full process is completely done, the system will provide the output/result.

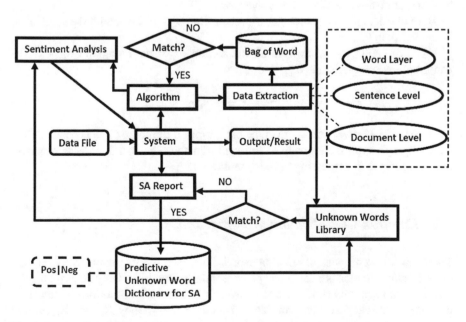

Fig. 1 Proposed model for the system

4 Proposed Method

In this paper, a model has been proposed for sentiment analysis using lexicon analysis technique in the paragraph level. Sentiment analysis in paragraph level is complex due to data complexity in different level. Solving them using one particular method is not sufficient always. So use of different technique is required for better analysis and accuracy. Therefore, this proposed model is containing multiple approach and algorithms.

4.1 Determining Positive or Negative Sentiment of the Paragraph

A paragraph contains multiple sentences and because of that the positivity or negativity depends on the score of the sentence. If the sentence represents positive score, then the probability of becoming positive paragraph will be high. So we can write this as

$$S_s = P_W - N_W \tag{1}$$

From Eq. (1), S_s is sentence score, P_W is positive word score, and N_W is negative word score. Each positive and negative word will get 1 point, and the total score will be some of positive words and negative words.

Here in Eq. (1), if the score of S_s is greater than 0, $S_s = PS_s$ and if the score of S_s is smaller than 0, $S_s = NS_s$.

From the collected score from all the sentences of the paragraph for a single paragraph, we can write the equation as:

$$S_P = PS_S + NS_S \tag{2}$$

Here in Eq. (2), S_P is the score of the paragraph. If the score of S_P is positive then the paragraph expressing positive sentiment. On the other hand, if S_P is negative, then the paragraph expressing negative sentiment.

4.2 Emotion Classification of Paragraph

There are different types of emotion expressions used by the people every day. So emotion classification is a very large part of sentiment analysis. Only determining the positivity or negativity cannot always provide a full scenario. Emotion classification can help us in that part. Let's consider a few emotions—happy E_H, joy E_J, excited E_E, sad E_S, angry E_A, and frustrated E_F.

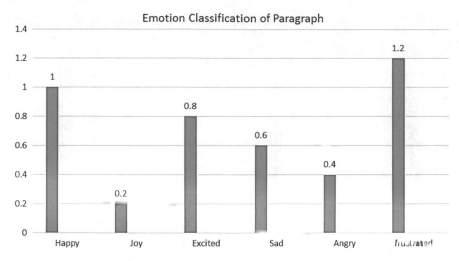

Fig. 2 Emotion classification of a paragraph

If we consider a paragraph from a document and apply an algorithm with bag of words, we will be able to find out individual score for the emotion classes.

For example, here $E_H = 5$, $E_J = 1$, $E_E = 4$, $E_S = 3$, $E_A = 2$, $E_F = 6$ scores for five sentences in a paragraph.

For the following result, we can calculate the paragraph score as,

$$\text{Paragraph score,} \quad P_S = \frac{\text{Emotion Score}}{\text{Number of Sentence}} \tag{3}$$

From the given Eq. (3), we get

$P_H = 1$, $P_J = 0.2$, $P_E = 0.8$, $P_S = 0.6$, $P_A = 0.4$, $P_F = 1.2$.

So, we can say that paragraph containing highest number of frustrated class words and lowest joy. On the other hand, if we consider the repeat of same word for an emotion class and count 1 for a single word using we get

$E_H = 3$, $E_J = 1$, $E_E = 3$, $E_S = 1$, $E_A = 2$, $E_F = 4$.

Applying Eq. (3) here we get,

$P_H = 0.6$, $P_J = 0.2$, $P_E = 0.6$, $P_S = 0.2$, $P_A = 0.4$, $P_F = 0.8$.

Here comparing both Figs. 2 and 3, we are getting better accuracy for the test data because there is no repetition of data in the calculation. Comparing both results, we can simply determine that the following paragraph is a frustrated-related paragraph.

4.3 Percentage of Positivity and Negativity

To determine if a paragraph is positive or negative, we can use the percentage method of positivity and negativity because greater percentage means greater chance of being

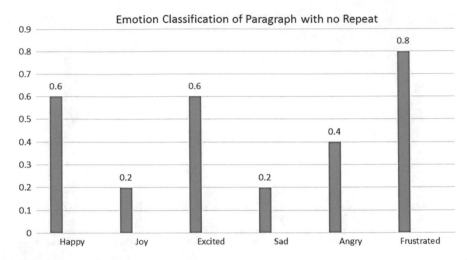

Fig. 3 Emotion classification of a paragraph with no repeat

that type of sentiment the paragraph is having. Bag of words method can be used for this approach also. Addition of positive and negative words will give a sentiment score and dividing it with positive and negative score, and finally multiplication of 100 will give us a clear percentage of each polarity class.

The equation can be written as,

$$S = NS_P + PS_P \tag{4}$$

$$\therefore P_S = \frac{PS_P}{S} \times 100 \tag{5}$$

$$\therefore N_S = \frac{NS_P}{S} \times 100 \tag{6}$$

Suppose, for a paragraph sum of emotion-related words (S) is 73, negative words (NS_P) is 26 and positive words (PS_P) is 47 by using Eq. (4). Applying that to the following Eqs. (5) and (6),

$$P_S = (47/73) \times 100 = 64.38\%$$

$$N_S = (26/73) \times 100 = 35.62\%$$

Here in Fig. 4, it is visible that maximum of the total 100% is positive words in the paragraph between all emotion-related words. So that it is possible that the given paragraph is a positive sentiment-related paragraph.

Fig. 4 Positive and negative word presentence

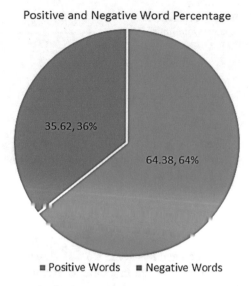

Positive and Negative Word Percentage

35.62, 36%

64.38, 64%

■ Positive Words   ■ Negative Words

4.4 Impact of the Word in the Sentence

Prediction of emotion can be done using impact of the word in the sentence. Higher impact value will create higher possibility for the paragraph. As other text analysis methods explained in this paper this one also can use lexicon-based bag of words for sentiment analysis.

For example "today is the best day of my life" containing 8 words and 1 of them is sentiment-related word.

So it can be written as, weight of the sentence, $S_W = 8$ and emotion weight of the sentence, $S_E = 1$ for the Eq. (7).

$$\text{Sentence emotion weight,} \quad S_{EW} = \sum_{S_{W>2}} \frac{S_E}{S_W} = (1/8) = 0.13 \qquad (7)$$

Here if the sentence is possible, then score is positive otherwise negative and the limit of the word must be greater than 3.

If there is multiple emotion-related words and both positive and negative word exist, the equation can be rewritten as,

$$S_{EW} = \sum_{W>2} \frac{S_{PE} - S_{NE}}{S_W} \qquad (8)$$

In this case, Eq. (8) is for only one single sentence. But if the sentiment weight for the paragraph is required, equation will be

Table 1 Positive and negative word library from the test data	Positive words	Negative words
	Drink, good	Hot, bad, painful, work

$$P_{EW} = \sum_{S_{W>2}} \frac{S_{EW1} + S_{EW2} + S_{EW3} + \cdots + S_{EWn}}{S} \qquad (9)$$

From the given Eq. (9), paragraph weight can be measured and value type will determine if the paragraph is positive or negative.

4.5 Adaptive Model for Unknown Words

Every sentence has some words that have no sentiment-related value but they play a direct role for the emotion. Those words can be used as an adaptive model and depending on the user. For example, a place can be liked by someone for a special memory, but for same reason someone can hate it. For the term "visit," we get both positive and negative sentiment. So visit can be used as an adaptive data. Data set can be trained for different user differently. But it is also found that the emotion-related words depend on the main verb of the sentence. So by considering those unknown words as training set it is possible to determine which sentiment is applicable for it. For example, consider some sentences: "It is a hot summer day and I am feeling bad. But by drinking water I am feeling good. It is painful to do any work at this time."

After preprocessing (remove noun, proposition, to be verb, article, etc.) data left——hot, feel, bad, drink, feel, good, painful, work, and so on. Depending on the sentence, it can be categorized as the positive and negative words as well as according to the sentence sentiment unknown words can be categorized as the Table 1. Therefore, the final outcome/training data set is built.

Only conflict happened in here was with feel, because it was in the both positive and the negative sentence. Because of that, both of them were shorted and library will wait until the next insertion of feel and depending on the emotion it will take place into one of those. Like this adaptive model will be able to change its value depending on the necessity.

5 Conclusion

Nowadays because of continuous increase of data made it complex to find out useful information. Researchers all over the world are trying to create methods to analyze data and extract information from it. As all data cannot be analyzed using same technique and method, different purpose need different algorithms. In this research work, a lexical method was proposed for paragraph-level sentiment analysis with step by step process. This method was proposed for text data and increase accuracy than the existing methods.

Accuracy of the algorithm depends on the data processing and logic applied to classify data. Therefore, our proposed method is not only considering the general text analysis technique. It is also focusing on the unknown words or words that are not related to the sentiments. This will increase the validity of this research method. On the other hand, it is holding uniqueness of the research as well as this proposed algorithm.

Acknowledgements We would like to thank DIU-NLP and Machine Learning Research LAB for all their support and help. Any error in this research paper is our own and should not tarnish the reputations of these esteemed persons.

References

1. Krishna, D.S., Kulkarni, G.A., Mohan, A.: Sentiment analysis-time variant analytics. Int. J. Adv. Res. Comput. Sci. Softw. Eng. **5**(3) (2015). ISSN. 2277 128X
2. Celikyilmaz, A., Hakkani-Tur, D., Feng J.: Probabilistic model based sentiment analysis of twitter messages. In: Spoken Language Technology Workshop (SLT), 2010, pp. 79–84. IEEE (2010)
3. Ferguson, P., O'Hare, N., Davy, M., Bermingham, A., Tattersall, S., Sheridan, P., Gurrin, C., Smeaton, A.F.: Exploring the use of paragraph-level annotations for sentiment analysis of financial blogs. In: WOMAS 2009—Workshop on Opinion Mining and Sentiment Analysis, Seville, Spain (2009)
4. Pamungkas, E.W., Putri, D.G.P.: Word sense disambiguation for lexicon-based sentiment analysis. In: 9th International Conference on Machine Learning and Computing, Singapore, Singapore, pp. 442–446 (2017)
5. Laura, C., José, O., Mathieu R., Pascal, P.: Dictionary-based sentiment analysis applied to a specific domain. In: Information Management and Big Data, pp. 57–68, Springer International Publishing (2017)
6. Ekawati, D., Khodra, M.L.: Aspect-based sentiment analysis for Indonesian restaurant reviews. In: International Conference on Advanced Informatics, Concepts, Theory, and Applications, Denpasar, Indonesia (2017)
7. Trilla, A., Alías, F.: Sentence-based sentiment analysis for expressive text-to-speech. IEEE Trans. Audio Speech Lang. Process. **21**(2), 223–233 (2013)
8. Sanguansat, P.: Paragraph2Vec-based sentiment analysis on social media for business in Thailand. In: 8th International Conference on Knowledge and Smart Technology, Chiangmai, Thailand (2016)
9. Nguyen, H., Nguyen M.L.: A deep neural architecture for sentence-level sentiment classification in twitter social networking. In: Conference of the Pacific Association for Computational Linguistics, abs/1706.08032 (2017)
10. Thelwall, M., Buckley, K., Paltoglou, G.: Sentiment strength detection for the social web. J. Am. Soc. Inf. Sci. Technol. **63**(1), 163–173 (2012)
11. Dutta, S., Roy, M., Das, A.K., Ghosh, S.: Sentiment detection in online content: a WordNet based approach. In: Panigrahi, B., Suganthan, P., Das, S. (eds.) Swarm, Evolutionary, and Memetic Computing. SEMCCO 2014. Lecture Notes in Computer Science, vol. 8947. Springer, Cham (2015)

An Efficient Sentiment Mining Approach on Social Media Networks

Al-Amin, Md. Amirul Islam, Sajal Halder, Md. Ashraf Uddin
and Uzzal Kumar Acharjee

Abstract In today's competitive environment, there is an essential need to collect and analyze data from social media, news, and other data streams that concern processing of huge amounts of data. A large number of posts, news, and blogs include opinions about product, service, and different issues. To accomplish an upper advantage, it is regularly important to listen and comprehend what individuals are saying in regard to contenders' item, benefit, and distinctive issues. We proposed a sentiment mining technique for social media analytics to identify influential opinions. Our aim to mine and to compress every one of the people surveys of an item as well as the polarity of subjective topics which isn't determine combined with opinion mining previously. Proposed task is performed in three steps: Firstly, mining item includes that have been remarked on by clients; secondly, recognize the supposition sentences in each survey and choosing whether every opinion sentence positive or negative or neutral; and finally, we summarize the results based on real datasets.

Keywords Social media · Text mining · Sentiment analysis · Opinion mining

Al-Amin (✉) · Md. A. Islam · S. Halder · Md. A. Uddin · U. K. Acharjee
Department of Computer Science and Engineering, Jagannath University,
Dhaka, Bangladesh
e-mail: alamin2293@gmail.com

Md. A. Islam
e-mail: sujan.jnu.05@gmail.com

S. Halder
e-mail: sajal@cse.jnu.ac.bd

Md. A. Uddin
e-mail: mdashrafuddin@students.federation.edu.au

U. K. Acharjee
e-mail: uzzal@cse.jnu.ac.bd

© Springer Nature Singapore Pte Ltd. 2019   451
A. Abraham et al. (eds.), *Emerging Technologies in Data Mining and Information
Security*, Advances in Intelligent Systems and Computing 814,
https://doi.org/10.1007/978-981-13-1501-5_39

1   Introduction

In recent years, we have watched the quick advancement of online networking, which has essentially changed the course in which people bestow and get. Microbloggings such as forums, blogs, online communities, and social media sites have emerged as new communication channels that link consumers with companies. These microbloggings, often called social media, directly influence marketing intelligence; consequently, organizations forcefully take an interest in social media correspondence and utilize it for advertising strategies and public relations. Web-based social networking likewise produces a huge volume of substance, for example, client audits, online articles, and tweets. Subsequently, investigating social media content is viewed as a chance to discover meaningful information and intelligence for business [3]. Specifically, opinion mining and sentiment examination are as often as possible utilized for such examination since they underscore the extraction of conclusion extremity and creator's feelings, for example, positive, negative, and impartial assumptions. For instance, if an organization can mine customer sentiments about items, administrations, mark picture, and notoriety by breaking down this information, the data would enable it to supervise bargains advancing and business strategy more effectively and productively. In this matter, opinion mining as a sub-train inside information mining and phonetics is alluded to as the computational method used to remove, group, realize, and evaluate the sentiment communicated in different online news sources, Web-based social networking remarks, and other client-produced content [10]. Sentiment investigation is frequently utilized as a part of thought excavation to discover sentiment, influence, subjectivity terms, and different enthusiastic states in the Web content. Therefore, assessment mining that includes assumption investigation can be characterized as a progression of procedures used to distinguish notion, subtlety, and the creator's perspective as showed up in content and can change this information into essential data for use in decision-making.

The remaining part of this paper is organized as follows: In Sect. 2, the related works are described. Our proposed method, sentiment mining approach, is discussed in Sect. 3. Experimental evaluation is shown in Sect. 4, and Sect. 5 concludes our work with future research directions.

2   Related Works

Web opinion mining generally implies the route toward expelling thing features and conclusions from review reports and condensing them using a graphical depiction. Su et al. [12] proposed a novel shared fortification approach to manage and deal with the component-level sentiment mining issue. Morinaga et al. [9] display a system for mining public opinions identified with item notoriety on the Internet. Kloptchenko et al. [5] use data and substance mining methods to analyze the printed some portion of an organization's budgetary report. Hung [4] uses grouping examination as an exploratory procedure to break down e-getting the hang of composing and imagines plans by social occasion sources that offer similar words, trademark regards, and coding rules. Sentiment investigation is the computational location and investigation of conclusions, notions, feelings, and subjectivities in writings [7]. As an uncommon use of content mining, slant examination is worried about the programmed extraction of positive or negative sentiments from writings. Bollen et al. [1] utilize opinion investigation to mine an expansive corpus of Twitter messages to decide the disposition of the Twitter residents on a given day. Duan et al. [2] utilize the estimation examination strategy to mine 70,103 online client audits posted in different online scenes from 1999 to 2011 for 86 lodgings in Washington, DC. Lee et al. [6] utilize the two information mining and assessment investigation strategies to break down the dataset gathered from MyStarbucksIdea, a standout among the most prominent online open advancement groups. In particular, they utilize estimation examination to isolate the opinion contained in every thought and remark accumulated from the MyStarbucksIdea site. Stieglitz and Dang-Xuan [11] recommend that organizations should give careful consideration to the examination of estimation identified with their brands and things in online networking correspondence.

3   Proposed Method

We propose to distinguish the main organizations specifically enterprises, for example, innovation, banking, retail and real estate, look at their online networking notices for focused investigation, and make industry-particular feeling mining for marketing knowledge and basic leadership. The information accumulation method basically includes utilizing freely accessible APIs from Web-based social media and if APIs are not accessible, creeping particular sites and parsing HTML as needed to gather survey comments. In the starting phase, we extract data from social media using text mining. At our second phase, we perform sentiment analysis using opinion lexicon-based methods. Finally, simulation and performance analysis will be accomplished to attest the better result of our proposed method. Proposed method is shown in Fig. 1.

In this paper, we have work with the Twitter Web-based social networking stage to find how tweets from the Twitter channel can be utilized to perform opinion investigation. As specified before, we performed opinion examination on three driving aircraft and R programming lingo has been generally used to play out this investigation.

3.1   Sentiment Analysis Method

The technique took after here is to calculate the positive and negative words in each tweet and set a sentiment score. Thusly, we can decide how positive or negative a tweet is. There are various approaches to count such scores; here is one formula to execute such counts.

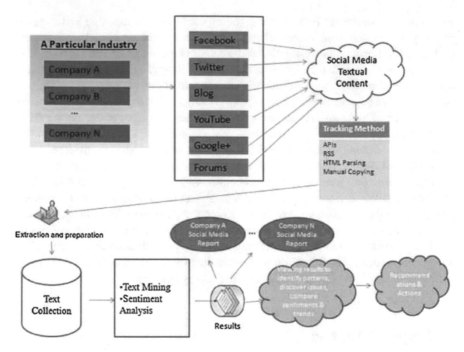

Fig. 1 Proposed methodology for text mining and sentiment analysis for microblogs

– Score = Total positive words – Total negative words
– If Score > 0, then tweet has 'positive sentiment'
– If Score < 0, then the tweet has 'negative sentiment'
– If Score = 0, then the tweet has 'neutral sentiment'

   To discover the number of positive and negative words, an opinion lexicon can be used.

3.2 Authorization Algorithm to Extract Tweets

TwitterR offers a simple approach to pluck tweets containing a given hashtag, word, or term from clients record. However, before stacking *TwitterR* library and utilizing its capacities, we have executed the following algorithm to set authorization to extract tweets.

Algorithm 1: Authorization algorithm to get access Tweets account.

Data: API key, API secret, Access token, Access token secret.
Result: Redirect for authentication;
1 **if** *API key= yourconsumerkey, API secret= yourconsumersecret, accesstoken = consumeraccesstoken AND access token secret= consumeraccesssecrettoken* **then**
2 | setup twitter authorization
3 **end**
4 Redirect for authetication;

3.3 Extracting Tweets with Hashtags

To represent the sentiment analysis, we examine tweets identifying with Delta, Jet-Blue, and United Airlines. Keeping in mind the end goal to extract particular tweets identifying these airlines, we inquire for Twitter for tweets with the hashtag Delta, JetBlue, and United.

Opinion lexicon: A list of English positive and negative sentiment words (around 6800 words). This list was accumulated over numerous years beginning from paper [8], and this list is shown in Table 1.

Algorithm 2: The Sentiment Score algorithm.

Input : Sentences, Positive Words (pos.word), Negative Words (neg.word), Progress (Initialize as none)
Output : Sentiment Score;
1 Begin
2 Remove Punctuation() /*Remove Punctuation from sentence*/
3 Remove Control Character() /*Remove control characters from sentence*/
4 Remove Digit() /*Remove digits from sentence*/
5 Convert LowerCase() /*Convert sentence into lower case*/
6 Split Sentence() /*Split sentence into words*/
7 Produce Vector() /*Produce single vector with all the atomic components (unlist)*/
8 Compare Words() /* Compare words to the lexicons of positive and negative terms*/
9 Get Value() /*Get positive and negative matched value as Boolean
10 Score = Difference from sum of positive matched value to sum of negative matched value
11 Returned calculated score
12 End;

Table 1 English positive and negative sentiment words

Positive words	Negative words
Accessable	Absence
Beloved	Banish
Capable	Cheat
Dedicated	Dangerous
Educated	Expensive

Algorithm 2 represents a framework to find and set the sentiment score. We get all positive and negative matched value as Boolean (Line 9) and then find out Score by calculating difference from sum of positive matched value to sum of negative matched value (Line 10).

4 Experimental Evaluation

R is a language and condition for measurable computing and designs. R gives a wide assortment of factual (direct and nonlinear displaying, established measurable tests, time-arrangement examination, order, grouping) and graphical procedures, and is very extensible.

4.1 Experimental Environment

R is a coordinated suite of programming facilities for information control, count, and graphical show. It incorporates:

- An effective information dealing and storage facility
- A suite of administrators for counts on exhibits, specifically matrices
- An expansive, intelligent, coordinated gathering of middle tools for data analysis
- Graphical facilities for information analysis and show either on-screen or on printed version
- An all-around created, basic and effective programming language which incorporates conditionals, loops, and client characterized recursive function, input, and output facilities.

4.2 Datasets

A dataset is a gathering of information. To check the validity of any algorithm, datasets are most important factor. There are many spatiotemporal datasets which have been classified into two categories: real datasets and synthetic datasets. We have used real data in our experiment analysis. We used three airlines' Twitter datasets got from the Twitter programming interface via looking through some question substances. The element terms and the comparing tweet include are recorded Table 2.

Table 2 Twitter datasets

Datasets	Number of tweets
Delta Airlines (@Delta)	5000; 10,000; 15,000
JetBlue Airlines (@JetBlue)	5000; 10,000; 15,000
United Airlines (@united)	5000; 10,000; 15,000

4.3 Result and Performance Analysis

In this section, we analyze performance in different experiments in different results of our proposed algorithm on the Twitter (three airlines) dataset.

4.3.1 Polarity Plot Customer Sentiments (Delta Airlines)

Table 3 depicts polarity. When Twitter users are more increased, positive, negative, and neutral comments by Twitter users gradually increase than previous number of Twitter users. So, for more Twitter users, we get better result.

4.3.2 Customer Sentiment Scores (Delta Airlines)

Figure 2 depicts Twitter user's sentiment score about Delta Airlines for 5000, 10,000, and 15,000 Twitter users, negative score denoted by the $(-)$ symbol, which shows despondency of clients with the aircraft, while the positive score implies that customers are content with the carrier. While, zero speaks to that Twitter clients are neutral.

4.3.3 Polarity Plot Customer Sentiments (JetBlue Airlines)

Table 4 represents polarity plot customer sentiment about the JetBlue Airlines.

Table 3 Polarity about Delta Airlines

Twitter users	Positive comments	Negative comments	Neutral comments
5000	1520 users	1100 users	2380 users
10,000	2450 users	2350 users	5200 users
15,000	3450 users	4100 users	7450 users

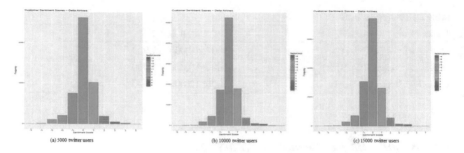

Fig. 2 Customer sentiment scores about Delta Airlines for different Twitter users

Table 4 Polarity about JetBlue Airlines

Twitter users	Positive comments	Negative comments	Neutral comments
5000	1750 users	550 users	2700 users
10,000	4650 users	2550 users	2800 users
15,000	6250 users	3850 users	4900 users

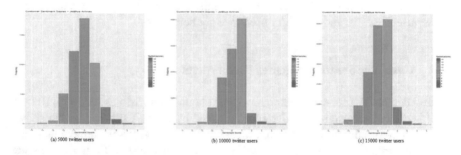

Fig. 3 Customer sentiment scores about JetBlue Airlines for different Twitter users

4.3.4 Customer Sentiment Scores (JetBlue Airlines)

Figure 3 depicts Twitter user's sentiment scores about JetBlue Airlines for 5000, 10,000, and 15,000 Twitter users.

4.3.5 Polarity Plot Customer Sentiments (United Airlines)

Table 5 represents polarity.

Table 5 Polarity about United Airlines

Twitter users	Positive comments	Negative comments	Neutral comments
5000	1450 users	1350 users	2200 users
10,000	1750 users	3450 users	4800 users
15,000	2450 users	5450 users	7100 users

Fig. 4 Customer sentiment scores about United Airlines for different Twitter users

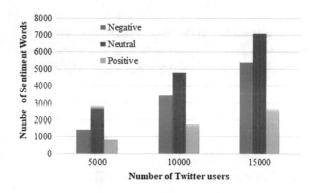

4.3.6 Customer Sentiment Scores (United Airlines)

Figure 4 depicts Twitter user's sentiment scores about United Airlines for 5000, 10,000, and 15,000 Twitter users.

4.4 Comparison Results

Now, we have summarized the overall positive, negative, and neutral scores. To put it in another way, we have created total count by adding positive, negative, and neutral sum. Additionally, we have calculated positive, negative, and neutral percentages (Tables 6, 7, and 8).

Table 6 Positive comments' comparison for three airlines

Name of airlines	Positive % score (5000) (%)	Positive % score (10,000) (%)	Positive % score (15,000)
Delta	30	24	23
JetBlue	35	46	42
United	29	18	16

Table 7 Negative comments' comparison for three airlines

Name of airlines	Negative % score (5000) (%)	Negative % score (10,000) (%)	Negative % score (15,000) (%)
Delta	22	24	27
JetBlue	11	26	26
United	27	34	36

Table 8 Neutral comments' comparison for three airlines

Name of airlines	Neutral % score (5000) (%)	Neutral % score (10,000) (%)	Neutral % score (15,000) (%)
Delta	48	52	50
JetBlue	54	28	33
United	44	48	47

5 Conclusions and Future Works

Sentiment analysis make able a business or company to comprehend shopper suppositions in connection to particular items or services. Besides, these experiences could be utilized to enhance their items and administrations by measuring shoppers' remarks and criticism utilizing sentiment analysis. Our proposed sentiment mining technique for social media analytics can identify influential opinions. Proposed method can also mine and to summarize all the people reviews of a product as well as polarity of subjective topics which isn't determine combined with opinion mining previously. In result section, we use real dataset for sentiment analysis. There are certain limitations in the paper. First of all, the tweets collected from the Twitter API are not as pure as required for the sentiment classification. Secondly, compared to the tweet data existing in Twitter, the dataset I collected and used in this paper is a very tiny part. For future research, we plan to examine the emotions on social media and sentiment analysis to decide the state of mind of an essayist or a speaker concerning a particular subject or the overall relevant polarity of an archive. We also plan to develop a tool so that general people can easily use it.

Acknowledgements This research was partially supported by Academic Innovation Fund, ICT Division, Government of the People's Republic of Bangladesh. Sajal Halder is the corresponding author.

References

1. Bollen, J., Mao, H., Zeng, X.: Twitter mood predicts the stock market. J. Comput. Sci. **2**(1), 1–8 (2011)
2. Duan, W., Cao, Q., Yu, Y., Levy, S.: Mining online user-generated content: using sentiment analysis technique to study hotel service quality. In: 2013 46th Hawaii International Conference on System Sciences (HICSS), pp. 3119–3128. IEEE (2013)
3. He, W., Harris, W., Yan, G., Akula, V., Shen, J.: A novel social media competitive analytics framework with sentiment benchmarks. Inf. Manage. **52**(7), 801–812 (2015)
4. Hung, J.: Trends of e-learning research from 2000 to 2008: use of text mining and bibliometrics. Br. J. Educ. Technol. **43**(1), 5–16 (2012)
5. Kloptchenko, A., Eklund, T., Karlsson, J., Back, B., Vanharanta, H., Visa, A.: Combining data and text mining techniques for analysing financial reports. Intell. Syst. Account. Finance Manage. **12**(1), 29–41 (2004)
6. Lee, H., Choi, K., Yoo, D., Suh, Y., He, G., Lee, S.: The more the worse? Mining valuable ideas with sentiment analysis for idea recommendation. In: PACIS, p. 30 (2013)
7. Li, N., Wu, D.D.: Using text mining and sentiment analysis for online forums hotspot detection and forecast. Decis. Support Syst. **48**(2), 354–368 (2010)
8. Liu, B., Zhang, L.: A survey of opinion mining and sentiment analysis. In: Mining Text Data, pp. 415–463. Springer, Berlin (2012)
9. Morinaga, S., Yamanishi, K., Tateishi, K., Fukushima, T.: Mining product reputations on the web. In: Proceedings of the Eighth ACM SIGKDD International Conference on Knowledge Discovery and Data Mining, pp. 341–349. ACM (2002)
10. Petasis, G., Spiliotopoulos, D., Tsirakis, N., Tsantilas, P.: Sentiment analysis for reputation management: mining the Greek web. In: SETN, pp. 327–340. Springer, Berlin (2014)
11. Stieglitz, S., Dang-Xuan, L.: Social media and political communication: a social media analytics framework. Soc. Netw. Anal. Min. **3**(4), 1277–1291 (2013)
12. Su, Q., Xu, X., Guo, H., Guo, Z., Wu, X., Zhang, X., Swen, B., Su, Z.: Hidden sentiment association in Chinese web opinion mining. In: Proceedings of the 17th International Conference on World Wide Web, pp. 959–968. ACM (2008)

A Crowdsource-Based Approach for Preparing Bangla POS Tagged Corpus

Shamim Ehsan, Sadia Tasnim Swarna and Sabir Ismail

Abstract Automated Parts of Speech Tagging plays a vital role in the natural language processing. For computational Bangla Language Processing, we do not have large-scale Parts of Speech tagged corpus. There are two basic approaches to implement a corpus, by written rules or automated. To implement a rule-based corpus, we need experts in Bangla linguistics and it is also time-consuming. And for the automated corpus, we need a trained corpus, which is currently not available. Crowdsourcing can be served a vital role to fulfill these two requirements. So, in this paper, we proposed a crowd source-based approach to building Bangla Parts of Speech tagged corpus. We have used a standard tag set for Bangla. Raw documents are collected from various newspapers, books, and online site. We first give some example of Parts of Speech and then provide data to people for crowdsourcing. Finally, we analyze the result of the data, and its accuracy is 95%.

1 Introduction

Crowdsourcing is the act of taking a job traditionally performed by a designated employee and outsourcing it to an undefined, generally large group of people in the form of an open call. So basically its like someone has a task to be done, but do not want to hire a specific person to do it, rather they put their task on the world wide web with a small amount of reward, let random people solve it and finally getting the task done. This is how crowd sourcing works.

Crowdsourcing can be used in Parts of Speech Tagging. All words in sentences are classified into some groups which have some common grammatical properties; they are called parts of speech. Generally, the words having same parts of speech

S. Ehsan (✉) · S. T. Swarna (✉) · S. Ismail
Shahjalal University of Science and Technology, Sylhet 3114, Bangladesh
e-mail: ehsanhrid@gmail.com

S. T. Swarna
e-mail: sadiatasnimswarna@gmail.com

S. Ismail
e-mail: sabir-cse@sust.edu

© Springer Nature Singapore Pte Ltd. 2019   463
A. Abraham et al. (eds.), *Emerging Technologies in Data Mining and Information Security*, Advances in Intelligent Systems and Computing 814,
https://doi.org/10.1007/978-981-13-1501-5_40

play similar roles inside the linguistic structure of sentences. Parts of Speech tagging means a system tags the parts of speech of a sentence automatically. It may be assumed that it can be easy for a computer by just looking into the dictionary for the appropriate parts of speech. But a word can fall into different parts of speech in the context of different sentences. Consider the sentences:

1. শরীরে বল নাই?
2. কথা বল।
3. বল খেলা শরীরের জন্য ভাল।

It is clearly seen that the word " বল " has three different meaning as well as different parts of speeches in the four sentences (adjective, verb, and noun). Unfortunately, there is no fixed rules to say " বল " belongs to which parts of speech in the sentences. To solve this ambiguity, we either need specific rules or we can train machine, to solve it using learning. Here comes the factor of data. Parts of speech tagging system needs a training dataset with all the sentences tagged with their corresponding parts of speech. But this kind of dataset is not available for Bangla.

So what to do? Then comes crowd sourcing with the solution. Crowdsourcing does not need very skilled workers or very sophisticated devices, rather a crowdsourcing system needs only a large number of normal people with no special talents/abilities.

Everyone with a minimum grammatical knowledge can tag sentence, and it does not take many skills. So tagging sentences using the general crowd can be a pretty good option to create a huge dataset of tagged sentences. It is pretty effective also in the fields of natural language processing like speech recognition, translation. Translation feature of Google Inc[1] is heavily depended on crowdsourcing from users all around the world, and they claim better accuracy than typical statistical machine translation. So we can say that crowdsourcing can solve the problem of automated parts of speech tagging.

2 Related Works

Crowdsourcing systems enlist a large number of humans to solve a wide variety of problems. Crowdsourcing was proposed to make efficient use of manpower and resources. In recent decade, numerous work has concentrated on various aspects of crowdsourcing, say for different performance analysis and computational techniques. Figure 1 shows a taxonomy of crowdsourcing.

Natural language processing is a work that may be difficult for automated process but comparatively easy for humans. In recent times, as a quick alternative of expert annotations, researchers found Mechanical Turk of Amazon [1–6]. Akkaya et al. [7] demonstrated that crowdsourcing would be good for subjectivity word sense comment. Callison-Burch and Dredze [8] demonstrated that they make information for discourse and dialect applications with a minimal effort. Gao and Vogel [3] showed that crowdsourcing laborers perform well on word arrangement assignments.

---

[1] https://translate.google.com/.

Fig. 1 A taxonomy of crowdsourcing

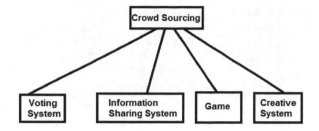

Jha et al. [4] demonstrated that an exact prepositional expression connection corpus can be developed by crowdsourcing specialists. Parent and Eskenazi [5] proposed a system to decompose a task for the meaning of dictionary words in MTurk. Skory and Eskenazi [6] discussed ways to calculate the quality of the results of MTurk workers tasks.

3 Raw Data Collection

First step is to select a group of sentence as our primary dataset. We cannot just choose some random sentence and ask users to tag them. Team Pipilikas[2] research team has shared their text corpus with us with 1000 documents in the corpus, all of which are collected from various newspapers. From there 330 documents are chosen. All the documents have passed through a validator then again manually checked to be selected for the dataset. The validator checks for too long and incomplete sentences. As user satisfaction should be the main priority in a crowdsourcing system, so if the user gets bored with very large sentences, that will prevent them from tagging more documents. After formatting, 330 documents have taken from PIPILIKA corpus, 35 documents from the novel " আমার বন্ধু রাশেদ " [9] By Dr. Muhammad Zafar Iqbal, and 35 from facebook articles from famous writers. So in total 400 documents have been chosen, each containing three sentences as our primary dataset. A brief summary is shown in Table 1.

Though there are five tags in Bangla grammar and eight parts of speech in English grammar, 12 tags are globally used in the parts of speech tagging research [10, 11]. They are:

1. বিশেষ্য(NN)
2. সর্বনাম(PR)
3. নির্দেশক(DM)
4. ক্রিয়া(V)
5. বিশেষণ(JJ)
6. ক্রিয়া-বিশেষণ(RB)

7. অনুসর্গ(PSP)
8. সংযোজক(CC)
9. অব্যয়(RP)
10. পরিমাণবাচক(QT)
11. যতি-চিহ্ন(PNC)
12. অন্যান্য(RD)

---

[2]Pipilika is the first Bangla search engine developed by the students of Shahjalal University of Science and Technology.

Table 1 Summary of the dataset

Total documents	400
Total sentences	1200
Total words	7657
Unique words	3082
Most frequent word	এ

A reference document is also used to help out the crowd workers. Niladri Sekhar Dashs [10] POS tagset for Bangla document is used as a reference document for crowd workers.

4 User Interface Implementation

4.1 Crowd Sourcing System

A system is called a crowdsourcing system if it depends on the opinions and answers of the general crowd. People who are participating in the crowdsourcing are called crowd users. Some crowdsourcing system sets some questions and takes opinion of the users, and most voted opinion is considered to be the correct answer. These systems are called Voting System. In some cases, there is no fixed answer to the question; users have to use their creativity to crowdsource. For example, Google Draw [12] ask users to draw an object within 20 s. Suppose a user is asked to draw a car, he has to draw it with his own creativity, these systems are called creative systems.

Our proposed crowdsourcing system is both a voting system and creative system. There will be sentences without any proper tags, users will tag the sentences, and with the response of the users, the model will tag the words.

4.2 Building the Interface

An interface is created to collect data from crowd workers. It's a web application which is built using PHP Laravel 5.3 framework. Laravel is chosen because it can handle concurrent requests smartly than most other frameworks. No pre-built web template is chosen, we tried to keep the site as simple as possible. Most of the templates have to load lots of JavaScript and CSS files and loading all of the files in every page makes the site a bit slower.

The user interface has register page, login page, and home page. Some people don't get motivated to enter when it requires too much information to log into a site,

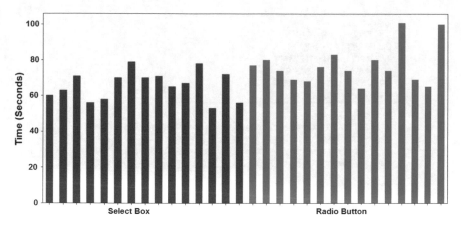

Fig. 2 Average time to tag a document in HTML select box and radio button

so a guest login feature where people get logged in with just a temporary username. After logging in, each user is assigned with a document and asked to tag the words of the documents.

We've planned to make use three different types of input fields of HTML (select box, radio button, and drag and drop). We successfully implemented first two and tested it to know which option will be suitable for the users. We select fifteen volunteers and assigned each of them to tag 50 sentences with solutions as fast as they can to check which method is faster and has less chance of selecting a wrong tag while in a hurry. Results are shown in Fig. 2.

It is seen that using select box, people take an average of 65 s and using radio button they take 76 s to tag a document. Using select box, people tag 98.45% words correctly while radio button has an accuracy of 91.45%. So we can have a conclusion that both chances of making silly mistakes and speed in radio button is higher than a select box. So we have excluded radio button from our site and take input from users with only HTML Select Box. When a user submits a document a confirmation window appears and the user is prompted with a view of the document with the corresponding parts of speech it has been tagged, if user noticed that he/she have tagged a word wrongly, he/she can move back to the previous page and correct it. In this case, the user doesn't have to select all the tags again, because if that happens, nobody wants to do that again. Instead, the user gets a view where all of his previous selected sentences is tagged and he just has to change the wrong tag and confirm. It is shown in Fig. 3.

4.3 Adding a LeaderBoard

As it is mentioned earlier, people only crowd-sourced when they get a handy reward or got tricked as they are competing or playing a game. So, a leaderboard is added where everyone can see who is at the top of data tagging. The user can see total

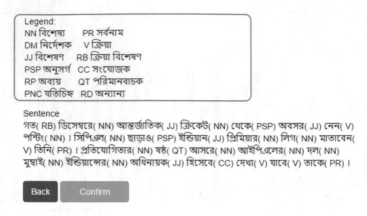

Fig. 3 Snapshot of Confirmation window

tagged documents region-wise. After adding the leaderboard feature, the average time to tag a document reduces more than 10% than the time takes before. Before adding leaderboard feature, the average time to tag a document by a user is 100 s, and after adding leaderboard, it reduced to 92 s.

5 Methodology

After the interface is created, data is collected from the crowd workers. Several people tag a sentence differently, and next step is to find out which tag should belong to which word in sentences, i.e., label the sentences. Like voting systems, we go for the majority. If the majority of people says that "ভাল" from "সে ভাল আছে" has the tag "বিশেষণ", then we label "ভাল" as "বিশেষণ" in the sentence "সে ভাল আছে". In case of a tie between two tags, we search for previous behavior of the word. Consider the sentence: "গত অর্থবছর থেকে আমদানিতে স্থবিরতা দেখা দেয়". Seven crowd workers have tagged the document and each of their tags have been shown in a 2-D matrix form. We can see that majority of people voted "অর্থবছর", "আমদানিতে" as "বিশেষ্য", "দেখা", "দেয়" as "ক্রিয়া" but confusion arises for the words গত and স্থবিরতা. In this case, we check previously in how many times "গত" has been tagged as "নির্দেশক" and as ক্রিয়া-বিশেষণ. Previous behavior breaks the tie. Matrix of the sentence with the most likely tag of: "গত অর্থবছর থেকে আমদানিতে স্থবিরতা দেখা দেয়" is shown in Fig. 4.

6 Finding Optimal Number of Users

As we have mentioned earlier, each document is tagged around 5–8 times. A question arises that how many times a document should be tagged? Five times should be enough? Is tagging a document eight times necessary? Or is it overkill? So the raw

	গত	অর্থবছর	থেকে	আমদানিতে	স্থবিরতা	দেখা	দেয়
বিশেষ্য	1	5	0	4	1	0	0
সর্বনাম	0	0	0	0	0	0	0
নির্দেশক	2	0	0	0	0	0	0
ক্রিয়া	0	0	0	2	0	7	6
বিশেষণ	1	2	0	1	3	0	0
ক্রিয়া-বিশেষণ	2	0	0	0	3	0	0
অনুসর্গ	0	0	5	0	0	0	1
সংযোজক	0	0	0	0	0	0	0
অব্যয়	1	0	2	0	0	0	0
পরিমাণবাচক	0	0	0	0	0	0	0
যতি-চিহ্ন	0	0	0	0	0	0	0
সাযায়া	0	0	0	0	0	0	0
Most Likely Tag	নির্দেশক	বিশেষ্য	অনুসর্গ	বিশেষ্য	বিশেষণ	ক্রিয়া	ক্রিয়া

Fig. 4 Most Likely Tags of a sample sentence

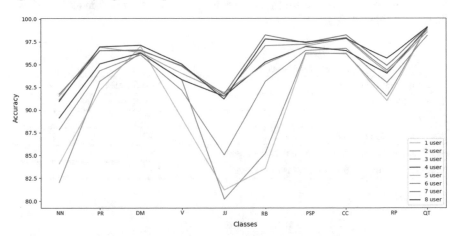

Fig. 5 Accuracy of different dataset to find optimal number of user to tag a document

dataset is analyzed again, and some tests are run. Eight data corpuses are created, each dataset is created based on the profile matrix of random n crowd workers response. If n exceeds the total number of responses to that document, all the responses to that document are taken. For the eight data corpuses, accuracy for each of the classes is calculated. The procedure of getting accuracy of dataset is discussed in Results section. The results are shown in Fig. 5. By observing the figure, we can see that when $n > 5$, accuracy for different parts of speeches is very slightly increasing, and in some cases, decreasing. So we can say that tagging a document by five different users should be enough.

7 Results

Total 170 crowd workers have been participated to build the tagged corpus, age
between 20 and 25, all of them are a university student. On average, each worker
tags 13 documents. Each document is tagged at least five times, and at most eight
times. A total of 2195 times the documents has been tagged by the crowd workers.
Each document contains three sentences, so in total 6585 times, the sentences are
tagged. Table 2 shows a brief summary in a tabular form.

We run a test to see how many words have been tagged by all the users the same
every time. Among the words which have been tagged same by all the users, a pie
chart is given to show which parts of speech have been tagged more correctly in Fig. 6.
The chart shows that tagging "বিশেষ্য" (Noun) and "ক্রিয়া" (Verb) are comparatively
easier than other documents.

For testing, 10% of the random document is chosen and sent to some Bangla
linguistic experts for checking. Then the accuracy is calculated for each of the classes.
Confusion matrix of the classes True Positive, True Negative, False Positive, False
Negative scores for the confusion matrix and the accuracy precision and recall are
shown in Tables 3, 4, and 5.

Table 2 Summary of tagged documents

Total documents tagged	2195 times
Total sentences tagged	6585 times
Crowd worker	170
Average document tagged	5.48
Maximum number of times a document is tagged	8
Minimum number of times a document is tagged	5

Fig. 6 Correctness of the parts of speeches

Table 3 Confusion matrix

	NN	PR	DM	V	JJ	RB	PSP	CC	RP	QT
NN	655	10	15	12	51	10	2	1	0	13
PR	5	221	0	0	15	0	0	0	0	0
DM	5	20	50	0	0	2	7	3	0	0
V	19	3	0	261	9	0	15	0	0	0
JJ	11	0	0	11	223	10	0	15	10	0
RB	0	0	0	0	15	26	0	0	0	0
PSP	0	0	0	0	0	0	24	0	10	0
CC	0	0	3	0	0	3	0	30	0	0
RP	13	0	0	0	15	0	5	15	40	0
QT	0	0	0	0	0	0	0	0	3	144

Table 4 True Positive, True Negative, False Positive, and False Negative table for confusion matrix

	TP	TN	FP	FN
NN	655	1019	114	53
PR	221	1453	20	33
DM	50	1624	37	18
V	261	1413	4	23
JJ	223	1451	57	105
RB	26	1648	15	25
PSP	24	1680	10	29
CC	30	1644	6	34
RP	40	1634	48	23
QT	144	1530	3	13

Table 5 Accuracy, Precision, and Recall of Test Dataset

	Accuracy	Precision	Recall
NN	90.93	85.18	92.51
PR	96.93	91.70	87.01
DM	96.82	57.47	73.53
V	96.04	85.02	91.90
JJ	91.18	79.64	67.99
RB	97.66	63.41	50.98
PSP	97.73	70.58	45.28
CC	97.67	83.33	46.88
RP	95.93	45.45	63.50
QT	99.05	97.96	91.72

A surprising thing is observed that the accuracy of the noun (বিশেষ্য) is a bit lower than the other classes. By common sense, we can assume that tagging noun is easier. But if we can look at the confusion matrix, we can see that many words have been falsely classified as a noun by the crowd workers, but they belong to some other class, Crowd workers tend to tag a document as noun when they are confused with the class of that word. That is the reason behind the lower accuracy of noun. The poor performance of Precision and Recall in some classes can also be explained; they seldom occur in the sentences, so a misclassification can damage more to Recall and Precision than Accuracy because in calculating accuracy, True Negative is also accounted. As all the True Negative values are comparatively larger, so a small value in False Positive and False Negative affects the accuracy less. The average accuracy for all the classes is 95.994%. So after calculating tags of every sentence according to our methodology stated above, our final tagged corpus is created. It is a table in MySQL with these rows:

- ID
- line
- tags
- category

Category field is kept for further research.

References

1. Quinn, A.J., Bederson, B.B.: Human computation: a survey and taxonomy of a growing field. In: Proceedings of the SIGCHI conference on human factors in computing systems. ACM (2011)
2. Gordon, J., Van Durme, B., Schubert, L.K.: Evaluation of commonsense knowledge with Mechanical Turk. In: Proceedings of the NAACL HLT 2010 Workshop on Creating Speech and Language Data with Amazon's Mechanical Turk. Association for Computational Linguistics (2010)
3. Gao, Q., Vogel, S.: Consensus versus expertise: a case study of word alignment with mechanical turk. In: Proceedings of the NAACL HLT 2010 Workshop on Creating Speech and Language Data with Amazon's Mechanical Turk. Association for Computational Linguistics (2010)
4. Jha, M., et al.: Corpus creation for new genres: a crowdsourced approach to PP attachment. In: Proceedings of the NAACL HLT 2010 Workshop on Creating Speech and Language Data with Amazon's Mechanical Turk. Association for Computational Linguistics (2010)
5. Parent, G., Eskenazi, M.: Clustering dictionary definitions using amazon mechanical turk.. In: Proceedings of the NAACL HLT 2010 Workshop on Creating Speech and Language Data with Amazon's Mechanical Turk. Association for Computational Linguistics (2010)
6. Skory, A., Eskenazi, M.: Predicting cloze task quality for vocabulary training. In: Proceedings of the NAACL HLT 2010 Fifth Workshop on Innovative Use of NLP for Building Educational Applications. Association for Computational Linguistics (2010)
7. Akkaya, C., et al.: Amazon mechanical turk for subjectivity word sense disambiguation. In: Proceedings of the NAACL HLT 2010 Workshop on Creating Speech and Language Data with Amazon's Mechanical Turk. Association for Computational Linguistics (2010)
8. Callison-Burch, C., Dredze, M.: Creating speech and language data with Amazon's Mechanical Turk. In: Proceedings of the NAACL HLT 2010 Workshop on Creating Speech and Language Data with Amazon's Mechanical Turk. Association for Computational Linguistics (2010)

9. Dr. Muhammad Zafar Iqbal, Rashed: My Friend, ISBN-984-437046-9
10. Niladri Sekhar Dash, POS tagset for Bangla Document, Microsoft Research India, Aug 2010
11. Categorizing and Tagging Words. http://www.nltk.org/book/ch05.html Cited 30 Aug 2017
12. Quick, Draw! https://quickdraw.withgoogle.com/, cited 30 Aug 2017

A Study of Feature Extraction Techniques for Sentiment Analysis

M. Avinash and E. Sivasankar

Abstract Sentiment analysis refers to the study of systematically extracting the meaning of subjective text. When analyzing sentiments from the subjective text using machine learning techniques, feature extraction becomes a significant part. We perform a study on the performance of feature extraction techniques, TF-IDF (term frequency-inverse document frequency) and Doc2vec (document to vector), using Cornell movie review datasets, UCI sentiment labeled datasets, stanford movie review datasets, effectively classifying the text into positive and negative polarities by using various preprocessing methods like eliminating stop words and tokenization which increases the performance of sentiment analysis in terms of accuracy and time taken by the classifier. The features obtained after applying feature extraction techniques on the text sentences are trained and tested using the classifiers logistic regression, support vector machines, K-nearest neighbors, decision tree, and Bernoulli Naive Bayes.

1 Introduction

In machine learning, a feature refers to the information which can be extracted from any data sample. A feature uniquely describes the properties possessed by the data. The data used in machine learning consists of features projected onto a high-dimensional feature space. These high-dimensional features must be mapped onto a small number of low-dimensional variables that preserve information about the data as much as possible. Feature extraction is one of the dimensionality reduction techniques used in machine learning to map higher-dimensional data onto a

M. Avinash · E. Sivasankar (✉)
National Institute of Technology Tiruchirapalli,
Tanjore Main Road National Highway 67, Near BHEL Trichy,
Tiruchirappalli 620015, Tamil Nadu, India
e-mail: sivasankar@nitt.edu

M. Avinash
e-mail: avinash.sai001@gmail.com

© Springer Nature Singapore Pte Ltd. 2019   475
A. Abraham et al. (eds.), *Emerging Technologies in Data Mining and Information Security*, Advances in Intelligent Systems and Computing 814,
https://doi.org/10.1007/978-981-13-1501-5_41

set of low-dimensional potential features. Extracting informative and essential features greatly enhances the performance of machine learning models and reduces the computational complexity.

The growth of modern Web applications like Facebook, Twitter persuaded users to express their opinions on products, persons, and places. Millions of consumers review the products on online shopping Web sites like Amazon and Flipkart. These reviews act as valuable sources of information that can increase the quality of services provided. With the growth of enormous amount of user-generated data, lot of efforts are made to analyze the sentiment from the consumer reviews. But analyzing unstructured form of data and extracting sentiment out of it requires a lot of natural language processing (nlp) and text mining methodologies. Sentiment analysis attempts to derive the polarities from the text data using nlp and text mining techniques. The classification algorithms in machine learning require the most appropriate set of features to classify the text as positive polarity and negative polarity. Hence, feature extraction plays a prominent role in sentiment analysis.

2 Literature Review

Bo Pang, Lillian Lee, and Shivakumar Vaithyanathan [1] have conducted a study on sentiment analysis using machine learning techniques. They compared the performance of machine learning techniques with human-generated baselines and proposed that machine learning techniques are quite good in comparison to human-generated baselines. Jain and Katkar [2] have performed sentiment analysis on Twitter data using data mining techniques. They analyzed the performance of various data mining techniques and proposed that data mining classifiers can be a good choice for sentiment prediction. Koprinska and O'Keefe [3] have done research on feature selection and weighting methods in sentiment analysis. In their research, they have combined various feature selection techniques with feature weighing methods to estimate the performance of classification algorithms. Albitar et al. [4] have a proposed an effective TF-IDF-based text-to-text semantic similarity measure for text classification. Le and Mikolov [5] introduced distributed representation of sentences and documents for text classification. Sanguansat [6] performed paragraph2vec-based sentiment analysis on social media for business in Thailand. Senturk and Bilgin [7] performed sentiment analysis on Twitter data using doc2vec. Analyzing all the above techniques gave us deep insight about various methodologies used in sentiment analysis. Therefore, we propose a comparison-based study on the performance of feature techniques used in sentiment analysis, and the results are described in the following sections.

The proposed study involves two different features extraction techniques TF-IDF and Doc2Vec. These techniques are used to extract features from the text data. These features are then used in identifying the polarity of text data using classification algorithms. The performance evaluation of TF-IDF and Doc2vec is experimented on five benchmark datasets of varying sizes. These datasets contain user reviews from various domains. Experimental results show that the performance of feature extraction techniques trained and tested using these benchmark datasets. A lot of factors affect the sentiment analysis of text data. When the text data is similar, it becomes difficult to extract suitable features. If the features extracted are not informative, it significantly affects the performance of classification algorithms. When the patterns lie closest to each other, drawing separating surfaces can be tricky for the classifiers. Classifiers like SVM aim to maximize the margin between two classes of interest.

3 Methodology

The performance evaluation of feature extraction techniques, TF-IDF and Doc2Vec, requires datasets of various sizes. The tests are conducted on the publicly available datasets related to movie reviews and reviews related to other commercial products. The performance is evaluated by varying the sizes of datasets, and the accuracy of each of the two feature extraction techniques is measured using the classifiers. The first dataset is a part of movie review dataset [8] consisting of 1500 reviews. The second dataset used is sentiment labeled dataset [9] consisting of 2000 reviews taken from UCI repository of datasets. The third dataset used is a part of polarity dataset v2.0 [10] taken from CS cornell.edu. The fourth dataset sentence polarity dataset v1.0 [11] is taken from CS cornell.edu. The fifth dataset used is large movie review dataset [8] taken from ai.stanford.edu. The reviews are categorized into positive and negative. We perform analysis on reviews in three stages, preprocessing, feature extraction, and classification. Figure 1 shows the experimental setup for sentiment analysis.

3.1 Preprocessing

We perform preprocessing in two stages—Tokenization and by eliminating stop words. Tokenization refers to the process of breaking the text into meaningful data that retains the information about the text. Text data contains some unwanted words that do not give any meaning to the data called as stop words. These words are removed from the text during feature extraction procedures.

3.2 Feature Extraction

Two techniques are used for feature extraction—TF-IDF and Doc2vec. Figure 2 shows the steps involved in extracting features using both the techniques.

Fig. 1 Experimental setup

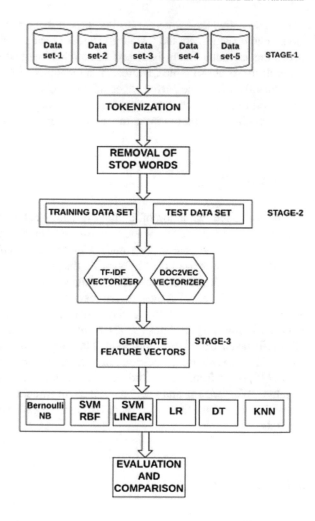

3.2.1　TF-IDF

TF-IDF is short form for term frequency-inverse document frequency. TF-IDF is one of the largely used methods in information retrieval and text mining. TF-IDF is a weight metric which determines the importance of word for that document.

TF

Term frequency measures number of times a particular term t occured in a document d. Frequency increases when the term has occured multiple times. TF is calculated by taking ratio of frequency of term t in document d to number of terms in that particular document d.

Fig. 2 Feature extraction procedure

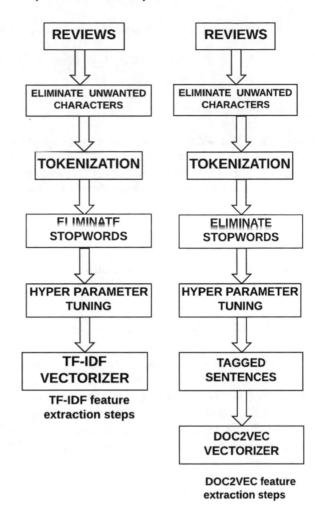

$$\text{TF}(t, d) = \frac{\text{Number of times term } t \text{ appears in a document } d}{\text{Total number terms in a document } d} \tag{1}$$

IDF

TF measures only the frequency of a term t. Some terms like stop words occur multiple times but may not be useful. Hence, inverse document frequency (IDF) is used to measure term's importance. IDF gives more importance to the rarely occurring terms in the document d. IDF is calculated as:

$$\text{IDF}(t) = \log_e \frac{\text{Total number of documents}}{\text{Total number of documents with term } t \text{ in it}} \tag{2}$$

The final weight for a term t in a document d is calculated as:

$$TF - IDF(t, d) = TF(t, d) \times IDF(t) \qquad (3)$$

3.2.2 Doc2Vec

An extended version to word2vec, doc2vec model was put forward by Le and Miklov [5] to improve the learning of embeddings from word-to-word sequences. doc2vec can be applied for word n-gram, sentence, paragraph, or document. Doc2vec is a set of approaches to represent documents as fixed-length low-dimensional vectors. Doc2Vec is a three-layer neural network with an input, one hidden layer, and an output layer. Doc2vec was proposed in two forms: DBoW and dmpv. In Word2vec, two algorithms continuous bag of words (CBOW) and skip-gram (SG) algorithms are implemented using deep learning, in Doc2vec these algorithms correspond to distributed memory (DM) and distributed bag of words (DBoW).

3.3 Classification

Classifiers are trained on training datasets with the features obtained from the feature extraction techniques TF-IDF and Doc2vec and with the corresponding output labels from the dataset. The test data is tested with the trained classifiers to predict the sentiments, and the accuracy is measured for test data. Classifiers logistic regression, K-nearest neighbors, decision tree, Bernoulli Naive Bayes, and support vector machines with linear and rbf kernels are used for sentiment analysis of the datasets.

3.3.1 Logistic Regression (LR)

In linear regression, we try to predict the value $y^{(i)}$ for the ith training sample using a linear function $Y = h_\theta(x) = \theta^T x$. This is clearly not a great solution to predict binary-valued labels ($y^{(i)} \in \{0, 1\}$). Logistic regression uses a different hypothesis that predicts the probability that a given example belongs to class "1" and the probability that the example belongs to class "0". Logistic regression is formulated as:

$$P(y = 1|x) = \frac{1}{1 + e^{-\theta^T x}} \qquad (4)$$

$$P(y = 0|x) = 1 - h_\theta(x) \qquad (5)$$

3.3.2   K-Nearest Neighbors (KNN)

In KNN, the input is represented in the feature space by k-closest training samples. KNN uses the procedure of class membership. Each object is classified by taking its neighbors votes, and the object is assigned to class which has got maximum number of votes. The training examples are represented as vectors in a multi-dimensional feature space with a class label. The training of the KNN algorithm involves storing training samples as feature vectors with their corresponding labels. k is decided by the user during classification, and the unlabeled data is assigned to class which is more frequent among k of the training samples nearest to that unlabeled sample. Euclidean distance is used to calculate the distance.

3.3.3   Naive Bayes (NB)

Bayes theorem forms the basis for Naive Bayes classifier. It assumes that features are strongly independent and calculates the probability. Multi-variate Bernoulli event model is one of the Naive Bayes classifiers used in sentimental analysis. If X_i is a boolean expressing the occurrence or absence of the ith term from the vocabulary, then the likelihood of a document given a class C_k is given by:

$$P(x|C_k) = \prod_{i=1}^{n} p_{k_i}^{x_i} (1 - p_{k_i})^{1-x_i} \tag{6}$$

3.3.4   Support Vector Machines (SVM)

The training samples are represented as points in the feature space. SVM performs classification by separating the points with a set of margin planes. The boundary hyperplane is chosen which maximizes the distance to the training samples. Support vectors are the points that determine the margin planes. Data which can be separated linearly is classified using linear kernel, and the data which is not linearly separable is classified using RBF kernel.

3.3.5   Decision Tree (DT)

Decision trees perform classification by using yes or no conditions. A decision tree consists of edges and nodes. The root node does not contain incoming edges, nodes which contain out going edges are called test nodes, and nodes with no outgoing nodes are called decision nodes. Edges represent the conditions. Each decision node holds a class label. When the unlabeled data samples are to be classified, they pass through series of test nodes finally leading to the decision node with a class label and the class label is assigned to that unlabeled data sample.

4 Experimental Results

The performance of TF-IDF and Doc2Vec is verified by experimenting the feature extraction techniques on five different datasets of varying sizes. First the experiment is done on a small dataset (700 positive and 800 negative reviews) of movie reviews taken from large polarity dataset 1.0 [8]. Secondly, the experiment is done on sentiment labeled dataset [9] (2000 reviews) of UCI data repository. The third experiment is done on a corpus (100 negative and 100 positive documents) made from the polarity dataset v2.0 [10] (1000 negative and 1000 positive documents) taken from the CS.cornell.edu. The fourth experiment is done on sentence polarity dataset v1.0 (5331 positive and 5331 negative reviews) taken from CS. Cornell.edu [11]. Fifth experiment is done on large movie review dataset [8] (25,000 positive reviews and 25,000 negative reviews) taken from the ai.stanford.edu. Using regular expressions and string processing methods, sentences and labels are separated from the datasets. The sentences obtained are fed into feature extraction techniques TF-IDF and Doc2Vec to generate vector(real numbers) features for each sentence. The split of training and testing samples is done by either hold out method where 50% data is used for training and 50% data is used for testing or by tenfold cross validation (CV) where nine folds are used for training and one fold is used for testing. Table 1 shows the methods used for splitting training and test datasets.

Each training and testing dataset obtained by any of the above methods are fed into feature extraction techniques TF-IDF and Doc2Vec to generate vectors. The TF-IDF vectorizer used is tuned with hyper parameters min_df (min number of times term t has to occur in all sentences), n_d = Number of sentences in a training or testing corpus, max_df (maximum number of times a term t can occur in all the sentences which is calculated as max_df * n_d), Encoding (encoding used), sublinear_tf (use of term frequency), use_idf (use of inverse document frequency), stop words (to eliminate stop words). The Doc2Vec vectorizer used is tuned with hyper parameters min_count (minimum number of times a word has to occur), window size (maximum distance between predicted word and context words used for prediction), vector size (size of vector for each sentence), sample (threshold for configuring which higher frequency words are randomly downsampled), negative (negative sampling is used for drawing noisy words), workers (number of workers used to extract feature vectors), dm (type of training algorithm used—0 for distributed bag of words and 1 for distributed model), Epochs (number of iterations for training). Tables 2 and 3 show hyperparameters tuned for both the techniques.

Table 1 Methods used for training and testing

Datasets	Dataset-1	Dataset-2	Dataset-3	Dataset-4	Dataset-5
Method used	Hold out	10 fold CV	10 fold CV	10 fold CV	Hold out
Training samples	800	9 folds	9 folds	9 folds	25,000
Testing samples	700	1 fold	1 fold	1 fold	25,000

Table 2 Hyper parameters tuned for TF-IDF Vectorizer

Parameters and datasets	Dataset-1	Dataset-2	Dataset-3	Dataset-4	Dataset-5
min_df	5	5	5	5	5
max_df	0.8 * n_d	0.8 * n_d	0.8 * n_d	0.8 * n_d	0.8 * n_d
encoding	utf-8	utf-8	utf-8	utf-8	utf-8
sublinear_df	True	True	True	True	True
use_idf	True	True	True	True	True
stop words	English	English	English	English	English

Table 3 Hyper parameters tuned for Doc2Vec Vectorizer

Parameters and datasets	Dataset-1	Dataset-2	Dataset-3	Dataset-4	Dataset-5
min count	1	1	1	1	1
window size	10	10	10	10	10
vector size	100	100	100	100	100
sample	1e-4	1e-5	1e-5	1e-5	1e-4
negative	5	0	5	5	5
workers	7	1	1	1	7
dm	1	0	0	0	1

5 Performance Analysis

To estimate the performance of TF-IDF and Doc2Vec on classification algorithms, we use accuracy measure as the metric. Accuracy is calculated as the ratio of number of reviews that are correctly predicted to the total number of reviews. Let us assume 2×2 matrix as defined in Table 4. Tables 5 and 6 show the accuracy of all classifiers for each feature extraction technique trained and tested on all the datasets.

Figure 3 shows the plot of accuracy for all classifiers using TF-IDF. Figure 4 shows the plot of accuracy for all classifiers using Doc2Vec. Based on the accuracy values from Tables 5 and 6,

1. For the first dataset, Doc2Vec performed better than TF-IDF for all classifiers. The accuracy is highest for logistic regression in case of TF-IDF, whereas SVM with rbf kernel achieved highest accuracy in case of Doc2Vec.

Table 4 Contingency matrix for sentiment analysis

		Predicted sentiment	
		+ve sentiment	−ve sentiment
Actual Sentiment	+ve sentiment	True Positive (TP)	False Negative (FN)
	−ve sentiment	False Positive (FP)	True Negative (TN)

Table 5 Accuracy of classifiers

Datasets	Dataset-1	Dataset-1	Dataset-2	Dataset-2	Dataset-3	Dataset-3
Classifiers	TF-IDF	Doc2Vec	TF-IDF	Doc2Vec	TF-IDF	Doc2Vec
Logistic regression	**83.428**	86.1428	81.950	**77.400**	98.864	**98.864**
SVM-rbf	75.142	**86.85714**	**82.350**	**77.400**	**98.887**	98.864
SVM-linear	78.000	85.71428	77.400	**77.400**	98.864	**98.864**
KNN ($n = 3$)	64.714	74.142	77.400	70.05	98.793	98.816
DT	65.571	70.285	76.950	64.850	97.633	97.065
Bernoulli-NB	80.714	83.71428	81.050	76.850	98.840	97.823

Table 6 Accuracy of classifiers

Datasets	Dataset-4	Dataset-4	Dataset-5	Dataset-5
Classifiers	TF-IDF	Doc2Vec	TF-IDF	Doc2Vec
Logistic regression	82.633	**99.98124**	**88.460**	86.487
SVM-rbf	**82.820**	**99.98124**	67.180	**86.760**
SVM-linear	82.596	**99.98124**	87.716	86.472
KNN ($n = 5$)	82.417	**99.98124**	68.888	74.592
DT	74.137	99.97186	71.896	67.140
Bernoulli-NB	79.330	**99.98124**	82.020	80.279

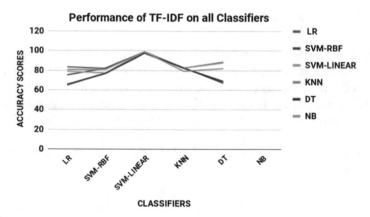

Fig. 3 TF-IDF performance

2. For the second dataset, TF-IDF performed better than Doc2Vec. The accuracy is highest for SVM with rbf kernel in case of TF-IDF whereas LR, SVM with rbf and linear kernels achieved highest accuracy in case of Doc2Vec.
3. For the third dataset, TF-IDF and Doc2Vec achieved similar accuracies. SVM with rbf kernel achieved highest accuracy in case of TF-IDF, whereas LR, SVM with rbf, and linear kernels achieved highest accuracy in case of Doc2Vec.

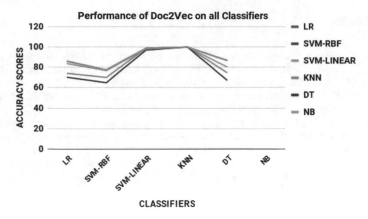

Fig. 4 Doc2Vec performance

4. For the fourth dataset, Doc2Vec performed better than TF-IDF for all classifiers. The accuracy is highest for SVM with rbf kernel in case of TF-IDF, whereas LR, SVM with rbf, linear kernels, and Bernoulli-NB achieved highest accuracy in case of Doc2Vec.
5. For the fifth dataset, TF-IDF performed slightly better than Doc2Vec. The accuracy is highest for LR in case of TF-IDF, whereas SVM with rbf kernel achieved highest accuracy for Doc2Vec.

Considering the performance analysis of all the classifiers on all datasets, we conclude that logistic regression and SVM with linear and rbf kernels perform better than all the other classifiers.

6 Conclusion

The purpose of this robust analysis is to provide deeper insight about the performance of feature extraction techniques TF-IDF and Doc2Vec. Based on the accuracy measures for all the datasets, Doc2vec and TF-IDF have achieved satisfactory performance for most of the datasets. But the accuracy scores of Doc2vec are better when compared to TF-IDF on most of the datasets. This work can be extended to test these techniques on unstructured data to analyze the performance of features extracted from both the methods.

References

1. Pang, B., Lee, L., Vaithyanathan, S.: Thumbs up? Sentiment classification using machine learning techniques. In: Proceedings of the ACL-02 Conference on Empirical Methods in Natural Language Processing, vol. 10, Series. EMNLP '02, pp. 79–86 (2002)
2. Jain, A.P., Katkar, V.D.: Sentiments analysis of Twitter data using data mining. In: International Conference on Information Processing (ICIP), pp. 807–810 (2015)
3. Koprinska, I., O'Keefe, T.: Feature selection and weighting methods in sentiment analysis. In: Proceedings of the 14th Australasian Document Computing Symposium, Sydney, Australia, pp. 67–74 (2009)
4. Albitar, S., Espinasse, B., Fournier, S.: An effective TF/IDF-based text-to-text semantic similarity measure for text classification. In: Benatallah, B., Bestavros, A., Manolopoulos, Y., Vakali, A., Zhang Y. (eds.) Web Information Systems Engineering WISE 2014. Lecture Notes in Computer Science, vol. 8786. Springer, Cham (2014)
5. Le, Q.V., Mikolov, T.: Distributed representations of sentences and documents. In: CoRR, vol. abs/1405.4053 (2014)
6. Sanguansat, P.: Paragraph2Vec-based sentiment analysis on social media for business in Thailand. In: 8th International Conference on Knowledge and Smart Technology (KST), pp. 175–178 (2016). https://doi.org/10.1109/KST.2016.7440526
7. Bilgin, M., Senturk, I.F.: Sentiment analysis on Twitter data with semi-supervised Doc2Vec. In: International Conference on Computer Science and Engineering (UBMK), pp. 661–666 (2017). https://doi.org/10.1109/UBMK.2017.8093492
8. Maas, A.L., Ng, A.Y., Potts, C., Huang, D., Pham, P.T., Daly, R.E.: Learning word vectors for sentiment analysis. In: Proceedings of the 49th Annual Meeting of the Association for Computational Linguistics: Human Language Technologies, vol. 1, Series. HLT '11, Portland, Oregon , pp. 142–150 (2011)
9. Kotzias, D., Denil, M., de Freitas, N., Smyth, P.: From group to individual labels using deep features. In: Proceedings of the 21th ACM SIGKDD International Conference on Knowledge Discovery and Data Mining, Series. KDD '15, Sydney, NSW, Australia, pp. 597–606 (2015)
10. Pang, B., Lee, L.: A sentimental education: sentiment analysis using subjectivity summarization based on minimum cuts. In: Proceedings of the 42nd Annual Meeting on Association for Computational Linguistics, Series. ACL '04, article no. 271 (2004)
11. Pang, B., Lee, L.: Seeing stars: exploiting class relationships for sentiment categorization with respect to rating scales. In: Proceedings of the 43rd Annual Meeting on Association for Computational Linguistics, Series. ACL'05, pp. 115–124 (2005)

Sentence-Based Topic Modeling Using Lexical Analysis

Shahinur Rahman, Sheikh Abujar, S. M. Mazharul Hoque Chowdhury, Mohd. Saifuzzaman and Syed Akhter Hossain

Abstract Data is not meaningful unless its information could be extracted. In every second in this world, we are generating millions of data over the Internet in different form. Most of them are in text format. Usually, data is written based on any topic, or sometimes on few topics. Following this, identifying topic of any text data is very important. Topic identification may help text summarization tools, text classification tool, etc. Machine learning applications may need less training on their data, only if once the topic of text is identified. Therefore, the demand of topic modeling is higher than ever right now. Data scientists are working day and night to make it more effective and accurate using different methods. Topic modeling focuses on the keywords that can express or identify the topic discussed in the document. Topic modeling can save a lot of time by releasing its user from page-to-page manual reviewing. In this paper, a model has been proposed to find out topic of a document. This model works based on the relations between most frequent words and their relation with sentences in the document. This model can be used to increase the accuracy of the topic modeling.

S. Rahman (✉) · S. Abujar · S. M. Mazharul Hoque Chowdhury · Mohd. Saifuzzaman
S. A. Hossain
Department of Computer Science and Engineering, Daffodil International University,
Dhanmondi, Dhaka 1205, Bangladesh
e-mail: shahinur3606@diu.edu.bd

S. Abujar
e-mail: sheikh.cse@diu.edu.bd

S. M. Mazharul Hoque Chowdhury
e-mail: mazharul2213@diu.edu.bd

Mohd. Saifuzzaman
e-mail: saifuzzaman.cse@diu.edu.bd

S. A. Hossain
e-mail: aktarhossain@daffodilvarsity.edu.bd

© Springer Nature Singapore Pte Ltd. 2019   487
A. Abraham et al. (eds.), *Emerging Technologies in Data Mining and Information
Security*, Advances in Intelligent Systems and Computing 814,
https://doi.org/10.1007/978-981-13-1501-5_42

1   Introduction

Nowadays with development and increasing use of the Internet, a reader not only reads the document but also contributes information publicly. From that large amount of information, it is quite difficult to sort a particular or desired word. So, it has become more important to compress and summarize data. From that huge amount of data, extracting information using manual method is actually incompetents [1].

In text mining, one of the most important problems is text summarization. In text summarization a lot of collection text are make smaller and neatly packed together summarized text represents the gist of main text. Text summarization helps to understand huge amount of text easily which saves a lot of time.

Text summarization is divided into two types. They are single document text summarization and multi-document text summarization. In a single text summarization, a large size of single text is summarized to another single document summary. In multi-text summarization, a set of documents are summarized to a single document summary. In both approaches, a large amount of data is summarized and stored in a single file.

In text summarization, extractive summarization is a common and mature technique which extracts important sentence then recombines them and generates a summary base of this sentence.

In topic modeling, latent Dirichlet allocation (LDA) model is used to explore topic firstly. First, from a document, we extract sentence in association with the most frequent word. Now it is possible to find out relations from high-scored words and high-length sentences.

2   Literature Review

Nowadays, there are lots of numbers studies about topic modeling but rarely have been a small number of studies in Bengali sentence-based topic modeling. In this paper, we use LDA model with lexical analysis to extract topic from a large collection of information.

LDA is one of the most common methods to extract topic modeling from different types of data examples that use auxiliary information which are the author-topic model [2], tag-topic model [3], and topic-link LDA [4]. All of this work has been done in English. Previously Geetanjali and Pushpak analysis Bengali Poet classification and Identification [5] and Amitava and Sivaji have done to analysis a document and find out Opinion base summarization [6] but There is no work-related about Topic modeling in Bengali.

In this current year, some good research has been done by different researchers all over the world. Jiang and Zhou (2017) worked on topic modeling based on the poison decomposition. In this work, they tried to find out statistical results based on multidimensional characteristics of the topic [7]. Now at the same time, Truică and his team worked on this same topic using contextual cause. They applied automatic term recognition system using contextual cause for topic modeling. Another researcher named Ruohonen (2017) tried to classify Web exploits using topic modeling [8]. Karami, Gangopadhyay, Zhou, and Kharrazi (2017) worked on fuzzy approach. Their target was health and media corpora topic modeling. Fuzzy approach was used to analyze medical document and extract information [9]. Work on probabilistic topic model has been done by Zhai (2017) for text data retrieval and analysis [10].

3 Proposed Method

The proposed process of sentence-based topic modeling is shown in Fig. 1. It has covered the steps of text document to extract topic, and it is following six different steps.

3.1 Data Preprocess

For text summarization, total twelve thousand data were collected from online news portal. To summarize, each summary consists of three parts and its beginning, body, and ending part. Before summarization, it is needed to preprocess the text document. Therefore, chapter segmentation, sentence segmentation, word segmentation occur then it remove stop words and finally stemming.

3.2 Chapter Segmentation

Basically, a text document consists of lots of chapters, and each chapter is dependent on the other chapter. First, it is needed to separate text into many chapters that is why chapter segmentation was used to divide the whole text. Chapter segmentation

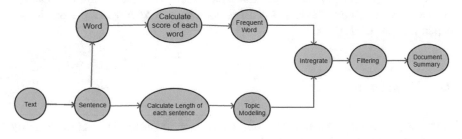

Fig. 1 Data flow of proposed model

Table 1 Sentences separated from document

Sentence number	Sentence
S1	মানুষ হিসাবে সবারই কিছু ইতিবাচক দিক রয়েছে
S2	আবার বিপরীত মেরুতেই রয়েছে বিভিন্ন দোষক্রটি
S3	নিজেরাই অনেক সময় ভুলগুলো নিয়ে সচেতন থাকি না।

is a process of dividing the text into the meaningful unit. From each unit, different meaning can be extracted to summarize data.

3.3 Sentence Segmentation

From a large collection of data, it is much difficult to make a meaningful summary. In a text document, each sentence disclosures a meaning. Sentence segmentation is splitting the text into a sentence. Whole text is exposed a set of sentences. We use NLTK toolkit in our work to separate sentence from given text, i.e., (Table 1).

S=মানুষ হিসাবে সবারই কিছু ইতিবাচক দিক রয়েছে। আবার বিপরীত মেরুতেই রয়েছে বিভিন্ন দোষক্রটি নিজেরাই অনেক সময় ভুলগুলো নিয়ে সচেতন থাকি না।

3.4 Word Segmentation

Word segmentation is referring to extract sentence as a set of independent words. Generally, in Bengali and some other language, using space is separator from one word to another word.

3.5 Remove Stop Words

Stop words are a set of commonly used words in any language which has no actual meaning. It is just used to make a sentence, but does not carry any tangible meaning. So, we need to remove the stop word from the text. To remove stop word in our paper, we use NLTK tools.

3.6 Stemming

In this step, clean data are collected and stored in a document. Data are now ready to be analyzed and further processed.

Table 2 Sentence scoring

Sentence number	Sentence	Length
S1	নদী আর নদী পাড়ের মানুষের জীবন নিয়ে এর আগে গৌতম ঘোষের পদ্মা নদীর মাঝি দেখেছিলাম	16
S2	মানিকের উপন্যাসের মত করে নয়, গৌতম নিজের মত করে পদ্মা পাড়ের জেলেদের জীবন বাস্তবতাকে ফুটিয়ে তুলেছিলেন	11

3.7 Calculate Sentence Score

In sentence segmentation, just separation of sentence happened and set of sentences were created. But basically no one has clear idea about the value of the word inside the sentence. The total number of word in each sentence represent the length of sentence. After counting the sentence length, all sentence were sorted on high score to low (Table 2)

S= {নদী আর নদী পাড়ের মানুষের জীবন নিয়ে এর আগে গৌতম ঘোষের পদ্মা নদীর মাঝি দেখেছিলাম। মানিকের উপন্যাসের মত করে নয়, গৌতম নিজের মত করে পদ্মা পাড়ের জেলেদের জীবন বাস্তবতাকে ফুটিয়ে তুলেছিলেন}

3.8 Calculate Word Score

To find out the most frequent word had to calculate the score of each word which is most important because after calculating the score of each word it is visible which group word is most used and which group of words has scored more than average. If a word is used in many sentences for ten times that means this word score is 10.

3.9 Topic Modeling

Latent Dirichlet allocation (LDA) is one of most popular topic modeling which is followed by data preprocessing. It helps to search the topic words related to a document. LDA obtains feature word list for each topic. For example, if in a collection of words, there are numbers of words like as "sports," "play" are used mostly, then consider that it becomes sports news. Consider the following equation

$$X = T_w - T_p - T_n - T_A - T_{TBV} \tag{1}$$

$$\text{Count} \rightarrow [W_1, W_2, \ldots\ldots\ldots\ldots W_n]$$

Here, X is the total value of valid word and T_w is the total word, T_p is total preposition, T_n is total nouns, T_A is total articles, and T_{TBV} is to be verbs. After calculating the total value of valid word, we find out the probability of each word following the equation.

$$P_{W1} = \frac{W_1}{X} \tag{2}$$

$$P_{W2} = \frac{W_2}{X} \tag{3}$$

$$P_{Wn} = \frac{W_n}{X} \tag{4}$$

If probability of P_{W1} is bigger then $P_{W2}, P_{W3} \ldots \ldots P_{Wn}$ then P_{W1} is respectively most probability to become the P_{W1} related topic. On the other hand, if the probability of P_{W2}, P_{W3}, or P_{Wn} are bigger, then the topic would be the biggest probability related word.

3.10 Filtering/Smoothing

In this stage, we are going to find out the score for the sentences and relations between the sentence and words. For that, we are proposing a method. As before, we explained that sentence length is the score for the sentence. This time sentence score will be the number of valid words were used in the sentence. To do that, it is necessary to remove determinate, nouns, articles, etc. Therefore, a set of clear data will be stored in the sentence. Now it is time to find out the expected topic of the document. We can write the process as,

$$S1 = [w1, \ w2, \ w3 \ldots \ldots wn]$$
$$S2 = [w1, \ w2, \ w3 \ldots \ldots wn]$$

Start
If
S1[i] = S2[i]
Match found
Else S2[i + +]
Go to Start

3.11 Document Summary

Suppose the word 'Bangladesh' got the highest value, and for this word, we got seven sentences. This time those sentences can be compared with each other using a two-dimensional array. This process will go on until next five top words to ensure maximum accuracy. After processing, most frequent matching will be taken as the best value and corresponding word will be the topic of the document.

4 Final Outcome

In this process, it is possible to find out best possible output topic for a document. This can be used for any language as well as Bangla. Using Bangla language, corpus data processing and other things can be done in much more efficient way.

5 Conclusion and Future Scope

In this era of technology, data brought us new opportunity as well as new complexity. Handling new data requires new method, sometimes new technology. Reading files one by one to find out topic of the document is one of the toughest tasks nowadays. If a simple system can solve this, why should not we use this to proper use our brain and time in other important tasks. Topic modeling is a very important sector of data science and requires very large amount of research work. This system can be a simple but efficient method to use topic modeling for different languages. Several adaptive algorithms may develop to understand the topic of any text based on context. Because, understanding the text based on context is very important. It may change the entire concept of what has been understood earlier.

Acknowledgements We would like to thank Daffodil International University and DIU NLP and Machine Learning Research LAB for all their support and help.

References

1. Gambhir, M., Gupta, V.: Recent automatic text summarization techniques: a survey. Artif. Intell. Rev. **47**(1), 1–66 (2017). https://doi.org/10.1007/s10462-016-9475-9
2. Rosen-Zvi, M., Griffiths, T., Steyvers, M., Smyth, P.: The author-topic model for authors and documents. In: Proceedings of the 20th Conference on Uncertainty in Artificial Intelligence, pp. 487–494 (2004)
3. Tsai, F.S.: A tag-topic model for blog mining. Expert Syst. Appl. **38**(5), 5330–5335 (2011)
4. Liu, Y., Niculescu-Mizil, A., Gryc, W.: Topic-link LDA: joint models of topic and author community. In: Proceedings of the 26th Annual International Conference on Machine Learning, pp. 665–672 (2009)
5. Rakshit, G., Ghosh, A., Bhattacharyya, P., Haffri, G.: Automated analysis of Bangla poetry for classifiation and poet identifiation. IITB-Monash Research Academy, India, IIT Bombay, India Monash University, Australia
6. Das, A., Bandyopadhyay, S.: Topic-based Bengali opinion summarization

7. Jiang, H., Zhou, R., Zhang, L., Zhang, Y.: A topic model based on poisson decomposition. In: Proceedings of the 2017 ACM on Conference on Information and Knowledge Management, Singapore, pp 1489–1498, November (2017)
8. Ruohonen, J.: Classifying web exploits with topic modeling. In: 28th International Workshop on Database and Expert Systems Applications, Lyon, France (2017). https://doi.org/10.1109/DEXA.2017.35
9. Karami, A., Gangopadhyay, A., Zhou B., Kharrazi, H.: Fuzzy approach topic modeling for health and medical corpora. Int. J. Fuzzy Syst. (2017)
10. Zhai, C.: Probabilistic topic models for text data retrieval and analysis. In: 40th International ACM SIGIR Conference, Shinjuku, Tokyo, Japan, pp. 1399–1401 (2017)

Anatomy of Preprocessing of Big Data for Monolingual Corpora Paraphrase Extraction: Source Language Sentence Selection

Vivek Kumar, Abhishek Verma, Namita Mittal and Sergey V. Gromov

Abstract In the scope of our work cross-language information retrieval, the ultimate goal is to develop an intelligent model using non-corresponding corpora of Hindi and English language. It consists of two phases which are Source Language Sentence Extraction (SLSE) and model building for translation. SLSE is the training data for the model, comprising the 70% of entire work. In this paper, we have proposed a novel pipeline for SLSE by creating first bilingual dictionary, N-grams, inverse term document index, etc. As mentioned, SLSE is being used as training data so it plays a very crucial role in building model so more attention has been paid to ensure the content richness. In this work, two non-corresponding English and Hindi corpora ranging from 60 GB of text have been constructed. Collecting data at this scale is even more tedious as it is highly unstructured, and the processing time for big data is also substantially large. To reduce the processing time, Hadoop was implemented throughout.

1 Introduction

Natural language processing (NLP) has long been one of the holy grails of computer science. Humans are no doubt better than machines when it comes to understanding the contextual meaning of natural language, while machines work more efficiently with large structured data. Although in reality, processing language and comprehend-

V. Kumar · S. V. Gromov
National University of Science and Technology-MISiS, Moscow 119049, Russian Federation
e-mail: vivekkumar@misis.ru

S. V. Gromov
e-mail: gromov@asu.misis.ru

A. Verma (✉) · N. Mittal
Malaviya National Institute of Technology, Jaipur 302017, India
e-mail: iamvermaabhishek@gmail.com

N. Mittal
e-mail: mittalnamita@gmail.com

© Springer Nature Singapore Pte Ltd. 2019
A. Abraham et al. (eds.), *Emerging Technologies in Data Mining and Information Security*, Advances in Intelligent Systems and Computing 814,
https://doi.org/10.1007/978-981-13-1501-5_43

ing the meaning are an extremely complex task. With the increasing advancement, natural language processing system is now capable of performing several tasks such as segmentation, parsing named entity recognition, tokenization, stemming, part of speech tagging. Traditionally, we use three methods for cross-language for paraphrasing—machine translation (MT) systems, where the query is automatically translated to a specified language; using multilingual dictionaries in which the query terms are replaced by the terms found in the thesaurus or dictionary [1]; and throughout analysis of parallel or comparable corpora, where term equivalences are automatically extracted [2, 3].

In cross-language information retrieval, a query which is training data of one natural language is matched against another language data. Practically, the problem of CLPA is more difficult when it comes to match a segment of text. It is necessary that text of one language resembles or matches to a segment of the text of equal content in another language which varies from many consecutive word, one sentence to multiple pages [4, 5].

The first problem is that the system must resolve how a term written in one language might be expressed in another. The second problem is to decide to keep which translation of the possible translations. The third problem is to decide how to weight the importance of translation alternatives in case of having more than one. The first two problems, how to make an interpretation of and how to prune choices, are additionally endemic to machine translation frameworks. The CUR framework, in any case, has the advantage of disposing of a few translations while holding others [6]. Holding equivocalness can be valuable in advancing review in information retrieval framework [7]. Assume that the underlying query contains two free test terms. In the occurrence that the first term can be interpreted in a wide range of conceivable ways, and if the second term can be expressed in just a single way, the retrieval framework attempts to not to give more weight to the first test word only in light of the fact that it has more translation choices [8–10].

2 Literature Survey

These various methods have been produced to adjust machine learning algorithms to work with vast data sets: Cases are new handling ideal models, for example, MapReduce, and distributed processing structures, for example, Hadoop [11]. Branches of machine learning including deep and online learning have likewise been adjusted with an end goal to beat the difficulties of machine learning with big data [12].

The first man who proposed the idea of cross-language text classification was N. Bel in 2003 [13]. The proposed CL text classification method has the potential that can move forward original single-language automatic text classification system to the more than one language framework without human intercessions. Contrary to the single-language text classification, the CL text classification method is opposed confronted with an issue that training sets of source language text and classification sets of aimed language text are in diverse language spaces.

Concerning query expansion techniques expect that extra terms that are connected with the central ideas in the query are supposed to be significant and that phrases in query extension by means of textual analysis and regional feedback can be utilized to decrease the errors related to automatic dictionary-based translation [14].

There are three classes of key structures on the cross-language text classification, which relies upon the techniques for exchange in numerous language areas: the CL classification primarily based on the parallel corpus, bilingual dictionary, and the CL classification based on machine translation [15]. Among these techniques, the CL classification based on machine translation does not require the parallel corpus and the bilingual lexicon. Therefore, it is it is financially savvy and spreads all the more regularly.

As an outcome, the projection of language area is the key reason for CL text classification. With a purpose to lead the "Japanese–English" CL text classification tests consolidated with the SVM algorithm, Y. Y. Li et al. used kernel canonical correlation analysis (KCCA), latent semantic indexing (LSI), and distinct techniques, free on interpretation sources, to finish the distance change of supply conversion of source languages and goal languages in lieu of the parallel corpus [16]. N. Bel led the terminology translation and the profile-based translation by proposing an area wording interpretation table on "English–Spanish" and were given text vectors, which have been utilized to conduct text classification experiments of Rocchio and Winnow. L. Xiao et al. introduced the CL text classification algorithm given the information bottleneck hypothesis and finished the undertakings of "Chinese–English" dialect area transformation with the aid of using Google Translator. They additionally utilized people in the broad segment of data sets of different languages to enhance the characterization work and affirm the productivity of this applied algorithm [17–20].

3 Modules of Source Language Sentence Extraction

3.1 Data Collection

It is imperative to pay attention to the quality of data and resources considered for forming the corpora. For result accuracy, this work needs data to be compiled in a manner so that it encompasses to all kinds of text. So, substantial time was invested in searching authentic data in the fields of politics, business, military, war, art, religion, contextual stories, sports, etc., to ensure richness of the text. Data for constructing the English and Hindi corpora was handpicked from Wikipedia, government Web sites, and rest was provided by the Malaviya National Institute of Technology, India. For this research work, two non-corresponding English and Hindi corpora ranging from 60 GB of text were utilized. The raw English and Hindi corpora are being shown in Fig. 1.

```
<doc id="12" url="https://en.wikipedia.org/wiki?curid=12" title="Anarchism">
Anarchism

Anarchism is a political philosophy that advocates self-governed societies based on voluntary institutions. These are often described as
stateless societies, although several authors have defined them more specifically as institutions based on non-hierarchical free associations.
Anarchism holds the state to be undesirable, unnecessary, and harmful. While anti-statism is central, anarchism entails opposing authority or
hierarchical organisation in the conduct of all human relations, including, but not limited to, the state system.
Anarchism draws on many currents of thought and strategy. Anarchism does not offer a fixed body of doctrine from a single particular world
view, instead fluxing and flowing as a philosophy. Many types and traditions of anarchism exist, not all of which are mutually exclusive.
Anarchist schools of thought can differ fundamentally, supporting anything from extreme individualism to complete collectivism. Strains of
anarchism have often been divided into the categories of social and individualist anarchism or similar dual classifications. Anarchism is
usually considered a radical left-wing ideology, and much of anarchist economics and anarchist legal philosophy reflect anti-authoritarian
interpretations of communism, collectivism, syndicalism, mutualism, or participatory economics.
The term "anarchism" is a compound word composed from the word "anarchy" and the suffix "-ism", themselves derived respectively from the
Greek , i.e. "anarchy" (from , "anarchos", meaning "one without rulers"; from the privative prefix áv- ("an-", i.e. "without") and , "archos",
commoncrawl    <a>       व्याख्या - कल जो गलतियां हुई हैं , उनसे सबक लेकर अपने आज को बेहतर करना ही उचित है ताकि हमारा आने वाला कल सुखद और सफलतादायक हो.
hwt2013 <s>    उन्होंने प्रदेश के अन्य मण्डलों की चीनी मिलों को 25 से 30 नवम्बर, 2013 तक शुरू कराने के भी निर्देश दिए हैं।
w2cweb <a>     Facebook on suhtlusportaal, mis ühendab inimesi nii oma sõprade kui ka kolleegide, koolikaaslaste ja ümbruskondsetega.
commoncrawl    <a>       पर्यावरणीय विज्ञान (
spiderling     <s>       फिल्म में मेरे हीरो अक्षय कुमार अपने काम को लेकर काफी सीरियस रहते हैं।
spiderling     <s>       मोदी के अनुसार जिस कांग्रेस का विरोध राम मनोहर लोहिया ने जिंदगी भर किया , उन्हीं का रास् ता छोड़कर नीतीश प्रधानमंत्री बनने का ख् वाब लिए कांग्रेस में शामिल होने के लिए
तैयार है।
commoncrawl    <a>       ↑ स्कूल जिला तेजी से वित्तपोषण की वैकल्पिक तलाश:
commoncrawl    <a>       चार बस में कम से कम होना चाहिए
w2cwiki <a>    अपने "रोतीफित्रतस" नामक लोकप्रिय उपन्यास में यह मध्ययुगीन रोमांस को आधुनिक मोड़ देने में सफल हुए हैं।
commoncrawl    <a>       समुन्दर के थोड़ा और करीब जाएं।
commoncrawl    <a>       जहां एक ओर व्यापारियों को बीच में से हटा दिया जायेगा , वहीं दूसरी और किसानों को उत्पादन की कम से कम कीमत भी नहीं प्राप्त हो पायेगी ।
spiderling     <s>       अगर इंसान से इतर शक्तियों के प्रति आस्थाविर्ष यूंहीं बनी रहेगी तो शिक्षा की कृपा से प्राप्त बौद्धिक पैमाने भी कट्टरता के रूप में मौजूद रहेगा । देश हित में हमें अपने जातीय / धार्मिक स्व
और उसकी सार्वजनिक स्वीकृति के अग्रह को खोने के लिए तैयार होना ही होगा ।
```

Fig. 1 Raw English and Hindi corpora

```
7 Anarchism does not offer a fixed body of doctrine from a single particular world view instead fluxing and flowing as a philosophy.
8 Many types and traditions of anarchism exist not all of which are mutually exclusive.
9 Anarchist schools of thought can differ fundamentally supporting anything from extreme individualism to complete collectivism.
10 Strains of anarchism have often been divided into the categories of social and individualist anarchism or similar dual classifications.
11 Anarchism is usually considered a radical leftwing ideology and much of anarchist economics and anarchist legal philosophy reflect
antiauthoritarian interpretations of communism collectivism syndicalism mutualism or participatory economics.
12 The term anarchism is a compound word composed from the word anarchy and the suffix ism themselves derived respectively from the Greek.
13 Authority sovereignty realm magistracy and the suffix or ismos isma from the verbal infinitive suffix.
14 Various factions within the French Revolution labelled opponents as anarchists as Robespierre did the Hébertists although few shared many
views of later anarchists.
16 There would be many revolutionaries of the early nineteenth century who contributed to the anarchist doctrines of the next generation such
as William Godwin and Wilhelm Weitling but they did not use the word anarchist or anarchism in describing themselves or their beliefs.
17 The first political philosopher to call himself an anarchist was PierreJoseph Proudhon marking the formal birth of anarchism in the
midnineteenth century.
```
```
8374494 ट्रांसपेरेंसी इंटरनेशनल की की सूची में भारत लगातार भ्रष्टतम देशों में शामिल किया गया है।
8374495 मुझे मेरे बड़े भैया बताते हैं कि जब मां के साथ जन्म के बाद जब मैंने पहली बार सफर किया तो वह तांगे में किया।
8374498 कल शाम ही देश के सामने इस बात का खुलासा हो गया था कि देश को धोखे में रखकर बाबा रामदेव ने किस तरह अपने उन्हीं की जुबान में एक करोड़ से दस करोड़ तक लोगों को अनशन में झोंक दिया था।
8374499 संगठन की राज्य महासचिव कविता श्रीवास्तव ने विभिन्न जिलों में हुई संगठित प्रगति की जानकारी देते हुए नए सदस्यों की सदस्यता पर बल दिया।
8374500 पांचवां बल्लेबाज कोई भी खेले खिलाड़ा विश्वसनीय नहीं ।
8374501 सट्टेबाज रमेश व्यास के बयान के आधार पर विंदु की गिरफ्तारी हुई है।
8374502 बैठक में बताया गया कि सरकार प्राथमिकता के आधार पर वह फंसे लोगों को निकाल रही है।
8374503 आम आदमी पार्टी स्टिंग का रॉ फुटेज देबना चाहती थी लेकिन मीडिया सरकार आम आदमी पार्टी को रॉ फुटेज दिखाने को तैयार नहीं है।
8374504 हरानाबू समझते बे धर्म साधना के कारण उनकी दृष्टि इतनी स्वच्छ व तेज हो गई है कि दूसरे सब लोगों का भला बुरा और झूठ सच वह सहज ही समझ सकते हैं।
8374505 बाकी बच्चे तो खुशी खुशी स्कूल जा रहे थे क्योंकि उन्हें अपने दोस्त मिलने वाले थे पर पप्पू को खुशी नहीं थी।
```

Fig. 2 Preprocessed English and Hindi corpora

3.2 Data Preprocessing

Preprocessing is a crucial undertaking and basic stride in natural language processing, text mining, and information retrieval. Data preprocessing is a data mining strategy that includes transforming raw data into a format suitable for experimental purposes. In fact, often raw data is inadequate, confusing, as well as confusing in terms of finding some trends for the use. It also contains numerous errors such as other languages text, alphabets. Data preprocessing is a demonstrated strategy for overcoming such issues. Data preprocessing stage is supposedly the most time-consuming of the whole knowledge discovery phase. The complexity of data preprocessing depends on the data sources used. It is necessary to remove all the special characters for hassle-free further processing. In preprocessing, all the new line characters are also removed to establish a coherency of data. As my focus is for Hindi to English translations, corpora containing other language input in them or in cross-fashion are also needed to be removed. The first stage preprocessed data is shown in Fig. 2.

Finally, each line of both the corpora is assigned with sentence number in increasing order for ease of identification and extraction. So, in this preprocessing of step, we have basically cleaned the big data in processable form by performing:

- Removal of special characters.
- Removal of stop words.
- Removal of newline characters.
- Removal of characters of any other languages or non-identifiable characters.
- Numbering the entire corpora in ascending order.

3.3 Stop Words

It is also a section of natural language. The intention behind removing stop words from the corpora is that it makes the data look less understandable, heavy for processing, and at some point less useful or important for analysis. The dimensionality of term space is reduced by removing stop words. The most common words in text documents are pronouns, articles, prepositions, etc., that does not contribute to adding meaning of the texts. These words are considered as stop words. Some examples of top words: a, an, the, in, with, etc. For the Hindi language, आदि आप इन इनका इन्हीं इन्हें इस इसका इसके इसमें इसी उन उनका उनकी उसे are some examples.

3.4 Inverse Term Document Indexing

In information retrieval, term frequency–inverse document frequency (tf–idf) is a numerical method which displays how vital a word is to a document in a corpus. It calculates the frequency of occurrence for a word in a file and relative value to the entire set of documents. In ML and NLP, it is used for building classifiers or predictive models. We modify weights primarily based on how frequently terms appear in a selected document set.

The tf–idf value rises correspondingly to the occurrence number of a word inside a file, which enables to modify for the fact that few words appear more regularly in widespread than others. Machine learning algorithm uses features for classification based on the weights of different words. It can also assure accuracy and quality of a classifier in the context of making many wrong predictions. For this work, each unique word of both corpora is indexed, following the sentence number of occurrence. Figure 3 shows the prepared index for English and Hindi corpora.

```
7 Anarchism does not offer a fixed body of doctrine from a single particular world view instead fluxing and flowing as a philosophy.
8 Many types and traditions of anarchism exist not all of which are mutually exclusive.
9 Anarchist schools of thought can differ fundamentally supporting anything from extreme individualism to complete collectivism.
10 Strains of anarchism have often been divided into the categories of social and individualist anarchism or similar dual classifications.
11 Anarchism is usually considered a radical leftwing ideology and much of anarchist economics and anarchist legal philosophy reflect
antiauthoritarian interpretations of communism collectivism syndicalism mutualism or participatory economics.
12 The term anarchism is a compound word composed from the word anarchy and the suffix ism themselves derived respectively from the Greek.
13 Authority sovereignty realm magistracy and the suffix or ismos isma from the verbal infinitive suffix.
14 Various factions within the French Revolution labelled opponents as anarchists as Robespierre did the Hébertists although few shared many
views of later anarchists.
16 There would be many revolutionaries of the early nineteenth century who contributed to the anarchist doctrines of the next generation such
as William Godwin and Wilhelm Weitling but they did not use the word anarchist or anarchism in describing themselves or their beliefs.
17 The first political philosopher to call himself an anarchist was PierreJoseph Proudhon marking the formal birth of anarchism in the
midnineteenth century.

8374494 ट्रांसपेरेंसी इंटरनेशनल की  की सूची में भारत लगातार भ्रष्टतम देशों में शामिल किया गया है।
8374495 मुझे मेरे बडे भैया बताते हैं कि जब मां के साथ जन्म के बाद जब मैंने पहली बार सफर किया तो वह भी तांगे में किया।
8374498 कल शाम ही देश के सामने इस बात का खुलासा हो गया था कि देश को धोखे में रखकर बाबा रामदेव ने किस तरह अपने उन्हीं की जुबान में एक करोड़ से दस करोड़ तक लोगों को अनशन में झोंक दिया था।
8374499 संगठन की राज्य महासचिव कविता श्रीवास्तव ने विभिन्न जिलों में हुई संगठित प्रगति की जानकारी देते हुए नए सदस्यों की सदस्यता पर बल दिया।
8374500 पांचवां बल्लेबाज कोई भी खेले फिलहाल विश्वसनीय नहीं ।
8374501 सट्टेबाज रमेश व्यास के बयान के आधार पर विंदू की गिरफ्तारी हुई है।
8374502 बैठक में बताया गया कि सरकार प्राथमिकता के आधार पर वहां फंसे लोगों को निकाल रही है।
8374503 आम आदमी पार्टी स्टिंग का रॉ फुटेज देखना चाहती थी लेकिन मीडिया सरकार आम आदमी पार्टी को रॉ फुटेज दिखाने को तैयार नहीं है।
8374504 हरानबाबू समझते थे धर्म साधना के कारण उनकी दृष्टि इतनी स्वच्छ व तेज़ हो गई है कि दूसरे सब लोगों का भला बुरा और झूठ सच वह सहज ही समझ सकते हैं।
8374505 बाकी बच्चे तो खुशी खुशी स्कूल जा रहे थे क्योंकि उन्हें अपने दोस्त मिलने वाले थे पर पप्पू को खुशी नहीं थी।
```

Fig. 3 Index prepared for English and Hindi corpora

793238:new york	214814:world war ii	86284:पहली बार	16743:भारॅत उत्तराखण्ड राज्य
438130:high school	182199:years age older	86228:उत्तर प्रदेश	16584:उत्तराखण्ड राज्य अन्तर्गत
406340:world war	181685:65 years age	84652:कई बार	16578:तहसील भारत उत्तराखण्ड
324330:age 18	174425:new york city	83105:अलग अलग	13404:नाम जाना जाता
292924:wikipedia articles	144488:every 100 females	70509:कभी कभी	10613:राज्य अन्तर्गत कुमाऊँ
285282:articles deletion	126821:per square mile	67214:दी गई	10613:अन्तर्गत कुमाऊँ मण्डल
222615:also known	93713:average household size	66139:किए गए	9641:मुलायम सिंह यादव
217016:median income	93125:children age 18	65873:माना जाता	9369:ईसा मसीह जन्म
215420:war ii	93116:average family size	64729:बार बार	9360:ईसा पूर्व वर्ष
213091:los angeles	93004:couples living together	64122:आ रही	9334:वर्ष पूर्व वर्ष
211825:new zealand	92896:married couples living	63391:आ गया	9327:ईसा पूर्व ईसा
210088:may refer	92702:age 18 living	61379:धीरे धीरे	9322:संवत सप्तर्षि संवत
206844:years age	92037:housing units average	60184:ईसा पूर्व	9321:कलियुग संवत प्राचीन
183751:65 years	92015:units average density	56293:जाता है	9321:संवत कलियुग संवत
183243:first time	90998:female householder husband	54392:इससे पहले	9320:आदि वर्ष आदि
182389:age older	90994:householder husband present	54143:पता चला	9320:ईसा पूर्व अनुसार
177182:york city	90545:wikipedia wikiproject spam	52429:चल रही	9320:अनुसार विक्रमी संवत
174579:united kingdom	90392:someone living alone	52138:नई दिल्ली	9320:वर्ष आदि निकाले
164717:two years	90356:living alone 65	50013:लिया गया	9320:सप्तर्षि आदि वर्ष
162043:19th century	90305:alone 65 years	47951:साल पहले	9320:प्राचीन सप्तर्षि आदि
156233:took place	89802:households made individuals	46314:आम आदमी	9320:सप्तर्षि संवत कलियुग
150406:air force	80872:wikiproject spam linkreports	45364:माता पिता	9320:विक्रमी संवत सप्तर्षि
145544:every 100	80687:hispanic latino race	43136:राज्य सरकार	9320:संवत प्राचीन सप्तर्षि
144655:100 females	76650:per capita income	41750:गया है	9319:पूर्व वर्ष पूर्व
144030:years later	75876:united states census	40630:जाने बाद	9318:पूर्व अनुसार विक्रमी
140827:prime minister	72866:males median income	40203:किए जाने	9311:पूर्व वर्ष प्रकार
138559:20th century	71856:females age 18	39890:बन गया	9311:वर्ष प्रकार प्रदर्शित
132559:wikipedia wikiproject	71854:100 females age	39262:उन्होंने बताया	9311:प्रकार प्रदर्शित जाता
131710:per square	71382:median income household	39027:जाने वाली	9301:ईसा जन्म अधार
129600:square mile	70957:median income family		
128488:well known			

Fig. 4 *N*-grams of Hindi and English corpora

3.5 N-Grams

An *N*-gram is defined as the adjoining arrangement of n items from a given progression of text.

The elements can be phonemes, syllables, letters, words, or base pairs according to the application. For $N = 1$ it is called unigram, for $N = 2$, Bigram, and so on. *N*-grams are used for developing features for supervised machine learning models, for instance, SVMs, decision tree, naive Bayes. *N*-gram models are also used for spelling correction, word breaking, and text summarization for search engines and experimental purposes. In Fig. 4, bigrams and trigrams are shown which are prepared from the English and Hindi corpora.

लालमा	landslips
नीलमंडल	phalsa
सहपक्ष	aileron
तुरहीनाद	fanfare|
जार्ज चतुर्थ	george iv
सिविलेजेशन	dehumanizing
इतर	perfume
आकारविहीन	shapeless
आकारविहीन	amorphous
आकारविहीन	formless
लड़ियाँ	threads

अनपकारी	harmless
चीरचर्म	deerskin
इकसिंगे	unicorn
परभक्षीपरजीवी	predatorparasite
यौवन	youth
यौवन	adulthood
अंबर	cloud
अवतारतत्व	incarnation
पेंट	paint
पेंच	screw

Fig. 5 N-grams of Hindi and English corpora

3.6 Bilingual Dictionary Preparation

A bilingual dictionary or translation dictionary is a specialized dictionary used to translate words or phrases from one language to another. In this step, we have created a bilingual dictionary for Hindi to English translation for further implementation of query extraction. In our work, the dictionary we have made is made from IIT-B Shabdanjali and bilingual dictionary. Figure 5 shows the prepared bilingual dictionary for this research work.

3.7 Query Generation

Queries from years from standard resources which is FIRE have been taken for preparing the list of a query for Hindi to English translations (Fig. 6).

Out of 200 queries from years 2008, 2010, 2011, and 2012, only these queries have been retained, in which at least one word is not present in the bilingual dictionary. The first sentence of the text is matched with the prepared dictionary, to search OOV words. For the queries (having OOV words after matching with the dictionary) are extracted for Source Language Sentence Selection. A total of 15 queries has been filtered along with the out of vocabulary words (OVV) [21–24] as shown in Fig. 7.

4 Proposed Methodology

For query-based Source Language Sentence Extraction, we went step by step as proposed in pipeline flowchart diagram. After constructing the Hindi and English corpora it was refined. So it was passed through the preprocessing stage, where

1 सिंगूर और नंदीग्राम भूमि विवाद

OVV   सिंगूर नंदीग्राम

2 हिजबुल्लाह गुरिल्लाओं के हमले

OVV हिजबुल्लाह

3 राम मंदिर को लेकर आडवाणी सिंघल   विवाद

OVV आडवाणी सिंघल

4 ग्रेग चैपल और ग्रेग चैपल के बीज सशंक विरामसंधि

OVV ग्रेग चैपल ग्रेग चैपल

5 शैक्षणिक संस्थानों में ड्रैस-कोड आरोपित करना

OVV ड्रैस-कोड

6 विश्वकप में जिंडेने जिडान हैड बटिंग दुर्घटना

OVV जिंडेने जिडान बटिंग

Fig. 6 FIRE filtered queries based on OOV

```
 50_query.txt  ×

1       सिंगूर नंदीग्राम भूमि विवाद
ovv      सिंगूर नंदीग्राम

सिंगूर या श्रीपेरमबदूर ।
कंदील सिंगूर में जल रही है।
यही स्थिति नंदीग्राम में थी ।
सिंगूर इसका ताज़ा उदाहरण है ।
नंदीग्राम का उदाहरण सामने है ।
रतन टाटा सिंगूर से सानंद तक।
सिंगूर इसका ताज़ा उदाहरण है ।
नंदीग्राम सिंगूर से सबक ले माकपा।
सिंगूर से टाटा की नैनो वापस लौटी ।
लेफ्ट के खाते में नंदीग्राम सिंगूर है।
सिंगूर कलिंगनगर में जनता लड़ रही है ।
सिंगूर में जमीन नहीं अधिग्रहीत करेगी ।
सिंगूर ने विस्थापन पर भी सवाल उठाए हैं ।
जो सिंगूर नंदीग्राम में हुआ वो भूलाने लायक है ।
सिंगूर में टाटा की परियोजना पर काम चल रहा है ।
सिंगूर में टाटा को मिली ज़मीन को विरोध हो रहा है ।
पार्टी ने बुधवार को पूर्वी मिदनापुर के नंदीग्राम में हुए इस नरसंहार के मामले में मुख्यमंत्री बुद्धदेब भट्टाचार्य से त्यागपत्र देने की भी मांग की है ।
अगर टाटा सिंगूर में यह परियोजना लगा पाता तो उसे दक्षिण एशियाई देशों में अपनी पहुंच बनाने में बहुत आसानी होती क्योंकि हल्दिया पोर्ट यहां से बहुत करीब है ।
उन्होंने नंदीग्राम के मुद्दे के साथ तसलीमा नसरीन का मुद्दा जो कि पूरी तरह से एक धार्मिक मुद्दा है को जोड़ कर किस तरह की दायित्व हीनता का परिचय दिया है ।
यह हमने सिंगूर कलिंगनगर नंदीग्राम कोयलकारो नेत्रहाट रायगढ़ जशपुर जगतसिंपुर लोहंडीगुडा में देखा है और सबसे हाल में लालगढ़ में हम यह होते देख रहे हैं ।
यह हमने सिंगूर कलिंगनगर नंदीग्राम कोयलकारो नेत्रहाट रायगढ़ जशपुर जगतसिंपुर लोहंडीगुडा में देखा है और सबसे हाल में लालगढ़ में हम यह होते देख रहे हैं ।
सेज़ की सेज पर नंदीग्राम में कई लाशे सोई पड़ी है तो विदर्भ में अपने पेट को घुटने से चिपकाकर अपनी भूख को मारता किसान आत्महत्या करने पर मजबूर हो रहा है ।
उद्योगपतियों से सांठगांठ का परिणाम नंदीगांव खेजरी सिंगूर जैसी जगहों पर बेघर किये जाने से नकसलियों के उकसाये में आदिवासियों ने प्रशासन के छक्के छुड़ा दिए ।
सेंसर बोर्ड ने एक बंगाली फिल्म को वर्तमान मुख्यमंत्री ममता बनर्जी के शपथ ग्रहण समारोह और सिंगूर भूमि पर हुए आन्दोलन की फोटो के प्रयोग पर रोक लगा दी है ।
दिसम्बर सिंगूर में अनिच्छुक किसानों की अधिग्रहित ज़मीन वापस लौटाने की मांग को लेकर ममता बनर्जी ने शहर में स्थित मेट्रो चैनल पर दिनों की भूख हड़ताल की ।
कम से कम नंदीग्राम सिंगूर मंगलकोट की घटनाओं में विपक्षी नेताओं के एक्शन को बाधित करने वाले लोगों को राज्य प्रशासन गिरफ्तार करके एक संदेश दे सकता था ।
कम से कम नंदीग्राम सिंगूर मंगलकोट की घटनाओं में विपक्षी नेताओं के एक्शन को बाधित करने वाले लोगों को राज्य प्रशासन गिरफ्तार करके एक संदेश दे सकता था ।
केन्द्र सरकार ने सिंगूर विवाद से पूरी तरह पल्ला झाड़ते हुए आज कहा कि यह मामला राज्य सरकार के अधिकार क्षेत्र का है और उसे ही इस समस्या का हल ढूंढना होगा ।
बेहतर होगा कि सरकार मुआवजे के लिए छिड़े इस आंदोलन को नंदीग्राम या सिंगूर जैसी लड़ाई में बदलने की भूमिका न बनाये और किसानों के पक्ष में माकूल फैसला करे ।
```

Fig. 7 50 sentences extraction from the OOV-based queries

alignment and removal of special characters and stop words were performed. We
need the entire corpus to be in a sequential manner so it was assigned to sentence
number. Out of 200 queries, 15 queries were having out of vocabulary words after
comparing with the bilingual dictionary. Now to do source language sentence we
used these queries and index. We have extracted sentences for each query in the

numbers of 50, 100, 200, 300, 400, and 500. So the methodology was to check which query has the OVV word by comparing it with the dictionary. All the 15 queries were stripped of the stop words and after that on the basis of which sentences are selected. The selection parameters were the length of sentences and frequency of occurrence. The first priority is to gather all the sentences which have the maximum number of the words of the query. For instance, if the query is "Army General Court Martial", which has four words, the objective is to find those sentences which have first all the four words, then three, and in subsequent decreasing orders. Out of all the extracted sentences, the average length is computed and sentences of below length were discarded. The two corpora are clustered on the basis of context.

5 Results

Both bilingual dictionary and filtered queries are used for sentence selection. In this step, files containing 50–500 extracted sentenced have been created separately. The selection of sentences has been done in a well-defined fashion.

First, 20% of the sentences have been extracted on the basis of the frequency of words except stop words of each generated query. For the rest of the 80% sentence extraction, the average length per word of the query is computed. After that, for each word, a certain number of sentences are extracted; this makes up to 80% of sentences putting together, for instance, if the average length of sentences for a word is seven words each. Then, the required number of sentences is extracted from sentences having seven words only. Overall, 90 text files from 50 to 500 sources were generated as the outcome. The result for some queries is shown in Fig. 7.

6 Future Work

In this work, we have extracted the source sentences in a designated manner. These sentences have a contextual similarity; as for every text file they were extracted from one query. This data is the training data for the experiment to achieve these objectives—first, cross-language plagiarism detection, cross-language information retrieval, and cross-language paraphrase detection. In the beginning, we want to train it from Hindi to the English language and then use it bidirectionally. We will implement the model for more than two languages depending upon the accuracy achieved in the first two cases.

References

1. Ballesteros, L., Croft, B.: Dictionary methods for cross-lingual information retrieval. In: Proceedings of the 7th DEXA Conference on Database and Expert Systems Applications, Zurich, Switzerland, Sept 1996, pp. 791–801
2. Berry Michael, W.: Automatic Discovery of Similar Words, in Survey of Text Mining: Clustering, Classification, and Retrieval, pp. 24–43. Springer, New York, LLC (2004)
3. Hearst, M.A.: Untangling text data mining. In: Proceedings of ACL'99: The 37th Annual Meeting of the Association for Computational Linguistics, University of Maryland, 20–26 June 1999
4. Jianming, C., Jianming, L., Zhouyu, L.: Research of Text Categorization Based on Support Vector Machine. Comput. Simul. **30**(2), 299–302 (2013)
5. Ballesteros, L., Croft, B.: Resolving ambiguity for cross-language retrieval. In: Proceedings of SIGIR'98, Melbourne, Australia, pp. 64–71, Aug 1998
6. Nguyen-Son, H.-Q., Miyao, Y., Echizen, I.: Paraphrase detection based on identical phrase and similar word matching. In: 29th Pacific Asia Conference on Language (2015)
7. Yin, W., Schütze, H.: Convolutional neural network for paraphrase identification. In: Human Language Technologies: The 2015 Annual Conference of the North American Chapter of the ACL, Denver, Colorado, pp. 901–911 (2015)
8. Liang, C., Paritosh, P., Rajendran, V., Forbus, K.D.: Learning paraphrase identification with structural alignment. In: Conference: IJCAI 2016, at New York
9. Lee, J.C., Cheah, Y.: Paraphrase detection using string similarity with synonyms. In: The Fourth Asian Conference on Information Systems, ACIS 2015
10. McCallum, A.: Information extraction: Distilling structured data from unstructured text. Queue **3**(9), 48–57 (2005)
11. Dean, J., Ghemawat, S., MapReduce: simplified data processing on large clusters. In: Proceedings of 6th Symposium on Operating Systems Design and Implementation, pp. 137–149 (2004)
12. Shvachko, K., Kuang, H., Radia, S., Chandler, R.: The hadoop distributed file system. In: Proceedings of the 2010 IEEE 26th Symposium on Mass Storage Systems and Technologies (MSST2010), pp. 1–10 (2010)
13. Li, Y., Shawe-Taylor, J.: Using KCCA for Japanese-English cross-language information retrieval and document classification. J. Intell. Inf. Syst. **27**(2), 117–133 (2006)
14. Crofts, N., Doerr, M., Gill, T., Stead, S., Stiff, M. (eds.): Definition of the CIDOC Conceptual Reference Model, Version 5.0 (2008)
15. Yue, L.: Research of Cross-Language Text Classification. Beijing Institute of Technology, Beijing (2011)
16. Bel, N., Koster, C.H.A., Villegas, M.: Cross-lingual text categorization. Lect. Notes Comput. Sci. **2003**(2769), 126–139 (2003)
17. Smith, J.R., Quirk, C., Toutanova, K.: Extracting parallel sentences from comparable corpora using document level alignment. In: NAACL/HLT, pp. 403–411 (2010)
18. Harris, Z.S.: Transformations in linguistic structure. Proc. Am. Philos. Soc. **108**(5), 418–422 (1982)
19. Lee, J.C., Cheah, Y.-N.: Paraphrase detection using semantic relatedness based on Synset Shortest Path in WordNet. In: International Conference on Advanced Informatics: Concepts, Theory, and Applications, 16–17 Aug 2016, Parkroyal Penang Resort
20. Harris, Z.S.: A Grammar of English on Mathematical Principles. Wiley, New York, USA (1982)
21. Ballesteros, L., Croft, B.: Phrasal translation and query expansion techniques for cross-language information retrieval. In: Proceedings of the 20th Annual International ACM SIGIR Conference on Research and Development in Information Retrieval (1997)
22. Gao, J., Nie, J., Xun, E., Zhang, J., Zhou, M., Huang, C.: Improving query translation for cross-language information retrieval using statistical models. In: Proceedings of the 24th Annual International ACM SIGIR Conference on Research and Development in Information Retrieval. ACM (2001)

23. Kumar, V., Kalitin, D., Tiwari, P.: Unsupervised learning dimensionality reduction algorithm PCA for face recognition. In: International Conference on Computing Communication and Automation (ICCCA), 5–6 May 2017, pp. 32–37
24. Tiwari, P., Mishra, B.K., Kumar, S., Kumar, V.: Implementation of n-gram methodology for rotten tomatoes review dataset sentiment analysis. Int. J. Knowl. Disc. Bioinf. (IJKDB) 7(1), 30–41. https://doi.org/10.4018/ijkdb.2017010103

Parallel Frequent Pattern Mining on Natural Language-Based Social Media Data

Shubhangi Chaturvedi and Sri Khetwat Saritha

Abstract Social media data on Web sites such as Twitter, Facebook, LinkedIn, YouTube, and Instagram is increasing tremendously because of their significant number of users. Newspaper, radio, television provide one-way communication, whereas social media provides many-to-many communication. Thus, analysis of social media data can produce many hidden information. Frequent patterns in social media can generate hidden information that can be useful. In this paper, we discussed how parallel frequent pattern mining algorithm is useful in finding patterns of natural language-based social media data. We present a process to retrieve frequent patterns (or rules) from a social media using thresholds of support and confidence. The parallel computing is achieved with the help of a scalable Apache Spark program. The retrieved patterns can be useful in making decisions related to social media.

Keywords Social media · Frequent pattern mining · Association rule

1 Introduction

Data mining on social media data is important because it is useful during its arrival. An example of this is the trending topics on social media that are valuable at the time of arrival and become obsolete after some time [1]. Increase in quantity of data or information in business, government, and science is called as data deluge [2]. Data mining extracts meaningful information, discovers hidden patterns, and mines knowledge [3]. The Knowledge Discovery in Databases (KDDs) [4] is another term used for data mining. Finding new or hidden information in a large dataset is the

S. Chaturvedi (✉) · S. K. Saritha
Computer Science and Engineering Department, Maulana Azad National
Institute of Technology, Bhopal 462003, MP, India
e-mail: chaturvedishubhangi51@gmail.com

S. K. Saritha
e-mail: sarithakishan@gmail.com

© Springer Nature Singapore Pte Ltd. 2019   507
A. Abraham et al. (eds.), *Emerging Technologies in Data Mining and Information
Security*, Advances in Intelligent Systems and Computing 814,
https://doi.org/10.1007/978-981-13-1501-5_44

main task behind data mining. Usually, social media data is too large and changes frequently because of the change in real-world social conditions.

1.1 Social Media

Social media data can be categorized into three: structured, semi-structured, or unstructured data. First, if data is stored in row and column, then it is termed as structured data. Second, XML and JSON are the examples of semi-structured data. Third, text and multimedia data such as audio, video, images, e-mail messages are categorized into unstructured data. The three kinds of data have several methods for their data preprocessing, which is followed by hidden information retrieval. The information can be in the form of pattern that can be useful for marketing of products, designing new strategies, and several other decision-making.

People or groups who have same interest are connected to each other in a network. In future, it is expected that entities such as software components, Web-based services, data resources, and workflows are connected using social media [2]. Users who are connected in a same network are known to each other in some of the way; i.e., they might have same hobbies, friends, career, and opinions. Different social media have different vocabularies; for example, Facebook uses "friends" and "status," Twitter uses "tweets" and "followers," and Google plus uses "circles" and "hangouts" [2].

Social media data is also helpful in health care where patients can give their views. It also provides a platform for advertising products, which influences a customer that may result in growth of business. For example, when a customer "likes" an organization's official page on Facebook, it may help to influence the customer's friends by showing the page on the friend's feed [5]. As described by Batrinca et al. [6], social media data mining has the following research challenges:

(1) *Scrapping*: Access to raw data of social media is becoming difficult because of their commercial value. For academia and research purpose, there are a very few sources available at affordable prize. Premium is charged by the news services like Thomson Reuters and Bloomberg for their data access. Twitter's public tweets and its historical data can be accessed by applying to Twitter's Data Grants.
(2) *Data Cleansing*: Unstructured text data which is real-time streaming have many research challenges.
(3) *Holistic Data Sources*: For research, various types of data such as customer data, real-time market, social media data, and geospatial data from different sources need to be combined.
(4) *Data Protection*: Security of big data resource has great importance. There is a chance that some unauthorized user may use all valuable data from database. Thus, different access levels should be provided to users.

(5) *Data Analytics*: Foreign words, errors in spelling, foreign languages, slang, and evolving natural language can cause challenge in social media data analysis for opinion mining.
(6) *Analytics Dashboards*: Non-programming interfaces are required to access raw data because most of the naïve user does not have skills to access raw data from social media platforms.
(7) *Data Visualization*: Representing data in graphical form can produce clear information; thus, visualization of data is also valuable.

Fan and Gordon [7] explained the process of social media in three steps. First step is "capture" in which data is gathered. Second step is "understand" in which noisy data is removed and advanced analytics (such as sentiment analysis, opinion mining, and trend analysis) is performed. Third step is "present," which summarizes, evaluates, and presents the extracted knowledge from previous steps.

1.2 Association Rule Mining

Association rule mining can be explained using a supermarket example where many items are present at same place. Each customer transaction has items purchased by a customer when they visit the supermarket. Items purchased on per transaction basis or over a certain period are known as basket data [8]. Barcode technology has helped to store this basket data.

There is requirement of making some business decision like: what coupons needed to be applied on which time, which products should be put on sale, which items to be placed near to each other so that profit can be increased. Such decision can be made based on the information generated by performing data analysis on past transactions; this can improve quality of business decisions.

Agrawal et al. [8] proposed an efficient algorithm for mining a large collection of basket data transaction to find decision-making information. Formally, $X => Y$ is an association rule, where X and Y are itemsets, X is antecedent, and Y is consequent. Consider a transactional database, which consists of customer record. Support of an itemset X is frequency of the itemset X given as

$$\text{Supp}(X) = |\{t \in T; X \subseteq t\}| / |T| \tag{1}$$

where T denotes set of transactions. Minimum support (minSup) is a user-defined threshold above which an itemset is used to make rules. Confidence of a rule ($X => Y$) is a conditional probability given as

$$\text{Conf}(X => Y) = \text{Supp}(X \cup Y) / \text{Supp}(X) \tag{2}$$

Minimum confidence (minConf) is a user-defined threshold above which a rule is generated as output. The association rule mining generates rules that have minimum support (s) and minimum confidence (c) in a transaction database. Generally, for large values of minSup and minConf, the association rule mining algorithm generates a few numbers of rule. Conversely, for small values of minSup and minConf, the association rule mining algorithm generates a large number of rules. Following example queries can be used on the retrieved association rules:

(1) Find the rules in which "table" is consequent.
(2) Find the rules in which "chair" is antecedent.
(3) Find all rules in which "fish" is antecedent and "aquarium" is consequent.
(4) Find "best" k rules in which "chair" occur as a consequent.

1.3 Parallel Frequent Pattern Growth Algorithm

Mining huge dataset using earlier algorithms like Apriori [9] and FP-Growth [10] can be resource intensive. The well-known FP-Growth algorithm [10] requires two scans of the database. In the first scan, it computes frequent itemset list sorted in descending order of their frequency. During its second scan, a FP-Tree is constructed. Then, for each of the items whose support is greater than threshold, mining is performed by constructing their conditional FP-Tree. The following resource challenges occur in FP-Growth:

(1) Storage: The constructed FP-Tree may be huge and cannot fit into main memory.
(2) Computation distribution: There can be parallelization of the steps of FP-Growth.
(3) Costly communication: Because of the interdependency between FP-Trees synchronization, thus parallel threads are needed frequently.
(4) Support value: Using correct minimum support (minSup) threshold value is very important. Usually, optimized minSup value is used for larger databases; otherwise, storage would be overflowed by the FP-Tree generation.

Li et al. [11] proposed an algorithm on distributed machines known as Parallel Frequent Pattern Growth (PFP-Growth) algorithm. Significant speedup can be achieved using the PFP-Growth algorithm. The idea behind this algorithm is to split the computation such that each of machines has its own mining tasks. Reduction in memory usage and computational cost is the greatest advantage of using parallel algorithms. Li et al. [11] performed experimentation using PFP-Growth algorithm to do query recommendations. The following are the PFP outline as described by the Li et al. [11]:

(1) Sharding: In the process of sharding, the database is divided into parts and these parts are stored in P different computers such that each part is referred as shard.
(2) Parallel counting: In this step, support values of all items of the database are counted using MapReduce and stored in F-list.
(3) Grouping items: Q groups are formed by dividing the |I| items on F-list. These groups are stored in a list known as group list (G-list).

(4) Parallel FP-Growth: One MapReduce passes in this step.

- *Mapper*: Each of the machines gets shard of database as input. It also reads G-list. Each mapper instance produces an output as one or more key-value pairs. Here, *key* is a group-id, whereas group-dependent transactions are *value*.
- *Reducer*: Using MapReduce grouping of group-dependent transaction is done corresponding to every group-id.

(5) Aggregation: This step aggregates the results of step 4 to generate the output.

The rest of the paper is organized as follows. Section 2 describes related works. Section 3 describes our approach for mining social media dataset. Section 4 demonstrates experiment that we performed on the natural language-based social media dataset. In the end, we conclude the paper in Sect. 5.

2 Related Work

Kovacs and Illés [12] investigated frequent itemset mining algorithms based on MapReduce; they provided a solution for porting Count Distribution (CD) algorithm and done improvement on it.

Social networks, computer vision, and bioinformatics are the main applications of frequent subgraph mining. Lin et al. [13] proposed two-step approach that gives an advantage of parallelism and the communication among worker. They used MapReduce in partitioning the collection of graphs among worker nodes, and then each of the worker nodes applies the filter step. Locally, frequent subgraphs are determined at this step. The input of a refinement step is the union of these locally frequent subgraphs. Those graphs, which are globally frequent, are retained at the refinement step.

For social media marketing, Fan et al. [14] have proposed a graph pattern association rules (GPARs) and algorithm for applying GPARs. Social graphs contain regularities between entities, and it can be discovered using GPARs. Various examples of association rules as social graph patterns were discussed in [14]. One of the examples is that if there are two friends living in same city and there are at least three French restaurants such that both the friends like in same city, then it can be said that "if one of them visits a new French restaurant, then other friend would also visit that place".

3 PFP-Growth for Natural Language-Based Social Media Data

A process for analysis of a dataset involves three steps as shown in Fig. 1. In first step, the dataset is preprocessed by using techniques like scrapping and cleaning.

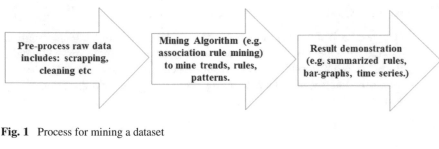

Fig. 1 Process for mining a dataset

Fig. 2 Process for mining social media dataset using PFP

The second step uses the preprocessed data to do the data mining algorithms such as association rule mining, sequential rule mining, or Apriori; this generates patterns, rules, and trends. The output generated in step two is then analyzed in the third step, and the results are demonstrated in graphs and time series.

Figure 2 explains the systematic process to mine a social media dataset. First step preprocesses the dataset according to the PFP-Growth algorithm. It involves merging of files, removal of duplicate words in a line, and removal of stop words. Then on the preprocessed data, we applied PFP-Growth mining algorithm (based on Spark [15, 16]), which produces various rules. We intentionally generated interesting rules [17], which are then used for further analyses and demonstration.

Before executing the PFP-Growth algorithm, we would need to preprocess our dataset. The preprocessing involves three steps.

3.1 Merging of the Social Media File to Make a Single Dataset

As shown in *Algorithm 1 Merge_Files*, we first merged numerous *text_files* into one single file to make a single dataset. For this, each of the lines of each *text_file* is read and then written in a newly created file named as *Merged_file*.

Algorithm 1 Merge_Files(text_files)

Create new file *Merged_file*
For each *text_file* from *text_files*
    Read *text_file*
    While line in *text_file* != null
        Write line into the *Merged_file*
    End while
End for
Return *Merged_file*

3.2 Removal of the Duplicate Entries from Each Line of the Dataset

In the *Algorithm 2 Remove_Duplicates*, we remove the duplicates from each of the lines in *Merged_file*. This algorithm takes *Merged_file* as input and converts each of the uppercase words to lowercase. It removes duplicates by storing the words into a *HashSet*. The unique words stored in *lset* are added into *line* by adding spaces between words, and then resultant is written into newly created file named *merge_without_duplicate*.

Algorithm 2 Remove_ Duplicates(Merged_file)

Read *Merged_file*
While *line* != null
    Replace symbols (.,:,;,\,(,),{,}) with ""   // means replace unnecessary symbols
                                           // with empty string
    Create variable *lset* ∈ HashSet   //HashSet has property that it always
                                           //keeps unique elements
        For each *word* of a *line*
            Convert uppercase to lowercase of each
            word and store in variable *result*
            Add *result* to *lset*   // duplicate result is automatically
                                          // discarded i.e. not added to the HashSet
        End For
    Create variable *s*[*lset*.size] ∈ set
        For *i* ∈ integer varies from 0 to *s*.length
                line[*i*] += *s*[*i*] + " "   // similar to the i = i +1 ➜ i += 1
                                             // where + is for string concatenation
        End For
    write *line* to the newly made file *merge_without_duplicate*
End while
Return *merge_without_duplicate*

3.3 Removal of the Stop Words from Each Line of the Dataset

The *Algorithm 3 Stop_Words_Removal* takes *merge_without_duplicate* file as input.
Firstly, a **getStopWords()** function is used to make a file, which consists of a list of
stop words, named as *stopWordFile*. The **RemoveStopWords()** function first read
the *stopWordFile* and input file, i.e., *merge_without_duplicate* file, and then matches
each of the words in given lines of *merge_without_duplicate* file with the list of stop
words in *stopWordFile*. For each of the words in a *line*, if the read word of the line is
not a stop word, then the word is concatenated to *writeline* using a space, and then
the *writeline* is stored in a new file named as *merge_without_duplicates_stopWords*.

Algorithm 3 Stop_Words_Removal(merge_without_duplicate)

Initialize *stopwords* as HashMap

Function **getStopWords**(File stopWordFile)
If (*stopwords* != null)
    Read the stopWordFile containing stopwords as list
    While end of the lines does not occur
    Remove leading and trailing spaces in *stopwords*
    End while
End if

Function **RemoveStopWords**(*merge_without_duplicates*,
                                                     *merge_wit
hout_dupli
cates_stop
Words*)

To construct a list of *stopwords,*
run the **getStopWords** function for a given stop word file
If (*stopwords* == null)   // note that *stopwords* is a global
Return

Read input file which is *merge_without_duplicates*
    While end of the *lines* does not occur
    Initialize *writeline* ∈ String
        For each of the *words* in a *line*
            If *word* in a line is not a *stopwords* then
                writeline = writeline + " "   // where + is for string concatenation
                writeline = writeline + word   //where + is for string concatenation
                Store *writeline* in the output file which is
                merge_without_duplicates_stopWords
            End if
        End For
    End while
End if
Return *merge_without_duplicates_stopWords*

After the preprocessing, the dataset is prepared for the rule mining using the PFP-
Growth algorithm. In the next subsection, we describe mining on the prepared social
media dataset (*merge_without_duplicates_stopWords*).

3.4   PFP-Growth on the Prepared Dataset of Social Media

Identifying relationships among groups can help in understanding social behavior and interactions. Efficiency is the main challenge in rule (or frequent pattern) mining. Parallel computing environment like MapReduce paradigm has a limitation that the given input and intermediate data have to work on key-value pair. Figure 3 shows how MapReduce works for counting of words in a text. The figure is straightforward to understand its mapping and reducing phase.

Various challenges occur while using MapReduce paradigm and during executing an algorithm. First challenge is that the algorithm is to be divided into two phases: Map and Reduce phase. Second challenge is that the result occurs only after the mapping computation and then the reduce phase aggregated the items to generate rules. For minimum confidence = 0.5, the following rules were generated:

[outside] => [think], 1.0	[think] => [i], 0.5	[what's] => [done], 1.0
[outside, think] => [box], 1.0	[think] => [box], 0.5	[box] => [outside], 1.0
[box, think] => [outside], 1.0	[outside] => [box], 1.0	[i] => [think], 1.0
[think] => [outside], 0.5	[done] => [what's], 1.0	[box] => [think], 1.0
[box, outside] => [think], 1.0		

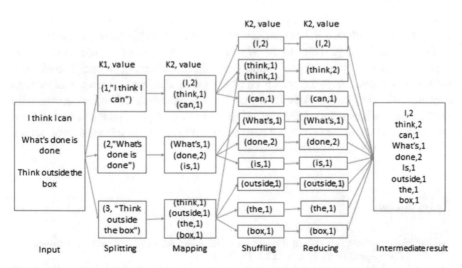

Fig. 3 An example of MapReduce process for counting words in a piece of text

4 Experiment on Natural Language-Based Social Media Data

We conducted experiments on IMDb movie review dataset collected from the link [18], which has a collection of positive and negative reviews about movies. We preprocessed it according to the algorithms in the Sects. 3.1, 3.2, and 3.3. Then, we executed PFP [11] on various support and confidence values. Numerous rules were generated on which refinement was done by removing irrelevant stop words.

We analyzed the positive and negative sentiment of the dataset. We discover positive words were frequently occurring in positive review *text files*. For example, with support $= 0.05$ and confidence $= 0.05$ [watch, best] together occurred 670 times and [great, see, movie] occurred 1018 times. Similarly, we found rule such as [character, story] $=>$ film with confidence 0.6937, which means when "character" and "story" occur together then "film" will also occur, as they are associated with each other. For support $= 0.1$ and confidence $= 0.2$: [movie, film] $=>$ [see], 0.4198; [movie, film] $=>$ [great], 0.3856 was generated. For minSup $= 0.15$ and minConf $= 0.15$ [great, movie] $=>$ [film], 0.528 was generated. The words used in the rules are positive feeling of audiences about movies in positive reviews. Similarly, we also retrieved rules for negative reviews, but we avoid its description to maintain brevity.

5 Conclusion

In this paper, we discussed how parallelism in FP-Growth has been implemented as PFP-Growth mining [11] algorithm. We also discussed how association rule mining is used in various approaches. Using association rule mining in social media can show many hidden aspects, which can be used in decision-making. Rather than executing the tasks sequentially, if they are executed parallelly, then performance can be increased. We concluded that performing parallel frequent pattern mining based on Spark for a social media dataset can produce rules that can generate information, which can be helpful in decision-making.

References

1. Zaharia, M., et al.: Discretized streams: an efficient and fault-tolerant model for stream processing on large clusters. HotCloud **12**, 10 (2012)
2. Wei, Tan, et al.: Social-network-sourced big data analytics. IEEE Internet Comput. **17**(5), 62–69 (2013)
3. Barbier, G., Liu, H.: Data mining in social media. In: Social Network Data Analytics, pp. 327–352 (2011)
4. Han, J., Pei, J., Kamber, M..: Data Mining: Concepts and Techniques. Elsevier (2011)
5. Balan, S., Rege, J.: Mining for social media: usage patterns of small businesses. Bus. Syst. Res. J. **8**(1), 43–50 (2017)

6. Batrinca, B., Treleaven, P.C.: Social media analytics: a survey of techniques, tools and plat-forms. AI Soc. **30**(1), 89–116 (2015)
7. Fan, W., Gordon, M.D.: The power of social media analytics. Commun. ACM **57**(6), 74–81 (2014)
8. Agrawal, R., Imieliński, T., Swami, A.: Mining association rules between sets of items in large databases. In: ACM SIGMOD Record, vol. 22, issue no. 2. ACM, New York (1993)
9. Agrawal, R., Srikant, R.: Fast algorithms for mining association rules. In: Proceedongs of the 20th International Conference on Very Large Data Bases, VLDB, vol. 1215 (1994)
10. Han, J., Pei, J., Yin, Y.: Mining frequent patterns without candidate generation. In: ACM SIGMOD Record, vol. 29, issue no. 2. ACM, New York (2000)
11. Li, H., ct al.: PFP: Parallel FP-Growth for query recommendation. In: Proceedings of the 2008 ACM Conference on Recommender systems. ACM, New York (2008)
12. Kovacs, F., Illés, J.: Frequent itemset mining on Hadoop. In: 2013 IEEE 9th International Conference on Computational Cybernetics (ICCC). IEEE (2013)
13. Lin, W., Xiao, X., Ghinita, G.: Large-scale frequent subgraph mining in mapreduce. In: 2014 IEEE 30th International Conference on Data Engineering (ICDE). IEEE (2014)
14. Fan, W., et al.: Association rules with graph patterns. Proc. VLDB Endowment **8**(12), 1502–1513 (2015)
15. Matei, Zaharia, et al.: Fast and interactive analytics over Hadoop data with Spark. USENIX Login **37**(4), 45–51 (2012)
16. Xiangrui, Meng, et al.: Mllib: machine learning in apache spark. J. Mach. Learn. Res. **17**(1), 1235–1241 (2016)
17. Zhang, Y., et al.: A survey of interestingness measures for association rules. In: International Conference on Business Intelligence and Financial Engineering, 2009. BIFE'09. IEEE (2009)
18. Accessed 1 Nov 2017. http://ai.stanford.edu/~amaas/data/sentiment/

Part V
Telecommunication
and Computer Network

Adaptive Smart Antenna Using Microstrip Array for Cellular Network

Anwesha Halder, Anupama Senapati and Jibendu Sekhar Roy

Abstract In future mobile communication, adaptive smart antenna is one of the important key technologies. In this paper, beamformation of smart antenna is investigated using microstrip antenna array. Least mean square (LMS) algorithm is used for adaptive beam formation. Linear antenna array of 30 elements and 40 elements is used for beam generation at 28 GHz. The radiation patterns at different angles are presented. Performances of microstrip array in beamforming of smart antenna system are presented here. Low-side lobe levels are achieved for 3-sector cellular network.

Keywords Smart antenna · Signal processing · Microstrip antenna array
Side lobe level

1 Introduction

Multiple antenna system in the form of an array with smart signal processing algorithm is used in smart antenna and is used to enhance the performance of a cellular network [3, 5]. In switched beam antenna system, antenna can generate beam toward some predefined fixed directions, but not very accurately. In adaptive smart antenna, based on the direction of arrival of desired signal, using signal processing algorithm, beam can be pointed exactly toward the user and null of the radiation pattern of the array can be generated [1, 8, 9] toward the undesired interferer. The smart antenna system provides efficient spectrum utilization, high security, and less power consumption in mobile communication. Commonly used signal processing algorithms

A. Halder (✉) · A. Senapati · J. S. Roy
School of Electronics Engineering, KIIT University, Bhubaneswar 751024, Odisha, India
e-mail: anwesha2406@gmail.com

A. Senapati
e-mail: senapati.anupama@gmail.com

J. S. Roy
e-mail: drjsroy@rediffmail.com

© Springer Nature Singapore Pte Ltd. 2019
A. Abraham et al. (eds.), *Emerging Technologies in Data Mining and Information Security*, Advances in Intelligent Systems and Computing 814,
https://doi.org/10.1007/978-981-13-1501-5_45

521

for adaptive beamforming are least mean square (LMS) and its variants, recursive least square (RLS) algorithm, sample matrix inversion (SMI) algorithm [1, 3, 5, 8, 9].

In most of the research papers on adaptive beamforming of smart antenna, isotropic antennas are used. In this paper, microstrip antenna array is used for the beamforming of adaptive smart antenna. Millimeter wave frequency band of 28 GHz is used for antenna design. This 28 GHz band is recommended by ITU for future 5G mobile communication [7]. Microstrip antenna is low profile, thin patch antenna, fabricated on printed circuit board and is useful for high-frequency applications [2, 6]. In this paper for beamforming of E-plane microstrip antenna array, adaptive signal processing algorithm, LMS algorithm is used. Performances of smart microstrip array based on beam direction, null direction, and SLL are reported.

2 Least Mean Square (LMS) Algorithm

For beamforming of adaptive smart antenna, LMS algorithm is used. LMS is an adaptive signal processing algorithm where to reduce error in computation weight vectors are updated by iterative process along with cost function to achieve desired performance of the system under consideration. Error $e(n)$ between desired signal $d(n)$ and array output $y(n)$ is minimized by using adaptive algorithm as [1, 9]

$$e(n) = d(n) - y(n). \tag{1}$$

Weight vectors in LMS algorithm is updated by the equation

$$w(n + 1) = w(n) + \mu x(n) e^*(n). \tag{2}$$

where μ is the step size parameter and $e(n)$ is the error, complex conjugate of $e(n)$ is $e^*(n)$ and $x(n) = [x_1(n), x_2(n) \ldots x_N(n)]$ is the signal received by the multiple antenna elements.

3 Adaptive Beamforming Using Microstrip Array

A linear antenna array of N number of antenna elements with uniform spacing 'd' is shown in Fig. 1. Antennas are fed by equal current (I_0) having progressive phase shift of 'α.'

Array factor for N element isotropic antennas is given by Balanis [4]

$$AF' = \sum_{n=1}^{N} A_n e^{j(n-1)\left(\frac{2\pi d}{\lambda} \cos\theta + \alpha\right)}. \tag{3}$$

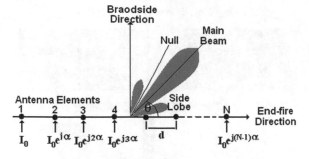

Fig. 1 Antenna array arrangement

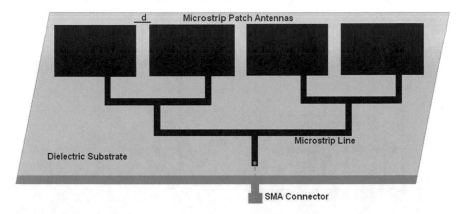

Fig. 2 Microstrip array

The resonant frequency of a rectangular microstrip antenna with dimensions L and W, substrate dielectric constant ε_r, substrate height 't' can be determined from the formula [2]

$$f_r = \left(c/2\sqrt{\varepsilon_e} \right) \sqrt{\left[\left(\frac{m\pi}{L} \right)^2 + \left(\frac{n\pi}{W} \right)^2 \right]}. \qquad (4)$$

Effective dielectric constant (ε_e) is defined as

$$\varepsilon_e = (1/2) \left\{ (\varepsilon_r + 1) + (\varepsilon_r - 1) \left(1 + \frac{12t}{W} \right) - 1/2 \right\}. \qquad (5)$$

Figure 2 shows the schematic diagram of a linear microstrip antenna array where antennas are fed by a coaxial SMA connector via microstrip lines.

Microstrip antenna array may be of E-plane array or H-plane array depending on the mutual coupling between the antenna elements in E-plane and H-plane. In this paper, E-plane microstrip array is used for beamforming and the array arrangement, for two microstrip elements, is shown in Fig. 3.

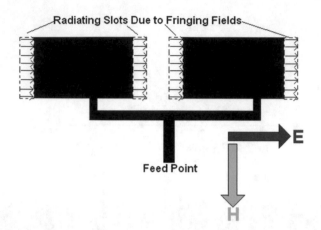

Fig. 3 E-plane microstrip array

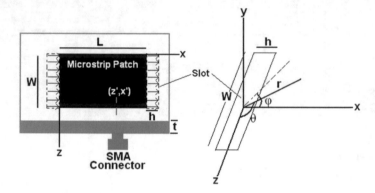

Fig. 4 Diagram to calculate radiation field of a microstrip antenna

Radiation field of microstrip antenna can be calculated using Fig. 4, by considering radiation of a microstrip antenna due to two slots (of width 'h') at the two edges of the patch.

For single radiating slot, radiation field at a distance 'r' from the origin of the source is [2]

$$E\varphi = -j2V_0 W k_0 \left(e^{-jk_0 r} \big/ 4\pi r\right) F(\ominus, \varphi). \tag{6}$$

$$E_\ominus = 0. \tag{7}$$

V_0 is the voltage across the slot and $V_0 = tE_y$. Here, E_y is calculated considering a rectangular microstrip antenna as a cavity and this microstrip cavity is excited at fundamental TM_{10} mode.

where

$$F(\ominus, \varnothing) = \frac{\sin\left(\frac{k_0 h}{2}\sin\theta\cos\varnothing\right)}{\frac{k_0 h}{2}\sin\theta\cos\varnothing} \frac{\sin\left(\frac{k_0 W}{2}\cos\ominus\right)}{\frac{k_0 W}{2}\cos\ominus} \sin\ominus. \tag{8}$$

for $\ominus = \pi/2$, $F(\varphi)$, E-plane pattern (x-y plane) is determined by considering microstrip antenna radiation due to two slots separated by a distance 'L' and can be expressed as [2]

$$F(\varnothing) = \frac{\sin\left(\frac{k_0 h}{2}\cos\varnothing\right)}{\frac{k_0 h}{2}\cos\varnothing} \cos\left(\frac{k_0 L}{2}\cos\varnothing\right). \tag{9}$$

For linear array of N element microstrip antennas, array factor is

$$AF = \sum_{n=1}^{N} F(\varnothing)\, e^{j(n-1)\left(\frac{2\pi d}{\lambda}\cos\theta + \alpha\right)}. \tag{10}$$

Normalized array factor is defined as

$$AF_{norm} = |AF|/|AF_{max}|. \tag{11}$$

where AF_{max} is the maximum value of AF.

For microstrip antenna array design, dielectric substrate RT/Duroid 5870 with $\varepsilon_r = 2.32$, thickness $(t) = 1.575$ mm, loss tangent $(\tan \delta) = 0.0005$ is considered. The dimensions of rectangular microstrip patch at frequency $f_r = 28$ GHz are length of the patch $(L) = 3.88$ mm, width of the patch $(W) = 3$ mm. A numbers of antenna elements (N) used in the linear microstrip array are 40 and 30.

The simulated results for adaptive beamforming, using 40 element E-plane microstrip array, at different desired beam direction (BD) and null direction (ND) are shown in Figs. 5 and 6.

Simulated results for adaptive beamforming, using 30 element E-plane microstrip array, are shown in Figs. 7 and 8.

Figures 9 and 10 show simulated square error plots for $N = 40$ and $N = 30$, respectively. The number of iteration in simulation is 10°.

The simulated results for smart antenna using E-plane microstrip array are provided in Table 1 for 40 element array and for 30 element array.

Nowadays, cellular networks are sectored using sector antennas. For practical use, most of the networks are 3-sectored, that is, 120° sectors. Therefore, information for side lobe levels (SLLs) is given in Table 1 only for 120° sectors. The sector under consideration is centered at 0° and covering −60° to +60°.

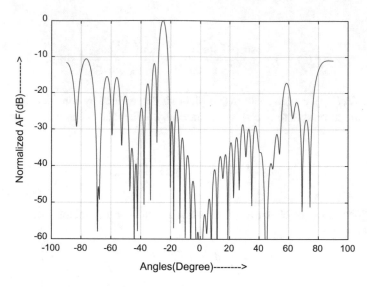

Fig. 5 Adaptive beamforming for BD $= -25°$, ND $= -20°$

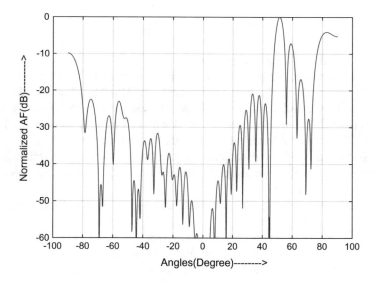

Fig. 6 Adaptive beamforming for BD $= 50°$, ND $= 45°$

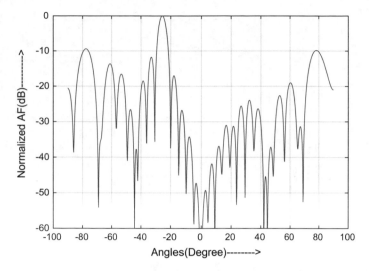

Fig. 7 Adaptive beamforming for BD = −25°, ND = −20°

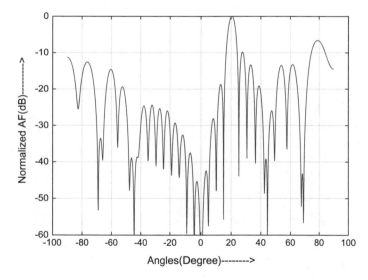

Fig. 8 Adaptive beamforming for BD = 20°, ND = 15°

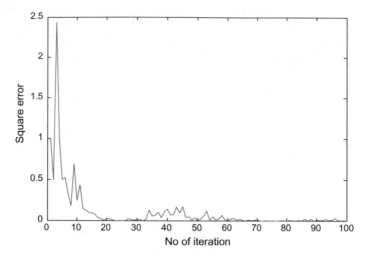

Fig. 9 Square error plot for $N = 40$ BD$= 40°$, ND$= 45°$

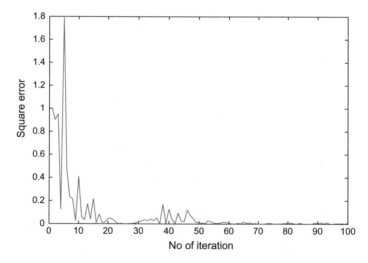

Fig. 10 Square error plot for $N = 30°$, BD$= 20°$, ND$= 15°$

Table 1 Simulated results for $d = 0.5\lambda$, step size $(\mu) = 0.002, f_r = 28$ GHz

Number of antennas	Desired beam direction	Obtained beam direction	Desired null direction	Obtained null direction	SLL$_{max}$ (dB)	HPBW
$N = 40$	$-25°$	$-25.2°$	$-20°$	$-20°$	-17	$3.4°$
	$50°$	$51.4°$	$45°$	$44.8°$	-7.5	$4.2°$
$N = 30$	$-25°$	$-25.6°$	$-20°$	$-19.8°$	-13.5	$4.2°$
	$20°$	$20°$	$15°$	$14.8°$	-9.6	$4.6°$

4   Conclusion

Beamforming of adaptive smart antenna of microstrip array using LMS algorithm is presented in this paper. Microstrip antennas are assumed to be excited at TM$_{10}$ mode. In MATLAB simulation, inter-element spacing in microstrip array is taken 0.5λ and therefore due to wide separation, mutual coupling is not taken into account. E-plane microstrip arrays are considered for smart antenna. In LMS algorithm step size of 0.002 is considered because with this value best result can be found. Results presented here are the best results obtained after simulation of 10 times for each case. The research work presented here may be useful for future millimeter wave cellular communication.

References

1. Ali, W.A.E., Mohamed, D.A.E., Hassan, A.H.G.: Performance analysis of least mean square and sample matrix inversion algorithms for smart antenna system. In: Loughborough Antennas & Propagation Conference IEEE Xplore, pp. 624–629 (2013)
2. Bahl, I.J., Bhartia, P.: Microstrip Antennas. Artech House, Dedham, MA (1980)
3. Balanis, C.A., Bellofiore, S.: Smart antenna system for mobile communication network., part-1: overview and antenna design. IEEE Antennas Propag. Mag. **44**(3), 145–154 (2002)
4. Balanis, C.A.: Antenna Theory—Analysis and Design, 3rd edn. Wiley, New York (2005)
5. Godara, L.C.: Application of antenna arrays to mobile communications, part II: beam-forming and direction-of-arrival considerations. Proc. IEEE **85**(8), 119–1245 (1997)
6. Lo, Y.T., Solomon, D., Richards, W.F.: Theory and experiments on microstrip antennas. IEEE Trans. Antennas Propag. **AP-27**, 138–145 (1979)
7. Rappaport, T.S., Sun, S., Mayzus, R.: Millimeterwave mobile communication for 5G cellular: it will Work. IEEE Access **1**, 333–349 (2013)
8. Sarkar, T.K., Wicks, M.C., Salazar-Palma, M.: Smart Antenna. Wiley-IEEE Press (2003)
9. Senapati, A., Roy, J.S.: Adaptive beamforming in smart antenna using Tchebyscheff distribution and variants of least mean square algorithm. J. Eng. Sci. Technol. (JESTEC) **12**(3), 716–724 (2017)

Social Energy-Based Techniques in Delay-Tolerant Network

Pushkar Jagtap and Lalit Kulkarni

Abstract Energy of nodes plays an important role in all kinds of networks whether it is wired network or wireless network, since most of the nodes are battery powered and it is not possible to recharge the battery whenever necessary. In most of the networks, the data cannot be sent completely when the connection is terminated from source to destination. Hence, the concept of delay-tolerant network (DTN) is introduced. In DTN, due to the mobility of nodes most of the nodes have sufficient amount of energy which is required for transferring of data. Energy is crucial issue in the concept of DTN, especially in disastrous like scenario, where recharging of battery becomes difficult. The proper use of energy for data transmission increases nodes lifetime and network lifetime situation. In this paper, some of the energy-saving techniques are explained to reduce the energy consumption of the node and also increase the packet delivery ratio.

Keywords Energy consumption · Data transmission · Store-carry-forward
DTN · Mobility

1 Introduction

Many of the routing protocols were introduced to limit the number of copies to be forwarded and increase packet delivery ratio, but few of the routing protocols are not energy aware. There are various factors which consume more energy such as mobility of nodes and speed of transferring data. In order to achieve the minimal usage of energy while maintaining the packet delivery ratio, it is necessary to develop highly autonomous system [1]. Autonomous system is a system which is independent

P. Jagtap · L. Kulkarni (✉)
Department of Information Technology, Maharashtra Institute of Technology, Pune, India
e-mail: lvkulkarni@gmail.com

P. Jagtap
e-mail: pushkarjagtap@gmail.com

© Springer Nature Singapore Pte Ltd. 2019
A. Abraham et al. (eds.), *Emerging Technologies in Data Mining and Information Security*, Advances in Intelligent Systems and Computing 814,
https://doi.org/10.1007/978-981-13-1501-5_46

from other systems in the network, and for data transmission, previous knowledge of system's behavior is not required.

The DTN is a type of network in which the connectivity within the nodes is not continuous so it is consider to be intermittent network. Due to the mobility of nodes, the connectivity within the nodes is not established. Energy is the main requirement for data transmission which is considered as most critical issue. Many of the routing protocols were used in DTN. Some of them are spray and wait, epidemic, Prophet, etc. [2]. These routing protocols do not consider the energy factor. In these protocols, they continuously forward the data packets to its destination for achieving high delivery ratio without considering energy consumption.

Representation

■■■■▶: Connection termination

■■■■▶: Connection establishment

DTN uses store-carry-forward mechanism for data transmission. In Fig. 1, when node A sends the data to node F, it will forward in n-hop mechanism. As the connection is terminated from node D to node E, the data is stored in node D. In Fig. 2, when the connectivity is established between D and E, the data is forwarded to the destination.

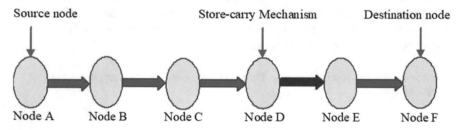

Fig. 1 Connection termination

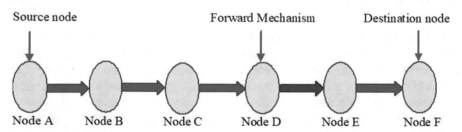

Fig. 2 Connection establishment

2 Issues in Delay-Tolerant Network

Compared to traditional Internet network and mobile ad hoc networks, delay-tolerant network has the following basic issues [3]

- Buffer space: In DTN, connection is not established all the time and follows "store-carry-forward" approach. Enough buffer space is required in the routers for storing all the messages, and these messages can be forwarded to the destination node when link failure issue or some other issues were solved. So, there is a requirement of enough buffer space in the routers.
- Energy: In the mobility of the node and the connectivity between the power stations, the nodes in DTN have low-energy level. For sending, receiving, and storing of data, sufficient amount of energy is required. Hence, it is necessary to design a routing protocol which is energy efficient.
- Security: Security is the major kind of problem in any network whether it is wired network or wireless network. Before message is to send to the destination it goes through the various intermediate node, and it is not guarantee that each node is legitimate and can transfer the data securely.
- Reliability: In DTN, for the reliable delivery of messages between source nodes and destination nodes, the routing protocol is used for the communication and has some acknowledgment, i.e., a receiver sent a response to sender that indicates successful delivery of messages. The successful delivery of messages between nodes is ensured by this acknowledgment [4].

3 Social Energy-Based Techniques

3.1 Social Energy-Based Routing

Li et al. [5] explained that the energy is generated when the nodes get encountered with other nodes in the community, and this energy is called social energy. This social energy is distributed and shared within the community and with the other nodes. Each node plays an important role in the community as each node has their own features and social energy levels in the network. Whenever the node gets encountered with other nodes very frequently, then it generates a higher social energy level. With the higher social energy level, the nodes are active within the community and in the network, due to which it provides high packet delivery to the destination.

3.2 Energy-Aware Social-Based Routing

Chilipirea et al. [1] explained for selection of the routing decision, and energy plays an important role. If energy consumption is taken into an account in a social-driven opportunistic network routing, the sink node becomes slow down and resources of that node cannot work properly. This kind of behavior of sink node makes the node unavailable and the advantages of that node disappear in the network. This kind of scenario is seen into an environment where recharging point for the battery of the nodes easily available for improving the lifetime of the battery. But in disaster-prone-like areas, recharging of the battery of the node becomes difficult and energy of data transmission is also limited. Hence, energy-aware bubble rap is explained below.

3.3 Energy-Aware Bubble Rap

Chilipirea et al. [1] explained energy-aware bubble rap is a combination of socially aware routing with energy consumption optimization. The main aim of energy-aware bubble rap is to balance the energy consumption of opportunistic network and also reducing the delivery cost and hop count and increases lifetime of the network.

4 Advance Energy-Saving Techniques

4.1 Energy and Memory Efficient Clone Detection

Zheng et al. [6] explained to reduce the energy consumption of the node and make the node memory efficient, and it is necessary to detect the clone in the network. To detect the clone the network, the set of nodes are selected randomly which are called as witness nodes. This witness node then helps to check the legitimacy of the nodes in the network. The confidential information of the source node such as id of the node, memory size of the node is shared with that witness node at the time of witness selection. This witness node is selected randomly from the network without announcing for the election of the witness node. This kind of selection of the nodes helps to select the nodes which are legitimate in the network. Whenever a source wants to send the data, it first sends the request to the witness nodes for the verification, then this witness nodes report the source node as a clone if the source node fails the certification.

To detect the clone in the network, witness selection and verification of the node should satisfy the two requirements:

- Witness node should be selected randomly,
- At least on witness node should receive message and can check the legitimacy verification of the node [6].

The first requirement is difficult for the unauthorized used to generate the duplicate verification message as the witness node is selected randomly. In the second requirement, at least one node should verify the id of the node and then detect whether the node is clone or not.

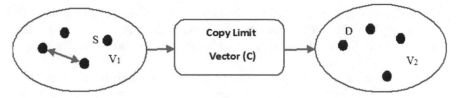

Fig. 3 Energy consumption with social selfishness

4.2 Transmission of Data with Social Selfishness

In an opportunistic mobile network, the mobile nodes can travel from one community to other with their own interest. For transmitting of data from one node to other node, the source node has to spend their own resources such as energy power which is one of the issues in DTN for transmitting within the same community [7]. Transmitting of data within the same community requires less energy as compared with node being placed outside the community. Due to the more energy consumption, the source node does not forward the data to other node outside the community in order to save their storage and energy. This behavior of the node is referred as social selfishness. To overcome the problem of social selfishness in the community, a copy limit vector is used for data transmission which is explained below in Fig. 3.

As shown in Fig. 3, C is represented as copy limit vector, V_1 and V_2 are network communities, S and D are source node and destination node, respectively [8].

Energy consumption for transferring of data is consider as directly proportional to the number of copies the source node sends from source to destination in their whole network lifetime. This energy consumption includes both transmitting energy from source node and the reception energy at the receiving node. Hence, for reducing the energy consumption the copy limit vector is placed in between the different communities. By placing the copy limit vector, only limited number of messages can be forwarded to the destination node [9].

4.3 Energy-Efficient Technique for Adaption of Interface Activation

For transmitting of data in the network, end-to-end path must be established in advance. But end-to-end connectivity is a major challenged in DTN which is proposed by Internet Research Task Force (IRTF) for developing a technique in which transmitting of data can be forwarded in "store-carry-forward" manner. The concept of bundle layer was introduced by IRTF which is placed in between transport layer and application layer. The bundle layer is a lock of raw data, which is stored in a mobile terminal (MT) and using routing scheme it transfers the data directly or through other mobile terminals (MT) to its destination. Generally, MT is battery powered due to which recharging of the battery becomes necessity which consumes more energy. Therefore, it is necessary to have communication interface I/F which usually inactivate the battery and reduces the energy consumption when the node not in use.

Izumikawa et al. [10] described energy-efficient mechanism called as DTN-oriented wireless interface activation mechanism based on radio fluctuation (DWARF) which satisfies the following requirement:

- More for consumed for searching the MT,
- Less infrastructure is sufficient for data transmission,
- No previous knowledge of the node is required for data transmission.

The DWARF mechanism has different approach for data transmission in which optimal interval in interface activation varies from situation to situation. It is necessary to keep the interface active for data dissemination which causes high order to achieve efficient activation.

5 Energy-Efficient Technique

5.1 Energy-Efficient Sparse Routing

In DTN, energy consumption is the major issue in the nodes due to which communication between the nodes rarely happens. In [11], spray and wait protocol is used which is similar to epidemic routing protocol. But, spray and wait protocol is used to restrict the number of copies of each message that nodes want to send. This can be done in two ways:

- Non-binary way: In this approach, a node sends limited number of copies of messages, say L. This approach is time-consuming because the source node can only send another data when the previous one is delivered to the destination node.
- Binary way: In this approach, L/2 messages are send to the destination node and these messages delivered to the destination node continuously. Here L indicated total number of packets.

By combining spray and wait protocol with Prophet protocol, a new routing protocol is introduced called as sparse routing protocol which can be used for energy consumption while achieving high packet delivery ratio.

5.2 Geographic-Based Spray-and-Relay (GSaR)

GSaR is an energy-consuming routing algorithm for delay-tolerant network. GSaR is only a routing algorithm which uses the historical record of encounter nodes and gets the nodes' information for making the routing decisions. By taking into account, the movement of destination node using encounter record, GSaR forwards the messages to the destination node by sending it by hop-to-hop approach, and it successfully delivers the message before the expiration deadline and also achieve high packet delivery ratio. Further, investigation in [12] indicates that the GSaR routing algorithm is efficient in terms of energy consumption for data transmission and also achieve high packet delivery ratio.

6 Conclusion

Various kinds of energy-saving techniques are available for routing in DTN. All the energy techniques discussed above takes energy consumption into an account for designing the routing protocol. By using these techniques, energy efficiency of the node will be increased by maintaining packet delivery ratio. All the techniques discussed in the paper are recently proposed techniques. Each technique has their own significance and gives different results. The main aim of all techniques is to reduce energy consumption with proper data transmission while maintaining high probability packet delivery ratio.

References

1. Chilipirea, C., Petre, A.C., Dobre, C.: Energy-aware social-based routing in opportunistic networks. In: 27th International Conference on Advanced Information Networking and Applications Workshops, Barcelona, 2013, pp. 791–796 (2013)
2. Hastings, M., Yang, S.: Energy-efficient sparse routing protocol for Delay Tolerant Networks. In: International Conference on Computing, Networking and Communications (ICNC), Santa Clara, CA, 2017, pp. 803–807 (2017)
3. Urunov, K., Vaqqasov, S., Namgung, J.I., Park, S.H.: Security issues for DTN mechanism of UIoT. In: 18th IEEE International Symposium on Consumer Electronics (ISCE 2014), JeJu Island, 2014, pp. 1–3 (2014)
4. Lu, Y., Wang, W., Chen, L., Zhang, Z., Huang, A.: Distance-based energy-efficient opportunistic broadcast forwarding in mobile delay-tolerant networks. IEEE Trans. Veh. Technol. 65(7), 5512–5524 (2016)
5. Li, F., Jiang, H., Li, H., Cheng, Y., Wang, Y.: SEBAR: social-energy-based routing for mobile social delay-tolerant networks. IEEE Trans. Veh. Technol. 66(8), 7195–7206 (2017)
6. Zheng, Z., Liu, A., Cai, L.X., Chen, Z., Shen, X.: Energy and memory efficient clone detection in wireless sensor networks. IEEE Trans. Mob. Comput. 15(5), 1130–1143 (2016)
7. Wang, W., Motani, M., Srinivasan, V.: Opportunistic energy-efficient contact probing in delay-tolerant applications. IEEE/ACM Trans. Networking 17(5), 1592–1605 (2009)
8. Wu, J., Zhu, Y., Liu, L., Yu, B., Pan, J.: Energy-efficient routing in multi-community DTN with social selfishness considerations. In: IEEE Global Communications Conference (GLOBECOM), Washington, DC, 2016, pp. 1–7 (2016)
9. Roy, A., Acharya, T., DasBit, S.: Energy-aware social-based multicast in delay-tolerant networks. In: IEEE 81st Vehicular Technology Conference (VTC Spring), Glasgow, 2015, pp. 1–5 (2015)

10. Izumikawa, H., Pitkänen, M., Ott, J., Timm-Giel, A., Bormann, C.: Energy-efficient adaptive interface activation for delay/disruption tolerant networks. In: 12th International Conference on Advanced Communication Technology (ICACT), Phoenix Park, 2010, pp. 645–650 (2010)
11. Hastings, M., Yang, S.: Energy-efficient sparse routing protocol for Delay Tolerant Networks. In: International Conference on Computing, Networking and Communications (ICNC), Santa Clara, CA, pp. 803–807 (2017)
12. Alone, S.V., Mangrulkar, R.S.: Implementation on geographical location based energy efficient direction restricted routing in delay tolerant network. In: International Conference on Power, Automation and Communication (INPAC), Amravati, pp. 129–135, (2014)

The Cryptographic Properties of Feistel Network-Based Quasigroups

Tajender Kumar and Sugata Gangopadhyay

Abstract Quasigroups defined by Feistel network are used in the designing of several cryptographic primitives, where Feistel network is created by a bijection map from $\mathbb{F}_2^n$ to $\mathbb{F}_2^n$. It is known that quasigroups are represented as vectorial Boolean functions and these functions with high nonlinearity and low differential uniformity are more important in cryptography. In this paper, we identify the relation between the cryptographic properties [nonlinearity, differential uniformity, and strict avalanche criteria (SAC)] of bijection map and Feistel network-based quasigroups.

1 Introduction

The quasigroup structures and their properties are applied to many areas like: authentication schemes, secret sharing schemes, DES block cipher, pseudo random number generators and cryptographic hash functions for some applications we refer to [2, 6, 9, 10]. Gligoroski et al. [7] defined quasigroups as Boolean functions, and Markovski and Mileva [14] generated huge quasigroups from small nonlinear bijections via extended Feistel network. Mihajloska and Gligoroski [15] proposed a technique for constructing cryptographically strong 4 × 4-bit S-boxes via quasigroups of order 4. S-boxes [4, 12] were also constructed by use of Feistel and MISTY structures. In [11], Leander and Poschmann classified all optimal 4-bit S-boxes on the cryptographic point of view. For preventing many attacks on the ciphers, S-boxes are required to satisfy certain cryptographic properties [13, 16, 18, 21], for example, having high nonlinearity, low differential uniformity, and strict avalanche criteria (SAC) etc.

T. Kumar (✉)
Department of Mathematics,
Indian Institute of Technology Roorkee, Roorkee 247667, India
e-mail: mathstezu@rediffmail.com

S. Gangopadhyay
Department of Computer Science and Engineering,
Indian Institute of Technology Roorkee, Roorkee 247667, India
e-mail: gsugata@gmail.com

© Springer Nature Singapore Pte Ltd. 2019
A. Abraham et al. (eds.), *Emerging Technologies in Data Mining and Information Security*, Advances in Intelligent Systems and Computing 814,
https://doi.org/10.1007/978-981-13-1501-5_47

539

We start by a brief introduction to the notion of quasigroups, vectorial Boolean functions and their properties. Then we give the representation of Feistel network-based quasigroups and its description as vectorial Boolean functions. In the end, we find the relationship between the cryptographic properties (nonlinearity, differential uniformity, and SAC) of bijection map from $\mathbb{F}_2^n$ to $\mathbb{F}_2^n$ and Feistel network-based quasigroups.

2   Preliminaries

2.1   Quasigroup

A pair $(Q, *)$ is called a quasigroup if the operation $*$ is closed on Q and the equations:

$$a * x = b,$$
$$x * a = b.$$

have unique solution for every $a, b \in Q$. Hence, the multiplication table of each quasigroup is equivalent to a Latin Square.

2.2   Vectorial Boolean Function

Let $\mathbb{F}_2^n$ be the n-dimensional vector space over $\mathbb{F}_2$, where $\mathbb{F}_2$ is a Galois field with two elements. A function $f : \mathbb{F}_2^n \longrightarrow \mathbb{F}_2$ is a Boolean function of n variables, and the set of n variables Boolean functions from $\mathbb{F}_2^n$ to $\mathbb{F}_2$ is denoted by $\mathscr{B}_n$. Let $f_0, f_1, \ldots, f_{m-1} \in \mathscr{B}_n$ and a Boolean map $F : \mathbb{F}_2^n \longrightarrow \mathbb{F}_2^m$ is defined as:

$$F(x) = (f_0(x), f_1(x), \ldots, f_{m-1}(x))$$

is called a vectorial Boolean function and $f_0, f_1, \ldots, f_{m-1}$ are called coordinate functions, where $f_k : \mathbb{F}_2^n \longrightarrow \mathbb{F}_2$ and $f_k \in \mathscr{B}_n$, $k = 0, 1, \ldots, m - 1$. Let $\mathscr{B}_{n,m}$ is the set of all vectorial Boolean functions $F : \mathbb{F}_2^n \longrightarrow \mathbb{F}_2^m$. For two vectors $u, v \in \mathbb{F}_2^n$, where $u = (u_0, u_1, \ldots, u_{n-1})$ and $v = (v_0, v_1, \ldots, v_{n-1})$, the canonical dot product is defined as:

$$u \cdot v = \sum_{i=0}^{n-1} u_i v_i.$$

2.3 Walsh Transform and Nonlinearity

Besides the coordinates, all linear combinations of the coordinates are involved for determining the cryptographic properties of a vectorial Boolean function. In this sense, the definitions are as follows:

Definition 1 Let F be a vectorial Boolean function from $\mathbb{F}_2^n$ into $\mathbb{F}_2^m$. The Boolean components of F are the n variables Boolean functions

$$F_\lambda : x \to \lambda \cdot F(x)$$

for any $\lambda \in \mathbb{F}_2^m$. The component corresponding to $\lambda = 0$ is called the zero (or trivial) component.

Definition 2 The nonlinearity $nl(F)$ of $F \in \mathscr{B}_{n,m}$ is the minimum nonlinearity of all the component functions $x \in \mathbb{F}_2^n \mapsto v \cdot F(x), v \in \mathbb{F}_2^m, v \neq 0$.

The walsh transform of a vectorial Boolean function $F \in \mathscr{B}_{n,m}$ is map $W_F : \mathbb{F}_2^n \times \mathbb{F}_2^m \longrightarrow \mathbb{Z}$, defined as:

$$W_F(a, b) = \sum_{x \in \mathbb{F}_2^n} (-1)^{a.x+b.F(x)}.$$

Moreover, the linearity [5] of F is given by

$$\mathscr{L}(F) = \max_{b \in (\mathbb{F}_2^m)^*} \mathscr{L}(F_b) = \max_{a \in \mathbb{F}_2^n, b \in (\mathbb{F}_2^m)^*} |W_F(a, b)|$$

and the nonlinearity [3] of F is given by

$$nl(F) = 2^{n-1} - (1/2)(\mathscr{L}(F))$$

or

$$nl(F) = 2^{n-1} - (1/2) \max_{a \in \mathbb{F}_2^n, b \in (\mathbb{F}_2^m)^*} |W_F(a, b)|.$$

2.4 Differential Uniformity

The derivative of the vectorial Boolean function $F \in \mathscr{B}_{n,m}$ in the direction of $a \in \mathbb{F}_2^n$ coincides with $b \in \mathbb{F}_2^m$ is defined [1] as:

$$D_F(a, b) = \{x \in \mathbb{F}_2^n | F(x + a) + F(x) = b\}.$$

Definition 3 [17] Let F be a vectorial Boolean function from $\mathbb{F}_2^n$ into $\mathbb{F}_2^n$. The derivative of F for differences pair (a, b) in $\mathbb{F}_2^n$ is defined as:

$$D_F(a \to b) = \{x \in \mathbb{F}_2^n | F(x \oplus a) \oplus F(x) = b\}.$$

The cardinality of the $D_F(a \to b)$ is corresponded to the entry at (a, b) in the difference table of F. It is denoted by $\delta_F(a, b)$.

In addition, the differential uniformity of F is given by

$$\delta(F) = \max_{a \neq 0, b} \delta_F(a, b).$$

Therefore, the differential uniformity of a vectorial Boolean function is always even or $\delta(F) \geq 2$.

Conditions for Differential uniformity:

The differential uniformity is achieved based on the value of m and n. The vectorial Boolean function $F : \mathbb{F}_2^n \mapsto \mathbb{F}_2^m$ has

Case-1 ([16], Theorem 3.2): For $n > m$ the minimum differential uniformity 2^{n-m} is reached iff $2m \leq n$ and n is even. For $n/2 < m < n$ the minimum differential uniformity is unknown.

Case-2 ([19], Sect. 3): For $n \leq m$ the minimum differential uniformity is 2. A function which reaches this bound is called almost perfect nonlinear (APN).

Case-3 ([18], Sect. 3): For $n < m$ the minimum differential uniformity is 2 and can be reached by simple modifications of APN functions.

Completeness ([8], Definition 2) For any positive integer n, $c_1^{(n)}, c_2^{(n)}, \ldots, c_n^{(n)} \in \mathbb{F}_2^n$ are defined as:

$$c_1^{(n)} = [0, 0, \ldots, 1]$$

$$c_2^{(n)} = [0, \ldots, 1, 0]$$

$$\vdots$$

$$c_n^{(n)} = [1, 0, \ldots, 0].$$

A function from $F \colon \mathbb{F}_2^n \to \mathbb{F}_2^m$ is complete if and only if

$$\sum_{x \in \mathbb{F}_2^n} F(x) \oplus F(x \oplus c_i^{(n)}) > (0, 0, ..., 0)$$

for all i ($1 \leq i \leq n$), where both the greater-than and the summation are componentwise over $\mathbb{F}_2^m$.

Avalanche effect ([8], Definition 3) A function from $F\colon \mathbb{F}_2^n \to \mathbb{F}_2^m$ shows the avalanche effect if and only if

$$\sum_{x \in \mathbb{F}_2^n} wt(F(x) \oplus F(x \oplus c_i^{(n)})) = m2^{n-1}$$

for all i ($1 \le i \le n$), where $wt()$ denotes the Hamming weight function.

2.5 Strict Avalanche Criteria

If a function $F : \mathbb{F}_2^n \to \mathbb{F}_2^m$ satisfies the following equations:

$$\sum_{x \in \mathbb{F}_2^n} F(x) \oplus F(x \oplus c_i^{(n)}) = (2^{n-1}, 2^{n-1}, \ldots, 2^{n-1})$$

for all i ($1 \le i \le n$). We say that F satisfies SAC ([8], Definition 4) or F is said to be a strong S-box. When a single bit of the input vector is complemented, then each output bits should be changed with 50%. Therefore, a strong S-box is complete and shows the avalanche effect.

3 Feistel Network-Based Quasigroups

Definition 4 Suppose Q is an additive group and I is the identity mapping on Q. Then, $\theta : Q \to Q$ is a complete mapping permutation if both θ and $\theta - I$ are permutations. A group Q is admissible if there is a complete mapping $\theta : Q \to Q$.

The following proposition is proved by Sade [20].

Proposition 1 ([20], 14) *Let $(Q, +)$ be an admissible group with complete mapping θ. And operation $* : Q \times Q \to Q$ is defined as:*

$$x * y = \theta(x - y) + y$$

where $x, y \in Q$. Then $(Q, *)$ is a quasigroup.

Suppose f is a bijective map on $\mathbb{F}_2^n$ to $\mathbb{F}_2^n$ and F is defined as:

$$F : \mathbb{F}_2^n \times \mathbb{F}_2^n \to \mathbb{F}_2^n \times \mathbb{F}_2^n$$
$$F(l, r) = (r, l + f(r)) \quad \forall \ (l, r) \in \mathbb{F}_2^n \times \mathbb{F}_2^n.$$

This is known as Feistel network, and it is shown in [14] that F is also bijective. Markovski and Mileva [14] have shown the outcomes of an F especially as relative to f, i.e., bijection of F does not depend on f but for complete mapping it is required that f should be bijective.

Therefore, $(\mathbb{F}_2^{2n}, +)$ is admissible with a complete mapping $F : \mathbb{F}_2^{2n} \times \mathbb{F}_2^{2n} \to \mathbb{F}_2^{2n} \times \mathbb{F}_2^{2n}$. Let $Q = \mathbb{F}_2^n \times \mathbb{F}_2^n$ and $* : Q \times Q \to Q$ be defined as a binary operation on Q

$$* : \mathbb{F}_2^{2n} \times \mathbb{F}_2^{2n} \to \mathbb{F}_2^{2n}$$
$$x * y = F(x + y) + y. \tag{1}$$

Then Q is also a quasigroup with respect to $*$ by Proposition 1.

Using that $x \equiv (x_l, x_r)$ and $y \equiv (y_l, y_r)$ and let G be its corresponding representation as vectorial Boolean function. Then, Eq. 1 gives

$$\begin{aligned}
x * y &= F((x_l, x_r) + (y_l, y_r)) + (y_l, y_r) \\
&= F(x_l + y_l, x_r + y_r) + (y_l, y_r) \\
&= (x_r + y_r, x_l + y_l + f(x_r + y_r)) + (y_l, y_r) \\
&= (x_r + y_l + y_r, x_l + y_l + y_r + f(x_r + y_r)),
\end{aligned}$$

or

$$G(x_L, x_R) = G(x_L \| x_R) = (x_r \oplus y_l \oplus y_r, x_l \oplus y_l \oplus y_r \oplus f(x_r \oplus y_r)).$$

4   Description of Quasigroup as Vectorial Boolean Function

For any quasigroup $(Q, *)$ of order $|Q| = 2^d$, we can define a bijection from the set of quasigroup to the set of binary strings of length d. Let α be a bijection and it is defined as:

$$\alpha : Q \to \mathbb{F}_2^d$$
$$q_1 \mapsto (x_1, x_2, \ldots, x_d).$$
$$*_\alpha : \mathbb{F}_2^d \times \mathbb{F}_2^d \to \mathbb{F}_2^d$$
$$(\alpha(q_1), \alpha(q_2)) \mapsto \alpha(q_1 * q_2).$$

Let $\alpha(q_1 * q_2) = (x_1, x_2, \ldots, x_d) *_\alpha (x_{d+1}, x_{d+2}, \ldots, x_{2d}) = (z_1, z_2, \ldots, z_d)$. Here we see that each z_i is presented as $2d-$ary Boolean function $z_i = f_i(x_1, x_2, \ldots, x_{2d})$, where $f_i : \mathbb{F}_2^d \to \mathbb{F}_2$ is determined by $*$, $i = 1, 2, \ldots, d$.

Lemma 1 ([6], Lemma 1) *For any quasigroup $(Q, *)$ of order $|Q| = 2^d$, let $\alpha :$ $Q \to \mathbb{F}_2^d$ be any bijection. Then there exist a unique vectorial Boolean function $*_\alpha$*

and $2d$-ary Boolean functions $f_1, f_2, \ldots, f_d$ uniquely determined by d. For all a, b, and c in Q, we have:

$$a * b = c.$$

$$\Downarrow$$

$$(x_1, x_2, \ldots, x_d) *_\alpha (x_{d+1}, x_{d+2}, \ldots, x_{2d}) = (f_1(x_1, x_2, \ldots, x_{2d}), f_2(x_1, x_2, \ldots, x_{2d}), \ldots,$$
$$f_d(x_1, x_2, \ldots, x_{2d})).$$

5 Cryptographic Properties of Feistel Network-Based Quasigroups

In this section, we give a detailed proof of the results, using Canteaut et al. [4] technique which shows the dependency of the cryptographic properties (nonlinearity, differential uniformity, and *SAC*) of the resulting G (Feistel network-based quasigroups) on the f (bijection map). The generalized results and their proofs are as follows:

5.1 Nonlinearity

Proposition 2 *Let f be the n-bit vectorial Boolean function and G be the $2n$-bit function defined by Feistel network. Then, we get:*

$$W_G((a, b) \parallel (c, d), e \parallel k) = \begin{cases} 2^{3n} \cdot W_f(b \oplus e, k) & \text{if } a \oplus k = 0, c \oplus e \oplus k = 0 \\ & \text{and } b \oplus d \oplus k = 0, \\ 0 & \text{elsewhere.} \end{cases}$$

for all a, b, c, d, e, and k in $\mathbb{F}_2^n$. Moreover, the linearity of f is

$$\mathscr{L}(f) = \max_{b \oplus e \in \mathbb{F}_2^n, k \in (\mathbb{F}_2^n)^*} \mid W_f(b \oplus e, k) \mid.$$

When $a \oplus k = 0$, $c \oplus e \oplus k = 0$ and $b \oplus d \oplus k = 0$, we get

$$\mathscr{L}(G) > \mathscr{L}(f)$$

and

$$nl(G) < nl(f).$$

Fig. 1 Blue values indicate linear masks

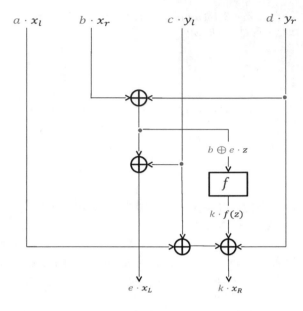

Proof The walsh transform of G is map $W_G((u_1, u_2), v) : (\mathbb{F}_2^{2n} \times \mathbb{F}_2^{2n}) \times \mathbb{F}_2^{2n} \to \mathbb{Z}$, defined as

$$W_G((u_1, u_2), v) = \sum_{(x,y)\in(\mathbb{F}_2^{2n}\times\mathbb{F}_2^{2n})} (-1)^{(u_1,u_2)\cdot(x,y)\oplus v\cdot G(x,y)}$$

or

$$W_G(u_1\|u_2, v) = \sum_{(x,y)\in(\mathbb{F}_2^{2n}\times\mathbb{F}_2^{2n})} (-1)^{(u_1\|u_2)\cdot(x\|y)\oplus v\cdot G(x\|y)}$$

Let $u_1 \equiv (a, b)$, $u_2 \equiv (c, d)$, and $v \equiv (e, k)$, the following result for the configuration depicted in Fig. 1 is given by

$$W_G((a, b) \| (c, d), e \| k) = \sum_{(x_l,x_r,y_l,y_r)\in(\mathbb{F}_2^n)^4}(-1)^{((a,b)\|(c,d))\cdot((x_l,x_r)\|(y_l,y_r))\oplus(e,k)\cdot G(x_L\|x_R)}$$

$$= \sum_{(x_l,x_r,y_l,y_r)\in(\mathbb{F}_2^n)^4}(-1)^{(a,b,c,d)\cdot(x_l,x_r,y_l,y_r)\oplus(e,k)\cdot(x_r\oplus y_l\oplus y_r,x_l\oplus y_l\oplus y_r\oplus f(x_r\oplus y_r))}$$

$$= \sum_{(x_l,x_r,y_l,y_r)\in(\mathbb{F}_2^n)^4}(-1)^{a\cdot x_l\oplus b\cdot x_r\oplus c\cdot y_l\oplus d\cdot y_r\oplus e\cdot x_r\oplus e\cdot y_l\oplus e\cdot y_r\oplus k\cdot x_l\oplus k\cdot y_l\oplus k\cdot y_r\oplus k\cdot f(x_r\oplus y_r)}$$

We set $x_r = y_r \oplus z$ and observe that, for any fixed y_r, z takes all possible values in $\mathbb{F}_2^n$ when x_r varies, implying that

$$W_G((a, b) \parallel (c, d), e \parallel k) = \sum_{x_l \in \mathbb{F}_2^n} (-1)^{(a \oplus k) \cdot x_l} \sum_{y_l \in \mathbb{F}_2^n} (-1)^{(c \oplus e \oplus k) \cdot y_l}$$
$$\sum_{y_r \in \mathbb{F}_2^n} (-1)^{(b \oplus d \oplus k) \cdot y_r} \sum_{z \in \mathbb{F}_2^n} (-1)^{(b \oplus e) \cdot z \oplus k \cdot f(z)}$$
$$W_G((a, b) \parallel (c, d), e \parallel k) = \sum_{x_l \in \mathbb{F}_2^n} (-1)^{(a \oplus k) \cdot x_l} \sum_{y_l \in \mathbb{F}_2^n} (-1)^{(c \oplus e \oplus k) \cdot y_l}$$
$$\sum_{y_r \in \mathbb{F}_2^n} (-1)^{(b \oplus d \oplus k) \cdot y_r} W_f(b \oplus e, k)$$

If $w \in \mathbb{F}_2^n$, we have

$$\sum_{u \in \mathbb{F}_2^n} (-1)^{u \cdot w} = \begin{cases} 2^n & \text{if } w = 0, \\ 0 & \text{elsewhere.} \end{cases}$$

Then, we obtain the following bound

$$W_G((a, b) \parallel (c, d), e \parallel k) = \begin{cases} 2^{3n} \cdot W_f(b \oplus e, k) & \text{if } a \oplus k = 0, c \oplus e \oplus k = 0 \\ & \text{and } b \oplus d \oplus k = 0, \\ 0 & \text{elsewhere.} \end{cases}$$

5.2 Differential Uniformity

Proposition 3 *Let f be n-bit vectorial Boolean function and G be the $2n$-bit function defined by Feistel network. Then, we get:*

$$\delta_G((a, b) \parallel (c, d), e \parallel k) = \delta_f(b \oplus d \to a \oplus b \oplus e \oplus k)$$

for all $a, b, c, d, e,$ and k in $\mathbb{F}_2^n$.

Proof The following result for the configuration depicted in Fig. 2 is given by
Then, (x_L, x_R) satisfies $G(x_L \| x_R) \oplus G((x_L \oplus u_1) \| (x_R \oplus u_2)) = e \| k$, where $u_1 \equiv (a, b)$ and $u_2 \equiv (c, d)$, if and only if

$$\begin{cases} x_r \oplus y_l \oplus y_r \oplus x_r \oplus b \oplus y_l \oplus c \oplus y_r \oplus d = e, \\ x_l \oplus y_l \oplus y_r \oplus f(x_r \oplus y_r) \oplus x_l \oplus a \oplus y_l \oplus c \oplus y_r \oplus d \oplus f(x_r \oplus b \oplus y_r \oplus d) = k. \end{cases}$$

$$\Leftrightarrow \begin{cases} b \oplus c \oplus d = e, \\ f(x_r \oplus y_r) \oplus f(x_r \oplus y_r \oplus b \oplus d) = a \oplus c \oplus d \oplus k. \end{cases}$$

equivalently

$$x_r \oplus y_r \in D_f(b \oplus d \to a \oplus c \oplus d \oplus k) \text{ or } x_r \oplus y_r \in D_f(b \oplus d \to a \oplus b \oplus e \oplus k)$$

or

$$y_r \in x_r \oplus D_f(b \oplus d \to a \oplus b \oplus e \oplus k).$$

Fig. 2 Red values indicate differences

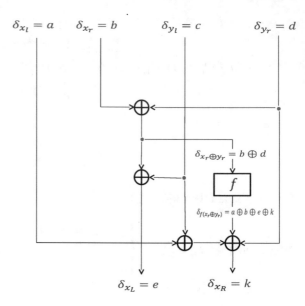

$$\delta_{x_l} = a \qquad \delta_{x_r} = b \qquad \delta_{y_l} = c \qquad \delta_{y_r} = d$$

$$\delta_{x_r \oplus y_r} = b \oplus d$$

f

$$\delta_{f(x_r \oplus y_r)} = a \oplus b \oplus e \oplus k$$

$$\delta_{x_L} = e \qquad \qquad \delta_{x_R} = k$$

Hence, for any fixed x_r in $\mathbb{F}_2^n$, a unique value of y_r is determined by above condition. Therefore, the number of (x_L, x_R) satisfying the differential is exactly $1 \cdot 1 \cdot 1 \cdot \delta_f (b \oplus d \to a \oplus b \oplus e \oplus k)$ or $\delta_f (b \oplus d \to a \oplus b \oplus e \oplus k)$.

5.3 Strict Avalanche Criteria (SAC)

Proposition 4 *Let f be n-bit vectorial Boolean function and G the 2n-bit function defined by Feistel network. Then even if f satisfies SAC, G can never satisfy SAC.*

Proof We have

$$G(x_L \| x_R) = (x_r \oplus y_l \oplus y_r, x_l \oplus y_l \oplus y_r \oplus f(x_r \oplus y_r)).$$

Here we see that the left part of G has n-coordinate functions and each coordinate function is either linear or affine. Kim et al. ([8], Theorem 1) proved that neither linear nor affine functions can satisfy SAC. Hence, the left part of G never satisfies SAC even the right part of G somehow satisfies. Therefore, G can never be satisfied SAC.

Acknowledgements We thank the Ministry of Human Resource Development, Government of India, for financial assistance.

References

1. Álvarez-Cubero, J.A., Zufiria, P.J.: Cryptographic criteria on vector boolean functions. In: Sen, J. (ed.) Cryptography and Security in Computing. ISBN: 978-953-51-0179-6. InTech (2012), Available from: https://www.intechopen.com/books/cryptography-and-security-in-computing/cryptographic-criteria-on-vector-boolean-functions
2. Bakeva, V., Dimitrova, V., Kostadinoski, M.: Pseudo random sequence generators based on the parastrophic quasigroup transformation. In: ICT Innovations 2014, pp. 125–134. Springer, Cham (2015)
3. Carlet, C.: Vectorial Boolean functions for cryptography. In: Crama, Y., Hammer, P.L. (eds.) Boolean Models and Methods in Mathematics, Computer science, and Engineering, 1st edn. pp. 398–469. Encyclopedia of Mathematics and its Applications 134, Cambridge university Press (2010)
4. Canteaut, A., Duval, S., Leurent, G.: Construction of Lightweight S-Boxes using Feistel and MISTY structures (Full Version). IACR eprint report 2015/711 (2015). https://eprint.iacr.org/2015/711.pdf
5. Canteaut, A.: Lectures Notes on Cryptographic Boolean Functions. Inria, Paris, France (2016)
6. Faugère, J.C., Ødegård, R.S., Perret, L., Gligoroski, D.: Analysis of the MQQ public key cryptosystem. In: International Conference on Cryptology and Network Security (CANS), vol. 6467, pp. 169–183 (2010)
7. Gligoroski, D., Dimitrova, V., Markovski, S.: Quasigroups as Boolean functions, their equation systems and Groebner bases. In: Gröbner Bases. Coding, and Cryptography, pp. 415–420. Springer, Berlin (2009)
8. Kim, K., Matsumoto, T., Imai, H.: A recursive construction method of S-boxes satisfying strict avalanche criterion. In: Conference on the Theory and Application of Cryptography, pp. 565–574. Springer, Berlin (1990)
9. Kościelny, C.: Generating quasigroups for cryptographic applications. Int. J. Appl. Math. Comput. Sci. 12(4), 559–570 (2002)
10. Lim, C.H.: CRYPTON: a new 128-bit block cipher. NIsT AEs Submission (1998)
11. Leander, G., Poschmann, A.: On the classification of 4 bit S-boxes. In: International Workshop on the Arithmetic of Finite Fields, pp. 159–176. Springer, Heidelberg (2007)
12. Li, Y., Wang, M.: Constructing S-boxes for lightweight cryptography with Feistel structure. In: International Workshop on Cryptographic Hardware and Embedded Systems, pp. 127–146. Springer, Heidelberg (2014)
13. Matsui, M.: New structure of block ciphers with provable security against differential and linear cryptanalysis. In: International Workshop on Fast Software Encryption, pp. 205–218. Springer, Heidelberg (1996)
14. Markovski, S., Mileva, A.: Generating huge quasigroups from small non-linear bijections via extended Feistel function. Quasigroups Relat. Syst. 17(1), 91–106 (2009)
15. Mihajloska, H., Gligoroski, D.: Construction of Optimal 4-bit S-boxes by Quasigroups of Order 4. In: The Sixth International Conference on Emerging Security Information, Systems and Technologies, SECURWARE (2012)
16. Nyberg, K.: Perfect nonlinear S-boxes. In: Workshop on the Theory and Application of Cryptographic Techniques, pp. 378–386. Springer, Heidelberg (1991)
17. Nyberg, K.: Differentially uniform mappings for cryptography. In: Workshop on the Theory and Application of Cryptographic Techniques, pp. 55–64. Springer, Heidelberg (1993)
18. Nyberg, K.: S-boxes and round functions with controlled linearity and differential uniformity. In: International Workshop on Fast Software Encryption, pp. 111–130. Springer, Heidelberg (1994)
19. Nyberg, K., Knudsen, L.R.: Provable security against a differential attack. J. Cryptol. 8(1), 27–37 (1995)
20. Sade, A.: Quasigroups automorphes par le groupe cylique. Can. J. Math. 9, 321–335 (1957)
21. Tan, Y., Gong, G., Zhu, B.: Enhanced criteria on differential uniformity and nonlinearity of cryptographically significant functions. Crypt. Commun. 8(2), 291–311 (2016)

Studies of Optimization of Throughput: Combining Receiver Diversity in Hybrid ARQ Scheme Over Fading Channel

Mayuri Kundu and Swarnendu K. Chakraborty

Abstract Hybrid automatic repeat request (HARQ) is one of the reliable schemes in wireless communication environment, and it is well established in this paper. To combat the fading process in wireless channels, antenna diversity is widely used. But hybrid ARQ scheme with antenna diversity is not studied so far. In this chapter, two important space diversity techniques, namely switch antenna diversity (SAD) and non-switch diversity (NSD), are discussed with HARQ scheme to achieve higher throughput. The mathematical analysis of HARQ is carried out, and the results show that the protocol may provide a higher value of throughput.

1 Introduction

The wireless communication technologies have become very popular because of its features like accessibility of network to users with anywhere/anytime services. The number of wireless devices is more than the population in the world. Due to this, wireless communication is having high data transmission rate and it is increasing day by day. In wireless communication, the main challenge is to maintain the huge data traffic with less error probability for a short interval of time [1]. Wireless network is cost-effective compared to aired network. But sometimes, the issues of fading may come and it makes constraint to achieve better throughput. The chances of packet loss get increased. This sporadic fading process runs down the wireless communication. Backward error correction (BEC) and forward error correction (FEC) are used to combat the fading process. BEC is implemented by automatic repeat request (ARQ). ARQ and FEC are the main strategies. In case of FEC, error correcting codes channel

M. Kundu (✉) · S. K. Chakraborty
Department of Computer Science Engineering, National Institute of Technology,
Yupia 791112, Arunachal Pradesh, India
e-mail: kundu.mayuri@gmail.com

S. K. Chakraborty
e-mail: swarnendu.chakraborty@gmail.com

© Springer Nature Singapore Pte Ltd. 2019   551
A. Abraham et al. (eds.), *Emerging Technologies in Data Mining and Information
Security*, Advances in Intelligent Systems and Computing 814,
https://doi.org/10.1007/978-981-13-1501-5_48

are used to retrieve the original packet from erroneous packet. In contrast, error detection codes are used in case of ARQ.

Here, the receiver attempts to find out errors in the received packet. And the result of the decoding procedure is informed to the transmitter through a reliable feedback channel. It may be done by sending acknowledgement (ACK) or negative acknowledgement (NACK) messages depending on success or failure of the decoding process. The solution is to send the same message again to the ARQ channel [2, 3]. Throughput is a major criterion to measure the performance of a wireless channel. Throughput of a channel can be defined as the ratio of correctly received information bits per channel to the total number of bits per channel (bpcu). To increase the channel throughput, decrement in retransmission is necessary and hence FEC is combinatorially used with ARQ. This new technique is known as hybrid automatic repeat request (HARQ). HARQ seems to be more efficient as it combines the retransmitted data packet of the same packet on failure decoding in the initial attempt to decode the data packet [4]. Hybrid ARQ or adaptive modulation technique and coding (AMC) is the advanced technique to combat with the fading problem without comprising the high data rate along with the system performance [5]. Here, two important space diversity techniques, i.e., (i) non-switch diversity (NSD) and (ii) switch diversity (SAD), are considered [6]. All branches are monitored for receiving a transmitted block correctly in NSD. SAD being a pre-reception technique, blocks are switched from one branch to another branch if instantaneous envelope of received signal crosses predefined threshold. NSD operates in block-by-block technique; after transmission, a correctly received block is chosen at the output end. This block-by-block operation technique keeps the SAD and NSD scheme conceptually closer. In this paper, the optimization of throughput for HARQ scheme is investigated. HARQ scheme is applied in switched and non-switched diversity with different combination to achieve the optimal throughput.

2   Diversity on HARQ Scheme

HARQ-based two different receiver diversity schemes are studied: (i) HARQ with SAD and (ii) HARQ with NSD.

In HARQ-SAD scheme, receiver has M number of antennas ($M=2$ in Fig. 2) instead of a single antenna. Here, only one antenna works at a time. When a received packet is faulty, it is ensured that the present transmission path has faced a deep fade; a receiver starts using another antenna. In the same way, the scheme can be implemented for M transmitter antennas. Herein, the throughput of HARQ-SAD is at least the same as basic SR scheme (Fig. 1) [7].

In HARQ-NSD, each receiver is associated with a dedicated antenna. After receiving the packet, it passes through a decision process to detect the correct reception over any of the diversity branches. The throughput of this scheme ensures higher rate that of the HARQ-SAD scheme. The HARQ scheme is combined with receiver diversity. Figure 3 shows the operation of the combined scheme. It is accepted when

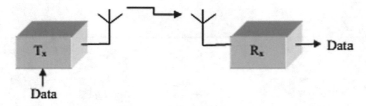

Fig. 1 Schematic representation of typical SR-ARQ scheme

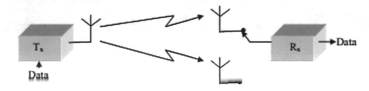

Fig. 2 Schematic representation of typical HARQ-SAD scheme

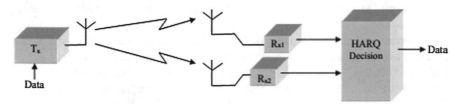

Fig. 3 Schematic representation of typical HARQ-NSD scheme

at least one of the copies of packets over a diversity branch is error-free. And the erroneous packet is buffered. If both the received copies are erroneous, both are saved in buffer, just in the same way as in the last case. It is studied that HARQ scheme provides a higher throughput than that of all other conventional schemes.

3 HARQ and Its Performance Analysis in Additive White Gaussian Noise (AWGN)

HARQ is classified into two parts: chase combining (CC) scheme and incremental redundancy (IR) scheme, where every retransmission uses the same packets. The receiver antenna utilizes maximal ratio combining (MRC) for received packets during ARQ retransmission. Multiple set of code words are used in IR-HARQ, where each corresponds to the same set of information bits. By puncturing the encoder output, a new set of code word from the previous transmission is generated with a different redundancy version [8]. Figure 4 shows that in CC-HARQ, all of the received copies of the same packet will be used during the decoding process. On failure decoding instead of dropping the packet, HARQ keeps the received information from the first

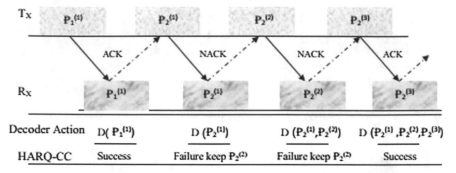

Fig. 4 Schematic representation of decoding of CC-HARQ transmission scheme. $P_n^{(k)}$ implies kth transmission attempt of nth packet

transmission attempt. The received information is buffered in terms of soft channel bits. This soft channel bits are used by soft MRC in CC or by soft combined decoding in IR [9].

In this case, a packet of N_b bits of information is to be conveyed to destination node (D), but it is only known to source node (S) at time zero. The encoded code word X of N_S symbol is $x_1, x_2 \ldots x_{N_S}$. C_{N_S} which is known to all nodes and guaranteed from random and independent symbols considering complex zero-mean unit variance Gaussian distribution [1]. Outage occurs when decoding fails at node D after almost K transmission attempts. As a performance analysis, outage probability and throughput are optimized. It is reported that decoding at receiver end, can have an arbitrarily less failure probability at the Kth attempt (for $1 \leq k \leq K$) when N_S is considerably large because of large value of $N_{s,k}$ [10, 11]. It happens when

$$R < \sum_{l=1}^{k} I(p_x(x).p_l(y|x)) \tag{1}$$

where

$R =$ tranmission rate,
$\sum_{l=1}^{k} =$ implies the summation of transmitted code word up to k,
l is the transmission block,
$p_x(x)$ is pdf of code generator,
$p_l(y|x)$ is the single letter transmission pdf.

For the transmission block, the channel capacity C_l is the maximum achievable $I(p_x(x), p_y(y|x))$. If the normalized accumulated mutual information (NACMI) is greater than 1, the outage is there. Otherwise, for asymptotically long sub-code words, the probability of unsuccessful decoding becomes considerably small.

Mathematically [12, 13],

$$P_r(\text{error}|I_k > 1) = 0, \tag{2}$$

$$P_r(\text{error}|I_k < 1) = 1. \tag{3}$$

And, I_k is the accumulated mutual information (ACMI) with asymptotically long code words after k transmission attempts [14, 15].

$$I_k \sum_{l=1}^{k} \frac{N_{s,l}}{N_{s,1}} C_l, \tag{4}$$

$$C_l = C(\gamma_l) = \log(1 + \gamma_l). \tag{5}$$

where γ_l is the SNR doing lth transmission. Decoding is successful when ACMI is greater than or equal to transmission rate [16].
i.e.,

$$I_k \geq R_1 = \frac{N_b}{N_{s,1}}, \tag{6}$$

hence,

$$1 \leq \sum_{l=1}^{k} \frac{N_{s,l}}{N_b} \times C_l, \tag{7}$$

or,

$$1 \leq \sum_{l=1}^{k} \rho_l \times C_l, \tag{8}$$

or,

$$1 \leq I_k, \tag{9}$$

hence,

$$I_k = \frac{I_k}{N_b}, \tag{10}$$

where I_k is the NACMI after kth transmission.
　　HARQ seems to be successful if and only if $I_k \geq 1$.

But, still a possibility may arise that there is an error while decoding the message at the receiver. This is defined as outage, and the probability of happening of an outage (P_{out}) is equal to the probability of $I_k < 1$.

Mathematically,

$$P_{\text{out}} = P_r\left(I_k^D < 1\right) \tag{11}$$

The throughput (η) calculation is followed [1].

$$\eta = \lim_{t \to \infty} b(t)/c(t). \tag{12}$$

where $b(t)$ is the correctly decoded bits up to time t.
$c(t)$ denotes the total number of bits, used by channel.

By using renewal-reward theorem, it can be written as,

$$\eta = \frac{\overline{N_b}}{\overline{N_s}}. \tag{13}$$

where $\overline{N_b}$ is the expected successfully received bits.
$\overline{N_s}$ is the transmitted symbols or number of channels used.

So, in kth attempt, the expected number of channel uses $\left(\overline{N_{s,k}}\right)$ is given by [17],

$$\overline{N_s} = \sum_{k=1}^{k} \left\{\overline{N_{s,k}}\right\}, \tag{14}$$

or,

$$\overline{N_s} = \sum_{k=1}^{K} E\{N_{s,k}\}. \tag{15}$$

where $E\{.\}$ is the expectation.

At the end of the process, correctly decoded bits may be zero with probability of P_{out} or N_b with probability $(1 - p_{\text{out}})$, and throughput changes to [18]:

$$\eta = \frac{N_b(1 - P_{\text{out}})}{\overline{N_s}}. \tag{16}$$

Considering fixed power criterion, the optimal throughput $\hat{\eta}$ can be calculated as:

$$\hat{\eta} = \max_{\pi} \eta(\pi), \tag{17}$$

$$\text{s.t. } P_{\text{out}}(\pi) \leq \epsilon. \tag{18}$$

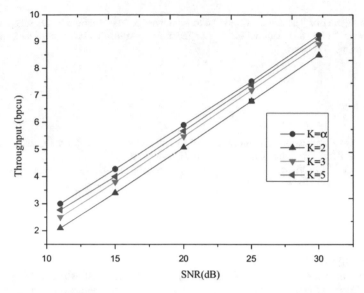

Fig. 5 Throughput versus SNR (dB) plot of HARQ scheme for different values of K

4 Result and Discussion

The HARQ (truncated) is considered here, and dynamic programming is used to find the optimal rate adaptation for optimizing the throughput while the transmission is outage-constrained. Rate-adaptive cooperative HARQ scheme significantly improves the value of the throughput. The result obtained is shown in the figure, and it is the optimal throughput for adaption rate (AD). Figure 5 corresponds to the throughput for rate adaption $\eta_{(AD-S)}$ for $K = 2, 3, 4$.

5 Conclusion

This work analyzes policies for fading channel to meet the higher throughput performance through HARQ transmission process. Diversity along with HARQ is proposed here. Two different scenarios are studied: First, where HARQ is implemented in SAD scheme, and second, NSD is merged with HARQ. These two scenarios show different results for throughput optimization. But both the cases provide better results than other conventional techniques (like SR-ARQ). Finally, numerical analysis of HARQ is investigated, and it is shown that in high signal-to-noise ratio regime, an optimal value of throughput can be achieved depending on K value.

Acknowledgements The author would like to acknowledge Visvesvaraya Ph.D. Scheme (MeitY, Government of India) and National Institute of Technology Arunachal Pradesh for their infrastructural and financial support.

References

1. Caire, G., Tuninetti, D.: The throughput of hybrid-ARQ protocols for the Gaussian collision channel. IEEE Trans. Inf. Theory **47**(5), 1971–1988 (2001)
2. Kundu, M., Chakraborty, K.S.: Mathematical comparison of throughput analysis of ARQ mechanism. IJRSI **4**(10), 85–88 (2017)
3. Bhunia, C.T.: ARQ techniques: review and modifications. J. IETE Tech. Rev. **18**(5), 381–401 (2001)
4. Lin, S., Costello, D.J.: Error control coding, vol. 2. Prentice Hall, Englewood Cliffs (2004)
5. Khosravirad, S.R., Szczecinski, L., Labeau, F.: Rate allocation for HARQ in relay-based cooperative transmission. In: Wireless Communications and Networking Conference, pp. 2757–2762, IEEE (2014)
6. Tellambura, C., Annamalai, A., Bhargava, V.K.: Unified analysis of switched diversity systems in independent and correlated fading channels. IEEE Trans. Commun. **49**(11), 1955–1965 (2001)
7. Chakraborty, S.S., Liinaharja, M., Ruttik, K.: Diversity and packet combining in Rayleigh fading channels. IEE Proc. Commun. **152**(3), 353–356 (2005)
8. Lott, C., Milenkovic, O., Soljanin, E. Hybrid ARQ: theory, state of the art and future directions. In: Information Theory for Wireless Networks, 2007 IEEE Information Theory Workshop, pp. 1–5 (2007)
9. Gopalakrishnan, N.: Achievable rates and rate selection algorithms for incremental redundancy (IR) hybrid ARQ (HARQ) wireless systems. Doctoral dissertation, Purdue University (2008)
10. Gopalakrishnan, N., Gelfand, S.: Achievable rates for adaptive IR hybrid ARQ. In: Sarnoff Symposium, pp. 1–6 (2008)
11. Malkamaki, E., Leib, H.: Coded diversity on block-fading channels. IEEE Trans. Inf. Theory **45**(2), 771–781 (1999)
12. Wu, P., Jindal, N.: Performance of hybrid-ARQ in block-fading channels: a fixed outage probability analysis. IEEE Trans. Commun. **58**(4), 1129–1141 (2010)
13. Byun, I., Kim, K.S.: Cooperative hybrid-ARQ protocols: unified frameworks for protocol analysis. CoRR, vol. abs/0812.2301 (2008)
14. Cheng, J.F., Wang, Y.P., Parkvall, S.: Adaptive incremental redundancy [WCDMA systems]. In: IEEE 58th Vehicular Technology Conference, vol. 2, pp. 737–741 (2003)
15. Malkamaki, E., Leib, H.: Coded diversity on block-fading channels. IEEE Trans. Inf. Theory **45**, 771–781 (1999)
16. Khosravirad, S.R., Szczecinski, L., Labeau, F.: Opportunistic relaying without CSI: optimizing variable-rate HARQ, submitted to IEEE Transactions on Vehicular Technology (2014)
17. Szczecinski, L., Khosravirad, S.R., Duhamel, P., Rahman, M.: Rate allocation and adaptation for incremental redundancy truncated HARQ. IEEE Trans. Commun. **61**(6), 2580–2590 (2013)
18. Khosravirad, S.R., Szczecinski, L., Labeau, F.: Rate adaptation for cooperative HARQ. IEEE Trans. Commun. **62**(5), 1469–1479 (2014)

Gray Hole and Cooperative Attack Prevention Protocol for MANETs

Sandeep S. Musale, Sandeep L. Dhende, S. D. Shirbahadurkar
and Anand S. Najan

Abstract A MANET has many wireless nodes that are arbitrarily moving and communicating each other. The communication is without the use of any central coordinate or base station. It is infrastructure-independent network. It has different unique characteristics that make it more complex in routing. The routing decision is made in a decentralized manner. Although many protocols have been proposed for wireless communication, the ADOV is most widely used. The intermediate node helps to transmit data packets from source to destination. The interference of intermediate nodes introduces some serious attacks in mobile ad hoc networks. Some of them are gray hole, black hole, flooding, and selfish node attacks. In this chapter, the gray hole and cooperative attack prevention method is discussed and the results of the same are presented.

Keywords MANET · Gray hole attack · AODV · Nodes · Routing

1 Introduction

A MANET is collection of wireless mobile nodes. It is the infrastructure-less network with dynamic topology. In this network, the path setup between two nodes is completed without the use of central coordinator or base station. In the absence of

S. S. Musale
MKSSS's COEW, Savitribai Phule Pune University, Pune, Maharashtra, India
e-mail: sandeepmusale@yahoo.co.in

S. L. Dhende (✉)
SCTR's PICT, Savitribai Phule Pune University, Pune, Maharashtra, India
e-mail: sldhende@pict.edu ; sandeep.dhende381@gmail.com

S. D. Shirbahadurkar
ZCOER Savitribai Phule Pune University, Pune, Maharashtra, India
e-mail: s_shir00@yahoo.co.in

A. S. Najan
PVG's COET, Savitribai Phule Pune University, Pune, Maharashtra, India
e-mail: najan.anand693@gmail.com

© Springer Nature Singapore Pte Ltd. 2019
A. Abraham et al. (eds.), *Emerging Technologies in Data Mining and Information Security*, Advances in Intelligent Systems and Computing 814,
https://doi.org/10.1007/978-981-13-1501-5_49

access point, the routing becomes complex task compared to infrastructure-based network. The resource management is on demand based and not allocated by any central coordinator. There are many routing protocols that can be used to route information from source to destination. Some of them are AODV, DSDV, and DSR. The AODV is most widely used protocol in MANETs. AODV is source-initiated routing protocol. The route setup is on demand when source has some data packets to send to destination. The route from source to destination is set up by sending route request packet to all the neighbors. It is done by the source node. The node that has route to the destination node sends an acknowledgment to the source node. This is done by the intermediate node or destination node. Once the route setup process is completed, the data packets are sent to destination. A MANET has some unique characteristics that help intruder to break the security [1]. The attack implementation is easy in MANETs compared to infrastructure-based network. The black hole and gray hole are some serious attacks that can be implemented in MANETs [2]. In these attacks, the data packets are intensely dropped without forwarding it to the destination [3]. When the data packets are dropped for particular node, the attack is known as gray hole attack. In cooperative attack, the attackers work together to create more serious problem in network. In this paper, the black hole, gray hole, and cooperative attacks are examined and the solution for the same is presented.

2 Literature Review

Al-Shurman et al. [4] propose a method that can be used to identify black hole node in MANETs. In this method, the authenticity of an intermediate or destination node is checked by source, and if the node is identified reliable for communication, the data packets are sent through that node. The negative of this method is large delay. The identity of the black hole node is confirmed using the second solution. In the second solution, the sequence number of the previously sent and received packet is used to confirm it. If there is mismatch in the sequence number, the node is identified as black hole node. In other case, the node will be normal node that can be used for sending data packets between source and destination.

Sun et al. [5] propose a method that uses neighbors' information to identify the black hole attack in MANETs. In this method, the sender requests intermediate or destination node to reply with neighbors' list. Then, the sender compares the neighbors' list of more than one replier, and if it identifies the larger difference between them, the node with less number of neighbors is identified as black hole node. The negative of this method is that it is costlier and less reliable.

Kurosawa et al. [6] propose a method that uses mean vector to identify the black hole attack in MANETs. To calculate mean vector, it uses number of route requests sent and received, respectively. It also uses the average difference between the sequence numbers of destination. This difference is calculated comparing the sequence number of destination received in route reply with the sequence number that is already in list of sender. Now, the calculated difference is compared with set value to identify black hole attack in MANETs. The black hole node is identified with larger difference in mean vector.

Deng et al. [7] propose a simple method that detects black hole attack in MANETs. The further route request and reply is used to confirm black hole node. This method is executed in two phases. In the first phase of detection, the route discovery process is initiated and the black hole node is detected. In this case, the first reply is assumed from the black hole node and hence the first replier is detected as black hole node. Now, to confirm this in the second phase of detection, the further route request is sent and the replies from the intermediate or destination nodes are obtained. Here, the first route replier is cross-checked with the first solution. If both of them are same, the node is detected as black hole node.

Tamilselvan et al. [8] propose a black hole attack detection method in MANETs. In this method, the next hop information is used to check the loyalty of replier node. The source node asks the replier to send reply with next hop information. Now, the source node sends further request to check the loyalty of replier. The loyalty of node is decided on number of rules. According to the next hop node, the disloyal node is detected as black hole node.

3   Methodology

Authors propose a solution to detect and prevent gray hole and cooperative attack in MANETs. This method is capable to prevent mentioned attacks. It is also capable to reduce packet loss in significant amount.

In order to communicate to destination node, source node initiates route discovery process and it is initiated by forwarding route request. The source node obtains the route replies from intermediate or destination nodes. These replies are maintained in table. Figure 1 shows the route discovery process executed by source node.

Fig. 1 Route discovery process

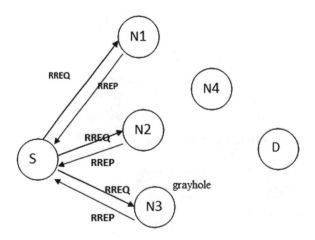

Table 1 Route reply table

Source	Replier	Destination
S	N3	D
S	N1	D
S	N2	D

Table 1 shows the route replies that are maintained by source node. The source node detects two or more route replies for the same destination. The first route reply is considered from the suspicious node, i.e., (N3). This node is marked as unreliable or suspicious node. Now, to send data packets to destination, the second route reply is used. It is assumed that the second and succeeding route replies are from the reliable nodes. When data communication is completed, the source uses feedback mechanism to confirm suspicious node. The detailed description of this process is given below.

To implement this, the existing AODV routing protocol is modified. The source node sends one-hop FB_REQ in the network. The nodes that are affected by it send the FB_REP. Now, the source node sends the FB_ACK to the replier node. Figure 2 shows the broadcasted one-hop FB_REQ. It is broadcasted by source node. It obtains the one-hop FB_REP, and it is stored in Table 2.

Now, the source node sends the two-hop FB_REQ and obtains FB_REP. Here also the source node sends the FB_ACK to the replier nodes. Figure 3 shows the two-hop

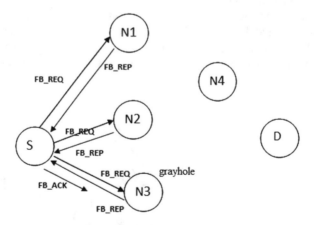

Fig. 2 One-hop feedback process

Table 2 One-hop FB_REP at source

Source	Feedback replier	Destination
S	N3	D

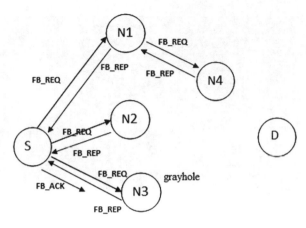

Fig. 3 Two-hop feedback process

Table 3 Two-hop FB_REP at source

Source	Replier	Destination
S	N3	D
S	N1	D

FB_REQ that is broadcasted by source node. Then, source node obtains FB_REPs, and it is stored in Table 3.

Now, the source node compares one-hop and two-hop FB_REP in tables. If the same node replies first in both the cases, the node is identified as suspicious node. Here, the node N3 sends first route reply in both the cases. Hence, it is detected as suspicious node. Such a node is discarded from the table, and it is prevented to be used for further communication.

4 Simulation Results

The simulation is performed using NS-2, and the results for the same are obtained. The output trace file is used to obtain these results. Figure 4 shows the simulation of 50 nodes.

Figure 5 shows the packet loss in the presence of full gray hole and selective gray hole nodes. It is observed that in both the cases the packet loss is about 100%. The nodes 48 and 41 are the full gray hole and selective gray hole, respectively. When these nodes are introduced in communication, the 100% packets are lost. Figure 5 also shows that the received packets are about 0%.

Figure 6 shows the analysis of packets in the network. This analysis is done in the presence and absence of gray hole node. There is some packet loss in transit. But

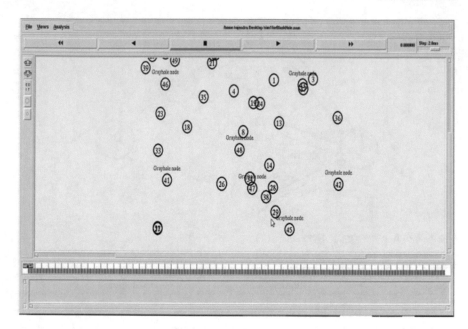

Fig. 4 Simulation scenario of 50 nodes

Fig. 5 Analysis of packet loss

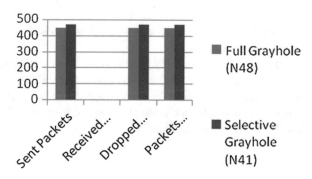

it is observed that in the presence of gray hole node the packets are lost in greater amount. As shown in Fig. 6, the route through the nodes 29 and 26 contains the gray hole node. Hence, there is larger packet loss. It is also observed that the route through 32 to 35 is safe and it is used after detection. When this route is used, the packet loss is reduced in greater amount.

Figure 7 shows the packet loss ratio in the network. It is observed that after gray hole attack detection the packet loss ratio is reduced from 22.14 to 20.11%.

Fig. 6 Analysis of packets sent

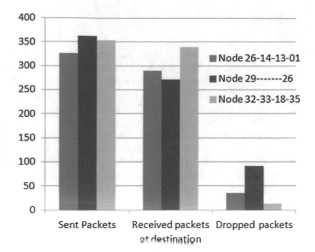

Fig. 7 Packet loss ratio

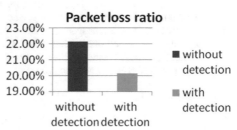

5 Conclusion and Future Work

The simulation results are already analyzed which demonstrate the effectiveness of the mechanism by saving the 4000 packets in the networks. Having simulated the gray hole attack, authors observed that the packet loss is 22.14% in the presence of gray hole node and 20.11% after the detection of gray hole node. The 20.11% of packets are lost because of the environmental condition and not because of malicious activity of node. The authors also analyzed that 100% packets are lost because of full and selective gray hole nodes. But selective gray hole node drops 100% packets for the particular source node and not for other node. The simulation results also show that 89% of packets get saved in case of full gray hole node and 96.04% of packets get saved in case of selective gray hole node. The author performs the simulation by modifying existing AODV routing protocol. In future, the authors are interested to simulate the same problem by using some other routing protocols in MANETs.

References

1. Kannhavong, B., Nakayama, H., Nemoto, Y., Kato, N.: A survey of routing attacks in mobile ad hoc networks. IEEE Wireless Commun. pp. 85–91 (October 2007)
2. Woungang, I., Dhurandher, S.K., Peddi, R.D., Obaidat, M.S.: Detecting blackhole attacks on DSR-based mobile ad hoc networks. In: IEEE Conference on Computer, Information and Telecommunication Systems (CITS), 14–16 May 2012, pp. 1–5
3. Bindra, G.S., Kapoor, A., Narang, A., Agrawal, A.: Detection and removal of co-operative blackhole and grayhole attacks in MANETs. In: IEEE Conference on System Engineering and Technology (ICSET), 11–12 Serptember 2012, pp. 1–5
4. AL-Shurman, M., Yoo, S.-M., Park, S.: Black hole attack in mobile ad hoc networks. In: ACMSE04, 2–3 April 2004, Huntsville, AL, USA (2004)
5. Sun, B., Guan, Y., Chen, J., Udo, W.: Detecting black-hole attack in mobile ad hoc network. UW Pooch, The institute of Electrical Engineers (2003)
6. Kurosawa, S., Nakayama, H., Kato, N., Jamalipour, A., Nemoto, Y.: Detecting blackhole attack on AODV-based mobile ad hoc networks by dynamic learning method. Int. J. Netw. Secur. 5(3), 338–346 (2007)
7. Deng, H., Li, W., Agrawal, D.P.: Routing security in wireless ad-hoc network. University of Cincinnati, IEEE Communications Magazine (2002)
8. Tamilselvan, L., Sankaranarayanan, V.: Prevention of black hole attacks in MANET. In: J. Clerk Maxwell (ed.) The 2nd International Conference on Wireless Broadband and Ultra Wideband Communications, A Treatise on Electricity and Magnetism, 3rd ed., vol. 2. Oxford, Clarendon, 1892, pp. 68–73 IEEE, New York (2007)

Threat Intelligence Analysis of Onion Websites Using Sublinks and Keywords

Tarun Trivedi, Vinod Parihar, Manas Khatua and B. M. Mehtre

Abstract With advances in dark web technology, cybercrimes are increasing. Onion websites are the main resources of unauthorized crime activities in the dark web. One of the main objectives of cyber threat intelligence (CTI) is to find out popular onion websites which are responsible for cybercrimes. It is imperative but cumbersome to monitor dark world and gather threat intelligence. Government and intelligence agencies manually look for hidden networks and their connections to dark world for building up threat intelligence. However, the existing onion websites use dynamic IP addresses which are difficult to trace. In this paper, we propose a *Threat iNtelligence Tool* (*TnT*) for automatic monitoring of onion websites and build up threat intelligence by predicting their popularity in the dark world. TnT is developed based on two parameters—number of sublinks and keywords—which are collected from every website. The proposed TnT is tested on a set of onion websites presently exist in the dark world. Our testing results extract the most popular onion sites which are the source of information and discussion platform about criminal activities and services in the dark web.

Keywords Threat intelligence · Tor network · Onion website · Dark web

T. Trivedi · V. Parihar
Sardar Patel University of Police Security and Criminal Justice, Jodhpur, India
e-mail: taruntrivedi20@gmail.com

V. Parihar
e-mail: vinodparihar121@gmail.com

M. Khatua (✉)
Indian Institute of Technology, Jodhpur, India
e-mail: manaskhatua@iitj.ac.in

B. M. Mehtre
IDRBT (RBI Institute), Hyderabad, India
e-mail: bmmehtre@idrbt.ac.in

© Springer Nature Singapore Pte Ltd. 2019
A. Abraham et al. (eds.), *Emerging Technologies in Data Mining and Information Security*, Advances in Intelligent Systems and Computing 814,
https://doi.org/10.1007/978-981-13-1501-5_50

1 Introduction

The Onion Routing (Tor) is a free open-source software which provides anonymity in the Internet [4, 16]. It protects users from network surveillance and traffic analysis attacks as well as physical location tracking. Similar difficulty arises in monitoring communications between a user and a server in the dark web as it encrypts data multiple times in a virtual network and bounces the communication through different distributed relays. Therefore, journalists, non-governmental organizations (NGO) and many private organizations use Tor network to improve privacy of their private communications. Military agents also use Tor network to mask the websites which need high privacy, and, thus, protecting them from any external harm. This is one side of the Tor network.

The other side of the network is a dark world. Several activities are performed in the dark world which includes hacking, cybercrime, cyber warfare, anonymity, politics, selling weapons and drugs [1]. Onion websites are provided for exporting the above illegal services which construct dark web in the dark world [12]. However, it is difficult to trace those onion websites in the Tor network as they use dynamic Internet Protocol (IP) address. Therefore, it is important but challenging to filter out those onion websites which are popularly involved in unauthorized activity in the dark web.

One of the main objectives of cyber threat intelligence (CTI) is to detect and prevent any cyber attack generated from dark web. Finding out any indication of such attacks a priori followed by their successful detection has immense importance. Intelligence agencies monitor the activity of hidden services in the dark web. Government agencies check all the services provided by onion websites and try to understand when, where and how cyber attacks happen. Intelligence agencies also monitor proxy-based traffic and data packets which are doubtful [16]. However, these kind of manual investigations are time consuming as well as inefficient. Some methods (e.g. [6, 9, 18]) have been proposed for automated monitoring of Tor networks. However, those methods are limited to recognize popular websites which are direct or indirect source of criminal activities in the dark web.

In this work, we propose a Threat intelligence Tool (TnT) for automated monitoring of onion websites and building up threat intelligence by recognizing the most popular onion websites in the dark web. The TnT finds out popularity of an onion website based on the number of sublinks as well as keywords related to a given objective. The TnT first extracts all the existing sublinks and the keywords present in the given URL, and, then, calculates the page score according to the number of intended keywords found on the onion site. Finally, it computes a score, namely "popularity index" corresponding to the given URL. We propose two versions of the score computation method, namely sublink-based analysis (SLBA) and sublink with keyword-based analysis (SLKBA). We performed experiment using both the methods as well as a benchmark method proposed by Zulkarnine et al. [18]. We observed that TnT can successfully extract the popular websites which are the main sources of criminal activities in the dark web.

2 Literature Survey

The onion sites provide illegal services such as selling of weapons, drugs, pornography and hacking tools in the dark world. According to a study performed by Cox [3], the percentage of hidden services provided by the onion websites of dark world clearly shows that a significant amount of unauthorized activities are going on in the dark world. Those illegal activities do affect the government and private organizations. Therefore, in this paper, our objective is to dynamically analyse the onion websites of the dark world for understanding the existence of illegal contents inside those websites which makes them popular.

2.1 Tor Network

The Tor network is a group of hidden servers that provide security and privacy of a user in a network. The Tor network is designed and implemented by Naval Research Laboratory in USA for providing privacy and security to government communications [4, 16]. Presently, Tor is open source for public use. The Tor hidden services hide the users' locations when they access instant messaging services (IMS) and web publishing services (WPS). Therefore, the Tor browsers are widely used by the journalists, bloggers, whistleblowers, military and hacktivist to hide their activity as well as identity [4, 11].

According to Defence Science Board Report in 2013, hidden web was 400–500 times bigger than the clear web or surface web [5]. The clear web or the surface web uses authorized extensions such as .net, .com, and .org. On the other hand, hidden web services use .onion extensions. Many dark sites provide users anonymity to a user when it does chatting with other users. In this situation, IP addresses cannot be tracked by an attacker who tries to eavesdrop a data packet [8].

2.2 Popularity Ranking

Many intelligence agencies such as Intelligence Bureau (IB) and Federal Bureau of Investigation (FBI) monitor and surveillance on the dark web for preventing and finding out cybercrime in the cyber world. For searching onion sites and their contents, dark web crawler is used as the existing normal crawlers do not work in the dark world [13]. The dark web crawler is an automated tool which can access HTML contents of a webpage. HTML contents provide text, hyperlinks, images and other links of an onion site. Thereafter, lexical analysis is performed on the extracted text for predicting meaningful information out of that [2].

Ghosh et al. [7] suggested keyword-based text analysis in which the keywords provide different categories and services of an onion site. The method of deep traffic inspection has been used for recognizing the pattern of anonymous networks [15]. Zulkarinne et al. [18] proposed a method for searching popular dark web on Tor network. This method is based on graph structure topology of dark web [17]. In this graph, each node represents a dark web and each edge represents a link between two different dark webs. In this work, in-degree and out-degree metrics are used for ranking the dark webs. In-degree represents total number of direct links coming from other dark web, and out-degree represents total number of direct link going out from this website to other website. There exist many other algorithms which can directly access the dark web through the combination of valid input [10]. The benchmark method in [18] tried to find out popularity of a single website. However, this method did not focus on functionality and structure of an onion website. On the other hand, our proposed method concentrates on a single onion website and its sublinks. We find out popularity of an onion website on the basis of functionality (using keywords) and structure (using sublinks) of the website.

3   Proposed Method

We propose a *threat intelligence tool* (TnT) for computing popularity rank of an onion site based on the number of keywords and sublinks available in the site.

3.1   Complete Workflow of TnT

A flowchart of the proposed method executed by the TnT is shown in Fig. 1. The proposed TnT is an advanced crawler which starts crawling process at user-specified website. Thereafter, it retrieves the HTML content of the pages corresponding to the given URL, analyse them, and compute the popularity page score for each webpage. According to the objective, keywords are selected a priori from the discussion and the HTML contents of the webpages. Whenever the TnT crawls on home page of an onion site, it extracts all sublinks corresponding to the intended keywords such as hacking, tracking, trafficking, attacking, and DoS [8, 18], and returns output to advanced filtering module. The objective of advanced filtering module is to get rid off of irrelevant and redundant content as well as the hyperlinks to other domain and images. It also removes duplicate sublinks. For example, any repeated sublinks, unintended webpage links, and others links associated to .css, .xml, .jpg, .jpeg, and .ico are filtered out. We did this filtering as we analyse each webpage and compute its score separately. The popularity of a URL depends on the types of product available in this domain but not in other domain. The TnT counts the number of sublinks corresponding to each intended keyword exist in each webpage. For performing this count, TnT applies text mining procedure as explained in [2]. Finally, we construct a network tree for the user-specified URL and compute a cumulative score corresponding to the URL for predicting its popularity ranking.

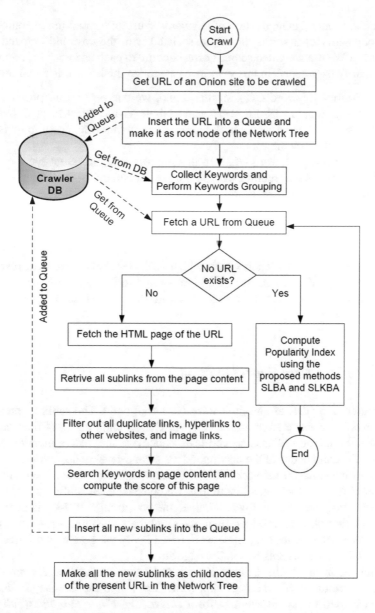

Fig. 1 Complete workflow of the proposed tool TnT

3.2 Grouping of Keywords

In this work, the keywords are the names of different "products" and "services" of an onion site. Those keywords are divided into three categories on the basis of their *costs* and *demands*. Categorization of keywords is useful for ranking the onion sites as the popularity of a site is directly associated with the cost and demand of its product [7]. We use weighted scores corresponding to each category for computing the popularity page score. In this work, we categorize the keywords as follows:

1. High Priority Keyword (K_h): In this group, we put very costly products such as weapons (e.g. FLIR LS32 Thermal Night Vision Black, FN SCAR 17S 16.25 7.62 × 51) and drugs (e.g. Bluegrass Armory Moonshiner Bullpup, Heroin, Cocaine, Flakka, Zombie).
2. Medium Priority Keyword (K_m): In this group, we put different types of shops in dark world such as mobile, safe email, hosting, drugs, weapons, malware and different discussion, chat, blogs.
3. Low Priority Keyword (K_l): In this group, we put normal keywords which are commonly used by onion sites such as bitcoin, comment, hack, anonim, market, onion, dark web, hidden and money.

We assign a value for each group which indicates their costs with respect to other products. For example, we give higher value to the keywords belong to the group K_h and lower value to the keywords in K_l. In brief, $\delta_h > \delta_m > \delta_l$ where $\delta_h + \delta_m + \delta_l = 1$.

3.3   Tree Structure Representation

We represent the URL of an onion website as root node. The sublinks present in the main onion website are represented by next-level child nodes of the tree. Subsequently, the sublinks of each second-level nodes (i.e. sublinks of main onion site) represent the child nodes of the corresponding second-level node, and so on. In this process, we construct the network tree having maximum N number of nodes in which the limit of N is user specified. Note that, we consider each sublink only one time for constructing the network tree, and, thus, the construction of loop is avoided. In brief, each sublink is represented by a node and the connection between two sublinks is represented by the edge of network tree. Finally, we get a network tree T and computes the score as explained in the Algorithm 1.

In the Algorithm 1, the *Extract_Keywords(U)* function is used for mining the existing keywords inside the page and form *U.text*, and the *Extract_Sublinks(U)* function is used to extract sublinks from *U.text*. The *Filter_Sub_links(Sub_links)* function is used for filtering repeated sublinks, unintended webpage links, and other links associated to .css, .xml, .jpg, .jpeg, and .ico. Finally, the algorithm returns a network tree T corresponding to the user-specified URL. Thereafter, we propose two different methods for analysing the constructed network tree T. First one is based on sublinks of the given onion website, namely *Sublink-based Analysis* (SLBA). In SLBA method, we increase the *Page_score* corresponding to each node by unity. The second method is based on both the keywords and their corresponding sublinks, namely *Sublink with Keyword-based Analysis* (SLKBA). In SLKBA method, page

---

Algorithm 1 Function Create Tree: *Create_Tree(.)*

Inputs: *URL* : URL of an onion website; Q : empty queue for storing N elements; three constants: $\delta_h, \delta_m, \delta_l$; K_h : list of hight priority keywords; K_m : list of medium priority keywords; K_l : list of low priority keywords.

Output: T : Tree of Onion Website.

1: *T is empty tree.*
2: *T.insert(URL)*   ▷ Add URL as a root node in tree T
3: *Q.insert(URL)*
4: **while** Q is not empty **do**
5:     *U = Q.delete()*
6:     *Fetch the home HTML page of U*
7:     **if** SLBA method is used **then**
8:         *Page_score = 1*
9:     **end if**
10:    **if** SLKBA method is used **then**
11.   *U.text = Extracts_Keywords(U)*
12:        **if** *any K_h exists in U.text* **then**
13:            *Page_score = Page_score + δ_h*
14:        **end if**
15:        **if** *any K_m exists in U.text* **then**
16:            *Page_score = Page_score + δ_m*
17:        **end if**
18:        **if** *any K_l exists in U.text* **then**
19:            *Page_score = Page_score + δ_l*
20:        **end if**
21:    **end if**
22:    *Sub_links = Extract_Sublinks(U)*
23:    *Sub_links = Filter_Sub_links(Sub_links)*
24:    *T.insert(U, Sub_links)*
25:    *T.U.Page_score = Page_score*
26:    *Q.insert(Sub_links)*
27: **end while**

---

score increment for each node depends on the type of keyword as explained in Algorithm 1. Finally, we compute the value of a new metric *Popularity Index* (M_F) following the method as discussed in the following subsection.

3.4 Popularity Index Computation

In both the versions of the proposed method for tree structure analysis, we count the number of leaf nodes, and, then, calculate popularity index corresponding to the constructed tree of the given onion site. Each leaf node computes cumulative page score which is the sum of individual page scores of its predecessor nodes including itself. Finally, Eq. (1) computes the cumulative score corresponding to the given URL by summing up the cumulative scores of all leaf nodes where L is the total number of leaf nodes in the tree, P_i is the total number of predecessor node of ith leaf node,

$Page\_score_i$ is the popularity page score of the ith leaf node, and $Page\_score_{i,j}$ is the popularity page score of the jth predecessor node of the ith leaf node. The result of this summation yields different values for different network trees depending on the structure of the network tree of an onion website. Therefore, there is a need of normalizing this cumulative score. We normalize the value of $F(T)$ in Eq. (2). For normalizing the value computed in Eq. (1), we divide the computed value of $F(T)$ by its upper limit which equals $\lceil \frac{N}{2}(\frac{N}{2}+1) \rceil$ where N is the number of nodes of the tree. We show the proof of this upper limit in Lemma 1.

$$F(T) = \sum_{i=1}^{L} \left(\text{Page\_score}_i + \sum_{j=1}^{P_i} \text{Page\_score}_{i,j} \right) \qquad (1)$$

$$\text{Popularity Index}(M_F) = \frac{F(T)}{\lceil \frac{N}{2}(\frac{N}{2}+1) \rceil} \qquad (2)$$

Lemma 1 *Let us consider that a tree T is an N-ary tree having L leaf nodes. In that tree, the ith leaf node has P_i predecessor nodes, and the total number of nodes of the tree equals N. We can show that the maximum value of $\sum_i^L (P_i + 1)$ equals $\lceil \frac{N}{2}(\frac{N}{2}+1) \rceil$.*

Proof According to the tree structure representation of a given URL, as discussed in Sect. 3.3, a node in the constructed tree T could have any number of child nodes. So, we consider the tree as N-ary tree where N is the total number of nodes of the tree. Let L is the total number of leaf nodes of the tree, and P_i is number of predecessor nodes for ith leaf node. According to the Algorithm 1, the maximum Page_score for a node equals unity. Therefore, we rewrite the value of $F(T)$ equals $(\sum_i^L P_i + L)$. Now, we would have the maximum value for $F(T)$ when all the P_i as well as the L will having maximum value. In other words, all the leaf nodes are at same as well as in highest level, and $P_i \approx L$. As we have N numbers of nodes in the tree, we could have max L as well as max P_i when L and P_i are approximately equal to $\frac{N}{2}$. In fact, the maximum possible values of L is $\lceil \frac{N}{2} \rceil$ for maximizing the $F(T)$. On the other hand, the maximum possible value of P_i is also $\lceil \frac{N}{2} \rceil$ for the same. Therefore, we can conclude that the maximum value of $F(T)$ is equal to $\lceil \frac{N}{2}(\frac{N}{2}+1) \rceil$. Hence, we conclude the proof.

4 Result and Analysis

4.1 System Configuration and Dataset

The proposed tool is implemented using a Windows machine with i5 processor, 4GB RAM and 500 GB hard disk. This tool is developed using Python programming language, and therefore, it works on any platform. In our analysis, we used a list of 4320 onion sites present in the dark world, specifically those sites are highly used for attacking purpose [14].

4.2 Evaluation Using the TnT

Zulkarinne et al. [18] proposed a method for finding out illicit and criminal content in dark web based on in-degree and out-degree. The number of in-degree and out-degree of an onion site decides its raking having criminal content. We have implemented this method to compare its performance with our proposed method. The in-degree and out-degree values for the given onion sites are calculated, and only the values of six most popular websites are shown in Tables 1 and 2, respectively. In-degree result, as shown in Table 1, shows highest ranking to a URL of category "services". This is because all the onion site owners would like to have a link to a website which

Table 1 Popular Onion website based on in-degree centrality

Website	URL	Category	$< K_{in} >$	Rank
Hidden answer 1	http://answersbbddrdcwo.onion/	Blogs and comments	0.0020	3
Hidden answer 2	http://answerstedhctbek.onion/	Blogs and comments	0.0010	4
0 day forum	http://qzbkwswfv5k2oj5d.onion/	Services	0.4445	1
The hidden wiki	http://wikitjerrta4qgz4.onion/	Hidden link	0.0226	2
The undernet directory	http://underdj5ziov3ic7.onion/	Link provider	0.0007	5
Exodus	http://exoduockgfq3ikf7.onion/	Hacking tool	0.0005	6

Table 2 Popular Onion website based on out-degree centrality

Website	URL	Category	$< K_{out} >$	Rank
Hidden answer 1	http://answersbbddrdcwo.onion/	Blogs and comments	0.0296	3
Hidden answer 2	http://answerstedhctbek.onion/	Blogs and comments	0.0131	4
0 day forum	http://qzbkwswfv5k2oj5d.onion/	Services	0.5556	2
The hidden wiki	http://wikitjerrta4qgz4.onion/	Hidden link	0.7474	1
The undernet directory	http://underdj5ziov3ic7.onion/	Link provider	0.0016	6
Exodus	http://exoduockgfq3ikf7.onion/	Hacking tool	0.0021	5

provides more services. But, it does not mean that a website connected to maximum number of websites is most popular. Similar analysis for out-degree result, as shown in Table 2, shows that the website having huge hidden links could get highest rank, but, does not indicate popularity either.

The TnT creates a tree while crawls on different webpages of the user-specified URL. Constructed tree shows the connectivity between the URL and its different levels of sublinks. While crawling to different sublinks, if any of the intended keywords is not present with a sublink then the *Page_score* of the sublink is considered zero in the SLKBA method. Otherwise, its score is computed following the computation method SLBA as discussed in Algorithm 1. In this evaluation, we used 4320 onion websites and computed the popularity score of each website using both the SLBA and SLKBA methods. Tables 3 and 4 show the popularity scores of six most popular websites. We observed that both the methods extract the onion website which belongs to "blogs and comments" category. This result is as per our expectation as blog websites are more popular in the dark world. This experiment shows that the TnT is more suitable for finding out popular websites in dark web. Further, we compare

Table 3 Popular onion websites based on SLBA

Website	URL	Category	M_F	Rank
Hidden answer 1	http://answersbbddrdcwo.onion/	Blogs and comments	0.1762	1
Hidden answer 2	http://answerstedhctbek.onion/	Blogs and comments	0.1092	2
0 day forum	http://qzbkwswfv5k2oj5d.onion/	Services	0.1007	3
The hidden wiki	http://wikitjerrta4qgz4.onion/	Hidden link	0.0876	4
The undernet directory	http://underdj5ziov3ic7.onion/	Link provider	0.0746	5
Exodus	http://exoduockgfq3ikf7.onion/	Hacking tool	0.0291	6

Table 4 Popular onion websites based on SLKBA

Website	URL	Category	M_F	Rank
Hidden answer 1	http://answersbbddrdcwo.onion/	Blogs and comments	0.0774	2
Hidden answer 2	http://answerstedhctbek.onion/	Blogs and comments	0.0934	1
0 day forum	http://qzbkwswfv5k2oj5d.onion/	Services	0.0269	5
The hidden wiki	http://wikitjerrta4qgz4.onion/	Hidden link	0.0274	4
The undernet directory	http://underdj5ziov3ic7.onion/	Link provider	0.0448	3
Exodus	http://exoduockgfq3ikf7.onion/	Hacking tool	0.0179	6

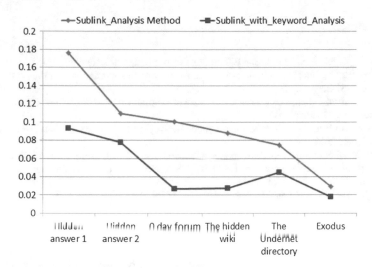

Fig. 2 Comparison between the SLBA and the SLKBA

the results of SLBA and SLKBA in Fig. 2. This comparison shows that both the SLBA and SLKBA methods are almost equivalently efficient for finding out popular website in the dark world.

5   Conclusion

In this work, we have discussed the challenges for identifying popular onion websites in the dark web, and, then, proposed a threat intelligence tool TnT which is useful for automatic collection of threat intelligence from the hidden dark web. The collected intelligence would be useful for the intelligence agencies such as FBI in USA and IB in India for close monitoring and/or blocking the websites if any illegal activity is observed. The TnT is developed following the basic idea which says that the popularity of a onion website increases if it provides multiple services to the clients. We capture this attribute by considering the functionality and structure of an onion website. We have tested the performance of TnT tool on 4320 onion websites of the dark web. The results show that we are able to successfully extract and identify the most popular onion websites which are acting as the source of information and discussion platform about criminal activities and services in the dark web.

References

1. Antonopoulos, A.M.: Mastering Bitcoin. O'Reilly Media, Inc., Newton (2015)
2. Barrio, P., Gravano, L.: Sampling strategies for information extraction over the deep web. Inf. Process. Manage. **53**(2), 309–331 (2017). (Elsevier)
3. Cox, J.: Study claims dark web sites are most commonly used for crime (February 2016). Accessed on 19 June 2017
4. Dredge, S.: What is Tor? A beginner's guide to the privacy tool (November 2013). Accessed on 19 June 2017
5. DSB: Resilient military systems and the advanced cyber threat, January 2013. [Online] Available in http://nsarchive.gwu.edu/NSAEBB/NSAEBB424/docs/Cyber-081.pdf. Accessed on 25 Nov 2017
6. Fu, T., Abbasi, A., Chen, H.: A focused crawler for Dark Web forums. J. Am. Soc. Inf. Sci. Technol. **61**(6), 1213–1231 (2010)
7. Ghosh, S., Porras, P., Yegneswaran, V., Nitz, K., Das, A.: ATOL: A framework for automated analysis and categorization of the Darkweb Ecosystem. In: Proceedings of the AAAI-17 Workshop on Artificial Intelligence for Cyber Security, San Fransisco, USA (February 2017)
8. Greenberg, A.: Hacker lexicon: What is the dark web? (November 2014). Accessed on 19 June 2017
9. Guitton, C.: A review of the available content on Tor hidden services: the case against further development. Comput. Hum. Behav. **29**(6), 2805–2815 (2013)
10. He, B., Patel, M., Zhang, Z., Chang, K.C.-C.: Accessing the Deep Web. Commun. ACM **50**(5), 94–101 (2007)
11. Johnson, A., Syverson, P., Dingledine, R., Mathewson, N.: Trust-based anonymous communication: adversary models and routing algorithms. In: Proceedings of the 18th ACM Conference on Computer and Communications Security
12. McMillan, R.: Definition: threat intelligence (May 2013). Accessed on 19 June 2017
13. Olston, C., Najork, M.: Web crawling. Found. Trends Inf. Retrieval **4**(3), 175–246 (2010)
14. Raghavan, S., Garcia-Molina, H.: Crawling the hidden web. In: Proceedings of the 27th International Conference on Very Large Data Bases
15. Shaikh, Z.A., Harkut, D.: An overview of network traffic classification methods. Int. J. Recent Innovation Trends Comput. Commun. **3**(2), 482–488 (2015)
16. Tor. Tor: Overview (September 2002). Accessed on 19 June 2017
17. Xu, J., Chen, H.: The topology of dark networks. Commun. ACM **51**(10), 58–65 (2008)
18. Zulkarnine, A.T., Frank, R., Monk, B., Mitchell, J., Davies, G.: Surfacing collaborated networks in dark web to find illicit and criminal content. In: Proceedings of IEEE Conference on Intelligence and Security Informatics (ISI), pp. 109–114, Tucson, AZ, USA (2016)

Controllability of Network: Identification of Controller Genes in a Gene–Gene Interaction Network

Anjan Kumar Payra, Anupam Ghosh and Pabitra Mitra

Abstract Lung cancer is the one of the most deadly diseases because of its constant change in biological behavior with respect to different infected patients. So, Identification of genes responsible for lung cancer is a major challenging task. This detection of genes can be done using gene–gene interaction network where each node represents a specific gene, whereas interconnection between them is considered as edges. Here, in this work, we have proposed a methodology which basically detects controller genes from gene–gene interaction network. Initially, our proposed work is started from gene expression data of lung cancer which consists of 7129 genes samples with their corresponding numeric values at different time slots both for normal and disease state. Out of 7129 genes samples, 3556 variant set of genes (i.e., the over or under expressive gene set) have been filtered out by the application of two-tailed t-test. Finally out of 3556 gene samples, 122 genes are identified as controller genes through the application of K-means clustering, finding the best cluster of the set of genes by counting maximum number of occurrences of functional groups or GO terms for each genes in clusters, formation of gene–gene interaction network (using String DB), and finally pruning of nonessential genes through node and edge weight. 122 predicted controller genes are also validated with the existing data of NCBI, and an accuracy of 91.8% has been obtained which will be discussed in the result and analysis section.

A. K. Payra (✉)
Department of Computer Science & Engineering, Dr. Sudhir Chandra Sur
Degree Engineering College, Dumdum, Kolkata 700074, India
e-mail: anjan.payra@gmail.com

A. Ghosh
Department of Computer Science &-Engineering, Netaji Subhash Engineering College,
Garia, Kolkata 700152, India
e-mail: anupam.ghosh@rediffmail.com

P. Mitra
Department of Computer Science & Engineering, Indian Institute of Technology,
Kharagpur 721302, India
e-mail: pabitra@gmail.com

© Springer Nature Singapore Pte Ltd. 2019
A. Abraham et al. (eds.), *Emerging Technologies in Data Mining and Information
Security*, Advances in Intelligent Systems and Computing 814,
https://doi.org/10.1007/978-981-13-1501-5_51

Keywords Node weight · Edge weight · Controller genes · Clustering
GO-attribute

1 Introduction

Lung cancer is difficult to diagnose if its symptoms cannot be understood completely.
Symptoms usually detect cancer when it is too late to cure. Lung cancer is so serious
that almost 1.3 million new cases have been registered for diagnose/year and 1.17
million died/year worldwide [1] out of which 58% are from developing countries
[2]. There are two categories of lung cancer non-small cell lung cancer (NSCLC)
(almost 80–85%) and small cell lung cancer (SCLC) (almost 15–20%). They are
opted based on histological differences, response to chemotherapy, susceptibility to
lung cancer, active smoking, pollution due to solid fuels, etc. [3].

There are multiple genetic factors that play an important part in lung cancer. So,
researches have been initiated to predict controller or candidate genes that conferred
risk of lung cancer. Recent trends of experiments have been also extended toward
function prediction of genes that actually control the entire network. Liu et al. [4]
introduced maximum matching approach in a directed network. While the concept
of minimum dominating set (MDSets) to control undirected network was proposed
by Nacher and Akutsu [5]. Jeong et al. [6] and Yu et al. [7] suggested the concept of denser network [8, 9], and their importance in complex involvement [10].
While similar properties like, degree centrality for counting the direct neighbors and
betweenness centrality for counting the shortest path in a network have been also
taken under consideration in certain works [11–13] for controllability of network.

Controllability of a network means control functionality, which is essential to
restrict the disease. The term "controller gene" has been introduced in this work for
observing the controllability of complex networks. Nodes that control other nodes in
its associated network are called as driver nodes or controller. A set of genes (driver
nodes or controller genes) which participate more frequently in an interacting dense
network have obviously higher participation scoring values in comparison with the
other genes. So, the other genes should be dependent on that set of controller genes
for different functionality. Hence, proper selection of controller genes will lead to
the control other genes in the network too. In a nutshell, it can be said that proper
observation and detection of controller genes are needed to persuade any deadlier
impact resulting from them.

2 Relevant Tools and Terms

The prediction of controller is a depending factor on the nature of data present in gene
expression dataset (GED) and gene–gene interaction network (GGIN). GED values
of human lung cancer are collected from web-based repositories. The variant set of

genes is obtained using two-tailed t-test [14]. Weka [15], R [16] are used to obtain similar sets of gene together from variant set of genes. Hard partitioning algorithm is used for grouping and validation of generated results, whereas FuncAssociate [17] is used to get enriched GO-attributes. It is really challenging to predict controller genes as all clusters show different functionality. Network statistics are calculated using String DB [18]. Controller genes are predicted clusterwise using edge and node weight to control GGIN and GED. The edge weight (W_{mn}) [19] is used to find the similarity between the nodes m and n in a network. It is rightly so that two connecting nodes can be present in same cluster if they are similar in activity. The similarity is measured using Jaccard's coefficient. In an interaction network (G_v), there are some nodes with degree 1 are often considered to be less essential nodes according to reliability measures [19]. So nodes with degree 1 and corresponding edges are eliminated from the network. The remaining subgraph of the network is marked as G'_v. The node weight (W_v) of node $v \in V$ in a network is the average degree of all nodes in G'_v

2.1 Gene–Gene Interaction Network

A gene interacts with another gene in order to perform some functionality. A gene—gene interaction network is defined as a collection of nodes (genes) connecting with each other with edges. Different controller genes and its corresponding different properties considered in this work are highlighted in Table 1 and Fig. 1. Genes RPL9, RPL8, RPL11, RPL12, RPL13, RPL19, RPL35, RPS9, RPS8, RPS16, and RPS21 are directly or indirectly dependent on controller gene RPL32 in Fig. 1.

Table 1 Different types of gene–gene interaction network properties like degree, edge–node weight, centralities are used in the above undirected sample network of our methodology

Properties	Values/Instant nodes		
Max degree	RPL32		11
Edge weight	<RPS9, RPL13, 0.2>, <RPL32, RPS9, 0.133> etc.		
Node weight	<RPL32, 1.077>, <RPL13, 1.0> etc.		
Controller set (using centralities)	Degree	RPL32	11
		RPS9, RPL8	3
	Betweenness	<RPL32, 100.0>	
	Closeness	<RPL32, 1>, <RPS9, 0.59>, <RPL8, 0.59>	
	Information	<RPL32, 2.89>, <RPS9, 1.73>, <RPL8, 1.73>	

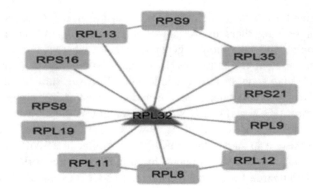

Fig. 1 Sample gene–gene interaction network, where rectangle boxes, solid straight lines, and triangle represent nodes, edges, and controller gene, respectively

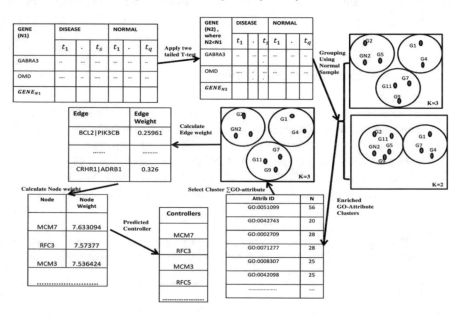

Fig. 2 Prediction of controllers (workflow) using edge–node weight

3   Methodology

Our entire work comprises of five distinct sections. These sections are applied on lung adenocarcinoma gene expression datasets of human. All these methodologies are outlined below as well as highlighted in Fig. 2.

Algorithm: contl_pred_ENweight
(Select genes from GED and GGIN whose are predicted as controller set of genes)

Input: $\bar{g} = \{\overline{g_1}, \overline{g_2}, .., \overline{g_i}, \ldots, \overline{g_n}\}$ is a set of n genes. D is dataset of sample collection of $\bar{g}$ genes.

$\bar{g}_i = \{\bar{g}_i^{dis}, \bar{g}_i^{nor}\} = \{d_{i1}, d_{i2}, .., d_{ij}, .., d_{im}\}$ where d_{ij} is jth gene expression data (GED) sample of ith gene in D.

$\bar{g}_i^{nor} = \{d_{i1}, d_{i2}, .., d_{ij}, .., d_{is}\}$ where total s numbers of normal sample collection of ith gene in D.

$\bar{g}_i^{dis} = \{d_{i1}, d_{i2}, .., d_{ij}, .., d_{iq}\}$ where, total q numbers of tumor/disease sample collection of ith gene in D.

Output: Rank list(R1) of controller genes

Start

//Preprocessing and two-tailed t-test:
Remove duplicate gene entry of $\overline{g_l}$ and obtain new dataset of D, where name of the gene $(\overline{g_i})$ is equal to gene $(\overline{g_l})$ in $\bar{g} = \{\overline{g_1}, \overline{g_2}, .., \overline{g_i}, \ldots, \overline{g_l}, \ldots, \overline{g_n}\}$.

Gene expression profile is represented as $n \times m$ or $n \times m$ microarray where n is number of genes and m is time variant samples of genes.

$m = (Tumour\ sample\ set(s) + normal\ sample\ set(q))$.
Calculate means of two samples m_1, m_2, respectively.
Calculate degree of freedom $(df) = (s + q - 2)$.
Estimate standard deviation (sd) and t value of a gene. $t = (m_1 - m_2)/sd$
Select the variant set of genes (R) with respect to critical value (t_{crit}) for specific probability cutoff value.

//Grouping or clustering:
Generate hard clustering using normal dataset of obtained genes in (R).
Validate clusters using standard criterion indices (like Dunn, DB index, and XB-index).

//Function Annotation and Selection:
Threshold parameter (p) value initially set to 5×10^{-6}.
Consider $c_1, c_2, c_3, \ldots\ldots, c_k$ to be the clusters with their respective enriched attribute counts $c_1^{count}, c_2^{count}, .., c_k^{count}$ where k is the number of clusters and $2 \le k \le n$. Consider $c_1^{count}, c_2^{count}, \ldots\ldots, c_k^{count} \ge p$.

Total count of enriched attribute is $C_k^{total} = \sum\limits_{i=1}^{m} c_i^{count}$.

Select cluster (C_{select}) by evaluating largest value of total count.
$C_{select} = Max(C_k^{total})$, where $2 \le k \le n$.

//Find Neighbors:
Consider w to be the total number of genes $(g_1, g_2, \ldots..g_w)$ present in selected cluster (C_{select}). An interaction of gene pair is considered where both genes should present in C_{select}. All interactions of the gene pairs are collected from String DB repository. When g_i interacts with g_j then $i! = j$ and $1 \le i, j \le w$. GGIN dataset (d_1) is also generated using gene pair (g_i, g_j).

//Controller of Clusters:
Calculate edge weight (W_{uv}) of the network using dataset (d_1). $W_{uv} = \frac{N(u) \cap N(v)}{N(u) \cup N(v)}$, where $N(u)$ and $N(v)$ are neighbors of nodes u and v, respectively. $N(u) \cap N(v)$ represents all common neighbors of u and, and $N(u) \cup N(v)$ represents all distinct neighbors of nodes u and v.
Find node weight (W_v) of node v using calculated edge weight (W_{uv}).
$W_v = \frac{\sum_{x \in V''} \deg(x)}{|V''|}$, where V'' is the set of nodes in G'_v. $|V''|$ is the number of nodes in G'_v. And $\deg(x) \ge 2$ is the degree of a node $x \in V''$ in W_v.
Gene rank list (R1) is obtained using descending value of node weight (W_v) based on threshold probability value $(\ge \alpha)$.
End

4   Results and Discussion

All preprocessed instances are used for clustering to find homologous set of collection for better usability with respect to functionality. There are ($s = 86$) time-variant sample of tumor and ($q = 10$) of normal sample with 7129 gene entries in the dataset. In preprocessing, almost 3% of repeated gene instances of entire datasets are eliminated. The important observation is to find out the set of genes which change significantly as compared to the rest of the genes in the dataset. Student distribution or t-test is used to find the variant set of genes. The results are given in Table 2. Almost 50% genes are selected as variant set of genes using two-tailed t-test approach. The filtered set of genes is used to calculate similarity measures using clustering methodologies. DBSCAN, K-means, and hierarchical clustering, etc., algorithms are applied on dataset. Both DBSCAN and hierarchical approaches have not given positive result for our dataset. The distribution of dataset is not suitable for DBSCAN since the unclustered points in DBSCAN remains unaltered with midpoint values. Results of both hierarchical and K-means are validated using standard criterion indices (see Table 3). The performance of K-means is best in comparable to the rest of the two considered algorithms. So, DBSCAN and hierarchical clustering algorithms are not selected to extend our work.

Table 2 Result of t-test, where p is probability value

Degree of freedom = 86 + 10 − 2 = 94	Gene ($p = 0.05$)	Gene ($p = 0.01$)	Gene ($p = 0.001$)	Two-tailed
Count	2471	1862	1337	3556

Table 3 Criterion values of hierarchical and K-means clustering algorithms

K	Algorithms	Ball Hall	Davies–Bouldin	Dunn Ratio	Xie-Beni
2	Hierarchical	149513391	0.637	0.00979	16.96
	K-means	171819717	0.500	0.02423	2.92
3	Hierarchical	100365351	0.691	0.00158	586.43
	K-means	139004118	0.593	0.00909	16.83
4	Hierarchical	104999431	0.485	0.00207	264.27
	K-means	106326013	0.657	0.00360	79.02
5	Hierarchical	84154207	0.513	0.00207	242.12
	K-means	85013543	0.418	0.00465	45.88
6	Hierarchical	70151085	0.477	0.00079	1643.50
	K-means	105400722	0.636	0.00204	213.28
7	Hierarchical	61284242	0.516	0.00079	1314.30
	K-means	90304862	0.185	0.00199	213.73
8	Hierarchical	53653414	0.515	0.00079	1287.82
	K-means	79515382	0.676	0.00174	250.36

From Table 3, it is clear that K-means algorithm outperforms the rest. So, simple K-means clustering algorithm using Euclidian distance is implemented to create clusters. All generated clusters are then evaluated using web-based tool Weka and R through the estimation of various cluster properties like centroids, percentages of instances. Sum of squared error variations are also computed during change of K in clusters to detect the best suitable one among them. Detailed analysis of generated approximate sum of squared errors is highlighted in Fig. 3.

All statistical information is collected after processing of Weka, but results are not satisfactory enough to select K-value of K-means algorithm with respect to sum square error. So, other different standard cluster criterion values [16] are also calculated which is given in Table 4.

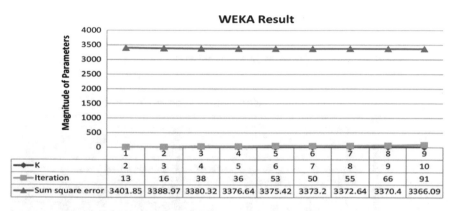

Fig. 3 Sum of squared errors of clusters for $K = 2, 3 \dots 10$

Table 4 Cluster criterion indices and best clusters, respectively

Indices	Best criterion value	Cluster (2–10)
Banfield–Raftery	−INF	6
C index	0.009	5
Calinski–Harabasz	6239.387	3
DB index	0.657	4
Det_Ratio index	−17.660	5
Dunn index	0.024	2
Gamma	0.997	2
McClain–Rao	0.066	2
Point-biserial	−2683.68	10
Ratkowsky–Lance	0.565	2
Silhouette	NA	NAN
Wemmert–Gancarski	0.933	2
Xie-Beni	2.929	2

The criterion values in Table 4 are also simultaneously not suitable as before for the best K-value in K-means algorithm. So, FuncAssociate is selected both for the computation and annotation of enriched GO-attribute and p value, respectively. Then, set of clusters having maximum number of functionality are selected depending on the constraint $p \geq 5 \times 10^{-6}$.

The results of FuncAssociate have been displayed in Fig. 4. From Fig. 4, it is clearly visible that for $K = 6$, the total number of generated GO-attributes is maximum if compared to the others, i.e., 181 with an average enriched GO of 30.17.

Depending on higher enriched GO-attribute values in Table 5, C_2, C_6 are selected. String DB gene–gene interconnection data are used to form GGIN for each of the selected cluster. Then the interactions in the GGIN are used to find the controller genes in each cluster using probability, edge weight, node weight, and pruning. In experimental lung cancer dataset, we figure out the set of controllers genes. Some of them are listed here MCM7, RFC3, MCM3, RFC5, PCNA, and FEN1. Predicted results are validated using existing NCBI lung cancer data repositories. The precision value of our prediction is 91.8% as given in Table 6.

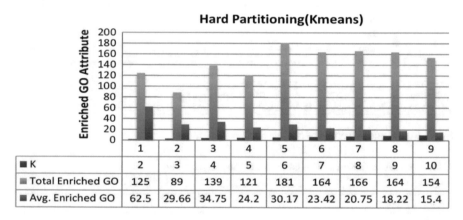

Hard Partitioning(Kmeans)

	1	2	3	4	5	6	7	8	9
K	2	3	4	5	6	7	8	9	10
Total Enriched GO	125	89	139	121	181	164	166	164	154
Avg. Enriched GO	62.5	29.66	34.75	24.2	30.17	23.42	20.75	18.22	15.4

Fig. 4 Bar chart of enriched GO-attributes (Tot-Total, Avg-Average, K-number of clusters)

Table 5 Network statistics of lung cancer dataset with $K = 6$, means $C = \{ C_1, C_2, ..., C_6 \}$

Lung cancer expression dataset ($K = 6$, $\sum$ GOAttribute $= 181$)

Cluster no.	Avg. node degree	Avg. local clustering coeff.	Total enriched GO-attribute	Selected cluster (GO-attribute $\geq \Omega$)
C_1	8.31	0.478	18	
C_2	10.2	0.557	8	
C_3	17.2	0.369	86	C_3
C_4	9.05	0.838	6	
C_5	1.2	0.567	2	
C_6	7.12	0.347	61	C_6

Table 6 Accuracy measurement of predicted controller gene sets using statistical measures

Generated statistics of our methodology	Scores/Values
Total genes count rank list	122
Responsible genes count in NCBI	122
Common genes	112
Precision	91.8%
Recall	91.8%
F score	0.91
G score	0.91

5   Conclusion

Both SCLC and NSCLC are alarming with higher risk for men and women. The nature of macromolecules (genes) can eradicate the aggressiveness of lung cancer. So, it is essential to detect controller genes in GGIN. Here, we initially consider GED of human lung cancer sample. Two-tailed t-test is applied on it, to figure out most variant set of genes. These variant sets of genes have been used to form cluster in order to find homogenous set of genes. The cluster results are functionally annotated to select clusters with higher enriched GO-attributes. Once annotation is done, and then GGIN has been formed for each of the selected clusters and hence, clusterwise network statistics has been also calculated for them using String DB. Finally, controller genes are predicted using edge and node weight in GGIN. In our proposed work, total 122 genes are predicted as controller genes out of 7129 genes. These 122 predicted controller genes are also validated against NCBI data from which an accuracy level of 91.8% has been obtained.

References

1. Ferlay, J., Shin, H.R., Bray, F., Forman, D., Mathers, C., Parkin, D.M.: Estimates of worldwide burden of cancer in 2008: GLOBOCAN 2008. Int. J. Cancer **127**, 2893–2917 (2010)
2. Barrera, R., Morales Fuentes, J.: Lung cancer in women. Lung Cancer Targets Ther. **48**, 79 (2012)
3. Larsen, J.E., Minna, J.D.: Molecular biology of lung cancer: clinical implications (2011)
4. Liu, Y.-Y., Slotine, J.-J., Barabási, A.-L.: Controllability of complex networks. Nature **473**, 167–173 (2011)
5. Nacher, J.C., Akutsu, T.: Analysis on controlling complex networks based on dominating sets. J. Phys: Conf. Ser. **410**, 12104 (2013)
6. Jeong, H., Mason, S.P., Barabási, A.L., Oltvai, Z.N.: Lethality and centrality in protein networks. Nature **411**, 41–42 (2001)
7. Yu, H., Kim, P.M., Sprecher, E., Trifonov, V., Gerstein, M.: The importance of bottlenecks in protein networks: correlation with gene essentiality and expression dynamics. PLoS Comput. Biol. **3**, 713–720 (2007)
8. Wuchty, S.: Controllability in protein interaction networks. Proc. Natl. Acad. Sci. USA **111**, 7156–7160 (2014)

9. Zhang, X.F., Ou-Yang, L., Zhu, Y., Wu, M.Y., Dai, D.Q.: Determining minimum set of driver nodes in protein-protein interaction networks. BMC Bioinform. **16**, 146 (2015)

10. Barabasi, A.-L., Oltvai, Z.N.Z.N., Barabási, A.-L.: Network biology: understanding the cell's functional organization. Nat. Rev. Genet. **5**, 101–113 (2004)

11. Yu, H., Braun, P., Yildirim, M.A., Lemmens, I., Venkatesan, K., Sahalie, J., Hirozane-Kishikawa, T., Gebreab, F., Li, N., Simonis, N., Hao, T., Rual, J.-F., Dricot, A., Vazquez, A., Murray, R.R., Simon, C., Tardivo, L., Tam, S., Svrzikapa, N., Fan, C., de Smet, A.-S., Motyl, A., Hudson, M.E., Park, J., Xin, X., Cusick, M.E., Moore, T., Boone, C., Snyder, M., Roth, F.P., Barabási, A.-L., Tavernier, J., Hill, D.E., Vidal, M., Yıldırım, M.: High-quality binary protein interaction map of the yeast interactome network. Science **322**, 104–110 (2008)

12. Freeman, L.C.: A set of measures of centrality based on betweenness. http://www.jstor.org/stable/3033543?origin=crossref (1977)

13. Vinayagam, A., Gibson, T.E., Lee, H.-J., Yilmazel, B., Roesel, C., Hu, Y., Kwon, Y., Sharma, A., Liu, Y.-Y., Perrimon, N., Barabási, A.-L.: Controllability analysis of the directed human protein interaction network identifies disease genes and drug targets. Proc. Natl. Acad. Sci. USA **113**(18), 4976–4981 (2016). https://doi.org/10.1073/pnas.1603992113

14. Kim, T.K.: T test as a parametric statistic. Korean J. Anesthesiol. **68**, 540–546 (2015)

15. Holmes, G., Donkin, A., Witten, I.H.: WEKA: a machine learning workbench. In: Proceedings of ANZIIS '94—Australian New Zealand Intelligent Information Systems Conference, pp. 357–361 (1994)

16. Venables, W.N., Smith, D.M.: R Core Team: an introduction to R. User Man. **2**, 99 (2015)

17. Berriz, G.F., King, O.D., Bryant, B., Sander, C., Roth, F.P.: Characterizing gene sets with FuncAssociate. Bioinformatics **19**, 2502–2504 (2003)

18. Franceschini, A., Szklarczyk, D., Frankild, S., Kuhn, M., Simonovic, M., Roth, A., Lin, J., Minguez, P., Bork, P., Von Mering, C., Jensen, L.J.: STRING v9.1: Protein-protein interaction networks, with increased coverage and integration. Nucleic Acids Res. **41** (2013)

19. Wang, S., Wu, F.: Detecting overlapping protein complexes in PPI networks based on robustness. Proteome Sci. **11**, S18 (2013)

Analysing MPLS Performance by SDN

Snehal Patil and Mansi S. Subhedar

Abstract Nowadays, MPLS has become the first choice for enterprises to connect remote branch offices as it offers several benefits to packet forwarding. MPLS-VPN combines the features of both overlay and peer-to-peer VPNs thus offering the most robust connectivity. However, MPLS-TE faces the problem of creating backup path immediately when the best path goes down or gets congested. At some point of time in near future, all IP networks will be converted into programmable networks. During this transition, there should be some sort of mechanism which will couple the MPLS network to software-defined networking (SDN). This paper analyses the performance of SDN when coupled with MPLS. By using some of the core features of SDN, MPLS performance is enhanced for the tunnel creation. It has been found that coupling SDN with MPLS offers better performance in terms of latency, response time and bandwidth utilization.

1 Introduction

Enterprises use VPN technologies for increasing their working efficiencies by connecting different branch offices to each other. MPLS is fast, but it derives its operations from interior gateway protocol (IGP). Most of the Internet service providers (ISP) use OSPF for their intra-autonomous system (AS) operations. The convergence of MPLS and path selection totally depends on OSPF convergence and its reliability. Companies are always concerned about cost-effectiveness, downtime, security, traffic shaping, traffic aggregation and extensibility. Multi-protocol label switching is the most efficient forwarding mechanism used by many service providers currently in the world. It offers robust, fast and secure connectivity for VPNs. Basically,

S. Patil (✉) · M. S. Subhedar
Department of Electronics and Telecommunication Engineering,
Pillai HOC College of Engineering and Technology, Rasayani,
Raigad 410207, Maharashtra, India
e-mail: snehalspatil12@gmail.com

M. S. Subhedar
e-mail: mansi_subhedar@rediffmail.com

© Springer Nature Singapore Pte Ltd. 2019   589
A. Abraham et al. (eds.), *Emerging Technologies in Data Mining and Information
Security*, Advances in Intelligent Systems and Computing 814,
https://doi.org/10.1007/978-981-13-1501-5_52

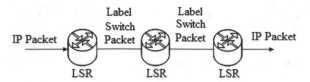

Fig. 1 MPLS mechanism

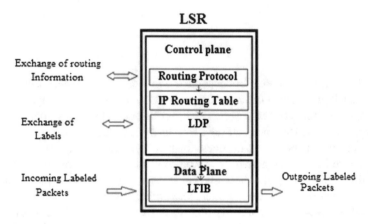

Fig. 2 MPLS operation in a router

MPLS divides the machine into two planes, control plane and data plane. MPLS creates a Label Forwarding Information Base (LFIB) table in data plane and forwards data based on the entries in this table. It does not take into consideration IP address unless the specified action is 'pop'. Data is forwarded based on the labels and not IP address. The label resides in between Layer 2 and Layer 3 headers. Since control plane lookup and data plane processing are avoided, minimal latency and transmission become secure. The use of MPLS-TE makes alternate tunnel creation very simple. Figure 1 shows MPLS switching mechanism that uses labels to forward packets. Usage of label allows edge routers to perform a routing lookup; all the core routers simply forward the packet based on labels assigned at the edge.

Figure 2 shows MPLS operation in routers. Label switch routers exchange routing information and labels in control plane. In data plane, LSR forwards packets. Exchanged routing information and labels are part of the control plane, while forward packets are part of the data plane.

2 Related Work

Variety of studies have been proposed in the literature for improving the performance of MPLS-TE tunnel creation. However, intelligence is provided on routers and SDN is not considered effectively in these studies. The minimum interface routing algorithm (MIRA) proposed in [1] uses Dijkstra's algorithm for calculating the best path. It uses residual bandwidth under consideration on alternative path calculations by identifying critical links which will interfere with future request of other ingress–egress pair. Earlier, people were using offline algorithm for providing alternating paths to IP-VPN. It came up with the solution of online computation of alternating path when actual path gets crowded. MIRA provides alternating path dynamically but convergence is slow. The online dynamic routing algorithm proposed in [2] uses two types of information models which gives low and high bounds so that better routing can be done. There are defects in scenario. The first defect gives inaccurate output, and the other defect applicable to complete information scenario results in loops and network becomes inefficient [3]. Hop-constrained adaptive shortest path (HCASP) provides bandwidth guaranteed routing of MPLS-based tunnels [4]. This method uses OSPF extension information, but it can be transmitted within same area only. They worked on MIRA algorithm to reduce the time required to calculate alternate path when best path goes down. The OSPF extension information traverses in area of origination only and because of that tunnel establishment is difficult [5]. Here, they have suggested storing ten LSA type information in LSDB of ABR and then advertising in cross regional. The main aim was to boost performance of CSPF. Even if we transmit all LSA types by using BGP extended communities between remote sites of customer, sham-link between the links needs to be provided to bring two customer sites in single area. Tu et al. discussed splicing of MPLS with open flow tunnel on SDN conception. A central controller, path translator and command installer are responsible for making MPLS-based routing policies and open flow entries of the available routers. Traffic is split into best effort traffic and diffserv traffic which needs QoS. When QoS data traffic increases, best effort delivery traffic rate decreases in CSPF. For best effort traffic, best path is longest—widest path. If all paths under consideration fail to fulfil above consideration, CSPF is considered. Authors have developed online algorithm based on offline optimal computations which increases best effort traffic rate even if QoS traffic increases as compared to CSPF. If aggregated traffic matrix provided for offline calculations is not accurate, decision goes to CSPF. In [6], an auto-QoS feature of Cisco is employed that provides alternate paths for MPLS VPNs by using RSVP. Cisco methodology takes around 2–3 s for new tunnel path establishments. They have proposed fast reroute algorithm (FRR) which calculates preferred tunnels and alternative tunnels. Bandwidth allocation of concern link, serving other alternative tunnels [9].

3 Software-Defined Networking

Figure 3 shows SDN architecture. The SDN-based centralized controller provides open interface which gives automatic control on entire network. The open SDN uses two types of open flow APIs, namely northbound API and southbound API. Controller uses the southbound interfaces to program devices. Northbound interfaces consist of applications to be loaded into controller and so that algorithms and pro-

Fig. 3 Software-defined
networking architecture

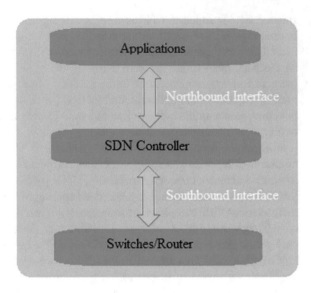

tocols can be provided to proceed network efficiently. When flow entry does not match, packet is forwarded to the controller, depending on the nature of application; it may even drop the packets [7]. SDN gives physical separation of the network control plane from data plane same as MPLS and even provides network virtualization. SDN architecture is accommodative, productive, feasible and profitable and is suitable for applications that require more bandwidth. Vigorous nature of SDN addresses the fact that static architecture of conventional networks is less suited for contempo data centres, campuses and carrier environments.

SDN devices constitute forwarding functionality for taking decisions on packet forwarding. Controller defines the flows which represents the data. A flow describes packets set forward from a particular set of endpoint to another endpoint. The packet forwarding actions belonging to a particular flow is described by one set of rule. Flows are represented by the term flow entry. A flow table consists of chain of flow entries and packet forwarding actions to be taken by network device. The flow entry is programming expression of control plane calculations done by the controller.

4 MPLS Provision in Different SDN Versions

First version of OpenFlow provides full VLAN support, and it also permits modification of a going tag whether it is MPLS header or virtual LAN. Level of VLAN requires energetic support for popping and pushing of various steps of tags and MPLS. A new tag is added by PUSH action. When PUSH is run, a new header of a specific type is added in front of ongoing outermost header. Current outermost tag is removed by POP action. A field containing newly added header is copied from existing field in current outermost header. If it not found, it is started to zero. By SET action, fresh values are then assigned to this outermost header. If contact with controller is lost, switch will be entered in either fail secure mode or fail standalone mode. The next version is uprated to maintain simultaneous connection with multiple controllers and the switch. The switch must ensure that it only sends messages to controller that corresponds to a command sent by that controller. Here, message is duplicated and a copy is sent to each controller. A controller may be one of the three following roles to a switch—Equal, Slave or Master. Equal and Master modes enable the controller to configure the switch. In case of Master mode, switch invokes that only one switch should be in Master mode and remaining in Slave mode. The multiple controller is for high availability requirements.

5   SDN and MPLS Operation

In open SDN environment, only controller is credible for all the operations of whole network. As controller itself contains a control plane, SDN switch forwards the packet matching with the presently defined flows. Only controller can make changes to flow table. As a network is not able to do any changes and controller can felt SOF (single point of failure). But here upon detecting a failed node by using distributed intelligence of network, it will reconfigure itself to overcome single point of failure problem. The network will use an alternate route which is calculated by controller using IGP extractions.

MPLS is not an encapsulation technique; it works on label swapping method for data forwarding. In architectural framework of MPLS, it decouples transport from services. Decoupling might be done by encoding instructions in packet headers. Instruction stacking is another important feature of MPLS. Stacking resembles to providing chain of instructions. SDN era started with OpenFlow protocol. OpenFlow works by assuming that there is a central controller software running as virtual machine or directly on host OS of server. There must be IP connectivity between controller and switches by using either out-of-band signalling or in-band signalling. Flow table is an important state of OpenFlow switch. Let one host send packet to another host. When packet arrives at a port connecting to the switch, switch realizes that this particular flow is not programmed on its flow table and then its control agent sends packet to controller. Controller had previously learnt location of the host and programs new accordingly on the switch. Afterwards, the switch can forward flow packets to destination host. Label edge routers (LERs) compute traffic engineering database (TED) according to distributed link state database. When it has TED view, controller performs path calculations and asks different LERs of the network to signal, resignal or tear down LSPs in the proper manner. To do this, controller can use a method of protocol abstraction to signal LSPs which is much more scalable and preferred way. Such protocols are called as Path Computation Element Protocol (PCEP). The Provider Edge (PE) function is distributed in service endpoints. The device with Virtual Forwarding and Routing or terminate psuedowire (PW) link becomes service endpoint.

In IGP domain, many path reservation techniques are available such as Resource Reservation Protocol (RSVP), Source Packet Routing in Networking (SPRING), Label Distribution Protocol (LDP). Four MPLS-TE builders are LDP, RSVP-TE, BGP and IGP. When two neighbours establish LDP sessions, they start to exchange label mapping messages which are associated with IPv4 prefixes to MPLS labels. It helps in forming label information base (LIB). RSVP-TE reserves resources along paths in Internet [8]. It is most powerful, robust and flexible MPLS signalling protocol. The constraint shorted path (CSPF) procedure initiated at very first router (head-end router) to create best path (LSP) by using RSVP [9]. For tunnel establishment, two types of RSVP messages are used, RSVP- PATH and RSVP- RESV. The very first router sends PATH message to last router (tail-end router) which contains tunnel creation information. Tail-end router then replies to head-end router by forwarding RESV message which follows similar reversed path to the head-end router. Bandwidth is the significant parameter to be considered while creating tunnel. If available bandwidth is not sufficient to fulfil requirement of tunnel, then PATH message will stop at immediately not reverting to source and ERROR message will be forwarded to last (tail-end) router. In such events, it does not create tunnel. To elect best path, CSPF uses IGP extractions [9]. The discovered shortest path may not be best to be used currently. Cisco's auto-bandwidth allocator is used to adjust bandwidth automatically. At every interval, applied flow data rate is monitored, and highest value is recorded and is used as a reference value to modify tunnel bandwidth of upcoming interval. Path alteration is done by first router. Whenever topology changes or congestion occurs, first router obtains message by RSVP that the path will not be preserved. Then, it removes faulty links and creates new TE database [10].

6 Our Contribution

The latency of plain-vanilla MPLS-VPN network that offers the fastest tunnelling method amongst currently available methods in the industry is observed. Figure 4 shows MPLS-VPN network. All routers are running OSPF, and MPLS is enabled on all the interfaces. Let there are three customers, viz. Customer A, Customer B and Customer C. Tunnels are reserved for all the customers for connecting their branch offices as shown in Fig. 4. RSVP is enabled on all interfaces for reserving tunnel path. When different load conditions are applied, time required for end-to-end packet delivery was found to be varying. R1-R2-R3 is the best path for Customer A which is currently being used if link R2-R3 is congested. When Customer B is forwarding traffic on its site using path R1-R2-R3-R6, controller will sense congestion on link

R2-R3 and notify R1 to forward traffic to path R1-R4-R5-R6-R3. Now, if link R4-R5 gets congested or damaged then controller will sense congestion through R4 and notify R1 to forward traffic via R1-R4-R7-R8-R9-R6-R3. With this, there is no latency as controller is already programmed with all possible routes to reach all customer sites, and therefore, path allocation is dynamic and random. Path reservation is not used. The said process is very fast as it saves the time required for creating MPLS tunnels, readjustment of MPLS tunnels, removing tunnels, path allocation by RSVP or some other reservation method.

The same scenario is implemented using OpenvSwitches and SDN controller and is depicted in Fig. 5. The SDN controller was preprogrammed with the flow entries, and it had reserved the resources for all customers with respect to the bandwidth requirements of the customers. Figure 6 exhibits the entire operation for controlling routers through SDN controller. Routers just keep the record of bandwidth utilization. But they are not concerned for backup path creation, as statistics collected from OSPF extension headers are calculated by SDN.

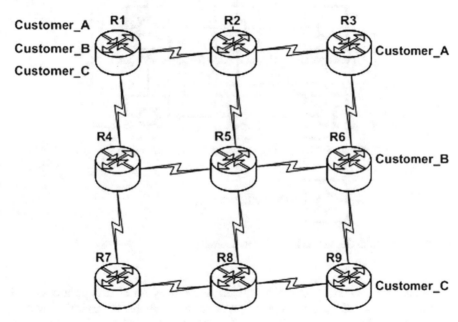

Fig. 4 MPLS-VPN network

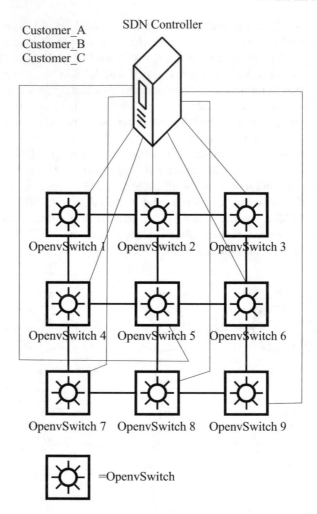

Fig. 5 Network with OpenvSwitches and SDN controller

Figure 7 shows comparison of bandwidth vs. latency for a SDN controller and MPLS. It proves that increase in latency reduces the available bandwidth in MPLS whereas bandwidth availability increases in SDN. However, if one particular link is shared by all the customers then efficiency of that link will be degraded. Our future work will propose a solution to this problem.

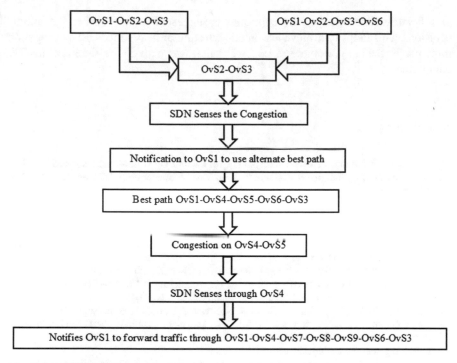

Fig. 6 Controlling OvS through SDN

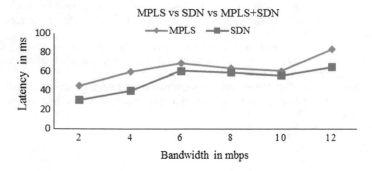

Fig. 7 Bandwidth versus latency comparison of SDN and MPLS

7 Conclusion

The active formation of MPLS tunnels using conventional method is fast, but when applied load increases, tunnel creation time gets affected. Also, increase in latency reduces the available bandwidth and the network requires high duration of time to manipulate the path of MPLS tunnel if modern path passes via some number of sources and/or there is heavy load over the network. To address these issues, use of software-defined networking controller is suggested to take packet forwarding decisions. This helps to achieve less latency in spite of heavy loads. SDN programming makes the network more responsive than it was with conventional MPLS-TE environment.

References

1. Tu, X., Li, X., Zhou, J., Chen, S.: Splicing MPLS and OpenFlow tunnels based on SDN paradigm. In: IEEE International Conference on Cloud Engineering (IC2E), pp. 489–493 (11–14 March 2014)
2. Hao, K., Jin, Z.: An on-line routing algorithm based on the off-line optimal computing in MPLS. In: 5th International Conference on Wireless Communications, Networking and Mobile Computing WiCom '09, 1–5 (24–26 September 2009)
3. Hasan, H., Cosmas, J., Zaharis, Z., Lazaridis, P., Khwandah, S.: Development of FRR Mechanism by Adopting SDN Notion
4. Elsayed, K.M.F.: HCASP: A hop-constrained adaptive shortest path algorithm for routing Bandwidth guaranteed tunnels in MPLS networks. In: Proceedings ISCC 2004 Ninth International Symposium on Computers and Communications, vol. 2 (July 2004)
5. Goransson, P., Black, C.: Software Defined Networks-A Comprehensive Approach
6. Cisco Systems. Advanced topics in MPLS-TEDeployment. [Online]. Available http://www.cisco.com/c/en/us/products/collateral/ios-nx-os-software/multiprotocol-label-switching-traffic-engineering/whitepaper_c11-551235.html
7. Wang, C., Hu, J.: Improvement of running tunnel based on OSPF TE. In: International Conference on Web Information Systems and Mining (WISM), vol. 2, 3–7 (23–24 October 2010)
8. Kar, K., Kodialam, M., Lakshman, T.V.: Minimum interference routing of bandwidth guaranteed tunnels with MPLS traffic engineering applications. IEEE J. Sel. Areas Commun. **18**(12), 2566–2579 (2000)
9. Hasan, H., Cosmas, J., Zaharis, Z., Lazaridis, P., Khwandah, S.: Creating and managing dynamic MPLS tunnel by using SDN notion. In: International Conference on Telecommunications and Multimedia (TEMU)(2016)
10. Cisco Systems. MPLS traffic engineering, constrained-base routing and operation in MPLS-TE. Cisco press [online]. Available http://www.cisco.com/articles/article.asp-p=426640&seqNumber=3

Serving New Interest in Named Data Network

Tanusree Chatterjee and Pritam Banerjee

Abstract Today's Internet has mostly become content-centric. The distribution of data securely and efficiently is the most important agenda today. Current IP network and content distribution network services are sometimes unable to fulfil all such requirements of data distribution. Named data network (NDN) has emerged as a promising candidate to cope with the use of today's Internet. Despite providing potential advantages over conventional network such as in-network caching, easier data forwarding, routing, it also leaves few issues unattended. One of such is the probability of finding the corresponding data for a request which has not been served yet by any router. In this paper, we propose a simple probabilistic approach towards it and present the modified algorithm for data forwarding accordingly. We also explain the method by suitable example. The proof of the same in support of our claim is left as a future proposal.

Keywords Named data network · Name prefixes · Routing · Forwarding
Probability

1 Introduction

Named data networking (NDN) [1–3] is a future Internet architecture inspired by years of empirical research on growing change of network usage and unsolved problems in contemporary Internet architectures like IP. NDN considers data as the first-class entity irrespective of its source or destination. Fast and secured data transfer is the prime requirement in today's Internet. Currently, the content providers, e.g.

T. Chatterjee (✉)
Techno India College of Technology, Kolkata 700156, West Bengal, India
e-mail: tnsr.chatterjee@gmail.com

P. Banerjee
Vinod Gupta School of Management, Indian Institute of Technology, Kharagpur 721302,
West Bengal, India
e-mail: pritam2506@gmail.com

© Springer Nature Singapore Pte Ltd. 2019   599
A. Abraham et al. (eds.), *Emerging Technologies in Data Mining and Information
Security*, Advances in Intelligent Systems and Computing 814,
https://doi.org/10.1007/978-981-13-1501-5_53

YouTube, Amazon, iTunes, occupy half of the world traffic. They operate on Content Distribution Network (CDN) to deliver their content to their end-users. But, mapping each content request from customers to the nearest CDN node serving the content still remains an issue. Unlike IP and CDN, by naming data instead of their locations, NDN transforms data into a first-class entity. This simple change in architecture has brought the potential advantages in seamless content distribution, reduced congestion, improved delivery speed through content caching, etc.

Unlike IP, an NDN router announces name prefixes of the content that the router can serve. The names are hierarchically structured; e.g., a video in YouTube may have the name/youtube.com/video/ndn_security.mpg, where '/' is not part of the name but it specifies a boundary between the name components. The names are not transparent to the network. The routers see the boundaries between the components; they do not know the meaning of a name.

1.1 Motivation

Communication in NDN is initiated from receiver end called consumer. Node which can serve the requested data is the sender, called producer. Here, the data transfer takes place through the exchange of two packets namely interest and data from/to consumer and producer. Each request is generated from a consumer in the form of an interest packet which carries a name of content. Subsequently, each reply generated from the producer of the corresponding request carries the same name while routed back to the consumer.

As NDN follows traditional link state routing, each router in NDN has the topology of entire network to forward an interest packet. Each router has the information about all the requests, i.e. name prefixes served by all other routers in the network in a form of a database. The database is called link-state database (LSDB). The packets in NDN do not carry source and destination addresses. So, upon receiving an interest, it is routed and forwarded in the network in a hop-by-hop manner until the corresponding data against a particular requested name is found. Each router uses Dijkstra algorithm to find the least cost path to reach to the router serving the requested name. But, due to in-network caching mechanism the data may be found in any intervening router and can be immediately sent back to consumer. Different routing protocols like OSPFN [4] and NLSR [5] have been developed in recent time to route and forward interest packets in NDN. The protocols address the problems of namespace mapping, security, trust management, routing and forwarding, etc. Now, let us assume an interest reaches to a router with such name which is not advertised by any routers yet. According to the above protocols in NDN, flooding is the only solution left to handle such an issue. But, flooding imposes a decent amount of overhead in terms of communications in the whole network. These motivate us to evaluate whether there exists any solutions other than flooding to forward the interest in case no router yet ready to serve such.

1.2 How the Names Are Used to Serve an Interest

Each NDN router stores three data structures such as forwarding information base (FIB), pending interest table (PIT) and content store (CS). These are required for data transmission in the network [1, 2]. Figure 1 which has been adapted from [1] and redrawn illustrates how NDN works. It shows how a requested content is forwarded to the consumer who sends an interest for the name */iiest.edu/cst/student.htm*. The figure depicts two producers (*Producer1* and *Producer2*) and one consumer (*Consumer*) connected through three routers 'A', 'B' and 'C'. It also shows the status of FIB, PIT and CS of each router after receiving the interest and data packet.

Router 'A' in Fig. 1 first searches whether the same interest has been served recently; i.e., it searches the data in CS. As the data is not found there, the router searches its PIT to check whether the request (interest) is already pending from any other interface. If the interest is not in PIT, a new entry is created in PIT. Then, 'A' performs the longest prefix match on its FIB and finds router 'C' as the next hop where the request should be forwarded. Again, as the data is not found in CS or in PIT of router 'C', a new PIT entry is created in 'C' and the interest is forwarded to *Producer1* according to FIB information. The data is found in CS of *Producer1*. Now, 'C' retrieves the data and sends back the data packet which is cryptographically signed by *Producer1*. The data follows the reverse path of the interest to reach *Consumer*.

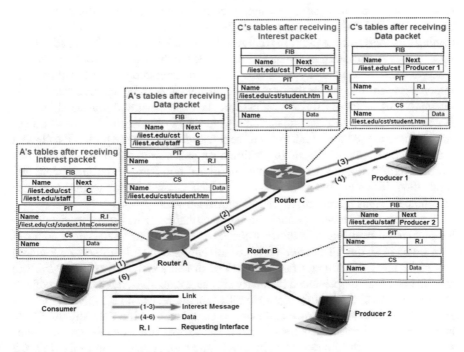

Fig. 1 Data forwarding in NDN according to names [4]

The content is stored in CS entry of each router while returning back. The collection of possible destinations in the network is also called link-state database (LSDB) in link-state routing method.

Now, let us assume an interest with name prefix/*iiest.ac/office/staff.htm* has been initiated by the consumer which is not served by any router according to LSDB. If flooding is not a desirable method in the network to forward the interest, then the following two points have to be taken into consideration.

- How to identify the router which may serve it and
- If identified one, then what is the probability to get the desired data there.

The rest of the paper is organized as follows. Section 2 discusses several some related works. Section 3 elaborates our proposed scheme to forward a new interest in NDN along with the algorithm and illustrative example based on a probabilistic approach. We conclude in Sect. 4 along with future research scope of our proposal.

2 Related Works

Open Shortest Path First for Named data (OSPFN) [4] is the first NDN routing protocol which is an extension of traditional routing protocol OSPF. It defines a new type of opaque link-state advertisement to carry name prefixes and compute name-based FIB. OSPFN does not support full-fledged dynamic and automatic multipath routing capability; unlike OSPF, a configured multipath feature has been added here. Also, it uses IP addresses as router id. So, the need to design a new routing protocol emerges which automatically calculates multipath forwarding choices without user intervention and also avoids the burden of managing underlying IP addresses. In this regard, NLSR protocol [6] has been evolved as a more matured routing algorithm. It uses interest/data to disseminate routing updates in the network. Dijkstra algorithm is used to produce a ranked list of next hops for each name prefix. It follows a hierarchical trust model for routing within a single administrative domain and uses Chronosync [5] protocol to replace the traditional flooding for routing update dissemination. However, this method uses too much message exchange during update dissemination thus increasing overheads on the network.

The authors in [7] have introduced an approach that can realistically scale variable-length name forwarding to billions of prefixes. The work has designed a compact data structure so that FIBs with a few million entries can still fit in Static Random Access Memory (SRAM) for fast lookup. The structure is named as binary Patricia trie. It shows that if represented by binary strings, a name-based forwarding table of several millions of entries even grows to billion in future, can be notably compressed and scalable. The performance and average complexity of the data structure have been analyzed which presents good result. But experimentations on several issues such as extending the data structure in other name-based functionalities like PIT, CS or investigating the data structure in speculative data plane are still remained.

The routing protocols and other routing and forwarding strategy have been developed and proposed in recent times to address several issues in routing [4], security [6, 5], namespace mapping [7, 8], etc., in NDN. Few of them are discussed above. However, none of the works address the issue of how to forward an interest packet in case there is no router yet ready to serve such a name prefix. So, after detail analysis of NDN architecture, routing and forwarding strategy we have decided to take into consideration the issue of serving a new interest in the network.

3 Proposed Scheme

In current NDN routing methodology [9], whenever a new interest reaches to a router, the request is flooded to the entire network to search who is the producer of the requested piece of data. Once the actual producer(s) is known for sure, the requested data can be retrieved and reach to the requesting consumer following the exact reverse path of the interest. Then update can be accordingly made in the FIB of the routers during the downstream path of the data packet. But, the demerit of such system is the initial flooding mechanism of the interest which results in congestion and imposing high communication overheads in the entire network. With the objective of reducing flooding and overheads in the network, we have tried to propose a better and feasible solution of the above problem.

3.1 The Scheme

The scheme uses a substring matching algorithm to find the probability of finding the corresponding data in a path even when only a small substring of the requested name prefix matches with any of the names.

When an interest is new, then a searching can be introduced to find a partial match between the new interest and all the name prefixes originated by different routers. If a certain degree of match is found with any of the existing name prefixes, then instead of flooding the request, the new interest can be forwarded towards originator router of the name prefix. The matching can be done in the following way. We start with the rightmost substring of the requested name prefix (S). If any partial match is found with any existing name prefixes of the LSDB, then a certain probability is assigned to get the requested name in the originator router of the existing name. When N is the number of substrings in the requested name, the probability can be calculated by $T = \frac{100}{N}$.

After the first iteration, we keep on picking the nth hierarchical level from the right of the requested name prefix and concatenate it with S_{n-1}, to form S_n. Again, S_n is tried against all the entries of LSDB to check for partial match. If indeed a partial match is found with the substring S_n, then the probability of forwarding the requested name prefix towards the destination router increases by $T\%$. This is repeated until and unless a match is no longer found against the selected substring from the requested name prefix.

It may happen that the last substring of a requested name prefix to have the best partial match with an entry in the LSDB served by a certain router (say *RTR-X*) is S_n; i.e., the requested name prefix can be forwarded to *RTR-X* with the probability of correct forwarding being $(n \times 16)\%$. However, another substring S_m (where $m < n$) of the requested name prefix had a partial match with an entry in the LSDB served by another router, say *RTR-Y*. Although $m < n$, meaning that the probability of correct forwarding towards *RTR-X* is more than that of *RTR-Y* [as $(n \times 16) > (m \times 16)$], still the information regarding *RTR-Y* is temporarily saved in the FIB, since it will be used in case forwarding towards *RTR-X* fails to retrieve the required data. Thus, information of every substring match is temporarily saved, so that it can be used, when forwarding to those routers having higher probability of correct forwarding, fails.

3.2 The Algorithm

The algorithm of the proposed scheme is presented in the following pseudocode. The following notations are used in the algorithm.

RTR_A—Router A, *CS_RTR_A*—CS of Router A, *PIT_RTR_A*—PIT of Router A, *FIB_RTR_A*—FIB of Router A, *Requesting_Face*—requesting Router/Node, *Origin_Router*—Origin Router of "Name_Prefix", S_i_LSDB—ith hierarchical level from the right of the "LSDB" entries, *N*—total hierarchical levels in "Name_Prefix", *Priority* [R_r]—Router arranged in terms of forwarding priority where Priority $R_r \propto P_r$.

INPUT: Name prefix requested by consumer (**Name_Prefix**)
OUTPUT: Serve data packet to consumer (**Data_Packet**)

BEGIN
1.   Request "Name_Prefix"
2.   "Name_Prefix" reaches "RTR_A"
3.   Search_CS (RTR_A)
4.   Serve "Data_Packet"
END

Search_CS (arg Router)
BEGIN
1.   if   (matching   entry   found   to
      "Name_Prefix" in "CS_RTR_A")
2.   Return "Data_Packet"
3.   else
4.   Search_PIT (RTR_A)
5.   End if
END

Search_PIT (arg Router)
BEGIN
1.   if   (matching   entry   found   to
      "Name_Prefix" in "PIT_RTR_A")
2.   Update   "Requesting_Face"
      against corresponding matched entry
      of "Name_Prefix"
3.   else
4.   Add "Requesting_Face" against
      the   corresponding   requested
      "Name_Prefix"
5.   End if
6.   Update "PIT_RTR_A"
7.   Search_FIB (RTR_A)
END

Search_ FIB (arg Router)
BEGIN
1.   if   (matching   entry   found   to
      "Name_Prefix" in "FIB_RTR_A")
2.   Forward "Name_Prefix" to the
      corresponding next destination
3.   else
4.   Consult "LSDB"

12.   if (partly matching entry to
      "Name_Prefix"   still   exists   in
      "LSDB")
13.   Modified_FIB_from_LSDB
      (Pr = 0)
14.   else
15.   for all (partly matching
      entry found to "Name_Prefix" in
      "LSDB")
16.   Set R_r ← origin
      routers of "Name_Prefix"
17.   End for

5.   Search   "Origin_Router"
      corresponding   to   requested
      "Name_Prefix"
6.   Search_LSDB
7.   End if
END

Search_LSDB
BEGIN
1.   if   (matching   entry   found   to
      "Name_Prefix" in "LSDB")
2.   Search   next   hops   to
      "Origin_Router"   and   their
      corresponding path cost
3.   Search_ FIB (RTR_A)
4.   else
5.   Set i: = 1
6.   S_1_Name_Prefix ← Rightmost
      hierarchical level of "Name_Prefix"
7.   Modified_FIB_from_LSDB (Pr
      = 0)
END

Modified_FIB_from_LSDB (arg Pr)
//obtain an updated FIB from LSDB
BEGIN
8.   if   ("S_i_Name_Prefix"   matches
      "S_i_LSDB")
1.   Compute $Pr \mathrel{+}= \left[\dfrac{100}{N} \times i\right]$
2. $i \mathrel{+}{+}$
3.   Select "S_i_Name_Prefix"
4.   Modified_FIB_from_LSDB (Pr)
5.   Else
6.   for (all listed routers)
8.   if ($i == 1$)
9.   Flood "Name_Prefix"
10.   else
11.   Discontinue (non-matching
      entry to "Name_Prefix" in "LSDB")
18.   Priority [R_r] = Descending
      [Pr]
19.   Update "FIB_RTR_A"
20.   Forward   "Name_Prefix"
      to the corresponding next destination
21.   Search_CS (RTR_B)
22.   End if
23.   End if
24.   End for
9.   End if
END

3.3   Illustrative Example

Figure 2, adapted and redrawn from [4], represents the sample network consisting of six NDN routers that we have considered to illustrate the proposed solution. The public IP and the tunnel address of each router are displayed in the figure. The cost of each link is also shown. The list of name prefixes is served by the routers; i.e., the LSDB is announced. Table 1 represents a part of the LSDB. Let us assume that each router is connected to a number of nodes, and each such connection is termed as interfaces (faces). We have presented Face 0 of RTR3, Face 1 of RTR2 and Face 2 of RTR6. Here, we discuss the example, where the requested name prefix is not present in the LSDB, to clearly depict the working of our proposed solution and how it is different from the existing scheme.

To start with, let us assume the data structures depicted in Tables 2 and 3 are presently maintained at RTR-3. Let us assume we are at RTR3 and a name prefix *MP/NITS/me/ndn/sports/soccer* is received from node-0 (face-0). Now, as per convention, at first the CS of RTR3 is searched, for a local cache hit. However, no data serving this name is present in the CS of RTR3. Next, the PIT of RTR3 is searched and no entry in the PIT matches with the requested name prefix. So, a new entry is made in the PIT of RTR3.

Now, while searching the LSDB to construct the FIB, it is found that there are no routers to serve this kind of a name prefix yet. So, no router can be designated as next hop in the FIB for the interest/*MP/NITS/me/ndn/sports soccer*. A close observation

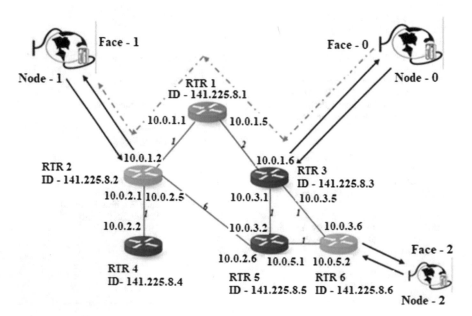

Fig. 2 A network topology

Table 1 LSDB

Router	Name prefixes
RTR2	WB/IIESTS/cst/ndn/sports/soccer WB/IIESTS/cst/ndn/sports/bb
RTR3	WB/IIESTS/cst/ndn/lifestyle/home WB/IIESTS/cst/ndn/lifestyle/cooking
RTR4	WB/IIESTS/cst/ndn/leisure/park WB/IIESTS/cst/ndn/leisure/theater
RTR5	WB/IIESTS/cst/ndn/travel/Michigan WB/IIESTS/cst/ndn/travel/Illinois

Table 2 RTR3 CS

Name	Data
WB/IIESTS/cst/ndn/sports/bb	Data-I
WB/IIESTS/cst/ndn/travel/Illinois	Data-II

Table 3 RTR3 PIT

Name prefix	Requesting face
WB/IIESTS/cst/ndn/leisure/park	Node—1, 3
WB/IIESTS/cst/ndn/travel/Michigan	Node—2

reveals that the requested data with the name prefix, *MP/NITS/me/ndn/sports/soccer* has a certain degree of match with the name prefix *WB/IIESTS/cst/ndn/sports/soccer* served by RTR2.

Hence, according to the proposed solution, we start matching with the rightmost hierarchical level of the requested name prefix, i.e. substring *soccer* (S_1). Now, S_1 is tried against all the entries of LSDB. It can be concluded that since S_1 (*soccer*) has a partial match with a name prefix served by RTR2. So, the requested name prefix can be forwarded towards RTR2, with a probability of correct forwarding being 16%. Next, the second substring (*sports*) is picked up and concatenated with S_1. Thus, *sports/soccer* is designated as S_2. Now, S_2 is tried against all entries of LSDB. After second iteration, since S_2 (*spots/soccer*) has a partial match with a name prefix served by RTR2 again, the requested name prefix can be forwarded towards RTR2, with a probability of correct forwarding being 32%. Proceeding this way, the next substring of the requested name prefix is S_3 (*ndn/sports/soccer*). It gets the best partial match with the name prefix *WB/IIESTS/cst/ndn/sports/soccer* served by RTR2. No more matches are found with the next substring of the requested name prefix. So, the interest can be forwarded towards RTR2 with the probability of match being 48%.

Other routers with which a partial match is found for the requested name prefix but with lower probability are also considered. So, all information regarding RTR6 is also saved temporarily in FIB. This temporary information is used when forwarding to the routers having higher probability of correct forwarding, fails. For now, RTR1 is the next hop in the least cost path, so the requested name prefix *MP/NITS/me/ndn/sports/soccer* is forwarded to RTR1. Thus, as per the convention,

Table 4 RTR1 FIB

Name prefix	Next hop	Cost
MP/NITS/me/ndn/sports/soccer	RTR3—10.0.1.5	2
MP/NITS/me/ndn/sports/soccer	RTR2—10.0.1.1	1

once again, the CS of RTR1 is searched at first. As no matching entry is found, next the PIT of RTR1 is searched and once again no match is found. So, a new entry is made into the PIT of RTR1. Then, the FIB (Table 4) is constructed, and as a result, the interest *MP/NITS/me/ndn/sports/soccer* is forwarded towards RTR2. Accordingly, the CS, PIT and FIB of RTR2 are also updated.

Finally, after consulting the FIB of RTR2, the request is sent to Node-1, who in this case is assumed to be the producer of data. Next, the data, with which the partial match is found, is sent back to RTR3 following the reverse path of the interest. Again, CS, PIT and FIB of all the nodes are updated. If the data satisfies the request, then the interest need not to be sent again. Otherwise, the interest is flooded in the network to find the producer.

P_r gives us the probability value of finding a correct match on forwarding the requested prefix to a partially matched router which is represented as follows.

$$P_r := \left[\frac{100}{N} \times i \right]$$

N is total no of hierarchical levels in name prefix, while i represents the ith hierarchical level from the right side.

The probabilistic approach is better than flooding in terms of handling network congestion. While in flooding the requested name prefix is forwarded to all the routers, leading to large overhead in the network communication, our approach, by assigning probability (P_r) against each matching routers ensures that the requested prefix is forwarded to the partially matched router, only one at a time, in descending order of the probability value. If forwarding to all the partially matched routers fails to satisfy the specific interest, then flooding is left as the only alternative method.

4 Conclusion

To conclude, it can be stated that the application-driven, experimental approach to NDN research has enabled progress and provided new depth to the original vision for NDN. But scientists till now have only been able to scratch the surface of research on namespace structure, trust models, scalable forwarding, routing, etc. [10].

In this paper, we have analyzed the routing and data forwarding method in NDN which is based on the hierarchical naming scheme. We have explored the issue of serving a new interest in NDN left unattended and so tried to develop a simple probabilistic approach to address it. Due to scarcity of space, we have not been able to provide any proof to show the effectiveness of the proposed scheme. In future, we would like to implement the approach in an NDN emulator like Mini-NDN [11] to evaluate the performance and adaptability of the method.

References

1. Zhang, L., Claffy, K., Crowley, P., Papadopoulos, C., Wang, L., Zhang B.: Named data networking. In: ACM SIGCOMM Computer Communication Review, vol. 44, no. 3, pp. 66–73. ACM, New York, USA (2014)
2. Moiseenko, I., Oran, D.: TCP/ICN: carrying TCP over content centric and named data networks. In: 2nd ACM International Conference on Information-Centric Networking, pp. 112–121. ACM, Kyoto, Japan (2016)
3. Xylomenos, G., Ververidis, C.N., Siris, V.A., Fotiou N., Tsilopoulos, C., Vasilakos, X., Katsaros, K.V., Polyzos, G.C.: A survey of information centric networking research. In: IEEE Communications Surveys and Tutorials, vol. 16, no. 2, pp. 1024–1049. IEEE (2013)
4. Wang, L., Hoque, A.K.M.M., Yi, C., Alyyan, A., Zhang, B.: Ospfn: an ospf based routing protocol for named data networking. NDN Technical Report NDN-0003, Tech. Rep., pp. 1–15 (2012)
5. Zhu, Z., Afanasyev, A.: Let's chronosync: decentralized dataset state synchronization in named data networking. In: 21st IEEE International Conference on Network Protocols (ICNP), pp. 1–10. IEEE, Goettingen, Germany (2013)
6. Lehman V., Hoque, A.K.M.M., Yu, Y., Wang, L., Zhang, B., Zhang, L.: A secure link state routing protocol for NDN. NDN Technical Report NDN0037, Tech. Rep., pp. 1–9 (2016)
7. Song, T., Yuan, H., Crowley, P., Zhang, B.: Scalable name-based packet forwarding: from millions to billions. In: ACM Conference on Information-Centric Networking, pp. 19–28. ACM, San Francisco, USA (2015)
8. Afanasyev, A., Yi, C., Wang, L., Z Zhang, B., Zhang, L.: SNAMP: secure namespace mapping to scale NDN forwarding. In: 18th IEEE Global Internet Symposium, pp. 1–6. IEEE, China (2015)
9. Yi, C., Abraham, J., Afanasyev, A., Wang, L., Zhang, B., Zhang, L.: On the role of routing in named data networking. In: ACM Conference on Information-Centric Networking (ICN), pp. 27–36. ACM, New York, USA (2014)
10. https://named-data.net/project/. Accessed on 9 Jan 2018
11. http://minindn.memphis.edu/. Accessed on 28 Nov 2017

A Survey on Miscellaneous Attacks and Countermeasures for RPL Routing Protocol in IoT

Akanksha Jain and Sweta Jain

Abstract The Internet of things is a worldview, where everyday objects can be equipped with distinguishing detecting, networking, systems administration, and handling capacities that will enable them to speak with each other and with different devices and services over the Internet to achieve some goals. At last, IoT devices will be ubiquitous and context-aware and will remain empower surrounding knowledge. Be that as it may, without solid security establishments, attacks and threats in the IoT will overweight any of its advantages. This overview paper examines how traditional methodologies addressed security issue in routing with respect to their limitations, threats to secure routing in IoT and techniques to enhance security in routing scenario as a future work.

1 Introduction

The IoT is outlined as an arrangement of heterogeneous systems interconnected with each other. It could be portrayed as the inescapable and worldwide network, which helps and gives a framework to the monitoring and control of the physical world through the gathering, preparing and investigation of produced information by IoT sensor devices [1]. Wireless sensor networks (WSNs), low-power wireless personal area networks (LOWPANs) and machine-to-machine (M2M) communication are supporting technologies in IoT [1]. With billions of devices associated with the system, a tough challenge is securing the system from different types of threats and attacks. Security is portrayed as one of the greatest issues, which is a keen interest area for research in IoT Routing and addressing are basic issues in IoT owing to the necessity of keeping up consistency in the way packets are routed amongst source

A. Jain (✉) · S. Jain
Department of Computer Science and Engineering, Maulana Azad National Institute
of Technology, Bhopal 462003, MP, India
e-mail: withakanksha19@gmail.com

S. Jain
e-mail: shweta_j82@yahoo.co.in

© Springer Nature Singapore Pte Ltd. 2019
A. Abraham et al. (eds.), *Emerging Technologies in Data Mining and Information
Security*, Advances in Intelligent Systems and Computing 814,
https://doi.org/10.1007/978-981-13-1501-5_54

and goal between IoT devices bridging over various versatile system topologies [2]. Routing protocol for low-power and lossy systems (RPL) designed for ensuring security in IPv6 was recently exhibited as the newly approved routing standard for the IoT. It reduces the general power utilization by limiting the traffic control, which is a noteworthy necessity for the resource-constrained devices imagined later on IoT [3]. RPL is lightweight protocol basically utilized as a part of a 6LOWPAN network. In spite of the fact that RPL characterizes some essential security modes, it stays helpless against different attack. Recent research reported that threats subsist only with the presence of a network and they include: botnets, malware, Web-based malware, denial-of-service (DOS) attacks on financial services and distributed denial-of-service(DDOS) attacks, android malware and spam are considered as IT security threats for 2013–15 [4]. The contributions of the paper are threefold: (1) present an outline of difficulties and the prerequisites to construct a secure IoT. (2) Provide categorization of various vulnerabilities and security protocols proposed for WSN and IoT as for routing in LLNs. (3) Finally, give a survey of continuous research activities in the field of security in the IoT. Our discussion proceeds as follows: In Sect. 2, RPL routing protocol is discussed that is the focus of our discussion together with its security requirements by discussing the vulnerabilities to IoT routing protocol, and this section also describes various methods for mitigating security issues in RPL along with their limitations followed by Sect. 3. In Sect. 4, comparative analysis of secure routing mechanisms is presented. Finally in Sect. 5, we conclude the survey with highlighted future research challenges.

2 Routing Protocol for Low-Power and Lossy Networks (RPL)

The routing convention for low-power and lossy system is a standout amongst other contender for guaranteeing routing in 6LOWPAN environments [5]. RPL is a distance vector IPv6 routing convention for low-power and lossy networks (LLNs) [6]. RPL is an open routing convention standardized by the IETF Roll working gathering in 2008 [7], and it is proactive separation vector in view of IPV6 routing convention for UNS that indicates how to construct a destination-oriented directed acyclic graph (DODAG) utilizing an objective function set of metrics as shown in Fig. 1 [6]. RPL depends on DODAG coordinated towards a particular node called root node. The sink node is associated with whatever is left of DODAG because of upward and descending routes or path which is set up independently [8]. To deliver a network path or route topology, each node chooses a set of guardians or parents that involves nodes with equivalent or better ways towards the sink. The node with the best route interface is picked as parent. RPL utilizes three sorts of control message keeping in mind the end goal to perform and oversee routing of data in the system: DODAG Information Object (DIO), DODAG Advertisement Object (DAO) and DODAG Information Solicitation (DIS). The development of the upward topology is performed through

Fig. 1 RPL basic DODAG
construction

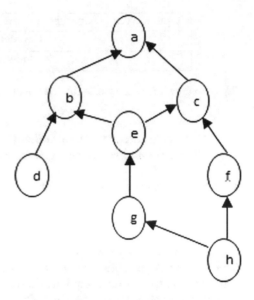

the dispersal of DODAG Information Object (DIO) messages which is a control
packet started by the DODAG roots. After coming to the DIO message, the node
ascertains its rank by utilizing the objective function determined by instance. To
assemble descending route which is not obligatory, a node must send the Destination
Advertisement Object (DIO) message. A node may join on existing system or network topology by dissemination of DODAG Information Solicitation message (DIS)
with a specific end goal to request DIO message from a RPL node. The periodicity of
these messages is resolved through the ticker clock/timer algorithm which depends
on the strength of network topology to progressively dynamically modify ID. Rank
is ascertained by the objective function with a specific end goal to find the position of
every node in DODAG. This rank can be traded off and compromised [9]. Diminish-
ing its rank, a compromised node could be erroneously arranged near root node and
could maliciously control a large traffic or movement moving through it. Expanding
its rank, a node may disrupt the network topology, reset the clock timer trickle and
trigger the sending of control message along these lines exhausting neighbour nodes
resources

2.1 Security in RPL

The information in the security field in Fig. 2 shows the level of security and cryptographic calculations utilized to process security of the message. Current RPL specification defines three security modes named as Unsecured, Preinstall and Verified
mode [6].

T	Reserved	Algorithm	KIM	Reserved	LVL	Flags
Counter						
Key Identifier						

Fig. 2 Security section of a secure RPL control message

2.1.1 Ensuring Integrity, Data Authenticity, Confidentiality, Semantic Security and Protection Against Replay Attack, Key Management

It guarantees integrity by the work of AES/CCM with 128-bit keys for generation of MAC, and by utilizing RSA with SHA-256 for advanced digital signatures, it guarantees integrity and additionally legitimacy [6]. The LVL (security level) field demonstrates the given packet security and permits to differing levels of validation, of data authentication and alternatively of information privacy. This security in RPL determination is guaranteed with the assistance of counter field, which might be utilized to transport a timestamp, as showed by T in Fig. 2. Privacy is guaranteed by utilizing cryptographic algorithms like AES with CCM mode and additionally conspire utilizing separate algorithms for encryption and validation [6]. The Key Identifier Mode (KIM) field in the security segment delineated in Fig. 2 speaks to whether the cryptographic key required to process security for this message might be resolved certainly and explicitly. This field supports protection of different granularity levels of packet [6]. Despite various security-enabled features RPL self-repenting or healing property shields it from internal threats or attacks; however, it stays helpless against external attacks [10].

3 Attacks on RPL Topology in IoT and Methods to Countermeasure

RPL does not have self-repenting mechanisms for different sorts of attacks that are compressed in Tables 1 and 2. They are portrayed as follows:

1. **Selective Forwarding Attack**: This attack happens by specifically forwarding packets. With this attack, denial-of-service (DoS) attack can be propelled. In RPL, an attacker node could forward all RPL control messages and drop whatever is left of the traffic movement [9].
2. **Sinkhole Attack**: In sinkhole attack, a malicious node publicizes useful way to attract many numerous adjacent nodes to route traffic through it.
3. **Wormhole Attack**: This attack can happen by making burrow between the two malicious nodes and transmitting the specific of all activities through it [9].
4. **Hello flood Attack**: In Hello flooding attack, an attacker node acquaints itself as a neighbour with numerous nodes, perhaps the whole system, however in some

Table 1 Summary of RPL attack that violates confidentiality and integrity and their effect on topology along with countermeasures

Attack	Effect on network parameters	Countermeasures	Remarks
Selective forwarding	Disrupt routing path and filter any protocol	Heartbeat protocol and end-to-end packet loss adaptation algorithm [9]	Both are attack detection techniques
Sinkhole	Compromising huge traffic passing through attacker node	SVELTE [11] and parent failover rank authentication technique [12]	IDS solution, parent failover detects the attack, and rank authentication technique avoids the attack
Wormhole	Disrupt the network topology and traffic flow	Markle tree authentication [13]	Attack prevention technique
Blackhole	Packet delay and control overhead, increased route traffic	SVELTE [11], monitoring of counter and parent failover technique	Attack detection techniques
Local repair attack	Control overhead, low delivery ratio, disrupt routing and traffic flow	IDS system	Attack detection mechanism
Rank attack	Packet delay, delivery ratio and generation of un-optimized path and loop	SVELTE [11], VeRA [14], TRAIL [15]	SVELTE is attack detection technique, while VeRA and TRAIL are prevention techniques
Version number	Control overhead, low packet delivery ratio, end-to-end delay	VeRA [14]	Attack prevention technique
Sybil and clone ID attack	Route compromise and disrupt traffic flow	Track of number of instances of each identity and geographical location; IPsec in tunnel mode is used	It helps to identify the cloned identities
Neighbour attack	False route, route disruption and resource consumption	TRAIL [15]	TRAIL prevents the attack

Table 2 Summary of RPL attack that violates availability and their effect on topology along with countermeasures

Attack	Effect on network parameters	Countermeasures	Remarks
Hello flooding attack	Route formation through attacker node, dissipation of sensor battery power	RPLs global and local repair	Mechanisms eliminate the attack existence
Blackhole	Packet delay and control overhead, increased route traffic	SVELTE [11], monitoring of counter and parent failover technique	Attack detection techniques
Rank attack	Packet delay, delivery ratio and generation of un-optimized path and loop	SVELTE [11], VeRA [14], TRAIL [15]	SVELTE is attack detection technique, while VeRA and TRAIL are prevention techniques
Denial of service	Make resources unavailable	IDS based solution	Framework is not compactable to general network architecture
DIS attack	Packet delay, resource consumption	TRAIL [15]	TRAIL prevents the attack
Neighbour attack	False route, route disruption and resource consumption	TRAIL [15]	TRAIL prevents the attack

of the nodes, conceivably the attacker nodes region, when attempting to join the malicious node or attacker, their message may get lost in the light of the fact that the attacker may be out of range [16]. This attack can be enhanced as sinkhole attack.

5. **Blackhole Attack**: In the blackhole attack, like an opening which sucks in everything, a malicious node or attacker drops all information packets quietly. Thus, all packets in the network routing through that node are dropped [10].

6. **Local Repair Attack**: In nearby repair, attacker with no issue with interface quality occasionally sends the local repair message. This causes the local repair around the nodes which hear the local repair message. Local repair attack makes more effect on delivery proportion than some other sort of attacks [17].

7. **Rank Attack**: By changing rank value, an attacker can attract in youngster or child node for choosing as parent or guardians or enhance some other metric and can attract vast activity going towards the root [18].

8. **Denial-of-Service Attack**: Denial-of-service or distributed denial-of-service attack is endeavouring to make assets inaccessible to its proposed client. In RPL, this attack can bring utilizing the IPv6 UDP packet flooding [10].

9. **Version Attack**: This attack happens by distributing the higher version number of DODAG tree. At the point when node gets the new higher adaptation or version number DIO message, they begin the development of new DODAG tree [10].

10. **Sybil Attack and clone ID Attack**: In clone ID attack, an malicious node copies the personalities of legitimate nodes onto another physical node, access to on large part of the system or so as to beat voting schemes [9]. In Sybil attack which is like clone ID attack, a malicious node utilizes a few consistent entities on the same physical node [10].

11. **DODAG Information Solicitation (DIS) Attack**: DIS message is utilized by new node to get the topology information before joining the RPL network. In this attack, vindictive node intermittently sends the DIS messages to its neighbours. At the point when the DIS messages disseminate by the malicious node, the recipient node after getting DIS message resets the DIO clock timer expecting something turned out wrong with the topology around it [17].

12. **Neighbour Attack**: In this attack, the vindictive or malicious node disseminates DIO messages that it got without including information of his own. The node who gets this kind of messages may believe that new neighbour node sends this DIO message. The victim node endeavours to choose the node which is not in the vicinity as parent node and change the path to out range neighbours [17]. This attack is like the wormhole attack with unique instance of just selectively forwarding of DIO message.

4 Comparative Analysis of Secure Routing Mechanisms

There are different outcomes of attacks like arrangement of un-optimized route or path, development of loop with no detection, existed enhanced path in the topology that can never be utilized. There is a decrease in packet delivery proportion with slight change in end-to-end delay when number of malicious nodes exceeded. In this section, an overview of the distinctive secure routing conventions proposed by the research fraternity and additionally a comparative analysis of all the routing mechanisms alongside their limitations are presented in Table 3.

1. **Version Number and Rank Authentication(VeRA)**: VeRA [14] abstains from malicious nodes from asserting lower rank than genuine rank by using one-way hash chain. One-way hash chain is progressive use of hash function, e.g. $h(h(h(h(x))))$, indicated $h^4(x)$ and along these lines every node computes $h^4(x)$ from $h(x)$, be that as it may, it cannot compute $h(x)$ from $h^4(x)$. Thus, one-way hash chain can be utilized to guarantee rank to be expanded entirely from the DODAG root to leaf nodes. Every node confirms neighbours' rank by figuring hash chain more than once. The DODAG root produces an irregular random number r and computes hash chain $h^c = h^R(r)$, where R is the most extreme rank an incentive in the DODAG. VeRA expects that every node knows the hash

Table 3 Comparative analysis of secure routing mechanisms

Protocols	Complexity	Techniques	Attack addressed	Limitations
VeRA [14]	High	One-way hash chain strategy	Version number and rank attacks	High complexity, vulnerable to hash chain forgery attack and replay attack
SVELTE [11]	Medium	Rank authentication technique	Sinkhole, blackhole and rank attack	High false detection rate
TRAIL [15]	Medium	Path validation	Rank, DIS and neighbour attack	Message overhead, suffers from scalability
Parent failover and rank authentication [12]	High	One-way hash chain strategy	Sinkhole and blackhole attack	High complexity, vulnerable to hash chain forgery attack
Markle tree authentication [13]	Medium	Estimates hash by using ID of a node and public key	Wormhole attack	Eliminates the existence of only warmhole attack
TSRF [19]	High	Direct and indirect trust metric systems	On-off, conflicting behaviour, selfish, badmouthing and collusion attack	High memory consumption, new rogue node identification evades detection

chain value $h^{\mathrm{R}}(r)$ and R since the DODAG root sends this incentive ahead of time. At the point when a node sends a DIO message, a node sends $h_{\mathrm{sender}} = h^{\mathrm{Rself}}(r)$ including its DIO message, where Rself is a rank of its own. Accepting a DIO message, every recipient node checks if $h^{\mathrm{R-Rsender}}(h_{\mathrm{sender}}) = h^{c}$, where Rsender is the sender's rank. If $h^{\mathrm{R-Rsender}}(h_{\mathrm{sender}})! = h^{c}$, then every node considers the sender as a vindictive or malicious node.

2. **SVELTE**: SVELTE [11] recognizes vindictive nodes by discovering rank irregularity in the DODAG root. The DODAG root judges the node as a malicious node if its rank is lower than its parent node since the rank of parent node must be lower than that of its descendent or child node. Along these lines, as the principal module, the DODAG root asks for each node to report its own particular rank and neighbour node positions or rank. After receiving a request, every node reacts with its neighbours' and parents' rank. As the second module, the DODAG root examines the gathered information and distinguishes malicious nodes. The DODAG root checks every node's rank irregularity by contrasting the rank that it demands and the rank that its neighbours report. The DODAG root judges that a node is a malicious node if the distinction between the ranks is larger than the pre-characterized threshold.

3. **Trust Anchor Interconnection Loop (TRAIL)**: TRAIL [15] distinguishes attacking or malicious node by discovering rank irregularity in each parent node rather than the DODAG root. A parent node judges its child node as a vindictive node if its child rank is lower than itself since the rank of parent node must be lower than that of its youngster or child node. A child node sends its rank to its parent node to check that its rank is straightforward. Each parent node checks whether the two conditions are fulfilled: (1) whether the rank in the message is higher than its own. (2) Whether the rank that sender advertises lies in the middle of the rank in the message and its own. When none of these condition is satisfied, the parent node considers its child node as a malicious node.

4. **Parent Failover and Rank Authentication Technique**: The rank validation method [12] depends on one-way hash strategy. The root starts to produce hash an incentive by picking arbitrary value and disseminate it in DIO message. All nodes compute the hash value utilizing previously received one and again disseminates it utilizing DIO message. Consider that vindictive node does not compute the hash value, it just disseminates received DIO message. Every node stores the hash value received by its parent alongside number of hops in the way. At the point when root node disseminates arbitrary random number safely, at that point node can check its parent rank utilizing that intermediates hop number. Parent failover technique utilizes unheard nodes set (UNS) field in DIO message showing that the nodes are in sinkhole compromised route. In the event that the node gets the DIO message containing its ID in UNS, it includes its parent in blacklist.

5. **Markle Tree Authentication**: Markle tree development [13] begins from leaf nodes to root. It utilizes ID of node and public key for estimation of hash. Each parent is recognized by its descendants or child. Validation of any node starts with the root node up to the node itself. In the event that any node failed to validate, at that point descendant nodes stay away from the wrong parent selection.

6. **A Trust-aware Secure Routing Framework(TSRF)**: Its outline intended for WSNs depended on trust determination which comprises of immediate and delayed indirect analysis of behavioural examples of sensor nodes with trust values amongst nodes spoke to in a range from 0 to 1. A 0 trust value means no trust exists amongst nodes and a 1 demonstrating a decent level of trust for the corresponding node.

5 Conclusion and Future Work

In this paper, we have reviewed various attacks on RPL network along with their highlighted strength and weaknesses that can be exploited by IDSs. This paper concludes that various mechanisms and countermeasures are proposed against the attacks on RPL network. In addition, a comparative analysis of the countermeasures is performed to protect communication on the IoT. The main aim of this paper is to highlight the importance of security in the RPL network and to provide grounds to the future researchers who plan to design and implement secure protocols and IDSs for LLNs consisting of resource-constrained devices.

References

1. Zhao, K., Ge, L.: A survey on the internet of things security. In: Proceedings of the 9th International Conferences on Computational Intelligence and Security (CIS), pp. 663–667 (2013)
2. Gubbi, J., Buyya, R., Marusic, S., Palaniswami, M.: Internet of Things (IoT): a vision, architectural elements and future directions. Future Gener. Comput. Syst. **29**, 1645–1660 (2013)
3. Winter, T., Thubert, P., Brandt, A., Hui, J., Kelsey, R., Levis, P., Pister, K., Struik, R., Vasseur, J., Alexander, R.: RPL: IPv6 routing protocol for low-power and lossy networks. IETF, RFC 6550, March (RPL: IPv6 routing protocol for low-power and lossy networks 2012)
4. Airehrour, D., Gutierrez, J., Ray, S.K.: Secure routing for internet of things: a survey. J. Netw. Comput. Appl. **66**, 198–213 (2016)
5. IEEE Standard for Local and Metropolitan Area Networks-Part 15.4: Low-Rate Wireless Personal Area Networks (LR-WPANs) Amendment 1: MAC Sublayer IEEE Std. 802.15.4e-2012 (Amendment to IEEE Std. 802.15.4-2011), pp. 1–225 (2012)
6. Thubert P. et al.: RPL: IPv6 routing protocol for low power and lossy networks, RFC 6550 (2012)
7. Ishaq, I.: IETF standardization in the field of the Internet of Things (IoT): a survey. Sens. Actuator Netw. **2**(2), 235–287 (2013)
8. Palattella, M., et al.: Standardized protocol stack for the internet of (important) things. IEEE Commun. Surv. Tutorials **15**(3), 1389–1406 (2013)
9. Wallgren, L., Raza, S., Voigt, T.: Routing attack and countermeasures in the RPL-based internet of things. Int. J. Distrib. Sens. Netw. 9(8) (2013)
10. Pongle, P., Chavan, G.: A survey: attacks on RPL and 6LOWPAN in IoT. In: International Conference on Pervasive Computing (ICPC). IEEE, New York (2015)
11. Raza, S., Wallgren, L., Voigt, T.: SVELTE: real-time intrusion detection in the internet of things. Ad Hoc Netw. **11**(8), 2661–2674 (2013)
12. Weekly, K., Pister, K.: Evaluating sinkhole defense techniques in RPL networks. In: Proceedings of 20th IEEE International Conference on Network Protocols (ICNP), pp. 1–6 (2012)
13. Khan, F.I., Shon, T., Lee, T., Kim, K.: Warmhole attack prevention mechanism for RPL based LLN network. In: Fifth International Conference on Ubiquitous and Future Networks (ICUFN). IEEE, New York (2013)
14. Dvir, A., Holczer,T., Buttyan, L.: VeRA-Version number and rank authentication in RPL. In: Proceedings of 8th IEEE International Conference on Mobile Ad-hoc and Sensor Systems, MASS 2011, pp. 709–714 (2011)
15. Perrey, H., Landsmann, M., Ugus, O., Schmidt, T.C., Wahlisch, M.: TRAIL: Topology Authentication in RPL. EWSN-16, arXiv:1312.0984 (2013)
16. Tsao T. et al.: A security threat analysis for routing over low power and lossy networks. Draft-ietf-roll-security-threats-2011, Routing Over Low-Power and Lossy Networks Internet-Draft (2011)
17. Le, A., Loo, J., Luo, Y., Lasebae, A.: The impacts of internal threats towards routing protocol for low power and lossy network performance. In: IEEE Symposium on Computers and Communications (ISCC) (2013)
18. Le, A., Loo, J., Lasebae, A., Vinel, A., Chen, Y., Chai, M.: The impact of rank attack on network topology of routing protocol for low-power and lossy networks. IEEE Sens. J. **13**(10), 3685–3692 (2013)
19. Hummen, R., Wirtz, H., Ziegeldorf, J.H., Hiller, J., Wehrle, K.: Tailoring end-to-end IP security protocols to the internet of things. In: 21st IEEE International Conference on Network Protocols (ICNP), pp. 1–10. IEEE, New York (2013)

Reduction of Web Latency: An Integrated Proxy Prefetch-Cache System Framework

Sirshendu Sekhar Ghosh, Moujhuri Patra and Aruna Jain

Abstract Present Web infrastructure is severely suffering from heavy congestion and enormous traffic burden upon the widely distributed client network around the globe due to proliferating growth in terms of overpopulated users along with day-by-day addition of latest complex and resource hungry Web applications, tools, and services. This leads to continuous increase in Web access delay (latency) faced by the users challenging the unprecedented popularity of the World Wide Web (WWW). Proxy caching is a renowned and current strategy for refining the distributed Web-based system performance particularly applicable for remote areas which severely suffers from poor network communication and limited bandwidth availability. Web prefetching is another very useful strategy which prefetches and keeps the closely related Web objects to be accessed by the users' in future into cache, based on their interest pattern. But still, there is a significant performance gap between the users' requests and the result provided. The intelligent and sophisticated integration of these two strategies along with one more performance enhancement technique, i.e., Web log mining, can provide an excellent and most successful solution to minimize the problem of latency and improve the overall Web quality of service. This paper provides an Integrated Proxy Prefetch-Cache System framework with the above benefits to satisfy maximum clients' requests with improved response time as well as with minimum latency. Also, the framework has been successfully incorporated into the highly scalable, hierarchical as well as distributed geographical region-based proxy server clusters with enhanced content co-operation, efficient metadata manage-

S. S. Ghosh (✉)
Department of Computer Applications, National Institute of Technology,
Jamshedpur 831014, Jharkhand, India
e-mail: ssghosh.ca@nitjsr.ac.in

M. Patra
Department of Master of Computer Applications, Netaji Subhash Engineering College,
Garia, Kolkata 700152, India
e-mail: moujhuri@gmail.com

A. Jain
Department of Computer Science & Engineering, Birla Institute of Technology,
Mesra, Ranchi 835215, Jharkhand, India
e-mail: arunajain@bitmesra.ac.in

© Springer Nature Singapore Pte Ltd. 2019   621
A. Abraham et al. (eds.), *Emerging Technologies in Data Mining and Information Security*, Advances in Intelligent Systems and Computing 814,
https://doi.org/10.1007/978-981-13-1501-5_55

ment, and low access delay. Using real-time Web proxy server network deployment with online trace-driven results analysis, the current framework guarantees improved object consistency between clustered proxy server caches and original Web servers.

Keywords Integrated Proxy Prefetch-Cache System Framework · Response time · Hit ratio · Distributed clustered network · Response time gain factor

1 Introduction

Current Web architecture is mostly dynamic and highly distributed around the globe where numerous Web objects are scattered in various different hosts. Any Web object request from a client travels through many routers, gateways, hubs, and several networking applications from one end of the world to million miles away its destination server. Also, due to multiple user requests for the same document in this scenario considerable amount of Web latency is experienced. For improving response time, the most popular software solution is Web caching. In this strategy, frequently requested (accessed) objects by users are stored locally within a cache (preferably at proxy) by which frequent object requests can be served effectively instead of contacting repeatedly the remote origin server. Among the three levels of Web caching deployment mechanisms, i.e., at client, proxy, and origin server level, proxy server (residing between client and Web server) acts a vital role in lessening the response time and thus saving significant network bandwidth. Each time whenever a client issues an object request, initially the proxy is contacted and if it contains the requested object within its local cache; this is considered as a cache hit; otherwise, a cache miss occurs and the proxy forwards the same request on behalf of the user to retrieve the object from a peer or parent proxy cache or still not found then finally contacts the origin server. If the requested object is cacheable (based on the information provided by the origin server or resolved from the URL), the proxy may or may not decide to add a copy of the object to its local cache. Proxy caching takes advantage of the Web object's temporal locality to reduce the user-perceived latency, minimizes network congestion, and bypasses bottlenecks at origin server and overall system performance improvement. To further improve the Web system performance, it can be integrated with another effective technique: Web prefetching which deduces user's future most likely requested objects into a local (proxy) cache populating at idle time or when bandwidth is available prior to an explicit user request. An intelligent and sophisticated integration of these two techniques can considerably reduce the Web latency by predicting and storing the most frequent and associated future requested objects to be accessed by users into a local (proxy) cache. The performance of Web proxies employing combined prefetching and caching strategies can further be enhanced with Web log mining technique, which is by deriving useful knowledge and extracting meaningful pattern from the Web access log analysis which are automatically stored on the server (origin or proxy) to predict current as well as future objects requests by the clients. Collectively, these techniques will reduce significant Web latency, minimize heavy network congestion and traffic, and improve reliability by effectively serving more users' requests lessening heavy workload from the origin Servers and also protecting them from "flash crowd" events. But systems with aggressive prefetching policies have serious drawbacks as some prefetched objects may not eventually be requested by users causing again the increase of network traffic as well as servers' loads. In order to overcome such limitation, high accuracy prediction models are being used.

In the current paper, we have implemented an integrated proxy prefetch-cache system framework intelligently combining both proxy caching and prefetching strategies. Also, we have incorporated a distributed intelligent agent (IA)-controlled approach into our integrated proxy framework. Cooperative, hierarchical, and distributed Web caching with prefetching have been combined with Web log mining techniques in our approach. Further, we have implemented a stepwise client request handling algorithm successfully justifying our combined strategy for widely distributed clustered network. The rest of the paper is prepared as follows. The background study and related research work carried out by eminent researchers in this area have been discussed in Sect. 2. In Sect. 3, we introduce in detail the Integrated Proxy Prefetch-Cache System Framework along with IA-controlled proxy-based combined prefetch-cache model in distributed clustered architecture. The detailed experimental setup along with comprehensive result analysis has been showed in Sect. 4. And finally, in Sect. 5 we conclude the paper explaining certain in-future work toward this invaluable research direction.

2 Related Work

Ghosh et al. [1] explained a generous survey on several different Web prefetching and caching strategies toward Web latency reduction. Griffioen et al. [2] presumed that caching and prefetching share the same cache space and presented that integrated Web caching and prefetching can significantly improve the performance of cache system architecture. Cao et al. [3] explained a model of integrated Web caching and prefetching on file system, and based on that, she illustrated that the integrated model could valuably reduce the elapsed times of the applications upto 50%. Teng et al. [4] implemented Integration of Web Caching and prefetching (IWCP) algorithm by intelligently integrating Web prefetching and caching which outperforms the previous LNC-R-W3-PC algorithm in terms of hit ratio and delay-saving ratio. Alexander [5] projected an active caching mechanism to support caching of dynamic contents within the Web proxies. Nair et al. [6] explained a dynamic prefetching strategy implemented at proxy server level by sophisticatedly integrating Web caching and prefetching and discovered that the cache hit ratio is interestingly increased to 40–75% and consequently object latency is reduced to 20–63%. Rodriguez et al. [7] examined Web caching architectures on the basis of load, connection time, and transmission time latencies. Tiwari et al. [8] showed distributed Web caching (DWC) scheme which supports successful metadata exchange among proxy servers periodically to maintain a complete network of metadata management. Garg et al. [9] implemented the distributed Web caching with clustering (DWCC) technique where they provided a solution aspect for scalability and robustness problem in Web caching scenario due to heavy load. Sengupta [10] sketched out the basic design principle of an adaptive caching framework for assisting cache management along with forwarding of Web query decisions. Francs [11] implemented an integrated prefetching technique into search engine, inspected the prefetching of search results, and demonstrated that using appropriate prefetching will successfully increase the cache hit ratio by almost 50%.

3   Integrated Proxy Prefetch-Cache System Framework

The integrated prefetch-cache system framework is deployed at proxy server level as it is built to serve a wide range of clients' requests. To diminish the overpopulated origin server overload, bottlenecks and repeated user access latency in today's extremely congested and complex network scenario, deployment of proxy server between Web clients and origin server is very much essential. Figure 1 illustrates our integrated system framework whose working functionalities are explained in detail as follows:

1. Web clients' issue object requests which are first of all sent to the Cache manager which authenticate the legitimate users by their saved user names and passwords by checking the stored preference list. The local (proxy) cache is logically partitioned into two portions: normal cache queue (with LFU) and prefetch cache queue (with LRU). The cache (replacement) manager internally issues the hybrid replacement algorithm that collects information of the requested object in proxy cache and makes the proper replacement decision.
2. Cache manager checks if the object is present in the proxy cache (either in normal cache or prefetch cache). If the object is found, it is considered as cache hit and sends it to the client in response with minimal latency; otherwise, a cache miss occurs and sends the underlying request to the origin server. It also prepares a list of accessed as well as missed URLs and transfers to the prediction engine.
3. The prediction engine maintains a hash table containing both the lists and their respective weight information. It also internally initiates and runs a prediction algorithm to predict the next user's object request and supply these predictions as valuable hints to the prefetch engine.
4. The prefetch manager within the prefetch engine maintains the hints, generates and stores interesting prefetching rules at prefetching rule depository by discovering users' object access patterns and browsing behavior by reading and analyzing the proxy server's access log periodically. Thus, the users' preference list of objects can be populated and supplied to the cache manager. It also decides whether to prefetch those objects from origin server or not depending on certain conditions like the available bandwidth or the idle time and automatically

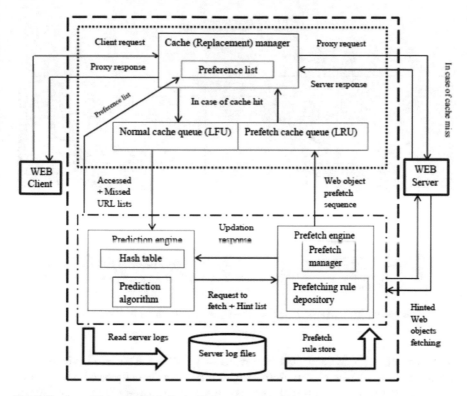

Fig. 1 Integrated Proxy Prefetch-Cache System Framework

resumes downloading the hinted Web objects from the origin server and sends as prefetch sequence queue of the prefetch cache. Also, it sends the objects updation information as a valuable response to the prediction engine.

The World Wide Web is entirely distributed around the globe and the proxy servers satisfying billions of clients' requests are grouped together into humongous number of several clusters accordingly as the clients' interests and behavior as well as geographical region based. At the highest level, origin servers (OS) are connected to PSs within several clusters via various form of communication medium. The proxies employ first level of cache storing frequently referenced objects by clients from OS and also provide cooperative document/page transfer. We have successfully incorporated our integrated framework into the current distributed clustered client network architecture. Our intelligent agent (IA)-controlled combined proxy prefetch-cache framework architecture has been depicted in Fig. 2. The framework consists of origin server (OS), proxy server clusters, IA-controlled local cache (LC) and clients. We have deployed in our scheme a second level of caching (a local cache, LC within the cluster located between proxies and clients controlled by some intelligent agents). This LC accumulates the most frequently accessed objects by clients from the upward (first)-level proxies. The LC contains two important modules: prefetch-

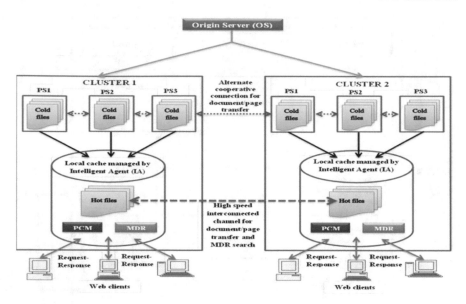

Fig. 2 Intelligent Agent (IA)-controlled combined proxy prefetch-cache framework

cache manager (PCM) and metadata repository (MDR). The PCM will monitor the prefetching of Web objects from proxies either periodically or with bandwidth availability or in idle time. It will also update the LC, provide suitable replacement policy and satisfy clients' requests. The frequently accessed objects stored in first level caches (i.e., proxies within cluster) from OS are named "cold files" while more frequently accessed objects those are stored into LC from upward proxies are named "hot files." The MDR contains information about documents (i.e., metadata) stored in LC as well as PSs within same and neighboring cluster servers (CS_{i-1} and CS_{i+1}).

First of all clients' requests are verified by this MDR whether it is in its own LC or within neighboring cluster's LC and it guides any request to the corresponding PS of same cluster or in neighboring clusters and if fails then finally the request transmitted to the highest level OS. By providing a high-speed document/page transfer among hot files of neighboring LCs, we can minimize the latency time of satisfying client requests without cooperative objects transfer among PSs. In our strategy, the IA located within the LC will entirely monitor and provide smart mechanism to fetch objects from PSs, populate the LC with hot files, update the cache with effective cache replacement policy, search MDR for any requested object and finally keep track of high-speed object transfer among LCs. Whenever a client issues an object request to the proxy server, firstly all the queue lengths of each cluster are verified and if it is under the permissible limit then further the queue lengths of corresponding PSs are verified. If the allowed no. of client limit of PSs has not surpassed, the client's request is served; otherwise, the request is forwarded to a relatively less loaded PS. This makes efficient load-balancing utilization in the overall network for proper handling of total client requests. This strategy will come into work in the following

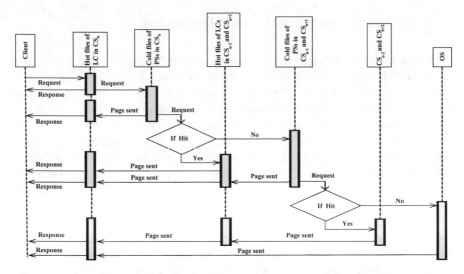

Fig. 3 Sequence diagram showing all possible phases of client's request

conceivable phases whenever one client requests some document/object/page from a proxy server (PS_i) of cluster n (CS_n) as depicted by the following sequence diagram shown in Fig. 3. We have used our own hybrid cache replacement policy [12, 13] to bring the cold files or evict the hot files in order to update the cache content into LC. This approach significantly improves hit ratio and reduces latency time by satisfying user requests from two hierarchies of cache, first prefetched from OS and second prefetched from PSs. This strategy also lessens the overall network cost since updating such a vast and complex metadata in most cases is relatively quick and cheaper than updating the (potentially much larger) requested Web objects. Since most proxy caching systems are installed with high-speed interconnect between cluster servers in order to meet the needs of todays' compute-intensive and large-scale applications, the hot files will be successfully migrated between LCs with tremendously high throughput and low latency.

Our distributed IA framework approach is a sophisticated combination of both hierarchical and cooperative Web caching strategies combined with prefetching technique that can easily be deployed in the current as well as future Web-clustered architecture. The scheme will alleviate the additional overhead of intricate metadata management of the PSs, enhance scalability, and also reduce the heavy network traffic as well.

4 Experimental Setup and Result Discussion

Our integrated proxy prefetch-cache system framework has been studied with the live proxy trace collected from the running Web proxy server of Birla Institute of Technology (BIT), Mesra, Ranchi, Jharkhand. The proxy is enormously popular among the various faculty members, administrative and office staffs, research scholars, and associates along with huge number of students of as many twenty-five departments along with various administrative sections, blocks, hostels, and quarters. In our experiment, we have taken the proxy traces refer to the period from 12-Feb-2012:07:45:03 to 18-Feb-2012:00:00:04, roughly of one week. In our experiment, we have taken into consideration the following four parametric scenarios in our system framework evaluation process: requested objects reverted from the origin server, requested objects reverted from proxy cache without verification, requested objects reverted from local (Proxy) cache after validating that they have not been changed, and requested objects reverted from the origin server by updating objects in cache. The improved proxy cache performance result analysis considering the above four cases with the chosen one-week proxy traces is revealed in Table 1.

Our thoroughly analyzed results show that around 38% of all clients' requests are returned from proxy cache without verification, which is considerably better of using our integrated proxy framework. Figure 4 depicts the pictorial representation of percentage of total requested objects according to four considerable parameters.

Table 1 Proxy cache performance result analysis

Parametric scenarios	Requested objects	Total requests (in %)	Total bytes
Requested objects reverted from the origin server	20,873	59.30	617.73 MB
Requested objects reverted from Proxy cache without verification	13,450	38.20	26.93 MB
Requested objects reverted from local (Proxy) cache after validating that they have not been changed	489	1.40	0.99 MB
Requested objects reverted from the origin server by updating objects in cache	59	0.10	14.62 KB
Information not available	354	1.00	49.74 KB
Total	35,225	100.00	645.72 MB

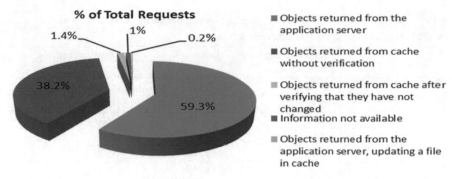

Fig. 4 Proxy cache performance analysis

In order to evaluate our integrated system framework, we have used two most appropriate performance metrics: response time (RT) and hit ratio (HR) which are most widely used parameters in evaluating the performance of integrated prefetch-cache system. RT is calculated as the ratio of the sum of transfer time of requested objects satisfied by the local (proxy) cache over the sum of total downloading time of objects, and HR is calculated as the percentage of presence of object requests that can be satisfied by the local (proxy) cache. We assume that, N is the total number of requested objects and $\delta_i = 1$, if the requested object i is found in the local (proxy) cache, while $\delta_i = 0$, otherwise. Mathematically, it is expressed as follows:

$$\text{Hit Ratio (HR)} = \frac{\sum_{i=1}^{N} \delta_i}{N} \tag{1}$$

$$\text{Response Time (RT)} = \frac{\sum_{i=1}^{N} t_i \delta_i}{\sum_{i=1}^{N} t_i} \tag{2}$$

t_i is the time to transfer the ith referenced Web object from origin server to the proxy cache. Also, the amount of gain in Web cache response time can be expressed using another important factor: Response time gain factor (RTGF) by comparing average Web object hit ratio and object response time.

$$\text{Response Time Gain Factor (RTGF)} = \frac{(\text{Time Without Cache} - \text{Time With Cache}) \times 100}{\text{Time Without Cache}} \tag{3}$$

Tables 2 and 3 summarize average network traffic at the beginning and after a certain time of consuming Web proxy cache along with the calculated RTGF parameter. We have monitored the overall HTTP requests made by individual client systems and calculated the summative number of hits applying all four strategies (cooperative, hierarchical, distributive, and combined) with a total of 512 MB of proxy cache.

An increment in HR designates enhanced user's satisfaction showing a better user servicing and a lower RT signifies an improved network performance and reduction of user-perceived latency. Figures 5 and 6 depict the hit ratio (HR) and response time (RT) of the integrated proxy prefetch-cache framework compared with other three

Table 2 Average network traffic at the beginning of using integrated proxy framework

At the beginning of using the integrated proxy framework

Requests	Avg. processing time (s)	Total bytes	Proxy HR (in %)
1924	133.00	3.61 GB	0.00
1708	126.40	9.75 MB	0.00
1654	127.10	8.33 MB	0.00
1927	115.20	8.42 MB	0.00
1731	121.10	8.41 MB	0.00
1713	121.60	8.51 MB	0.20
1642	126.60	8.36 MB	0.00
1835	128.70	8.48 MB	0.00
1859	126.30	9.13 MB	0.00
2455	135.10	11.50 MB	0.00

Table 3 Average network traffic after a certain time with the integrated proxy framework along with the calculated RTGF

After a certain time of using the integrated proxy framework				RTGF
Requests	Avg. processing time (s)	Total bytes	Proxy HR (in %)	
6297	58.90	2.94 GB	1.00	55.7142
6157	61.30	37.91 MB	0.00	51.5031
5954	62.30	36.27 MB	0.00	50.9834
6004	58.00	36.19 MB	0.00	49.6527
5934	57.90	36.12 MB	0.00	70.1156
5901	58.30	35.80 MB	0.00	50.1187
5899	57.30	35.80 MB	0.00	54.6318
6249	54.50	38.34 MB	2.80	57.6543
6014	57.60	36.69 MB	0.30	54.3942
7212	52.00	40.92 MB	12.00	69.7113

strategies taken individually. Our integrated framework outperforms the other three strategies taken individually with the increase in user loads and size of the cache for both the performance metrics.

We have calculated the HR with respect of variable local (proxy) cache size and the RT in milliseconds (ms) with respect to increasing user load (in %). It is observed that for cooperative and hierarchical strategies, the HR is almost identical in all cache sizes. In contrast, for RT upto 25% load level all the four strategies provide similar response time. This is because upto that much load level, the local proxy cache is not fully populated with objects from proxies and after reaching the specified load level the local proxy cache is started updating using our own hybrid cache replacement policy [12]. The comprehensive result analysis indicates that the

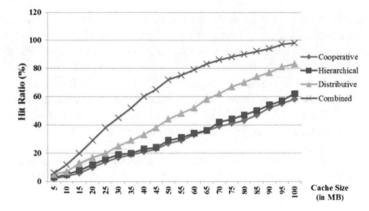

Fig. 5 Hit ratio (HR)

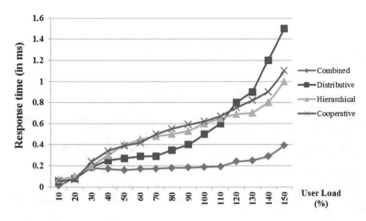

Fig. 6 Response time (RT)

combined framework has improved the overall performance in terms of two measures under consideration, i.e., HR increment up to 69.13% and RT minimization up to 48.33% compared to other individual strategies, i.e., cooperative, hierarchical, and distributive.

5 Conclusion

In the current paper, we have presented An Integrated Proxy Prefetch-Cache System Framework for Distributed Clustered Client Network Architecture. The main objective is to satisfy most of the clients' requests through a local proxy cache populated from various distributed proxies within the same or neighboring clusters such that a very few requests are sent to the remote origin server. The IA provides intelligent mechanism for retrieving requested objects within the local cache of respective or neighboring clusters, implement prefetching and caching of objects in two levels and also clients' optimum load balancing through a relatively less loaded cluster. Also, the framework can easily be adopted and deployed in current dynamic Web infrastructure. We wish to outspread our approach in future by including more diversified clusters and many more distributed features such as encrypted file handling, cookies management, optimal load balancing, and dynamic document updation within proxies residing in clusters as per in-future coming requirements.

Acknowledgements We would like to acknowledge all the Institute as well as campus Server and Network support staffs from Birla Institute of Technology Mesra, Ranchi, Jharkhand, for providing the secure and voluminous proxy workload traces for our research and development purpose. Also, they were always very supportive throughout the entire work done in establishing and maintaining our experimental proxy server cluster network setup along with necessary hardware and software support.

References

1. Ghosh, S.S., Jain, A.: Web latency reduction techniques: a review paper. Int. J. Inf. Technol. Netw. Appl. **1**(2), 12–18 (2011)
2. Griffioen, J., Appleton, R.: Reducing file system latency using a predictive approach. In: Proceedings of the USENIX Summer Conference, pp. 197–207 (1994)
3. Cao, P., Felten, E.W., Karlin, A.R., Li, K.: A study of integrated prefetching and caching strategies. In: Proceedings of the ACM SIGMETRICS Conference on Measurement and Modeling of Computer Systems, pp. 188–197 (1995)
4. Teng, W.G., Chang, C.Y., Chen, M.S.: Integrating Web caching and Web prefetching in client-side proxies. In: Proceedings of the IEEE Transactions on Parallel and Distributed Systems, vol. 16, no. 5, pp. 444–455 (2005)
5. Alexander, H., Khalil, I., Cameron, C., et al.: Cooperative web caching using dynamic interest-tagged filtered bloom filters. In: Proceedings of the IEEE Transactions on Parallel and Distributed Systems, vol. 26, no. 11 (2015)
6. Nair, A.S., Jayasudha, J.S.: Dynamic web pre-fetching technique for latency reduction. In: Proceedings of the International Conference on Computational Intelligence and Multimedia Applications, pp. 202–206. IEEE (2007)
7. Rodriguez, P., Spanner, C., Biersack, E.W.: Analysis of web caching architectures: hierarchical and distributed caching. IEEE/ACM Trans. Netw. **9**(4) (2001)
8. Tiwari, R., Khan, G.: Load balancing in distributed web caching: a novel clustering approach. In: Proceedings of the International Conference on Methods and Models in Science and Technology, vol. 24, pp. 341–345 (2010)
9. Garg, L., Tiwari, R.: Robust distributed web caching scheme: a dynamic clustering approach. Proc. Int. J. Eng. Sci. Technol. **3**(2), 1069–1076 (2011)
10. Sengupta, A., Amuru, S.D., Tandon, R., et al.: Learning distributed caching strategies in small cell networks. In: Proceedings of the 11th International Symposium on Wireless Communications Systems (ISWCS). IEEE (2014)
11. Francs, G., Bai, X., Cambazoglu, B.B., et al.: Improving the efficiency of multi-site Web search engines. In: Proceedings of the 7th ACM International Conference on Web Search and Data Mining. ACM (2014)

12. Ghosh, S.S., Kumar, V., Jain, A.: A novel hybrid policy for web caching. Int. J. Eng. Res. Develop. **6**(11), 15–22 (2013)
13. Ghosh, S.S., Jain, A.: Hybrid cache replacement policy for proxy server. Int. J. Adv. Res. Comput. Commun. Eng. **2**(3), 1527–1532 (2013)

Disease Prediction on the Basis of SNPs

Jyotiprakash Panigrahi, Bhabani Shankar Prasad Mishra
and Satya Ranjan Dash

Abstract In DNA and RNA, five types of nitrogenous bases are present; these are adenine (A), uracil (U), guanine (G), thymine (T), and cytosine (C). Sometimes, these arrangements are altered which is known as single nucleotide polymorphism. These polymorphisms appear due to two causes (i) mutation and (ii) disease. By classifying these two types of SNPs, we can conclude a disease-causing SNP. In this paper, we describe genetic analysis of simple and complex disease evaluation by various methods. Here, we describe Apriori algorithm, genetic algorithm, machine learning approach like support vector machine (SVM). An approach of Apriori-Gen algorithm is also discussed which evaluates statistical interaction between several SNPs to find association among them. Univariate marginal distribution algorithm (UMDA) and support vector machine (SVM) are also used to find disease identification. USVM is used not only for its redundancy feature but also it solves parameters selecting problem of SVM.

Keywords Single nucleotide polymorphisms · SVM · UMDA · USVM
RF · GA · Apriori algorithm

1 Introduction

Genetic mapping provides a powerful approach to identify the presence of SNP in gene. These single nucleotide polymorphisms are present in DNA which is very similar to mutation. But these SNPs are corrupted human DNA, which causes a major

J. Panigrahi · B. S. P. Mishra
School of Computer Engineering, KIIT, Deemed to be University, Odisha 751024, India
e-mail: jyoti.p.80386@gmail.com

B. S. P. Mishra
e-mail: bsmishrafcs@kiit.ac.in

S. R. Dash (✉)
School of Computer Applications, KIIT, Deemed to be University, Odisha 751024, India
e-mail: sdashfca@kiit.ac.in

© Springer Nature Singapore Pte Ltd. 2019
A. Abraham et al. (eds.), *Emerging Technologies in Data Mining and Information Security*, Advances in Intelligent Systems and Computing 814,
https://doi.org/10.1007/978-981-13-1501-5_56

635

disease in future. In this paper, we describe and discuss some approach by which we will able to identify the presence of SNP and its location in a DNA. SNPs are known as single nucleotide polymorphisms where we evaluate DNA by its nitrogenous bases. These nitrogenous bases are of adenine (A), uracil (U), guanine (G), thymine (T), and cytosine (C) which combined by hydrogen bond to form DNA. Sometimes, nitrogenous bases or nucleotide are altered in gene (in a part of DNA) which looks like mutation but in reality it is SNP [1]. These SNPs are of two types nonsynonymous and synonymous, where only nonsynonymous SNPs affect amino acid. Nonsynonymous SNPs are categorized as missense and nonsense. Here, association studies are taken to determine whether a variant is associated with any disease or trait.

There are many approaches present for disease identification till date. Here, we are discussing some approaches like USVM, Apriori algorithm, logistic regression, Bay's network, and genetic algorithm. For genetic approach, we evaluate these SNPs in high linkage disequilibrium (LD). LD is non-random association of alleles present in different loci. When association frequency of alleles is higher or lower, then co-responding loci are in linkage disequilibrium. LD has investigated four million SNPs in population [2]. This is an investigation of LD which produces haplotype map that may assist process of design and analysis of gene association for complex disease analysis [3]. To assist SNP presence, we need clustering similar to polypeptides present in DNA, which is known as gene clustering. There are many effective gene algorithms like HPM and fuzzy clustering present for clustering human gene [4]. By this approach, we can effectively identify disease-causing SNP location in DNA.

Some classical approaches like Markov chain Monte Carlo (MCMC) are not efficient for calculating convergence rate in huge model space. For major drawback of classical and some deterministic approach, we move toward soft computing. In soft computing algorithm like evolutionary computing, artificial neural network, fuzzy logic, and swarm optimization technique give better result than other classical approaches.

Evolutionary algorithms (EAs) provide an alternative for this solution because EA evolves a population of current solutions, which make it safer to explore in search space [5]. EA approach is developed by inspiration of genetic processes and has ability to find global solutions in complex optimization problems [6].

In [7], AIC method is proposed to identify accurate disease-causing SNPs. Here, a hybrid approach of logistic regression and genetic algorithm is used where logistic regression model is used to determine relation of disease with SNPs. On the other hand, genetic algorithm is used to optimize a large number of SNPs relation set.

Another approach of Apriori algorithm is also taken in [8]. In this approach, a graph filter is used to reduce unwanted data for gene data set. After getting the clean data, Apriori algorithm is evaluated to find genetic disease with respective SNPs. Apriori's support, confidence, and risk factor are evaluated to find their result.

USVM is approached in [9] where univariate marginal distribution algorithm is used with SVM to reduce redundancy and parameter selection problem of support vector machine. In this approach, USVM classifies SNPs set which causes disease.

In [10], a combination of machine learning algorithm is proposed to analyze disease-causing SNPs, where ANN, SVM, and graph-based algorithm are evaluated. These algorithms are tested on sensor data set for result analysis.

Bayesian networks approach is taken in [11] where epistatic interactions are detected by this. It uses branch and bound method for learning purpose. In this paper, experimental result proves that this approach is more effective than other approaches like Markov blanket-based method. This approach is more effective when numbers of sample SNPs are very less.

2 Literature Survey

2.1 Evolutionary Algorithm for Gene Mapping

EAs approach is adaptive procedures which evolve a population of potential solutions for individuals. The fittest individual [6] is concluded in this method. In [12], the author describes a process which evolves a population of models consisting of logic tree to find the fittest combination of SNPs in gene. The evolution is running by iterating over many generations, each generation consisting of flowing process.

(1) Selection
(2) Mutation
(3) Crossover.

2.1.1 Fitness Function

In [12], author describes two approaches for fitness function. First one is GLM framework, and another is *minimum description length (MDL)* which is very similar to machine learning approach. The score functions [12] are converted into a fitness function. To avoid scaling problem, we set highest fitness of tree as 1, the next fit value is 2, and so on. The function of scaling assigns values in which scaled value of a logic tree is proportional to its inverse square root value of its rank.

2.1.2 Evolutionary Algorithm

Step-1
Initialize trees and fitness generation= 0
Step-2
Start SELECTION step to produce a new generation
Step-3
Perform the step for MUTATION on tree models.

Step-4
```
Perform the step CROSSOVER on tree models.
```
Step-5
```
If convergence not found then Repeat steps 1 to Step 3.
```

2.2 Apriori-Gen Algorithm

In this approach, gene graph is used to reduce redundancy of data and Apriori algorithm is used to find most disease-causing SNPs from whole data set [8].

2.2.1 Graph Filter

Graph filtering process or construction of gene graph encoded data into numeric forms. If SNP is bi-allelic, then value will be taken as 0. Or if it is a majority value, then it may be taken as 1. If haplotypes are homogeneous, value is set to 0 or 1. For heterogeneous haplotype, value is set to 2. After assigning values, a N*M matrix is prepared, where N is number of individual and M is number of SNP in data set. A graph X = {H,G} is constructed, where H is known as haplotype and G is known as genotype. In graph, two types of edges are present: (i) case and (ii) control.

As we know SNPs set are very large. So in [8], SNPs are reduced to tag SNP which represent small number of information.

2.2.2 Association Rule

A database contains some tuples where each tuple contains some item set. Here, importance of item is evaluated by support and confidence.

```
Item I={ $I_1, I_2, I_3....,I_m$ }

      Tuples T={ $t_1, t_2, t_3,....,t_k$ }

   Here M→N ,where M,N & I
```

$$M \cap N = \phi$$

```
Support = Percentage of tuples in data base that M U N

Confidence = Ratio of tuples contain M U N w.r.t M
```

Large set is also computed by an algorithm.

2.2.3   Apriori-Gen Algorithm

Input:

I   //item set

L   //large item set

S   // support factor

α   //confidence

Output:

R   //Association rules Satisfying S and α

Algorithm:

R=Φ

For I∈ L

    For $x \subset L \Rightarrow x \neq \Phi$

      If $\dfrac{SUPPORT(I)}{SUPPORT(x)} \geq \alpha$ then

$$R = R \cup \{x \Rightarrow (I - x)\}$$

    End

  End

Risk ratio is calculated as ratio between case and control. And odd ratio is calculated as ratio between odd numeric counts of an event.

$$\text{Risk ratio} = \frac{p_{case}}{q_{control}}$$

$$\text{Odd ratio} = \frac{p/(1-p)}{q/(1-q)} \text{ where p as case and q as control}$$

Here, case is known as disease-causing item set and control is known as non-disease item set. In [8], Crohn's disease-causing SNPs are analyzed where confidence is 60% and support is 50% obtained by this approach.

2.3 USVM Approach

In [9], a new hybrid method is applied known as USVM or univariate marginal distribution algorithm and support vector machine. By mixing those parameters, selecting problem of SVM is reduced. It will select its kernel dynamically to optimize its result.

2.3.1 Univariate Algorithm

This approach is used for selecting best entity of data from large population. It takes selection and population as input parameter and gives best entity. In [9], sample SNPs set are denoted as 0, 1, and 2 and 0 and 1 for homozygous type and 2 for heterozygous type. To optimize the disease-causing SNPs, classification algorithm p_e is applied.

$$\max(p_e) = f(R_s)$$

```
Here  R= {r_{s1}, r_{s1}, ...., r_{sD}},  D<n
r = is SNPs set
s = is sample
D = sample set
n = number of sample set
```

Here, problem of optimal subset is divided into two subproblems.

(1) Optimal subset selection
(2) Evaluation of subset.

2.3.2 Algorithm Description

Here, combination UMDA and SVM algorithm is used to optimize the solution. In step-1, data set is divided into two parts known as train set and test set. Train set is used to classify test set. After this operation, all values are decoded to select SVM kernel parameter type. In feature extraction phase, all subsets are marked as 1 or 0 where 1 is for selected entity and 0 for non-selected entity. In [9], fitness value is evaluated as

```
Fitness Value``` $= \frac{1+q}{2}$, where q = set of fitness value evaluated by algorithm.

A genetic operation is also taken to find higher fitness value and estimate probability distribution of selected population to generate new individual.
```

2.3.3 USVM Algorithm

Step-1

```
R=1
```

$$D_{train} \subset D$$
$$D_{test} \subset D \quad // \ D \ is \ sample \ set$$

Step-2

Generate a population p_0 randomly.

Step-3

While (~ termination)

Start SVM operation and find feature set

Classifier= D_{train}

Fitness Evaluation set= D_{test}

      Choose elite set//A elite set is taken for
                genetic evaluation and generate
                new individual.

```
r=r+1;
```

End while

3 Result Analysis

Data Sets: Here, we take data set which contains Crohn's disease SNPs sample with 144 cases and 243 controls.

    Evaluation: Here, we evaluate data with fivefold cross to check efficiency of algorithms. We take CN, SVM, USVM, Apriori, and random forest [13] to evaluate the same data set. By evaluating this algorithm, we find accuracy. These accuracy values are given in Table 1.

    Here, this graph shows the accuracy of SVM, USVM, random forest, CN, and Apriori-Gen algorithm. USVM approach shows highest accuracy among all. Here,

Table 1 Accuracy of algorithms

Cross-validation	Classifier	Accuracy (%)
Fivefold	USVM	89.6
Fivefold	Random forest	74.4
Fivefold	Apriori-Gen algorithm	53.1
Fivefold	SVM	63.6
Fivefold	CN	54.6

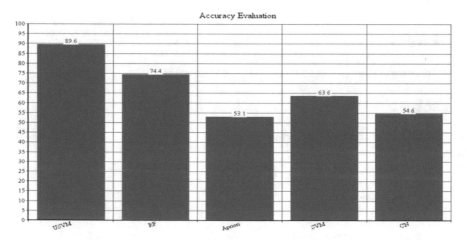

Fig. 1 Accuracy bar

we concluded that USVM approach is most efficient for disease association study (Fig. 1).

4 Conclusion

In this paper, we discuss various algorithms for predicting disease on basis of SNPs. Here, we take SVM, USVM, random forest, Apriori-Gen algorithm and get there accuracy with same data set. We found USVM gives most efficiency among all. In our future work, we are going to enhance accuracy and minimize time consumption by the algorithms.

References

1. Crawford, D.C., Akey, D.T., Nickerson, D.A.: The patterns of natural variation in human genes. Annu. Rev. Genomics Hum. Genet. **6**, 287–312 (2005)
2. International HapMap Consortium: A haplotype map of the human genome. Nature **437**(7063), 1299 (2005)
3. Balding, D.J.: The impact of low-cost, genome-wide resequencing on association studies. Human Genomics **2**, 78–81 (2005)
4. Eisen, M.B., et al.: Cluster analysis and display of genome-wide expression patterns. Proc. Nat. Acad. Sci. **95**(25), 14863–14868 (1998)
5. Fogel, G.B., Corne, D.W.: An introduction to evolutionary computation for biologists. Evol. Comput. Bioinf. 19–38 (2003)
6. Mitchell, M.: An introduction to genetic algorithms. MIT press (1998)
7. Nakamichi, R., Imoto, S., Miyano, S.: Case-control study of binary disease trait considering interactions between SNPs and environmental effects using logistic regression. In: Proceedings Fourth IEEE Symposium on Bioinformatics and Bioengineering BIBE 2004. IEEE (2004)
8. Mao, W.: The Application of apriori-gen algorithm on the interaction analysis of SNPs in genetic disease. In: 2009 WRI World Congress on Computer Science and Information Engineering, vol. 3. IEEE (2009)
9. Wei, B., et al.: USVM: selection of SNPs in diseases association study using UMDA and SVM. In: 2010 4th International Conference on Bioinformatics and Biomedical Engineering (ICBBE). IEEE (2010)
10. Kourou, K., et al.: Machine learning applications in cancer prognosis and prediction. Comput. Struct. Biotechnol. J. **1**(3), 8–17 (2015)
11. Han, B., Chen, X.: Detecting SNPs-disease associations using Bayesian networks. In: 2010 IEEE International Conference on Bioinformatics and Biomedicine (BIBM). IEEE (2010)
12. Clark, T.G., De Iorio, M., Griffths, R.C.: An evolutionary algorithm to find associations in dense genetic maps. IEEE Trans. Evol. Comput. **12**(3), 297–306 (2008)
13. Mao, W., Lee, J.: A combinatorial analysis of genetic data for Crohn's disease. In: 2007 The 1st International Conference on Bioinformatics and Biomedical Engineering ICBBE. IEEE (2007)

Further Reading

14. Porto, W.F., Franco, O.L., Alencar, S.A.: Computational analyses and prediction of guanylin deleterious SNPs. Peptides **69**, 92–102 (2015)
15. Zeng, T.-T., et al.: Influence of SCARB1 gene SNPs on serum lipid levels and susceptibility to coronary heart disease and cerebral infarction in a Chinese population. Gene **626**, 319–325 (2017)

Temporal Signature Mining for Network Intrusion Detection Using TEMR

Tanjila Mawla, Sharmishtha Dutta and Md. Forhad Rabbi

Abstract Network intrusion detection is being a big challenge to combine the network security with the growing speed of data transmission. Traditional data mining approaches were widely used to detect intrusions though having some drawbacks. But there remain some needs of the application of temporal data mining techniques on network intrusion detection as the flow of network data is continuous over time. This paper proposes the idea of applying a framework which is based on doubly sparse convolutional matrix that is closely related to nonnegative matrix factorization to identify the hidden patterns of events. One-sided convolution of two nonnegative matrices shows the convergence of the framework. There is an observation about applying the algorithm on network intrusion detection (NID), and this is how β-divergence increases and decreases with the consideration of a different number of data. The successful minimization of β-divergence can lead to categorizing attack and normal activities found in a dataset containing a packet of network data.

1 Introduction

In the realm of the Internet, the information security issue regarding network intrusion requires fast detection of unusual behavior from large-scale continuous Web log data. A lot of methods using data mining have been applied to detect intrusion of attacked network data. But still there are problems in the area of detecting network intrusion. Nowadays, a lot of useful information and patterns are hidden in the rapidly growing

T. Mawla (✉)
North East University Bangladesh, Telihaor, Sheikhghat Sylhet 3100, Bangladesh
e-mail: tani109.bd@gmail.com

S. Dutta
Metropolitan University, Zindabazar, Sylhet 3100, Bangladesh
e-mail: shoron.dutta321@gmail.com

Md. F. Rabbi
Shahjalal University of Science & Technology, Kumargaon, Sylhet 3114, Bangladesh
e-mail: post2rabbi@gmail.com

© Springer Nature Singapore Pte Ltd. 2019   645
A. Abraham et al. (eds.), *Emerging Technologies in Data Mining and Information Security*, Advances in Intelligent Systems and Computing 814,
https://doi.org/10.1007/978-981-13-1501-5_57

data. Having a vast number of datasets, we are drowning in information but starving for knowledge discovery.

Traditional data mining is important to process data and discover information but is less effective when it comes to handling data with temporal information. While working with temporal data mining, data should be represented as temporal sequences of events and need to be treated differently than normal data. A datum is not replaced by its later version in a temporal database. All versions of data are stored, and we can have a different perspective on finding interesting information.

There have been only a few works that dealt with intrusion detection taking the temporal aspect of data into account [1]. The framework in [2] presents a different approach for representing temporal information. It is called temporal event matrix representation (TEMR). It offers a matrix-to-image analogy as images are easier to visually understand the events in a temporal sequence. We intended to apply the framework for the well-known knowledge discovery in databases (KDD) cup 99 dataset.

At first as beginners, to understand how to work with network data, we applied a popular traditional algorithm—support vector machine (SVM) on the KDD cup dataset. This experiment helped us understand the behavior of network data. It will also help in comparing both the results. Then, we moved on to applying the framework to our dataset. The rest of the paper is designed as follows: The literature review is in Sect. 2; experiments and results are presented in Sect. 3. We discuss our work in Sect. 4 and draw the conclusion in Sect. 5.

2 Literature Review

2.1 Network Intrusion Detection

Network intrusion detection is an important task in the recent era because communication through the Internet has become so much popular. Many techniques using traditional data mining have been proposed to solve the network attacking issues. The intrusion detection system should be designed in such a way so that it can provide high accuracy and detection rate and low false alarm rate [3]. The data mining techniques including clustering and classification are introduced to deal with vast amount of network data. Techniques using traditional data mining are proposed to detect network intrusion in many research works [3–5].

2.2 Cyber-Attacks and Intrusion Detection

Intrusion detection is needed to prevent external attacks. These attacks are usually called cyber-attacks. The most common cyber-attacks are denial of service (DoS), distributed denial of service (DDoS), IP spoofing, man-in-the-middle, privilege mis-use, and so on. There are some embedded intrusion detection systems.

There are two different ways to detect intrusion—signature-based intrusion detection and anomaly-based intrusion detection [6, 7]. Signature-based intrusion detection is the most commonly and commercially used intrusion detection [8]. This technique can detect the known attacks. It mainly matches patterns found in the continuous flow of the network data with the known patterns and raises an alarm when finding one [8]. In this approach, intrusions must be hand-coded by analysts and experts and newly discovered signatures of attacks cannot be identified. Because matching is not possible as the unknown attacks are not being stored in the database [9, 10].

2.3 Data Mining Approaches for Intrusion Detection

Classification [11–14], clustering [11–14], link analysis [9], sequence analysis [9] are data mining techniques for identifying normal data from abnormal data. Datasets of live and real network data are important to detect intrusion. Air Forces Research Laboratory (AFRL) in Rome along with MIT used network traffic as a simulated network from their networks [9]. Bayesian classification model was used to detect the type of network data [15]. There [9] KDD dataset was used for training and testing. DARPA off-line intrusion detection evaluation was available to researchers after processing of data available. DARPA and KDD are the datasets that were used as very important datasets for evaluating various authors proposed models [9].

To merge data mining and network intrusion detection, researchers use misuse detection system that only detects known attacks. This process can only recognize known attacks stored in the database. The major drawbacks of traditional misuse detection system are that it cannot identify newly developed attacks.

3 Experiments and Results

3.1 Traditional Data Mining (Support Vector Machine)

To extract any hidden information from event sequences, we tried traditional data mining process on our experimental dataset of KDD cup. We thought about unsupervised clustering, but there is no importance of this clustering. Unsupervised clustering may divide the data into two clusters, but it is not possible to identify which cluster is for which type of data. Then coming back to our target, we worked with support vector machine (SVM). By applying SVM, we could be able to partially identify attack data in the dataset.

3.1.1 Dataset

The KDD cup dataset contains network data of 41 different features. Among them, 34 features are numeric. While applying SVM on the dataset, we used all the 41 features to train the classifier. In our dataset, there are 23 types of attack data are available. We trained the classifier as they could identify the type of attack. We did not bother about the normal data because they were easily identified because normal data are free from any kind of attack features. In real life, various types of attack data flow when any device deals with network issues. Our dataset has the attack types such as *back, teardrop, loadmodule, neptune, rootkit, phf, satan, buffer_overflow, ftp_write, land, spy, ipsweep, multihop, smurf, pod, perl, warezclient, nmap, imap, warezmaster, portsweep, guess_passwd*.

3.1.2 Parameters for Applying SVM

We tried different values of gamma and C for the classifier and settled for gamma $= 0.01$ and $C = 100$ which gave us the best results. Here, gamma means the influence while running a single training. The low values of gamma define the model as far from the correctness, and the high values of gamma define closeness to the classification, and the value of gamma is selected by the model as support vectors. Any misclassification occurred by the model for training examples is defined by c. The low and high values of c mean the decision surface smooth and aim of classification of training examples, respectively. In each experiment, we split the dataset into half for training and testing.

3.1.3 Result of Applying SVM

When we applied SVM on our dataset, we observed that it can detect which data are normal. So, we did not bother about the normal data of the test dataset. We only considered the attack data to compare the test data and output after applying SVM with the training dataset. The following table shows the outcome of the result after applying SVM. The number of mismatches increased with the increase of the number of data points in the dataset (Table 1).

Table 1 Performance details of SVM on KDD cup dataset

Dataset size	Number of attacks	Number of mismatches
10,000	1913	30
25,000	5955	101
50,000	21,864	260
100,000	114,596	453

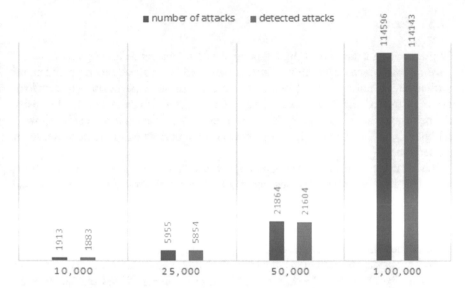

Fig. 1 The number of attacks present in the dataset, along with correctly detected attacks

The chart below shows the performance of the classifier in correctly detecting the attacks for each dataset (Fig. 1).

3.2 Temporal Event Signature Mining

In [2], Fei Wang proposed a framework for mining signatures from healthcare data. We reviewed the outcome of the application of this framework in healthcare data.

3.2.1 Methodology

The basic methodology of the framework is to use nonnegative matrix factorization on the dataset to split it into two sets of matrices and then reconstruct one matrix using one-sided convolution over the time axis. The reconstructed matrix is updated using a threshold value on the beta-divergence between the original matrix and the reconstructed one.

3.2.2 Features of Dataset

We work on KDD cup dataset, and our dataset has data of 41 dimensions among which 34 dimensions are for features represented by numeric data.

3.2.3   Results

We prepared eight datasets for eight types of attacks. These attacks are *neptune, satan, teardrop, back, smurf, pod, warezclient and ipsweep*. The datasets are prepared using normal: attack: normal = 1:2:1 ratio. The outcome of our experiment is to visualize the reconstructed matrix Y in such a way that attacks and normal data can be easily distinguished. Representation of Y for all these datasets except ipsweep is showed in Figs. 2, 3, 4, 5, 6, 7, and 8. Ipsweep shows similar characteristics to neptune, satan and teardrop.

It is clearly visualized that for *neptune, satan, and teardrop*, the attack data (from 50 to 150) have very lower values than normal data, and they can be easily distinguished. Also, for back and warezclient, the values of attack data are significantly higher than that of normal data. This visualization helps us to identify attacks and normal data. But, in case of *smurf* and *pod,* the attack data and normal data have values in the same limits. So, it is difficult for us to detect attacks only by visualizing them. As next step, we prepared datasets with multiple sets of attack data and normal to find out their behavior. We designed the datasets to check if *back* can be detected along with other attacks (*ipsweep, neptune, satan, teardrop*) with the dataset proportion normal: back: normal: other attack = 1:2:1:1. When a combination is generated consisting of normal data with *back* and other attacks (whose values were seen lower than that of normal data), it can be distinguished. The first 50 data are *normal*; next 100 data are *back* followed by 50 *normal* data. This is finally followed by 50 attack data of four kinds—*ipsweep*, *neptune*, *satan,* and *teardrop*. In each case, back has

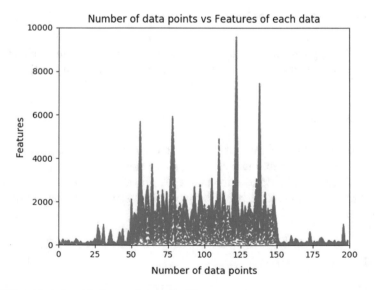

Fig. 2 Reconstructed matrix for attack type 'back'

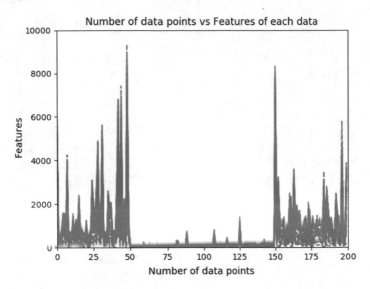

Fig. 3 Reconstructed matrix for attack type 'teardrop'

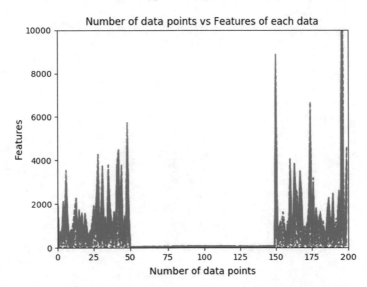

Fig. 4 Reconstructed matrix for attack type 'neptune'

the highest value, normal data have values lower than back and the rest of the four attacks have the lowest values of all.

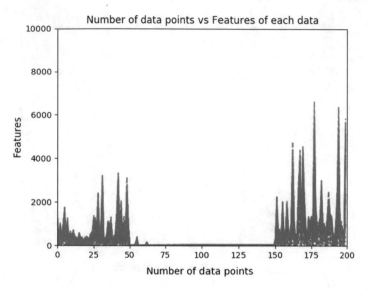

Fig. 5 Reconstructed matrix for attack type 'satan'

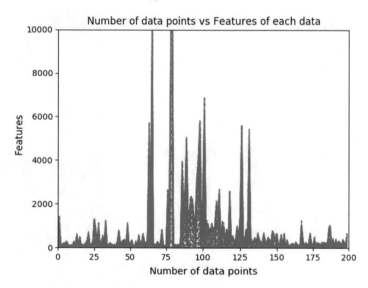

Fig. 6 Reconstructed matrix for attack type 'warezclient'

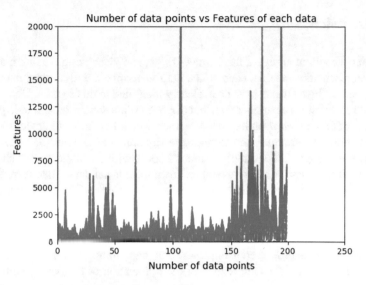

Fig. 7 Reconstructed matrix for attack type 'pod'

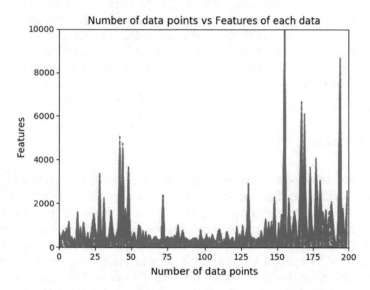

Fig. 8 Reconstructed matrix for attack type 'smurf'

4 Discussion

In our experiment of applying the framework [2] and visualizing it using matrix-to-image representation, at first, each attack data were mixed with normal data to find their signature. Five attacks have been clearly identified in this section. Here, *back* is identified as its values are higher than that of normal data. On the other hand, *ipsweep, neptune, satan*, and *teardrop* had fewer values which helped us to detect them in a flow with normal data. *Back* had an opposite signature to the other four attacks. But this was not enough. When multiple attacks come with normal data, we should be able to distinguish each one of them apart from the normal data. Therefore, the next experiment was preparing dataset with *normal* with *back* and the other four attacks. In these four experiments also, the signatures remained true to the previous ones. As for the other three attack types—*smurf, pod*, and *warezclient*, the values were like normal data and the attack data had no consistency among them. This can indicate that as the dataset is processed using traditional data mining technique, it may have considered an attack data as normal. But the application of temporal mining could detect it.

So, overall the task was successful as this approach can be used to transform the incoming Web log data into reconstructed matrix Y, and then, abnormalities can be visualized easily. A visualization helps us to detect attack data. Applying the framework, the normal data flow can be monitored and data packets resembling significantly higher or lower values than it can be detected as attacks remained true to the previous ones.

5 Conclusion

In this paper, we applied temporal signature mining to detect network intrusion. We reached a decision that traditional data mining approaches are not suitable for intrusion detection of network data. In this work, we analyzed a proposed methodology previously applied to healthcare data. Supervised machine learning is applied to detect an attack and normal data. We got the result which can detect attack data but cannot identify the type of the data and their shape to find hidden patterns. We studied temporal data mining and got a framework from [2] and modified some parameters to detect signatures. We assumed that the modified framework can be applied for network intrusion detection. We experimented our dataset with the learning framework and got something fruitful to detect some attacks accurately which was showed in the experiment and result section and discussed in the discussion section.

References

1. Hogo, M.A.: Temporal analysis of intrusion detection. In: 2014 International Carnahan Conference on Security Technology (ICCST), pp. 1–6. IEEE (2014)
2. Wang, F., Lee, N., Hu, J., Sun, J., Ebadollahi, S., Laine, A.F.: A framework for mining signatures from event sequences and its applications in healthcare data. IEEE Trans. Pattern Anal. Mach. Intell. **35**(2), 272–285 (2013)
3. Wankhade, K., Patka, S., Thool, R.: An overview of intrusion detection based on data mining techniques. In: 2013 International Conference on Communication Systems and Network Technologies (CSNT). IEEE (2013)
4. Dokas, P., et al.: Data mining for network intrusion detection. In: Proceeding NSF Workshop on Next Generation Data Mining (2002)
5. Tiwari, K.K., Tiwari, S., Yadav, S.: Intrusion detection using data mining techniques. Int. J. Adv. Comput. Technol. **2**(4), 1–5 (2013)
6. Caulkins, B.D., Lee, J., Wang, M.: A dynamic data mining technique for intrusion detection systems. In: Proceedings of the 43rd annual southeast regional conference, vol. 2, pp. 148–153. ACM (2005)
7. Sivaranjani, S., Pathak, M.R.: Network Intrusion Detection using Data Mining Technique
8. Youssef, A., Emam, A.: Network intrusion detection using data mining and network behaviour analysis. Int. J. Comput. Sci. Inf. Technol. **3**(6), 87 (2011)
9. Helali, R.G.M.: Data mining based network intrusion detection system: A survey. In: Novel Algorithms and Techniques in Telecommunications and Networking, pp. 501–505. Springer (2010)
10. Münz, G., Li, S., Carle, G.: Traffic anomaly detection using k-means clustering. In: GI/ITG Workshop MMBnet (2007)
11. Antunes, C.M., Oliveira, A.L.: Temporal data mining: an overview. In: KDD Workshop on Temporal Data Mining, vol. 1, p. 13 (2001)
12. Ye, N., Li, X.: A scalable clustering technique for intrusion signature recognition. In: Proceedings of 2001 IEEE workshop on information assurance and security. Citeseer, pp. 1–4 (2001)
13. Lin, W., Orgun, M.A., Williams, G.J.: An overview of temporal data mining. In: Proceedings of the 1st Australian data mining workshop, pp. 83 (2002)
14. Shahnawa, M., Ranjan, A., Danish, M.: Temporal data mining: an overview. Int. J. Eng. Adv. Technol. **1**(1), 2249–8958 (2011)
15. Altwaijry, H.: Bayesian based intrusion detection system. In: IAENG Transactions on Engineering Technologies, pp. 29–44. Springer Netherlands (2013)

High-Capacity Downlink for Millimeter Wave Communication Network Architecture

Abdullah Al-Mamun Bulbul, Md. Tariq Hasan, Mohammad Ismat Kadir,
Md. Mahbub Hossain, Abdullah Al Nahid and Md. Nazmul Hasan

Abstract With the explosive growth in the demand for higher bandwidth, more new technologies are emerging. Frequencies 10 GHz will be fully occupied within few years by communication channel. The millimeter-wave (mm-wave) frequency band that ranges from 30 to 300 GHz is a new frontier for fifth generation (5G) mobile communication. The mm-wave frequencies suffer from very high attenuation in free space and through objects that limit the signal propagation range. In this paper, the downlink of 5G network architecture has been proposed in order to increase the data throughputs and coverage. The free space channel has been characterized by the Rayleigh fading channel. Orthogonal frequency-division multiple access (OFDMA) have been utilized in the downlink. The proposed network uses 16-quadrature amplitude modulation (QAM) which will ensure greater data throughputs above 5 Gbps. Also, using adaptive beam-forming antennas, the network is expected to provide increased coverage of about 2 km.

A. A.-M. Bulbul (✉) · Md. T. Hasan · M. I. Kadir · M. Hossain · A. A. Nahid · M. N. Hasan
Electronics and Communication Engineering Discipline, Khulna University,
Khulna 9208, Bangladesh
e-mail: bulbulmamun@yahoo.com

Md. T. Hasan
e-mail: mdthasan@gmail.com

M. I. Kadir
e-mail: iskadir@hotmail.com

Md. M. Hossain
e-mail: mahbub.eceku@yahoo.com

Md. N. Hasan
e-mail: nazmul.apu08@gmail.com

A. A. Nahid
School of Engineering, Macquarie University, Sydney 2109, Australia
e-mail: abdullah-al.nahid@students.mq.edu.au

© Springer Nature Singapore Pte Ltd. 2019
A. Abraham et al. (eds.), *Emerging Technologies in Data Mining and Information Security*, Advances in Intelligent Systems and Computing 814,
https://doi.org/10.1007/978-981-13-1501-5_58

657

1 Introduction

Coping with the globalization and digitalization, digital gadgets like smartphones, laptops, personal digital assistants (PDAs), smart watches etc. are being produced in a large volume to meet the demand for increasing people. By 2020, it is predicted that the demand for increased system capacity will be 1000 times more than the present [1]. Since the microwave band occupancy is about to reach the saturation, a new unutilized spectrum is the only option to mitigate the uprising demand for capacity and the millimeter-wave (mm-wave) band is expected to be that new alternative [2–6]. But due to inherent high frequency at mm-wave band, it suffers from high attenuation and gaseous absorption loss, which reduces the range and signal strength. This results in a reduction of the network coverage area.

Date throughput upto 2.1 Gbps [7] and only 200 m of coverage [1, 8] can be provided by the existing mm-wave networks. The unlicensed spectrum of the 60 GHz band suffers from less path loss than other parts of the mm-wave band. In this paper, E-band has been utilized to minimize signal propagation loss of the proposed network. The structure and performance of orthogonal frequency-division multiple access (OFDMA) is a potential options for multiple access [9] which makes it a appealing candidate for the downlink multiple access technique in this proposed mm-wave communication network. Besides, adaptive multiple-input multiple-output (MIMO) antennas have been used to increase the coverage of this proposed network.

In this paper, the existing 5G networks are first explored. The propagation properties of mm waves are taken into account and limitations of networks are being exhibited. The downlink of mm-wave communication (5G) network architecture is proposed. The simulation results, future scopes and colclusion are discussed in the subsequent sections.

2 Literature Review

Wireless network analyst and desiners are focusing on the higher stages of frequencies to cope with the increasing demand for network capacity and higher data throughputs [10, 11]. A large unused spectrum is available at mm-wave ranging from 30 to 300 GHz [2–6]. Researches have been carried out on the application of this spectrum at 5, 28, 38 and 60 GHz bands [1, 8, 12–14]. According to [6], a noteable amount of absorption loss is present at mm-wave bands. Though 60 GHz band offers large swath of spectrum, it undergoes about 15 dB/km O_2 absorption loss that accompanied by acute free space loss [15, 16]. As O_2 absorption loss is meaningfully minor above 70 GHz [15], it makes the 70 GHz band, the E-band, the most appropriate candidate for 5G Communication. The E-band, consisted of 81–86 GHz and 71–76 GHz, is liscesed by FCC [17].

Though mm-band offers large spectrum, mm-wave application in wireless interface still requires to undergo a large number of challenges. Antenna desing for mm-wave application is the major issue. A number of explorations have been accomplished in designing small sized array antenna operating at mm-wave band. The antenna designers have carried out such small sized antenna that amass of antenna array is possible on chip or printed-circuit-boards (PCBs).

An mm-wave systems operating at 28 and 38 GHz is proposed in [1] proposing the utilization of MIMO antenna. But the network suffers from poor coverage of only 200 m. Beside modulation and multiplexing techniques is not specified. Another mm-band system has been presented in [18]. The system uses massive MIMO and is expected to provide theoretical data rate of about 10 Gbps which does not include any practical or simulation evidence. Both 38 and 60 GHz has been utilized in [14] which includes 7 beam-width antennas. The system is expected to provide 265 m coverage. But data rate has not been analyzed in the paper.

Since E-band is not critically suffered from atmospheric absorption loss, this band has been utilized in the proposed network. To achieve high coverage along with reduced interference, adaptive beam-forming antennas have been used. To achieve high data rates and higher spectral efficiency, the proposed downlink utilizes 16-quadrature amplitude modulation (QAM) modulation and OFDMA.

3   Downlink Design

Between 81–86 and 71–76 GHz, the higher portion of E-band (i.e., 81–86 GHz) has been proposed in the downlink and the remaining portion is kept for the application in the uplink as power consumption is a key challenge for uplink mobile station. MATLAB v.2015a has been used as the simulation platform for the proposed network. Besides 16-QAM, network performance has also been investigated for quadrature phase shift keying (QPSK) for comparison purposes.

3.1   OFDMA Transmission

As the orthogonal frequency are closely spaced, OFDMA provides high spectral efficiency [19]. It divides the available large bandwidth into a number of narrowband subcarriers that are orthogonal to each others. There subcarriers are utilized to carry data in parallel format by splitting the message into the parallel branches [19].

The input digital message signal is first transformed into parallel format. 16-QAM modulation scheme has been utilized to modulate the parallel message bits which is followed by N-point IDFT. The input to the IDFT resulting from the modulation with N orthogonal parallel subcarriers is shown in Eqs. 1 and 2.

$$s_m(t) = \sum_{m=0}^{N-1} R\left[(a_m + jb_m)\exp(j2\pi f_m t)\right] \tag{1}$$

$$= \sum_{m=0}^{N-1} [a_m \cos(2\pi f_m t) - b_m \sin(2\pi f_m t)] \tag{2}$$

Here a_m and b_m denote the data that are assigned to the in-phase and quadrature-phase of the m-th sub branch and f_m is the frequency of that branch.

To avoid the requirement for distinct oscillators, N-IDFT is applied on the signal. For each user, $512 (= N)$ subcarriers has been assigned. IDFT results in the following Eq. 3.

$$s_n = \frac{1}{\sqrt{N}} \sum_{m=0}^{N-1} \left[a_m \cos(\tfrac{2\pi mn}{N}) - b_m \sin(\tfrac{2\pi mn}{N})\right] \tag{3}$$

where: $n = 0, 1, 2, 3, ..., N-1$

The resulting IDFT signal is then fed to a LPF. Then cyclic prefixing (CP) is performed to reduce inter symbol interference (ISI). The length of CP, L, is chosen as 64. CP is followed by a digital to analog converter (DAC) to convert it in a form convenient for wireless transmission.

3.2 Signal Transmission Using Adaptive Beam-Forming Antenna from BS

The resulting analog signal at mm-band is transmitted using adaptive beam-forming antenna. This antenna is compitent to estimate the signal of interest and reject interference. Least-square-spatial (LMS) algorithm has been used to simulate and investigate the antenna performance. Due to its adaptive mechanism, the antenna is highly directional and expected to maintain long distance communication and provide high coverage.

3.3 Channel and Path-Loss Model Design

To design the wireless interface in MATLAB, Rayleigh fading channel has been utilized. Signal propagating at free space experiences a loss that is proportional to the distance and signal frequency [20] as shown in Eq. 4.

$$\text{Loss}_{\text{FSL}} = [92.4 + 20\log(f) + 20\log(R)]\text{dB} \tag{4}$$

where:

f = signal frequency (in GHz)
R = distance between BS and MT (in km)

Other losses like foliage loss, rain absorption loss, atmospheric gaseous loss etc. are taken into account during signal propagation. According to [21], the atmospheric gaseous loss is 0.12499 dB/km for E-band at sea level. Rain also aids a loss of 0.2899 dB/km during average rainfall of about 0.25 mm/h [21, 22]. Overall path loss estimation is done based on the following Eq. 5.

$$\text{Loss}_{\text{Total}} = [\text{Loss}_{\text{FSL}} + \text{Loss}_{\text{AGL}} + \text{Loss}_{\text{FL}} + \text{Loss}_{\text{Rain}}] \tag{5}$$

3.4 Signal Reception and Demodulation at MS

Mobile terminal (MT) receives the signal from base station (BS). The approximate signal strength at the MT (RSL) is carried out based on the following Eq. 6.

$$\text{RSL} = [P_{\text{TX}} + G_{\text{TX}} + G_R - \text{Loss}_{\text{Total}}]\text{dBm} \tag{6}$$

where:

P_{TX} = transmitting power in dBm
G_{TX} = transmitter antenna gain in dBi
G_R = receiver antenna gain in dBi

MT process the data in opposite mechanism as it was processed in BS. After converting the received signal into a base band signal, analog to digital transformation is performed. CPs are removed and data is converted from serial to parallel form. Then N-DFT, constellation demapper and parallel to serial conversion is accomplished in a subsequent manner. The block representation of the overall downlink is shown in the Fig. 1.

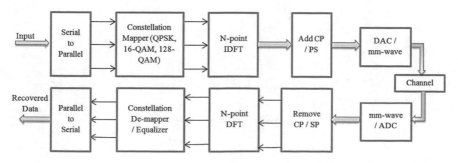

Fig. 1 Downlink of the proposed mm-wave network

4   Results and Discussion

In this section, the performance of the designed mm-wave communication network for optimizing the energy consumption and achieving a high data rate has been investigated with the aid of simulation results.

4.1   Path Loss Analysis

Free space loss (FSL) is directly proportional to the distance and operating frequency. So an increase in frequency and dintance will also increase FSL. As gaseous loss is high at 60 GHz band compared to E-band (around 70 and 80 GHz), the overall path loss is much higher at 60 GHz band as found in Fig. 2.

4.2   BER Performance

The bit-error rate (BER) is slightly higher in 16-QAM compared to QPSK as found in Fig. 3. This is due to the reason that 16-QAM is a higher order modulation as it has 4 bits per symbol whereas QPSK has only 2 bits per symbol. This slight increase in BER will be waged by the increased system capacity.

Fig. 2 Overall path-loss comparison for different mm-wave bands

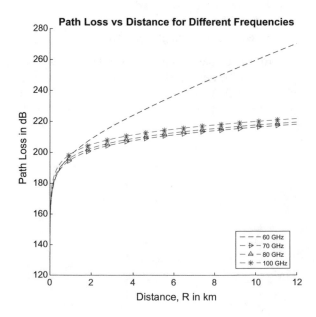

Fig. 3 BER comparison between uplink and downlink

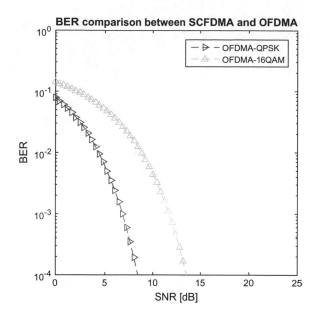

4.3   Comparison of RSL

The RSL is measured based on Eq. 6. It is found from simulation in Fig. 4 that the proposed network maintains a good signal strength at a distance of 2 km. The signal strength is about −80 dBm for 16-QAM and −87 dBm for QPSK at two kilometers distance.

4.4   System Capacity Estimation

The capacity of the proposed network has been shown in Fig. 5. System capacity is estimated using Shannons capacity theorem [23]. The proposed 16-QAM provides data rates of 5.4 Gbps for 5 GHz bandwidth whereas data rates is 2.7 Gbps for QPSK. This is evidence of the high capacity of this designed mm-wave communication network.

The following Table 1 consists of the features of the proposed downlink.

5   Results and Discussion

The E-band has been chosen as the operating frequency band for the downlink of the proposed mm-wave communication (5G) network. OFDMA and 16-QAM have been

Fig. 4 RSL comparison between 16-QAM and QPSK

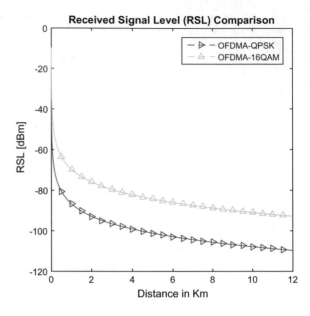

Fig. 5 Capacity estimation for the proposed network

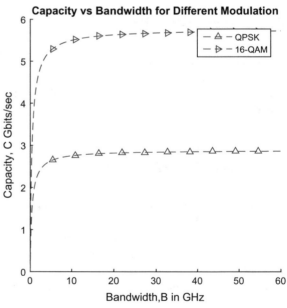

Table 1 Summary of contributions

Contribution fields	Existing network	Proposed network
Data rate (Bandwidth: 5 GHz)	Up to 2.5 Gbps	Up to 5.4 Gbps
Coverage	200 m	2 km
RSL at 2 km	Greater than −107 dBm	Less than 100 dBm

proposed as the multiple access technique and modulation technique respectively to ensure spectral efficiency and achieve high data rates. The proposed network is capable to provide data rate of about 5.4 Gbps which is much higher compared to existing mm-wave networks. Besides the network also ensures an increased coverage of above 2 km.

Only the downlink of the mm-wave communication has been analysed in this paper. In the preceeding work, a complete uplink and downlink architecture for 5G will be proposed and key challenges such as power consumption, peak-to-average power ratio (PAPR), performance of the proposed network for 128-QAM and interoperability of the proposed network with existing cellular networks will be investigated.

References

1. Rappaport, T.S., Sun, S., Mayzus, R., Zhao, H., Azar, Y., Wang, K., Wong, G.N., Schulz, J.K., Samimi, M., Gutierrez, F.: Millimeter wave mobile communications for 5G cellular: it will work!. IEEE Access **1**, 335–349 (2013). https://doi.org/10.1109/ACCESS.2013.2260813
2. Lockie, D., Peck, D.: High-data-rate millimeter-wave radios. IEEE Microw. Mag. **10**(5), 75–83 (2009). https://doi.org/10.1109/MMM.2009.932834
3. Rappaport, T.S., Murdock, J.N., Gutierrez, F.: State of the art in 60-GHz integrated circuits and systems for wireless communications. Proc. IEEE **99**(8), 1390–1436 (2011). https://doi.org/10.1109/JPROC.2011.2143650
4. Khan, F., Pi, Z.: mmWave mobile broadband (MMB): unleashing the 3–300 GHz spectrum. In: Proceeding of 34th IEEE Sarnoff Symposium, pp. 1–6 (2011) https://doi.org/10.1109/SARNOF.2011.5876482
5. Pi, Z., Khan, F.: An introduction to millimeter wave mobile broadband systems. IEEE Commun. Mag. **49**(6), 101–107 (2011). https://doi.org/10.1109/MCOM.2011.5783993
6. Pietraski, P., Britz, D., Roy, A., Pragada, R., Charlton, G.: Millimeter wave and terahertz communications: feasibility and challenges. ZTE Commun. **10**(4), 3–12 (2012)
7. Huang, K.C., Wang, Z.: Millimeter Wave Communication Systems. Wiley **29** (Sections 1.1–1.2) (2011). ISBN 1-118-10275-4
8. Rappaport, T., Gutierrez, F., Ben-Dor, E., Murdock, J.N., Qiao, Y., Tamir, J.I.: Broadband millimeter-wave propagation measurements and models using adaptive-beam antennas for outdoor urban cellular communications. IEEE Trans. Antennas Propag. **61**(4), 1850–1859 (2013). https://doi.org/10.1109/TAP.2012.2235056
9. Myung, H.G.: Introduction to single carrier FDMA. In: 15th European Signal Proceeding Conference (EUSIPCO 2007), pp. 2144–2148. IEEE, New york (2007)
10. Wang, P., Li, Y., Yuan, X., Song, L., Vucetic, B.: Tens of gigabits wireless communications over E-Band LoS MIMO channels with uniform linear antenna arrays. IEEE Trans. Wirel. Commun. **13**(7), 3791–3805 (2014). https://doi.org/10.1109/TWC.2014.2318053

11. Guo, Y.J., Liu, D., Bird, N.C.: Guest editorial for the special issue on antennas and propagation aspects of 60–90 GHz wireless communications. IEEE Trans. Antennas Propag. **57**(10), 2817–2819 (2009). https://doi.org/10.1109/TAP.2009.2032587
12. Zhao, X., Kivinen, J., Vainikainen, P., Skog, K.: Propagation characteristics for wideband outdoor mobile communications at 5.3 GHz. IEEE J. Sel. Areas Commun. **20**(3), 507–514 (2002). https://doi.org/10.1109/49.995509
13. Rajagopal, S., Abu-Surra, S., Malmirchegini, M.: Channel feasibility for outdoor non-line-of-sight mmwave mobile communication. In: Proceeding of IEEE Vehicle Technology Conference (VTC Fall), pp. 1–6 (2012) https://doi.org/10.1109/VTCFall.2012.6398884
14. Rappaport, T., Ben-Dor, E., Murdock, J.N., Qiao, Y.: 38 GHz and 60 GHz angle-dependent propagation for cellular and peer-to-peer wireless communications. In: Proceeding of IEEE International Conference on Communications (ICC), pp. 4568–4573 (2012) https://doi.org/10.1109/ICC.2012.6363891
15. Madhow, U.: Networking at 60 GHz: the emergence of multigigabit wireless. In: Proceeding 2nd International COMSNET, pp. 1–6 (2010) https://doi.org/10.1109/COMSNETS.2010.5431983
16. Yong, S.K., Chong, C.C.: An overview of multigigabit wireless through millimeter wave technology: potentials and technical challenges. EURASIP J. Wirel. Commun. Netw. **2007**(1), 78907-1-78907-10 (2007) https://doi.org/10.1155/2007/78907
17. Federal Communication Commission: Allocation and service rules for the 71–76 GHz, 81–86 GHz and 92–95 GHz bands. FCC Memorandum Opinion and Order, FCC **03-248** (2003)
18. Chimeh, J.D.: 5G Mobile communications: a mandatory wireless infrastructure for Big data. In: Proceeding of International Conference on Advances in Computing, Electronics and Electrical Technology (CEET), vol. 2015 (2015) https://doi.org/10.15224/978-1-63248-056-9-29
19. Ancora, A., Bona, C., Slock, D.T.: Down-sampled impulse response least-squares channel estimation for LTE OFDMA. In: IEEE International Conference on Acoustics, Speech and Signal Processing-ICASSP, vol. 3, pp. 293–296 (2007) https://doi.org/10.1109/ICASSP.2007.366530
20. Roddy, D.: Satellite Communication. 4/e, McGraw-Hill, New York (2006). ISBN 0-07-146298-8
21. CCIR Doc. Rep. 719-3.: Attenuation by Atmospheric Gases. ITU (1990)
22. Flock, W.L.: Propagation Effects on Satellite Systems at Frequencies Below 10 GHz: A handbook for satellite systems design. NASA Doc.1108(02) (Chaps. 3, 4 and 9 passim) (1987)
23. Lovsz, L.: On the Shannon capacity of a graph. IEEE Trans. Inf. Theory **25**(1), 1–7 (1979). https://doi.org/10.1109/TIT.1979.1055985

Part VI
Wireless Communication

Review on Intrigue Used for Caching of Information in View of Information Density in Wireless Ad Hoc Network

Saleha I. Saudagar, Sonika A. Chorey and Gayatri A. Jagnade

Abstract Node-to-node networking is completely decentralized or distributed architecture where every node has a workload having same capabilities and act as both client and server. Data caching approach barters information item in a node-to-node fashion. In Data caching where each intermediatory node when receive information, evaluates the cache drop time of that information and bring to a resolution which content to keep and which content to remove. These evaluations are taken based on the propinquity of the node, and at the same time, information cloning is avoided; hence, it evaluates an intrigue where nodes independent of each other check the legality of caching data.

Keywords Data caching · Node to node (N2N)
Ad hoc wireless network (AWN)

1 Introduction

In AWN, data impartment is carried out in N2N fashion. The computing devices could be conventional computer, laptop, smart phones or routing network or various embedded processors as sensors placed in sensor network. Caching is a technique for storing data in memory, and as when needed, it should directly access instead of regenerating the data. The ethical placement of data to intermediatory node diminishes the whole outlay of information admittance and is aggravated by below two defining characteristics of node-to-node networks. First, N2N networking is completely decentralized or distributed architecture not having any admittance system.

S. I. Saudagar · S. A. Chorey (✉) · G. A. Jagnade (✉)
Prof. Ram Meghe Institute of Technology and Research, Badnera, Amaravati 444701, India
e-mail: sachorey@mitra.ac.in

G. A. Jagnade
e-mail: gajagnade@mitra.ac.in

S. I. Saudagar
e-mail: sisaudagar@mitra.ac.in

© Springer Nature Singapore Pte Ltd. 2019   669
A. Abraham et al. (eds.), *Emerging Technologies in Data Mining and Information
Security*, Advances in Intelligent Systems and Computing 814,
https://doi.org/10.1007/978-981-13-1501-5_59

Thus, offsite location access to information is classically carried out via node-to-node routing, and it can highly promote from caching and thus improve access timings. Second, the network is mostly node depended with regards to channel bandwidth or power in the nodes. In such case, network burden can be lessen with the help of information caching; this leads to lessen bandwidth use, as well as battery energy. The optimum data caching becomes severe when network nodes have a limited memory capacity while caching information chunks.

Elaborating proficient intrigue for caching in node cluster with memory limitations can be achieved here. In particular, these intrigues can enact in sensor network, node-to-node networks and mesh networks, World Wide Web and making more generalize AWN. Nevertheless, the current implemented approaches are quite "ad hoc" in nature and observed that is depends, and not having any strong systematic foundation. However the theoretically it is flourish while analyzing that optimality properties of information caching and replicated chunk allotment problems, on the other hand, the distributed implementation of these techniques and their performances in complex network setting is not being scrutinized.

Our goal is to develop an intrigue which will be systematically discernible along with this incontestable recital in a central setting and in addition should be implemented to natural distributed environment. As in this model we focus on N2N system, it contains multiple information chunks at agreed frequency. Every node in the system accesses data items and picks to cache in its memory; doing this, it lessen the access amount. Essentially, here we expand proficient intrigue that picks information chunk for caching at each node. Using simulations of centralized approximation algorithm and localized distributed algorithm, we show through broad conduct test on ns2 that our projected intrigue performs a good deal better than a foregoing approach over a expansive assortment of parameter value. We strive here to represent the solution of caching and cache residency of multiple information contents in a node having limited memory capabilities.

These evaluations are taken based on the propinquity of the peer, and at the same time, duplication of data is avoided; hence, it evaluates an approach where nodes independent of each other check legality of caching data. Considering limited memory constraints, the efficient caching intrigue is developed which stores newly received information in parallel uphold content distribution system recital. Under both conditions, each peer takes a decision based on to its discernment of what nearby users may store in their caches with the aim of differentiating its own information chunks to be cached from other nodes. We strive here to represent the solution of caching and cache residency of multiple information content in a node having limited memory capabilities which results in the creation of the content diversity within clustering of nodes in N2N so that the requesting users probably find the desired information nearest. This simulator shows that presented caching intrigue is capable of creating the desired content diversity in different N2N networking scenario showing that our solution succeeds which in turn gives resource proficient information right of entry.

2 Problem Definition

N2N system could yield benefit, when implemented as in juxtaposition with cellular networks, to encourage information chunks exchange using huddle among mobile user. For highly popular content, N2N distribution can, indeed, remove bottleneck by pushing the distribution from the core to the edge of the network.

Nevertheless, cooperative caching techniques which are currently implemented would swap node memory ability with unneeded data that were chose meanwhile hence found to be unfeasible, as its few intrigues of proficiently information caching in AWN.

3 Related Work

In this section, we parpended several content caching and its replacement in node-to-node (N2N) way of communication. However, all mentioned caching methodology is based on cooperative caching (Fig. 1).

3.1 Cooperative Caching

Cooperative caching techniques which are currently implemented for the N2N are defined as the following message-oriented, directory-oriented, hash-oriented. The distributed routing algorithm also known as CARP is the largest part well-known

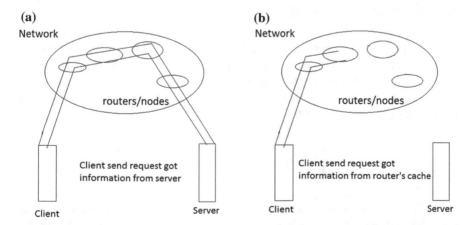

Fig. 1 **a** Detail about how the process flows in traditional existing systems. In such system every time we need to travel the data from given source to destination. **b** The process flow in the proposed system

hash-oriented cooperative caching mechanism. In cache array routing protocol, the Web proxy cache arrays have adopted hash-oriented routing and rely on load distribution. A mobile node in N2N might not be aware the information source that exists, or if the mobile node might not aware of information source availability, hybrid cache might not be a good option [1]. Besides, node information chunk for caching away from the route of the node asking for information and the information source might not be sharing cache information with the requesting node.

3.2 Caching in Small-Sized Memory

Information caching reduces burden on AWN, but while caching in small-sized cache memory, cache replacement is anticipated. In [2], it uses cache table which is to be maintained in AWN like to be maintained and updated like routing table.

3.3 Content Diversity

In AWN, creating content diversity with estimation of content availability improves network efficiency. Cooperation of nodes is mandatory in creating such content diversity where nodes of common interest and ad hoc pattern are grouped together and add improvement in cache hit rate [3]. Contrarily, such scenario needs additional exchange of messages and continuous feedback from nearby nodes moving in AWN.

3.4 Information Cloning

As we parpend various issues, various intrigues to information cloning are pertinent to caching information which is being specified. In [4], the author specifies the way of reducing information cloning, but it desires control messages for node cooperation and information accessing frequency. The method in [1] is about information cloning by analyzing its convergence time.

4 Proposed System

We aim to simulate a scenario that escorts the resource proficient information admittance. The intrigue of caching information in AWN where nodes switch over information chunks in a N2N fashion. This caching intrigue lets the information caching depend on the conscious occurrence of the information chunks in the node propinquity; its inference grounds less operating cost to the information distribution system.

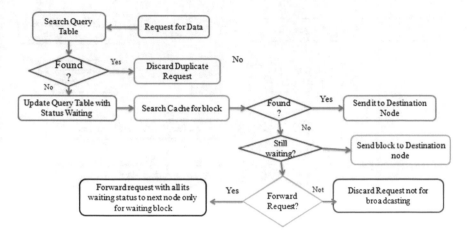

Fig. 2 Flow of request in node-to-node network

4.1 Request Flow

See Fig. 2.

4.2 Information Flow

However you look it at the result withdrawn here eventually creates virtual network of 'intermediatory' nodes, and which is suitable to stagnant inter-connected networks with steady associations among nodes. We are presenting a scaffold that can retrieve information and achieve caching of that information intrigue in mobile AWN, and it puts up on a core routing protocol and necessitates the manual locale of a network-wide "Caching expanse" (Fig. 3).

In density-based caching intrigue, if intermediate peer or router contains the required data, then it is directly transmitted to required destination decreasing the accessing time and thus helps to improve efficiently.

4.3 Large-Capacity Cache

In this pencil case, nodes are capable of storing a large fraction of the available information chunks. Efficient and compact memory acquirance is always enviable (if not required) provision, for the reason that the similar memory may be accessed by different application executing on node. During large capacity cache, caching intrigue consent to the caching based on the alleged "presence" of the content in

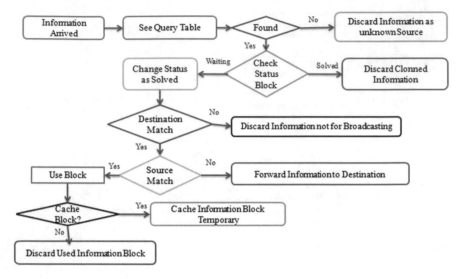

Fig. 3 Flow of message in node-to-node network

the node propinquity, its inference grounds less operating cost to the information distribution system and the sole purpose of diminish the memory acquirance and not affecting on the whole information retrieval performance can be achieved.

4.4 Small-Capacity Cache

In small capacity scenario, nodes memory are limited and dedicated where it could store a tiny percentage (10%) of the information particular node retrieve. The information caching result gets converted to a cache replacement intrigue which in turn chooses and drops the cached information chunk just arrived. As the Hamlet intrigue is evaluated in diverse AWN, where nodes communicate through N2N way. And all these results will assure high request-query pledge ratio, whereas balance the traffic load very low, even for scarcely popular content, constantly next to various network connectivity and mobility scenarios.

5   Simulation Results

The selected files of source node are then transmitted across AWN through intermediate nodes provided source and destination nodes should be in AWN range (Figs. 4, 5, 6 and 7).

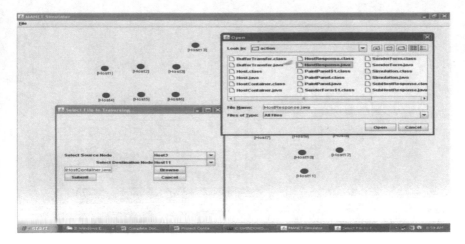

Fig. 4 How files of node are being selected for transmitting across AWN

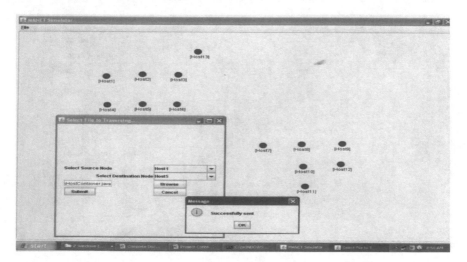

Fig. 5 Successful transmission from source to destination node in AWN range

Both successful and unsuccessful data transmissions are maintained with intermediate node receivers.

6 Conclusion

In AWN nodes exchange information items in N2N fashion as N2N Networking is completely decentralized or distributed architecture without a central base station. While considering data caching where upon receiving requested information deter-

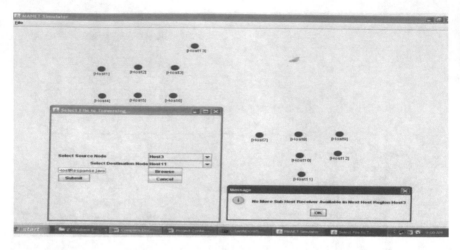

Fig. 6 Error in the transmission due to node not in AWN range

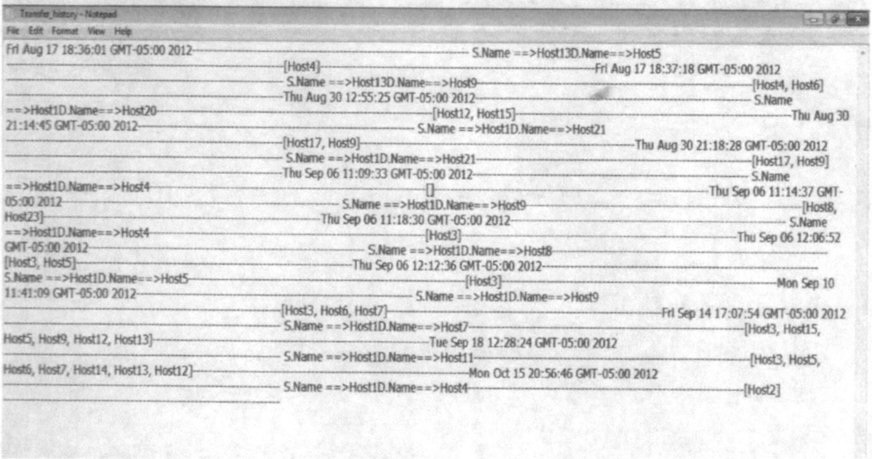

Fig. 7 Transfer history

mines access latency of information chunks or chooses the content is omitted for recently received information chunks. Here, intrigue created a paradigm of information caching way to support competent information access in AWN. In fastidious, we consider memory restraint of the devices in Ad hoc Wireless Network and analyze efficient information caching algorithms to resolve near most favorable cache custody to make the most of drop in access cost yields message cost hoard and hence, it gives improvised bandwidth use and power savings.

Nonetheless, the mock-up shows the recital of the caching algorithm for ample number of network as well as the application set of facts. Presented a distributed accomplishment fully relying on an presumption algorithm for problem of cached chunk residency under memory restraint. These evaluations are taken based on the propinquity of the node, and at the same time, information cloning is avoided; hence, it evaluates an intrigue where nodes independent of each other check legality of caching data. Here simulation of caching algorithm in diverse AWN and evaluate it with further caching intrigues, it is explaining when this implemented can create the desired information diversity, as a consequence yields competent information access.

7 Future Work

We will develop an intrigue which aims to design scenario that yields to resource proficient information admittance. Data caching intrigue for AWN in which nodes exchange information chunks in a N2N way. The solution succeeds in creating the expected copies of information items and creates resource-efficient information access.

References

1. Hara, T.: Effective replica allocation in ad hoc for improving data accessibility. In: IEEE INFOCOM (2001)
2. Chow, C.Y., Leong, H.V., Chan, A.T.S.: GroCoa: group based peer to peer cooperative caching in mobile environment. In: IEEE S. jel Communication (2007)
3. Ko, B.-J., Rubenstein, D.: Distributed self stabilizing placement of replicated source in emerging network. In: IEEE ACM (2009)
4. Varshney, N., Roy, T., Chaudhary, N.: Vampire attack detection I wireless sensor network. In: International Journal of Innovative Research in Computer and Communication Engineering (2014)
5. Nayak, S., Narvekarand, M., Mukhopadhayay, D.: Cooperative caching technique in peer to peer mobile environment. In: International Conference on System Modeling & Advancement in Research Trends (SMART) (2016)
6. Singh, S., Agrawal, S.: VANET Routing Protocols Issues and Challenges., published in IEEE, USA (2014)
7. Varshney, N., Roy, T., Chaudhary, N.: Security protocol for VANET by using digital certification to provides security with low digital bandwidth. In: International Coference on Communication and Signal processing (2014)
8. Elias, C.E., Zhang, S., Liu, E.: Vehicular ad hoc network (VANETs): current state, challenges, potential and way forward. In: Internatinal Conference on Automation and Commputing, Uk (2014)
9. Mahardhika, G.I., Mat, M.: Multi-criteria vertical handover decision in heterogeneous network. In: Wireless Technology and Applications IEEE (2012)
10. Zeadally, S., Hunt, R., Chen, Y.-S., Irwin, A., Hassan, A.: Vehicular ad hoc networks (VANETs): status, results, and challenges. In: Telecommunication Systems (2012)
11. Sumra, I.A., Ahmad, I., Hasbullah, H., AbManan, J.B.: Classes of attacks in VANET. In: International Conference on Wireless and Optical Communications Networks (2013)

12. Al-kahtani, S.B.A., Al Kharj: Survey on security attacks in Vehicular Ad hocNetworks (VANETs). In: International Conference on Signal Processing and Communication Systems (2012)
13. KadamMegha, V.: A Security analysis in VANETs: a survey. In: International Journal of Engineering Research and Technology (IJERT) (2012)
14. Al-Qutayri, M., Yeun, C., Al-Hawi, F.: Security and privacy of intelligent VANETs. In: Computational Intelligence and Modern Heuristics, book edited by Al-Dahoud Ali (2010)
15. Tang, B., Gupta, H., Das, S.: Benefit based data caching replacement policies for cooperative caching in mobile ad hoc network. In: IEEE Transaction Mobile Computing (2008)

Secure Encryption in Wireless Body Sensor Networks

A. Sivasangari, Suvam Bhowal and R. Subhashini

Abstract The transmission of the patient's data is transmitted through wireless medium. It is therefore necessary to provide adequate security. A lightweight selective encryption (LWSE) scheme is proposed in this work. The proposed security mechanism of WBAN ensures security parameter such as data confidentiality and authentication and designed to provide the security in three levels of communication in WBAN. The security of medical data either while stored in WBAN or during their transmission is an important issue concern. The sensor nodes are energy constrained. It is necessary to use lightweight security mechanism for WBAN. A LWSE scheme is proposed for minimizing transmission overhead.

Keywords Morphological operators · Symmetric encryption
Skipjack algorithm

1 Introduction

The sensor nodes participating in a WBAN can be classified as implanted or worn sensor nodes. The transmission range of these nodes is 2 m. The sensor nodes gather all the necessary physiological parameters of the human body such as ECG, pulse, and blood pressure. The SH collects the information from the sensor nodes and relays it to the hospital. This information collection process happens by using the Internet through an intermediate device. This type of network employs the collection of sensors on the human body for sensing certain physiological factors of a per-

A. Sivasangari (✉) · S. Bhowal · R. Subhashini
Sathyabama Institute of Science and Technology, JeppiaarNagar,
Rajiv Gandhi Salai, Chennai 119, Tamil Nadu, India
e-mail: sivasangarikavya@gmail.com

S. Bhowal
e-mail: suvambhawal180@gmail.com

R. Subhashini
e-mail: subhaagopi@gmail.com

© Springer Nature Singapore Pte Ltd. 2019   679
A. Abraham et al. (eds.), *Emerging Technologies in Data Mining and Information
Security*, Advances in Intelligent Systems and Computing 814,
https://doi.org/10.1007/978-981-13-1501-5_60

son and forwards the sensed data to the monitoring hospital or medical center. The WBAN communication architecture is divided into three levels. Level 1 intra-WBAN communication refers to communication between the sensors and communication between the sensors and personal server (PS). Level 2 inter-WABN communication refers to communication between the PS and BS. In level 3 beyond WBAN, the medical application, database, and healthcare service providers are employed for providing the service to the remote users including doctors, nurse, caretakers of patients, and researchers.

Besides this, medical data is highly confidential with privacy being preferred. Data privacy should also be assured to the patients. Yet, WBANs follow distributive architecture and may be subjected to data leakage. The issue of data leakage can be addressed by incorporating cryptography and access control mechanism for providing data privacy. However, establishing security and privacy mechanisms in WBAN is a challenging task, due to power or energy constraints. Incorporation of security mechanism must not have an impact on the performance and efficiency of the system. Hence, WBAN expects the security mechanism to be of lightweight that minimizes the encryption time, FAR, FRR, HETR values.

2   Related Works

Security and privacy are ensured by biometrics-based security mechanisms. However, this process involves an expensive key distribution as mentioned by Venkatasubramanian and Gupta [2]. Still, there are limited biometrics-based securities in the literature. The reason for limited usage of biometrical concept in achieving security is the computational overhead and the consumption of energy. Power consumption of sensor nodes in WBAN must preferably be limited in order to ensure a long lifetime for the network. Inter-pulse interval (IPI) is exploited as the biometric feature in the work proposed by Poon et al. [1]. However, these works expect time synchronization and lead to higher communication overhead.

Data transfer can be done between the sensors by taking such physiological values into account. ECG is one of the examples of physiological quantity. The ECG signal is used for feature extraction process. Both the sender and the receiving sensors sense the ECG signal, which is followed by the process of hashing or watermarking technique. The transformed signal is exchanged for generating keys as suggested by Ali and Khan (2010) and Venkasubramanian et al. (2008). The cluster-based key agreement protocol was presented by Ali et al. (2013). The cluster is formed, and the keys are generated by physiological signals. Venkatasubramanian et al. [2] have proposed a secure scheme for inter-sensor communication, which makes use of the fuzzy vault for physiological signal-based key agreement.

The work proposed by Poon et al. [1], the key values, is derived from ECG signal. The IPI values are calculated from ECG signal. Upmanyu et al. [3] have proposed a secure blind biometric authentication protocol. This protocol combines biometric mechanism with cryptography algorithm. The protocol provides a secure data transmission under different types of threats. The sensor nodes are loaded with polynomial shares.

The PSKA scheme proposed by Heinzelman et al. (2002) locks the key value in the vault of sender side and unlocks it in the vault by using features of physiological signal at the receiver side. It is based on physiological values between the sensor nodes. It provides two types of keys for sensor nodes. The individual key is shared with sensor and personal wireless hub. A pairwise key is used between the sensor nodes. A new node when added to the WBAN tries to locate the neighboring nodes by sending hcllo messages. After receiving the acknowledgement message from the neighboring nodes, it calculates the pairwise key by using polynomial share. The individual key is updated periodically by performing XOR with the medical information.

The PSKA allows use of key values generated from the physiological signal by communication of neighboring nodes. The sender generates the random key value from the physiological signal and hides use of physiological feature vector. The hidden key is sent to the receiver. The receiver obtains the key using its feature vector. The pre-deployment of keys is not needed. The additional extra chaff points lead to unnecessary communication overhead. The features of physiological signal are derived by using the discrete wavelet transform (DWT). Sivasangari et al. [7] proposed ECG-hummingbird encryption algorithm ensures security of medical information by using biosignal. Hummingbird algorithm is applied for encryption of medical information. The extracted ECG feature values are substituted in polynomial equation.

3 LWSE Procedure

In the three levels of data transfer, this work suggests an approach that will securely propose a security model that concentrates on three levels of communication. This section presents a secure system that provides privacy, authenticity, availability, confidentiality, and integrity.

The proposed work concentrates on three levels of data communication. In initial state, sensor nodes perform the registration operation with sensor head for avoiding miscommunication. Biosensors detect the medical parameters. The detected QRS peaks from ECG signal are encrypted by using the skipjack algorithm. The sensor nodes attach the time stamp and sensor id with encrypted information and transfer to SH. In the second phase, the registration is performed between SH and base station. The symmetric key encryption is used for encryption of the medical information. Watermarking is applied on this encrypted information. Both the encrypted and embedded data are forwarded to BS by SH. When the physician needs to access the data, the access request is sent to the BS. The BS then sends the decryption key which is encrypted by the physician id. The physician has to decrypt the key, in order to gain access to the data. In this work, focus has been rendered to extract QRS signals and to provide security through all the three different phases.

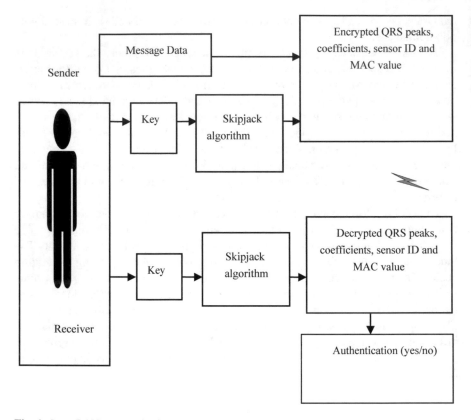

Fig. 1 Intra-BAN communications

The overall flow of the proposed inter-sensor communication work is shown in Fig. 1.

The algorithms are described below for data transmission between SH and BS and data transmission between BS and physician.

1. SH registers itself with BS.
2. BS verifies the ID and forwards the mk.
3. SH forwards D_R, sh_{id} to the BS.
4. BS verifies D_R and forwards k_{ms}, k_{mk}, ts.
5. SH encrypts data with k_{sym}.
6. Embed the encrypted data in a cover image.

//Data transmission between BS and physician

1. Physician forwards the Acc_R to the BS.
2. BS encrypts the decryption key with phy_{id}.
3. Extract data from the cover image.
4. Decrypt the data.

3.1 Data Transmission Between Biosensors and SH

The ECG is a waveform that represents the electrical and muscular functions of the heart. It consists of five waves such as P, Q, R, S, and T waves. The QRS complex waveform is the important waveform in ECG that represents the simultaneous activity of right and left ventricles. The doctor analyzes the heart rate analysis using QRS complex values reflecting the electrical activity within the heart during ventricular contraction. The proposed work exploits dilate and erodes operators. It performs the reduction of peak values and widening of the valleys in the ECG signal. The width of the peaks and valleys is determined by structuring element. The peak values are quantized and converted into a binary string.

3.2 Data Transmission Between SH and BS

The coefficient values are transferred from the sender to the receiver, and all the sensors in WBAN know the order of the polynomial. The coefficient values are pre-deployed in all the sensor nodes. The SN constructs a polynomial equation by using the feature values. The SH receives the coefficients and finds the value of the polynomial equation. An encrypted message is sent along with the coefficients and MAC value of the message to the SH which tries to recover the key value by constructing the polynomial. It decrypts the message by using key values. The peak values extracted from the ECG signal generate the key value. The MAC value is recalculated for verification. The QRS peaks are encrypted by using the skipjack algorithm. This transmission consists of the encrypted QRS peaks, coefficients, SN id, and the MAC value.

The embedding procedure is described in the following steps.

1. Consider the cover image size is $M \times N$.
2. The encrypted data bytes are arranged in a set.
3. LWT decomposes the image into different subband coefficients such as approximate coefficients, vertical coefficients, horizontal coefficients, and diagonal coefficients.
4. Separate the horizontal and vertical coefficients.
5. The horizontal and vertical subbands are used to embed the encrypted data.
6. Let $P = \{p1, p2 \dots pn\}$ represent the set of all neighboring pixel pairs in the image.
7. Let $W = \{w1, w2, \dots wn\}$ represent the encrypted data values. The difference between the pixel pair values is calculated. The encrypted data embedded directly a difference of pixel pairs.
8. If a pair is not suitable, the next pixel pair is checked and the process is repeated till all the bytes are embedded.

3.3 Data Transmission Between BS and Remote Users

Every doctor in a health care is needed to register their names and id in the server. Every user attribute is verified by the BS. The BS sends the information only to users whose attributes are satisfied. After successful verification, the decryption key is sent by the BS which is encrypted by the id of doctor. The embedded information is extracted from the cover image.

4 Performance Analysis

The security of the proposed algorithm is analyzed in this section. Encryption and decryption are performed in the sender and the receiver side. The proposed algorithm encrypts all the data prior to the data transmission for protecting the data from the adversary. The proposed work is analyzation takes place by the performance metrics such as false acceptance and rejection rate, half total error rate, and key generation time. MATLAB is utilized for carrying out the proposed research work. This work is evaluated with MIT/BIH standard ECG database, and the least FAR and FRR are achieved. All the sensors transmit the messages to the SH, which is the most powerful node.

The proposed algorithm is evaluated with the existing algorithm fuzzy vault. Tables 1, 2, and 3 show FAR, FRR, and HETR performance analysis. Analysis is based on degree of polynomial. But, the tolerance value is set to 4. When the s value increases, the probability of mismatching information is reduced. People are living nearer to each other. The message passed by sensors may transfer to other SH also possible in communication.

The encryption time of the existing system is compared with the proposed LWSE method. The present research work takes less time to generate the key. Analysis of the experimental results clearly shows the satisfactory working of the system for several performance measures. The key generation is time is much lesser than other existing work. FAR, FRR, and HETR values also reduced when compared to other approaches. The main contribution of the present research is to provide a

Table 1 Comparison of FAR

Degree (s)	Fuzzy vault scheme	LWSE model
6	0.012	0.01
7	0.005	0.004
8	0.003	0.002
9	0.002	0.001
10	0.001	0
11	0	0
12	0	0

Table 2 Comparison of FRR

Degree (s)	Fuzzy vault scheme	LWSE model
6	0.09	0.08
7	0.13	0.12
8	0.25	0.20
9	0.38	0.26
10	0.49	0.30
11	0.6	0.35
12	0.7	0.39

Table 3 Comparison of HETR

Degree (s)	Fuzzy vault scheme	LWSE model
6	0.051	0.0450
7	0.067	0.062
8	0.126	0.101
9	0.191	0.130
10	0.245	0.150
11	0.300	0.175
12	0.350	0.195

secure system that strives to provide access control, security, and privacy. This system achieves all these security measures by incorporating symmetric key encryption and watermarking technique.

5 Conclusion

In the proposed work, a lightweight symmetric key-based encryption along with the watermarking is exploited. In this work, a secure system that is based on symmetric key encryption and watermarking technique is presented. This work promises access control, data integrity, and confidentiality. The results of this work are satisfactory, when analyzed with standard performance metrics. The experimental results show that the proposed scheme is lightweight and reduce the amount of data transmission. The performance of the proposed system is analyzed by FAR, FRR, and HETR. On analyzing the experimental results, it is concluded that the proposed work consumes the least time for carrying out encryption than any other algorithms.

References

1. Poon, C., Zhang, Y., Bao, S.: A novel biometrics method to secure wireless body area sensor networks for telemedicine and M-health. IEEE Commun. Mag. **44**(4), 73–81 (2006)

2. Venkatasubramanian, K.K., Gupta, S.K.S.: Physiological value-based efficient usable security solutions for body sensor networks. ACM Transactions (2010)
3. ManeeshUpmanyu, Anoop M., Namboodiri, Kannan Srinathan, Jawahar, C.V.: Blind authentication: a secure crypto-biometric verification protocol. IEEE Trans. Inf. Forensics Secur. 5(2), 255–268 (2010)
4. Malasri, K., Wang, L.: Design and implementation of a secure wireless mote-based medical sensor network. Sensors 9(8), 6273–6297 (2007)
5. Chen, M., Gonzalez, S., Vasilakos, A., Cao, H., Leung, V.C.: Body area networks: a survey. Mobile Netw. Appl. 16(2), 171–193 (2011)
6. Chiu, C.T., Haodong, W., Sheng, Z., Qun, L.: Body sensor network security: an identity-based cryptography approach. In: Proceedings of the ACM conference on Wireless network security, pp. 148–153 (2008)
7. Sivasangari, A., Martin Leo Manickam, J.: A Light weight cryptography analysis for wireless based healthcare applications. J. Comput. Sci. 10(5), 2088–2094 (2014). ISSN: 1549-3636
8. He, Daojing, Chan, Sammy: A novel and light weight system to secure wireless medical sensors networks. IEEE J. Biomed. Health Inf. 18(1), 316–326 (2014)
9. Hong, D., Sungm, J., Hong, S., Lim, J., Lee, S., Koo, B.S., Lee, C., Chang, D., Lee, J., Jeong, J., Kim, H., Kim, J., Chee, S.: HIGHT: a new block cipher suitable for Low-Resource device. In: Proceedings of CHES 2006, vol. 4249, pp. 46–59. Springer (2006)
10. Hu, J., Bao, S.: An approach to QRS complex detection based on multi scale mathematical morphology. In: Proceeding of 3rd IEEE International Conference on Biomedical Engineering, pp. 725–729 (2010)
11. Lee, P., Lee, D.H.-J.: Secure health monitoring using medical wireless sensor networks. In: Proceedings of 6th International Conference on Networked Computing and Advanced Information Management, pp. 491–494 (2010)

An Effective Privacy-Preserving Reputation Scheme for Mobile Crowdsensing

Bayan Hashr Alamri, Muhammad Mostafa Monowar
and Suhair Alshehri

Abstract Mobile crowdsensing is an emerging technology in which mobile carriers collect and contribute valuable data for different applications. This technology enables a broad range of sensing applications by leveraging mobile objects world wide to improve people's quality of life. Protecting the privacy of participants and the trustworthiness of the sensor data are conflicting objectives, yet are key challenges in this area. Privacy issues arise from the disclosure of user-related context information of data provider such as participants' identities. The trustworthiness of sensed data is important to provide confidence to its end users. This work addresses the data trustworthiness of the sensed data and the privacy preserving of participants by designing an effective reputation scheme that preserves the privacy and developing a general architectural schema for mobile crowdsensing applications. Detail security analysis to prove the effectiveness of our approach in terms of its resistance to different kinds of attacks is presented.

1 Introduction

Mobile crowdsensing (MCS) is an emerging technology in which mobile carriers including participants, mobile objects, or data providers are able to collect and contribute valuable data for different applications in domains such as smart cities, road transportation, and health care [1, 2]. This technology enables a broad range of sensing applications by leveraging mobile objects worldwide to improve peoples quality of life [3]. Sensing applications can be classified into two categories based

B. H. Alamri · M. M. Monowar (✉) · S. Alshehri
Faculty of Computing and Information Technology, King AbdulAziz University,
Jeddah, Saudi Arabia
e-mail: mmonowar@kau.edu.sa

B. H. Alamri
e-mail: balamri0052@stu.kau.edu.sa

S. Alshehri
e-mail: sdalshehri@kau.edu.sa

© Springer Nature Singapore Pte Ltd. 2019   687
A. Abraham et al. (eds.), *Emerging Technologies in Data Mining and Information
Security*, Advances in Intelligent Systems and Computing 814,
https://doi.org/10.1007/978-981-13-1501-5_61

on what events being monitored: personal or community sensing. In personal sensing applications, the events pertain to a person such as movement pattern of individuals such as running, jogging. On the other hand, in community MCS applications, the events pertain to large-scale events. This category is classified further based on the involvement of mobile carriers in sensing actions to participatory or opportunistic. In participatory sensing applications, participants volunteer to collect data from the environment using mobile phones, so they are actively involved in the sensing action. In opportunistic sensing application, participants are not involved in the sensing actions [4]. Although many applications have been gained benefits from MCS, there are two important issues that are considered as obstacles for MCS evolution. First, protecting the privacy of these data providers [5], and second, ensuring the trustworthiness of the sensing data [6]. These two issues are two key challenges yet conflicting objectives in MCS applications [3]. In MCS, privacy challenges arise from the disclosure of user-related context information of data providers such as participants identities, IP addresses, and locations [3, 5]. Moreover, there are common attacks against the participant's privacy such as traffic analysis, monitoring, eavesdropping, and collusion. On the other hand, the trustworthiness of sensing data is important in MCS applications to provide confidence to its end users including clients, queries, and data requesters who used these sensing data. Data trustworthiness has common attack models including erroneous or malicious contributions, collision, and Sybil attacks [3, 6, 7]. Applying trust and ensuring the privacy of participants incur additional overhead such as computational power or additional battery consumption. This is often critical for different entities in MCS systems, especially for participants who carry the smartphones with limited capabilities [8]. The goal of this work is to design and develop an effective MCS system that preserves the privacy of participants while ensuring the trustworthiness of sensing data and reducing the associated overhead. Given the lack of work in this area considering the above-mentioned issues, the main contributions of this paper are: (1) the design of a centralized reputation distributed trust (CRDT) based on the collaborative path hiding concept (CPH) for MCS applications, (2) a method for computing the local trust for participants without exposing their privacy based on the feedback received from their neighbors, their past behavior, and the quality of service provided by these participants, (3) security analysis of the proposed schema in term of it is resistant to the different type of attacks.

The rest of this paper is organized as follows: Sect. 2 describes the related work. Section 3 presents the necessary preliminaries and defines the threat model. Section 4 introduces the proposed schema. Section 5 provides a security and privacy analysis showing the resistance of the proposed schema against numerous attacks. Finally, Sect. 6 concludes the paper and outlines the future work.

2 Related Work

Several techniques have been proposed to ensure the trustworthiness of sensing data [9, 10]. Though several works have studied the issues related to the privacy of participants [11, 12], little work has been done considering both problems [7, 8, 13–15]. In these cases, they did not consider the common attacks on the sensing systems such as Sybil attacks [8, 14] or the additional overhead including the computational, traffic, or energy consumption overhead [13].

Reputation systems were proposed by Huang et al. [9] and Mousa et al. [10] to evaluate the trustworthiness of the contributed data in the participatory sensing systems. Huang et al. use the Gompertz function to compute the reputation score of the contributed data. The limitation of this system is the accumulative feature of the reputation scores that may lead to de-anonymize the participants and compromise their privacy by linking the trust value associated with multiple contributions from the same mobile device. Mousa et al. compute the reputation and trust scores based on the contribution evolution module. This system depends on a dynamic trusted set of providers and that will make it easy to identify malicious users. However, this system did not address the Sybil attacks or how it will reduce the overhead.

A novel participatory sensing (PS) framework architecture that guarantees the security and accountability were proposed by Gisdakis et al. [11]; SPPEAR architecture in addition to preserving user privacy it provides incentives to the participants in PS. It uses group signature to enable the participants to anonymously authenticate themselves. But, the process to defend against the malicious user is not straightforward to evicting intelligent. Also, there is a trade-off between the overhead and the use of multiple pseudonyms. Reusing one pseudonym to protect more than one reports trade-off participants privacy for overhead. A very similar approach is taken by Gao et al. [12]; they present TrPF, which is an enhanced trajectory privacy-preserving framework. It is based on trusted third party server (TTPs) that will serve as a privacy-preserving agent by providing a certificate to the ligament participants and unlink the identity of those participants from their spatial-temporal information.

According to Christin et al. [14], TrustMeter scheme is proposed to assess the participants' contributions in a privacy-preserving mechanism. This scheme utilizes the CPH concept [16], to unlink between the participants' real identities and the spatiotemporal context of the reported data. It also prevents collusion attacks through maintaining a list of partners on the server. It also has minimum traffic overhead during the exchange of triplets and reputation values. On another hand, it does not prevent multiple identities attacks, i.e. Sybil attacks. applications that was proposed by Michalas and Komninos [8]. This system aims to provide privacy of participants, anonymity, accountability as well as supporting users reputation. Moreover, it prevents the collision attack through verifying the credential of users, but it fails to address the profiling of user attacks and how to defend against malicious attacks in trust evaluation when the identity of user is preserved. IncogniSense proposed by Christin et al. [13] is an anonymous reputation framework that is based on the blind signature to achieve periodic pseudonyms as well as secure reputation transfer between these periodic pseudonyms. This framework prevents Sybil attacks by enabling reputation and pseudonym manager (RPM) to sign blindly only one pseudonym for each client per period. Thought this framework is an energy efficient schema, it acquires additional communication and management overhead.

3   Preliminaries and Threat Models

This section reviews the necessary background to pave the way to the discussion of the proposed scheme and presents the threat model and assumptions.

3.1   Collaborative Path Hiding (CPH) Concept

Mobile devices utilize the concept of collaborative path hiding (CPH) to protect their privacy. Participants ensure their privacy themselves without the need to trust an application to protect their privacy. It is based on exchanging information between them physically in an opportunistic fashion. This breaks the association between the participants' real identity and the spatiotemporal information at which the sensor reading was taken. The exchange process can be done based on three strategies: 1. random-fair, 2. random-unfair, and 3. realistic [14].

3.2   Centralized Reputation Distributed Trust (CRDT) Schema

To manage the trust and reputation values, we utilize the centralized reputation distributed trust (CRDT) schema. CRDT is used in network environments to assess the similarity of multiple information about the same event from different sources using two values—first, the local trust (LT) value that is stored locally at each participant, which reflects his opinion for other participants that he has received information from, and second, global reputation (GR) value that is stored in central reputation manager (CRM) unit in the server, which reflects all other participants' opinion on a particular participant [17].

3.3   Threat Model

Our schema assumes that attackers follow the Dolev-Yao threat model [18]. Also, due to the nature of the CPH concept, participants can become attackers by attempting to access the triplet content during the exchange process. Participants can give whatever feedback during local trust computation and vote whatever value during reputation computation process. We also assume malicious participants can collude with each other to launch collusion attacks by giving each other positive feedback. Malicious participants can also attempt to create a fake ID and lunch Sybil attack (adopting multiple identities) in order to increase their reputation level by self-promotion. Also, there is multiple attacks model that can affect the accuracy of the reputation and trust model such as bad mouthing attack, on-off attack, and conflicting behavior attack [19, 20]. In the bad mouthing attack, the attacker gives dishonest feedback to decrease the trust and reputation of a good participant and the opposite for the malicious one. In on–off attack, the attacker acts good and bad alternatively, but in conflicting behavior attack, the attacker acted differently to different participant in order to decrease the trust and reputation of a good participant.

4 The Proposed Scheme

This section provides an overview of the proposed scheme and details the underlying mechanisms.

4.1 Overview

Figure 1 shows the framework for the proposed scheme. Our framework consists of the application server (App.Server) and participants (client, mobile devices). The

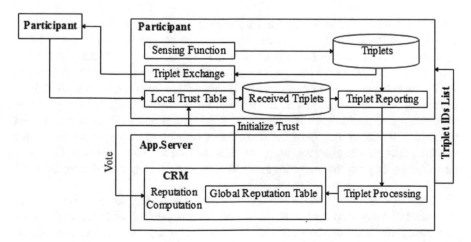

Fig. 1 The proposed framework

participants collect a sensing reading from their surroundings and prepare them to be exchanged with another participant using CPH [14]. Then each participant reports the exchanged and their own triplet to the server periodically. The trust and reputation compute based on CRDT [17] manner. Specifically, the trust level computes locally at each participant but the reputation value computes globally at the server in the CRM.

4.2 Triplet Exchange and Reporting

A participant collects triplets (time, location, and sensor reading) and stores it in his mobile device until he encounters another participant. The triplet is encrypted using the public-key of the server to prevents others from accessing its content, and it has an ID which is unique for each triplet. The triplet ID is a one-way hash function (SHA-252) for its participant ID, the triplet timestamp, and a random number. After verifying the trustworthiness of the opposite party, each participant starts exchanging his triplets with the other party using one of exchange strategies mentioned in Sect. 3.1. The exchanged and collected triplets report to the server to be processed. Then, participants request a list of the triplets IDs delivered to the server since their last connection. The main reason of this list is to remove the successfully delivered triplets from the participant eternal storage. Each triplet has delivered timeout, and if it is expired before it reaches the server, it is considered as lost and demand the participant to exchange it again.

4.3 Local Trust Update

The LT value is stored locally in the trust table in the participant device. LT reflects the participant's opinion for other participants that he has received information from. Each participant examines the trust level of the opposite party by looking at the LT value in the table. The examination is based on the difference between the LT values for the exchanging peers of participants. If this difference is less than a predefined threshold, they are considered as trusted and the exchange is started using one of the exchange strategies as mentioned in Sect. 3.1. If it is greater than the predefined threshold, the connection between these peers drops and they start to search for another participant to exchange with. If this is the first encounter between these two participants, they ask the server for each other reputation value to initialize the local trust.

A participant should always use LT value instead of GR value to evaluate the trust level of the other participants. Such a distributed mechanism in estimating trust minimizes the traffic overhead. In other words, each participant would use the LT value to estimate the opposite party trust level without the need to contact the server

Fig. 2 Process for updating
the local trust

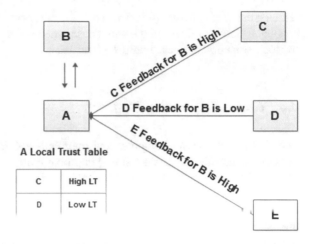

for trust information except in some cases such as to initialize the LT for participants
that communicate for the first time.

After the exchange of triplets between two participants, each one of them updates
their corresponding LT value. For example, if participant A exchanged triplet with
participant B, A updates the LT value for B such that A asks the other neighbors
participants to give their feedback for B. Local trust feedback ($LT F$) is the feedback
that a participant gets from the nearest neighbors about a specific participant at the
range of 10 m. These feedbacks weight based on a trust factor (K) that range between
[0, 1], which reflects how much A trusts these neighbors as depicted in Fig. 2.

The feedback from the participant with the highest local trust at the requester
table is more trusted. For example, C gives a high feedback about B and he has a
high LT value in A's local trust table, so A gives the feedback from C a high weight
and the opposite is done in the case of feedback form D. In case of E which does not
have any LT value in A's local trust table, the K value will be 0 as if E does not give
any feedback for B. Quality of service (W) is the ration of the number of delivered
triplets to the total number of exchange triplets as shown in Eq. (1).

$$W = \frac{DeliveredTriplets}{ExchangedTriplets} \tag{1}$$

If W is low that means B does not deliver all the triplets that were exchanged with
A to the server. We can get this information when A reports the sensing data to the
server and requests a list of the triplets IDs delivered to the server since their last
connection. A then can compute the LT value of B as follows:

$$LT_A(B) = [\Sigma_{C_i}(K \times LT F)] + LT_{old} + W \tag{2}$$

where C_i is all the neighbors in the vicinity of 10 m and gives a feedback to B. The
past behavior, Old Local Trust (OLT), of B that is already stored in A's local trust

table or the GR value that is retrieved from the server in case that this is the first encounter between them. In the worst case when no one gives a feedback about B, the local trust updates based on OLT and W.

4.4 Global Reputation Update

The CRM of the server computes and maintains a GR value for each participant in the system. These values are stored in global reputation table. CRM periodically updates these values by the asynchronously voting process. Here, each participant vote contains two values: (1) the LT for the participant that the server wants to compute its reputation, and (2) the number of time (F) the voter gives feedback to this participant in order to weight its vote. So, the reputation is based on the vote from all the participants that interact with the specific participant. In case that the server does not have a global reputation value for this participant, new participant, the server initializes the reputation with R0.

For example, if the server wants to compute the reputation of participant B, it requests a vote from all the participants that interact with B which is denoted as (S_B), and if A is one of the voters, then A submits to the server the local trust that he maintains for B, $(LT_A(B))$ and the number of times A gives feedback to B, (F_A^B). We set an upper limit for F_A^B which is (F_L), to prevent any attacks from targeting the global reputation by attempting to overwhelm it by his own votes. Also, we use the GR of A as $GR(A)$, so each vote should be weighted based on the reputation of corresponding voter. Based on the above parameters, we estimate the GR value, as in [17] based on the following equations.

$$GR_{new}(B) = \frac{\Sigma_{A \in S_B}(min\{F_A^B, F_L\} \times GR(A) \times LT_A(B))}{\Sigma_{A \in S_B}min\{F_A^B, F_L\} \times GR(A)} \qquad (3)$$

$$GR_{final}(B) = \alpha \times GR_{new}(B) + (1 - \alpha) \times GR_{old}(B) \qquad (4)$$

where GR_{final} is the final updated reputation, and GR_{old} is the reputation that already stored in the reputation table. GR_{final} replaces GR_{old} in the global reputation table. Since the reputation is the aggregate of the participant's stored reputations, we considered a factor α that ranges between [0, 1], which integrates the influence of past reputation on reputation update. This factor makes the weight of the old reputations less than the weight of the reputation that is made recently to reflect the participant's current behavior.

5 Security and Privacy Analysis

This section analyzes the behavior of the proposed scheme when different types of attacks take place based on two assumptions. First, we assume that this scheme is suitable for mobile sensing applications without real-time data delivery constraints. Second, each participant must authenticate himself to the server and the CRM, and both the server and CRM are well protected.

5.1 Server Cannot Correlate the Participant Identity with His Original Sensing Report

Using the CPH concept, each time a participant wants to report sensing data to the server, he reports a combination of his own triplet and the triplet of exchanging partner, so the server is not able to determine the exact participant who collects these triplets. Moreover, the process of exchanging information between participants physically breaks the association between the participant identity and the spatiotemporal information at which the sensor reading was taking.

5.2 Participants Cannot Collude Together to Give Each Other Positive Feedback Without Being Detected

One of the contributions of our scheme is to provide two mechanisms to defend against colluding participants. The server keeps a list of the partners for each participant that he exchanges with. Thus, an irregular rate of high-frequency exchange between the same participants can be identified. Also, if two participants keep exchanging information with each other and give positive feedback to each other, this only increases their mutual local trust, but it does not affect the local trust feedback given by other participants. Using centralized reputation in the proposed scheme ensures that the opinions of all participants are going to be reflected on a specific participant as we mentioned in Sect. 4.4. Thus, the level of global reputation value is not affected in the case of colluding participants.

5.3 Attempt to Alter the Computed Reputation or Report Incorrect Information on Behalf of Another Participant

The server and the CRM are protected against any unauthorized access by using standard cryptographic primitives that guarantee that no one can manipulate the reputation values. We assume that each client has a unique ID and public-/private key pair for authentication. Our scheme protects trusted participants against this attack by enforcing participants to be authenticated to the server. So, this attack cannot be launched without accessing the private key of the target participant.

5.4 Replay and Sybil Attacks

When the same triplets reported to the server at abnormal rates, the server considers them as spam. Our scheme does not prevent attackers from attempting Sybil attacks, but the attackers need to build a reputation first before launching any attacks. Because each time a new participant enters the system, their reputation is initialized with 0. Moreover, as illustrated in the example in Sect. 4.3, if a participant E gives a feedback about participant B and the requester A does not have any LT value for E, it is as if E does not give any feedback. Even in the voting process, we set an upper limit for F_B^A which is (F_L). Hence, no one can overwhelm the GR value by his vote through creating multiple identities as elaborated in Sect. 4.4 in Eq. (3). So it is difficult for any participant to create a fake identity in order to destroy the trust or the reputation process.

5.5 Voting Process Subject to Bad Mouthing Attack

The local trust update process takes into account the local trust of the participants who provide these feedback as in Eq. (2). We weight each feedback using K factor which reflects how much the requester trusts these feedback. Also, in the reputation update process, all votes are weighted based on the reputation level of corresponding voters as explained in Sect. 4.4. So, we give a high value to the feedback and vote from a participant who is highly trusted.

5.6 On–Off Attack and Conflicting Behavior Attack

We give more weight to the recent reputations as compared to old reputations, and this could be a key factor to defend against on–off attacks. So we utilize a factor α that ranges between [0, 1], to integrate the influence of past reputation on reputation update and reflecting the participant's current behavior as illustrated in Eq. (4). If the participant misbehaves after some good behavior, his reputation value collapses and to increase it a lot of good behaviors is required, so that it can deter the participant from launching this attack. The local trust update process is based on four parameters as mentioned in Eq. (2). In computing LT value, we do not only rely on other participant's feedback but also the past behavior of the participant and the quality of the service that he provides. So, even if a participant behaves differently to different participants to disrupt the feedback process, other aspects are considered in computing the trust besides the feedback recommendations. Moreover, in the feedback process to update the LT, these feedbacks are weighted according to how much the requester trusts the neighbors who provide these feedback. Hence, it is difficult for the malicious participant to know which one is highly trusted by the targeting one.

6 Conclusion

In this paper, we propose a privacy-preserving reputation scheme for effectively preserving the privacy of the participants and ensure their trustworthiness in MCS. This scheme preserves the privacy of participants by leveraging the path hiding concept and ensures the trustworthiness of the contributed data by CRDT schema. In addition, we prove the resistance of our scheme under different types of attacks using theoretical analysis. For our future work, we are going to simulate the proposed schema to measure the computational performance of the local trust and the global reputation update and to measure the communication overhead and computational overhead that might impose on the participant side. Also, we are planning to evaluate the performance of the proposed schema in a real-world scenario and compare it with existing approaches.

References

1. R.C. Borcea, C., Talasila, M.: Mobile Crowdsensing. Chapman and Hall/CRC, UK (2017)
2. Pournajaf, L., Xiong, L., Garcia-Ulloa, D.A., Sunderam, V.: Tech. Rep. TR-2014–002 (2014)
3. He, D., Chan, S., Guizani, M.: User privacy and data trustworthiness in mobile crowd sensing. IEEE Wirel. Commun. 22(1), 28–34 (2015)
4. Ganti, R.K., Ye, F., Lei, H.: Mobile crowdsensing: current state and future challenges. IEEE Commun. Mag. 49(11), 32–39 (2011)
5. Krontiris, I., Dimitriou, T.: In: 2013 IEEE International Conference on Distributed Computing in Sensor Systems, pp. 249–257 (2013). DOI https://doi.org/10.1109/DCOSS.2013.31
6. Mousa, H., Mokhtar, S.B., Hasan, O., Younes, O., Hadhoud, M., Brunie, L.: Trust management and reputation systems in mobile participatory sensing applications: a survey. Comput. Netw. 90, 49–73 (2015)
7. Wang, X.C., Cheng, W., Mohapatra, P., Abdelzaher, T.: Enabling reputation and trust in privacy-preserving mobile sensing. IEEE Trans. Mobile Comput. 13(12), 2777–2790 (2014)
8. Michalas, A., Komninos, N.: In: 2014 IEEE Symposium on Computers and Communications (ISCC), pp. 1–6 (2014). DOI https://doi.org/10.1109/ISCC.2014.6912480
9. Huang, K.L., Kanhere, S.S., Hu, W.: In: Proceedings of the 13th ACM International Conference on Modeling, Analysis, and Simulation of Wireless and Mobile Systems, MSWIM '10, pp. 14–22. ACM, New York, NY, USA (2010). https://doi.org/10.1145/1868521.1868526. http://doi.acm.org/10.1145/1868521.1868526
10. Mousa, H., Benmokhtar, S., Hasan, O., Brunie, L., Younes, O., Hadhoud, M.: In: 2017 14th IEEE Annual Consumer Communications Networking Conference (CCNC), pp. 829–834 (2017). https://doi.org/10.1109/CCNC.2017.7983241

11. Gisdakis, S., Giannetsos, T., Papadimitratos, P.: In: Proceedings of the 2014 ACM Conference on Security and Privacy in Wireless & Mobile Networks, WiSec '14, pp. 39–50. ACM, New York, NY, USA (2014). https://doi.org/10.1145/2627393.2627402. http://doi.acm.org/10.1145/2627393.2627402

12. Gao, S., Ma, J., Shi, W., Zhan, G., Sun, C.: TrPF: A trajectory privacy-preserving framework for participatory sensing. IEEE Trans. Inf. Forensics Secur. **8**(6), 874–887 (2013)

13. Christin, D., Rokopf, C., Hollick, M., Martucci, L.A., Kanhere, S.S.: Pervasive and Mobile Computing **9**(3), 353 (2013). https://doi.org/10.1016/j.pmcj.2013.01.003. http://www.sciencedirect.com/science/article/pii/S1574119213000382. Special Issue: Selected Papers from the 2012 IEEE International Conference on Pervasive Computing and Communications (PerCom 2012)

14. Christin, D., Pons-Sorolla, D.R., Hollick, M., Kanhere, S.S.: In: 2014 IEEE Ninth International Conference on Intelligent Sensors, Sensor Networks and Information Processing (ISSNIP), pp. 1–6 (2014). https://doi.org/10.1109/ISSNIP.2014.6827614

15. Huang, K.L., Kanhere, S.S., Hu, W.: In: 37th Annual IEEE Conference on Local Computer Networks, pp. 10–18 (2012). https://doi.org/10.1109/LCN.2012.6423585

16. Christin, D., Guillemet, J., Reinhardt, A., Hollick, M., Kanhere, S.S.: In: 2011 IEEE Eighth International Conference on Mobile Ad-Hoc and Sensor Systems, pp. 341–350 (2011). https://doi.org/10.1109/MASS.2011.41

17. Wang, X., Govindan, K., Mohapatra, P.: In: 2011 8th Annual IEEE Communications Society Conference on Sensor, Mesh and Ad Hoc Communications and Networks, pp. 395–403 (2011). https://doi.org/10.1109/SAHCN.2011.5984923

18. Dolev, D., Yao, A.: On the security of public key protocols. IEEE Trans. Inf. Theor. **29**(2), 198–208 (1983)

19. Sun, Y.L., Han, Z., Yu, W., Liu, K.J.R.: In: 2006 40th Annual Conference on Information Sciences and Systems, pp. 1461–1466 (2006). https://doi.org/10.1109/CISS.2006.286695

20. Hoffman, K., Zage, D., Nita-Rotaru, C.: A survey of attack and defense techniques for reputation systems. ACM Comput. Surv. **42**(1), 1 (2009)

Quantum PSO Algorithm for Clustering in Wireless Sensor Networks to Improve Network Lifetime

P. Kanchan and D. Pushparaj Shetty

Abstract Clustering is done in wireless sensor networks (WSN) to conserve the energy of sensor nodes in the network. The network lifetime of WSN can be defined as the duration for which the network remains operational. It is a critical design issue in WSN's since once a node is deployed, it may not be feasible to replace or recharge the sensor nodes. In this paper, we proposed a quantum PSO algorithm for improving network lifetime called quantum PSO clustering algorithm to improve network lifetime(QPCINL). The QPCINL uses quantum bits. A quantum bit can exist in '0' state, '1' state or a linear superposition of '0' and '1' states, unlike the binary bit which can exist in only '0' state or '1' state. We define a factor called *network lifetime factor(NLF)* which allows us to compare various algorithms. We test our algorithm by giving different values to the number of sensor nodes and cluster heads, varying the base station position, etc. Then, we compare our results to existing algorithms and demonstrate the superiority of our algorithm.

1 Introduction

A wireless sensor network (WSN) is a network of nodes deployed within an area in random positions. These nodes interact with each other to collect data, process it, and then communicate with a base station (BS). Most often, these sensors are battery operated. Once the sensors are deployed in a specific field, say, in the deep sea or in the battlefield, it is difficult to either replace the battery or supply additional energy. Therefore, the main issue is to conserve the energy in WSN's. Various mechanisms have been used by researchers for conserving energy, and one of the most effective methods is clustering. In clustering, the network is divided into various clusters.

P. Kanchan (✉) · D. Pushparaj Shetty
Department of MACS, National Institute of Technology Karnataka,
Surathkal 575025, India
e-mail: pradeepis_2000@yahoo.com

D. Pushparaj Shetty
e-mail: prajshetty@gmail.com

© Springer Nature Singapore Pte Ltd. 2019
A. Abraham et al. (eds.), *Emerging Technologies in Data Mining and Information Security*, Advances in Intelligent Systems and Computing 814,
https://doi.org/10.1007/978-981-13-1501-5_62

Each cluster has a cluster head (CH) which collects the data from all member nodes (sensors) of its cluster. Then, the CH aggregates the data and sends it to the base station (BS). The BS processes the data as per the requirement and performs actions like notifying an event.

Network lifetime can be defined as the time span from the deployment to the instant when the network stops working. The instant at which the network stops working depends on the application. It can be, for example, the instant at which the first node dies or the instant at which a certain predefined percentage of nodes runs out of energy or the instant at which the last node dies.

In this paper, we present an algorithm in which we perform position updates using quantum PSO and CH selection using PSO-ECHS [1] and find the *network lifetime factor(NLF)*. We label the algorithm as quantum PSO clustering algorithm to improve network lifetime (QPCINL). We compare it with existing algorithms like LEACH [3] and PSO-ECHS [1]. The following are our contributions:

- Quantum PSO-based position updates followed by PSO-based CH selection
- Deriving a weight function for cluster formation
- Finding the network lifetime factor(NLF) by varying number of nodes and CH's
- Demonstrating the efficiency of our proposed algorithm over existing algorithms through simulation.

The rest of the paper is organized as follows. Section 2 contains a review of literature related to the field of interest. Preliminaries like introduction to PSO and QPSO along with the network model and energy model are explained in Sect. 3. Section 4 consists of derivation of the fitness function, position update using QPSO, selection of cluster head, and formation of clusters. Section 5 contains discussion of the simulation results. The conclusion is explained in Sect. 6.

2 Review of Related Works

2.1 Nature-Inspired Approaches for Clustering

Tilett et al. [9] have proposed a novel method called particle swarm optimization (PSO) which uses the concept of a swarm for clustering problem. The disadvantage here is that the CH nodes are selected based only on the distance, which may cause energy imbalance in the network. In [10], Guru et al. proposed PSO-based cluster formation. There may be some residual energy in the sensor nodes which is ignored in this paper. In [2], Latiff et al. use the PSO-C algorithm which is used to perform energy-aware cluster head selection. Its disadvantage is that during cluster formation, sensor nodes which are not CH nodes are assigned to the nearest CH and this may cause the network to become inefficient in terms of energy. This also has one more disadvantage that it decreases lifetime of the network. In [1], Rao et al. discuss an algorithm which performs cluster head selection while maintaining the energy efficiency (PSO-ECHS). Here, the PSO algorithm performs cluster head selection. In cluster formation, a weight function is calculated and used by non-CH nodes to join their CHs.

2.2 Heuristic Approaches for Clustering

LEACH [3] is a very popular algorithm used for clustering. Here, clusters are formed by the nodes with one node performing as the cluster head (CH). Those nodes that are not CH (non-CH nodes) perform the job of transmitting data to CH and the CH receives the data. The CH has to collect the data received and transmit it to a remote base station (BS). The disadvantage of this algorithm is the possibility of a CH with low energy being selected. This may cause it to stop functioning, affecting the network. In PEGASIS [4], each node communicates only with a node that is its close neighbor and takes turns transmitting to BS. As far as energy efficiency is concerned, PEGASIS is better than LEACH, but it may become unstable when there are networks of large size. TL-LEACH [5] introduces the concept of using a two-level hierarchy. Here, local cluster base stations (primary CHs and secondary CHs) are rotated randomly. The disadvantage is that electing secondary CHs causes extra overhead. In addition, there may be an energy imbalance in the network because non-CH nodes are assigned to CHs based on the distance only. M-LEACH [6] and LEACH are almost similar except that M-LEACH forwards data to the next hop CH node; it does not send directly to the BS. The disadvantage is that it does not take into account the cluster formation phase. In V-LEACH [7], some CHs are selected as vice CHs. These vice CHs become CHs when the main CHs die. Disadvantage is the additional energy required for selecting vice CHs. In E-LEACH [8], nodes having more residual energy are made as cluster heads for the next round. This can extend the life of the network.

2.3 Quantum Computing Based Algorithms

In [11], Jun Sun et al. proposed a quantum-based PSO. They consider an individual particle of a PSO system moving in a quantum multidimensional space. In [12], Zhen-Lun Yang et al. proposed improved quantum-behaved PSO with elitist breeding (EB-QPSO). Here, elitist breeding strategy is used to escape from local optima and perform efficient search by the swarm. In [13], Jun Sun et al. perform an analysis of QPSO. A particles behavior is influenced by a parameter called Contraction Expansion (CE) coefficient. The upper bound of CE coefficient, within which the value of the CE coefficient selected can guarantee the convergence of the particle position, is found by the authors. In [14], Millie Pant et al. develop a variant of QPSO which they call Q-QPSO. Their method uses interpolation-based recombination operator. In [15], Yin et al. use a quantum PSO in which quantum rotation gates do the updation of quantum bits and mutation is done by quantum non-gates. Their algorithm performs better than GA and classical PSO. The review enabled us to find a research gap. QPSO requires only one parameter as opposed to PSO which requires two.

2.4 Algorithms on Network Lifetime

In [21], Chen et al. proposed a general formula to find the network lifetime without depending on the network model. In [22], Rahman et al. proposed an efficient algorithm which uses PSO to locate an optimal sink position. Their approach saves 40% energy and prolongs the network lifetime. In [23], Yetgin et al. proposed a two-stage network lifetime maximization technique. They proposed exhaustive search algorithm (ESA) and single-objective genetic algorithm (SOGA). In [24], Dietrich and Dressler conduct a review of the various definitions of lifetime, differences between the various methods, advantages and disadvantages.

3 Preliminaries

3.1 PSO Introduction

Real-life swarms inspire particle swarm optimization (PSO) [16]. An example is a flock of birds searching for food and shelter. Each bird is analogous to a particle. In PSO, N denotes the total number of particles. A position vector and a velocity vector define each particle P. The position vector $X = (x_1, x_2, \ldots, x_D)$ signifies a solution and velocity vector $V = (v_1, v_2, \ldots, v_D)$ is responsible for exploration of search space. Here, D denotes dimensionality of the search space. It has the same value for all particles. A fitness function, which depends on the application for which the algorithm is being used, is used to evaluate each particle. The objective of PSO is to find that position of a particle which results in the best value of fitness function. In the initialization stage of PSO, a position and velocity are assigned to each particle. During each iteration, the best/personal best of a particle denoted as P_{best_i} and global best of the swarm denoted as G_{best} are calculated. The aim is to reach a global best solution. To attain this, the position of the particle, X, and velocity, V, are updated by the following equations:

$$V_{i,d}(t+1) = \omega V_{i,d}(t) + C_1 r(X_{P_{best,d}} - X_{i,d}) + C_2 R(X_{G_{best}} - X_{i,d}) \qquad (1)$$

$$X_{i,d}(t+1) = X_{i,d}(t) + V_{i,d}(t+1) \qquad (2)$$

Here,

ω	Inertia weight ($0 < \omega < 1$)
C_1, C_2	Acceleration coefficients ($0 \leq C_1, C_2 \leq 2$)
r, R	Random numbers uniformly distributed in the interval (0,1)
d	Dimension component ($1 \leq d \leq D$)
i	Particle number.

The updation process is repeated until an acceptable value of G_{best} is generated. P_{best_i} and G_{best} are calculated as follows:

$$P_{best_i} = P_i, if \; [Fitness(P_i) < Fitness(P_{best_i})]$$
$$= P_{best_i}, otherwise \tag{3}$$

$$G_{best} = P_i, if \; [Fitness(P_i) < Fitness(G_{best})]$$
$$= G_{best}, otherwise \tag{4}$$

3.2 QPSO Introduction

Clerc and Kennedy [17] have stated that convergence of the PSO occurs when each particle converges to its local attractor p_i, which is defined as follows:

$$p_{i,d}^t = \psi_d^t * P_{best_{i,d}^t} + (1 - \psi_d^t) * G_{best_d^t} \tag{5}$$

where $\psi_d^t = C_1 r \,/\, (C_1 r + C_2 R)$
Here, C_1, C_2, r, and R are as in (1).
Quantum PSO (QPSO) [11] is developed based on the above analysis.
Each particle in QPSO is assumed to be a spinless particle which moves in quantum space. The probability of a particle appearing at a position denoted by x_i^t in the iteration t is determined from a probability density function [18].
Using the Monte Carlo method, a particle will fly according to the following formulas:

$$X_{i,d}^{t+1} = p_{i,d}^t + \alpha |x_{i,d}^t - m_{best_d^t}| ln(1/u_{i,d}^t), if \; (randv \geq 0.5)$$
$$= p_{i,d}^t - \alpha |x_{i,d}^t - m_{best_d^t}| ln(1/u_{i,d}^t), if \; (randv < 0.5) \tag{6}$$

Here

α	Contraction–Expansion (CE) coefficient
$u_{i,d}^t$	and randv are uniformly distributed random numbers in the range [0,1]
m_{best}	Mean best.
m_{best}	is calculated as follows:

$$m_{best_d}^t = (1/N) \sum_{i=1}^{N} P_{best_{i,d}}^t \tag{7}$$

where

N Swarm size.

The CE coefficient is controlled by using a time-varying decreasing method [13] as follows:

$$\alpha = \alpha_1 + ((T - t)(\alpha_0 - \alpha_1)/T) \tag{8}$$

where

α_0 Initial value of α
α_1 Final value of α
T Number of maximum possible iterations
t Current iteration number.

The QPSO algorithm is a probabilistic optimization algorithm. A major advantage of QPSO algorithm over PSO is that in QPSO, only the position vector is required for calculations, whereas in PSO, both position and velocity vectors are needed.

3.3   The Energy Model Used

The energy model used is the same as the one used in LEACH [3] and PSO-ECHS [1]. The energy consumed by a node to transmit a l-bit data packet is:

$$E_{Transmit} = lE_{elec} + l\varepsilon_{fs}d^2, \text{ if } d < d_0$$
$$= lE_{elec} + l\varepsilon_{mp}d^4, \text{ if } d \geq d_0 \tag{9}$$

where

l No. of bits in the data packet
E_{elec} The energy dissipated / bit for running the transmitter / receiver circuit
ε_{fs} Amplification energy (using free space method)
ε_{mp} Amplification energy (using multipath model)
d Propagation distance
d_0 Threshold distance.

Similarly, energy consumption by the receiver to receive l-bit of data is given by:

$$E_{Receive} = lE_{elec} \tag{10}$$

Several factors like digital coding, modulation, filtering, and signal spreading influence the E_{elec}.

Overall energy which is spent by the sensor node in order to transmit data and receive data is given as follows:

$$E_{Total} = E_{Transmit} + E_{Receive} \tag{11}$$

3.4 The Network Model

The sensors are deployed randomly and once deployed are assumed to be stationary. A single node can operate as both sensor as well as cluster head (CH). Each node performs sensing at regular intervals of time. A node sends data to its CH or BS. Usually, the number of sensors will be greater than the number of CHs. The sensors (CH or BS) use different levels of transmission power depending upon the distance to which data will be sent. An assumption is that sensor nodes are homogeneous and have equal capacity for processing and communication.

3.5 Network Lifetime

Network lifetime has been defined in various ways by different authors. In our work, we define it as the time from deployment of the node till the first node runs out of energy. For a single node, lifetime has been defined as follows in [1] and [22].

$$L = \frac{E_{initial}}{E_{total}} \tag{12}$$

where

$E_{initial}$ Initial energy of a single sensor node
E_{total} Total energy spent by single node for transmitting and receiving data
$E_{initial}$ will be initially 2 J for all the nodes, and E_{total} is calculated as in (11).

We introduce a new term called network lifetime factor (NLF) which we define as follows:

$$NLF = \frac{E_{NW\,initial}}{E_{NW\,total}} \tag{13}$$

where

$E_{NW\,initial}$ Sum of initial energies of all the nodes in network
$E_{NW\,total}$ Sum of final energies of all nodes in network.

The algorithm which gives us better NLF can be said to be better than the other algorithms.

4 Proposed Algorithm

The quantum PSO clustering algorithm to improve network lifetime (QPCINL) is a nature-inspired algorithm, which combines the good features of both PSO and quantum computing. The main advantage of our proposed algorithm over existing algorithms is that in the position-updating phase, only one parameter, the position vector, is required. The quantum PSO clustering algorithm to improve network lifetime (QPCINL) has three steps:

1. Position updating
2. CH selection
3. Formation of clusters

The position updating is done using QPSO [12], and selection of CH is done using PSO-ECHS [1]. The advantage of using QPSO for position updating is that many parameters are not needed as in [1]; only the position vector is needed. In the CH selection phase, the sensor nodes send information, which indicates their location and residual energy to the BS. Only those nodes, which meet a threshold energy, become eligible to be a CH. The CH selection algorithm is run at the BS. Then, cluster formation phase starts. In cluster formation, a weight function is derived based on factors like distance, node degrees of CHs, and energy as in [1].

4.1 Deriving the Fitness Function

The fitness function is derived in the same way as in [1]. f_1 is dependent on the average distance among clusters and average distance between CH and BS. Our aim is to *minimize f_1*.

$$f_1 = \sum_{j=1}^{m} (1/l_j) \sum_{i=1}^{l_j} (dist(s_i, CH_j + dist(CH_j, BS)) \tag{14}$$

where

m	No. of CH's
l_j	No. of sensor nodes in cluster j
$dist(s_i, CH_j)$	Distance between sensor s_i and its selected cluster head CH_j
$dist(CH_j, BS)$	Distance between cluster head CH_j and the BS.

f_2 is defined as the reciprocal of total energy of all selected CHs. Our aim is to *minimize f_2*.

$$f_2 = \frac{1}{\sum_{j=1}^{m} E_{CH_j}} \tag{15}$$

where

E_{CH_j} is the energy of cluster head CH_j.

$$Fitness = \alpha f_1 + (1 - \alpha)f_2, 0 < \alpha < 1 \tag{16}$$

Our aim is to minimize the fitness.

4.2 Position Update

Position update is done using QPSO [11]. It is based on quantum computing. Here, the particle can appear with a certain probability at any position in the search space, and it has better fitness value. Therefore, QPSO is superior to PSO. The state of the particle can be described with only a position vector unlike the PSO [1] where we need position vector and velocity vector. The updating of the position of particle is done by (6).

4.3 Cluster Head Selection

Cluster head selection is done as in PSO-ECHS [1]. The CHs are selected from the normal sensor nodes based on the energy efficiency. The fitness function is designed such that total energy consumption is minimized.

4.4 Cluster Formation

We perform cluster formation using a weight function as in [1]. Sensor nodes use a weight function to join a CH which is denoted as CH_{weight}. It is defined as follows:

$$CH_{weight}(s_i, CH_j) = L * Energy factor \tag{17}$$

where

$$Energy factor = \frac{E_{residual}(CH_j)}{dist(s_i, CH_j) * dist(CH_j, BS) * deg(CH_j)} \tag{18}$$

Here

L	is a constant, assumed to be 1
$E_{residual}(CH_j)$	Residual energy of cluster head CH_j
$dist(s_i, CH_j)$	Distance between sensor s_i and CH_j
$dist(CH_j, BS)$	Distance between CH_j and BS
$deg(CH_j)$	Degree of node CH_j.

During cluster formation, each sensor node calculates the CH_{weight} using (17) and joins the CH with highest weight value. The overall algorithm is explained in Algorithm 1.

Algorithm: *Quantum PSO Clustering algorithm to Improve Network Lifetime (QPCINL)*

Input: Sensor nodes $S = [s_1, s_2, ..., s_{sen}]$

Size of swarm: N

Number of dimensions of particle, D = No. of CH's

Output: Network lifetime factor (NLF)

Steps:

1. Initialize the particles
2. Calculate fitness using (16)
3. for $i = 0$ to No. of rounds

    a. Update position using QPSO
    b. Calculate fitness using (16)
    c. Find personal best, P_{best} and global best, G_{best}
    d. Use P_{best} and G_{best} for CH selection
    e. Form clusters

    end for
4. Calculate network lifetime factor (NLF) at the end of predefined of rounds
5. Stop

Algorithm 1. Pseudo code of QPCINL

5   Performance Evaluation

5.1   Simulation Environment

Our algorithm was tested using C (Dev C++), and the results are plotted using MAT-LAB (R2015a) on Windows 8 Pro Platform. For our simulation, the number of sensor nodes was varied from 300 to 700 and CHs from 15 to 50. Each sensor node is initialized with energy of 2J. The sensing field is of $200 \times 200 \, m^2$ (Table 1). In the test environment, comparison of the total energy consumption is done while varying the number of sensors from 300 to 700 and CHs from 15 to 50. For the first case of 300 sensors and 15 CHs, the BS was placed at (100, 100), (200, 200) and (300, 300) and the results were obtained. Later, the number of sensors was changed to 400, 500 and 700. The number of CHs was also varied as 35, 40, 50.

The PSO parameters used as in [1] are given in Table 2.

Table 1 Network parameters

Parameters	Value
Area	200×200 m^2
Base station	(100, 100), (200, 200), (300, 300)
No. of sensors	300–700
No. of CH's	15–50
E_{elec}	50 nJ/bit
ε_{fs}	10 pJ/bit/m^2
ε_{mp}	0.0013 pJ/bit/m^4
Packet length	4000 bits
Message size	500 bits

Table 2 PSO parameters

Parameters	Value
No. of particles	30
C_1	2.0
C_2	2.0
α	0.3
ω	0.7
D	15–50
No. of iterations	100

5.2 Performance Metric Used

The performance metric used is *network lifetime factor(NLF)*, a term we have introduced in this paper. We have considered the NLF for a particular the number of rounds (in our case 5000). The NLF decreases as the number of rounds increases. Then, LEACH, PSO-ECHS, and QPCINL algorithms are ranked by comparing the NLF values while varying the number of sensors from 300 to 700 and CHs from 15 to 50. It can be observed from the graphs that the NLF values for QPCINL are better than those of LEACH and PSO-ECHS. The NLF values obtained using QPCINL are approximately 3 times more than PSO-ECHS. For the first case of 300 sensors and 15 CHs, the BS was placed at (100, 100), (200, 200), and (300, 300) and the results are obtained. Similarly, values of sensors and CH's were varied, and results are obtained. QPCINL outperforms both LEACH and PSO-ECHS by a large margin. Similar results were obtained when BS was placed at (200, 200) and (300, 300) with QPCINL outperforming LEACH and PSO-ECHS. QPCINL got better results when the number of sensors was varied from 300 to 700 and the number of CHs was varied from 15 to 50.

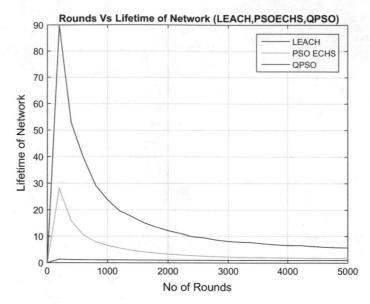

Fig. 1 Scenario S1

The scenario S1 consists of 300 sensors, 15 cluster heads, and base station at (100, 100), which is plotted in Fig. 1. The scenario S2 consists of 400 sensors, 45 cluster heads and base station at (100, 100), which is plotted in Fig. 2. The scenario S3 consists of 500 sensors, 50 cluster heads and base station at (100, 100), which is plotted in Fig. 3. The scenario S4 consists of 700 sensors, 35 cluster heads, and base station at (100, 100), which is plotted in Fig. 4.

6 Conclusion

In this paper, we have proposed an algorithm called quantum PSO clustering algorithm to improve network lifetime(QPCINL). The proposed algorithm uses only one parameter, the position vector, whereas PSO requires both position and velocity vectors. The simulation results of QPCINL algorithm are compared with existing algorithms LEACH and PSO-ECHS. The experimental results show that our algorithm performs better than LEACH and PSO-ECHS in terms of network lifetime factor (NLF).

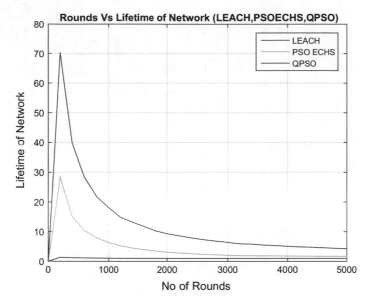

Fig. 2 Scenario S2

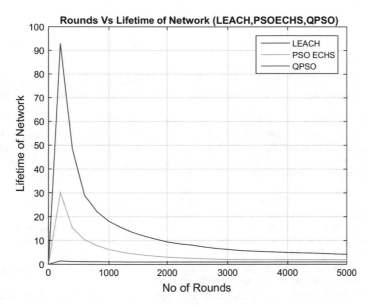

Fig. 3 Scenario S3

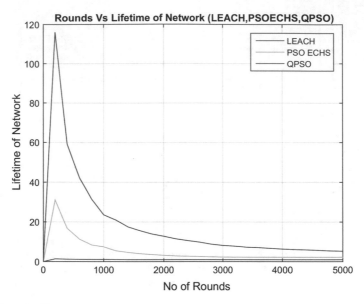

Fig. 4 Scenario S4

References

1. Rao, P.S., Jana, P.K., Banka, H.: A particle swarm optimization based energy efficient cluster head selection algorithm for wireless sensor networks. Wireless Networks **23**(7), 2005–2020 (2017)
2. Latiff, N.A., Tsimenidis, C.C., Sharif, B.S.: Energy-aware clustering for wireless sensor networks using particle swarm optimization. In: IEEE 18th International Symposium on 2007 Personal, Indoor and Mobile Radio Communications, PIMRC 2007. Sep 3, pp. 1–5 (2007)
3. Heinzelman, W.B., Chandrakasan, A.P., Balakrishnan, H.: An application-specific protocol architecture for wireless microsensor networks. IEEE Trans. Wireless Commun. **1**(4), 660–670 (2002)
4. Lindsey, S., Raghavendra, C.S.: PEGASIS: Power-efficient gathering in sensor information systems. In: Aerospace conference proceedings, vol. 3, pp. 3–3. IEEE (2002)
5. Loscri, V., Morabito, G., Marano, S.: A two-levels hierarchy for low-energy adaptive clustering hierarchy (TL-LEACH). In: 62nd Vehicular Technology Conference, VTC-2005-Fall, vol. 3, pp. 1809–1813. IEEE (2005)
6. Xiaoyan, M.: Study and design of clustering routing protocols of wireless sensor networks. Ph.D. Dissertation, Zhejiang University, Hanzhou (2006)
7. Yassein, M.B., Khamayseh, Y., Mardini, W.: Improvement on LEACH protocol of wireless sensor network (VLEACH). Int. J. Digit. Content Technol. Appl. (2009)
8. Xiangning, F., Yulin, S.: Improvement on LEACH protocol of wireless sensor network. In: International Conference on Sensor Technologies and Applications, SensorComm, pp. 260–264. IEEE (2007)
9. Tillett, J., Rao, R., Sahin, F.: Cluster-head identification in ad hoc sensor networks using particle swarm optimization. In: IEEE International Conference on Personal Wireless Communications, pp. 201–205. IEEE (2002)

10. Guru, S.M., Halgamuge, S.K., Fernando, S.: Particle swarm optimisers for cluster formation in wireless sensor networks. In: Proceedings of the 2005 International Conference on Intelligent Sensors, Sensor Networks and Information Processing Conference, pp. 319–324. IEEE (2005)
11. Sun, J., Feng, B., Xu, W.: Particle swarm optimization with particles having quantum behavior. In: Congress on Evolutionary Computation, CEC2004, vol. 1, pp. 325–331. IEEE. (2004)
12. Yang, Z.L., Wu, A., Min, H.Q.: An improved quantum-behaved particle swarm optimization algorithm with elitist breeding for unconstrained optimization. In: Computational intelligence and neuroscience, p. 41 (2015)
13. Sun, J., Fang, W., Wu, X., Palade, V., Xu, W.: Quantum-behaved particle swarm optimization: analysis of individual particle behavior and parameter selection. Evolutionary Comput. **20**(3), 349–393 (2012)
14. Pant, M., Thangaraj, R., Abraham, A.: A new quantum behaved particle swarm optimization. In: Proceedings of the 10th annual conference on Genetic and evolutionary computation, pp. 87–94. ACM (2008)
15. Yin, O., Li, W., Zhang, X., Huo, F.: Continuous quantum particle swarm optimization and its application to optimization calculation and analysis of energy-saving motor used in beam pumping unit. In: 2010 IEEE Fifth International Conference on Bio-Inspired Computing: Theories and Applications (BIC-TA), pp. 1231–1235. IEEE (2010)
16. Kennedy, J.: Particle swarm optimization. In: Encyclopedia of machine learning, pp. 760–766. Springer US (2011)
17. Clerc, M., Kennedy, J.: The particle swarm-explosion, stability, and convergence in a multidimensional complex space. IEEE Trans. Evolutionary Comput. **6**(1), 58–73 (2002)
18. Liu, J., Sun, J., Xu, W.: Quantum-behaved particle swarm optimization with adaptive mutation operator. In: Advances in Natural Computation, pp. 959–967 (2006)
19. Tsai, C.W., Kang, C.T., Chiang, M.C.: A quantum-inspired evolutionary algorithm based clustering method for wireless sensor networks. In: 2015 Seventh International Conference on Ubiquitous and Future Networks (ICUFN), pp. 103–108. IEEE (2015)
20. Calle Torres, M.G.: Energy consumption in wireless sensor networks using GSP. Doctoral dissertation, University of Pittsburgh (2006)
21. Chen, Y., Zhao, Q.: On the lifetime of wireless sensor networks. IEEE Commun. lett. **9**(11), 976–978 (2005)
22. Rahman, M.N., Matin, M.A.: Efficient algorithm for prolonging network lifetime of wireless sensor networks. Tsinghua Sci. Technol. **16**(6), 561–568 (2011)
23. Yetgin, H., Cheung, K.T.K., El-Hajjar, M., Hanzo, L.: Network-lifetime maximization of wireless sensor networks. IEEE Access **3**, 2191–2226 (2015)
24. Dietrich, I., Dressler, F.: On the lifetime of wireless sensor networks. In: ACM Transactions on Sensor Networks (TOSN), 5(1), p. 5 (2009)

Secure Remote Patient Monitoring with Location-Based Services

Mainak Sen and Gautam Mahapatra

Abstract Remote patient monitoring (RPM) refers to a wide variety of technologies designed to manage and monitor a range of health conditions. It gives chance to doctor's to monitor patient's behavior even when they are not within conventional medical attention and take necessary actions accordingly. Such system requires a seamless integration of hardware and software with information stored in Internet such that everyone including the doctor and the patient party can access and analyze data according to their need. An alarm service is generally integrated to make people aware of any unwanted situation. The current work proposes a monitoring ambience for people who meet with an accident or for some situation, and after surgery, some positions of patients are strictly prohibited as it might cause some serious damage to their health.

1 Introduction

World Health Organization (WHO) states that cardiovascular disease (CVD) is one of the prime causes of death. Enrichment of medical science has given humans a long life. Nursing homes are not large enough in number to tackle the huge load of elderly person (majority). Also doctors, nurses cannot reach to every corner of a city for checkup or vice versa. In that case, remote monitoring plays a vital role as patient party can have the luxury of an Ambient Assisted Living (AAL) environment that nearly acts as a nursing home. In a remote monitoring, the main focus is on the areas like health records, monitoring, personalized treatment, remote care, decision support system.

M. Sen (✉)
Department of Computer Science & Engineering, Techno India University,
EM-4, Sector V, Kolkata 700091, West Bengal, India
e-mail: mainaksen.1988@gmail.com

G. Mahapatra
Department of Computer Science, Asutosh College,
92, Shyama Prasad Mukherjee Road, Kolkata 700026, West Bengal, India
e-mail: gsp2ster@gmail.com

© Springer Nature Singapore Pte Ltd. 2019   715
A. Abraham et al. (eds.), *Emerging Technologies in Data Mining and Information Security*, Advances in Intelligent Systems and Computing 814,
https://doi.org/10.1007/978-981-13-1501-5_63

The AAL constitutes of different medical plantable or implantable sensors, radio frequency identification (RFID) readers and tags, gateways, smart phones, tablets, software application with analysis ability, Internet connections, databases, etc. Such complex heterogeneous communication system is presently known as Internet and embedded objects which are sharing their physical parameters are generally called Internet of things (IoT). The progression of wired and wireless communication system has made it possible to build a communication network among these embedded intelligent objects so that they can share their information easily to each other and these whole structures may create a integrated manageable information system.

MobiHealth [1] allows patient to be in fully free mode so that they can roam around easily and uses the facility of Universal Mobile Telecommunication System (UMTS) and General Packet Radio Service (GPRS) networks. Developing these kind of applications are extremely crucial for modern health care and its quite feasible today using the advancement in communication technology. HealthGear [2] is a wearable real-time health monitoring system developed for analyzing physiological signals connected to a mobile phone via Bluetooth, eWatch [3] is a wristwatch that provides both audio and visual notification during sensing.

But still trust management [4] that is how much trust one node can have is still a concern. Also privacy and Quality of Service (QoS) [5] must be incorporated into healthcare applications along with security. As wireless monitoring sometimes become error prone because of inefficient patient identification, here wired sensors have been used for the experiment.

IEEE 802.15.6 classifies the different job being performed by a node in a wireless body area network (WBAN) and as per that different naming convention is used as follows [6].

Implant Node Node planted either underneath the skin or inside the tissue.
Body Surface Node Either placed on the surface or 2 cm away from the human body.
External Node A few centimeters to 5 m away from the human body.

The organization of the paper is as follows. After this brief introduction, the necessary background for the current work has been described in Sect. 2. The problem with existing technology which initiates the present work has been stated in Sect. 3. Section 4 describes the proposed solution supported by the experimental results that are shown in Sect. 5. Finally, Sect. 6 concludes the discussion.

2   Background

In this section, the existing WBAN technology and its security issues are briefed. Also, the different forms of cryptosystem have been discussed here.

2.1 Existing WBAN Architecture

The existing WBAN follows the three-tier architecture as described below.

2.1.1 Tier-1

The tier-1 comprises of different sensors worn by the patient. Sensors like EMG for muscle monitoring, EEG for brain activity monitoring, ECG for heart condition, and breathing sensor for respiration monitoring, position sensors for patient position monitoring are used. In this model, patient position sensor is used. Each sensor nodes gets an initialization command either from Tier-2 or from some predefined event. The master node manages all the operations following the topologies like Star, Mesh as per the need [7].

2.1.2 Tier-2

Information received from WBAN coordinator to a local access point like PDA, PC or mobile phone that constitutes a personal server which acts as the gateway to transfer data to the Internet. The personal server communicates by means of a network coordinator following IEEE 802.15.4 (Zigbee) or IEEE 802.15.1 (Bluetooth) standards.

2.1.3 Tier-3

Information received by any tier-2 device is transferred to tier-3 for analysis. The medical practitioner, nurse or even the well-wisher of the patient can take some necessary action depending on the need. These captured data can also be used by different users for different purpose like recommendation of treatments, patient condition monitoring, short message service, e-mail or chat-based alarms, data mining.

In Fig. 1, the block diagrammatic structure of the RPM is shown. The encircled modules are forming an integrated patient wearable system that senses different physical parameters of the patient, stores them in local memory system integrated with microcontroller-based system and then with the help of wire/wireless connection, it exchanges data to the Home PC, Laptop or PDA system connected with wired/wireless Internet services available even in remote area. This local computing system will process the sensing data and stored it in plain text format. This comma-separated formatted, uploaded text data file, with the help of import facilities available in the DBMS software is stored in the Web database system.

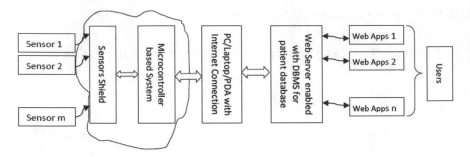

Fig. 1 Three-Tier Architecture [8]

2.2 Security and Privacy Requirements in WBAN

People with different perspective on security may define it in various manner but none can deny the need of security. Many threats have been identified which has been classified broadly into two categories—active and passive [9] attacks as mentioned in Table 1. To prevent misuse of medical data, a medical service provider must follow HIPPA rules [10]. But still the question is who owns medical data and how to control the access to these data as medical data are very sensitive suggested by the European Union Data Protection Directive [11] and to counter that proper encryption of data is in need. Ubimon [12] was developed to solve the issues against wearable, implantable sensors for distributed mobile monitoring for elderly person.

Table 1 Types of attacks

Types of attack
1. **Active attacks**
a. **Monitoring on patient**
Sometimes an attacker with a powerful receiver antenna may forcefully pickup messages from network and may change some of the information in the message and after getting the address of the patient, attacker may physically hurt him/her [13].
b. **Location threats**
Medical sensors support user mobility as the patient may change location. So, proper location of patient is also needful in healthcare application which is generally based on radio frequency, received signal strength indicator, etc. [14].
2. **Passive attack**
Threats during transmission
While the message containing the information of the patient is in transit, a third party can intercept the message. This is known as interception attack and in the modification attack after getting the information the attacker alters some of the information contained in the message and re-transmit it to the receiver side. Then while analyzing the message, a doctor or nurse would get these altered data and suggest some treatment which might endanger the patient.

Table 2 Forms of cryptosystem

1. DNA cryptosystem
The research on biotechnology and DNA computing combines cryptology with molecular biology initiated by Adleman. DNA shotgun sequencing [15] is currently the technology that attempts to combine DNA fragments into original long chain of DNA.
2. Location Dependent Cryptosystem
The 768-bit modulus was cracked and so, an extra means of protection is needful to ensure security of data which may be provided by geo-location that is latitude and longitude [16].
3. Embedded Cryptosystem
Devices that we use today such as mobile phones, PDA, sensors used in health care are all embedded systems that include an embedded processor. Different cryptographic algorithms are nowadays implemented for some specific applications (ASIC) or on re-configurable platform (FPGA) [17].

2.3 Types of Different Cryptosystem

Researchers are nowadays working on many different types of cryptographic techniques that can be incorporated with the existing patient monitoring system explained in Table 2. Except those, a new type of cryptosystem is in development phase that uses the concept of non-particle physics.

2.4 Location-Dependent Cryptosystem in Remote Monitoring

Here the concept of location-based cryptosystem has been used to work along with the existing remote monitoring concepts to encounter the existing security threats. But still location-dependent services need to be carefully used [18].

3 Problem Statement

Nuclear family size, long-distance work is very common in today's era. Road accident is very common to go along with the other disease. After an operation, patients are generally prescribed to be under safe eye throughout 24X7. But due to the limited number of hospitals, nursing home and their accommodating ability, number of doctor's, nurse's, etc, it is very hard to keep patient in the hospital for long time. On this regard, remote patient monitoring is very helpful that provides medical practitioner to watch patient from distance locations. Also, data security, privacy is a major concern in remote patient monitoring and so secure data transmission must be included in this system.

4 Proposed Work

The current work is to monitor the positions of a patient along with the temperature for whom some positions are highly restricted or very much risky. As it is not possible to continuously monitor the position of a patient manually due to so many reasons, so an automated system may be used which can be used 24 h for continuous monitoring. The monitoring system samples data using sensors in a certain interval and using data fusion technique change of positions are detected and immediately are uploaded in the Web-based/data-based system using integrated Global System for Mobile (GSM)-based wireless sensor network service. Stored data is regularly being analyzed by a Web-based application software, and any kind of abnormal positions is immediately communicated to the health organization or health service provider for proper care and immediate action.

4.1 Different Positions of Patient

The different position that can be sensed by our proposed architecture is shown in Fig. 2.

4.2 Operational Flow Diagram

When the patient comes to hospital, the location of the patients' relative and their mobile number, e-mail id, etc., are stored in the hospital database. Then doctor after checking the patient suggests how he or she should spend the day even during sleep hours. And in the system, mark positions which are safe or dangerous. Whenever the patient moves to those marked dangerous position, an alarm via a text message goes to nursing authority.

The proposed system is shown in Fig. 3.

Fig. 2 Different patient positions

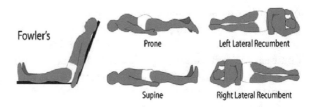

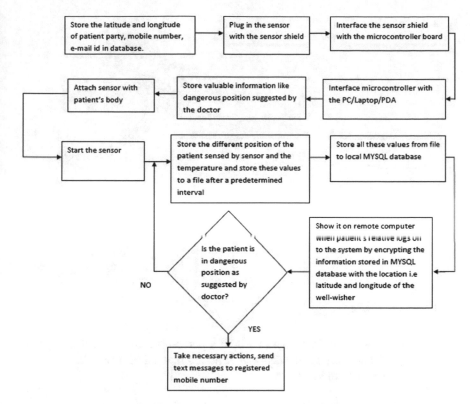

Fig. 3 Operational flow diagram

4.3 Components Used

4.3.1 Arduino Uno Microcontroller Board

Here, Arduino Uno System Development Kit (AUSDK) has been used which is integrated with PIC 8-bits microcontroller system. The AUSDK has built in A/D and D/A converters, USB connectors, etc. In this experiment, AUSDK is directly connected with the desktop computer enabled with USB port.

4.3.2 E-Health Sensor Shield

Along with AUSDK, a shielding system has been used with which the positioning sensor and temperature sensors are connected through wire.

Fig. 4 Position sensor

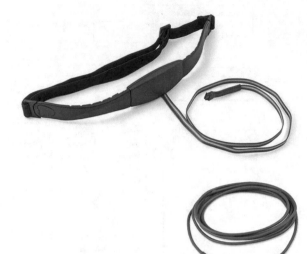

Fig. 5 Temparature sensor

4.3.3   Patient Position Sensor

This sensor can sense the position in which a person wearing the sensor is in as shown in Fig. 2, and these sensed data can be stored in a log file for further analysis. The sensor is depicted in Fig. 4.

4.3.4   Body Temperature Sensor

Figure 5 depicts the sensor that senses the temperature of a patient.

With this hardware configuration, an interfacing software system has been created. This software captures patient position data and then stores it in a text file. When a particular time interval expired, then the file is uploaded to the Web database system using FileZilla software system.

5   Results

To study the effectiveness and applicability of the proposed system with newly incorporated security features, several experiments were organized. In this segment, a patient is presented to illustrate the usefulness of the proposed system and how it provides advantage in security over typical existing system. Informed consents were obtained from all individual participants for whom only the identifying information is included here.

Fig. 6 Relative's location   **Lat/Long: 22.6050889,88.4290119**
Address: BD/241, Gauranganagar, Kestopur,
Kolkata, West Bengal 700101, India

5.1 Case Study

Ridam, an UG student of Techno India University travels by bike. After an accident, he was rescued to a hospital and he had one major shoulder surgery. Then, he was in the hospital for 10–12 days and doctor prescribed him to always stay either in flower or in left lateral recumbent position. But then he was released from the hospital and came back to his home, there was no one to look after him. So Ridam, as per the instruction of the hospital authority took the service of remote monitoring in such a way that the mentioned positions are marked as dangerous and except those he can stay in all other positions like supine, prone, and right lateral.

The proposed monitoring system stores the patient position and body temperature after some predefined interval and also when he unintentionally goes into any of those mentioned dangerous position, an alarm that is a text message goes to the registered family member and hospitals authorities mobile number.

When he was admitted to the hospital, the hospital authority took the spatial location that is the latitude and longitude of the address of Ridam's relatives as shown in Fig. 6.

5.2 Discussion

The above framework is very much useful for remote monitoring where number of hospitals nowadays becoming very limited as average age of people is growing up. The hospital authority or doctor can log into their own account and look into their patients activity same as patient party. Figure 7 shows the log file of the patient stored with temperature and position at different time. Here the information has been stored after 10 s.

But when he moved to any other position, an alarming message was sent to the registered mobile number and also to the hospital authority as shown in Fig. 8.

The following chart shown in Fig. 9 gives the idea how much our patient Ridam used to move in bed. The X-direction shows the different position a patient and Y-direction indicates the number of times the patient was in that position during the time span.

Fig. 7 Log File

DATE	TIME	POSITION	TEMPERATURE(F)
24-Aug-17	10:46:32	SUPINE	97.58
24-Aug-17	10:46:20	RIGHT LATERAL	97.95
24-Aug-17	10:46:06	FOWLER	94.58
24-Aug-17	10:45:54	SUPINE	97.55
24-Aug-17	10:45:43	SUPINE	97.5
24-Aug-17	10:45:32	RIGHT LATERAL	94.88
24-Aug-17	10:45:21	PRONE	95.47
24-Aug-17	10:45:07	LEFT LATERAL	97.81
24-Aug-17	10:44:57	PRONE	95.74
24-Aug-17	10:44:37	PRONE	97.31
24-Aug-17	10:44:21	LEFT LATERAL	94.81
24-Aug-17	10:44:06	LEFT LATERAL	94.07
24-Aug-17	10:43:55	LEFT LATERAL	95.37

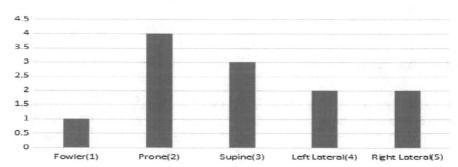

Mon, 8/21/2017

Patient in Danger

Recipient SIM1 Receive

14:04

Fig. 8 Alarming message

Fig. 9 Graph

6 Conclusion

In this paper, a remote patient monitoring system (RPM) has been proposed. The proposed system will be useful for monitoring patients who are critically ill and their positioning even in sleeping positions could affect treatment. The proposed system logs patients movement in terms of changing positions in bed around the hour. This log is accessible by the doctor or patient party from any remote location connected by Internet and/or mobile network. The system proactively generates alarm every time the patient changes into some position that is prohibited. The system has been experimentally deployed and verified alongside with location-based services to provide security against passive attack in remote health care.

While much has already been accomplished technologically, a lot is yet to be done so that patients and service providers have trust for such solutions. The proposed solution needs rigorous testing and quality control particularly around aspects of safety, security, privacy, accuracy, and integrity of their data. Having used wired sensors for this work, the patient has to be within the range of the wire and bidirectional communication would have made the system more acceptable to people.

Acknowledgements It is a pleasure to thank Prof. Nabendu Chaki, Department of Computer Science & Engineering, University of Calcutta, for his constant support and encouragement of this research. Also, the financial support from the Technical Education Quality Improvement Programme (**TEQIP** Phase-II) of Government of India with the assistance of World Bank through the University of Calcutta is being acknowledged.
Informed consent was obtained from all individual participants included in the study.

References

1. http://www.mobihealth.org/
2. Oliver, N., Flores-Mangas, F.: HealthGear: A real-time wearable system for monitoring and analysing physiological signals. In: Proceedings - BSN: international Workshop on Wearable and Implantable Body Sensor Networks, pp. 61–64 (2006). https://doi.org/10.1109/BSN.2006.27
3. Maurer, U., Rowe, A., Smailagic, A., Siewiorek, D.P.: eWatch: a wearable sensor and notification platform. In: Proceedings BSN: International Workshop on Wearable and Implantable Body Sensor Networks, pp. 142–145 (2006). https://doi.org/10.1109/BSN.2006.24
4. Rocker, C., Zie e, M.: E-Health, Assistive Technologies and Applications for Assisted Living: Challenges and Solutions, In: Holzinger, A., Kushniruk, A., Carayon, P., Patrick Rau, P.L., Beale, R., Helander, M.G., Hollan, M.D., Saaki Lahlou, S.B. (eds.) E-Health, Assistive Technologies and Applications for Assisted Living: Challenges and Solutions. Medical Information Science Reference, Hershy, New York, pp. 25 (2011)
5. Movassaghi, S., Abolhasan, M., Lipman, J., Smith, D., Jamalipour, A.: Wireless body area networks: a survey. IEEE Commun. Surveys Tutorials **16**, 1658–1686 (2014). https://doi.org/10.1109/SURV2013.121313.00064
6. Yazdandoost, K.Y.: Channel model for body area network (BAN), IEEE p. 802.15-08-0780-12-0006. Networks. (2009). IEEE P802.15-02/490r1-SG3a
7. Warren, S., Lebak, J., Yao, J., Creekmore, J., Milenkovic, A., Jovanov, E.: Interoperability and security in wireless body area network infrastructures. In: Annual International Conference of the IEEE Engineering in Medicine and Biology- Proceedings. Department of Electrical and Computer Engineering, Kansas State University, Manhattan, KS, United States, pp. 3837–3840 (2005)
8. Otto, C., Milenkovic, A.: System architecture of a wireless body area sensor network for ubiquitous health monitoring. J. Mobile Multimedia **1**(4), 307–326 (2006). 10.1.1.77.2522
9. Ng, H.S., Sim, M.L., Tan, C.M.: Security issues of wireless sensor networks in healthcare applications. BT Technol. J. **24**, 138–144 (2006). https://doi.org/10.1007/s10550-006-0051-8
10. Hipaa, H.H.S.-O., for C.R.: Medical privacy - national standards to protect the privacy of personal health information. federal register **65**(2002), (2002)
11. Wong, R.: The data protection directive 95/46/EC: idealisms and realisms. Int. Rev. Law, Comput. Technol. **26**, 229244 (2012). https://doi.org/10.1080/13600869.2012.698453

12. Saadaoui, S., Wolf, L.: Architecture concept of a wireless body area sensor network for health monitoring of elderly people. In: 2007 4th Annual IEEE Consumer Communications and Networking Conference, CCNC (2007) https://doi.org/10.1109/CCNC.2007.147
13. Dimitriou, T., Ioannis, K.: Security issues in biomedical wireless sensor net- works. In: First International Symposium on Applied Sciences on Biomedical and Communication Technologies (ISABEL 08). Aalborg, Denmark, pp. 15 (2008)
14. Redondi, A., Tagliasacchi, M., Cesana, M., Borsani, L., Tarrio, P., Salice, F.: LAURA Localization and Ubiquitous monitoring of patients for health care support. In: IEEE 21st International Symposium on Personal, Indoor and Mobile Radio Communications Workshops, pp. 218–222 (2010). https://doi.org/10.1109/PIMRCW.2010.5670365
15. Motahari, A.S., Bresler, G., Tse, D.N.C.: Information theory of DNA shotgun sequencing. In: IEEE Transactions on Information Theory, pp. 6273–6289 (2013). https://doi.org/10.1109/TIT.2013.2270273
16. Raper, J., Gartner, G., Karimi, H., Rizos, C.: Applications of location based services: a selected review. J. Location Based Services **1**, 89111 (2007). https://doi.org/10.1080/17489720701862184
17. Itoh, K., Takenaka, M., Torii, N.: Fast implementation of publickey cryptography on a DSP TMS320C6201. Ches 61–72 (1999). https://doi.org/10.1007/3-540-48059-5-7
18. Basiri, A., Moore, T., Hill, C., Bhatia, P.: The non-technical challenges of Location Based Services markets: Are the users concerns being ignored? In: Proceedings of International Conference on Localization and GNSS, ICL-GNSS. Institute of Electrical and Electronics Engineers Inc. (2016) https://doi.org/10.1109/ICL-GNSS.2016.7533866

Implementation of Mobility-Assisted Uncertainty Reduction Based on MCR-O Scheme in MANET

K. Pushpalatha, G. Karthikeyan and S. Umarani

Abstract The nodes of mobile ad hoc networks (MANET) allow flow of information through router/relay whenever the network uses multihop communication in a decentralized cluster architecture by which power utilization is minimized. A multisencar (mobile sensor)-based MANET is considering for data aggregation to mitigate potential attacks and ensure that the network is secure, reliable, and energy-efficient. The reliability of information received from the cluster head node to sencar is determined by using MCR-O method. This paper indicates that all the nodes in the system are in belief or disbelief states. The uncertainty nodes are considered as disbelief nodes based on cluster region. The system sets various boundaries whether the node has an accessible neighbor and the percentage of the belief communication in the network on the basis of degree distribution, location probability, and neighborhood distribution.

1 Introduction

Trust is an important role everywhere when sharing exposes major impact. Trustable sharing executes reliable, flexible, quick communication with an improved throughput having less overhead and a reasonable delay. It is a major classification for providing security to the system for improving network lifetime in the distributed dense network. The trust addresses issues on various application areas such as peer-to-peer networks, military fields, commercial communications, social networks. So the trust can be viewed either in a social or a cognitive manner. The trustworthiness in

K. Pushpalatha (✉) · S. Umarani
SRM Institute of Science and Technology, Ramapuram, Chennai 600 089, Tamil Nadu, India
e-mail: pushpalathakrishnan@gmail.com

S. Umarani
e-mail: ravania@gmail.com

G. Karthikeyan
Sona College of Technology, Salem 636 005, Tamil Nadu, India
e-mail: gkarthikeyan@yahoo.com

© Springer Nature Singapore Pte Ltd. 2019   727
A. Abraham et al. (eds.), *Emerging Technologies in Data Mining and Information Security*, Advances in Intelligent Systems and Computing 814,
https://doi.org/10.1007/978-981-13-1501-5_64

both approaches may have effect at any ends such as normal nodes, heads of groups if nodes are in cluster or at sink nodes.

Scarcity of trust is more challenging on decentralized nodes with dynamic topology which may also have group communication exchange. The trust measures in the proposed system for individual or group communication with the sencar are based on node degree and node reputation. Securing inter-cluster and intra-cluster communication provides local trust and global trust effectively. So the trust would successfully apply on individual entity as well as group entities, but complexities are even higher in group of nodes than with an individual. Application of on real-time applications is a challenging task, which would react on time on different dimensions. Due to various properties such as dynamicity, asymmetry, subjective, context-free nature and also non-transitive recommendations would even make system to function even harder to incorporate trustworthiness. The researches already made in existing system are discussed in the following section.

The previous work by Pushpalatha et al. [9] has been showing receiving data from the cluster heads which may or may not a trusted entity. Here, the researcher proposes an improved, reliable, trust-based, and energy-efficient data aggregation for wireless sensor networks. The system uses a proactive mechanism for estimating reliable communication in advance to avoid overheads, whereas the reactive technique has security credentials after the occurrence of overhead from stale the nodes.

Out of various proactive techniques, the hierarchical scheme is considered as best suited to the proposed system. The proposed two-way system initially calculates the degree of the distribution of cluster head nodes and also checks whether the node is bounded in the cluster region using MCR technique. Secondly, the cluster head node is authorized by neighborhood node with its available reputation for reliable transmission. It combines reputation system, residual energy, link availability, and a recovery mechanism to improve secure data aggregation and ensure that the network is secure, reliable, and energy-efficient.

2 Related Work

Various findings were made by many researchers on secure communication for trusted entities for mitigating risk. Many ideas were implemented to approach trust improvements in the unreliable network arising out of mobility. Literature survey on trust solutions has been viewed as distributed probability theory, fuzzy approach, and game-theoretic approach resulting from various attacks like misbehaviors, unknown or miscellaneous node arising from newly joining, routing, service deny, collusion. The metrics used in the existing system in various research categories are delay, trust level, throughput, overhead, lifetime, success rate, etc., for addressing issues on authentication due to topology diversion, key exchange, uncertainty reduction, resource sharing, intrusion detection, and so on, to secure the communication.

The researcher in [1] has analyzed subjective and objective trusts for indicating the differences between trust and trustworthiness. The term trusts can also be referred to as reputation as mentioned in [2], but the representation of reputation is passive in general [3]. Trust is the belief level of a node relating to other nodes that had been its relay in the network cluster. Tracing positive trust evidence is based on probability of belief communication through self-estimation and the opinion about reputation information of node through the neighbor advertisement. Solhaug et al. [4] assessed the authentication and trust cooperation on unreliable occasions in decentralized network. The term reputation is a part of trust management taxonomy and is classified as establishment, update, and revocation.

The initial part of trust management taxonomy such as establishment has been classified into two categories: (1) a reactive- or behavior-based and (2) a proactive- or certificate-based which are represented in [5]. The behavior-based method collects information from other nearer nodes as recommendations in a reactive manner based on assumptions by its built-in mechanism. The certificate-based proactive technique receives the relationship between nodes and rates the trust value either independently or mutually with others through the third-party certificate vendor. In [6], the trust can be rated through assumptions in either classification that would compute positive, negative, or its combination in order to have reputation management. But the network should be organized in distributed or hierarchical architecture mentioned in [5].

The policy on signed credential approach for binary decision and reputation approach for numerical computation methods have been suggested by Yunfang [7]. As part of the trust establishment in [8], the trust evidence can be evaluated either through direct observation or via secondhand information. Direct observation is possible when the nodes can interact between each other. The node gets trust information on various matters from the immediate neighbors on its track as numerical computations during multicast connection and aggregates the secondhand opinion and disseminates if direct coverage is not under communication range. The multiple data aggregation is discussed in [9]. In [10], evaluation of trust management is mitigation of the various potential attacks say active, passive, inter-cluster due to outsider, intra-cluster due to insider, new node arrival, poor authentication on frequent node mobility, intrusion on decentralized nature, etc.

Various effective methods are available for estimation of the trust value to take decision accurately. However, the trust values would fluctuate due to various characteristics discussed earlier. Any small fluctuation found in real-time applications like commercial activities and military fields would create serious risks in tracing the appropriate solution that would help in avoiding irrevocable losses. The author in [11] has proposed two classifications: The first technique gathers trust based on node id, authentication key (public/private depends on the cryptography method), evidence by self or via relay, or address known as evidence-based; the second technique is further subdivided into direct and indirect assessments known as monitor-based. The direct observation technique calculates trust values on the basis of neighbor's gentle or evil nature that creates risks on message reliability, jamming network via needless resource wastage through flooding, service deny, and selfishness on battery exhaustion upon processing irrelevant tasks. The indirect observation has recommendations and reputations for trust rating that are purely from accessible neighbors, and the trust information should forward to other nodes if needed as it receives.

The trust system has emerged as risk protection, and the authors [2, 11–13] have found some route to identify and execute sly present in dense network. The solution proposed by the researcher is based on two dimensions to know the neighbor's additional information and the security credentials from neighbor about the intended recipient. The detection of a suspicious node could be traced on evidence-based or monitory-based scheme using either cooperative or non-cooperative decision. The decisions on tricky communication made effectively in order to neutralize the energy consumption. These unforeseen situations produce no prior awareness of nodes behavior. Neutralizing power consumption is possible for a scaled network, although it would defer practically for some extent.

The researcher in [14] has proposed an accurate neighborhood with the maximum existence of a neighbor. This mechanism guarantees a healthy environment using a grouping algorithm for stable nodes, but this is not feasible for mobility conditions.

Torkestani and Meybodi [15] has proposed a centralized and distributed algorithm using automata tuning machine for distributed learning toward loosely connected ad hoc nodes, say cluster head and cluster members available in the entire network in each cluster. The authors have developed another algorithm in the year 2011 for localized learning of distributed networks which support trusted clustering in a network. The researcher's work [9] proposes multicast aggregation at cluster group head from multiple sources for improving throughput and minimizing blocking rate, but it needs trust information to perform secure communication.

3   Motivation

The proposed system discusses secure group communication on multicast transmission for hierarchical network. Stipulating security is a multidimensional approach that depends on specific applications. Though various ways are possible to provide security, the system has been motivated on the trust technique. Aggregating data from multiple sources for dynamically distributed dense network has computation overhead that affects system performance. Including complex security technique say public key cryptography to such scenario induces overburden to each node that leads poor performance. Application of trust management to the proposed system requires feasible outcome to maintain overall network performance in a better manner. The nodes in network may compromise due to various reasons. Energy depletion is the major factor where the node becomes a stale user. It is also possible to inject incorrect data. In this work, the researcher focuses on the reliability to cluster head nodes since the transmission to sencar are only from the valid heads of clusters.

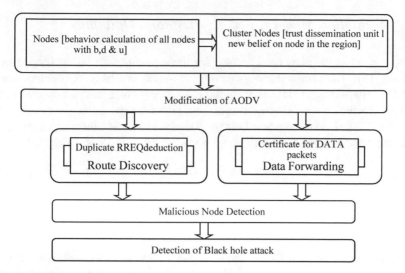

Fig. 1 Architectural diagram

4 Methodology

This work strengthens the secure transfer to the sencar from cluster group heads of valid sources. For ensuring the trust level, the system initiates degree distribution of connected nodes to identify the position of cluster group heads which is within a belief or a disbelief range. The system sets various boundaries and tests whether the node has an accessible neighbor only if the head located in positive area and the probability of proximity are in the bounded region. The head could perform communication to sencar by achieving successful test conditions to become a belief entity and analyze the percentage of the belief communication in the network on the basis of the degree of distribution, location probability, and neighborhood recommendations.

Trust predictions are costly task in finding and analyzing the contribution of every mobile node on the category of each security to enable reduction in false positive. The projected work is shown in Fig. 1, and a description thereof is provided in the sections that follow.

5 Mobility-Assisted Uncertainty Reduction Scheme

The basic ideas of checking trustworthiness are based on personal experience of truster toward trustee which are taken as firsthand opinion. The truster observes opinion about the trustee through its reliable neighbor which is known as secondhand opinion.

5.1 Reputation Model Based on Firsthand Opinion

The proposed system has been chosen for the reputation model developed by Li and Wu [16]. Uncertainty is the focus of [17] model as a basic element to support evidence. Therefore, a node uses a triplet $b, d, u \in \{0, 1\}^3$ to represent a node's opinion (belief, disbelief, and uncertainty, respectively) toward a trustee/neighbor such that $b + d + u = 1$. The beta distribution Beta (α, β) helps to estimate the renown reputation value out of differences from false recommendations due to uncertainty 'u'.

5.2 Uncertainty Reduction Trust Information Dissemination Based on Secondhand Opinion

Assuming consistency in the behavior of the nodes, the hierarchical uncertainty reduction scheme in [18] is assisted with probabilistic predictions, and the algorithm represented in Fig. 2 gives the data flow for random renown test based on actual observations using Bayesian analysis.

The Bayesian method determines the opinion α, β values relating to the intended recipient through its neighbor. There is no static approach available to follow for collecting node reputation through its peers. Assessment of the node reliability is based on some evidence disclosed by some other nodes in the cluster as be trustworthy or untrustworthy. Hence, the extracted value would also be suspicious.

The trusted threshold is also an assumption with some random value due to frequent topology on node mobility. The system is a trusted one if the assumed threshold value is valid and feasible. The system would be considered if the threshold is invalid and unbounded.

5.3 MCR-O Method

The MCR-O method uses a simple and fast technique which is based on firsthand and secondhand opinions. It does dual filtrations where the trusted values are bounded if a node has a distribution of acceptable degree, else the distribution is unbounded and the node is said to be untrustworthy. The bounded nodes are into the second filtration where the reputation of recipient node given by its neighbor using following parameters such as node id, probability value, delay, trust level, throughput, overhead, lifetime, and success rate is maintained by a sencar and the neighbor cluster head nodes. The node gets accepted for the performance of data dissemination to the sencar if reputations are positive, else the node gets isolated.

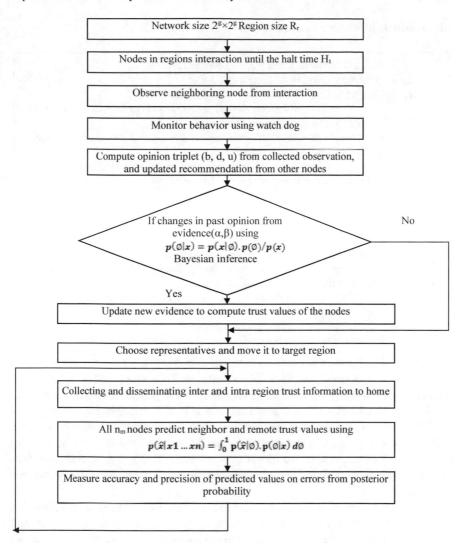

Fig. 2 A mobility-assisted uncertainty reduction scheme

Algorithm 1. Identification of trusted node based on MCR-O

Step 1: Assume Network size $2^g \times 2^g$ and Region size R_r
Step 2: Observe Cluster Head Nodes of Sencar
Step 3: Check the Cluster Head Node is belief, or disbelief
       Step 3.1: Calculate belief, or disbelief of cluster head nodes position using degree distribution.
       Step 3.2: Check whether the cluster head position is in bounded region using MCR technique
              Step 3.2.1: Locate the area of cluster head node and the function of area is positive
              Step 3.2.2: Set the upper bound, lower bound and side values of the cluster region.
              Step 3.2.3: If the probability of x_0 is greater than or equal to Y and the cluster head position is bounded with lower and upper limit, then go to Step 3.3, otherwise go to step 3.4.
       Step 3.3: If cluster head has accessible neighbor in bounded region then the node is belief node.
       Step 3.4: Otherwise select another node from cluster group and go to Step 2.
Step 4: Collect and disseminate data to sencar.

Mobile ad hoc networks compute trust opinion that relates various source nodes obtained by its neighbor node in sufficient volume. This type of network has uncertainties that are more prone to estimate neighborhoods' precise trust appraisal calculated in a hierarchical affiliation in cluster region [19]. Uncertainty can bring down proactive collection of trust information and disseminate the trust value to the entire nodes in the region until the completion of measurement of the new trust value. This process leads to increase in the overheads of communication, cost, and trust convergence time when looping up collection and dissemination tasks due to varying network size.

The proposed work focuses on a hierarchical scheme-based trust value estimation by gathering evidence for overcoming any stale opinion at the nodes on the network and dissemination thereof on cyclic intervals using posterior probabilistic prediction with Bayesian technique instead of frequent process [20]. All the nodes vote for the movement of a node which has the largest belief and the smallest uncertainty nodes after considered adequate pause time.

The researcher offers a description of how the data being observed, how the data data modifies the subjective views on the reputation of a node through posterior distribution and finally how the posterior extraction is used for building a predictive distribution for future values of reputation. The results are important for supporting the performance of the mobility-assisted uncertainty-reduced scheme for ensuring security and collaboration by formulating and evaluating trust among nodes in MANETs.

6 Results and Discussion

6.1 Sample Network

The network consists of 1500×1500 m^2 with 350 m as each region size in many clusters grouped together on the basis of node density in the decentralized ad hoc network. The nodes keep on sensing some target that depends on application. The proposed system has various clusters with each cluster having a cluster group head (CGH) for aggregating information sensed by cluster members. All nodes in network have the ability to move with battery supported for retaining residual energy; based on the requirements the trusted third-party sencars are fixed in various places across the network may have communication from multiple cluster group head. The location where sencars are placed is known as the polling point, and the system should have only minimum sencars possible for the simulation network. The sencars gain two activities, normally aggregation and dissemination to and from CGH. The ideas behind the sencar distribution in the dense network are energy consumption and multiple data aggregation. The regular monitoring system would always perform monitory activity and may often be drained its battery.

Sencars extend its support on data aggregation from the CGH whose battery is likely to drain faster and would have a very smaller residual energy. One of the classical ideas such as cluster formation is also used initially once the nodes entered into the network formation tasks for saving energy. Since the cluster is not the solitary solution for energy consumption, it also focuses on the simplification of the communication path at intra-regional, intra-cluster, and inter-cluster connections via CGH. Despite being a good technique, the sample network needs a secured trusted third agent as destination (sencar). The CGH transfers the content immediately when there is the likelihood of its draining with extremely small chance of changing CGH after network formation made. There are no security credentials made by the trusted sensor to authenticate the sources in [9] as shown in Fig. 3.

Figure 3 simulates the simple setup for a network having 30 nodes in the network. The sencar sojourn for data aggregation with multiple sources and those nodes are the cluster group heads which in turn receives data from its cluster members via cluster heads represented as SN in diagram as shown below. The node 32 which acts as sencar can receive cluster member information through the CGH numbered as 7 and 10 which act as sources from groups 2 and 3.

Each cluster group consists of variable number of nodes that may allow to join or leave the network at anytime. Each group has maximum of 8 nodes in this simulation setup. The sencar 32 will transfer all received information to the base station number as 0. The base station collects at the end of each iteration.

The checking process is needed at the time of CGH request toward sencar for data dissemination. The main aim of this work under this scenario from multiple sources needs reliability, in order to disable the sencar from receiving information to the base station from the fake sources. Testing the trust level of sources is needed for

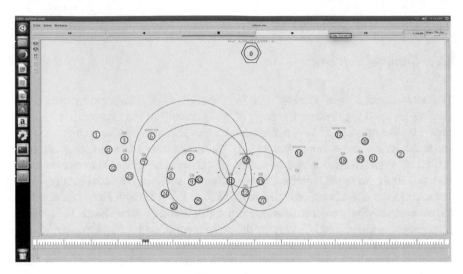

Fig. 3 Simulation setup of single sencar for trust integration

understanding its belief level. Knowledge of belief values of each source is required for calculating its location by using degree distribution as the initial process.

Later, the sencar would be tested to ascertain the identified location is regionally bounded. It is required to first test the area function and set the upper, lower, and sides' boundary values if area is identified positive. The data aggregation by sencar begins if the probability between upper and lower bounds is known and has an accessible neighbor using trust certificate. This is shown in Fig. 4. The figure describes the two clusters, a base station and a sencar at polling point $P1$. The diagram helps in understanding each CGH node maintains its trust information in the form of a certificate table. The sencar also maintains the table with entries about all possible CGH nodes under its coverage. The red circled nodes denote CGH, the blue circled nodes are other head nodes, and the white nodes are cluster members.

The sencar checks the neighbor opinion before accepting data communication from a sender. Various parameters are required for checking the neighborhood, normally node id, location probability, delay, trust level, throughput, overhead, lifetime, and success rate. This process is explained under neighbor node opinion.

6.2 Determining Degree Distribution of Each Node

The nodes in a network are distributed and also follow hierarchical discipline for node arrangement in some Euclidian distance. The degree distribution can be calculated for both dense and sparse networks. The proposed system uses power law for calculating node similarity through link density. The nodes in the network for the proposed

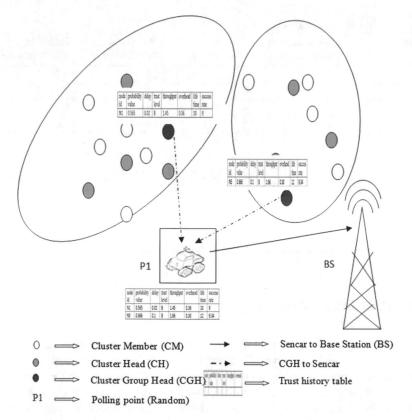

Fig. 4 Sample network diagram

scenario have a small number of links connected to the nodes in a few cases and highly connected nodes in few others, say CGH and sencar, respectively. The system is used for scale-free network using cluster-based connections. Hence, the nodes should know their similarity on the basis of similar interest, belief and the closeness or proximity between each other to the clusters to the sencar located at a polling point. The degree distribution decreases as the number of similarity increases.

However, the link density is based on the possible prediction for links in near future. The density is the difference of the ratio of available edges to the future predictions and can be denoted as $d = \frac{e}{0.5 \times f \times (f-1)}$.

Let d be the degree of a CGH node and k be the degree of its neighbor. If the sencar tries to access the multiple CGH as its sources, it can be made easily at least if the similarity on proximity is satisfied. The degree distribution of CGH node with degree k can have the probability to the stochastic neighbor CGH with degree d that can

Table 1 Distribution of sources in clusters

S. no	Sencar at polling point number (S_i)	Number of clusters (x_i)	Number of Sources (n_i)	Probability distribution $p(x_i)$
1	S1	1	25	0.10
2	S2	3	47	0.19
3	S3	4	38	0.15
4	S4	5	12	0.05

$$\text{be denoted as } n(d|k) = \frac{e(k+2)}{kd(d+1)} \left[1 - \frac{\left(\dfrac{2e+2}{e+1}\right)\left(\dfrac{k+d-2e}{d-e}\right)}{\left(\dfrac{k+d+2}{d+2}\right)} \right]. \text{ Since the proposed}$$

network has hierarchical clustering mechanism, it can be denoted as $c(k) = k^{-1}$.

6.3 Distribution Probability

The probability for distribution of stochastic or random process in a MANET scenario makes continuous observations. The observations are made on the position of CGH range from a lower bound to an upper bound region. The probability of a single cluster is logically zero for any specific location and 1 for all clusters in a region. The probability density function of a CGH located under coverage range as $x1$ and $x2$ of positive region is denoted as $\int_{x1}^{x2} f(x)dx f(x) \geq 0$. The graph representation has measures of the probability function known as key parameters driven at values for the curve can be denoted as $f(x) = xe^{-x} x \geq 0$. If the probability of cluster head node located between $x1$ as lower bound and $x2$ as upper bound, it is denoted as $f(x2) - f(x1) x2 > x1$.

Table 1 shows the probability distribution of sources in clusters connected to a selective polling point for a specific instance. For example, a shopping complex at a city has various divisions such as home appliances' section, beauty section, electronic section, fabric section. Customers purchase their required items of choice. When 25 customers buy only one item, the probability for distribution is 0.10. If 47 customers buy 3 items, then the probability is 0.19. The comparison is given in the graph in Fig. 5. In the technique proposed by researcher, the identification of source nodes available to a sencar distributed in a stochastic manner is between range $x1$ and $x2$ as mentioned above. The sources are available in the cluster located in coverage region of the network that can be shown in Fig. 6 from the reference of MCR technique.

The probabilities of the location of all cluster group nodes predictable in a bounded region produce the outcomes between 0 and 1. If the source node location is not predicted when the node changes its location or away from the network region, the outcome becomes negligible, say, 0. For instance, the drawing tool tending to draw an object beyond the window region of the monitor screen in multimedia

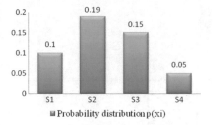

Fig. 5 Probability distribution of source nodes

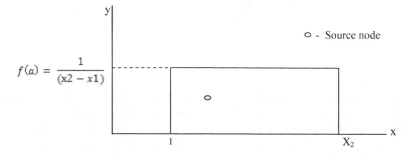

Fig. 6 Distribution of source nodes

application is not possible. Another example say flying across country without a passport is impossible. Even if necessarily needed, this would need a trust certificate in the form of visa approval for their commute, type of visa, permitted duration, purpose of transfer, etc. Trust system is most important in any application where group communication and integrity are crucial.

6.4 Bounded Region Identification

The mobile ad hoc network that uses the air medium for communication must cover a certain angular distance of coverage for communication. This subsection is derived from the above concept where it identifies the node availability in a network region.

Identification of the CGH nodes locations are bounded in the coverage region is an important activity in the initiation to initiate communication with the available sources. Figure 7 shows the proposed representation of distributed cluster head node that is presently some of the source nodes of the sencar at different instances. The green node and yellow node in figure are located in lower bound and upper bound, whereas the red node is completely away from the network region marked as rect-angular area of the figure. The lower bound is between x_1 and x_2 as minimum and maximum limits, respectively. Y_0 is the upper bound in the region. The arc shows the coverage level of the sencar. The source is located at two independent variants

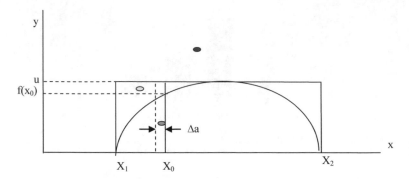

Fig. 7 Distribution of CGH in bounded region

f_1 and f_2 for each CGH. The numeric variates are distributed in a uniform manner as indicated by frequency distribution. The x_0 and y_0 values can be computed as $x_0 = f_1(x_2 - x_1)$ and $y_0 = u(f_2)$. If the y_0 value is less than or equal to the probability density function $f(x_0)$, the node is identified as a bounded node set positive belief value. The system has the ability to allow the node to do the trust evidence steps as discussed in Algorithm 1 for identifying trust from its neighbor. This is explained in the following section.

6.5 Neighbor Node Opinion

The sequence of steps begins with finding the degree distribution of nodes up to identification of the status of its bounded region required for all possible nodes ready for data distribution under a sencar. The sencar is needed to accept the requested source node as reliable entity. It needs the recommendations from the neighbor on the basis of the evidence of the node through observation. The personal experience about the source node through various interactions can be passed on to the sencar upon request as recommendations. The positive response is based purely on the success rate of interactions, belief value, and opinion from its neighbor. The positive response is would be a flag set to on or off if prior conditions are accepted or rejected, respectively.

7 Conclusion

The transmission of data present in previous work performs unreliable communications between any two entities. Those transmissions would not support for valid communication via trust authentication. The security boundaries of MANET were under analysis over several years to optimize its goal for shaping the protection mechanism on MANET. This type of network has uncertainties that are more prone to estimate neighborhoods' precise trust appraisal calculated in a hierarchical affiliation in cluster region. This work focuses on the sencar-received information whether from the trusted node or not by using degree distribution and collects the opinion from the neighbor of another cluster group head.

References

1. Gambetta, D.: Can we trust trust? Trust: Making and Breaking Cooperative Relations, pp. 213–237. Basil Blackwell, Oxford (1990)
2. Li, H., Singhal, M.: Trust management in distributed systems. Computers 40(2), 45–53 (2007)
3. Liu, J., Issarny, V.: Enhanced reputation mechanism for mobile ad hoc networks. In: Proceedings of 2nd International Conference of Trust Management (iTrust 2004), Oxford, UK (2004)
4. Solhaug, D.E., Stölen, K.: Why trust is not proportional to risk? In: Proceedings of 2nd Int'l Conference on Availability, Reliability, and Security (ARES'07), Vienna, Austria, pp. 11–18 (2007)
5. Aivaloglou, E., Gritxalis, S., Skianis, C.: Trust establishment in ad hoc and sensor networks. In: Proceedings of 1st Int'l Workshop on Critical Information Infrastructure Security, Lecture Notes in Computer Science, Samos, Greece, 31, vol. 4347, pp. 179–192. Springer, Berlin (2006)
6. Adams, W.J., Hadjichristofi, G.C., Davis, N.J.: Calculating a node's reputation in a mobile ad hoc network. In: Proceedings of 24th IEEE Int'l Performance Computing and Communications Conference, Phoenix, AX, vol. 7–9, pp. 303–307 (2005)
7. Yunfang, F.: Adaptive trust management in MANETs. In: Proceedings of 2007 Int'l Conference on Computational Intelligence and Security, Harbin, China, pp. 804–808 (2007)
8. Li, J., Li, R., Kato, J.: Future trust management framework for mobile ad hoc networks: security in mobile ad hoc networks. IEEE Commun. Mag. 46(4), 108–114 (2008)
9. Pushpalatha, K., Karthikeyan, G., Umarani, S.: An efficient sencar based multiple data aggregation in MANET. Int. J. Adv. Res. Dyn. Control Syst. 9(2) (2017) ISSN 1943–023X
10. Liu, J., Issarny, V.: Enhanced reputation mechanism for mobile ad hoc networks. In: Proceedings of 2nd Int'l Conference of Trust Management (iTrust 2004), Oxford, UK (2004)
11. Khalil, I., Bagchi, S.: Stealthy attacks in wireless ad hoc networks: detection and countermeasure. IEEE Trans. Mob. Comput. 10(8), 1096–1112 (2011)
12. Adelantado, F., Verikoukis, C.: A non-parametric statistical approach for malicious users detection in cognitive wireless ad-hoc networks. In: Proceedings of 2011 IEEE International Conference on Communications. ICC, Kyoto (2011)
13. Li, C.T., Yang, C.C., Hwang, M.S.: A secure routing protocol with node selfishness resistance in MANETs. Int. J. Mobile Commun. (IJMC) 10(1), 103–118 (2012)
14. Konstantopoulos, C., Gavalas, D., Pantziou, G.: Clustering in mobile ad hoc networks through neighborhood stability-based mobility prediction. Comput. Netw. 52, 1797–1824 (2008)
15. Torkestani, J.A., Meybodi, M.R.: Clustering the wireless ad hoc networks: a distributed Learning Automata Approach. J. Parallel Distrib. Comput. 70, 394–405 (2010)
16. Li, F., Wu, J.: Uncertainty modeling and reduction in MANETs. IEEE Trans. Mobile Comput. 9(7) (2010)
17. Pavai Madheswari, S., Suganthi, P.: An M/G/1 retrial queue with second optional service and starting failure under modified bernoulli vacation. Transylvanian Rev. 24(10), 1602–1621 (2016)
18. Pavai Madheswari, S., Suganthi, P., Josephine, S.A.: Retrial queuing system with retention of reneging customers. Int. J. Pure Appl. Math. 106(5), 11–20 (2016)
19. Ahmed, A., Bakar, K.A., Channa, M.I., Haseeb, K., Khan, A.W.: A survey on trust based detection and isolation of malicious nodes in ad-hoc and sensor networks. Front. Comput. Sci. 9, 280 (2015) ISSN 2095-2228

20. Wei, Z., Tang, H., Yu, F.R., Mason, P.: Trust establishment based on Bayesian networks for threat mitigation in mobile ad hoc networks. In: Military Communications Conference (MILCOM) 2014 IEEE, pp. 171–177 (2014)

IoT-Based Robot with Wireless and Voice Recognition Mode

Md. Amirul Islam, Nishat Halim Sharif, Md. Shahriar Parvez Tameem, S. M. Mazharul Hoque Chowdhury, Moriom Chowdhury Kumu and Md. Fokhray Hossain

Abstract Nowadays, robots are playing a very important role in industry level and also out of the industry. Dependency on robots is increasing for their fast and reliable working speed and accuracy. Considering that, the demand of robots is increasing every day. This research was conducted focusing on the necessity of robots in our daily life. This paper proposes a system where a robot can be controlled in different ways like voice, wireless, and full automatic mode. The prototype was built and tested. The robot prototype will be able to receive voice commands from short distance. In case of long-distance communication, the user will be able to connect through the Internet using IoT.

Keywords Arduino · Communication · Internet · IoT · Microcontroller
Robotic arm · Sensor · Servo · Voice recognition module · Wi-Fi

Md. A. Islam (✉) · N. H. Sharif · Md. S. P. Tameem · S. M. M. H. Chowdhury
M. C. Kumu · Md. F. Hossain
Department of Computer Science and Engineering, Daffodil International University,
Dhanmondi, Dhaka 1205, Bangladesh
e-mail: amirul2267@diu.edu.bd

N. H. Sharif
e-mail: nishat2330@diu.edu.bd

Md. S. P. Tameem
e-mail: tameemshahriar@diu.edu.bd

S. M. M. H. Chowdhury
e-mail: mazharul2213@diu.edu.bd

M. C. Kumu
e-mail: Kumu2439@diu.edu.bd

Md. F. Hossain
e-mail: drfokhray@daffodilvarsity.edu.bd

© Springer Nature Singapore Pte Ltd. 2019   743
A. Abraham et al. (eds.), *Emerging Technologies in Data Mining and Information
Security*, Advances in Intelligent Systems and Computing 814,
https://doi.org/10.1007/978-981-13-1501-5_65

1 Introduction

Nowadays, the high standards of living have encouraged automation to come in the van and be an integral part of home design. At the same time, the environmental relation has ensured that energy-efficient housing models and appliances are used. So, we proposed the system called Homey, a robot which can do tasks on behalf of a person at home.

Our project's aim is to develop a robot which can recognize the voice and perform tasks according to the command. It can be controlled remotely by Android smartphone application or through any Internet-enabled device, so that it can be used by anyone and especially can be helpful for physically disabled persons.

2 Literature Review

IoT is the concept which creates a relationship between user and system remotely. It also creates interconnection between devices. There are three C's on IoT, communication, control and automation, and cost savings [1]. We have tried to implement IoT so that user can control and communicate with the robot within a low budget. The user also can observe the environmental condition by using sensors.

To connect with the Internet, Wi-Fi module with its low-cost effectiveness can be very useful and easy to use. Through Internet, a user can control the robot and its action remotely [2].

In 2017, J. Chandramohan, R. Nagarajan, K. Satheeshkumar, N. Ajithkumar, P. A. Gopinath and S. Ranjithkumar worked with ESP8266 Wi-Fi module [3]. They worked with Arduino UNO and ESP8266 Wi-Fi module to control home appliances remotely.

The user can control it without Internet by using voice command [4]. There is a voice recognition system attached to the robot. There is a difference between recognizing voice and speech [5]. It depends on what type of recognition module and which algorithm will be implemented to the robot to recognize the voice of a user. The robot can only recognize the voice and cannot differentiate the users.

In 2015, K. Kannan and J. Selvakumar worked on the Arduino-based voice-controlled robot in India. They used EasyVR module as voice recognizer device [6]. They also used Arduino Mega 2560 as microcontroller.

3 Component Requirements

To make the system fully functional, different components were used. Some important components used in this system are Arduino Mega 2560, Arduino Nano, voice recognition module (V-3), ESP8266 Wi-Fi module, LM35 temperature sensor, ultrasonic sensor, servo motor, liquid crystal display (LCD), jumper wires, and lithium polymer battery. The Arduino was programmed by 'AVR C' to process and control different logical statements to input and to get outputs. This system can be modified using other sensors and components; for this, the user just needs to change the program only.

4 System Development

The Homey robot was built based on Arduino Open Source Platform. As the robot can be controlled by online command, to ensure the IoT connectivity, Wi-Fi module was used along with Arduino Mega. The robot connects itself with the Internet through Wi-Fi Shield (Fi250) V1.1 module. The user can give command using Web application or Android application. The IoT system can send command and also receive data from the robot. The user can observe sensor value from the robot.

Another way of giving the instruction is voice command. Voice recognition module (VR-3) was used to receive voice command along with Arduino Nano. Arduino Nano receives voice command and processes it. Then, it generates specific instruction for each voice command and sends it to Arduino Mega through TX–RX serial communication pins [7]. Then, Arduino Mega performs according to the command. Voice training sets are stored in Arduino Nano. Therefore, when it receives voice command, it processes and compares the voice with training dataset.

4.1 Block Diagram

Figure 1 shows the data flow of the prototype.

4.2 Voice Input and Output

Voice commands are stored in two different groups as given in Tables 1 and 2, and each group has their unique voice commands. The voice recognition module takes input command as group. Each group can hold six commands. Maximum 255 commands can be trained with the voice recognition module.

The commands for both groups are given below.

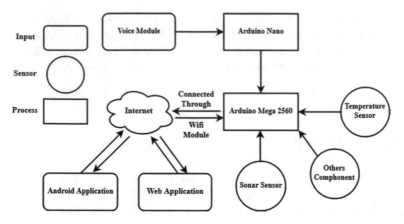

Fig. 1 Block diagram of the system

Table 1 Voice input and output in group 1

Voice command (Input)	Arduino command	Work (output)
Stop	1	Stop
Auto	2	Auto
Forward	3	Move forward
Piche	4	Move backward
Left	5	Move left
Right	6	Move right

Table 2 Voice input and output in group 2

Voice command (Input)	Arduino command	Work (Output)
Blink	7	Blink
Light on	8	Light on
Lightning	9	Light off
Temperature	T	Temperature
Handshake	P	Handshake
Night mode	N	Night mode

4.3 IoT Input and Output

A Web application and an Android application were developed to control the robot remotely through the Internet. The Android application interface is given below.

The robot can be controlled by Android application as Fig. 2 through the Internet using Wi-Fi module. While user presses any button on the application, corresponding commands are sent via Internet to the robot. The commands are in the form of ASCII character. The Arduino on the robot then compares the received command with its

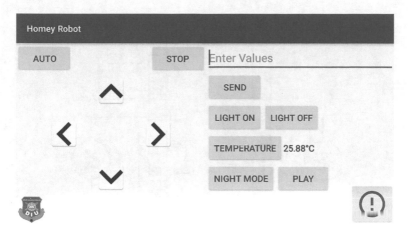

Fig. 2 Android application

pre-defined commands and control the servo motors, gear motors, sensors, and other peripherals to move forward, backward, left, right, stop, move hands, fingers, or measuring sensor values.

The temperature button sends command to Arduino to check temperature sensor value, and the value is sent back to the user. The temperature value is also displayed on the liquid crystal display attached to the front side of the robot. If the user presses the auto button or gets disconnected from the Internet, the robot can move anonymously. During the anonymous mode, sonar sensor helps it to avoid obstacles.

The sonar sensor HC-05 helps to measure the distance from the obstacle in front of the robot. The sonar sensor generates ultrasonic sound using trigger. Then, it reads the echo pin and returns the sound wave travel time in microseconds. The distance is calculated using the following rule:

$$\text{Duration} = \text{Distance}/\text{Speed} \qquad (1)$$

$$\text{Distance} = \text{duration} * 0.034/2 \qquad (2)$$

From Eqs. (1) and (2), the speed of sound in air is 0.034, and it travels twice to return, so divided by 2.

The Arduino Mega microcontroller gets the distance value and checks the condition when it should stop and search for another long-distance path.

Motor driver 1329d was used to control motors' speed [8]. There are two gear motors attached in the lags of the robot. The motors are connected with motor driver so that the speed can be controlled by the microcontroller. There are four digital inputs for two motors in motor driver. The diagram of motor driver makes it clear (Fig. 3).

The robot was built in the shape of a human. Different body parts like hand, head, and fingers are controlled by servo motors. It also can handshake with human using

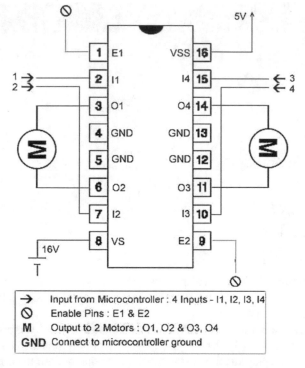

Fig. 3 Circuit diagram of motor driver [9]

Fig. 4 Functional robotic arm

its robotic arms and pick objects as shown in Fig. 4. As the structure is not that much strong, therefore at present, it only can pick lightweight objects.

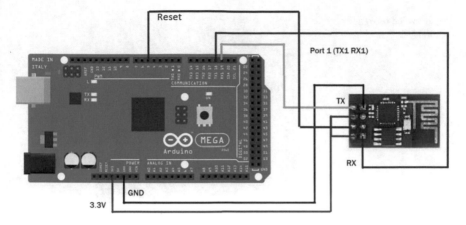

Fig. 5 Arduino mega 2560 with Wi-Fi module

4.4 Arduino

Arduino is an open-source platform. The hardware and software are open source [10]. There are various versions of Arduino in terms of processing speed, number of i/o pins, number of data communication pins, etc. Arduino Mega 2560 has been used as the main control board of the robot.

The Arduino Mega 2560 is built with ATmega2560 microprocessor [11]. The ATmega2560 on the Mega 2560 comes preprogrammed with a bootloader that allows to upload new code to it without the use of an external hardware programmer. It communicates using the original STK500 protocol. The Arduino IDE is the easy way to upload program on it [12] (Fig. 5).

4.5 ESP8266 Wi-Fi Module

ESP8266 Wi-Fi module is the best and cheapest Wi-Fi module for Arduino [13]. It is very easy to program and connect. The ESP8266 Wi-Fi module has built-in memory. So that it can be programmed and can process the data coming from the server.

4.6 Voice Recognition Module

This product is a speaker-dependent voice recognition module. It supports up to 255 voice commands in all. Maximum six voice commands can work at the same time. Any sound can be trained as command. Users need to train the module first before letting it recognizing any voice command.

Fig. 6 Voice recognition module V-3

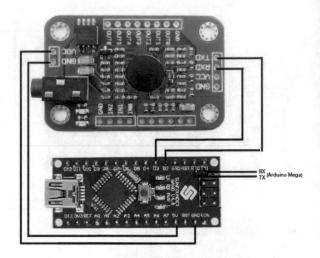

Fig. 7 Fully functional robot prototype

Voice recognition module (V-3) board as shown in Fig. 6 has two controlling ways: serial port (full function) and general input pins (part of function) [14]. General output pins on the board could generate several kinds of waves while corresponding voice command was recognized.

5 Final Prototype

Figure 7 shows the final visibility of the prototype.

6 Future Scope

This prototype can be modified and used for different purposes. The IP camera can be installed on the robot so that user can monitor and control the robot remotely. Many necessary sensors like gas sensor and humidity sensor can be added to measure atmospheric readings of the particular particles and so on. The robot can be used for the educational purpose also. The students can learn basic microcontroller hardware and software interaction and logical operations by experimenting with the robot.

7 Conclusion

During the development of the project, we studied the working principle of Arduino microcontroller, especially the Arduino Mega 2568, Arduino Nano, etc. We also studied that Arduino can be functional with modules. We used voice recognition module to take command input by voice. We designed a mobile app also to operate the robot. The Bluetooth module is used to connect the robot to the app.

The proposed system is developed with the combination of above-stated tools which will lead the users to hassle-free use of the robot and enjoy the features.

Acknowledgements This robot was developed as a final year project of Daffodil International University. We will specially thank Mr. Fahad Faisal, Lecturer, Department of Computer Science and Engineering, Daffodil International University, for his special guidelines. We feel thankful to Ms. Samia Nawshin, Lecturer, Department of Computer Science and Engineering, Daffodil International University, for her motivations. We will also like to thank Mr. Hafizul Imran, Founder and Chairman, Bangladesh Robotics Foundation, for his technical advice. Any error in this research paper is our own and should not tarnish the reputations of these esteemed persons.

References

1. Gawli, K., Karande, P., Belose, P., Bhadirke, T., Bhargava, A.: Internet of things (IoT) based robotic arm. Int. Res. J. Eng. Technol. **04**(03), India (2017)
2. Sehgal, T., More, S.: Home automation using IoT and mobile app. Int. Res. J. Eng. Technol. **04**(02), India (2017)
3. Chandramohan, J., Nagarajan, R., Satheeshkumar, K., Ajithkumar, N., Gopinath, P.A., Ranjithkumar, S.: Intelligent smart home automation and security system using Arduino and Wifi. Int. J. Eng. Comput. Sci. **6**(3), India (2017)
4. Kamdar, H., Karkera, R., Khanna, A., Kulkarni, P., Agrawal, S.: A review on home automation using voice recognition. Int. Res. J. Eng. Technol. **4**(10), India (2017)
5. Street Directory, http://www.streetdirectory.com/travel_guide/139545/technology/key_differences_between_speech_recognition_and_voice_recognition.html
6. Kannan, K., Selvakumar, J.: Arduino based voice controlled robot. Int. Res. J. Eng. Technol. **02**(1), India (2015)
7. Serial Communication, https://www.arduino.cc/reference/en/language/functions/communication/serial

8. Components101, https://components101.com/l293d-pinout-featuresdatasheet
9. Dual H-bridge Motor Driver—L293D IC, http://www.robotplatform.com/howto/L293/motor_driver_1.html
10. Arduino Computing platform, https://www.arduino.cc/en/Guide/Introduction
11. Arduino Computing platform, www.arduino.cc/en/Main/arduinoBoardMega
12. Arduino Computing platform, https://www.arduino.cc/en/Guide/Environment
13. Electro Schematics, http://www.electroschematics.com/11276/esp8266-datasheet
14. Voice Recognition Module vr3 Manual, https://www.scribd.com/document/311781701/Voice-recognition-module-vr3-Manual

Development of Joint Intelligent Millimeter wave Sensing and Communication

M. Chakraborty, B. Maji and D. Kandar

Abstract With the ever-growing data rate of 1 TB or more in vehicular sensing and communication in intelligent transportation system, millimeter wave is a promising candidate for meeting the requirement of joint operation of radar and communication. In millimeter wave, the high path loss may be compensated by adaptive beam formation. Because of very short wavelength, it supports the use of massive antenna arrays at both the vehicle in motion and base stations at roadside unit. Millimeter-wave radars are commonly used nowadays in autonomous driving, traffic efficiency, road safety. The same hardware may be shared with communication. This sharing of hardware effectively reduces cost/size and makes efficient spectrum usage. On one side, IEEE 802.11ad mmW is a better candidature for radar with good ranging and parameter estimation, and on the other side, the target information can directly be used for communication. In our work, we have demonstrated a comprehensive development of intelligent millimeter-wave sensing and communication working simultaneously. We have shown the beam tracking by the variation of beam steering angle with the distance and the resulting decreased BER.

Keywords Millimeter wave · OFDM-MIMO · V2V · V2X

M. Chakraborty (✉)
School of Engineering and Technology, Adamas University, Kolkata 700126, India
e-mail: mithunchakraborty03@gmail.com

B. Maji
Department of ECE, NIT, Durgapur 713209, West Bengal, India
e-mail: bmajiecenit@yahoo.com

D. Kandar
Department of IT, NEHU, Shillong 793022, India
e-mail: kdebdatta@gmail.com

© Springer Nature Singapore Pte Ltd. 2019 753
A. Abraham et al. (eds.), *Emerging Technologies in Data Mining and Information Security*, Advances in Intelligent Systems and Computing 814,
https://doi.org/10.1007/978-981-13-1501-5_66

1 Introduction

Existing dedicated short-range communication technology supports some of the safety applications and toll collection at 5.9 GHz but with low latency and a very low data rate. But in the advanced intelligent transportation system, the vehicles would be flooded with data from the radar sensors fitted at all the sides of the car, data from lider, data from other car sensors along with the communication data with the roadside unit, data from satellite, etc.. All these require a technology that can handle gigabytes or tera bytes of data. Millimeter-wave sensing and communication is deemed to be a candidate for meeting the ever-existing demand for V2V sensing and V2X communication [1, 2]. Because of the highly directional beam at 77 GHz, mmW radar can give range estimation accuracy by few centimeter. In millimeter-wave communication due to the inherent high attenuation loss, directional adaptive beam-forming and combining should be applied to guarantee high enough signal-to-noise ratio (SNR) for detection at communication receivers. Hence, accurate channel estimation is also an important issue in this aspect. The channel estimation is limited by scatterings due to large attenuation loss, high absorption loss, etc., at mmW [3]. The beam formation [4, 5] and combination are constrained by few RF blocks [6] in millimeter-wave communication transmitter and receiver containing high-speed A/D converters which consume measurable power as well as incur high costs. The authors have proposed here an intelligent transportation system in road scenario, whereby a vehicle containing radar with motion detects two moving cars as targets using 77 GHz mmW radar sensor, estimates their velocity and distance information, and communicates the sensed information to the fixed roadside unit with multiple-input single-output (OFDM-MIMO) technology. The proposed system is intelligent in the sense that the system is capable of continuously tracking the target car in motion for sensing and the RSU for communication while moving with speed.

2 System Model

2.1 RADAR System Model

The radar in our model is considered as FMCW radar as this can easily be embedded in vehicle for its less power requirement, small size. The radar transmits a continuous wave. Base band for modulation is used as sinusoidal signal. The received echo is the same as transmitted wave delayed in time by Δt which is related to the range. Because the signal is always sweeping through a frequency band, the frequency difference at any moment during the sweep remains constant and is usually called the beat frequency which can be translated to range.

Target modeling:

The target vehicles are modeled as car 1 and car 2 having 100 and 125 m^2. The distance of target car 1 is 150 m from the vehicle containing the radar with a speed of 90 km/h, whereas the distance of target car 2 is 100 m with a speed of 150km/h. The channel for the radar imaging is taken as free space.

Radar transmitter:

Uniform linear array consists of 12 elements, each of the elements is isotropic radiator, and spacing between any two is lambda/2. RADAR is installed on a vehicle having platform velocity. We have assumed the radar is a mono-static frequency modulated continuous-wave radar, operating at 77 GHz. The radar can distinguish two targets at distances of 300 and 500 m and estimates the relative velocity and Doppler.

Radar receiver:

Receiver collects the reflected target car signatures. De-chirp operation is performed on the echo signal and buffered for each sweep of the transmitted signal. This de-chirped signal helps to find the beat frequency and Doppler shift resulting in an estimation of range and speed of each target. The difference between transmitted signal frequency and received signal frequency is termed as beat frequency.

Range of beat frequency $f_r = \frac{|f_{bu}+f_{bd}|}{2}$ and

Doppler frequency $f_d = \frac{|f_{bu}-f_{bd}|}{2}$

where f_{bu} and f_{bd} are up-chirp and down-chirp frequencies, respectively.

Target distance and velocity can be estimated as,

$$R = \frac{CTf_r}{2B}$$

$$V = \frac{Cf_d}{2f_c}$$

Maximum unambiguous radar range $R_{unamb} = \frac{CT}{2}$

Estimated velocity and distance information of the target cars may be utilized by the vehicle containing radar for its automatic cruise control for effective collision avoidance.

2.2 Communication System Model

The communication system is configured as a mobile unit (moving car) enabled with MISO transmit beam formation and a (fixed) roadside unit (RSU) enabled with single antenna and working as a receiver. The transmitter contains multiple antenna (12 elements) with a transmitter power of 9 W and gain −8 dB. The onboard communication

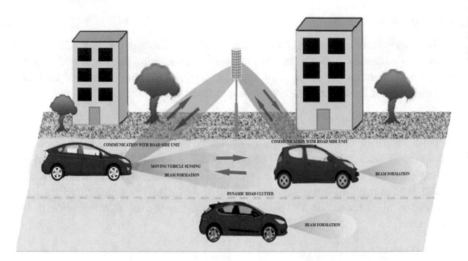

Fig. 1 Typical ITS on road scenario

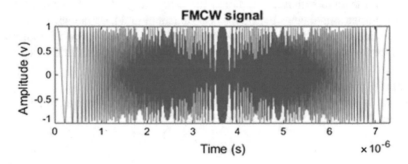

Fig. 2 Generated FMCW waveform for transmission

transceiver communicates with the fixed RSU and other vehicles for getting traffic information, road condition, vehicular infotainment, etc. The communication serial data is converted to parallel for OFDM frame generation to increase the effective data rate. As in the mobile scenario the channel contains high multipath, the channel is modeled here as a fading channel and OFDM is preferred to increase the system data rate canceling the interferences. Figure 1 shows the typical joint operation.

3   Simulation Results

The performance of the proposed model is demonstrated here. Table 1 describes the simulation parameters.

Table 1 System simulation parameters

Parameter name	Value
Carrier frequency (fc)	77 GHz
Radar sensor	FMCW long-range radar
Range resolution	1 m
Maximum allowed velocity of the vehicle with radar	230 km/h
Velocity of the vehicle with radar	150 km/h
Distance of target car 1	150 m
Velocity of target car 1	90 km/h
Car 1 radar cross section of target	100 m^2
Distance of target car 2	100 m
Velocity of target car 2	120 km/h
Car 2 radar cross section of target	125 m^2
OFDM-MIMO communication transmitter power	9 W
Communication transmitter gain	−8 dB
Initial transmitter range	2050/1000 m
Initial mobile angle	1°
Later mobile angle	1.36°
No of array antenna element	12

Initially without applying the transmit beam-forming, communication is established between the moving car and RSU, but because no beam formation is applied, BER becomes very high; after the application of directional beam-forming with a steering angle and generating the antenna weight vector for the individual antenna element, communication BER is greatly improved. The system is working like a phase-locked loop where initially there is maximum error generated due to non-alignment of the transmitter antenna from the roadside unit base station, but then the error is minimized by the application weight vector generation (Figs. 3 and 4).

The entire system simulated for two different distances between the communication transmitter and receiver such as 2050 and 1500 m with two different beam steering angles such as 1° and 1.36°, respectively, with the BER recorded 0.09 and 0.03% with the constellation diagrams received at the receiver is shown in Fig. 5.

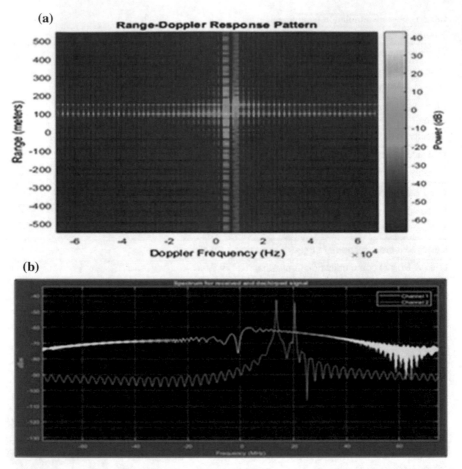

Fig. 3 **a** Two cars (120 and 150 m) with different velocities and different RCS detected by the radar sensor with their Doppler shift and **b** their Doppler spectrum

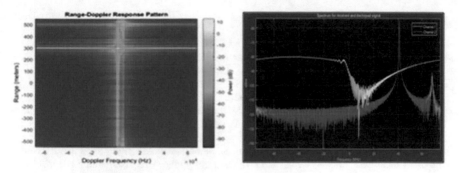

Fig. 4 Two cars (300 and 500 m) with different velocities detected by the radar sensor with their Doppler spectrum

(a)   **(b)**

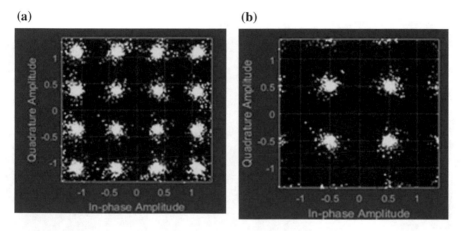

Fig. 5 a Rx constellation at a distance of 2050 mt and **b** 1500 mt

4   Conclusion

The simulation results show that the radar sensor can successfully differentiate between two vehicles 30 m apart as well as 200 m apart in their range Doppler diagram, and for the communication, beam formation is of great help to reduce the BER from the BER value 8.39% for 30,714 no. of bits with 2578 no. of bit errors to BER value 0.00% for the same no. of bits with no bit errors. Also from car-to-RSU communication as the car approaches the RSU, beam steering angle increases for maintaining the beam formation and BER decreases. Hence, we propose our mmW system for effective V2V sensing and WLAN communication to provide a joint framework of vehicular communication and radar technologies at 77 GHz.

References

1. Boccardi, F., Heath Jr., R.W., Lozano, A., Marzetta, T.L., Popovski, P.: Five disruptive technology directions for 5G. IEEE Commun. Mag. **52**(2), 74–80 (2014)
2. Andrews, J.G., Buzzi, S., Choi, W., Hanly, S.V., Lozano, A., Soong, A.C.K., Zhang, J.C.: What will 5G be? IEEE J. Sel. Areas Commun. **32**(6), 1065–1082 (2014)
3. Akdeniz, M.R., Liu, Y., Samimi, M.K., Sun, S., Rangan, S., Rappaport, T.S., Erkip, E.: Millimeter wave channel modeling and cellular capacity evaluation. IEEE J. Sel. Areas Commun. **32**(6), 1164–1179 (2014)
4. Jayaprakasam, S., Ma, X., Choi, J.W., Kim, S.: Robust Beam-Tracking for mmWave Mobile Communications. 1089–7798 (c) 2017 IEEE. https://doi.org/10.1109/lcomm.2017.2748938, IEEE Communications Letters
5. Zhang, C., Guo, D., Fan, P.: Tracking angles of departure and arrival in a mobile millimeter wave channel. IEEE ICC 2016—Communications Theory (2016)
6. Alkhateeb, A., Mo, J., Prelcic, N.G., Heath Jr., R.W.: MIMO precoding and combining solutions for millimeter-wave systems. IEEE Commun. Mag. **52**(12), 122–131 (2014)

Wi-Fi Optimization Using Parabolic Reflector and Blocking Materials in Intrusion Detection Systems

Sumanta Kumar Deb, Ankan Bhowmik, Biswajit Maity, Abhijit Sarkar and Amitava Chattopadhyay

Abstract Intrusion detection system (IDS) is a software application which monitors the system, network activities, finds vulnerabilities if present, and protects digital data in a safe manner. An IDS monitors network traffic and data, features selection, analysis and action or detection, and also alert generation during life cycle. Firewall technique is one of the system-based protection techniques which is used to protect the private network from the public network. The areas where IDSs are used are in financial, healthcare, technical fields like MANET, cloud computing and its security, data mining. There are three types of intrusion detection systems—HIDS, NIDS, and APIDS. HIDS is based on sensors, where it can obtain data from operating system. HIDS can also tell attacker's activity by analyzing network. NIDS is also based on network sensors. NIDSs can collect network information and can audit network attacks, while packet is moving across the network. APIDS works based on behavior and event of the protocol. IDS prevents various attacks based on OSI layer like DoS or DDoS attack, eavesdropping, spoofing, U2R, logon abuse, application-based. IDS system can be affected by the signal strength where in this paper the objective is to enhance detection rate of intrusion detection system based on Wi-fi optimization where we have tried to show that how some basic materials can affect intrusion detection system through Wi-fi signal strength.

S. K. Deb (✉) · A. Bhowmik · B. Maity · A. Sarkar · A. Chattopadhyay
Department of Computer Application,
Institute of Engineering & Management, Kolkata 700091, India
e-mail: sumanta.deb@iemcal.com

A. Bhowmik
e-mail: ankan.bhowmik@iemcal.com

B. Maity
e-mail: biswajit.maity@iemcal.com

A. Sarkar
e-mail: abhi41001@gmail.com

A. Chattopadhyay
e-mail: amitava.chattopadhyay@iemcal.com

© Springer Nature Singapore Pte Ltd. 2019
A. Abraham et al. (eds.), *Emerging Technologies in Data Mining and Information Security*, Advances in Intelligent Systems and Computing 814,
https://doi.org/10.1007/978-981-13-1501-5_67

Keywords IDS · Vulnerabilities · Security · MANET · HIDS
NIDS · Parabolic reflector · Wi-fi

1 Introduction

IDS can be categorized into some types like—signature-based IDS, anomaly-based IDS, hybrid IDS.

In anomaly-based IDS, it is of two types—static detector and dynamic detector. Anomaly detection is used to detect attacks like misuse of protocol and service ports, DoS or DDoS, buffer overflow, and other application payload. In anomaly IDS, two types of statistical models are proposed, operational model or threshold metric and Marker model or Markov process [1–3]. In signature-based IDS, there are two techniques used in misuse detection, i.e., expression matching and state transition analysis. Specifically, NIDS techniques are statistical-based, knowledge-based, and machine learning-based. There are many tools used as IDS like Snort, OSSEC-HIDS, fragroute, Honeyd, and Kismet [4, 5] (Figs. 1, 2, and 3).

Some works have been proposed in machine learning methods for intrusion detection with respect to parameters like accuracy, algorithm complexity, and time complexity. In this work, we saw how different sensors are connected wirelessly using radio waves which are a type of electromagnetic radiation. Electromagnetic spectrum is a combination of known frequencies (f) and number of waves per second (t) and their wavelengths (λ) [6].

$$f = 1/T \quad \text{and } f = c/\lambda$$

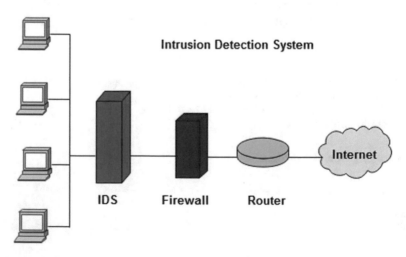

Fig. 1 Intrusion detection system

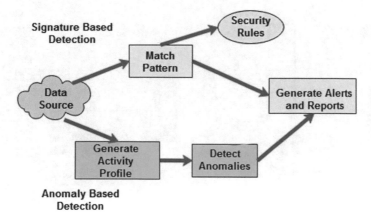

Fig. 2 Signature-based IDS and anomaly-based IDS

We have analyzed how Wi-fi signal [7–9] strength can affect the different parameters in intrusion detection system (IDS), and we have experimented that the Wi-fi signal strength can be enhanced by using the parabolic reflector theory.

2 Existing Work

2.1 Background Study on IDS

Many existing works in the area of IDS have been evaluated by using KDDCup' 99 data as standard dataset [10, 11]. In a military network environment, various intrusions simulated for several weeks are included in the dataset. Four main categories of KDDCup' 99 attacks are: remote to local (R2L), user to root (U2R), probing, and denial of service (DOS) [10, 11]. As indicated, Fuzzy Intrusion Recognition Engine (FIRE) is an irregularity based system launched after the DDoS attacks of February 2000 [2, 12]. The IDS alone cannot prevent the attack but provides us with valuable security information administrator about the type of activity. FIRE uses fuzzy logic to determine the similar network attacks. FIRE consists of three components: Network data collector (NDC) reads raw network packets off the wire and stores them in disk; network data processor (NDP) condenses the information in chose classes, performs the data mining, and contrasts the information and authentic mined information. Fuzzy inputs are the data fetched, which was not present in any historical mined data; fuzzy threat analyzer (FTA) analyzes the fuzzy outputs from NDP and sends it to the security administrator for response. The data collection interval in FIRE is 15 min [12]. To create an aggregate key consisting of the IP destination, IP source, and destination port fields [12], data mining techniques are used in FIRE. This new field is known as the sdp. The sdp represents the status of a TCP service channel

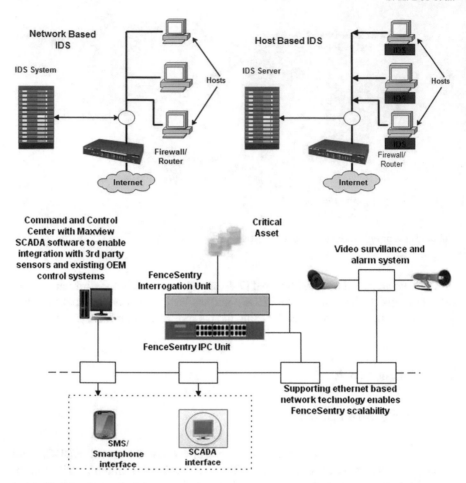

Fig. 3 NIDS and HIDS and PIDS

(whether successful or not) between two IP end points. The NDP then tabulates and records the number of packets seen. This table is updated and revised by NDP over time. Generally, only sdps with matching destination service are recorded. FIRE was first used at Iowa State University where it was connected to unprotected networks. Initial phase of data collection went for 3 weeks, and it was able to locate 9 distinct TCP port scan and 4 ICMP. Two primary forms of NIDS are: misuse detection and anomaly detection; misuse detection system works on previous database and past attack patterns but is unable to identify any new attack. On the other hand, anomaly detection system applies statistical measures and artificial intelligence, but its disadvantage is that it uses extensive data for artificial learning algorithms and is quickly overloaded [12].

2.2 Background Study on Parabolic Reflector

Light is a wave which is converted into electric and magnetic fields. As an example, most of the emitted electromagnetic energy of visible stars is converted as light. Our eyes sense only tiny portion of the spectrum. Wavelength correlates with energy so the color of a star tells how hot it is, red stars are coolest, the coldest stars emit hardly any visible light at all, blue stars are hottest, and they can only be seen with infrared telescopes [9]. Electromagnetic spectrum can be vitiated and get disturbed by various objects. If we think of the visible light in electromagnetic spectrum, it gets obstructed by some of the materials like wood or metal which would completely block the visible light, while tinted glass or water would partially block it. The question which arises is that will the same thing happen for radio waves like Wi-fi signals? In this particular analysis, it would be studied that which material would block the signal or reduce its strength at what measure [6, 8, 9]. Wireless signal is measured using decibel-milliwatts (dBm). Positive dBm values have a power more than 1 mW, and negative dBm values have a power less than 1 mW. Spectrum of electromagnetic radiation consists of known frequencies and wavelength, where frequency is measured in hertz, so one hertz is one cycle per second. Wi-fi signals percolate at 2.5 or 5 GHz [6] (Figs. 4 and 5).

Electromagnetic *radiation* can be compared with a flow of *photons*, where they are roaming like a waveform, consisting energy and travelling with the speed *same as light*. The unique differentiating factor that separates *gamma rays, visible light*, and *radio* waves are the energy carried by the photons. Radio waves have photons with the lowest energies [6, 13]. *Microwaves* have somewhat more vitality than radio waves. Among visible *X-rays*, *ultraviolet*, and gamma rays, the infrared has more energy. The energy of a photon causes it to act similar like a wave or like a particle. This is known as "wave–particle duality" of *light* [6, 14, 15]. It is vital to comprehend that we are not discussing a distinction in what light is, yet by the way it acts. Low-energy photons (such as radio photons) carry on more like waves, while higher energy photons (such as *X*-rays) act more like particles. The electromagnetic

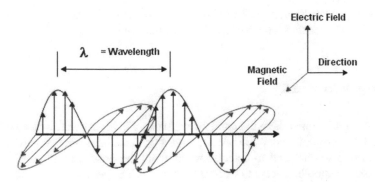

Fig. 4 Relationship between light and magnetic fields

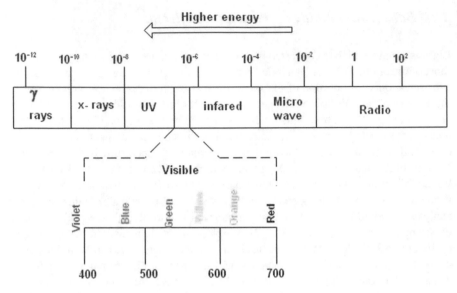

Fig. 5 Energy distribution of electromagnetic waves

spectrum can be communicated in terms of energy, *frequency*, or *wavelength*. Every state of mind about the EM range is identified with the others in an exact numerical manner. Researchers speak to wavelength and recurrence by the Greek letters lambda (λ) and nu (ν) [6, 13]. Using those symbols, the relationships between wavelength, energy, and frequency can be written as wavelength equals to the speed of light is divided by the frequency, or $\lambda = c/\nu$ [16] and energy equals to the Planck's constant times is multiplied with the frequency, or $E = h \times \nu$

where

- λ is the wavelength,
- ν is the frequency,
- E is the energy,
- c is the speed of light, $c = 299,792,458$ m/s (186,212 miles/s),
- h is Planck's constant, $h = 6.626 \times 10^{-27}$ erg-seconds.

3   Proposed Work

We have experimented the wifi signal strength using Wifi analyzer which can block wifi signal to properly work IDS: Aluminum foil; Cardboard (Figs. 6 and 7).

We have also analyzed to detect how to avoid the problem and enhance Wi-fi signal using parabola concept to work IDS properly. The point of a parabola consists of some focusing signals for better wireless connection [16] (Fig. 8).

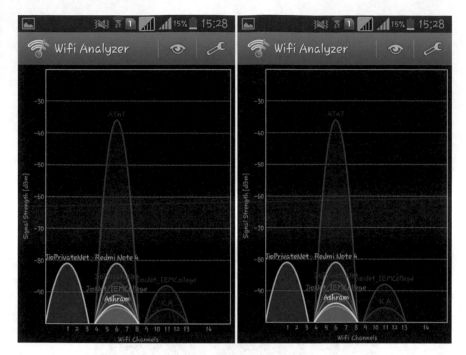

Fig. 6 Wi-fi signal strength in normal case and using aluminum

Material	Case 1	Case 2	Case 3	Average case	Attenuation (dBm)
In general	−37	−37	−37	−37	0
Aluminum foil	−45	−45	−45	−45	8
Cardboard	−38	−38	−38	−38	1

The reflecting properties of a parabola make parabolic reflectors useful in many practical applications. Reviewing the law of reflection which expresses that the angle of incidence is equivalent to the point of reflection measured forms the normal and that the typical is perpendicular to the surface; the reflective property of the parabola has various functional applications. Since light is a wave, if a light source is set at the focal point of a paraboloid, the outcome will be a beam of light emerging outward along the direction of the axis. This is how flashlights, headlights, and searchlights work [7–9, 14, 15] (Fig. 9).

The opposite is also true. Any parallel beam of light or wave, when incident on a parabola, converges on the focus after reflection [6].

Parabolic Reflector Theory

Here, total length A1 + A2 is equal to B1 + B2, and this means that the phase integrity

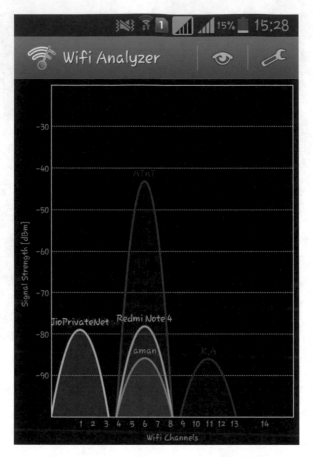

Fig. 7 Wi-fi signal strength obstacle by using cardboard

of the system is retained. Incoming waves add at the focal point, and outgoing waves produce a single wave front moving in parallel away from the reflector [13–15] (Fig. 10).

4 Conclusion

In this paper, we have analyzed that IDS specially NIDS does not properly work through some signal blocking materials by which vulnerabilities may reach in any HIDS, NIDS, or PIDS. Here, we have also measured how to avoid the situation and enhance the signal strength by implementing parabolic reflector property. In future, other methods are been looked into to generate different ranges of input and to detect

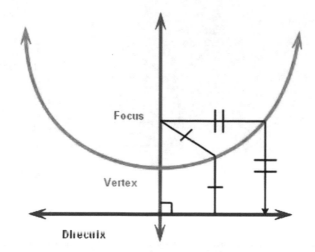

Fig. 8 Directrix and focus are equidistant from any point on a parabola

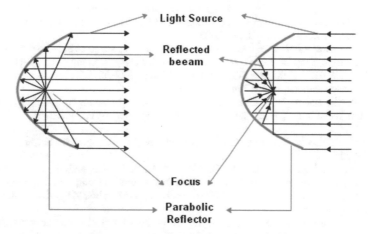

Fig. 9 A parabolic reflector uses energy of the sun to heat a liquid. The heated liquid is ultimately used to generate electrical power

a greater diversity of intrusions. Providing the log-in information, email and Web server activity are the areas looked into.

References

1. Vijayarani, S., Maria Sylviaa, S.: Intrusion detection system—a study. Int. J. Secur. Priv. Trust Manag. (IJSPTM) **4**(1) (2015)

Fig. 10 Parabola reflection
principle

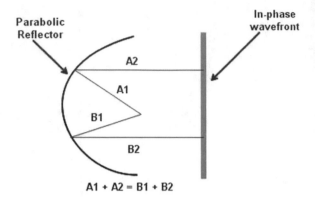

A1 + A2 = B1 + B2

2. Dickerson, J., Dickerson, J.: Fuzzy network profiling for intrusion detection. In: Annual Conference of the North American Fuzzy Information Processing Society—NAFIPS, pp. 301–306 (2000). https://doi.org/10.1109/nafips.2000.877441
3. Aydın, M.A., Zaim, A.H., Ceylan, K.G.: A hybrid intrusion detection system design for computer network security. Comput. Electr. Eng. **35**, 517–526 (2009). ISSN:0045-7906 https://doi.org/10.1016/j.compeleceng.2008.12.005
4. García-Teodoro, P., Díaz-Verdejo, J., Maciá-Fernández, G., Vázquez, E.: Anomaly-based network intrusion detection: techniques, systems and challenges. Comput. Secur. **28**(1–2), 18–28 (2009). ISSN 0167-4048, https://doi.org/10.1016/j.cose.2008.08.003. (http://www.sciencedirect.com/science/article/pii/S0167404808000692)
5. Goodall, J.R., Lutters, W.G., Komlodi, A.: Developing expertise for network intrusion detection. Inf. Technol. People **22**(2), 92–108 (2009). https://doi.org/10.1108/09593840910962186
6. Erickson, J.: Planck's Equation. http://www.csun.edu/~jte35633/worksheets/Chemistry/52PlancksEq.pdf
7. Tavallaee, M., Bagheri, E., Lu, W., Ghorbani, A.A.: A detailed analysis of the KDD CUP 99 data set. In: Proceedings of the 2009 IEEE Symposium on Computational Intelligence in Security and Defence Applications (CISDA 2009) pp. 53–58. ISBN: 978-1-4244-37634. http://dl.acm.org/citation.cfm?id=1736481.1736489
8. Poole, I.: Radio-electronics.com, resources and analysis for electronics engineers, parabolic reflector antenna theory. IEEE 802.11 Wikipedia, Channel and frequencies Electromagnetic spectrum Wikipedia
9. Salunkhe, U.R., Mali, S.N.: Security enrichment in intrusion detection system using classifier ensemble. J. Electr. Comput. Eng. 2017, Article ID 1794849, 6 pages (2017). https://doi.org/10.1155/2017/1794849
10. Scullion, F.: Atomic theory, finding the energy of a photon. www.scribd.com/document/169352643/Energy-of-Foton
11. Astronomer's toolbox.: The Electromagnetic Spectrum—National Aeronautics and Space Administration. https://imagine.gsfc.nasa.gov/science/toolbox/toolbox.html
12. Salunkhe, U.R., Mali, S.N.: Security enrichment in intrusion detection system using classifier ensemble. J. Electr. Comput. Eng. 2017, Article ID 1794849 (2017)
13. Parabola Reflector Antenna Theory.: http://www.radioelectronics.com/info/antennas/parabolic/parabolic-reflector-antenna-theory.php
14. Patcha, A., Park, J.-M.: An overview of anomaly detection techniques: Existing solutions and latest technological trends. Comput. Netw. **51**, 3448–3470 (2007). ISSN:1389-1286, http://dx.doi.org/10.1016/j.comnet.2007.02.001, https://doi.org/10.1016/j.comnet.2007.02.001
15. Stapel, E.: Conics: Parabolas: Introduction http://www.purplemath.com/modules/parabola.htm

16. Telagarapua, P., Lakshmi Prasanthib, A., Vijaya Santhic, G., Ravi Kirand, B.: Design and analysis of parabolic reflector with high gain pencil beam and low side lobes by varying feed. Int. J. Adv. Netw. Appl. **03**(02), 1105–1115 (2011)

Lifetime and Transport Delay Optimization in Presence of Delay in WSN

Bhushan Jichkar, Deepak Mehetre and S. Emalda Roslin

Abstract The wireless sensor network is fast-emerging field nowadays. The small sensor nodes in WSN compacts with functionalities sensing, processing, and communication. This technology has many applications such as home automation, health monitoring, traffic, security monitoring. The primary function of sensor nodes is to detect and capture the information and send to sink node. This transmission of data consumes minimum amount of energy to maximize lifetime of WSN. For reducing delays, we discuss the different approaches to maximize the life of WSN. This paper implements the BCMN/A protocol to achieves energy and delay efficiency both in intracluster and intercluster. In the proposed system, we make system more secure by sending data in the encrypted format using encryption algorithm and detect the delayed attack. This method improves the energy and delay efficiency of the network and enhances the network lifetime.

1 Introduction

A WSN is a collection of sensors connected by wireless channels where each sensor node is a small device that can collect data from its designated surrounding area. With this data, computations are carried out and communicate with other sensor nodes or controlling authorities in the network. In ad hoc network, various routing protocols have been proposed and due to scalability issues, and these protocols are not precisely suitable for WSN. WSN includes large range of low-power sensor nodes. WSN is operated in harsh environment. When WSN compared with the traditional sensor networks, WSN provides a flexible proposition. The nature of the WSN permits

B. Jichkar (✉)
Departement of Computer Engineering, K. J. College of Engineering
and Management Research, Pune 411048, Maharashtra, India
e-mail: bhushanjichkar@gmail.com

D. Mehetre · S. Emalda Roslin
Department of Computer Engineering, Sathyabama University,
Kamaraj Nagar, Kanchipuram, Chennai 600119, Tamil Nadu, India
e-mail: dcmehetre@gmail.com

© Springer Nature Singapore Pte Ltd. 2019   773
A. Abraham et al. (eds.), *Emerging Technologies in Data Mining and Information Security*, Advances in Intelligent Systems and Computing 814,
https://doi.org/10.1007/978-981-13-1501-5_68

many sensor nodes to be randomly placed in inaccessible terrains. Also, the nodes can perform functions like data processing and routing.

Routing is challenging as it is difficult to distinguish networks from other wireless networks. Because of lot of number of sensor nodes, it is impossible to build a global addressing scheme which results in traditional IP-based protocols may not be applied to WSNs.

Data yield is addressed by applying a data delivery protocol based on ARQ. With ARQ, a receiver sends acknowledgment packets (ACKs) for successfully received packet and transmitter retransmits the packet upon timeout. They are variants of ARQ protocols, as well as other data delivery protocols. Protocols differ by their data delivery reliability properties as well as their energy efficiency. Aggregation of all received data performed by cluster head and then forwards it to the sink in a multi-hop manner. Due to the unreliability of WSN, it shows an large energy consumption and it gives poor network performance. At each hop, more network energy is consumed. Thus, it is necessary to design a new protocol to guarantee QoS in WSNs such as lifetime, end-to-end reliability, and delay. Send-and-wait automatic repeater quest (ARQ) protocol (SW-ARQ) is commonly used to ensure the reliability by employing multiple retransmissions. Because of the complexity of cluster-based networks, there is a need of little research efforts in achieving all the networks reliability, transport delay, and lifetime optimization.

2   Literature Survey

This paper [1] obtains the energy consumption and delay under SW-ARQ protocol with theoretical analysis to obtain the optimal cluster radius r, which provides a theoretical guidance. In this paper, the author proposed the advanced BCMN/A protocol, BCMN/A protocol broadcasts intracluster and returns multi-ACK.

In this paper [2], wireless sensor network used delay- sensitive transport protocol. This protocol provides timely and reliably transport event.

This work investigates how LEDBAT [4], a delay-based less-than-best-effort protocol, would behave in a wireless sensor network. LEDBAT is used to produce TCP. At the time of sharing a bottleneck link, simulations showed in certain conditions, A fair share of wireless bandwidth is obtained by LEDBAT.

Author presents an energy-saving routing scheme [5] for mobile WSN. Our proposed routing algorithm adapts dynamically to network node's energy level.

In this paper [7], author proposed delay aware network structure for WSNs with in network data fusion. The proposed network structure organized wireless sensor nodes into multiple single-layer clusters.

In this paper, Shi et al. [8] propose a solution to the scheduling problem in clustered wireless sensor networks (WSNs). It provides network-wide optimized time-division multiple access (TDMA) schedules and that achieve high power efficiency, zero conflict, and reduced end-to-end delay.

In this paper [9], author proposes a distributed scheduling method with an upper bound for delay time slots and study the problem of distributed aggregation scheduling in WSN .

The scope of the work in [10] is to review the effects of replication attack in WSN exploitation. Comparative analysis of replication attack for each protocol is taken under consideration.

In [11], author presents the secure and QoS-aware routing protocol (SQRP) that takes into consideration the trust and QoS parameters pertaining to link's quality to select an optimized end-to-end route.

In [12], author simulates and analyzes black hole and DoS by Hello Flooding attack in wireless sensor networks. A generic WSN model has been created in Qual-Net and necessary changes have been done in the code library to simulate the attacks.

3 Implementation Details

3.1 Problem Definition

Develop a secure and efficient data routing scheme in presence of delay attack, to route the sensing information from all sensor nodes to base station. Also, maintain the security of data with cryptographic technique.

3.2 Proposed System Overview

The architecture view of proposed system is depicted in Fig. 1. It presents the overall workflow of the system. Initially, a network is built with node deployment. To make efficient routing, clustering approach is used. In this, all number of nodes are divided into clusters. Cluster members to base station through respective cluster heads. Also, security and integrity of the data are maintained through cryptographic and hashing algorithm, respectively. The details description of each functionality is described in Sect. 3.3. In contribution, the system detects the delayed attack and makes the system more secure.

3.3 System Functionality

1. **Clustering**: To improve the data routing efficiency, clustering approach is used. It reduces the data routing time. In our system, once the nodes are deployed in network, and they are divided into four clusters. The cluster formation is based on geographical location. For all clusters, one cluster head (CH) is allocated, which

Fig. 1 System architecture

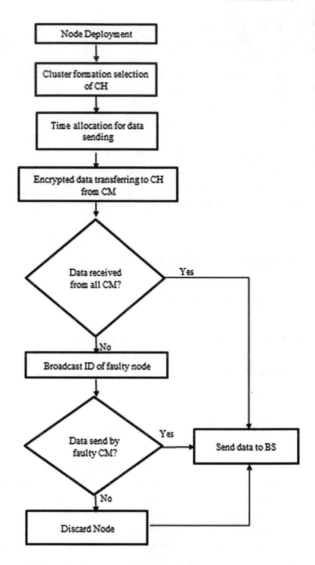

collects the data from its cluster members and forward to the base station in hop-by-hop manner. For cluster head selection, BS computes the energy and distance to base station for each node. The node with maximum energy and minimum distance to base station is selected as the CH for that particular cluster

CH selection criteria = E/D,

Where,

E = Energy of Node

D = Distance to Base Station.

2. **Data Security**: To secure the data during routing, cryptographic technique is used. For this, base station makes use of ECC algorithm. In this, BS generated key pair (private key, public key) for each cluster member and distributes to all CMs. After receiving this key pair, CMs encrypt their data with the help of public key and forwarding to the respective CHs. CH cannot decrypt this data, because it does not have the private key. Only BS can decrypt this data by using the private key.

3. **Data Integrity Checking**: After receiving the data from CMs, each CH performs data integrity checking. At the time of data sending, CM encrypts its data and computes the hash of this encrypted data by using SHA1 Algorithm. CM sends this hash along with encrypted data to CH. At CH side, CH receives the encrypted data and hash and computes the new hash of receiving data. If received hash is equal to the new computed hash, CH assumes that the data is safe and it forwards to the base station.
If $Hash_{ReceivedfromCM} == Hash_{ComputedatCH}$
Data integrity Passed
Otherwise fails.

4. **Delay Attack Detection and Prevention**: For data sending, CH broadcasts the time to all CM. During this time, CMs should transmit their data to respective CHs. If CH found that the data is not received from some CM, then CH broadcast the ID of that specific CM. After receiving the broadcast message, the CM should resend their data. Even if the data is not received, CH found that the CM is attacker node and it discards this node and its data. Otherwise, the data is forwarding from CH to BS in hop-by-hop manner.

5. **Data sending from CH to CM**: To reduce the energy consumption, data is sending from CH to BS hop by hop. In this, one CH sends their data to its nearest CH to reach to the BS. For this, BS computes the path of data routing in descending order. BS computes the distance of each CH, and the CH with minimum distance is the last node in path.

For example, suppose we have four cluster heads, CH1, CH2, CH3, and CH4 and their distance to base station is 500, 400, 600, and 300 m, respectively. Then, BS identifies the path as CH3-CH1-CH2-CH4. That is, CH3 collects its data and forwarding to CH1, CH1 collects its data and data from CH3 and combines forward to CH2. In this way, data is reached to the base station.

3.4 Algorithm

3.4.1 Algorithm 1: Proposed System Algorithm

Input: Sensed Data
Output: Secure Data Collection at Base Station
Process:

1. Create a network graph as Graph g(v,e) where V are vertices/nodes and E are edges.
2. Cluster formation within a network.
3. Cluster head (CH) selection on the basis of energy, distance and number of neighborhood nodes attached to parameter within an each cluster.
4. Time allocation while data sending.
5. Encrypt data using ECC encryption algorithm.
6. CMs send encrypted data to the CH.
7. CH check is data received from all the CM?
8. If data is received from any CM, then CH forwards the data to the base station (BS).
9. If data is not received from any CM, then CH broadcasts the CM id within the cluster.
10. Still if the data is not received the CM, it is declared as a delay attacker and discarded that node from the network.

3.4.2 Algorithm 2: CH Order for Data Sending

Input: Encrypted Data from CMs
Output: Data Sending to BS
Process:

1. Data collection at CH
2. Compute the distance from each CH to BS
3. Call Euclidean distance formula
4. Compute x, y coordinates of each CH, say (x1, Y1)
5. Compute x, y coordinates of each BS, say (x2, y2)
6. Compute the distance D(BS, CH),
$$D(BS, CH) = \sqrt{(X2 - X1)^2 + (Y2 - Y1)^2}$$
7. Arrange all CH in path in descending order
   D(CH1) $\gg$ D(CH2) $\gg$ D(CH3) $\gg$ D(CH4) to BS.
8. Collect and send the data through selected path to BS.

3.5 Stepwise System Description

1. **Home Screen**: At home screen (Fig. 2), we can enter number of nodes. These nodes are further processed either with existing and proposed system. Here, we enter the 50 nodes and go with the option of proposed system.
2. **Network Deployment**: The network is deployed with 40 nodes as shown in below screen. The nodes are scattered in system and each node has ID for their unique identification. The nodes are represented by red-colored circles (Fig. 3).

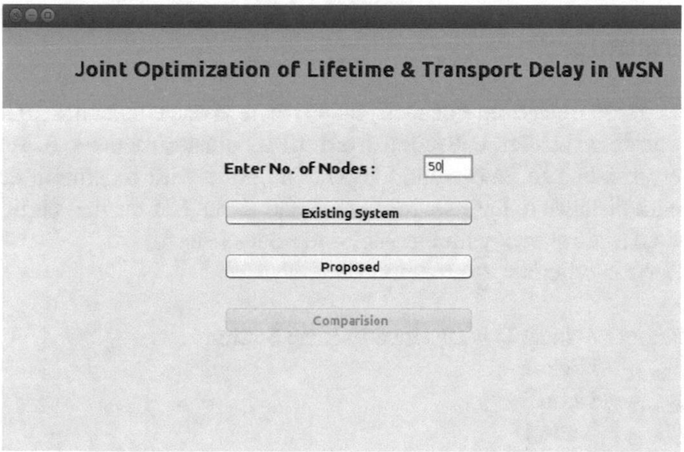

Fig. 2 Enter number of nodes and select system

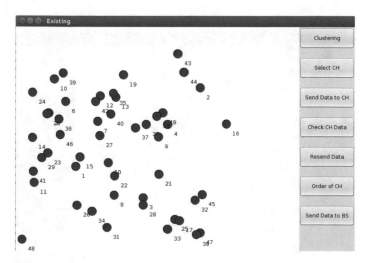

Fig. 3 Network deployment

3. **Cluster Formation**: Nodes are divided into four clusters based on their location. All clusters contain following nodes. The nodes in one cluster are represented by the unique colored circle. Following is the set of nodes available in each cluster (Fig. 4).
   Cluster 1 = Magenta-Colored Nodes = 1, 12, 47, 21, 2, 26, 27, 37, 7, 22, 48, 45, 25
   Cluster 2 = Yellow-Colored Nodes = 39, 10, 41, 34, 8, 27, 50, 24, 6, 35, 44, 36, 42, 29
   Cluster 3 = Cyan-Colored Nodes = 14, 33, 9, 20, 46, 19, 23, 43, 15, 37, 4
   Cluster 4 = Pink-Colored Nodes = 5, 23, 11, 30, 49, 28, 13, 18, 40, 16, 31, 38
   There is one node is separately assigned as the base station node, represented by green-colored circle.

4. **Cluster Head Selection**: For each cluster, there is one cluster head (CH), which is responsible for data collection from all its cluster member (CM) and collectively sending to base station (BS). The node with maximum energy and minimum distance to base station is selected as the CH for that particular cluster. All CHs are shown with blue-colored nodes (Fig. 5.)
   CH selection criteria = $\frac{E}{D}$
   where,
   E = Energy of Node, D = Distance to Base Station
   $CH_{Cluster1}$ = Node 1
   $CH_{Cluster2}$ = Node 39
   $CH_{Cluster3}$ = Node 19
   $CH_{Cluster4}$ = Node 11

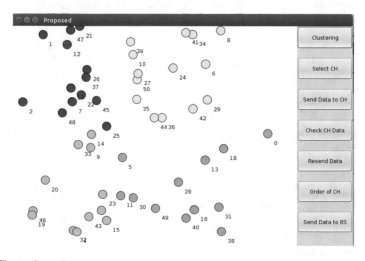

Fig. 4 Cluster formation

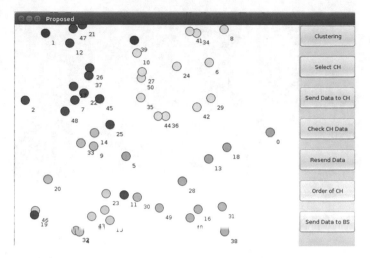

Fig. 5 CH selection

5. **Data sending form CM to CH**: All CMs collected the information. In data sending stage, all cluster members send their data to their respective cluster members. For security purpose, data is encrypted before sending. And during data sending, its hash is also sending to CH. So that each CH performs data integrity checking for all its CMs. If the hash is not matched, CH discards that data, otherwise forwarding to BS (Fig. 6).

6. **Process of delay attack detection and prevention**: After data receiving, each CH checks that from which node data is not received. Following screen shows that nodes 37, 44, 20, and 5 does not send their data to CH of cluster 1, 2, 3, and 4, respectively (Fig. 7).

   After this, all CHs requesting to each CMs, those not sending data, to resend their data. CH sets some time limit. During this time limit, CMs should resend their data. This will reduce the delay time. Here, all CHs waiting for 5 s to receive the data (Fig. 8).

   After this waiting, again CH checks that from which node data is not received. Following screen shows that node 37 in cluster 1 does not resend their data to CH within 5 seconds. So that the CH1 discards that node for further computation. But other nodes resend their data within specified time. The discarded nodes are represented by black-colored nodes, which are reflected in next screen (Fig. 9).

7. **Data sending from CH to BS**: To reduce the energy consumption, data is sending from CH to BS in hop-by-hop manner. In this, one CH sends their data to its nearest CH to reach to the BS. For this, BS computes the path of data routing in

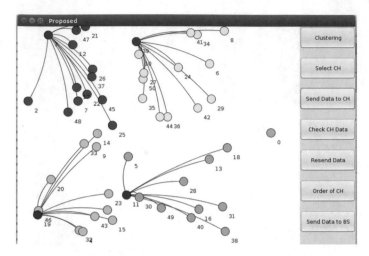

Fig. 6 Data sending from CM to CH

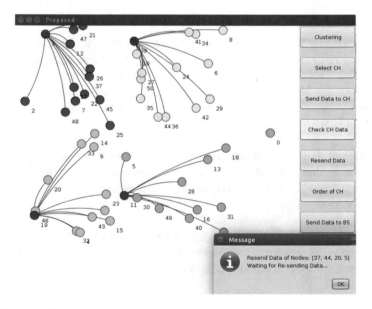

Fig. 7 Data resending request

descending order. BS computes the distance of each CH, and the CH with minimum distance is the last node in path. In the following screen, path for sending data from CHs to BS is computed as: [19, 1, 39, 11].

That is, CH3 ≫ CH1 ≫ CH2 ≫ CH4. This path is based on the distance between CH to BS (Figs. 10 and 11).

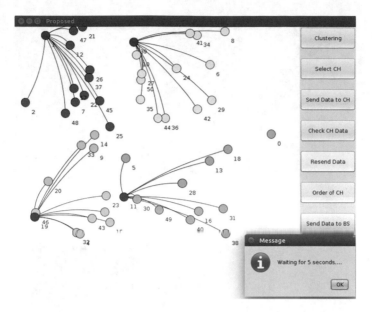

Fig. 8 CHs waiting for 5 s

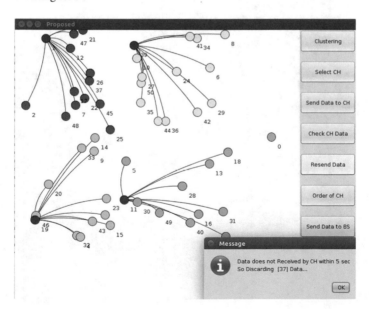

Fig. 9 CM discarding

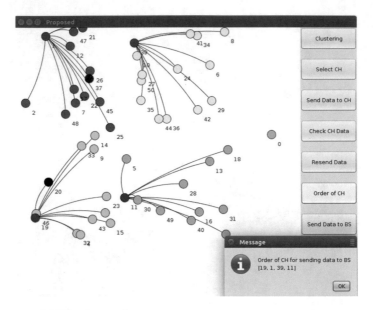

Fig. 10 Order of CH for data sending from BS

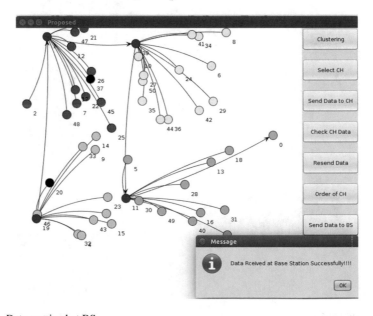

Fig. 11 Data received at BS

4 Result and Discussion

4.1 Experimental Setup

The proposed system are built using Java Framework version JDK 6 and Windows Platform are Used. Netbeans version 6.9 is used as a development tool. For network creation, system used the Jung tool. The system does not require any hardware for running it; any standard machine can run the application.

4.2 Evaluation Result

The existing system does not detect the delay attack. In the proposed system, we detect delay attack as well as prevent it by discarding that node. This makes system more secure and consumes less energy, transport delay and enhances the network lifetime. The performance of the system is measured in terms of following:
- Delay
- Energy consumption
- Throughput
- Network lifetime

1. Delay :
   Table 1 states the comparison of existing and proposed system based on time efficiency. Time efficiency of the system is based on the detection of delay attack on the node. The proposed system is more time-efficient than existing one because it manages the delay attack very well.
   The performance of system is tested with the different sized network. The system is tested with five different networks consisting of 40, 50, 60, 70, and 80 nodes, respectively.
   Following Fig. 12 depicts the time efficiency comparison graph of the proposed system with the existing system. The time required for total data routing in the existing system is more than the time required for the proposed system.
   Table 2 shows comparison between proposed and existing system in the basis of attack, security, time parameters, and delay.
2. Energy consumption:
   Table 3 depicts the comparison of energy consumption in existing and proposed system. The system is evaluated five times with different size of network. The size of network is node with 40, 50, 60, 70, and 80 nodes. From this table, it is clear that, as the network size increases, energy consumption also increases.
   Fig. 13 represents the graphical comparison of energy consumption in existing and proposed system. The proposed system consumed less energy that existing one, because in proposed system, path of data sending from CH to BS is already selected. Also, the CH discards the data of attacker node, so that the energy is saved. The energy is measured in terms of joules.

Table 1 Time efficiency comparison

Number of nodes	Existing system	Proposed system
40	160.993	104
50	130.686	65.996
60	149.748	104.438
70	118.730	100.23
80	155.976	76.105

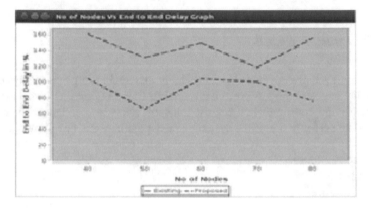

Fig. 12 Time efficiency comparison graph

Table 2 Comparative analysis

Parameters	Existing system	Proposed system
Delay attack	Not detected and prevented	Detected and prevented
Transport delay	More	Less
Security for data	Data not secured	Data secured with ECC
Time required	More	Less

Table 3 Energy comparision

Number of nodes	Existing system	Proposed system
40	520	333
50	507	450
60	650	545
70	745	699
80	822	750

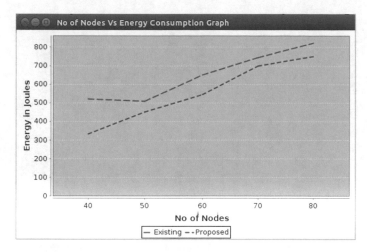

Fig. 13 Energy comparison

Table 4 Throughput comparison

Number of nodes	Existing system	Proposed system
40	11.07	39.25
50	13.33	40.00
60	14.22	33.99
70	13.77	23.74
80	15.55	33.27

4. Throughput:

Table 4 represents the throughput comparison of proposed and existing system. Throughput is the number of packets delivered in unit time. The throughput increases as the size of network size increases.

Figure 14 represents the graphical comparison of proposed system and existing system in terms of throughput. As the proposed system, it allows CMs to resend their data, and throughput increases. The throughput is measured in terms of percentage.

4. Network lifetime:

Table 5 represents the network lifetime comparison of proposed and existing system. As the energy consumption is reduced in proposed system, lifetime of network is also increased.

Figure 15 represents the graphical comparison of proposed system and existing system in terms of network lifetime. For network lifetime measurement, we calculate the residual energy in joules.

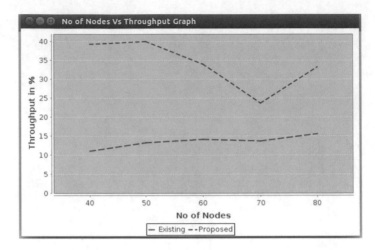

Fig. 14 Throughput comparison

Table 5 Network lifetime comparison

Number of nodes	Existing system	Proposed system
40	32,887	45,333
50	35,693	44,363
60	46,888	56,635
70	66,332	62,747
80	63,931	78,010

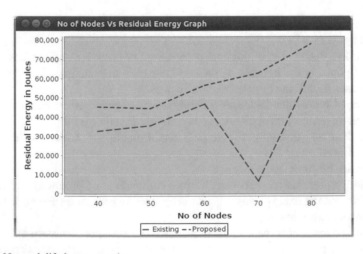

Fig. 15 Network lifetime comparison

5　Conclusion

In this paper, an efficient data routing scheme is proposed which optimized the delay and improved the network lifetime in the presence of delay attack. This system can efficiently detects and prevents the delay attack is network. Delay attack is identified at cluster heads. Also, system performs the data security and data integrity checking throughout the data routing procedure. The ECC and SHA1 algorithms are implemented to make the data secure and to chcck the integrity of data, respectively. The experimental results prove that the proposed system is better than the existing one in terms of delay attack prevention, minimum transport delay, more secure and time efficiency.

References

1. Ota, K., Dong, M.: Joint optimization of lifetime and transport delay under reliability constraint wireless sensor networks. IEEE (2015)
2. Gungor, V.C., Akan, O.B.: DST: delay sensitive transport in wireless sensor networks. In : International Symposium on Computer Networks, pp. 116–122. Istanbul (2006)
3. Patel, B.D., Patel, A.D.: A trust based solution for detection of network layer attacks in sensor networks. In: 2016 International Conference on Micro-Electronics and Telecommunication Engineering (ICMETE), pp. 121–126. GHAZIABAD, India (2016)
4. Montes, I., Tiglao, N., Ocampo, R., Festin, C.: Delay-based end-to-end congestion control for wireless sensor networks. In: 2015 Seventh International Conference on Ubiquitous and Future Networks, pp. 497–502. Sapporo (2015)
5. Alwakeel, S., Prasetijo, A., Alnabhan, N.: An adaptive energy-saving routing algorithm for mobile wireless sensor networks. In: 2015 International Conference on Electrical and Information Technologies (ICEIT), pp. 104–108. Marrakech (2015)
6. Albath, J., Thakur, M., Madria, S.: Energy constraint clustering algorithms for wireless sensor networks. Ad Hoc Netw. **11**(8), 2512–2525 (2013)
7. Cheng, C., Leung, H.: A delay aware network structure for wireless sensor networks with in network data fusion. IEEE Sens. J. **13**(5), 1622–1631 (2013)
8. Shi, L.Q., Abraham, O.F.: TDMA scheduling with optimized energy efficiency and minimum delay in clustered wireless sensor networks. IEEE Trans. Mobile Comput. **9**(7), 927–940 (2010)
9. Xu, X.H., Li, M., Mao, X.F., et al.: A delay efficient algorithm for data aggregation in multihop wireless sensor networks. IEEE Trans. Parallel Distrib. Syst. **22**(1), 163–175 (2011)
10. Sachan, R.S., Wazid, M., Singh, D.P., Katal, A., Goudar, R.H.: Misdirection Attack in WSN: Topological Analysis and an Algorithm for Delay and Throughput Prediction (ISCO), pp. 427–432. India (2013)
11. Shimpi, B., Shrivastava, S.: A modified algorithm and protocol for replication attack and prevention for wireless sensor networks. In: 2016 International Conference on ICT in Business Industry & Government (ICTBIG), pp. 1–5. Indore (2016)
12. Ahmed, A., Kumar, P., Bhangwar, A.R., Channa, M.I.: A secure and QoS aware routing protocol for Wireless Sensor Network. In: 11th International Conference for Internet Technology and Secured Transactions (ICITST), pp. 313–317. Barcelona (2016)
13. Moon, A.H., Shah, N.A., Khan, U.I., Ayub, A.: Simulating and analysing security attacks in wireless sensor networks using QualNet. In: 2013 International Conference on Machine Intelligence and Research Advancement, pp. 68–76. Katra (2013)
14. Umrao, S., Verma, D., Tripathi, A.K.: Detection and mitigation of node replication with pulse delay attacks in wireless sensor network. In: IEEE International Conference in MOOC, Innovation and Technology in Education (MITE), pp 390–392. Jaipur (2013)

Types of Sensor and Their Applications, Advantages, and Disadvantages

Manish Pandey and Gaurav Mishra

Abstract This chapter investigates the various types of sensor on the basis of what they measure as well as on their area of application. Brief descriptions of the sensors are given along with their advantages and disadvantages. A sensor is an electronic device which measures changes in a quantity, for example, voltage, temperature, pressure, and humidity. Classification occurs, therefore, on the basis of the property that is measured by a sensor. A temperature sensor measures changes in temperature, namely hot or cold weather, or minute changes in its surroundings. This chapter further investigates the limitations and disadvantages of each type of sensor and discusses why a particular sensor is not deployable in some applications or locations.

Keywords Sensors · Types · Features · Comparison

1 Introduction

A sensor is an electronic device designed to monitor and measure changes in the environment, convert those changes to electrical signals and communicate that information to other electronic devices for further computing and analysis [1].

M. Pandey (✉) · G. Mishra
Department of Computer Science and Engineering, MANIT, Bhopal 462003,
Madhya Pradesh, India
e-mail: contactmanishpandey@yahoo.co.in

G. Mishra
e-mail: gauravmalak@gmail.com

© Springer Nature Singapore Pte Ltd. 2019

791

A. Abraham et al. (eds.), *Emerging Technologies in Data Mining and Information Security*, Advances in Intelligent Systems and Computing 814,
https://doi.org/10.1007/978-981-13-1501-5_69

1.1 Classification of Sensors Is Based on the Following Criteria

Primary input quantity-The quantity that is measured, e.g., voltage.
Transduction principles-Making use of physical as well as chemical effects.
Material and Technology-Capacitors, transistor etc.
Property-Temperature, pressure, distance, and motion etc.
Application-Industrial or non-industrial.

2 Property-Based Classification

2.1 Optical Sensors

Optical sensors are categorized on the basis of the nature of their transduction, manufacturing technology of transduction quantity measure etc. [2]. According to their nature of transduction, optical sensors are divided into two categories:

- **Intrinsic** Optical sensors in which the optical waveguide is used to carry out the transduction by revising, in some way, their essential characteristics of transmission subject to the magnitude being measured.
- **Extrinsic** These are optical sensors in which an interaction between light its measurement is produced in an external optical device.

2.1.1 Applications

Optical sensors can be employed in many fields, e.g., in biomedicine the analysis of breath is performed using a tunable diode laser and for optical heart rate observation using an optical sensor based liquid level indicator in computers and mobiles. Optical sensors are routinely employed in the automotive industry to measure different quantities, and are also used in the robotics industry. They can be used in places and components in which nuclear radiation would cause a danger to our health etc.

Advantages are their capability of operating over large distances, their low cost, and small size etc. Disadvantages are their short-lived stability.

2.2 Temperature Sensors

The various types of temperature sensors are:

- **Thermocouples** Made by connecting the ends of two dissimilar metals. In the case of a temperature difference across a junction, a minor voltage is generated called an electromotive force (EMF) and this is used to indicate temperature [3]. Advantages are their wide temperature range, low cost, wide availability, and robustness.

- **RTDs** (Resistance Temperature Detectors) This type of sensor applies a constant current, computes the resulting voltage, and the RTD resistance. Advantages are the narrow range of temperatures, high degree of accuracy, and high resolution.
- **Thermistors** Just like RTDs, thermistors show a highly nonlinear resistance versus temperature curvature. Advantages are the narrow temperature range, peak, and resolution.
- **Semiconductor sensor** These are categorized into various types, e.g., voltage, current, and digital output. Advantages are their high linearity and high accuracy.

2.2.1 Applications

Regulated industries, for example, food processing, chemical industry, calibration labs, data centers, commercial applications in the home in devices such as refrigerators, air conditioners, and ovens [3]. They are also used in the mobile industry in phones and tablets, the medical industry in medicine and blood storage room and labs, the mechanical industry in machine parts and heating control [1], in the robotics industry to regulate working environments, and in stores, e.g., Wal-Mart.

Advantages of temperature sensors are their accuracy, greater reliability, flexibility, and sensibility. Disadvantages are resistance error, vibration, and a high response time.

2.3 Position Sensors

Position sensors are used to measure the distance traveled by a body from its start position. Examples are BWD sensors, and motorcraft ignition pickups.

They are classified on the basis of the various sensing methods they employ, as follows:

- **Potentiometric Position Sensors** These measure a voltage drop when a resistive path is slide by an electrical contact, in other words the position is in direct proportion to the output voltage. Advantages are very low cost, low weight, and compact build. Disadvantages are vibration, attire, high temperature etc.
- **Optical** Light is sensed by a photodetector when a beam is shone over a grating followed by the generation of a position signal. Advantages are a good degree of accuracy when mounted correctly, good resolution, and widespread availability. Disadvantages as fatal failure without any cautioning shock, extreme temperature, alien matter etc.
- **Magnetic** When a magnet enters the effect field of a magnetic detector, the magnetic field changes in proportion to the relative displacement. Advantages are their robust nature and the fact that most liquids have a negligible or no effect

on them. Disadvantages are hysteresis effects, temperature, poor impact/shock performance, and the requirement for precision mechanical engineering.

- **Magneto restrictive** An unusual phenomenon in ferromagnetic substances is known as 'magnetostriction'. This is a quality which causes them to vary their shape and/or dimensions during the procedure of magnetization. Advantages are that the percentage degree of accuracy rapidly increases with length, robustness, and capability of operating in high pressures. Disadvantages are susceptibility to temperature effects and shock, and a rather high cost etc.
- **Capacitive** Usually two conductive plates separating an insulator. The circuits' capacitance along its axis changes when another conductive plate moves across the other two, representing the relative situation of the two parts. Advantages are their compact build, narrow installation tolerances, and low power usage. Disadvantages are sensitive, humidity coefficients.

2.3.1 Applications

Position sensors have many applications in the gaming industry, e.g., in game controllers and Blu-ray disc stamper thickness measurement, and in the mobile industry as part of accelerometers in phones, tablets, and robotics [1]. In the automobile industry they are used in precision motor and shaft run out measurement, spindle metrology and spindle run out measurement, and can be further used in pedometers, tracking devices, in the military and in hospitals, for semiconductor wafer surface sensing, for security purposes like intruder alarm systems, and in tracking, location identification, object movement, and direction guidance etc.

Advantages include their security, portability, low cost, and reliability. Disadvantages are that they can be affected by environmental changes, are less accurate in the dark, and have a limited coverage area etc.

2.4 Image Sensors

Image sensors are used to capture images in 2D to be further processed and used in numerous applications. All image sensors are constructed with an array of photodetectors called pixels, which gather photons (single light particles). The photons in pixels are then transformed into electrons [4]. The collected electrons are moved to a signal conditioning and processing circuit. Types of image sensors are described below:

- **Charged Coupled Device (CCD)** CCDs are produced using a technique which enables transportation of charge across a chip with no distortion, resulting in high-quality sensors with reliability and photosensitivity.
- **Complementary Metal Oxide Semiconductor (CMOS)** This is an integrated circuit technology that has been adapted to capture images. Chip output is in the form

of digital bits. Advantages are a fast image processing speed and a low sensitivity meaning that every pixel of a CMOS sensor can be distinguished separately.

2.4.1 Applications

Image sensors have many application areas, such as in cameras, phones, tablets, military uses such as surveillance, in the capturing of location data, uses in the robotic industry, such as biometrics, face detection systems, and traffic monitoring. They are also employed in vehicle parking and automatic car systems, ambient noise seismic imaging in wireless networks, real-time imaging, health care systems, security purposes, surveillance, measuring light intensity, investigating structures under the surface of the earth, for more reliable use of image sensor combine with radar sensor like real-time imaging system for millimeter wave synthetic aperture radar sensors, ambient light sensors are used in the prediction of changes in functional health, and in commercial uses, such as capturing images and video.

Advantages are security, reduced energy consumption, accuracy, sensitivity, and for digital lock. Disadvantages are susceptibility to distortion, heating, and their slow speed.

2.5 Humidity Sensors

Humidity is defined as the amount of moisture present in the surrounding air. Humidity sensors measure air temperature as well as moisture [5].

Types of humidity sensors:

- **Capacitive** Sensors measure relative humidity by placing a thin strip of metal oxide between two electrodes.
- **Resistive** Sensors measure the electrical impedance of atoms.
- **Thermal** Thermal sensors use electric phenomena to measure humidity in their surroundings.

2.5.1 Applications

Humidity sensors have a broad range of applications, for example, in meteorology centers to predict and report weather, in commercial applications, such as air conditioners and ventilating system [1], in the textile industry, in greenhouses, in an inkjet printer chip-less RFID (radio frequency identification) sensor, in the automobile industry, in the robotic industry to maintain temperatures and in robots that are used for testing and trials. They are also employed in environment control, agricultural processing, storage, industrial production, and clinical trials.

Advantages are their accuracy, low cost, and high performance etc. Disadvantages are that they can be affected by environmental changes and contamination, e.g., a gas that they are measuring can affect the performance of the sensor [5].

2.6 Radar Sensors

A radar sensor is an obstacle or object recognition system which uses radio waves to examine the angle, range or velocity of objects. Radio waves move outward from an antenna until they hit an object and travel back to the antenna. Radar sensors can be combined with other sensors to provide improved security, reliability, and accuracy in the detection and sensing of objects [6].

2.6.1 Applications

Some applications of radar sensors are in home automation (automatic on/off of a device over a network or through the use of a remote control device), in threat detection systems, intruder detection applications, driverless cars, traffic monitoring systems that controlling traffic signals, in collision avoidance, distance measurement, Level, Measurement in Refinery, in the automatic detection of humans outdoors, and in the sports industry. Further uses are in the agricultural industry for monitoring weather (remote sensing using radar), in the space program, for air and road traffic control, in the provision of directions, and in human gesture and motion recognition [1].

Advantages are that the radar signal can penetrate through objects, hence can be used in any environmental conditions, can distinguish between moving and still objects, count passing people, and is accurate and reliable. Disadvantages are that it cannot distinguish between very close objects, does not recognize the color of objects or objects that are placed behind a conductive sheet.

2.7 Laser Sensors

These sensors are used where very tiny objects or precise positions are the subject of detection [7].

Types of laser sensor:

- **Light intensity based laser sensor** A light beam is emitted from the sender and travels to the receiver. It is used to examine very small objects.
- **Position-based laser sensor** This type of sensor detects the position of a target. This is achieved by using a triangulation system (the change in the distance to the target affects the position of light focused on the CMOS detecting element) or a

time-measuring structure (distance is measured depending on the time, or in other words, the total time taken by the emitted laser beam to strike the target and return back to the sensor.

2.7.1 Applications

Areas of applications are the robotics industry for determining local coordinate frames attached to objects in industrial robots, workshops [7], in navigation software systems to avoid obstacles in dynamic environments, in aerial and unmanned aerial vehicles (UAVs), in sports to measure the depth of water on a track, displacement measurement, drill pipe run out, electronic manufacturing, thickness measurement, train and railroad inspection, weld gap detection, crash test sensors, manufacturing, data storage, commercial/military aviation, the automotive industry, and hospitals etc.

Advantages are their sensitivity, high resolution, reliability, and wide measurement range. Disadvantages are that they must be clean and free from dirt and other foreign materials otherwise accuracy will be affected. There is also a narrow range of operating temperatures.

2.8 Infrared Sensors

Infrared transmission requires a transmission medium comprised of either a vacuum, atmosphere, or an optical fiber. Band-pass filters can be applied to limit spectral response. Infrared detectors are used to detect focused radiation [4].

Types of Infrared Sensors:

- **Thermal Infrared sensor** These use infrared energy as heat to detect objects which radiate heat. Disadvantages are the slow response time and low detection capabilities.
- **Quantum Infrared sensor** These having the advantage of a higher detection performance and a faster response time.

2.8.1 Applications

Infrared sensors are used in astronomy in studies of astronomical objects which are visible in infrared radiation, for the tracking of vehicles, people [1], and aircraft as infrared is radiated strongly by hot bodies, in rail safety to detect cracks in rail tracks, by the military to measure chemical composition, in real-time combustion control and pollution monitoring, in industrial in process control, gas leak and fire detection, for flame monitoring, moisture analysis, gas analysis, obstacle detection systems, and in night vision devices.

Advantages are that infrared passes through both living and non-living objects and is effective in detecting defects. Disadvantages are that it cannot detect multiple objects with a small temperature difference, and the sensors are expensive [4].

2.9 Ultrasonic Sensors

These sensors emit high-frequency sound waves/pulses and measure how long it takes for that frequency to echo or bounce back. One opening of the sensor sends ultrasonic waves while the other opening receives them [8].

It uses the following equation:

$$\textbf{Distance} = \textbf{Time} \times \textbf{Speed of Sound}/2$$

Types of Ultrasonic Sensor:

- **Ultrasonic Proximity Detection** An object that is passing within range will be detected and an output signal will be generated. Advantages are that it works independently of target size, material, and reflectivity.
- **Ranging Measurement** (Reflective sensor) Precise/accurate measurement of the distance an object moves from the sensor is calculated via the time intervals between sent and received bursts of ultrasonic waves.

2.9.1 Applications

Ultrasonic sensors are used in commercial industry on production lines to streamline the production process, in distance measurement, in the robotics industry for robotic sensing, in the automobile industry for vehicle detection in car washes and on automotive assembly lines, in the measuring and control of liquid levels, to detect boom height on agriculture machinery, for anti-collision detection in aerial work platforms, in tension control, in wire or thread break detection, and in the detection of people etc.

Advantages are their high sensitivity, high frequency, high penetrating power, the ease with which they detect external and deep objects, their accuracy, simple interface with a microcontroller or any type of controller, low power utilization, and low cost. Disadvantages are that some materials can distort an ultrasonic sensor's reading, density, and consistency.

2.10 Motion Sensors

A motion sensor is a device that detects moving objects. It is an important part of home security and office security etc. [3].

Motion sensors are of two types:

- **Active Detectors** Energy generated within the sensor system is beamed outwards, and the fraction returned is measured to identify moving objects, as in radar.
- **Passive Detectors** Energy leading to radiation received comes from external sources, e.g., the sun.

2.10.1 Applications

Motion sensors are used in several applications, such as intruder detectors and security lighting. Commercial applications include automatic door opening, entry way lighting, automated sinks/toilet flushers, and hand dryers [3]. In the railway industry they are used in automatic ticket gates and in lift lobbies, multi-apartment complexes, and parking etc. In the mobile industry they are used in phones and tablets to recognize motions and gestures, and in the robotics industry to detect the movement of components. In the military field they are used to detect false motion and for surveillance, and in the automobile industry for vehicle and shaft motion, and vehicle washing. They are also now used in energy efficiency and control systems in the home and in home automation mechanisms.

Advantages are that they provide greater security, i.e., when a sensor is tripped, the signal is sent straight to your security systems, reducing crime. They are also cheap and save energy. Disadvantages are the small area of coverage, the fact that they can be affected by environmental changes, and if installed near a light source can result in the unwanted triggering of motion detectors.

2.11 Touch Sensors

Touch sensor are used to detect touch, especially from the human body.

Types of touch sensor are:

- **Capacitive touch sensor** Use of the electrical properties of the human body to detect the force or body part used to touch a particular sensor.
- **Resistive touch sensor** The pressure applied causes the screen to respond accordingly.

2.11.1 Applications

Touch sensors are widely used in the mobile industry in phones and tablets, in touch-pads on laptops, in biometric systems for fingerprint or hand detection systems, in traffic control and navigation system, for security in ATM machines and locks etc., in the gaming industry for game and motion control, in the robotics industry to develop different components, and in automotive and industrial applications, remote controls, and control panels.

Advantages are their reliability, low cost, no physical contact, speed, ease of maintenance, and design flexibility. Disadvantages are a low accuracy if the object is very small, and the possibility of a malfunction due to scratches on the screen.

2.12 Proximity Sensors

Proximity sensors detect nearby objects that are within the range of sensor without requiring a physical touch from an object.

Types of proximity sensor:

- **Inductive proximity sensor** This is a device which generates an output when an object is within its sensing region or enters into the sensing region of a sensor from any direction output signals or electrical signals.
- **Capacitive proximity sensor** This sensor uses the charge of the capacitor to detect metals, but can also detect resins, liquids, and powders.
- **Magnetic proximity sensor** A magnetic effect is used to measure the distance of an object and it can work on DC, AC, AC/DC, DC.

2.12.1 Applications

Proximity sensors are used in the mobile industry on industrial production lines to check for leakage, in design, counting of pieces, for machine protection by measuring distances and avoiding collisions. They are employed in scientific laboratories to detect materials, measure distances, and monitor tools, in the automobile industry for velocity measurement, gear checking, direction motion, and obstacle detection etc. [1]. In robotics they provide accurate measurements, and are used in roller coasters, anti-aircraft warfare to detect nearby objects, in the detection of liquid levels, in object detection, and for computer displays etc. [3].

Advantages are their accuracy, rapid switching rate, ability to work in tough environmental conditions, stability, and ease of operation. Disadvantages are that their operating range may be limited, they can be affected by nature and surface of objects, their noisy interface, and the fact that there can be a blind zone when in close proximity to objects.

2.13   Acoustic and Sound Sensors

Acoustic sensors sense sound by means of microphones or other filters, and use sound waves to communicate. They are widely used in wireless sensor networks (WSNs) [9].

Types of acoustic sensor:

- **Microphone** An acoustic sensors for air waves in the audible range.
- **Hydrophone** An acoustic sensor for liquid waves.

2.13.1   Applications

Acoustic sensor are used in the characterization of thin-film materials, for determining film thickness, in real-time analysis and monitoring of material modification, corrosion as well as diffusion, in the characterization of liquid properties, namely density and viscosity, in electrochemical studies and sonoelectrochemistry, in acoustoelectric interactions to probe electrical properties of solutions, transitions temperatures, storage and loss moduli, dissemination and infusion, in chemical and biological sensing of vapor, in biological determination (concentration, identification), and in biomedical diagnosis [9, 10].

Advantages are their accuracy and the fact that they are potentially more sensitive than bulk wave devices, such as TSM resonators, as acoustic energy get stuck close to the surface. Disadvantages are their higher operating frequency and inexpensive etc..

2.14   Pressure Sensors

Pressure sensors are used to measure the pressure of gases or liquids. They are classified based on their operating temperature ranges, the range of pressures they measure, and the type of pressure they measure [11].

Absolute Pressure Sensor These sensors measure pressure relative to a perfect vacuum.

Gauge Pressure Sensor These sensors measures pressure relative to atmospheric pressure.

Vacuum Pressure Sensor These sensors measure pressures below atmospheric pressure. In other words they show the difference between a low pressure and atmospheric pressure.

Differential Pressure Sensor These sensors measures the difference between two pressures.

Sealed Pressure Sensor This type of sensor is similar to a gauge pressure sensor except that it measures pressure relative to a fixed pressure rather than the ambient atmospheric pressure.

2.14.1 Applications

Pressure sensors are widely used and have applications in aircraft, automobiles, and any machinery which has implemented pressure functionality, in altitude sensing for aircraft, satellites, rockets, and weather balloons (using changes in pressure relative to altitude) [11], in flow sensing, namely the pressure variance between two sections that is in direct proportion to flow rate through Venturi tube like water flow at particular point of tube etc.. Other uses are in leakage testing in the plastics industry and manufacturing sector, in the mobile sector to measure pressure in different areas of mobiles and tablets, in the military sector for mine detection, and in pressure measurement and other real-time applications.

Advantages are that there is no parallax inaccuracy, they are easy to rearrange, have consistent readings even in operating areas subject to high vibration, and the fact that operator interpretation is not required etc. Disadvantages are full scale and trend is not easy to see, they require power, and errors may occur as a result of oscillating values.

2.15 Electric and Magnetic Sensors

These are transducers which operate on the basis of magnetic field changes and are used to measure current, displacement, speed, and position [12]. There are three types of sensor.

Low Field Sensors These can sense very low values of magnetic field less than 1 μGauss. They are used in medical and nuclear applications.
Earth Field Sensors These can sense values of magnetic field from 1 μGauss to 10 Gauss. They make use of the earth's magnetic field in many applications, for example, navigation and vehicle detection.
BIAS Magnetic Field Sensors These sense large magnetic fields of more than 10 Gauss and are used in the industrial sector in read switches, hall devices, and giant magneto resistance sensors (GMRs).

2.15.1 Applications

Electric and magnetic sensors are used in the automobile industry in engine systems, anti-lock braking systems, steering angle recognition, force and torque detection, in commercial industry for current sensing, hydraulic motors, mobile and computer memory, peristaltic pumps, printers, graphic arts, in military fighting vehicles, in Crusader self-propelled howitzers, amphibious assault vehicles, in biological/medical centrifuges, in aerospace for steering mirror actuators, valve actuators in the railway industry for engine control, in turbines, valve position sensing, in agriculture for diesel engines, engine-powered compressors, and engine-powered generators, in aviation for brake control, and in jet engines and turbochargers [12].

Advantages are their high sensitivity, reduced power consumption, high switching speed, temperature stability, low noise, high performance, and immunity to light changes. Disadvantages are that they can be disturbed by magnetic metals and the field size can affect the trip point etc.

3 Conclusion

There are very many sensors available, covering a wide range of qualities, for example, mechanical, electrical, pressure, and optical sensors. They can be used in many different applications to improve reliability, increase production, and reduce error rates compared to those of humans. Sensors affect the day-to-day life of all of us, from phones to cars, and AC to machines etc. They provide a smart means of controlling different processes in an improved manner. Furthermore, a combination of different sensors can be employed to provide improved results and performance, for example, mobile phone sensors, proximity sensors, temperature sensors, gyrometer sensors, barometer sensors, and IR blasters, which can be combined to enhance the working output of devices.

References

1. Suh, C., Ko, Y.: Design and implementation of intelligent home control systems based on active sensor networks. IEEE Trans. Consum. Electron. **54**(3), 1177–1184 (2008)
2. Byun, J., Hong, I., Lee, B., Park, S.: Intelligent household LED lighting system considering energy efficiency and user satisfaction. IEEE Trans. Consum. Electron. **59**(1), 70–76 (2013)
3. Han, D., Lim, J.: Smart home energy management system using IEEE 802.15.4 and ZigBee. IEEE Trans. Consum. Electron. **56**(3), 1403–1410 (2010)
4. Sobrino, J.A., Del Frate, F., Drusch, M., Jiménez-Muñoz, J.C., Manunta, P., Regan, A.: Review of thermal infrared applications and requirements for future high-resolution sensors. IEEE Trans. Geosci. Remote Sens. **54**(5) (2016)
5. Chen, Z., Lu, C.: Humidity sensors: a review of materials and mechanisms. Sens. Lett. **3**(4), 274–295 (2005)
6. Amin, M.G.: Radar for Indoor Monitoring. CRC Press, Boca Raton (2017)
7. SICK AG. LMS 200 Laser Measurement Manual. Auto Indent, pp. 10–11 (2000)
8. Jang, Y., Shin, S., Lee, J.W., Kim, S.: A preliminary study for portable walking distance measurement system using ultrasonic sensors. In: 2007 29th Annual International Conference of the IEEE Engineering in Medicine and Biology Society, pp. 5290–5293. IEEE, Piscataway (2007)
9. Plinge, A., Jacob, F., Haeb-Umbach, R., Fink, G.A.: Acoustic microphone geometry calibration: an overview and experimental evaluation of state-of-the-art algorithms. IEEE Signal Process. Mag. **33**(4), 14–29 (2016)
10. White, R.M.: Acoustic sensors for physical, chemical and biochemical applications. In: 1998 IEEE International Frequency Control Symposium (Cat No 98CH36165) FREQ-98

11. Mannsfeld, S.C.B., et al.: Highly sensitive flexible pressure sensors with microstructured rubber dielectric layers. Nat. Mater. **9**, 859–864 (2010). https://doi.org/10.1038/NMAT2834
12. Blitz, J.: Electrical and Magnetic Methods of Non-destructive Testing, vol. 3. Springer Science & Business Media (2012)
13. Robben, S., Englebienne, G., Kr'ose, B.: Delta features from ambient sensor data are good predictors of change in functional health. IEEE J. Biomed. Health Inform. **21**(4) (2017)
14. Sabatini, A.M.: Estimating three-dimensional orientation of human body parts by inertial/magnetic sensing. Sensors **11**(2), 1489–1525 (2011)

Mamdani and Sugeno Fuzzy Inference Systems' Comparison for Detection of Packet Dropping Attack in Mobile Ad Hoc Networks

Alka Chaudhary

Abstract Mobile ad hoc networks (MANETs) are very vulnerable to attacks due to their complex properties such as dynamic topologies, restricted bandwidth, and battery powers of mobile nodes. One of the very popular attacks in MANETs is packet dropping attack. In this paper, intrusion detection systems are developed for detecting the packet dropping attack using Mamdani-type and Sugeno-type fuzzy inference systems. This paper presents the basic difference between Mamdani-type and Sugeno-type fuzzy inference systems. For this purpose, the Qualnet simulator is used to create the scenario of packet dropping attack and MATLAB toolbox is used to analyze the performance of the developed systems. The simulation results demonstrated the performance comparison of both the developed intrusion detection systems in respect to true positive rate and false positive rate and also discussed the advantages of using Sugeno-type model over Mamdani-type model.

Keywords Fuzzy inference system (FIS) · Mobile ad hoc networks (MANETs)
Intrusion detection system (IDS) · MANETs' security issues
Mamdani-type fuzzy inference system · Sugeno-type fuzzy inference
system and packet dropping attack

1 Introduction

Mobile ad hoc network (MANET) is a temporary network which facilitates to form an ad hoc network in war situations and disaster fields because MANET does not require any pre-infrastructure. This flexibility makes MANET attractive for many applications such as rescue operations, education applications, and virtual conferences. Mobile nodes play the role of hosts as well as routers.

A. Chaudhary (✉)
Department of IT, Manipal University Jaipur, Dehmi Kalan,
Jaipur-Ajmer Expressway, Near GVK Toll Plaza, Jaipur 303007, Rajasthan, India
e-mail: alka.chaudhary@jaipur.manipal.edu

© Springer Nature Singapore Pte Ltd. 2019
A. Abraham et al. (eds.), *Emerging Technologies in Data Mining and Information Security*, Advances in Intelligent Systems and Computing 814,
https://doi.org/10.1007/978-981-13-1501-5_70

In dropping data packet attack, malicious node(s) prevents data packets to forward other mobile nodes and after that drops these all packets [1]. Due to MANET's characteristics, prevention-related techniques are not the best solution for ad hoc networks. So intrusion detection system is an important solution for the security of MANETs.

This paper presents the comparison between our developed fuzzy logic-based approaches using Mamdani-type and Sugeno-type inference systems to detect the packet dropping attack through malicious nodes in MANETs [3, 4] and showed the advantages of using Sugeno-type model over Mamdani-type model. The proposed solutions are able to detect data dropping attack in a distributed manner by each node.

The rest of this paper is organized as follows: Sect. 2 describes the concept of fuzzy inference system. Section 3 presents the developed fuzzy inference systems for packet dropping attack. Section 4 defines the scenarios and simulation parameter by Qualnet simulator 6.1 and discusses and analyzes the results of both the proposed fuzzy inference systems, and finally, conclusion is given in Sect. 5.

2 Fuzzy Inference System

Fuzzy logic can be able to deal with uncertainty so that it is applied in intrusion detection before since 90s [5]. Intrusion detection features can be viewed with the help of fuzzy variables or linguistic terms and able to make the decision on normal and abnormal activities in the network [6, 7].

Fuzzy rules which are based on if–then–else rules are used to define all situations in the network for detecting the attacks. The fuzzy rule-based system is known as fuzzy inference system (FIS) that is responsible to take decision. Many of the few types of fuzzy inference systems are proposed in the literature [8].

This paper selected two-input single-output-based Mamdani fuzzy inference system for making the decision which is shown in Fig. 1. There is an example of Mamdani fuzzy inference system with two rules which can be given as

$$\text{if } x \text{ is } A1 \text{ and } y \text{ is } B1 \text{ then } z \text{ is } C1,$$
$$\text{if } x \text{ is } A2 \text{ and } y \text{ is } B2 \text{ then } z \text{ is } C2,$$

where A1, A2 and B1, B2 are the membership functions of inputs for A and B fuzzy sets, and C is the fuzzy set for output, respectively. But, it is a time-consuming procedure.

Takagi et al. suggested an approach to render the fuzzy rules from the dataset which is depicted in Fig. 2. There is an example of Sugeno fuzzy inference system with two rules which can be given as

$$\text{if } x \text{ is } A1 \text{ and } y \text{ is } B1, \text{ then } z1 = p1x + q1y + r1,$$
$$\text{if } x \text{ is } A2 \text{ and } y \text{ is } B2, \text{ then } z2 = p2x + q2y + r2,$$

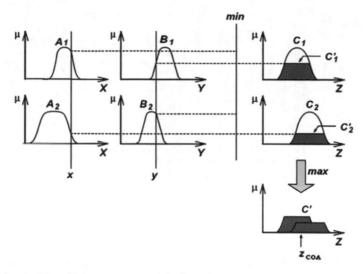

Fig. 1 Mamdani fuzzy inference system with min and max operators

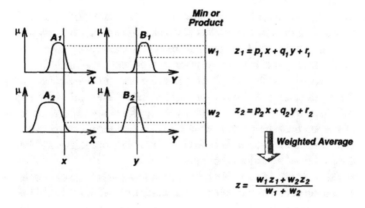

Fig. 2 Fuzzy reasoning of Sugeno model

where A1, A2 and B1, B2 are the membership functions of inputs for A and B fuzzy sets, and z1, z2 is the crisp function for output, respectively.

3 Proposed Mamdani-Type and Sugeno-Type Fuzzy Inference Systems

The developed fuzzy inference systems for the evaluation of packet dropping attack consists of two-input parameters: Data Packet Forwarded Ratio and Average Data Packet Dropped Rate. The systems have one output that indicates the verity level to

Table 1 Rule for fuzzy system

S. No.	Data packet forwarded ratio	Average data packet dropped rate	Verity level
I.	High	Low	High
II.	High	Medium	High
III.	High	High	Low
IV.	Medium	Low	Medium
V.	Medium	Medium	Medium
VI.	Medium	High	Low
VII.	Low	Low	Low
VIII.	Low	Medium	Low
IX.	Low	High	Low

check behavior of node on the bases of input parameters, i.e., malicious or normal. For both FISs, the proposed system rule base [9] is presented in Table 1 for evaluating the behavior of a node. On the bases of optimum performance, membership functions [9] are selected for input and output parameters which are depicted in Fig. 3. First rule of rule base in FIS is interpreted as follows: If Data Packet Forwarded Ratio is high and Average Data Packet Dropped Rate is low, then verity level is high.

Accordingly, the ith rule of rule base interprets as Mamdani fuzzy-based inference system. Verity level of each node is calculated to its direct neighbor nodes' table behalf on the input parameters' membership functions. The value of verity level lies within 0–10. As a result, the low value of verity level shows the more malicious behavior of a node than the normal behavior of neighbor nodes. So verity level of value 0 indicates the particular node behavior completely malicious and 10 represents the normal behavior of a particular node.

A threshold of verity level is already set for comparing the calculated value of verity level to determine the behavior of the node in MANETs. Here, the threshold value set is 5.2. If the calculated verity level is greater than the threshold value, the node is not malicious node, otherwise it is.

4   Results and Discussion

The following results are depicted during the simulation of both types of fuzzy inference systems in terms of detection of packet dropping attack on the bases of true positive rate and false positive rate that are the metrics [10] to evaluate the performance of intrusion detection system. Table 2 depicts the parameters of Qualnet simulator that are used to create the scenario of packet dropping attack. MATLAB toolbox is used to evolve the Mamdani-type and Sugeno-type fuzzy inference systems. Particularly, Table 3 shows that the performance of detection rate of Mamdani-type and Sugeno-type fuzzy inference systems at the verity threshold value is equal to 5.2.

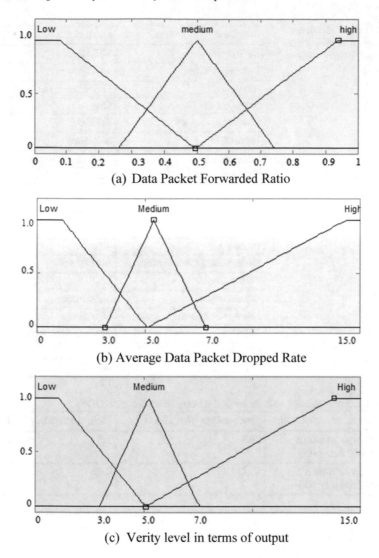

(a) Data Packet Forwarded Ratio

(b) Average Data Packet Dropped Rate

(c) Verity level in terms of output

Fig. 3 Membership functions for fuzzy system

From the results, performance of both the proposed inference systems shows quite similar. Defuzzification criteria of Sugeno-based FIS make it more suitable than the Mamdani FIS.

Table 2 List of simulation parameters used during data extraction for fuzzy system

Simulator	Qualnet 6.1.
Radio type	802.11b
Routing protocol	AODV
Antenna	Omnidirectional
No. of channels	One
Channel frequency	2.4 GHz
Packet size	512 bytes
Simulation area	1500 m × 1500 m
Simulation time	200 s
Path loss model	Two ray
Mobility speeds	1–25 mps
Mac Type	IEEE 802.11
Pause time	30 s
Battery charge monitoring	Interval 60 s
Traffic type	CBR
Battery model	Linear model
Number of nodes	30 nodes
Mobility	Random way point
Malicious nodes	3
Energy model	Generic

Table 3 Detection rates of both the proposed fuzzy inference-based IDSs

Inference system-based IDSs	True positive rate (TPR) (%)	False positive rate (FPR) (%)
Mamdani-type inference system-based fuzzy IDS	98.3	1.3
Sugeno-type inference system-based fuzzy IDS	98.1	0.09

5 Conclusion

In this paper, a comparison has been made between the Mamdani-type and Sugeno-type fuzzy inference systems based on intrusion detection systems for detecting the packet dropping attack in MANETs. The performance of both fuzzy inference systems is quite similar although designing of both fuzzy inference systems is same but the Sugeno FIS works better in terms of defuzzification process. Mamdani FIS uses defuzzification module that converts the fuzzy values to crisp values in terms of output. But, it is a time-consuming procedure. Sugeno-type FIS is more efficient toward computation and also more suitable with adaptive techniques. Here, defuzzification is done by using weighted average.

References

1. Wu, B., Chen, J., Wu, J., Cardei, M.: A survey of attacks and countermeasures in mobile ad hoc networks. In: Department of Computer Science and Engineering, Florida Atlantic University
2. Zhang, Y., Lee, W.: Intrusion detection in wireless ad hoc networks. In: Proceedings of the 6th Annual International Conference on Mobile Computing and Networking (MobiCom'00), pp. 275–283 (2000)
3. Chaudhary, A., Tiwari, V.N., Kumar, A.: A reliable solution against Packet dropping attack due to malicious nodes using fuzzy logic in MANETs. In: International Conference on Optimization, Reliability and Information Technology (ICROIT), IEEE (2014)
4. Chaudhary, A., Tiwari, V.N., Kumar, A.: Design an anomaly based fuzzy intrusion detection system for packet dropping attack in mobile ad hoc networks. In: 2014 IEEE International Advance Computing Conference (IACC), IEEE (2014)
5. Shanmugam, B., Idris, N.B.: Anomaly intrusion detection based on fuzzy logic and data mining. In: Proceedings of the Postgraduate Annual Research Seminar, Malaysia (2006)
6. Wahengbam, W., Marchang, N.: Intrusion detection in Manet using fuzzy logic. In: 3rd IEEE National Conference on Emerging Trends and Applications in Computer Science (NCETACS), pp. 189–192. ISBN: 978-1-4577-0749-0, Shillong, 30–31 Mar (2012)
7. Verma, A.K., Anil, R., Jain, O.P.: Fuzzy logic based revised defect rating for software lifecycle performance prediction using GMR. In: Bharati Vidyapeeth's Institute of Computer Applications and Management (2009)
8. Jang, J.S.R., Sun, C.T., Mizutani, E.: Neuro-Fuzzy and Soft Computing—A computational Approach to Learning and Machine Intelligence, 1st edn. Prentice Hall of India (1997)
9. Ross, T.J.: Fuzzy Logic with Engineering Applications. McGraw Hill International Editions
10. Sen, S., Clark, J.A.: Intrusion detection in mobile ad hoc networks. In: Chapter 17, Guide to Wireless Ad Hoc Networks, Springer (2008)

A Novel Transfer Learning-Based Missing Value Imputation on Discipline Diverse Real Test Datasets—A Comparative Study with Different Machine Learning Algorithms

Jit Gupta, Sayak Paul and Anupam Ghosh

Abstract Intelligent systems are dependent on data. If they are fed with wrong and noisy data, the results are erroneous too. Most of the cases, the data is not present in a favorable structured format and hence predictive models have been introduced. Missing values and noisy data is one of the problems that are tackled very frequently during data analysis. In this chapter, we propose a transfer learning-influenced missing value imputation technique based on the concept of artificial neural network that establishes the novelty of the work. The hyperparameters of the neural network are tuned with evolutionary search-based technique. The algorithm is demonstrated on three benchmark datasets, viz. Mammographic Mass dataset, Retail Chain Customer Demographic dataset, and Credit Card Approval dataset. The performance of the proposed algorithm is compared with some earlier investigations, viz. KNN-based missing value imputation and support vector regressor (SVR). The results are appropriately validated using RMSE and MAE scores. It has been shown that our proposed algorithm performs better with respect to other existing techniques.

Keywords Artificial neural networks · Transfer learning · Hyperparameters
Missing value · EvolutionarySearchCV

J. Gupta (✉)
Johnson Controls India Engineering, Pune 411006, India
e-mail: dotgupta@gmail.com

S. Paul
TCS Research and Innovation, Pune 411013, India
e-mail: sayak.p@tcs.com

A. Ghosh
Department of Computer Science & Engineering, Netaji Subhash Engineering College,
Kolkata 700152, West Bengal, India
e-mail: anupam.ghosh@rediffmail.com

1   Introduction

We are probably living in the most fascinating period of computing history, a period, when computing moved from large mainframes to PCs and subsequently to cloud, a period, when the machines are intelligent enough to compute on their own with or without any human intervention. A variety of approaches have turned into real solutions which are impacting our lives in a fruitful way. Machine learning algorithms constitute a major portion of this impact. The algorithms can be classified into three major types—supervised learning, unsupervised learning, and reinforcement learning [1].

Learning algorithms have evolved from the field of pattern recognition. These algorithms provide scope for the computer to learn from errors instead of the programmer explicitly detailing out every step to be executed. There are generally two datasets involved in the creation of these models—training dataset and testing dataset. As their names suggest, training dataset is used to train the model so as to increase its efficiency by reducing the error rate, while testing dataset is used for testing the finally trained model. A third dataset, known as the validation dataset, may be used to tune the parameters so as to produce a more generalized and not data-specific model [2].

Missing data tends to be present in almost all forms of real datasets. The presence of these missing values in the training data often reduces the accuracy of a model or leads to a biased model, leading to inaccurate predictions. This happens because of the incorrect analysis of the behavior and relationship with other variables present in the dataset [3, 4].

In many datasets (a dataset is the foundation of any predictive modeling application), useful attribute values can be missing. For example, in the Titanic dataset, many instance values are missing in the attributes like Age, Cabin [5]. Missing values result in problems in image datasets also. For example, in satellite images, several pixels often remain missing from the images due to transmission problems, which make it really difficult for the intelligent systems to detect a particular image [6]. In fact, data can be missing in the biological datasets which make it troublesome for even biologists and doctors when they analyze the datasets for diagnosis purpose. Handling missing values in these datasets is one of the burning topics of the hour [4].

Julián et al. suggested a radial basis function-based method for imputing missing values [5].

In this chapter, we have developed a transfer learning-based approach using tuned hyperparameters, which not only performs efficiently on data with missing values at a single-dimensional level but also gracefully scales down to missing value problems at multi-dimensional level. The model has been tested on datasets of three different fields, and the results have been compared with preexisting algorithms such as grid search and random search for the tuning process and k-nearest neighbor (KNN) as well as support vector regressor (SVR) for the imputation process. The work is organized in several sections: Sect. 1 presents a brief introduction toward the contribution of this paper. Related works are briefed in Sect. 2. Section 3 shows methodology of the proposed work. In Sect. 4, all the experimental results have been shown and discussed. Section 5 gives the conclusion.

2 Related Works

Substantial amount of work has been done in the field of soft computing for efficient imputation of missing values. Several researchers have used statistical methods for the purpose, while some have used neural networks for the same. Some researchers have also used tools like NIP. Raquel et al. proposed a predictable model which imputes missing values from information that may be of low quality when it is the only possibility [6]. Julián et al. suggested a radial basis function-based method for imputing missing values. They have also incorporated a method called event covering along with radial basis function networks to improve the results [7]. Saravanan et al. proposed a fuzzy-possibilistic C means method for data imputation which they optimized using genetic algorithm with support vector regression. Their proposed system considered both membership function and typicality of the data [8].

Esther et al. worked with multi-layer perceptrons with three specific learning rules, viz. Levenberg–Marquardt, BFGS quasi-Newton, and conjugate gradient Fletcher–Reeves update with a very less number of hidden nodes. Their work proved to be performing better than classical imputation approaches, especially with the datasets that have categorical variables [9]. Thomás et al. worked with statistical method like metric matching, Bayesian bootstrap, and regression-based minimal square for data imputation for its use in artificial neural networks [10]. Fulufhelo et al. presented a comparative study of two missing value imputation techniques, viz. expectation–maximization and auto-associative neural networks combined with genetic algorithms. According to their study, it was inferred that, when there is little or no interdependency between the input variables expectation-maximization is suitable, whereas the auto-associative neural network and genetic algorithm combination is suitable when there is some inherent nonlinear relationships between some of the given variables [11].

One of the major causes of discrepancies in data analytics is the presence of empty fields or missing values in the given dataset. This may result in amplified errors throughout the data and the analysis. Common ways to handle these missing values include finding the mean or median of values or even replacing the missing value with its nearest neighbors. These results in high error rates, and the methods that provide the better solutions are applicable to genre-specific datasets. This acted as the major impetus for us to explore and develop a novel method to combine the advantages of both types of imputation algorithms and keep the disadvantages in check.

3 Methodology

The methodology is divided into four major sections. The first section concerns preprocessing the given dataset into numerical data so as to ease the performance of operations. This is done using label-encoding [1]. This is followed by the process of normalizing which scales data of every feature vector within a certain range, [0,1] in this case [2]. The third division deals with tuning of hyperparameters which uses the EvolutionarySearchCV. Most machine algorithms choose values for parameters based on a hit-and-miss approach. This algorithm however optimizes the values for the parameters used in 3.5 so as to result in lower error rates in the model [4]. The final part uses a much used concept to provide a solution for a problem that is not related to the field that it is usually applied on. The resultant model is called the transfer learning imputer, and as illustrated further in following sections, it is applicable on missing data irrespective of its genre and field. The flowchart of the proposed model is depicted in Fig. 1.

3.1 Description of the Symbols Used

$X_i = [x_{i1}; x_{i2}; x_{ij} \ldots x_{im}]$, where x_{ij} signifies the jth data point for ith feature
σ = Normalization range
γ = Value to denote missing value
ε = Epoch
α = Hidden unit range
η_R = Learning rate range
ρ = Initial population
μ = Mutation probability
λ = Crossover probability
Ω = Number of generations
error = Error matrix

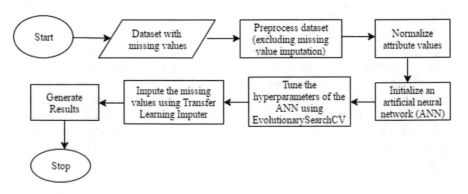

Fig. 1 Flowchart of the proposed method

3.2   Preprocessing

In this step, all the non-numeric instance values in the respective dataset have been converted to numeric ones because of ease in calculation. This conversion is done using label-encoding [1].

3.3   Normalizing

The ranges of the values for different features in the dataset may differ to a great extent. Due to this, a small change in the value of one feature instance may not affect the other which is not desirable. The datasets that are used in this experiment were prone to this problem. To check with it, the datasets were normalized using min–max scaling [2].

3.4   Tuning of Hyperparameters

EvolutionarySearchCV is used for this purpose. Following algorithm describes its methodology.

    EvolutionaryAlgorithmSearch(α, η_R, ρ, μ, λ, Ω)

1. Populate matrix Y of size ρ with values from α and η_R
2. A Fitness Function is defined to classify if the population is good or bad according an evaluation metric i.e. RMSE.
3. The good ones are selected for mating to produce their off-springs.
4. The Selection method is built on top of Roulette Wheel Selection method.
5. For building the wheel, m partitions are taken, where m is the number of Chromosomes in the population. The area occupied by each chromosome is proportional to its fitness score. Based on these scores, the wheel is built.
6. After the wheel is built, it is rotated and the region of wheel which comes in front of the fixed point is chosen as the parent. For the second parent, the same process is repeated.(Stochastic Universal Selection method).
7. for i=1 to genΩ

    7.1. for i=1 to ρ do

        7.1.1. for j=1 to ρ do

            7.1.1.1. Crossover between Y(i) and Y(j) to produce offspring (having properties of both parents) using μ and λ for fitness function evaluation.

            7.1.1.2. Calculate fitness score of offspring.

            7.1.1.3. Add offspring to Y

8. Display offspring A with best fitness score.

9. Display execution time.

The above algorithm is used for tuning the hyperparameters of the artificial neural network which is later used in the transfer learning imputer (3.5). Hyperparameters cannot be directly learned from the training process. These parameters express "higher-level" properties of the model such as the number of hidden units, learning rate, activation functions. They denote the expressivity of the network.

In the above algorithm, first an initial population is defined followed by a fitness function to classify if the population is good or bad based on a certain evaluation metric called RMSE. Those, which are classified as good ones, are further mated to produce their offsprings, and these offsprings replace the bad ones from the population. This process repeats [12, 13].

After the tuning of hyperparameters, transfer learning impute method is incorporated.

3.5 Transfer Learning Imputer

Transfer learning [14, 15] happens to be a widely talked about machine learning problem where the knowledge gained from solving a certain conundrum is used in solving another related problem which may or may not be different to the initial one. In this given algorithm, transfer learning has been applied for imputing [16] values at every step and even at a multi-dimensional level. A regression [17] neural network [17, 18], after being fed by the hyperparameters from 3.4, is trained using a matrix consisting of fields with no missing values for no features in the given dataset. Using logistic regression, the neural network imputes values for fields with a single missing value for each feature at a time. The matrices containing the imputed values are concatenated to the training matrix, and the resultant matrix is used as the training set for fields containing multiple missing values [19–21].

The algorithm is stated below:

TransferLearningImputer(D, σ, γ)

1. Load the dataset is loaded into matrix B and normalize it within the range of σ.
2. Replace every missing value in the dataset by a value γ that is much larger than the existing range of values.
3. for i=1 to number of vectors in B, do

    3.1. If the vector has only 1 missing value (i.e., γ) in column i then add it to vector X_i.
    3.2. If the vector has no missing values in column i then add it to matrix **train**.

4. Create an Artificial Neural Network MOD that performs having one hidden layer, with parameters from offspring **A** produced by Sect. 3.4 and using gradient descent as a training function, tansgmoid as the activation function and mean square error as the performance function.
5. Feed matrix **train** to MOD to train it.

6. for i=1 to number of vectors in B, do

    6.1. Using **train(i,:)** as the target value, test MOD with vector $\mathbf{X_i}$

    6.2. Add imputed vector $\mathbf{X_i}$

7. for i=1 to number of vectors in B, do

    7.1. If vector has multiple missing values in column i then

        7.1.1. Let the missing value columns be $p_1...p_j$

        7.1.2. Train MOD with matrix **train[$1....p_1$-$1,p_1$+$1.....p_2$-$1,p_2$+$1...p_j$-1]**

        7.1.3. for q=1 to j, do

            7.1.3.1 Using **train(p_j,:)** as the target value, test MOD with column p_j of B. Store error in **error(j)**

        7.1.4. Find min of **error**. Test MOD with corresponding column.

        7.1.5. Add imputed column to **train**

8. Copy **train** to matrix **imputed**.

In the given algorithm, each missing value is substituted with a common value that is outside the normalization range. This is followed by separating rows with single missing values from rows with multiple missing values. The artificial neural network MOD is trained and tested on single missing value; i.e., each feature vector with missing values is imputed, such that no other feature has missing values for that field. Each of these vectors is added to the training matrix, so as to result in a new training set which is used for testing on multiple missing values [4, 17, 18].

The new training set is tested on each missing value in a row, and the RMSE value is compared; the one with the lowest error is imputed first and added to the training set, and this step is repeated until all values are imputed.

3.6 Analysis of Complexity

The time complexity of EvolutionarySearchCV (Sect. 3.4) is heavily problem-specific because the fitness function, population volume, and the number of generations change from problem to problem. Primarily evolution strategies depend on the following points:

- The fitness function that is to be optimized
- The number of population
- The number generation

A stochastic selection will require sorting the population, and it runs in the time of $O(N \log N)$. And the process is followed by transforming the population with crossover and mutation operation which run in $O(NL)$. Therefore, the basic complexity becomes

$$O(N \log N) + O(NL) \tag{1}$$

According to the Big O notation, where N is the number of population and L is the number of generations [6].

The given transfer learning imputer (Sect. 3.5) algorithm has a complexity of

$$O(f A_n) + O(n A_n) \tag{2}$$

where f signifies the number of features present in the dataset, A_n signifies the time complexity taken by the artificial neural network taken to converge, and n signifies the number of multiple missing fields present in the dataset.

Like every machine learning algorithm, the time complexity of the algorithm presented in this work is dependent on the dataset and also on the time taken by the model to converge while fed by training data. Therefore, it is highly dynamic and may be subject to change depending on the various factors influencing by the dataset.

4 Results

In the subsequent sections, viz. 4.1, 4.2 and 4.3, the datasets' description, a detailed analysis of the results, and validation of those results have been discussed, respectively.

4.1 Description of the Datasets

The first dataset that is used is Mammographic Mass dataset which is multivariate in nature. It has 176 missing values in total spanning across 5 features (BI-RADS assessment, age, shape, margin, and density) [22].

The second dataset is a Credit Card Approval dataset which is also multivariate in nature. It has 15 attributes and 690 numbers of instances in total [23].

The final dataset is a Customer Demographic dataset. It has a total of 2117 number of missing values out of a total of 8993 instances [24].

4.2 Analysis of the Result

Table 1 presents the results that were obtained from the experiments performed with the proposed methodology on the three datasets. Values of the learning unit and number of the hidden units (hyperparameters) were learned by the Evolutionary-SearchCV, and the final RMSE and MAE scores were generated by the transfer learning imputer. With Table 1 results, it can be inferred that the hybridization of EvolutionarySearchCV and transfer learning imputation performs consistently across different datasets which further makes it not data-specific.

Table 1 Values of RMSE scores and learned hyperparameters

	Mammographic mass dataset	Credit card approval dataset	Customer demographic dataset
RMSE score	0.03288	0.04	0.0720
MAE score	0.3170	0.2049	0.2069
Hyper-tuned learning rate	0.047	0.043	0.04
Hyper-tuned hidden units	12	6	7

4.3 Validation of the Result

All the imputation scores are validated using root mean square error (RMSE) and mean absolute error (MAE) scoring [2]. Both are standard regression model evaluation error metrics. Figure 2 represents the RMSE score comparison between Grid-SearchCV, RandomizedSearchCV, and EvolutionarySearchCV, and the respective execution time comparison between the same is shown in Fig. 3.

From the above two figures, it can be seen that EvolutionarySearchCV takes very less time while compromising the RMSE score slightly compared to the other two techniques. It is quite vivid that EvolutionarySearchCV surpasses the other two methods if both execution time and RMSE score are taken into account.

In Figs. 4 and 5, the result shows how transfer learning imputer maintains its consistency across different datasets while KNN imputer and SVR fail to do so in terms of RMSE score and MAE score, respectively [1].

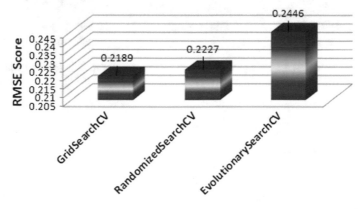

Fig. 2 Comparative study between RMSE scores between GridSearchCV, RandomizedSearchCV, and EvolutionarySearchCV

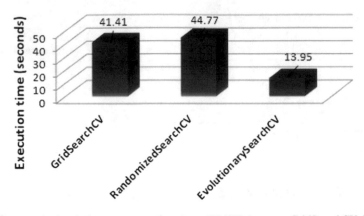

Fig. 3 Comparative study between execution times (RMSE) between GridSearchCV, Random-izedSearchCV, and EvolutionarySearchCV

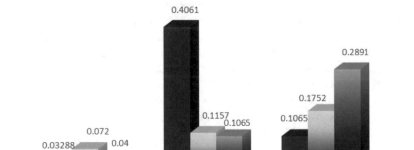

Fig. 4 RMSE scores of KNN imputer, SVR, and transfer learning imputer for three different datasets, respectively

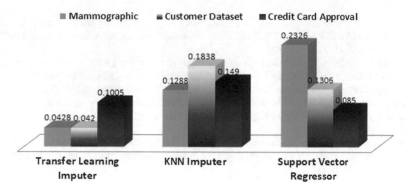

Fig. 3 MAE scores of KNN Imputer, SVR, and transfer learning imputer for three different datasets, respectively

5 Conclusion

The proposed work is a hybrid of evolutionary searches and neural networks, and it has been applied in the context of transfer learning. Hyperparameters play a vital role in modeling the performance of all the machine learning algorithms including neural networks. In this work, these hyperparameters (of the neural network) have been tuned using EvolutionarySearchCV (an extension of evolutionary search). The values of these hyperparameters have been further used to construct the architecture of the neural network used in the transfer learning imputer.

The process of improving the performance of the transfer learning imputer is one of the challenges that can be addressed in the future work as it involves multiple trainings of neural networks. This can act as a pillar for future reference so as to build on the work presented in this work and increase the efficiency of the overall algorithm in general.

References

1. Bishop, C.M.: Pattern Recognition and Machine Learning (1st edn.) Information Science and Statistics, pp. 23. Springer, New York, Inc. (2006)
2. Han, J., Kamber, M., Pei, J.: Data Mining: Concepts and Techniques (3rd edn.), pp. 56. Morgan Kaufmann Publishers Inc. San Francisco (2011)
3. Saha, S., Bandopadhyay, S., Ghosh, A.: An ensemble based missing value estimation in DNA microarray using artificial neural network. In: Second International Conference on Research in Computational Intelligence and Communication Networks (ICRCICN), pp. 279–284 (2016)
4. Ghosh, A., Dhara, C.B., De, K.R.: Selection of genes mediating certain cancers, using neuro-fuzzy approach. Neurocomputing **133**, 122–140 (2014)
5. Julián, et al.: A study on the use of imputation methods for experimentation with radial basis function network classifiers handling missing attribute values: the good synergy between RBFNs and event covering method. J. Neural Netw. **23**, 406–418

6. Martinez, R., Cadenas, M.J., Garrido, C.M., Martinez, A.: Imputing missing values from low quality data by NIP tool. In: 2013 IEEE International Conference on Fuzzy Systems (FUZZ-IEEE), pp. 2141–2148 (2014)
7. Luengo, J., García, S., Herrera, F.: A study on the use of imputation methods for experimentation with radial basis function network classifiers handling missing attribute values: the good synergy between RBFNs and EventCovering method. J. Neural Netw. **23**, 406–418 (2010)
8. Saravanan, P., Sailakshmi, P.: Missing value imputation using fuzzy possibilistic C means optimized with support vector regression and genetic algorithm. J. Theor. Appl. Inf. Technol. **21**, 34–39 (2015)
9. Silva-Ramírez, E., Pino-Mejías, R., López-Coello, M., Cubiles-de-la-Vega, M.: Missing value imputation on missing completely at random data using multilayer perceptrons. J. Neural Netw. **24**, 121–129 (2011)
10. López-Molina, T., Pérez-Méndez, A., Rivas-Echeverría, F.: Missing values imputation techniques for Neural Networks patterns. In: 12th WSEAS International Conference on SYSTEMS, Greece, pp. 290–295 (2008)
11. Nelwamondo, F., Mohamed, S., Marwala, T.: Missing data: a comparison of neural network and expectation maximization techniques. Current Sci. **93**(11), 1514–1521 (2007). Retrieved from http://www.jstor.org/stable/24099079
12. Sivanandam, N.S., Deepa, N.S.: Principles of Soft Computing (2nd edn.), pp. 373–374. Wiley Publishers (2011)
13. Roy, S., Chakraborty, U.: Soft Computing (1st edn.), pp. 201–203. Pearson Publishers (2013)
14. Torre, L., Shavlik, J.: Transfer Learning, pp. 1–3. University of Wisconsin, Madison, USA
15. Ruder, S., Plank, B.: Learning to select data for transfer learning with Bayesian optimization. In: Proceedings of the Conference on Empirical Methods in Natural Language Processing, pp. 372–382, Copenhagen, Denmark (2017)
16. Specht, D.F.: A general regression neural network. IEEE Trans. Neural Netw. **2**, 568–576 (1991)
17. Gheyas, I.A., Smith, L.S.: A neural network-based framework for the reconstruction of incomplete data sets. Neurocomputing **73**, 3039–3065 (2010)
18. Junninen, H., Niska, H., Tuppurainen, K., Ruuskanen, J., Kolehmainen, M.: Methods for imputation of missing values in air quality datasets. Atmos. Environ. **38**, 2895–2907 (2004)
19. Lapuerta, P., Azen, S.P., Labree, L.: Use of neural networks in predicting the risk of coronary artery disease. Comput. Biomed. Res. **28**, 38–52 (1995)
20. Mitchell, T.M., Machine Learning, pp. 70. WCB–McGraw-Hill. ISBN 0-07-042807-7 (1991)
21. Duchi, J., Hazan, E., Singer, Y.: Adaptive subgradient methods for online learning and stochastic optimization. J. Mach. Learn. Res. **12**, 2121–2159 (2011)
22. Mammographic mass dataset. URL: http://archive.ics.uci.edu/ml/machine-learning-databases/mammographic-masses/mammographic_masses.names
23. Credit card approval dataset. URL: http://archive.ics.uci.edu/ml/datasets/credit+approval
24. Customer demographic dataset. URL: http://sci2s.ugr.es/keel/dataset.php?cod=163

Machine Learning in Astronomy: A Case Study in Quasar-Star Classification

Mohammed Viquar, Suryoday Basak, Ariruna Dasgupta, Surbhi Agrawal and Snehanshu Saha

Abstract We present the results of various automated classification methods, based on machine learning (ML), of objects from data releases 6 and 7 (DR6 and DR7) of the Sloan Digital Sky Survey (SDSS), primarily distinguishing stars from quasars. We provide a careful scrutiny of approaches available in the literature and have highlighted the pitfalls in those approaches based on the nature of data used for the study. The aim is to investigate the appropriateness of the application of certain ML methods. The manuscript argues convincingly in favor of the efficacy of asymmetric AdaBoost to classify photometric data. The paper presents a critical review of existing study and puts forward an application of asymmetric AdaBoost, as an offspring of that exercise.

1 Introduction

A quasar is a *quasi-stellar radio source*, which was first discovered in 1960. They emit electromagnetic radiation in the frequency bands corresponding to radio waves, visible, ultraviolet, infrared, X-rays, and gamma rays. They are many light-years away from the Earth and the radiation from a quasar could take billions of years to reach us and may carry signatures of the early stages of the universe. This information

M. Viquar · S. Basak · S. Agrawal · S. Saha
PESIT Bangalore South Campus, Karnataka 560100, India
e-mail: viquar27x4@gmail.com

S. Basak
e-mail: suryodaybasak@gmail.com

S. Agrawal
e-mail: surbhiagrawal@pes.edu

S. Saha
e-mail: scibase.snehanshu@gmail.com

A. Dasgupta (✉)
Calcutta University, Kolkata 700106, West Bengal, India
e-mail: dasguptaariruna@gmail.com

© Springer Nature Singapore Pte Ltd. 2019
A. Abraham et al. (eds.), *Emerging Technologies in Data Mining and Information Security*, Advances in Intelligent Systems and Computing 814,
https://doi.org/10.1007/978-981-13-1501-5_72

gathering exercise and subsequent physical analysis of quasars pose strong motivation for the current study. It is difficult for astronomers to study quasars by relying on telescopic observations with template manual matching alone since quasars are difficult to distinguish from stars due to their great distance from Earth. Hence, in this paper, we present methods which can be scaled up to semi-automated or automated techniques to distinguish quasars from stars.

Machine learning (ML) [3] is a subfield of computer science which relies on statistical methods for predictive analysis. *Supervised* ML algorithms rely on a representative sample of data to make predictions of class belongingness for new or incoming data. The methods we elucidate in this paper use supervised machine learning approaches with proper bias handling in the data. The ML algorithms that we have tried are support vector machines (SVM), SVM and K nearest neighbor hybrid (SVM–KNN), AdaBoost, and asymmetric AdaBoost. Of these four methods, SVM and SVM–KNN have been previously tried for quasar-star classification, but we improve upon the performance (and the justification for using them) by introducing methods of bias handling. To the best of our knowledge, AdaBoost and asymmetric AdaBoost have not been previously tried to solve this problem. To contrast the effects of bias that arise due to the imbalance in the data, we have performed the experiments on naturally imbalanced as well as artificially balanced data sets.

The outcome of this research is two-fold. The first, to assert appropriate models for the separation of stars and quasars; and the second, to provide a solid reasoning for selecting these models, and consequently establishing a set of best practices for data scientific research in astronomy.

2 Literature Survey

Support vector machine (SVM) is one of the most widely used and powerful ML methods. The authors in [6] attempted to solve the quasars-stars classification problem by using SVM to classify the star and quasar samples that are present in the Sloan Digital Sky Survey (SDSS) database. Elting et al. [4] used SVM for classifying stars, galaxies, and quasars. Both of them use nonlinear radial basis function (RBF or Gaussian) kernel. Although the accuracies reported were high, a justification of selection of the RBF kernel was not forthcoming. The authors in the present manuscript have performed a linear separability test on the data set, discussed in Sect. 4, which clearly shows that the data are mostly linearly separable, and hence, a linear SVM can be used. Peng et al. [12] used an SVM–KNN method which is a combination of SVM and KNN. SVM–KNN improves the performance of SVM by using KNN to better classify the samples which occur near the boundary (hyperplane) constructed by the SVM learner. In other works, decision tree classifiers are also used for star-galaxy separation [10].

If data are linearly separable, then SVM may be implemented using a linear kernel. The absence of linear separability may justify SVM implementation in conjunction with the RBF kernel. In [4, 6, 12], such an exploration is not reported. Moreover, the class dominance was ignored by [4, 6, 12]. Class dominance must be considered; otherwise, the accuracy of classification obtained will be biased by the dominant class, and it will always be numerically very high. We have performed *artificial balancing* of data to counter the effects of class bias; the process of artificial balancing has been elaborated in 4.1. In addition to using previously tried ML models with improvements on bias handling, in this paper, we explore asymmetric AdaBoost, which is a method designed to handle imbalanced datasets.

3 Data Acquisition

The Sloan Digital Sky Survey (SDSS) is the most extensive redshift survey of the universe, whose data collection began in 1998. Data release (here on, just *DR*) 6 [2] comprioos of the complete imaging over the northern Galactic cap. As a part of this survey, about 287 million objects are registered, over 9583 deg^2. More than 1.27 million spectra are available from this survey in the u, g, r, i, and z bands. DR7 [1], released in 2009, covers 11,663 deg^2 of the sky. The DR7 was the end of the SDSS-II phase. This catalog contains the same five bands of data as in the DR6, but of 357 million distinct objects. All of the data that are released by SDSS is made available over the Internet [11]. The SkyServer provides interfaces for querying and obtaining data as per a user's needs. Using the available interfaces, spectral data, as well as images, can be obtained [7]. The data are available for non-commercial use only, without written permission. From this data, we make use of the classes of quasars and stars and extend the work done by [4, 6, 12].

4 Method

4.1 Artificial Balancing of Data

Artificial balancing of data needs to be performed such that the classes present in the dataset used for training a model do not present a bias to the learning algorithm. The ratio of the number of quasars to the number of stars is 7:1, and hence, either class is not equally represented to the classifier. In quasar-star classification, the stars' class dominates the quasars' class. This causes an increase in the influence of the stars' class on the learning algorithm and results in a higher accuracy of classification. In artificial balancing, an equal number of samples from both the classes are taken for training the classifier. This eliminates the class bias and the data imbalance.

Without artificial balancing, the dataset used for analysis uses a larger number of samples belonging to the stars' class as compared to the number of samples in the quasars' class. The samples that are classified as belonging to the stars' class are more when compared to the number of samples classified as belonging to the quasars' class as the voting for the dominating class increases with imbalance and results in a higher accuracy of classification. Hence, the voting for the stars' class was found to be 99.41% which is higher than the voting of quasars, which is 98.19%, by [12]. The accuracy claimed is doubtful as data imbalance, and class bias is prevalent.

4.2 Separability Test

A *separability test* is used to determine the nature of the separability of data. In particular, if the data are not linearly separable, certain classifiers may not work well or may not be appropriate.

The *convex hulls* of different classes in the dataset provides us with an indication of separability: The convex hull of a given set of points is the smallest *n*-dimensional polygon which can adequately envelope all the points in the respective set. In general, if the convex hull of at least any two classes of any data set intersects or overlaps, then it may be concluded that the classes in the data are not linearly separable.

In the existing literature on quasars-stars classification, a strong justification is not provided for the use of an RBF kernel. However, in Fig. 1 is observed that the

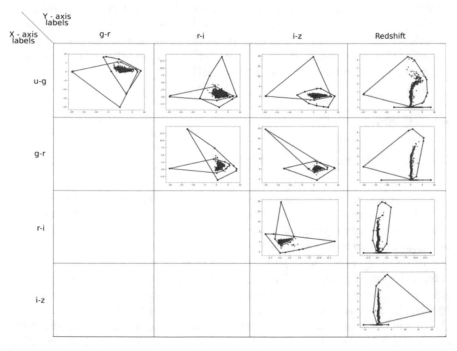

Fig. 1 Convex hulls across every pair of features show that the two classes can be approximately wrapped into two separate, non-overlapping polygons when considering redshift as a feature. The data points belonging to the class of quasars are plotted in red, and those belonging to the class of stars are plotted in blue

majority of the data belonging to the class of stars are not present within the convex hull of the class of quasars. Thus, the two classes in the dataset are mostly linearly separable, and SVM can be used here. Since the data exhibits linear separability, an RBF kernel need not be used.

4.3 Support Vector Machine

An SVM classifier requires the data to be separable so that it is possible to yield a hyperplane separating both the classes. Consider a set of n samples from the data set and two classes C_1 and C_2 corresponding to quasars and stars, or vice versa. Let x be the input matrix with n rows corresponding to the n data points and an array y with n elements, where the jth element of y is the class label of the jth row in x. Out of the set of n points, a pair of points, created by taking one from either class, is used to create a support vector S. Each point is then added to the support vector S. The position of samples from both the classes is determined in a five-dimensional support vector (the five dimensions being $u - g$, $g - r$, $r - i$, $i - z$, and Z); any points which are geometrically present on the wrong side of the hyperplane by virtue of their class belongingness are added to a vector V such that $S = S \cup V$. If any coefficients are negative due to the addition of V to S, then such points are pruned.

4.4 SVM–KNN

The K-nearest neighbor (KNN) classifier is a simple method for algorithmic classification, based on geometric similarity of the K closest training samples in the feature space. When a previously unobserved sample is fed for classification to the KNN classifier, it searches the feature space for the K samples which are closest to the test sample. The K closest samples may belong to different classes; the learning algorithm selects the class to which the majority of the K nearest samples belong and determines it to be the class to which the test sample belongs. Here, the parameter K needs to be fed as an input and often depends on the data being explored. However, in practice, a value of K between 7 and 11 works well [8].

4.5 AdaBoost

Adaptive Boost or *AdaBoost* [5] is a general ensemble learning approach that makes use of the results of multiple weak learners to make a strong prediction. AdaBoost works in multiple rounds by incrementally training weak learners, where each successive weak learner tries to classify the misclassified samples of the previous learner, with increased weights on the misclassified samples. AdaBoost can be used on any learning algorithm, but the most popular learners for AdaBoost are short *decision trees* or *decision stumps* [3]. In the current study, the weak learners over which AdaBoost was used are decision trees with one level.

4.6 Asymmetric AdaBoost: Handling the Data Imbalance Problem Mathematically

The asymmetric AdaBoost algorithm [9] aims to incorporate initial costs of misclassification in order to make the AdaBoost algorithm more sensitive to biases.

Consider a set of n training samples $(\mathbf{x}_i, y_i; i = 1, 2, \ldots, n)$ where $\mathbf{x}_i$ and y_i are the feature vector, and class label of the ith sample, respectively. Without loss of generality, it can be assumed that, the first m examples have class label $y_i = 1; i = 1, 2, \ldots, m$, and the remaining $n - m$ examples has class label $y_i = -1; i = m + 1, m + 2, \ldots, n$, corresponding to the classes of quasars and s-tars, respectively. Here, m = 74,463 and $n - m$ = 430,827. Let us define a weight distribution $D_t(i); t = 1, 2, \ldots; i = 1, 2, \ldots, n$ over the whole training set where the index t denotes the tth iteration of the AdaBoost algorithm, and the total number of iterations is equal to 1000. The weak learner selects the best classifier according to the weight distribution. In regular AdaBoost, the initial weights are usually assigned as $D_t(i) = 1/n, \forall i$. After each iteration, the weight distribution is modified in such a way that misclassified samples get a higher penalty than the correctly classified samples: This is similar to regular AdaBoost. However, an asymmetric behavior is observed in AdaBoost: While updating weights in successive iterations, it treats the misclassification of positive samples and negative samples equally. But there may be situations where misclassification of a positive sample may be more expensive than that of a negative sample, which introduces an asymmetry to the problem. Asymmetry can also be introduced when the number of samples belonging to one class dominates over that of the other. The classification power of regular AdaBoost diminishes as such asymmetry increases.

5 Results

5.1 Results Obtained Using the Unbalanced Data Set

The ROC curves of SVM, SVM–KNN, and AdaBoost on an unbalanced dataset are shown in Fig. 2a–c, respectively. The accuracies of these methods are 98.6, 98.86, and 97.2% respectively, as shown in Table 1. Notably, the difference between the sensitivity and specificity of SVM and SVM–KNN is approximately 9%.

Table 1 Results of classification of unbalanced dataset: F-score is an essential measure of the performance of any classifier applied to an unbalanced dataset, which has been ignored in the available literature

Methods	Accuracy (%)	Sensitivity	Specificity	Fscore
SVM	98.6	0.9150	0.9937	0.9551
SVM–KNN	98.86	0.9159	1	0.9159
AdaBoost	97.2	0.9012	0.9129	0.9406
Asymmetric AdaBoost (new contribution)	99.99	1	1	1

Table 2 Results of classification of balanced dataset: the accuracy of classification drops when data is balanced

Methods	Accuracy (%)	Sensitivity	Specificity
SVM	96.92	0.9576	0.9808
SVM–KNN	97.87	0.9575	1
AdaBoost	96.54	0.9663	0.9645

5.2 Results After Artificially Balancing the Data Set

The ROC curves of SVM, SVM–KNN, and AdaBoost after artificial balancing are shown in Fig. 2a–c, respectively. The accuracies of these methods are 96.92, 97.87, and 96.54%, respectively, as shown in Table 2. Notably, the difference between the sensitivity and specificity of all the models is negligible; in the case of AdaBoost, both the sensitivity and specificity are about 5% higher compared to the values attained with an unbalanced dataset. In this case, there is no requirement to report the F-score as it is a metric that should be used in the case of unbalanced or biased datasets. The method of artificial balancing does well to reduce the effects of bias, as seen from the small difference between sensitivity and specificity.

5.3 Results of the Asymmetric AdaBoost Classifier

The entire dataset was split into training and testing sets. Weights were assigned to both the classes: The stars' class was assigned a weight of 0.10, and the weight of the quasar class is kept constant, and equal to 1 (these numbers were selected based on iterative experimentation with different values of initial weights). The mean accuracy of classification was 99.9995% after running the asymmetric AdaBoost classifier for 1000 iterations. The ROC curve of this method is shown in Fig. 2g.

Simply put, an appropriate weight initialization arrives at the best weight distribution for a given number of estimators faster than equal initial weights. The ROC

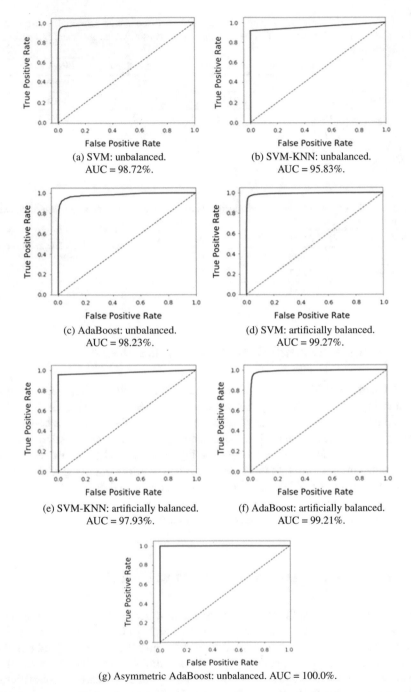

(a) SVM: unbalanced.
AUC = 98.72%.

(b) SVM-KNN: unbalanced.
AUC = 95.83%.

(c) AdaBoost: unbalanced.
AUC = 98.23%.

(d) SVM: artificially balanced.
AUC = 99.27%.

(e) SVM-KNN: artificially balanced.
AUC = 97.93%.

(f) AdaBoost: artificially balanced.
AUC = 99.21%.

(g) Asymmetric AdaBoost: unbalanced. AUC = 100.0%.

Fig. 2 ROC curves of the different methods explored: all the values for *area under the curve* (AUC) are provided with the plots, for the different cases. Note how the values of AUC are more for the balanced cases, as compared to the unbalanced cases

Table 3 Comparison of accuracies of classification achieved by Gao et al. (2008), Elting et al. (2008), Peng et al. (2013) before and after artificial balancing: accuracy drops after balancing

Methods	Accuracy before balancing (%)	Accuracy after balancing (%)
Gao et al. (2008)	97.55	96.92
Elting et al. (2008)	98.5	96.92
Peng et al. (2013)	98.85	97.87

curve plotted for asymmetric AdaBoost is shown in Fig. 2g. Asymmetric AdaBoost tends to classify positive samples more carefully when compared to negative samples as it corrects the misclassification. Its precision and recall values are found to be equal to 1. The value of F-score is also equal to 1, as shown in Table 1.

In the design of any experiment, there exists an inherent trade-off between good results and time of execution. Using asymmetric AdaBoost improves the time of execution, while best-preserving accuracy (Table 3).

6 Discussion

We implemented the methods for quasar-star classification which are already reported in the literature, with and without artificial balancing. An accuracy of 98.6% was obtained for SVM and 98.86% accuracy for the SVM–KNN method without artificial balancing. The artificial balancing of the dataset was accomplished by considering an equal number of quasar and star samples for classification (which is equal to the number of quasars in the dataset, as the quasars' class has lesser number of samples). The accuracy of classification of artificially balanced data drops from 98.6 to 95.8% for SVM with linear kernel and 98.86 to 97.05% for the SVM–KNN method. This is shown in Tables 1 and 2.

The choice of classifiers has further been verified by exploring the separability of the data. The data are not separable across the axes of $u - g$, $g - r$, $r - i$, and $i - z$: Using any of these four features alone, it is very difficult to discern between the two classes, as the majority of the data points are overlapping, or very close to each other in the feature space, along these axes. However, when considering the redshift (Z), we can observe that the data are considerably separable based on this feature. There is a slight overlap, near the edge or corner points of the quasars' class (this can be observed by inspecting the corner points of the convex hull of the quasars' class): Since the overlap is very little, SVM is an appropriate method to be explored as a classifier. However, the slight overlap results in accuracy of 96.92% by SVM (Table 2) and not 100%. On the other hand, tree-based classifiers work by *multiple recursive partitioning* of the feature space, and hence, in general, are the choice of classifiers for datasets which are mostly linearly inseparable. Since the overlap is not much, with the appropriate initial weights and with the cumulative effect of the remaining features, asymmetric AdaBoost resulted in an accuracy which is near perfect (Table 1)!

7 Conclusion

Asymmetric AdaBoost is endowed with greater computational efficacy compared to SVM. Given high accuracy, fast speed and easy modulation of parameters in contrast to SVM, asymmetric AdaBoost is a good choice as a classifier as specified in Tables 1 and 2.

The approaches explored in this paper can be used to solve the star-quasar classification problem in particular, and other problems in astronomy in general. These classifiers can be used to classify multi-wavelength astronomical data sources and pre-select quasar candidates for large surveys. The paper is firmly focused on scientific correctness and algorithmic relevance. Different ML approaches have been discussed and should be interpreted in that light, not as a suite of trial and error approaches to pick the better ones.

References

1. Abazajian, K.N., Adelman-McCarthy, J.K., et al.: The seventh data release of the sloan digital sky survey. Astrophys. J. Suppl. (2009). https://doi.org/10.1088/0067-0049/182/2/543
2. Adelman-McCarthy, J.K., Agüeros, M.A., et al.: The sixth data release of the sloan digital sky survey. Astrophys. J. Suppl. (2008). https://doi.org/10.1086/524984
3. Basak, S., Saha, S., et al.: Star galaxy separation using adaboost and asymmetric adaboost (2016). https://doi.org/10.13140/RG.2.2.20538.59842
4. Elting, C., Bailer-Jones, C.A.L., Smith, K.W.: Photometric classification of stars, galaxies and quasars in the sloan digital sky survey DR6 using support vector machines. In: AIP Conference Proceedings (2008). https://doi.org/10.1063/1.3059095
5. Freund, Y., Schapire, R.E.: Experiments with a new boosting algorithm. In: Saitta, L. (ed.) Proceedings of the Thirteenth International Conference on Machine Learning (ICML 1996), pp. 148–156 (1996)
6. Gao, D., Zhang, Y., Zhao, Y.: Support vector machines and kd-tree for separating quasars from large survey data bases. Mon. Not. R. Astron. Soc. **386**, 1417–1425 (2008). https://doi.org/10.1111/j.1365-2966.2008.13070.x
7. Hambly, N.C., Irwin, M.J., MacGillivray, H.T.: The SuperCOSMOS Sky Survey II. Image detection, parametrization, classification and photometry. Mon. Not. R. Astron. Soc. **326**, 1295–1314 (2001). https://doi.org/10.1111/j.1365-2966.2001.04661.x
8. Hassanat, A.B., Abbadi, M.A. et al.: Solving the Problem of the K Parameter in the KNN Classifier Using an Ensemble Learning Approach (2014). Available via arXiv.https://arxiv.org/abs/1409.0919
9. Landesa-Vázquez, I., Alba-Castro, J.L.: Shedding light on the asymmetric learning capability of AdaBoost. Pattern Recogn. Lett. **33**(3), 247–255 (2012). https://doi.org/10.1016/j.patrec.2011.10.022
10. Miller, A.A., Kulkarni, M.K., et al.: Preparing for advanced LIGO: a stargalaxy separation catalog for the Palomar transient factory. Astron. J. **153**(2), 73 (2017)
11. O'Mullane, W., María, N.L., et al.: Batch is back: CasJobs, serving multi-TB data on the Web. Available via Microsoft's website (2005). https://www.microsoft.com/en-us/research/wp-content/uploads/2005/02/tr-2005-19.pdf
12. Peng, N., Zhang, Y., Zhao, Y.: A SVM-kNN method for quasar-star classification. Sci. China Phys. Mech. Astron. **56**(6), 1227–1234 (2013). https://doi.org/10.1007/s11433-013-5083-8

A Supervised Method to Find the Relevance of Extracted Keywords Using Deep Learning Approaches

Rajesh Kumar, Gopichand Agnihotram, Pandurang Naik and Suyog Trivedi

Abstract Keyword relevancy refers to how relevant, or vital, certain keywords or phrases are to any kind of task driven by the users. From the documents, we derive the keywords and its relevancy to predict the context and importance of the documents. If user tries to target for too many keywords, it is difficult to analyse them and important information will be missed out. As we have to identify right keywords based on the context of the documents, in this paper, we will discuss the extracted keywords' relevancy in limited knowledge on known relevant keywords using deep learning approaches. We applied this in the domain of resume database where we have thousands of resumes and we want to arrive technical skills which will be the relevant keywords in this case from those resumes. This will help the recruiters to automate the job description of different roles and also in identifying the resumes of candidates with the key skills of hiring. Here, we used convolutional neural network (CNN) method, a deep learning approach to identify the technical skills and its relevancy from the resume documents. We trained the CNN model with thousands of documents with limited known technical skills and known non-technical skills from the history of resume database. This is a supervised approach to predict the new technical skills' relevancy from the new resume database.

Keywords CNN (Convolutional neural networks) · Skill relevancy
Word embeddings · Supervised learning · Training model · Job description

R. Kumar (✉) · G. Agnihotram · P. Naik · S. Trivedi
Wipro Technology Limited, Wipro CTO Office, Wipro, Bangalore 560100, India
e-mail: rajesh.kumar133@wipro.com

G. Agnihotram
e-mail: gopichand.agnihotram@wipro.com

P. Naik
e-mail: Pradeep.naik@wipro.com

S. Trivedi
e-mail: strivedi2505@gmail.com

© Springer Nature Singapore Pte Ltd. 2019   837
A. Abraham et al. (eds.), *Emerging Technologies in Data Mining and Information
Security*, Advances in Intelligent Systems and Computing 814,
https://doi.org/10.1007/978-981-13-1501-5_73

1   Introduction

Keyword relevance is an important topic for contextually driven information. We have multiple documents and want to infer the keywords relevant in the document to obtain the context of the documents. The individuals will run the topic models or frequency-based models to derive the relevant keywords. These are the unsupervised methods to derive the relevant or important keywords from the documents. Here, we are addressing the keyword relevance in supervised way where we know some keywords which are relevant and some keywords which are non-relevant. We have created training dataset using these two labels (relevant and non-relevant) and trained model with deep learning approaches. The training model will predict for the new datasets whether the extracted keyword is relevant or non-relevant. This is a generic problem in identifying the extracted keyword relevancy.

We have applied this approach to automate the job description for hiring using history of resume database for each role. Hiring is an integral part of different organizations which involve a lot of manual effort. To automate many stages involved in hiring like sorting relevant resumes and ordering them requires a program to understand and parse resumes properly. In this process, the system requires to identify the relevant technology skills from resume database on each role. Since resumes are not having any predefined template where user can easily identify the technology skills sections to extract the technology skills. These resume documents are unstructured in nature, and designing a generic skill extractor becomes very difficult. In the same way, rule-based approaches take a lot of human effort in analysing and making rules for all resumes of different templates and they do not work as and when new template arrives. Here, we address this application using deep learning models as it takes entire resume database as input and able to predict the technology skills for the new resumes.

2   Literature Survey

Named-entity recognition (NER) has been widely studied in NLP using deep learning approaches along with supervised machine learning algorithms like conditional random fields (CRF) models, hidden Markov models (HMM). With increase in computational power, deep learning models like LSTM, RNN are also being used for named-entity recognition.

Sequential models like LSTM and RNN are extensively used for sequence tagging task in text, speech, and images. Wang et al. have used a bidirectional LSTM and word embeddings for POS tagging task [1]. Huang, Yu, and Xu have also used LSTM and bidirectional LSTM with a CRF layer for POS and NER tagging task [2]. Using LSTM, they are able to capture long-term dependencies between output and input feature vectors. BI-LSTM models can find correlation between both past and future samples for tagging. To use sentence-level tags, they have added a CRF on top of LSTM. But LSTM and RNN models are sequential in nature, so to do prediction about next sample in the sequence, you need to wait for prediction for previous sample. So, it cannot be parallelized. Alternatively, CNN has also been used in tagging task. Advantage of using CNN comes from its parallelizability, because in the model prediction for samples in the sequence it can be computed independently. Strubell et al. [3] have used iterated dilated convolutional neural network (ID-CNN) to bypass this limit, and using iterated dilated CNN, they can capture large contexts in the documents. Kim [4, 5] has also used CNN for different NLP tasks like sentiment analysis and question classification and achieved state-of-the-art results. Author had shown that word-vector embedding computed by Mikolov et al. [6] is universal and can be used for tasks other than what they are trained for. The authors have also described using pretrained and task-specific word-vector embedding in multiple channels to use both for prediction. In another article [7], the authors have used CNN for POS tagging task.

As per the literature, the classical machine learning techniques and deep learning approaches describe the named-entity recognition (NER) problems in NLP and also some unsupervised methods such as LDA, TFIDF methods to compute keyword relevance based on their scores with some cut-off. In our paper, we propose a supervised method using deep learning approach to predict the keyword relevancy of the documents, and with this, user can be easily get to know the context of the documents.

3 Keyword Relevancy Models

Deep learning is a class of machine learning algorithms and is having good number of applications in different domain problems. Here, we will use convolutional neural networks as part of deep learning approach to predict the keyword relevancy. The details about the algorithm are as given below.

3.1 Convolutional Neural Networks

Convolutional neural network has been widely used in computer vision applications such as image classification, image segmentation. In natural language processing, Long short-term memory (LSTM) and recurrent neural network (RNN) models are preferred because of sequential nature of the problem. Advantage of using LSTM and RNN comes from the ability of these models to capture long-term dependencies in sequences. But these models are hard to train and require more computational power. CNN models are 10x faster and perform better to capture short-term dependencies of input sequence. The training architecture is given in Fig. 1.

In this work, we describe a method to use CNN to capture context features of a target word (a keyword to find the relevancy) and later use this context along with

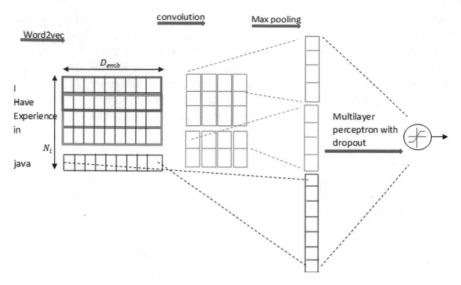

Fig. 1 Training architecture using convolutional neural network

vector embedding of the word to make the prediction about relevancy of the keyword. We show that model can distinguish between different contexts and is able to predict different relevancy accurately in different contexts for same target word. For each document in corpus after preprocessing, tokenizing, and stemming, it can be represented as list of tokens $[T_1, T_2, \ldots, T_n]$ where T_i is ith token in document. We replace all tokens with their vector embedding obtained in Sect. 3.1.2 and get sequence $[T_1, T_2, \ldots, T_n]$ where $T_i \in \mathbb{R}^m$, where m is dimension of vector embedding, is vector embedding of ith token in the sequence. Note that symbol T_i has been used for both—a word and its embedding, interchangeably as they are uniquely mapped. For $k = 1, 2, \ldots, n - w + 1$, we construct the samples as given in (1):

$$X_k = \left[T_k, T_{k+1}, T_{k+2}, \ldots, T_{k+w-1}\right]$$

$$Y_k = \begin{cases} 1, & T_{k+w-1} \text{ is a relevant keyword} \\ 0, & \text{otherwise} \end{cases} \tag{1}$$

Now, we can write the tokens' probability values as given in (2)

$$P\left[T_{k+w-1} \text{ is relevant } |T_k, T_{k+1}, \ldots, T_{k+w-1}\right]$$
$$= P\left[Y_k = 1 | T_k, T_{k+1}, \ldots, T_{k+w-1}\right]$$
$$= P\left[Y_k = 1 | F(T_k, T_{k+1}, \ldots, T_{k+w-2}), T_{k+w-1}\right] \tag{2}$$

where $F(T_k, T_{k+1}, \ldots, T_{k+w-2})$ represents semantic context of the target word (a keyword to find the relevancy). We train the model to predict the probability of a word being a relevant keyword given its context.

Model Architecture

The architecture of the model is described below—same as depicted in Fig. 1

a. **Embedding layer**—This layer converts a word into its vector embedding. It stores trained vector embeddings of the words in the vocabulary. Its dimensions are number of words in the vocabulary * dimension of the vector embeddings. Input to this layer is N_w-dimensional vector of words, and output of this layer is sequence of vector embeddings of those words which is a matrix of dimension $N_w * m$.

b. **Convolutional filters**—Convolutional filters are used to find new features from input. A convolutional filter operation can be written as given in (3):

$$c - f(W \quad X \mid b) \tag{3}$$

where

W matrix of dimension $h * m$, m is size of embedding dimensions, and h is number of words used in convolution.

X concatenation of m-dimensional vector embeddings of h words producing a matrix of dimension $h * m$.

b a scalar representing bias term.

f a nonlinear function (ex—hyperbolic tangent).

c output of convolution.

We are using a total of N_C number of convolutional filters with varying values of h. This layer helps in capturing abstract semantic meaning of context which helps in identifying if target word is relevant or not.

c. **Max pooling**—Each convolutional filter of size $h * m$ produces a feature vector of dimension $N_w - h + 1$ where N_w is size of context used in the CNN. We use max pooling to extract most relevant feature of this vector [4]. Using N_C number of convolutional filters, we get N_C-dimensional feature vector after max pooling.

d. **Regularization**—To avoid overfitting, we are using dropout [8] after extracting features by max pooling.

e. **Classifier**—Features obtained from max pooling layer is concatenated with vector embedding of target word and used as final feature vector for classification task. It is input to fully connected layers which are followed by SoftMax layer for binary classification. We use rectified linear unit (ReLu) as activation function for fully connected layers. Output of SoftMax layer gives the probability of a word being relevant which is the relevancy score of the target word.

The training model involves preprocessing the data, creating a word embedding for each token, training data preparation, and finally training the model using CNN approach. These steps will be explained in the following sections as given below.

3.1.1 Preprocessing

We will create a corpus using different documents, and documents in the corpus contain different types of noises. The documents need to be preprocessed before processing them to the next steps; that means we need to remove these noises. This step may vary on type of input documents, type of application, and keywords we want to extract. We convert all the documents in the corpus to lower case and remove noises like header, footer, cardinals, special characters, stop words such as "is", "the", "are" from them. Finally, we tokenize the words and perform stemming. Stemming helps us in reducing vocabulary size by recognizing different forms of verbs and nouns as same keyword. For example, after stemming "fishing", "fished", and "fisher" to the root word, "fish", "running" and "runs" both will be mapped to "run". After preprocessing each document, they are transformed to sequence of tokens which are the keywords of our interest.

3.1.2 Computing Semantically Similar Word-Vector Embedding

Word2vec is nothing but vector embeddings of tokens. There are many types of embeddings used in different types of NLP tasks like Tf-idf, continuous bag of words (CBOW), or skip-gram vector embeddings. Tf-idf or one hot encoding vectors are sparse in nature, but CBOW and skip-gram are continuous in nature and have been shown to capture semantic meaning of words. It has been shown that CBOW word embeddings of words with similar meaning are closer in embedding space. We will use these embeddings as input so that model will give similar output even if synonyms are used. Also, by using vector embeddings we are computing extended features for existing features. We compute vector embeddings for words in our vocabulary using continuous bag of words model (CBOW) [6]. It is entirely unsupervised method. We also use GloVe [9] vector embeddings as they are trained on larger corpus (100 billion words from Google news) and hence more robust. If a vector embedding of a word is present in GloV, we use it; otherwise, we use embeddings learned on our own corpus using CBOW. For CNN training, we initialize the embeddings with these CBOW embeddings. We will further fine-tune word embedding during training.

3.1.3 Training Data Preparation

To train the model using supervised method, we prepare a labelled training dataset. In this classification task, we will have two classes as output—relevant and Non-relevant. Relevant keywords are those keywords in the document which are of our interest and vice versa for non-relevant keywords. We already have a small vocabulary of relevant and non-relevant keywords. We build a vocabulary of all keywords in the corpus and label them manually as relevant or non-relevant using known relevant keywords and non-relevant keywords and make two vocabularies—$V_{Relevant}$ and $V_{nonRelevant}$. For each document, after preprocessing we get a sequence of tokens and replace the elements in sequence with their word embeddings and these are the extended features of words. So, we get a new sequence $[T_1, T_2, \ldots, T_n]$ where T_i is vector embedding of ith token in the sequence. We use a shifting window of size W (W is fixed number. It can vary from 1, 2, 3.., r; $r < n$) to make samples from this sequence. We take elements from k to $(k + W - 1)$th index to construct a sample of size W, and its label is decided based on target word. Target word is the word at the last index $(T_k + W - 1)$ of the sample. If target word is element of $V_{relevant}$, then its label is taken as 1. If target word is element of $V_{nonRelevant}$, then label is made to be 0. If target word is neither in $V_{Relevant}$ nor in $V_{nonRelevant}$, we ignore such samples.

For $k = 1, 2, \ldots, n - w + 1$, we construct samples as given in (4)

$$X_k = \begin{bmatrix} T_k, T_{k+1}, T_{k+2}, & \ldots, T_{k+w-1} \end{bmatrix}$$
$$Y_k = \begin{cases} 1, & T_{k+w-1} \text{ is a Relevant Keyword} \\ 0, & \text{otherwise} \end{cases} \qquad (4)$$

To separate these samples into training and validation sets given in (5), we randomly split $V_{Relevant}$ and $V_{nonRelevant}$ in the ratio of 80:20 and get four vocabularies $V_{Relevant/train}$, $V_{Relevant/val}$, $V_{nonRelevant/train}$, and $V_{nonRelevant/val}$. For compact representation, let us define

$$V_{train} = V_{nonRelevant/train} \bigcup V_{relevant/train}$$
$$V_{val} = V_{relevant/val} \bigcup V_{nonRelevant/val} \qquad (5)$$

Now using V_{train} and V_{val}, we will divide above dataset into two sets —training (S_{train}) and validation (S_{val}). Now for kth sample $\begin{bmatrix} T_k, T_{k+1}, T_{k+2}, \ldots, T_{k+w-1} \end{bmatrix}$, if target word T_{k+w-1} belongs to V_{train}, then we add it to training set S_{train}, and if it belongs to V_{val}, we add it to validation set S_{val}. We train the model using S_{train} and use S_{val} for evaluation.

3.1.4 Training

Finally, the model is trained using convolutional neural networks (CNNs) with extended features of dataset which we prepared in the above sections to minimize the prediction loss over training data. Performance of the model varies with the choice of loss function. We are using cross-entropy loss for binary classification. To train this model, back propagation is used. In this binary classification problem, we face problem of unbalanced dataset where number of samples in one class is more than that of in another class and if trained improperly model may get biased towards one class. We need to employ proper data balancing technique while training the model. Also, assessment of the model should be based on F-score instead of accuracy (K-fold cross-validation). For data balancing, while training with mini-batch gradient descent, we are using balanced mini-batches in each iteration. To generate a balanced batch of size N, we use randomly sampled αN positive samples and $(1 - \alpha)N$ negative samples, where $0.4 \leq \alpha \leq 0.6$.

After training the model, we will have new documents and keywords of those documents for predicting the relevancy of those keywords. Using CNN method, we can able to derive the relevancy of keywords in better way with the limited knowledge about the known keywords' relevancy. Once you know the relevant keywords in the documents which are the highly important keyword in the documents, this will help user to know the context of documents without going through those documents.

4 Proposed Methodology with Experiments

In this section, we will use the method described above for skill extraction from a large corpus of resume database available with the Indian IT industry. In this case, skills are relevant keywords and non-skills are non-relevant keywords. Using the method described earlier, we compute relevancy of each word in the document, which represents the likelihood of word being skill. We will collect the words which are more than predefined predicted relevancy threshold (let say 0.6) and those are skills (technical skills) from the resume document. These technical skills from resume documents will help the recruiters those who are hiring for different roles (such as Java developer, Dot Net developer, data scientists, big data analytics) in the company to create a job description to those roles. This will also help to obtain the resumes of candidates with the key skills of hiring. We have collected 150 thousand resumes for different roles hired in the company. The documents are of type .doc, .docx, pdf, and we convert them into text format. We perform following steps and train the model to extract skills from new documents which will help in generating job description automatically.

STEP 1: Preprocessing of resume documents:

Text-converted resumes contain different forms of noises, and we need to remove them. The preprocessing will defer from documents to documents. We remove non-ASCII characters, special characters ('|',':'), numbers, bullet characters, etc. We replace multiple contagious spaces with single space character and multiple contagious newline character with single newline character. We will tokenize the document using NLTK tokenizer [10] and perform stemming to reduce the vocabulary size. Since keywords of our interest contain alphanumeric characters, we remove all the tokenized keywords which do not contain any alphanumeric keyword. We remove all general stop words ("is", "was", "have", "i", etc.) which are present in the documents.

STEP 2: Creating Word embedding—Extended features of the words:

To obtain the vector embeddings of words in the vocabulary, we compute CBOW vector embedding of the documents. We follow the method described in Sect. 3.1.2

Table 1 Processing input text in each step

Steps	Stepwise result
Input text	I am a software developer. I have expertise in java and html
Preprocessing	software, develop, expertise, java, html
Manual labels by the user	(software, 0), (develop, 0), (expertise, 0), (java, 1), (html, 1)
Learning word embedding	$T_{\text{software}}, T_{\text{develop}}, T_{\text{expertise}}, T_{\text{java}}, T_{\text{html}}$ Where T_i is 100-dimensional vector embedding of i. *for example* $T_i = [0.01, 0.02, -0.05, \ldots,]_{1 \times 100}$ for $i = 1$
Preparing samples for training set	$(x_1, y_1) = ((T_{\text{software}}, T_{\text{develop}}, T_{\text{expertise}}), 0)$ $(x_2, y_2) = ((T_{\text{develop}}, T_{\text{expertise}}, T_{\text{java}}), 1)$ $(x_3, y_3) = ((T_{\text{expertise}}, T_{\text{java}}, T_{\text{html}}), 1)$

to get the vector embeddings. We keep the size of vector embeddings 100, size of context window $= 5$. We train the model for 10 iterations. Along with CBOW vectors trained on this corpus, we also use GloVe vector embeddings as described in Sect. 3.1.2.

STEP 3: Preparing training data using documents:

After preprocessing all resume documents in the corpus, we are creating a vocabulary V_{ref} by keeping only words with minimum term frequency f_{min} ($=100$) resume document frequency. To handle new words, which are not in V_{ref}, we add one more word UNKN in the vocabulary. We obtain word-vector embedding for all tokens in V_{ref} using method described in Sect. 3.1.2. To create labelled samples, we need a reference vocabulary of skills (relevant keyword) and non-skills (non-relevant keywords). We use a semi-automatic method to prepare this vocabulary. We first extract all the potential relevant keywords which follow a set of patterns, and an expert goes through them to label them as relevant or not. We use heuristics such as skills are separated by comma and usually they come after phrases—"experienced in" and "developed using". We extract all the words following these patterns and go through it manually and label them as skill and non-skill. All skills are added to the vocabulary V_{skill}, and non-skills are added to vocabulary V_{nonSkill}. Then to further extend the vocabularies, we label all words in V_{ref} as skill or non-skill. We augment existing vocabularies V_{skill} and V_{nonSkill} with these new labelled words by adding skill words to V_{skill} and non-skill words to V_{nonSkill}. Now for each document in the corpus, after preprocessing the document we get a list of tokens. Then, we use shifting window of size N_w ($=15$) to create samples from the document (discussed in Sect. 3.1.3). For each sample, label is assigned 1 if target word is a skill and 0 if it is not a skill. If target word is neither in skill vocabulary nor in non-skill vocabulary, then we ignore such samples.

For example, consider text "i am a software developer. i have expertise in java and html". Here after labelling manually, the relevant keywords/skills are java, html and non-relevant keywords/non-skills are software, develop, expertise. For $N_w = 3$ (shifting window size), the stepwise processing of the sentence is given in Table 1.

To separate these samples into training and validation sets, we randomly split V_{skill} and V_{nonSkill} in the ratio of 80:20 and get four vocabularies $V_{\text{skill/train}}$, $V_{\text{skill/val}}$, $V_{\text{nonSkill/train}}$, and $V_{\text{nonSkill/val}}$. For compact representation, let us define $V_{\text{train}} = V_{\text{nonSkill/train}} \bigcup V_{\text{skill/train}}$ and $V_{\text{val}} = V_{\text{skill/val}} \bigcup V_{\text{nonSkill/val}}$. Now using V_{train} and V_{val}, we will divide above dataset into two sets—training (S_{train}) and validation (S_{val}). Now for kth sample $\left[T_k, T_{k+1}, T_{k+2}, \ldots, T_{k+w-1}\right]$, if target word T_{k+w-1}, which is last word in sample, belongs to V_{train}, then we add it training set S_{train} and if it belongs to V_{val}, we add it to validation set S_{val}. Using above procedures, we have created a training dataset consisting of 0.6 million positives (represents skill/relevant keyword) and 36 million negative samples (represents non-skill/non-relevant) for a corpus consisting of 0.15 million documents.

STEP 4: Training the data:

We have kept the shifting window size $N_w = 15$. Two types of convolutional filter of sizes 3 and 5 with counts 50 each are used, and after max pooling, we get 100-dimensional feature vector layer. For training, we are using binary cross-entropy as loss function. For optimization, we are using Adam optimization [11] with parameters $lr = 0.001$, $\beta_1 = 0.9$, $\beta_2 = 0.999$, $\varepsilon = 1e - 08$, decay $= 0.0$. We are using a batch of 600 samples in each iteration. To avoid overfitting, 0.8 regularization has been used after max pooling. We trained our CNN model on training set S_{train} and obtained training accuracy 86% and validation accuracy of 82% on S_{val}.

4.1 Testing on New Resume Documents

In testing phase, to extract skills from a new resume documents first we convert it to text followed by preprocessing. Then, we create a set of samples using the procedure discussed in Sect. 3.1.3. For each sample s, we find the label for s by passing it through the trained CNN model. If the predicted label is 1 (skill), we take the target word (a keyword to find the relevancy) from the sample. Note that vector embeddings of words used during testing is same as the one during training. We extract all such target words to get the skills from the resumes which will help the recruiters to automate the job description of each role and in identifying the resumes of candidates with the key skills of hiring. **For example:** For $N_w = 3$, the model input and output are given in Table 2.

Table 2 Model output for testing document

Steps	Stepwise result
Input text	I have worked at Oracle for 5 years. I have hands-on experience in Oracle, MySQL, Spring
Model output	Oracle, MySQL, Spring—skills worked, years, experience, hands—non-skills

5 The References Section

The following section shows a sample reference list with entries for journal articles [5, 8], conference proceedings [1–4, 6, 7, 11], and URLs [9, 10].

References

1. Wang, P., Qian, Y., Soong, F.K., He, L., Zhao, H.: Part-of-speech tagging with bidirectional long short-term memory recurrent neural network. arXiv preprint arXiv:1510.06168 (2015)
2. Huang, Z., Xu, W., Yu, K.: Bidirectional LSTM-CRF models for sequence tagging. arXiv preprint arXiv:1508.01991 (2015)
3. Strubell, E., Verga, P., Belanger, D., McCallum, A.: Fast and accurate entity recognition with iterated dilated convolutions. In: Proceedings of the 2017 Conference on Empirical Methods in Natural Language Processing, pp. 2660–2670 (2017)
4. Kim, Y.: Convolutional neural networks for sentence classification. arXiv preprint arXiv:140 8.5882 (2014)
5. Bengio, Y., Ducharme, R., Vincent, P., Jauvin, C.: A neural probabilistic language model. J. Mach. Learn. Res. **3**, 1137–1155 (2003)
6. Le, Q., Mikolov, T.: Distributed representations of sentences and documents. In: Proceedings of the 31st International Conference on Machine Learning (ICML-14), pp. 1188–1196 (2014)
7. Yu, X., Faleńska, A., Vu, N.T.: A general-purpose tagger with convolutional neural networks. arXiv preprint arXiv:1706.01723 (2017)
8. Srivastava, N., Hinton, G.E., Krizhevsky, A., Sutskever, I., Salakhutdinov, R.: Dropout: a simple way to prevent neural networks from overfitting. J. Mach. Learn. Res. **15**(1), 1929–1958 (2014)
9. GloVe: Global Vectors for Word Representation. https://nlp.stanford.edu/projects/glove/
10. nltk.tokenize package. http://www.nltk.org/api/nltk.tokenize.html
11. Kinga, D., Adam, J.B.: A method for stochastic optimization. In: International Conference on Learning Representations (ICLR) (2015)

An Efficient Traffic Sign Recognition Approach Using a Novel Deep Neural Network Selection Architecture

Sourajit Saha, Md. Saiful Islam, Md. Asif Bin Khaled
and Suraiya Tairin

Abstract Traffic sign classification is an important aspect of autonomous driving systems. A slight improvement on classification performance can potentially lower the rate of car accidents. In view of this, we propose three different deep convolutional neural networks in a hierarchical pattern, yet not convoluted among themselves for classifying traffic sign. A very popular and reliable traffic sign dataset called GTSRB is used to train our proposed networks. In our work, we present a novel approach to classify images. Furthermore, we modify all three convolutional neural networks over some of the existing neural nets. While modifying the networks, we redesign them based on specific requirements which may also prove handy for other datasets. Along with the new methods, we are able to reduce the computational complexity as well. On top of the new architecture, we achieve a notably higher accuracy in performance of 99.92% surpassing the state-of-the-art performance of 99.81%. In a nutshell, we trained an artificial intelligence (AI) model that learns to chose between two different AI models while classifying an image.

Keywords Deep learning · Convolutional neural networks · Image classification
Computer vision

S. Saha (✉) · Md. S. Islam · Md. A. B. Khaled · S. Tairin
Department of Computer Science and Engineering,
BRAC University, 66 Mohakhali, Dhaka 1212, Bangladesh
e-mail: sourajit13301004@gmail.com

M. S. Islam
e-mail: md.saiful.islam@bracu.ac.bd

M. A. B. Khaled
e-mail: mdasifbinkhaled@gmail.com

S. Tairin
e-mail: suraiya@bracu.ac.bd

© Springer Nature Singapore Pte Ltd. 2019
A. Abraham et al. (eds.), *Emerging Technologies in Data Mining and Information
Security*, Advances in Intelligent Systems and Computing 814,
https://doi.org/10.1007/978-981-13-1501-5_74

1   Introduction

While driving a vehicle, it is sometimes very likely for drivers to misread traffic signs due to the dearth of attention, reflection of light, anonymous obstacles, and many other factors which can lead to abysmal accidents. Drawing attention to these appreciable problems, autonomous vehicles and advanced driver assistance systems (ADASs) took on the noble task of classifying traffic signs for human [2, 14]. Conventional machine learning and computer vision techniques were popular for classifying traffic signs; however, the industry now heavily relies on deep learning-based classifiers [11]. With the boon of Graphical Processing Unit (GPU), it is now more convenient and efficient to accomplish the task of traffic sign classification through deep learning and convolutional neural network-based computer vision models. These models try to learn the important features corresponding to each class through a network that is trained by biologically inspired neural structures. Nonetheless, these models do not quite mimic the complex neural network behavior of the human brain. This automated traffic sign recognition task is extremely challenging since the traffic sign images on which the model is trained are neither nice looking nor amiable to learn features all the time. The different camera sensors mounted over vehicles constantly take pictures and pass onto the system to classify the traffic sign. Often, extracting features from pictures becomes difficult using image processing as there are motion blur, light reflection, distortion, and many possible factors that come into play. There are existing work on traffic sign recognition (TSR) that uses boosting [17], Bayesian classifiers [15], and some ensemble classifiers with K-d trees and support vector machines (SVM) [12].

With all these approaches, it is highly likely to get stuck on local optimum and often end up training an overfitted model that works well on training images however fails to classify correctly when given a new image. A lot of convolutional neural network (CNN) methods have been applied on different traffic sign datasets that used handcrafted features like histogram of oriented gradient or HOG [5] and scale invariant feature transform or SIFT. However, using these handcrafted features are not useful for generic purpose as testing so many filters for every image and having a lot of images nesting into several subclasses can be very time-consuming. Consequently, there has to be a better technique that automatically learns the features for every class. Thus, adding a number of layers stacked inside a convolutional neural network gives this networks the feasibility to learn features itself and correct the errors while back-propagating. Another important term for TSR is classification accuracy which is the ratio of correctly classified testing images and total testing images expressed in percentage. Therefore, it goes without saying, the higher the accuracy, the better the model is. In our approach, a very reliable traffic sign data set called the German Traffic Sign Recognition Benchmark also known as GTSRB [20] is used for training and testing TSR and classification model. Figure 1 shows an overview of grayscale road traffic sign images in the GTSRB [20] dataset. In this paper, we make the following set of contributions:

Fig. 1 Overview of the GTSRB dataset

- We propose a new hierarchical multi-neural network approach using deep learning to classify image of road traffic sign.
- Besides, our proposed methodology also includes two newly customized neural network architectures which we utilize for efficient detection of traffic sign.
- Finally, our experimental evaluation confirms a significantly higher performance in accuracy for traffic sign detection on the GTSRB dataset.

2 Related Works

In the GTSRB competition [20], the winning team [4] obtained the highest accuracy of 99.46% exceeding the best human performance [21] of 98.84%. In their work, a total of 25 different CNN with three convolutional layers and two fully connected layers were used that made the network learn more than 88 Million parameters. Furthermore this method used data augmentation which does not ensure a reliable classification accuracy in general for unknown data which are the two most disadvantages this method holds. A recent research conducted [6] on GTSRB that uses a modified version of GoogLeNet [22] with batch normalization and spatial transformer layer [8] surpassed the state-of-the-art performance by achieving 99.81% accuracy.

Along with the deep learning methods, in [9], the authors use a number-oriented speed limit classifier which has been tested on 2880 images. This classifier achieves an accuracy of 92.4% for 1233 images. Though, this process cannot clarify that whether images of similar traffic sign samples are partitioned among sets. Clearly, the accuracy rate did not climb up the ladder until neural nets were applied and ever since the performance analysis between different models has been focused on the reliability of the network in terms of computational efficiency and test set classification accuracy rate. Consequently, another research [13] was conducted with a proposed CNN-based clustering algorithm to divide classes into k classes followed by a hierarchical CNN training the $k + 1$ classification CNNs that achieved 99.67% accuracy yet the model was a touch expensive in terms of computation.

Another research [16] held at the same year used max-pooling position (MPP) as classifier instead of multi-layer perceptron (MLP) and achieved 96.95% accuracy. They demonstrated why MPP is a better classifier in general though did not contribute much to the accuracy improvement on GTSRB dataset. A year prior to that another research team [24] trained GTSRB on a CNN for feature extraction nonetheless, for classifying they replaced fully connected layers with an extreme learning machine (ELM) which was fed with the CNN features previously learnt and gained 99.4% accuracy. This was a good performance given their gossamer computational complexity though they did not surpass the state-of-the-art performance.

3   Proposed Methodology

In this section, we discuss our proposed methodology and underlying considerations for devising our methodology. In our work, we propose a total of three distinct deep neural networks. The GTSRB dataset has a total of 31,367 training images, 7842 validation images, and 12,630 testing images. The images are distributed over a total of 43 classes. The classes are divided into a group of six subclasses in terms of similarity. All the images are extracted from a sequence of videos that are recorded at different time of the day under different lighting conditions, and the video is also recorded from a moving vehicle which takes care of the motion blur issue. We elaborate our approach for preprocessing the images of traffic sign in the next subsection.

3.1   Image Preprocessing

The images have huge amount of distortion in them due to different light setting thus some of the images are not even likely to be recognized by human. Since most of the images are over or under exposed and the backgrounds are mostly dark or bright, an effective solution to get rid of this flaw is applying histogram equalization to all the images before training, validating, and testing [1]. We evenly spread the most frequent intensity values through all the images. The images after histogram equalization as shown in Fig. 2 increases important spatial correlation among the pixels, though it results in a huge cumulative noise rise in the background.

Fig. 2 Image after
histogram equalization

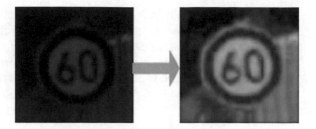

3.2 Deep Learning to Analyze Features

In this subsection, we present how our two distinct neural nets are made to analyze features. We constitute a total of three different networks in our work. The first one is inspired by both GoogLeNet [22] and VGG net [19]. This network has 13 layers. We modified the inception module of GoogLeNet. We also mounted a tiny VGG net-based neural network into our inception module. Unlike GoogLeNet, there is only six inception layers in our model. The max-pooling path of our inception is similar to VGG net in terms of layer depth. Consequently, as an image goes deeper through the max-pooling path of an inception module, we spread out the features through the spatial dimension. This setting ensures that the very first layer emphasizes only on the local features and the last layers get more knowledge about those features. To lessen the internal covariate shift of the network, we also use a batch normalization layer [7] right before the third inception module.

Figure 3 shows our modified inception module architecture. The second network $N2$ is a VGG like deep neural network with nine layers. Instead of starting from 64 unit deep kernels, our network starts from 32 unit depth and ends at 128 unit instead of 512 unit. Our first model works better for traffic signs which are more convenient for human eyes to mis-classify. Besides, $N2$ responds better to some other classes than the first network does on those. The trade-off here is that our first network works better overall, however has a higher number of parameters compared to the

Fig. 3 Modified inception
module architecture

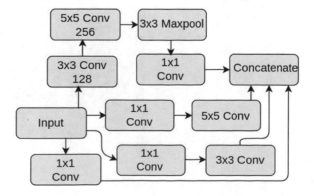

second one. On the contrary, $N2$ does not classify as well as the first network overall, nonetheless works significantly better on some specific classes. In a nutshell, a way to combine these two networks can provide significant improvements on accuracy even though that increases the number of parameters cumulatively. In the next subsection, we illustrate our approach to efficiently combine these two networks without added parameter upsurge.

3.3 Deep Learning Model for Network Selection

In this subsection, we illustrate our third network that chooses between the first two networks. At this point, it seems inevitable to keep the networks separate and the primary concern shifts to a completely different decision-making process. It is certain that, for images of certain classes, it is efficient to take the path to network 1 and for the other images network 2 is the better choice as shown in Fig. 4. This is where a third deep neural network comes into picture which decides if an image should take on the first network or the second one. We augment the dataset with only two classes where class 1 and class 2 correspond, respectively, to the first and second network. Our third deep learning model learns features from the images that correspond to the probabilistic decision of picking any one of the two networks that serve that belonging class of the image the best.

Our third deep convolutional neural network model has a structure which is very much like the VGG net. However, the working of this eight-layer network is different as after each pooling layer, there is a softmax classifier that backpropagates the error in earlier layers. These softmax classifiers are necessary since in most of the cases this network is only interested in learning the shape and color and some other anonymous features of the bounding box of traffic signs instead of looking at what is inside the sign. Nonetheless, the other two networks are behooved to perform that job. That is

Fig. 4 Deep neural network selection architecture

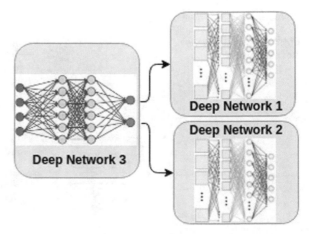

why we put the softmax classifiers in earlier stages to check if the network needs to learn furthermore or not about that image. Initially, after image preprocessing, we train the images on the first two networks. Sequentially, we train those images on the third network with data augmentation to learn which network to compute while predicting a class for any new image. Clearly, this is a new approach to classify images that have never been used before.

4 Experimental Evaluation

In this section, we discuss our network architectures, training, and testing specifications. The images vary in dimension by big margin, and also, there are major issues like occlusion due to obstacles, inconsistency in viewpoint due to motion. Addressing this problem, we resize both the testing and training images to 128×128 in the first place. After image preprocessing, we divide the training set into "train-only set" and "validation set". Both image preprocessing and all three neural network architectures are implemented, trained, and tested on deep learning library keras [3]. For processing, we have used NVIDIA GeForce 940MX which has a memory of 2 gigabyte. Sequentially, we feed the training images to our first and second networks, respectively. Then through applying the test images, we calculate the accuracy matrix for each network. From the accuracy matrix, we constitute the sub- class accuracy matrix of 6×1, since there are six subclasses. Then, we train our third network with augmented training images.

4.1 The First Network

In our first network, we inject a modified inception module where 1×1 convolutional layers are used before 3×3 and 5×5 convolutions to reduce dimension explosion in the concatenation layer. We reduce the max-pool path's steadily increasing convolutional pooling layer's dimension by stacking another 1×1 convolution on top of that. This way, our inception module itself can learn between four choices. The first choice is keeping the input features unchanged and not updating the features. The second and third options are going through 3×3 and 5×5 convolutions, respectively. The fourth choice which we modified in this research is to learn fewer features firstly and then if necessary learn more features and pool the most significant ones from within. Eventually, we concatenate all these feature maps to work as a single layer for the next phase of the network. Figure 5 shows the architecture of this network. Here, "Avg pool" stands for average pool and "Incept" stands for modified inception layers.

As the images go deeper through the network, we increase the kernel depth and the network becomes more confident on its learning. We use the softmax 0 layer to update parameters in backpropagation and softmax 1 layer to update parameters as

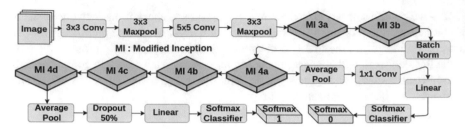

Fig. 5 Deep network 1 architecture

well as classifying images. For training, we split the images into batch of size 16 and utilize SGD with momentum 0.9 and consider weight decay of 0.002359. The initial learning rate is 0.0001, and we increase it by 10% after every 10 epochs. We use rectified linear unit (ReLU) [10] as activation function. Each batch has 32 images with 30 epochs in total. In Table 1, 3×3 reduce and 5×5 reduce refers to the amount of 1×1 filters we use before 3×3 and 5×5 filters respectively. Also, 3×3 pool reduce symbolizes the number of 1×1 filters we use after the 3×3 max-pooling layer. This 13-layer network gives an accuracy of 99.65% and learns a total of 11 million parameters.

4.2   The Second Network

In the second network, we maintain the idea of keeping the kernel size fixed as VGG net does. We also multiply the kernel depth by two after every pooling layer. We only use 3×3 kernels throughout the entire network to emphasis on correlation of one pixel to its eight other adjacent neighboring pixels. In the deeper layers, we provide our network more depth to better understand the features. Unlike VGG net, before expanding the kernel depth, we apply a dropout to avoid redundancy as well. The network therefore can learn features which are only relevant to pass onto the next phase of the network.

We design each batch with 16 images and run 30 epochs in total. We use SGD with momentum 0.9 to update the network parameters and rectified linear unit (ReLU) [10] for activation. We set the weight decay to 0.00001 and initial learning rate to 0.01 which we update every 10 epochs. The network architecture is shown in Fig. 6. Table 2 demonstrates the kernel dimensions and parameter numbers. This nine-layer model gives an accuracy of 98.92% which is prior to human performance of 98.84%. This model learns a total of 1.35 million parameters.

Table 1 Details of deep network 1

Type	Filter-size	Output size	1 × 1	3 × 3	3 × 3 reduce	5 × 5	5 × 5 reduce	3 × 3 pool-reduce	Parameter
Conv	3 × 3	128 × 128 × 32	0	0	0	0	0	0	0.9K
Maxpool	3 × 3	64 × 64 × 32	0	0	0	0	0	0	0
Conv	5 × 5	32 × 32 × 48	0	0	0	0	0	0	38.5K
Maxpool	3 × 3	16 × 16 × 48	0	0	0	0	0	0	0
Incept 3a	0	16 × 16 × 300	24	64	24	128	24	84	253K
Incept 3b	0	16 × 16 × 436	64	116	64	200	128	56	603K
Incept 4a	0	8 × 8 × 512	128	96	200	256	224	32	856K
Incept 4b	0	8 × 8 × 512	96	128	256	240	380	48	1.4M
Incept 4c	0	4 × 4 × 1024	256	256	196	256	320	256	3M
Incept 4d	0	4 × 4 × 1024	280	148	524	356	400	240	4.9M
Avg pool	4 × 4	1 × 1 × 1024	0	0	0	0	0	0	
Linear	0	1 × 1 × 43	0	0	0	0	0	0	0

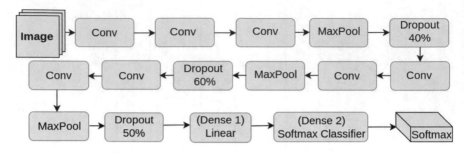

Fig. 6 Deep network 2 architecture

Table 2 Details of deep network 2

Type	Kernel size	Output size	Parameter
Conv	3 × 3	64 × 64 × 64	1.7K
Conv	3 × 3	48 × 48 × 32	3.5K
Conv	3 × 3	46 × 46 × 32	9.2K
Maxpool	2 × 2	23 × 23 × 32	0
Conv	3 × 3	23 × 23 × 64	18K
Conv	3 × 3	21 × 21 × 64	36K
Maxpool	2 × 2	10 × 10 × 64	0
Conv	3 × 3	10 × 10 × 128	74K
Conv	3 × 3	8 × 8 × 128	.14M
Maxpool	2 × 2	4 × 4 × 128	0
Dense 1	0	1 × 1 × 512	1.05M
Dense 2	0	1 × 1 × 43	22K

4.3 Accuracy of the Two Networks

In this section, we discuss how the 43 classes which are numbered from 0 to 42 are divided into six subclasses. The first subclass is called "Blue" and contains eight classes including class 33 to class 40. The second one is called "Danger" and contains 15 classes including class 18 to class 31 and 11. The third one is called "End-of" and contains four classes including class 6, 32, 41, 42. The fourth one is called "Speed" and contains eight classes including class 0 to class 5 and 7, 8. The fifth one is called "Red-other" and contains four classes including class 9, 10, 15, 16. The sixth one is called "Spezial" and contains four classes including class 17, 12, 13, 14. As we observe, the first network ends up working better on subclass 2, 3, 6, and the second network on subclass 1, 4, 5. Table 3 contains details of subclass accuracy from the two networks.

Table 3 Subclass accuracy of proposed method and existing methods

Method	Blue (%)	Danger (%)	End-of (%)	Speed (%)	Red other (%)	Spezial (%)
Deep network 1	99.52	99.93	99.98	99.03	99.89	99.94
Deep network 2	99.96	99.17	97.19	99.92	99.95	98.27
Haloi [6]	99.72	99.89	99.95	99.86	100	99.87
Committee of CNNs [4]	99.89	98.03	94.44	98.61	99.87	98.63
Human performance [21]	99.72	98.67	98.89	97.63	99.93	100

4.4 Third Network with Data Augmentation

In this section, we focus on how we augment the dataset and feed into our third neural network. For the newly augmented dataset, we label images belonging to class 6, 11–14, 17–32, 41, 42 as 1. Subsequently, we label images belonging to the remaining classes as 2. Therefore, network 1 better predicts on images labeled as 1 and network 2 better predicts on images labeled as 2. The third network has to learn which network to choose when an image is given. Our eight-layer network which is our third network has a total of three softmax classifiers. We divide this network into three phases where each phase consists of two 3×3 convolutions and a 2×2 max-pool followed by the next phase of the network and a softmax layer, except for the last phase which only has a softmax layer. Like our second network, we utilize SGD considering momentum 0.9 and decay 0.0005. We set our initial learning rate to 0.0001 and increase by 10% every four epochs. This networks compute very fast in terms of updating parameters due to the addition of extra softmax layers. Figure 7 shows the network architecture that we implemented. Dividing each batch with 32 images and running a total of 30 epochs, we gain 99.94% accuracy. This means 9994 out of every 10,000 images know which of the two networks is best for them. Since this architecture is similar to our second model, the number of parameters are almost same.

This model learns a total of 1.37 million parameters. Figure 8 demonstrates features learned through different layers of the second network.

4.5 Comparison with the Existing Solutions

Our research has exceeded the current state-of-the-art performance. The current state-of-the-art performance method [6] uses about 10.5 million parameters and has

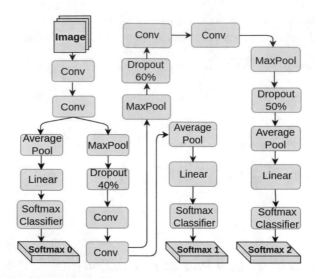

Fig. 7 Deep network 3 architecture

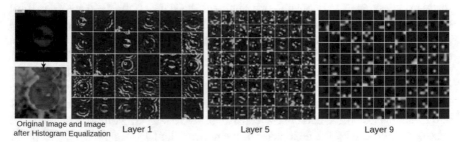

Original Image and Image
after Histogram Equalization   Layer 1   Layer 5   Layer 9

Fig. 8 Features learned in different layers of the second network (N2)

a 99.81% accuracy with three spatial transformer layer mounted in the network, whereas in our research, we have reached an accuracy of 99.92%. Combining the three deep neural networks, our model learns as much as 13.72 million parameters to accomplish the image classification task. Although our method holds two extra deep networks and around 3 million extra parameters than the current best solution; however, with this trade-off we have reached a new height in terms of performance that autonomous vehicles systems shall find helpful.

Table 4 shows comparison among different solutions. Furthermore, the committee of CNNs [4] have used 25 distinct deep neural networks with data augmentation such as rotation, extension, translation applied to the dataset. However, in this work we have not done any such, thus eliminating the risk of data dependency and generalization failure.

Table 4 Classification accuracy comparison with state-of-the- art models

Method	Accuracy (%)
Our proposed method	99.92
Haloi [6]	99.81
Committee of CNNs [4]	99.46
Human performance [21]	98.84
Multi-scale CNNs [18]	98.31
Random forests [23]	96.14

5 Conclusion

In this paper, we presented a new technique to classify traffic sign images achieving increased accuracy in performance. The other existing approaches to traffic sign detection have shown a series of steps with multiple deep neural networks stacked together and nested deep networks. Either way that results in a colossal nested network that is prone to learn a huge number of redundant features. However, with our approach of a master network planted before the two other networks and modified architecture of VGG and GoogLeNet, we reduce the number of parameters all while improving the accuracy outperforming the previous solutions. The most exciting part of our work is the third network that learned itself how to choose between the first two networks, which has further research potentiality. Furthermore, we also combined multiple neural networks and avoided parameter upsurging successfully. Finally, due to the short number of parameters learned, our method is also scalable.

References

1. Acharya, T., Ray, A.K.: Image enhancement and restoration. In: Image Processing: Principles and Applications, pp. 105–129 (2005)
2. Braunagel, C., Kasneci, E., Stolzmann, W., Rosenstiel, W.: Driver-activity recognition in the context of conditionally autonomous driving. In: 2015 IEEE 18th International Conference on Intelligent Transportation Systems (ITSC), pp. 1652–1657. IEEE (2015)
3. Chollet, F., et al.: Keras: Deep Learning Library for Theano and Tensorflow. https://keras.io/k (2015)
4. CireşAn, D., Meier, U., Masci, J., Schmidhuber, J.: Multi-column deep neural network for traffic sign classification. Neural Netw. **32**, 333–338 (2012)
5. Dalal, N., Triggs, B.: Histograms of oriented gradients for human detection. In: IEEE Computer Society Conference on Computer Vision and Pattern Recognition CVPR 2005. , vol. 1, pp. 886–893. IEEE (2005)
6. Haloi, M.: Traffic sign classification using deep inception based convolutional networks. arXiv preprint arXiv:1511.02992 (2015)
7. Ioffe, S., Szegedy, C.: Batch normalization: accelerating deep network training by reducing internal covariate shift. In: International Conference on Machine Learning, pp. 448–456 (2015)

8. Jaderberg, M., Simonyan, K., Zisserman, A., et al.: Spatial transformer networks. In: Advances in Neural Information Processing Systems, pp. 2017–2025 (2015)

9. Keller, C.G., Sprunk, C., Bahlmann, C., Giebel, J., Baratoff, G.: Real-time recognition of us speed signs. In: Intelligent Vehicles Symposium, 2008 IEEE, pp. 518–523. IEEE (2008)

10. Krizhevsky, A., Sutskever, I., Hinton, G.E.: Imagenet classification with deep convolutional neural networks. In: Advances in Neural Information Processing Systems, pp. 1097–1105 (2012)

11. Le, T.T., Tran, S.T., Mita, S., Nguyen, T.D.: Real time traffic sign detection using color and shape-based features. In: Asian Conference on Intelligent Information and Database Systems, pp. 268–278. Springer (2010)

12. Maldonado-Bascon, S., Lafuente-Arroyo, S., Gil-Jimenez, P., Gomez-Moreno, H., López-Ferreras, F.: Road-sign detection and recognition based on support vector machines. IEEE Trans. Intell. Transp. Syst. **8**(2), 264–278 (2007)

13. Mao, X., Hijazi, S., Casas, R., Kaul, P., Kumar, R., Rowen, C.: Hierarchical CNN for traffic sign recognition. In: Intelligent Vehicles Symposium (IV), 2016 IEEE, pp. 130–135. IEEE (2016)

14. McCall, J.C., Trivedi, M.M.: Video-based lane estimation and tracking for driver assistance: survey, system, and evaluation. IEEE Trans. Intell. Transp. Syst. **7**(1), 20–37 (2006)

15. Meuter, M., Nunn, C., Gormer, S.M., Muller-Schneiders, S., Kummert, A.: A decision fusion and reasoning module for a traffic sign recognition system. IEEE Trans. Intell. Transp. Syst. **12**(4), 1126–1134 (2011)

16. Qian, R., Yue, Y., Coenen, F., Zhang, B.: Traffic sign recognition with convolutional neural network based on max pooling positions. In: 2016 12th International Conference on Natural Computation, Fuzzy Systems and Knowledge Discovery (ICNC-FSKD), pp. 578–582. IEEE (2016)

17. Ruta, A., Li, Y., Liu, X.: Robust class similarity measure for traffic sign recognition. IEEE Trans. Intell. Transp. Syst. **11**(4), 846–855 (2010)

18. Sermanet, P., LeCun, Y.: Traffic sign recognition with multi-scale convolutional networks. In: The 2011 International Joint Conference on Neural Networks (IJCNN), pp. 2809–2813. IEEE (2011)

19. Simonyan, K., Zisserman, A.: Very deep convolutional networks for large-scale image recognition. arXiv preprint arXiv:1409.1556 (2014)

20. Stallkamp, J., Schlipsing, M., Salmen, J., Igel, C.: The german traffic sign recognition benchmark: a multi-class classification competition. In: The 2011 International Joint Conference on Neural Networks (IJCNN), pp. 1453–1460. IEEE (2011)

21. Stallkamp, J., Schlipsing, M., Salmen, J., Igel, C.: Man vs. computer: benchmarking machine learning algorithms for traffic sign recognition. Neural Netw. **32**, 323–332 (2012)

22. Szegedy, C., Liu, W., Jia, Y., Sermanet, P., Reed, S., Anguelov, D., Erhan, D., Vanhoucke, V., Rabinovich, A.: Going deeper with convolutions. In: Proceedings of the IEEE Conference on Computer Vision and Pattern Recognition, pp. 1–9 (2015)

23. Zaklouta, F., Stanciulescu, B., Hamdoun, O.: Traffic sign classification using kd trees and random forests. In: The 2011 International Joint Conference on Neural Networks (IJCNN), pp. 2151–2155. IEEE (2011)

24. Zeng, Y., Xu, X., Fang, Y., Zhao, K.: Traffic sign recognition using extreme learning classifier with deep convolutional features. In: The 2015 International Conference on Intelligence Science and Big Data Engineering (IScIDE 2015), Suzhou, China, vol. 9242, pp. 272–280 (2015)

Kohonen's Self-organizing Map Optimizing Prediction of Gene Dependency for Cancer Mediating Biomarkers

Partho Mallick, Oindrila Ghosh, Priyanka Seth and Anupam Ghosh

Abstract Microarray gene expression data sets are huge in number. Hence, we have devised a procedure to optimize and simplify such multi-dimensional data to represent it in lower dimensions. Furthermore, we also have demonstrated an important test to determine whether a test sample of genes is cancerous or not. By mapping the test sample with a few cancer-affected gene samples, we have grouped the optimized data set to check the number of clusters that will help us to identify the affected genes. Therefore, this work reports an easy optimization technique to identify the cancer-affected genes. Kohonen's self-organizing map along with entropy, symmetrical uncertainty has been employed to develop the software tool to detect the cancer-affected gene in a fuzzy framework. The proposed work performs best with respect to some existing models, e.g., ANN, SVM-RFE, Apriori, and FP growth in predicting the true positives.

Keywords Fuzzy logic · Kohonen's self-organizing map · Symmetric uncertainty · Entropy · Information gain

P. Mallick (✉) · O. Ghosh · P. Seth
Techno India University, Kolkata, West Bengal, India
e-mail: partho.mallick@gmail.com

O. Ghosh
e-mail: della.priya@gmail.com

P. Seth
e-mail: priyankaseth1003@gmail.com

A. Ghosh
Department of Computer Science and Engineering, Netaji Subhash Engineering College, Garia, Kolkata 700152, West Bengal, India
e-mail: anupam.ghosh@rediffmail.com

© Springer Nature Singapore Pte Ltd. 2019
A. Abraham et al. (eds.), *Emerging Technologies in Data Mining and Information Security*, Advances in Intelligent Systems and Computing 814,
https://doi.org/10.1007/978-981-13-1501-5_75

1   Introduction

Cancer being one of the most challenging diseases in the present century, numerous methods have been employed for the diagnosis of cancer-affected patients. Cancer occurs when genes within a normal cell start functioning abnormally and get damaged and mutated. Extensive research works in molecular genetics opened new possibilities to find a series of primary genes which are responsible for the genesis of various types of cancers. Most genes do not influence the performance of the classification task. Culturing such genes during classification task. Culturing such genes during classification problem and poses computational difficulties, bringing unnecessary complexity in the process. It has been found that the self-organizing map which was first invented by the Finnish scientist Tuevo Kohonen's has been applied in a number of applications like automatic category generation for text documents by self-organizing maps [1], predicting stock prices using a hybrid Kohonen's self-organizing map [2], meteorological day type identification, applications in medicine, localization using self-organizing map.

In this context, using Kohonen's self-organizing map [3], we have optimized a high-dimensional data set of gene expressions, responsible for the cancer disease, into a much lower dimension which reduces the complexity of computation, cost, and time. We also have put efforts in easy diagnosis of cancer disease and identification of the type of cancer affecting the patient. Kohonen's self-organizing map has been found very effective in our work where we have optimized a huge data set containing 7129 genes rated for 86 different intensity values with respect to time. We have implemented Kohonen's SOM. Also we have classified the optimized data set into clusters which largely helped us to distinguish between a normal gene and a cancerous gene. Furthermore, it is also possible to identify the type of cancer with the help of this clustering.

2   Methodology

2.1   Prediction of Cancer by Applying Kohonen's SOM

We chose five sets of data matrix from the data set of genes which are cancerous. After applying fuzzy logic [4–6], entropy [4], conditional entropy [4], symmetric uncertainty [4], and information gain [4] on these data sets [6–8], we took the first row from each of the matrices and made another matrix pattern with only the first rows of each of the five sets. In this way, we made five different pattern matrices such that first pattern matrix contains only the first rows of the five matrices, second pattern matrix contains only the second rows of the five matrices above, and soon. Now, we took a sample data set of genes from a patient, and after applying all the concepts like fuzzy logic, entropy, conditional entropy, and symmetric uncertainty [6, 8, 9], we call that result matrix as the test matrix. Our intention is to find whether we get two different clusters of matrices or one singular cluster. If our test sample maps into any one of the five different samples taken, then we conclude that our test matrix falls into a single cluster, and hence, the patient as cancer. Else if we can conclude that the test matrix does not fall into the categories of cancer-affected genes, then it must be falling into another category, which means we will have two clusters of data; thus, we can conclude that the patient is normal. So in this way, we can detect a sample of genes whether it is cancerous or not. If the genes are cancerous, then we can also know what type of cancer has affected those genes. We have applied self-organizing map (SOM) to test our test matrix [10, 11], i.e., sample with the five samples.

By using the fuzzy C-means algorithm (FCC) [5], the prediction and automatic feature selection is accomplished. In FCC, the measure of confidence in feature selection may be obtained by assigning the values to fuzzy membership.

2.2 Algorithm of Self-organizing Map (SOM)

Step 1: Initialization

Random values are chosen for the initial weight vectors W_j (0). Given that $I =$ count (intensity of genes present in the lattice), $W_j(0)$ must not be equal for $1 < j \leq I$. The weights should be of lower magnitude.

Step 2: Sampling

A sample x is drawn from the input space with a certain probability; the vector x represents the activation pattern that is applied to the lattice. Dimension (vector x) = m.

Step 3: Matching the Similarity

The intensity of the gene which has matched the best, denoted by $i(x)$ at the nth step is obtained by the minimum Euclidian distance formula: $i(x) = \arg_j \min \|x(n) - W_j\|, j = 1, 2, \ldots$.

Step 4: Updating

The synaptic weights of all the gene intensities are adjusted by using update formula:

$$W_j(n+1) = W_j(n) + \eta(n) h_{j,i(x)}(n)(x(n) - W_j(n))$$

where $\eta(n)$ is the learning rate parameter, $h_{j,i(x)}(n)$ is the neighborhood function centered around the winning gene intensity $i(x)$, both $\eta(n)$ and $h_{j,i(x)}(n)$ are varied dynamically during learning for best results.

Step 5: Continuation

Step 2 is continued again until no noticeable change in the feature map has been observed.

Fig. 1 Flowchart

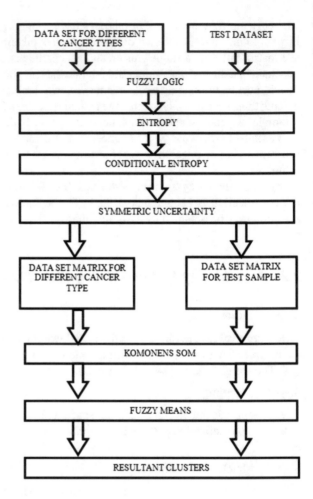

3   Result

3.1   Description of the Data Set

For testing purpose, we have taken 7129 different genes, which were rated for 86 different intensity values with respect to time. Now, we have applied self-organizing map, invented by Kohonen et al. in the year 1998 on our test material. We found that the genes having the similar intensity values are grouped in a straight horizontal line, and as a result of competition between the genes, the straight line, which is longest of all, shows the winner genes which are highly affected. Due to the cooperative phase of self-organizing map, the lines appearing above and below the winner genes show the values of intensities of those genes which are nearer to the winner genes. Hence, our data set of 7129 genes has been optimized to a reduced number of winner genes

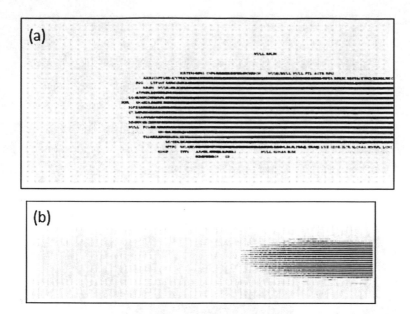

Fig. 2 **a** Training map output after 5000 iterations and **b** training map output after 10,000 iterations

which also show the neighboring genes which are affected due to the mostly affected winner genes. So as we will increase our iterations, the data set will get converged to an optimized number which will help us in detection of only those genes which got affected, neglecting those which are located far away from the winner genes (Fig. 1).

3.2 Analysis of Result of Our Proposed Model

After applying the concepts like fuzzy logic, entropy, conditional entropy, and symmetrical uncertainty on any randomly collected five sets of data from the sample data, we have obtained the five matrices which are shown in the tables below.

After applying self-organizing map technique on our data set of 7129 genes rated for 86 different intensity values with respect to time, we got the following output as shown in Fig. 2, where the huge data set of genes suspected to get affected by cancer has been optimized and converged to a lesser number of data set, facilitating the ease of detection of only the cancer-affected genes.

Our outputs are obtained after applying SOM on each of the five sample matrices to map with the test sample matrix to find out whether there is only single cluster or two different clusters (Fig. 3).

Here, we found we have two categories—1 and 0. So our sample test matrix did not match with the affected samples; hence, they got divided into two clusters. So the test sample is cancer-free.

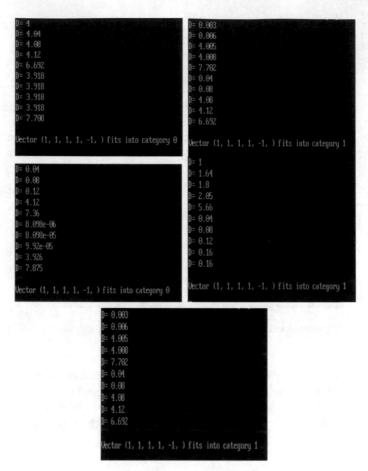

Fig. 3 Outputs after testing

3.3   Validation Using TOP-k Genes

In this work, we have extracted top 100 genes from the prediction graph that have been changed quite significantly from normal stage to carcinogenic stage. We have further validated our results from the TOP-k genes of NCBI database. Here, we have taken the value of k as 25, 50, 75, and 100. Moreover, we have compared the proposed model with some existing models based on artificial neural network (perceptron learning), SVM (recursive feature elimination), Apriori, and FP growth. It has been shown that our proposed model performs better with respect to other existing models in predicting the number of true positive genes (depicted in Fig. 4).

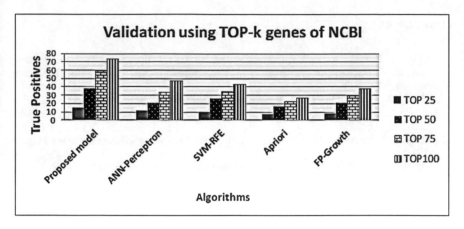

Fig. 4 Validation using TOP-k genes of NCBI

4　Conclusion

It is important to understand that all cancers are genetic, meaning that they result from the unnatural function of one or more genes. Abnormal mutations of cells caused by damaged genes, which itself may be due to various other factors, result in cancer. Moreover, when doctors decide how to diagnose a cancer, they are arming themselves with a wealth of knowledge about the molecular and genetic makeup of their patient's tumor. As a final result of this work, we can conclude that Kohonen's self-organizing map not only helped us optimizing our data set from a higher dimension to a much lower dimension, but we can also apply self-organizing map (SOM) for detection of cancer which obviously will help greatly in reducing the complexity in detection of the cancer-affected genes, and it gives a clear vision to the doctors about the treatment of the cancer-affected patients.

References

1. Yang, H., Lee, C.: Automatic category generation for text documents by self-organizing maps. In: IEEE-INNS-ENNS International Joint Conference on Neural Networks, Como, Italy, paperIII-581 (2000)
2. Afolabi, M., Olude, O.: Predicting stock prices using a hybrid Kohonen self organizing map (SOM). In: 40th Hawaii International Conference on Systems Science, Binghamton, NewYork (2007)
3. Kohonen, T., Somervuo, P.: Self-organizing maps of symbol strings. Neurocomputing **21**, p. 19 (1998)
4. Sarkar, S., Mallick, P.: CIIT International Journal, (July 2012). Gene Prediction graph: a novel graphical machine learning approach showing gene. Dependency For Cancer Prediction
5. Maji, P., Pal, S.: Fuzzy–rough sets for information measures and selection of relevant genes from microarray data. IEEE Trans. Syst. Man Cybern.—Part B: Cybern. **40**(3) (2010)
6. Wang, Z.: Neuro-fuzzy modeling for microarray cancer gene expression data. Thesis (2005)
7. Yu, L., Liu, H.: Redundancy Based Feature Selection for Microarray Data. Department of Computer Science and Engineering Arizon State University, Technical Report (2004)
8. Paul, T., Iba, H.April-: Prediction of cancer class with majority voting genetic programming classifier using gene expression data. IEEE/ACM Trans. Comput. Biol. Bioinf. **6**(2), 353 (2009)
9. Ghosh, A., Dhara, B.C., Rajat, K.: Selection of genes mediating certain cancers, using neuro-fuzzy approach. Neurocomputing **133**, 122–140 (2014). https://www.researchgate.net/publication/260429254_Selection_of_genes_mediating_certain_cancers_using_a_neuro-fuzzy_approach
10. Ghosh, A., Rajat, K.: Development of a fuzzy entropy based method for detecting altered gene-gene interactions in carcinogenic state. J. Intell. Fuzzy Syst. **26**, 2731–2746 (2014). https://content.iospress.com/articles/journal-of-intelligent-and-fuzzy-systems/ifs942
11. Ghosh, A., Rajat, K.: Fuzzy correlated association mining: Selecting altered associations among the genes, and some possible marker genes mediating certain cancers. Appl. Soft Comput. **38**, 587–605 (2016). https://www.sciencedirect.com/science/article/pii/S1568494615006596
12. Vuori, V.: Clustering writing styles with a self-organizing map. In: IEEE 8th International Workshop on Frontiers in Handwriting Recognition. Laboratory of Computer and Information Science, Helsinki University of Technology, Finland (2002)
13. Hung, C., Huang, L.: Extracting rules from optimal clusters of self-organizing maps. In: 2nd International Conference on Computer Modeling and Simulaion. Chung-Li, Department of Information Management, Chung Yuan Christian University, Taiwan (2010)

Key Feature Extraction and Machine Learning-Based Automatic Text Summarization

Vivek K. Verma, Anju Yadav and Tarun Jain

Abstract Text summarization is the way to produce important sentence from the original set of sentences. Summary of text sentences can be created in many ways and means, and various methods are being developed. Automatic mechanism can create a logical summary using parameters like text syntax, style and length. In this paper, the proposed approach is independent summarizer which is generalized and can work for any text domain. Some of the major domains are used for the benchmark testing of the system such as sports, business, education, Twitter article, medical.

Keywords Text summarization · Semantic similarity · Machine learning
Sentence segment extraction

1 Introduction

Day-to-day data sets are increasing very wisely. Hospitals, office or other working place are dealing very large number of data every day, and it is impossible to read all the data sets as it is, so here comes text summarization to make retrieve the main sentences from different datasets. Text summarization can be done in many ways—frequency-based, extraction-based, abstraction-based. Frequency-based scores are given by number of repeating key words in the text. In extraction-based, sentence and words are reused to find the scores. In abstraction-based, sentence and words are regenerated from extracted one [1, 2], and abstraction-based is complex and needs generation and machine learning techniques.

V. K. Verma · A. Yadav (✉) · T. Jain
SCIT, Manipal University Jaipur, Jaipur 303007, India
e-mail: anju.yadav@jaipur.manipal.edu

V. K. Verma
e-mail: vermavivek123@gmail.com

T. Jain
e-mail: tarunjainjain02@gmail.com

© Springer Nature Singapore Pte Ltd. 2019
A. Abraham et al. (eds.), *Emerging Technologies in Data Mining and Information Security*, Advances in Intelligent Systems and Computing 814,
https://doi.org/10.1007/978-981-13-1501-5_76

In our proposed approach, we use extraction-based text summarization, key words from every sentences are checks similarity between other keywords from other sentences. For checking semantic similarity, we used Wu and Palmer [3] (wup) similarity method, and scores are given according to the sematic similarity of one sentences with others. Highest score is considered as more important. Here every sentence and words are compared with other sentences and other words using Wu and palmer approach by using WordNet online dictionary. Similarity checks are carried out for every sentences among others using wup_similarity to provide scores for sentences in the text.

2 Related Work

This section will focus on studies carried out on text summarization and the related preprocessing for accurateness of the information analysis [4]. On the past few years, there has been increased research on text summarization due to its ever-increasing applications in many other tasks.

Mohd et al. [2] proposed a work in which sentence in the text is taken and checked how the information is semantically same. Lesk algorithm will deal with every word in the text using WordNet online dictionary. In WordNet, all the words are in semantically rather alphabetical order. Some words have two or more senses in WordNet; this provides ambiguity in the result. Lesk algorithm is simplified and used to avoid the ambiguity. WordNet and lesk algorithm are used to provide scores for each sentence [5]. Key words will be extracted from every sentence, and each and every key words will be compared to other key words in other sentences and value will be given for all key words. Summation of values of key words for particular sentence is the score of that sentence.

Saggion et al. [6] proposed a work in which text-to-text similarity rank is given for different corpus, knowledge-based data sets for finding automatic short answer generator. Finding the correct answer for the questions, we need to process text similarity between question and all stored answers in the corpus or data sets. Wu and Palmer methodology is used for finding the similarity between two texts.

Wu and Palmer measure and calculate similarity by considering the depths of the two concepts in the UMLS (it computes semantic relatedness of two concepts), along with the depth of the LCS.

Formula used for calculating scores in Wu and Palmer method is

$$\text{Score} = 2 * \text{depth}(\text{LCS})/(\text{depth}(s1) + \text{depth}(s2))$$

$0 < \text{score} \leq 1$: score will not reach zero because depth of LCS never becomes zero. If both the concept is same, then score becomes 1.

Ferreira et al. [7] proposed a work in which text semantic similarity for a corpus is calculates similarity using longest common subsequence (LCS) association with WordNet for selecting most matching sentences. Instead of selecting and giving scores for each individual words, select two sentences or two short paragraph and check the similarity and give the score. It compares two sentences; one sentence will be selected and compared with all other sentences and scores will give for it modified LCS with WordNet is used for comparing sentences.

3 Proposed Model

Text summarization can be done in many ways. Summary of the text is based on scores. For giving a score semantic-wise, we used Wu and Palmer method.

Phase 1 Load the corpus and read the text using text replace functions; replace all the short-form words and apostrophe words into their original word. Delete all the punctuations except full stop from the text. This preprocessing is done because scores cannot be calculated for punctuations, so to avoid ambiguity in the result, we are removing all punctuations and apostrophe words.

Input: (T1, T2, punctuation)

1.　　　Load & read T1

2.　　　Call replace ()

    a.　　　Set T1=T1.replace(s, n)

3.　　　cancel_punctuation()

    a.　　　Set T3 = T1

    b.　　　Get punctuation =''' !()-[]{}:;'"∧,<>?@#$%^&*_~" '

    c.　　　From X= 0 till T3

        If X in punctuation then,

        Set X = " "

    d.　　　End loop

Phase 2 Splitting the text in sentence-wise and word-wise using tokenizer. For calculating the scores, first sentence will be selected and compared using WordNet dictionary and check similarity with every other sentence using Wu and Palmer similarity. The summation of similarity score given by all other sentence is the main score for the selected sentences. Once the score is being allocated to the first sentence, then in sequence it selects other sentences and scores will be given for those sentences.

Input: (T2[], T1, word1[], filter1[], word2[], filter2[], sum1, sum)

T1: holding text

1. Set T2=T1.split(.).
2. From j = 0 till T2
   a. From I = 0 till T2
      i. IF I not in j then,
         1. Set word1 = word_tokenize(T2[j])
         2. Set word2 = word_tokenize(T2[i])
         3. For w = 0 till word1
            a. IF w not in stop_words then
            i. Set fillter1[] =w
         4. End loop
         5. For w = 0 till word2
            a. IF w not in stop_words then
            i. Set filter2[] =w
         6. End loop
         7. From I =0 till filter1
            a. From j = 0 till filter2
               i. Set p = i[1]. wup_similarity(j[1])
               ii. Set sum = sum + p
            b. End loop
         8. Set sum1=sum1+sum
         9. End loop
      ii. End IF
   b. End loop

Phase 3 Selecting important sentences will be based on scoring, after scoring is been allocated for every sentence in the text sorting to be done based on sentence and their scoring and position of the sentence in the corpus. Sorting can be done in matrix-based, graph-based and array- or list-based sorting, and many other. In our method we are creating a tupple with parameters of scoring, sentence, and index as their positions. At first tupple will be sorted in descending order by scores. Then according to the percentage of summary tupple will be divided and second part of the tupple will be removed. For arranging the sentence in their position-wise again sorting we have done according to their position. And finally from tupple sentences will be stored on other text file.

Input: (T2[], index[],score[],sent_str[],sort2,sort3,summary,tupple[: :])

T2 : holding sentence
1. For i=0 till T2

    a.   Set index [] =i

2. End loop

3. Set tupple= [T2, score,indcx]

4. Sort2=sorted (tupple,key=lambda tup: tup[1])

5. Get i

6. Set a=length(T2)

7. Set b=a/i

8. Sort2=sort2(: b)

9. Sort3=sorted (sort2, key= lambda tup:tup[2])

10. From i=0 till sort3

    a.   Summary=str(i[o])

11. From j=0 till summary

    a.   Sent_str+=str(i)+"."."

12. Set sent_str=sent_str[:-1]

13. Store sent_str in text file

Phase 4 Final goal of the proposed work is to select a few components as a summary. All the feature vectors selected in previous phases need to train using a strong a machine learning techniques for an efficient independent summarizer. Naïve Bayesian classifier has selected for the feature vector classification with neural network as an inductive learning machine learning technique [8–11].

$$P(\text{class}|\text{pred.}) = \frac{P(\text{pred.}|\text{class})P(\text{class})}{P(\text{pred.})}$$

4   Result Analysis

This system generates summary of electronic documents retaining important points of a document. Sentence Selection and Wu and Palmer algorithms are used to calculate priority of sentences and text summarization [12]. This algorithm has been tested on various kinds of text categories like legend personalities, different technical reports, and different newspaper articles on sports, politics, and short stories. Also the length

Table 1 Results of confusion matrix

Text type	TP	FP	TN	FN	Accuracy	Precision	Sensitivity	F-score
Newspaper	171	15	10	7	0.8768	0.9194	0.9448	0.9319
Magazine	150	10	8	5	0.8960	0.9375	0.9494	0.9434
Blog articles	140	11	8	4	0.8834	0.9272	0.9459	0.9365
Journal articles	110	9	6	3	0.8828	0.9244	0.9483	0.9362

of texts taken are different from each other in order to check the accuracy of the system, and all the texts taken as input are in English as the dictionary used here WordNet is in English. At first, the texts are manually summarized by two persons and parallel we did the same via this system. Then both the results are compared using the famous parameters—*PRECISION (P) AND RECALL(R)*. The parameters are expressed in the following way:

$$P = C/(C + W),$$

$$R = C/(C + M),$$

where

C the number of sentences extracted by both the systems.
W the number of sentences extracted by this system but not by the online summarizer.
M the number of sentences extracted by the online summarizer but not by this system.

Precision is a measure of result relevancy, and sensitivity is an amount of which truly significant results are returned. Precision is defined as the number of true positives over the number of true positives (T_p) plus the number of false positives (F_p). Recall (R) is defined as the number of true positives (T_p) over the number of true positives (T_p) plus the number of false negatives (F_n) (Table 1).

In both the cases, the texts are summarized to 50% of the original text, so the number of sentences in both the outputs is same. Hence, the number of the sentences missed and wrong for every testing corpus is same as shown in the table above. This algorithm gives good results for large texts as well as for small texts. It is also tested for 25% summarization, and results obtained are satisfying. It is because the sentences included in summary are semantically relevant. But there are some drawbacks of this system as well, for example, texts with punctuation marks like commas and apostrophe won't get processed. Secondly, it only works for English language. On analysis of the proposed approach, it is noticed that it produces very efficient results for articles with less number of named entity because such articles have more meaningful words in the sentence. As a result, a more number of words are available to be intersected with the text. Hence, the weightage of the sentence is calculated more effectively.

5 Conclusion

In this work, we have got a satisfactory result by summarizing 75% from the original text. Preprocessing is carried out to make summarization more generalized for all types of text. The best summary of the text can be given only by human, and we have compared this text summarization against manually summarized text using precision and recall technique. Ten different corpuses are tested, and our best accuracy result is 92%; least result is 73% and average is 82%.

References

1. Ferilli, S., et al.: A similarity-based abstract argumentation approach to extractive text summarization. In: Conference of the Italian Association for Artificial Intelligence. Springer, Cham (2017)
2. Mohd, M., et al.: Sumdoc: a unified approach for automatic text summarization. In: Proceedings of Fifth International Conference on Soft Computing for Problem Solving. Springer (2016)
3. Wu, Z., Martha, P.: Verbs semantics and lexical selection. In: Proceedings of the 32nd Annual Meeting on Association for Computational Linguistics. Association for Computational Linguistics (1994)
4. Gambhir, M., Gupta, V.: Recent automatic text summarization techniques: a survey. Artif. Intell. Rev. **47**, 11–66 (2017)
5. Shah, C., Anjali, J.: Literature study on multi-document text summarization techniques. In: International Conference on Smart Trends for Information Technology and Computer Communications. Springer, Singapore (2016)
6. Saggion, H., Thierry, P.: Automatic Text Summarization: Past, Present and Future. Multi-Source, Multilingual Information Extraction and Summarization, pp. 3–21. Springer Berlin, Heidelberg (2013)
7. Ferreira, R., et al.: Assessing sentence scoring techniques for extractive text summarization. Expert Systems with Appl. **40**(14), 5755–5764 (2013)
8. Chopra, S., Auli, M., Rush, A.M.: Abstractive sentence summarization with attentive recurrent neural networks. In: Proceedings of the 2016 Conference of the North American Chapter of the Association for Computational Linguistics: Human Language Technologies 2016
9. Chen, Q., et al.: Distraction-based neural networks for document summarization. arXiv preprint arXiv: 1610.08462, 2016
10. Arnulfo, R., et.al.: Text summarization by sentence extraction using unsupervised learning. In: Mexican International Conference on Artificial Intelligence, pp. 133–143, 2008
11. Gupta, V.: Hybrid algorithm for multilingual summarization of Hindi and Punjabi documents. Mining Intelligence and Knowledge Exploration, pp. 717–727, 2013
12. Fattah, M.A.: A hybrid machine learning model for multi-document summarization. Appl. Intell. **40**(4), 592–600 (2014)

Prediction of Gold Price Movement Using Geopolitical Risk as a Factor

Debanjan Banerjee, Arijit Ghosal and Imon Mukherjee

Abstract Accurate prediction of commodity prices by using machine learning techniques is considered as a significant challenge by the researchers and investors alike. The main objective of the proposed work is to highlight that geopolitical risk is a potentially relevant feature when it comes to predicting the gold prices. The current work utilizes the geopolitical risk index that has been developed based upon the risk associated with positive or negative political events as highlighted in the major newspapers around the world. The geopolitical risk index is measured by counting major negative events such as terrorist attacks, riots, violent conflict, nuclear missile tests in case of the Korean Peninsula and the Middle East and North Africa region. Many major newspapers from around the world such as the Guardian, the Independent, the New York Times, the Washington Post, the Boston Globe, the Daily Mail, the Jerusalem Post are chosen for this particular purpose.

Keywords Geopolitical risk · Machine learning · Random forest · Support vector machine

D. Banerjee
Department of Management Information Systems,
Sarva Siksha Mission Kolkata, Kolkata 700042, West Bengal, India
e-mail: debanjanbanerjee2009@gmail.com

A. Ghosal (✉)
Department of Information Technology, St. Thomas'
College of Engineering and Technology, Kolkata 700023, West Bengal, India
e-mail: ghosal.arijit@yahoo.com

I. Mukherjee (✉)
Department of Computer Science and Engineering,
Indian Institute of Information Technology, Kalyani 741235, West Bengal, India
e-mail: imon@iiitkalyani.ac.in

© Springer Nature Singapore Pte Ltd. 2019
A. Abraham et al. (eds.), *Emerging Technologies in Data Mining and Information
Security*, Advances in Intelligent Systems and Computing 814,
https://doi.org/10.1007/978-981-13-1501-5_77

1 Introduction

In the contemporary world, the aftermath of many spectacular events such as the burst of the subprime bubble in the USA, the 2008 financial crisis has witnessed significant volatility in the financial markets. The commodity gold is used as an investment hedge against any supposed upward or downward tendency in the market prices. Current work applies machine learning techniques such as the random forest and logistic regression using various features, specifically geopolitical risk to predict whether the price of the gold will go down or go up. The current work uses historical data from the Indian markets from January 2010 to December 2017 from the Web resource in.investing.com to make predictions whether the price of gold would increase or decrease.

2 Related Works

The first major challenge of predicting the movement of gold prices is to find a set of features that will be the optimum set for predicting the gold price movement. Agarwal et al. [1] have discussed various state-of-the art techniques that do impact the price of the stocks. Banerjee et al. [2] have used discretization procedure for gold price movement prediction. Logistic regression technique has been used by Bilberry et al. [3] to help users select from a set of exchange-traded funds (ETF) based upon a probability score of possibility of the growth of the stocks. Caldara and Iacovello [4] have created the geopolitical risk index and discussed the relations between the GPR and the gold price. Ciner [5] has observed with the help of Tokyo Stock Exchange historical data has become somewhat ambiguous from the late twentieth century.

Grudnitski and Osburn [6] have observed that applying neural networks increases the potentiality of features such as currency exchange rates and interest rates for accurate price prediction. Logistical regression has been used by Hargreaves and Hao [7] in the context of Australian stock markets to create a stock selection strategy for stock users. Imandoust and Bolandraftar [8] have utilized predictive machine learning techniques for the equity data from stock markets situated in different parts of the world. The concept of random forest has been utilized for predicting the direction of the market prices by Khaidem et al. [9]. In the work of Li [10], artificial bee colony algorithm was utilized for gold price prediction. In the work of Potoski [11], various financial market commodity prices including index prices were used as a set of features from different commodity exchanges across the world. Various machine learning techniques have been compared by Shah [12] and Shen et al. [13] in their respective works to predict the gold price accurately. Sjaastad [14] has observed that there are strong relations between major exchange currency blocks and the gold price. According to that work, any abrupt disruption in any major exchange currency blocks, for example, the US dollar could lead to instability in the gold price direction.

The second major challenge is to find out the appropriate machine learning technique for obtaining the accurate prediction. In the work of Potoski [11], logistic regression and SVM have been utilized by the authors with mostly continuous numerical features. Stock market data from Saudi Arabia has been utilized by Zaidi and Amanat [15] to predict the potential movement for a particular stock in a certain direction. These experiments have shown that with an optimum amount of features the logistic regression technique produces almost similar accuracy to that of a neural network model. In this work, technical indicators such as relative strength index (RSI), stochastic oscillator, on-balance volume were introduced by the authors as features.

3 Procedure for Predicting Gold Price Direction

The present approach for predicting the direction of gold price can be described in the following steps. (a) Feature definition: This is the very first step whereby the necessary features for the work are identified and collected. This work utilizes and compares the role of geopolitical risk to that of the other features. (b) Application of machine learning techniques: The current work uses machine learning techniques such as logistic regression, SVM, random forest, and neural network for the purpose. (c) Performance measurement: The performance of the classification is measured using metrics such as accuracy and precision. This work uses standard procedures for measuring these metrics.

3.1 Feature Definition: Geopolitical Risk as a Feature

Geopolitics has had a much diversified and wide-ranging definition over the centuries. Most practitioners of geopolitics define the term to explain the complicated relations between states vying for influence over geographical borders. The geopolitical risk is a phenomenon which can be defined as a negative event or action or the possibility of the occurrence of the same, the occurring in a particular geographical location. These events can be a terrorist attack, large-scale ethnic riots, incidents of hostile militaries clashing with each other, civil war, tension between large powers such as the recent one between the USA and China over the North Korean atomic bomb Korean peninsula or India–Pakistan tensions over Kashmir. The current work utilizes the monthly geopolitical risk index as a feature by incorporating the geopolitical risk index developed by Caldera and Iacoviello [4]. This particular geopolitical risk index has been developed by taking into consideration political events in each month from the source of eleven prominent global newspapers such as The Boston Globe, Chicago Tribune, The Daily Telegraph, Financial Times, The Globe and Mail, The Guardian, Los Angeles Times, The New York Times, The Times, The Wall Street Journal, and The Washington Post.

The current work uses linear regression technique to analyze influence of the feature over the gold price with the help of the following statistical parameters.

1. Slope: 3.368.
2. Probability: 2.345e−16.
3. R-squared error: 0.898.
4. Adj. R-squared error: 0.889.
5. F-statistic probability value: 3.238e−16.
6. The null hypothesis we consider here is that there is no influence of geopolitical risk over the gold price.
7. If our probability value is more than 0.05 and F-statistic p value is also more than 0.05 and R-squared error figures are less than 0.05, then we can say that null hypothesis is accepted.
8. From the above results, we can observe that both probability value and F-statistic probability value are less than 0.05 and both the R-squared errors are well above 0.05; thus, we can reject the null hypothesis.
9. After rejecting the null hypothesis, we can accept that the geopolitical risk does have influence over the gold price.
10. Slope coefficient of geopolitical risk is 3.368, so we can say that when geopolitical risk increases, gold price will also increase.

This work uses the following equation to compute geopolitical risk.

$$\text{Geopolitical risk} = (N/P) \times T \tag{1}$$

1. The total number of all negative events in a month as observed in the news publication sources is indicated by N.
2. The total number of all positive events in a month as observed in the news publication sources is indicated by P.
3. The total number of all events in a month as observed in the news publication sources is indicated by T.

3.2 Applying Machine Learning Techniques

In this work, four different types of machine learning algorithms have been utilized. These are (a) logistic regression, (b) support vector machine, (c) random forest, and (d) neural network.

The Logistic Regression The logistic regression always provides an output value between 0 and 1. Once we derive the values, a threshold (if the value ≥ 0.5) is applied. If we observe that the value is ≥ 0.5, thus we replace that value with 1, else we replace that value with 0. The logistic regression has been implemented by using the R-package glm.

Support Vector Machine The support vector machine algorithm that we use in this work has been utilized since adopting a nonlinear approach will provide greater accuracy.

1. In this work, we use *PythonCanopy* software and *scikit* package for implementing support vector machine.
2. We use the rbf kernel and the variables c and gamma whose values we use as a range from 0.1 to 1000.

Random Forest We explore the *R*-package Random Forest for our present work. The parameters that we have used here are maximum number of leaf nodes and maximum number of splits. The current work achieves best performance when we select maximum number of leaf nodes as 15 and maximum number of splits as 12.

The Neural Network The current work utilizes avNNet function from the caret package in the *R* framework for using neural network. For our work, we achieve the best results by using activation function as tanh or tan hyperbolic, solver as lbfgs, hidden layer size as 3, and maximum iteration as 100.

3.3 Performance Measurement

In this phase, the performance of the models is validated using techniques such as accuracy measurement. The standard procedures have been introduced to measure the various metrics of the work.

Accuracy Measurement The accuracy measurement is calculated by taking a percentage of the grand total of all the true predicted high values as well as all the true predicted low values from the set of all high values and all low values.

4 Experimental Results

The work applies logistic regression, SVM, random forest, and neural network techniques, respectively, over both non-discretized and discretized features. The accuracy measurements are also noted down. The three selected features that are utilized in this regard are crude oil price, silver price, copper price and compare the performances of these features to those of the geopolitical risk as a feature (Tables 1 and 2).

The improvement in terms of accuracy that the current work has achieved by utilizing geopolitical risk factor as a feature can also be observed from the ROC curves with area under curve (AUC) as drawn in the below images.

Table 1 Comparison between geopolitical risk and other features

Algorithm	Feature	Accuracy (in %)
Logistic regression	Geopolitical risk	81.82
Logistic regression	Silver future price	70.15
Logistic regression	Copper future price	70.57
Logistic regression	Crude oil future price	63.37
SVM	Geopolitical risk	83.58
SVM	Silver future price	67.75
SVM	Copper future price	66.57
SVM	Crude oil future price	68.37
Random forest	Geopolitical risk	82.32
Random forest	Silver future price	71.15
Random forest	Copper future price	72.57
Random forest	Crude oil future price	70.37
Neural network	Geopolitical risk	81.32
Neural network	Silver future price	67.15
Neural network	Copper future price	65.57
Neural network	Crude oil future price	61.37

Table 2 Comparative analysis of proposed work with other works

Algorithm	Feature	Accuracy (in %)
Potosky [11]	SVM	69.32
Banerjee et al. [2]	SVM	70.432
Proposed approach	SVM	83.58112
Potosky [11]	Logistic regression	69.5362
Banerjee et al. [2]	Logistic regression	70.8132
Proposed approach	Logistic regression	81.7832

4.1 Comparison with Previous Work

The current work performs better when being compared with previous work that was performed by Potosky [11]. Potosky after making use of her own set of features was able to obtain a percentage of 69% accuracy, whereas the current work after making use of the geopolitical risk as a factor has ensured that it is possible to obtain an accuracy of over 81% while applying four separate types of machine learning techniques (Fig. 1).

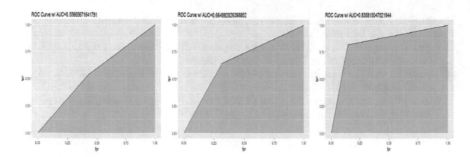

Fig. 1 ROC curve comparison between Potosky's work, Banerjee et al.'s work, and the current work

5 Conclusion

This can be observed from the experimental results that using geopolitical risk as a feature gives results comparable to other features such as the crude oil, copper or silver future prices. Since we did utilize four separate types of algorithm techniques, we could observe that the performance of the geopolitical risk as a feature is fairly secular given the choice of the machine learning technique.

References

1. Agarwal, J.G., Chourasia, V.S., Mitra, A.K.: State-of-the-art in stock prediction techniques. Electron. Instrum. Eng. **2**(4), 3–5 (2013)
2. Banerjee, D., Ghosal, A., Mukherjee, I.: Prediction of gold price movement using discretization procedure. In: International Conference on Computational Intelligence in Data Mining (Accepted) (2017)
3. Bilberry, J.K., Riley, N.F., Sams, C.L.: Short-term prediction of exchange traded funds (ETFs) using logistic regression generated client risk profiles. J. Finance Accountancy **3**(5), 5–6 (2015)
4. Caldara, D., Iacoviello, Matteo.: Measuring geopolitical risk. In: Working Paper, Board of. (2016)
5. Ciner, C.: A note on the long run relationship between gold and silver prices. Glob. Finance J. **12**(2), 299–303 (2000)
6. Grudnitski, G., Osburn, L.: Forecasting S and P and gold futures prices: an application of neural networks. J. Futures Markets **13**(6), 631–643 (1993)
7. Hargreaves, C., Hao, Y.: Prediction of stock performance using analytical techniques. J. Emerg. Technol. Web Intell. **5**(2), 136–143 (2000)
8. Imandoust, S.B., Bolandraftar, M.: Forecasting the direction of stock market index movement using three data mining techniques: the case of Tehran stock exchange. International. J. Eng. Res. Appl. **4**(6), 106–117 (2014)
9. Khaidem, L., Saha, S., Dey, S.R.: Predicting the direction of stock market prices using random forest. Appl. Math. Finance **1**(5), 1–20 (2016)
10. Li, B.: Research on WNN modeling for gold price forecasting based on improved artificial bee colony algorithm. Comput. Intell. Neurosci. **1**, 2–10 (2014)
11. Potoski, M.: Predicting Gold Prices, vol. 1, issue no 1, pp. 2–4. Stanford University 2013 publication (2013)

12. Shah, V.: Machine learning techniques on stock prediction. Found. Mach. Learn. **1**(1), 6–12 (2007). Spring, Berlin
13. Shen, S., Jiang, H., Zhang, T.: Stock Market Forecasting Using Machine-Learning Algorithms, vol. 1, issue no 1, pp. 26–32. Stanford: Stanford University (2012)
14. Sjaastad, L.A.: The price of gold and the exchange rates: once again. Resour. Policy **33**(2), 118–124 (2008)
15. Zaidi, M., Amanat, A.: Forecasting stock market trends by logistic regression and neural networks. Int. J. Econ. Commer. Manage. **4**(6), 4–7 (2016)

Author Index

Printed in the United States
By Bookmasters